Sector Field Mass Spectrometry for Elemental and Isotopic Analysis

**New Developments in Mass Spectrometry**

*Editor-in-Chief:*
Professor Simon J Gaskell, *Queen Mary University of London, UK*

*Series Editors:*
Professor Ron M A Heeren, *FOM Institute AMOLF, The Netherlands*
Professor Robert C Murphy, *University of Colorado Denver, USA*
Professor Mitsutoshi Setou, *Hamamatsu University School of Medicine, Japan*

*Titles in the Series:*
1: Quantitative Proteomics
2: Ambient Ionization Mass Spectrometry
3: Sector Field Mass Spectrometry for Elemental and Isotopic Analysis

*How to obtain future titles on publication:*
A standing order plan is available for this series. A standing order will bring delivery of each new volume immediately on publication.

*For further information please contact:*
Book Sales Department, Royal Society of Chemistry, Thomas Graham House, Science Park, Milton Road, Cambridge, CB4 0WF, UK
Telephone: +44 (0)1223 420066, Fax: +44 (0)1223 420247
Email: booksales@rsc.org
Visit our website at www.rsc.org/books

# *Sector Field Mass Spectrometry for Elemental and Isotopic Analysis*

Edited by

**Thomas Prohaska**
*University of Natural Resources and Life Sciences Vienna, Tulln, Austria*
*Email: thomas.prohaska@boku.ac.at*

**Johanna Irrgeher**
*University of Natural Resources and Life Sciences Vienna, Tulln, Austria*
*Email: johanna.irrgeher@boku.ac.at*

**Andreas Zitek**
*University of Natural Resources and Life Sciences Vienna, Tulln, Austria*
*Email: andreas.zitek@boku.ac.at*

**Norbert Jakubowski**
*Federal Institute for Material Research and Testing, Berlin, Germany*
*Email: norbert.jakubowski@bam.de*

ROYAL SOCIETY OF CHEMISTRY

THE QUEEN'S AWARDS FOR ENTERPRISE: INTERNATIONAL TRADE 2013

New Developments in Mass Spectrometry No. 3

Print ISBN: 978-1-84973-392-2
PDF eISBN: 978-1-84973-540-7
ISSN: 2044-253X

A catalogue record for this book is available from the British Library



Published by The Royal Society of Chemistry,
Thomas Graham House, Science Park, Milton Road,
Cambridge CB4 0WF, UK

Registered Charity Number 207890

For further information see our web site at www.rsc.org

# Foreword

Ralph Sturgeon, Principal Research Officer with Measurement Science and Standards of the National Research Council of Canada in Ottawa

The present volume, *Sector Field Mass Spectrometry for Elemental and Isotopic Analysis*, edited by Thomas Prohaska, Johanna Irrgeher, Andreas Zitek and Norbert Jakubowski, provides for the launch of a Royal Society of Chemistry series on new developments in mass spectrometry. Elemental mass spectrometry enjoys a long history and the cost-effective availability of modern instrumentation coupled to a significant publication database encompassing a plethora of applications in disparate scientific disciplines support its multidisciplinary use in diverse laboratories. Recent developments in mass spectrometers used for isotopic and elemental analysis have yielded improvements in critical performance metrics, pushing the frontiers of imagination, expectation and deliverable results into new and exciting domains ranging from revised atomic weights and isotopic abundances to geochronology and fundamental transport phenomena. Many excellent reference texts comprising such general subject areas already exist, such as those most recently issued by J.H. Gross (*Mass Spectrometry, A Textbook*, 2011) and C. Dass (*Fundamentals of Contemporary Mass Spectrometry*, 2007); along with older volumes such as those by Barshick *et al.* (*Inorganic Mass Spectrometry: Fundamentals and Applications*, 2000) and Adams *et al.* (*Inorganic Mass Spectrometry*, 1988), not to mention specific treatises on ICP-MS

New Developments in Mass Spectrometry No. 3
Sector Field Mass Spectrometry for Elemental and Isotopic Analysis
Edited by Thomas Prohaska, Johanna Irrgeher, Andreas Zitek and Norbert Jakubowski

Published by the Royal Society of Chemistry, www.rsc.org

proffered by Montaser, Nelms and Hill more than a decade ago. While these yet remain classic texts and excellent monographs, there is need and room for another textbook of this nature, which critically serves to bring a rapidly developing field up to date.

In contrast to these earlier citations that have attempted to comprehensively encompass all mass-spectrometry platforms (quadrupole, sector field, time-of-flight, ion trap, ICR, *etc.*), this volume is devoted only to sector field instrumentation, delineated on the basis of differences in sample ionisation strategies. The topic is covered in detail, utilising a generalised template that is consistently applied to each instrumental technique, giving the feel of a single author treatise while taking full advantage of the expertise provided through a multiauthored effort arising from contributions from talented, internationally recognised experts/users in each of the subject areas. Thus, a more coherent and homogeneous result is achieved than for most other volumes that are merely edited collections of individually written chapters. In all cases, coverage starts with a general description of the instrumentation followed by a thorough, fundamental technical consideration of each component in a logical fashion, beginning with sample introduction and moving through the ion source, sampling interface, vacuum system, ion optics, magnetic and electric sectors, entrance and exit slits, additional ion optics and detection system(s). A consistent use of general terminology and abbreviations has been followed in accordance with the recommendations of IUPAC. Ample selections of schematics and figures serve to complement these technical descriptions. A further useful inclusion is a comprehensive summary of instruments/manufacturers that further details unique specifications and characteristics of all marketed devices. Also presented with each instrumental technique is a set of applications relevant to life and environmental sciences, archaeometry and forensics, industrial and materials research as well as geological, cosmological and radionuclide research. In this manner, ICP-MS (both single and multicollector), GDMS, TIMS, SIMS and IRMS can be logically and comprehensively discussed following the fundamental detail presented earlier. Interestingly, and uniquely for treatises such as this, the subject of metrology (the science of measurement) is also introduced and its general relevance to the data generated using modern SFMS is outlined. This includes discussion of appropriate terminology for clarity, the concept of measurement uncertainty and a general comment on reference materials and traceability. The book concludes with a comprehensive summary chapter as well as a consideration of future developments expected to be realised in this exciting field of instrumentation.

The goal of this book is enunciated in the Introduction – "to discuss the main features, principles of operation and resulting applications of these different sector field mass spectrometers"; this has been achieved by providing a good balance between essentials and excess detail that would otherwise overburden the reader. Nevertheless, reference to a comprehensive set of additional reading options is provided for the interested bibliophile. This book should integrate well into the curriculum for graduate courses in inorganic mass spectrometry and equally belongs on the shelf of every practicing mass spectrometrist by providing succinct reading for newcomers and expert practitioners alike.

# Contents



New Developments in Mass Spectrometry No. 3
Sector Field Mass Spectrometry for Elemental and Isotopic Analysis
Edited by Thomas Prohaska, Johanna Irrgeher, Andreas Zitek and Norbert Jakubowski

Published by the Royal Society of Chemistry, www.rsc.org

**Fundamentals**

# *Acknowledgements (in Alphabetical Order)*

The following people are highly acknowledged for their support in the editing and proof-reading process: Maria Badrian, Johannes Draxler, Naida Dzigal, Ondřej Hanousek, Monika Horsky, Anastassiya Tchaikovsky and Florian Träger.

The following people are highly acknowledged for their valuable comments and contributions during the writing and reviewing process of the various sections.

(***Note***: the authors who contributed to chapters other than those they were involved as authors are acknowledged here):

Daniel Abou-Ras, Günter Allmaier, Yetunde Aregbe, Ian Bowen, Willi A. Brand, Stefan Bürger, Emmanuel de Chambost, David Donohue, Charles Douthitt, Johannes Draxler, Manuela Göbelt Steve Guilfoyle, Manfred Gröning, Tamara Gusarova, Ondřej Hanousek, Gary Hieftje, Takafume Hirata, Volker Hoffmann, Monika Horsky, Kevin McKeegan, Urs Klötzli, Stefanie Konegger-Kappel, Michael Latzel, Stephan Richter, Torsten Rissom, Lothar Rottmann, Laure Sangely, Sebastian Schmitt, Anastassiya Tchaikovsky, Staf Valikiers, Cornel Venzago, Thomas Walczyk and Andreas Zitek.

We acknowledge Klaus Heumann for his valuable comments to the book.

We would like to thank all contributing companies and their representatives for providing us information, data, figures and photos along with technical details of their products:

New Developments in Mass Spectrometry No. 3
Sector Field Mass Spectrometry for Elemental and Isotopic Analysis
Edited by Thomas Prohaska, Johanna Irrgeher, Andreas Zitek and Norbert Jakubowski

Published by the Royal Society of Chemistry, www.rsc.org

Bruce Godfrey (Australian Scientific Instruments)
Christian Le Pipec, Marion Chopin, Paula Peres, François Horréard, Alex Merkulov and Philippe Saliot (CAMECA)
Mike Seed, Kyle Taylor (Isoprime)
Ekbal Patel (Mass Spectrometry Instruments Ltd.)
Ian Bowen, Steve Guilfoyle (Nu instruments)
Laura Bishop (SerCon)
Zenon Palacz, Steve Locke (IsotopX Ltd)
Dirk Wissmann, Marco Buchholz, Dirk Adelt and Willi Barger (Spectro Analytical Instruments)
Lothar Rottmann, Gerhard Zinsberger and Charles Douthitt (Thermo Fisher Scientific)

The editors acknowledge especially the patience of their partners, families and friends during the last two and a half years of editing this book.

This book is dedicated
to the memory of Stefan Bürger,
a friend and competent colleague.

He was a passionate scientist and a passionate climber.
He passed away with his passion during the final stage of this book.

# *Authors (in Alphabetical Order)*

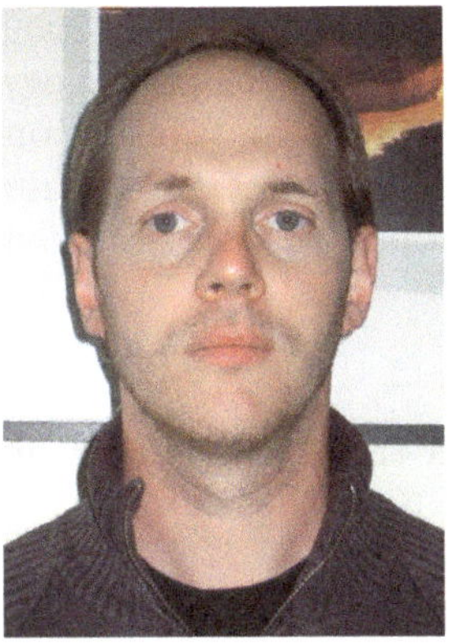

**Jérôme ALEON,** Centre de Sciences Nucléaires et de Sciences de la Matière (CSNSM), University Paris Sud laboratory in Orsay, France
(Photo Copyright: *Aléon Jérôme*)

Jérôme Aléon is a research scientist at the Centre de Sciences Nucléaires et de Sciences de la Matière (CSNSM) a joint CNRS/ University Paris Sud laboratory in Orsay, France. He studied geology at the National School of Geology in Nancy, France and received his PhD from the Lorraine University in isotope geochemistry by SIMS in 2001. After a postdoctoral stay in the SIMS group of the Earth, Planetary and Space Sciences department at the University of California Los Angeles, USA during which he shifted his research interests from geochemistry to cosmochemistry, he was hired in 2003 as a CNRS researcher in the SIMS group of the Centre de Recherches Pétrographiques et Géochimiques, Nancy, France. In 2005, he spent one year in the NanoSIMS group at the Lawrence Livermore National Laboratory, USA, before joining the Solid State Astrophysics group at CSNSM in Orsay in 2006. His research interests focus on the *in situ* study of the isotopic composition of light elements in the oldest rocks and solids of the solar system by SIMS and NanoSIMS to unravel the origin and conditions of formation of rocky material and planets in our solar system.

**Sergey ASSONOV**, International Atomic Energy Agency, Vienna, Austria

(Photo: *Sandor Tarjan*; Copyright: *Sergey Assonov*; Courtesy of: *Sergey Assonov*)

Sergey Assonov is a specialist in isotope ratio mass spectrometry (IRMS). After having graduated from the Moscow State University, Russia, he worked for the Russian Academy of Sciences in Moscow from 1985–2000. Later, he was a research scientist at the Max Planck Institute in Mainz, Germany until 2005. Then he worked for the European Commission Joint Research Centre in Geel, Belgium, as a research officer until 2009 when he became a lecturer and research scientist at the Institute of Geology at the University of Cologne, Germany. From 2013 he has taken on the challenge for the characterisation of new reference materials within the IAEA where he is currently working as an isotope reference material specialist.

**Jean-Nicolas AUDINOT**, Centre de Recherche Public - Gabriel Lippmann (CRP-GL), Luxemburg

(Photo: *Jean-Nicolas AUDINOT*; Copyright: *Jean-Nicolas AUDINOT*; Courtesy of: *Jean-Nicolas AUDINOT*)

Jean-Nicolas Audinot obtained his PhD on the Solid Oxide Fuel Cell at the Institute for Condensed Matter Chemistry of Bordeaux affiliate at the University of Bordeaux, France, in 1999. In 2000, he joined the Centre de Recherche Public Gabriel Lippmann (CRP-GL), Luxembourg, as a researcher in the Laboratory of Analysis of Materials (LAM) directed by Prof. H.-N. Migeon. In charge of the SIMS activities, he developed new analytical procedures for the analysis of any type of material. Since 2007, he has been the project leader in the Science and Analysis Materials (SAM) Department of the CRP-GL. He is in charge of the research projects involving the characterisation of the biological samples with sophisticated instruments, such as the (Nano)-SIMS, TEM and AFM.

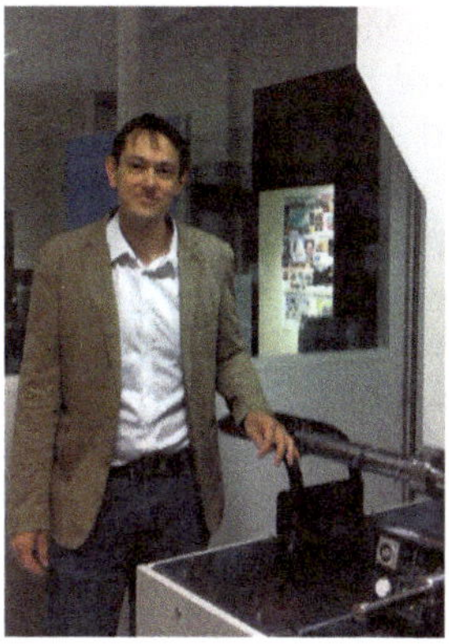

**Jean-Paul BARNES,** Commissariat à lénergie atomique et aux énergies alternatives, LETI, France

(Photo: *Jean-Paul BARNES*; Copyright: *Jean-Paul BARNES*; Courtesy of: *Jean-Paul BARNES*)

Jean-Paul Barnes received the M.Eng. and D.Phil. degrees in metallurgy and science of materials from University of Oxford, Oxford, UK, in 2000 and 2003, respectively. Since 2004, he has been with the CEA-LETI (Grenoble, France) in the characterisation service working on electron and ion spectrometry techniques. He is currently responsible for the ion beam analysis group. His current interests include ion beam analysis of microelectronic materials, ToF-SIMS of organic materials and correlative atom-probe and electron tomography for semiconductor devices.

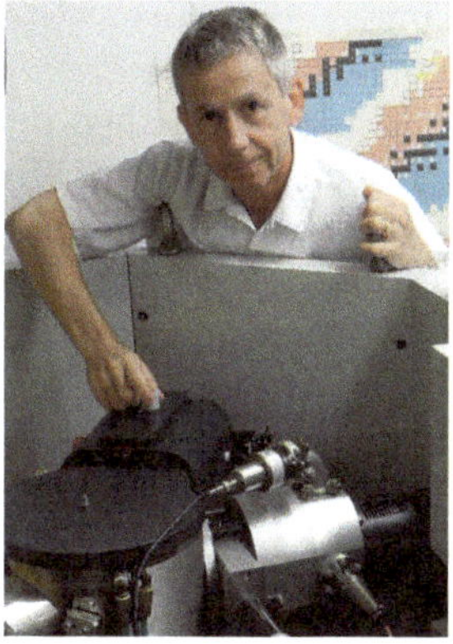

**Philippe BIENVENU,** Commissariat à lénergie atomique et aux énergies alternatives, France

(Photo Copyright: *CEA*; Courtesy of: *CEA*)

Philippe Bienvenu is a scientific researcher at the Commissariat à l'Energie Atomique et aux énergies alternatives (CEA) in the Cadarache center. After a PhD in geochemistry, he spent about 20 years in the field of analytical chemistry and radiochemistry. Most of his work was focused on the development of analytical methods using atomic and mass spectrometry techniques for the measurement of long-lived radionuclides in nuclear waste materials. In 2011, he joined the LECA hot laboratory to contribute to research programs on nuclear fuel development and characterisation. He is now in charge of SIMS applications on irradiated fuel samples, working with a shielded CAMECA SIMS *IMS6F R*.

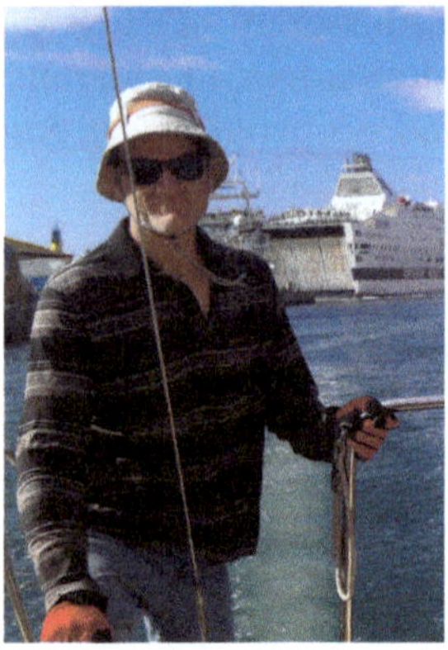

**Bernard BOYER,** Université Montpellier 2, France
(Photo Copyright: *Bernard Boyer*)

Bernard Boyer obtained his diploma in Electronics at Montpellier, then after two years in scientific bases in sub-Antarctica and Antarctica, he joined CAMECA for about twenty years. He contributed to delivering the first prototype *IMS5F* and later *IMSWF*. CAMECA allowed him to obtain very important know-how in mechanical and vacuum techniques. He is now an Instrumental engineer for the French CNRS (Centre National de la Recherche Scientifique) at University Montpellier 2, in charge of a material analysis platform using a CAMECA SIMS *IMS4F* and a CAMECA EPMA *SX100*.

**Willi A. BRAND**, Max-Planck-Institute for Biogeochemistry, Jena, Germany
(Photo: *Jürgen M. Richter*; Copyright: *W. Brand*)

Willi A. Brand has built up and established the Stable Isotope Mass Spectrometry facility at the Max-Planck Institute for Biogeochemistry in Jena, Germany, which he is heading. He obtained a Diploma degree in Chemistry from the University in Bonn in 1979 and finished with a PhD (with honours) in 1981. After three years as a postdoctoral fellow at the University of North Carolina and back at the University of Bonn, he joined the development team at Finnigan MAT in Bremen, where he was involved in developing several new stable isotope ratio mass spectrometers (*Delta S*, *MAT 252*, *on-line Delta*, *Delta Plus*, *Delta Plus XL*) and new automated sample preparation devices (*e.g.* Trapping box line, Kiel Carbonate systems, ConFlo, PreCon). During his time at Finnigan, he served as a project engineer, project manager, program manager, development laboratory manager and head of stable isotope developments. In 1998, he decided to leave Finnigan MAT and join the newly founded Max-Planck Institute in Jena. He continued developing tailored solutions for high-precision stable isotope systems and published numerous scientific papers about stable isotope techniques and referencing strategies. From 2009 to 2013 he acted as chair of the IUPAC Commission on Isotope Abundances and Atomic Weights (CIAAW).

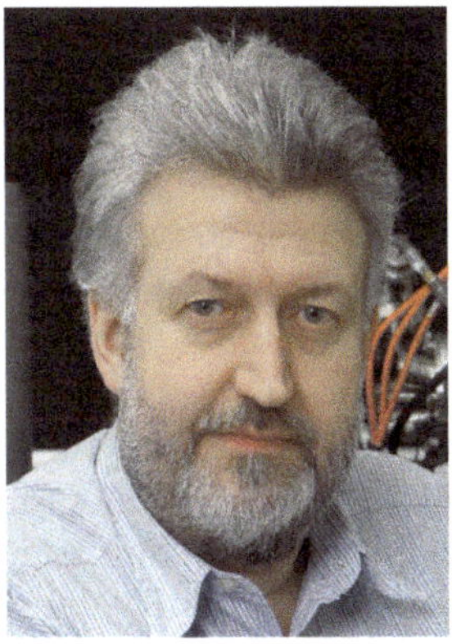

**Uwe BREUER,** Bundesinstitut für Risikobewertung, Germany

(Photo: *Ralf-Uwe Limbach*; Copyright: *Forschungszentrum Jülich, Germany*; Courtesy of: *Forschungszentrum Jülich*)

Uwe Breuer is responsible for Secondary Ion Mass Spectrometry (SIMS) and Atom Probe Tomography (APT) in the ZEA-3 of the Research Centre Jülich, Germany. He studied Physics at the RWTH-Aachen, Germany, where he also completed his PhD with works on Plasma-SNMS in 1992. In 1992 he joined the Research Centre Jülich working with a new installed VG-electronbeam-SNMS along with a CAMECA-4f. In 2003 a ToF-SIMS (ION-TOF SIMS IV) was installed. He is working on a broad variety of material and layered systems. Since November 2013 he is establishing the Atom Probe Tomography (APT) technique for applications in the framework of the Helmholtz Energy Materials Characterization Platform (HEMCP).

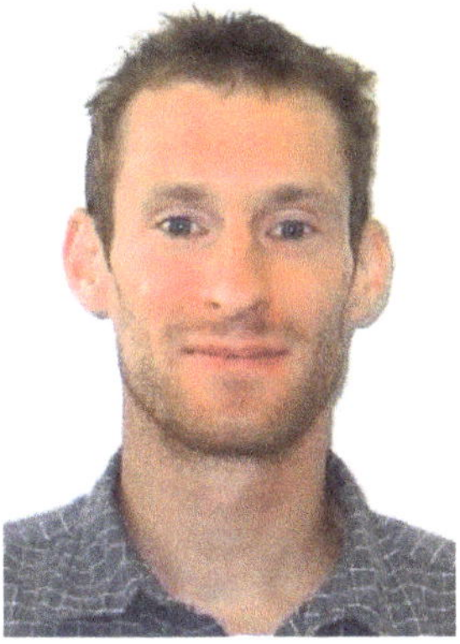

**Stefan BÜRGER,** International Atomic Energy Agency, Safeguards Analytical Services, Department of Safeguards, Vienna, Austria

Stefan Buerger studied physics at Mainz University and received his PhD with magna cum laude in 2005. He worked at the Max-Planck-Institute for Chemistry, Mainz, Germany, at Oak Ridge National Laboratory and at New Brunswick Laboratory, USA before he became a staff scientist at the Office of Safeguards Analytical Services of International Atomic Energy Agency (IAEA). During his relatively short but very fruitful scientific career Stefan made an important contribution to inorganic mass spectrometry and to metrology in analytical chemistry. In particular, his research on nuclear metrology, certification of reference materials, quality assurance and quality control for nuclear forensics and safeguards analysis has been a valuable driving force for the further progress in these fields. Stefan had a great passion for sport, for research and for teaching. He was a member of the German National League and German Champion for long-distance swimming. In his research work and teaching activities he was very attentive and always available for an unbiased opinion, which, practically always, offered a valuable step forward to resolving the issues under discussion. Stefan Bürger passed away in a fatal accident during mountaineering in the Austrian Alps on August 16, 2013. We remember Stefan Buerger as a great scientist, outstanding personality, and an extremely valuable colleague and teacher.

(by Sergei Boulyga, International Atomic Energy Agency, Austria)

**Emmanuel DE CHAMBOST,** formerly CAMECA, France

(Photo: *Michelle de Chambost;* Copyright: *This file is licensed under the Creative Commons Attribution-Share Alike 3.0 Unported license* ***(CC-BY-SA),*** Courtesy of: *n/a*)

Emmanuel de Chambost, born in 1948, studied physics and electrical engineering at the *École Supérieure d'Électricité*, Paris. He was first involved in e-beam lithography instruments and neural network projects in the frame of the Thomson-CSF *Laboratoire Central de Recherches* and joined Cameca in 1989 as *IMS 1270* project leader and was eventually in charge of the R&D department. After his retirement in 2009, he is currently involved in historical investigation and development of social housing.

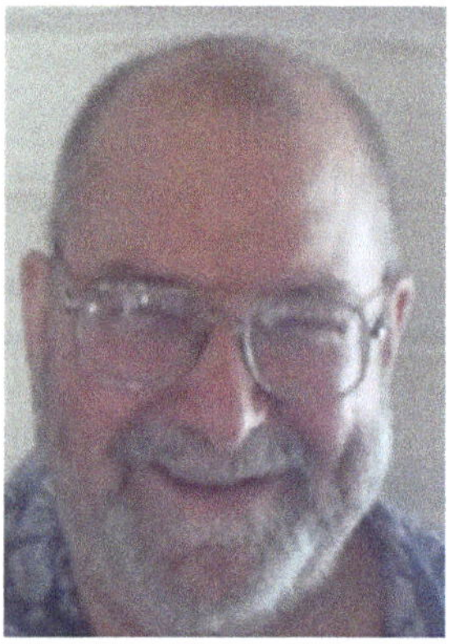

**Charles B. DOUTHITT,** Thermo Fisher Scientific, US

(Photo Courtesy of: *Hurricane Car Wash*)

Charles Douthitt has been selling mass spectrometers designed and manufactured in the Bremen, Germany factory of Thermo Fisher Scientific (formerly Finnigan MAT) since 1986. He studied isotope geochemistry under Professor Sam Epstein at the California Institute of Technology, working on the geochemistry of the stable isotopes of silicon. At Thermo, in addition to working with IRMS, he has been involved in bringing to market a variety of isotope ratio technologies, including multicollector and high-resolution ICP-MS. He has, for many years, been documenting the commercial history and global distribution of all aspects of isotope mass spectrometry.

**Francois FOUREL,** Laboratoire de Géologie de Lyon, CNRS-UMR 5276, Université Claude Bernard Lyon 1, Ecole Normale Supérieure de Lyon, France

(Photo: *Francois Fourel*; Copyright: *Francois Fourel*; Courtesy of: *Francois Fourel*)

Francois Fourel has held a position as research engineer in the "Laboratoire de Géologie de Lyon – Terre, planètes, environnement" directed by Prof. Francis Albarede, since 2010. He obtained his D.E.A. (master's equivalent) in geology (option geochemistry-petrology) in 1985 and his PhD in the field of U–Pb and Sm–Nd analyses using TIMS and ion probe techniques in the "Laboratoire de Géochimie-Cosmochimie de l'Institut de Physique du Globe de Paris" directed by Claude Allègre in 1988. Then he worked as a research associate at the California Institute of Technology (CALTECH), Pasadena, California in the course of his military duty. From 1990 to 1996 he was the manager of the stable isotope department for EUROFINS Scientific in Nantes, France. In 1996 he joined Micromass as a development and application scientist for stable isotope mass spectrometers and was appointed product manager for the stable isotope group of the inorganic division in 1998. After the inorganic products division split up from Micromass to operate as "GV Instruments", he remained in his position there until he resigned in 2004. From 2004 to 2010 he worked in the laboratory PaléoEnvironnements et PaléobioSphere PEPS at the Université Claude Bernard Lyon 1, directed by Pr. Christophe Lécuyer, where he was in charge of the stable isotope laboratory. His research interests are paleoenvironment, phosphate analyses and new isotopic techniques.

**Ondřej HANOUSEK,** University of Natural Resources and Life Sciences Vienna (BOKU), Department of Chemistry, Division of Analytical Chemistry, VIRIS Laboratory for Analytical Ecogeochemistry, Tulln, Austria

(Photo: *Ondřej Hanousek*; Copyright: *Ondřej Hanousek*, Courtesy of: *Ondřej Hanousek*)

Ondřej Hanousek is currently a PhD student at the University of Natural Resources and Life Sciences, Vienna (BOKU) and a member of the VIRIS Laboratory for Analytical Ecogeochemistry in Tulln. He studied Synthesis of Pharmaceuticals (BSc) and Analytical Chemistry (MSc) in the Czech Republic and Germany and graduated from the Institute of Chemical Technology, Prague in 2012. At the BOKU, his research focuses on analysis of stable isotopes of sulfur.

**Monika HORSKY,** University of Natural Resources and Life Sciences Vienna (BOKU), Department of Chemistry, Division of Analytical Chemistry, VIRIS Laboratory for Analytical Ecogeochemistry, Tulln, Austria

(Photo: *Gerlinde Grabmann*; Copyright: *Monika Horsky*)

Monika Horsky is currently a PhD student at the University of Natural Resources and Life Sciences, Vienna and works at the VIRIS lab in Tulln, Austria. In 2010 she graduated in chemistry from the University of Vienna, where she had specialised in analytical, inorganic and physical chemistry. Her PhD thesis deals with strontium isotope ratio analysis of wood using MC ICP-MS and provenance determination of archaeological findings.

**Trevor IRELAND,** The Australian National University, Australia

(Photo Courtesy of: *Trevor Ireland, ANU*)

Professor Trevor Ireland is head of the SHRIMP group at The Australian National University, Canberra. He studied Geology and Physics in New Zealand before starting his PhD studies in Canberra on the newly completed SHRIMP I ion microprobe at ANU. After completing his PhD studies he worked at Washington University with a Cameca IMS 3f, worked on the newly installed Cameca 1270 in UCLA and installed the SHRIMP RG at Stanford University before returning to Canberra in 2000. He has interests in geochemistry and cosmochemistry as well as instrument development.

**Johanna IRRGEHER,** University of Natural Resources and Life Sciences Vienna (BOKU), Department of Chemistry, Division of Analytical Chemistry, VIRIS Laboratory for Analytical Ecogeochemistry, Tulln, Austria

(Photo: *Andreas Zitek* Copyright: *Johanna Irrgeher*)

Johanna Irrgeher is currently a postdoctoral researcher in the field of analytical chemistry at the VIRIS Laboratory for analytical ecogeochemistry at the University of Natural Resources and Life Sciences Vienna (BOKU). She holds a master degree in biotechnology from BOKU Vienna and obtained her PhD granted by the Austrian Academy of Sciences in 2013 with honours for her work on stable strontium isotope ratio analysis by (LA)-MC ICP-MS. During her PhD she was a visiting researcher at the Institute of Isotopes in Budapest (Hungary), the National Cheng Kung University in Tainan (Taiwan) and the National Research Council Canada in Ottawa (Canada). Since 2014 she has been an associate member of the IUPAC committee on isotope abundances and atomic weights as well as the IUPAC subcommittee on isotope abundance measurements. Her current research focuses on analytical method development for elemental and isotopic analysis in the field of analytical ecogeochemistry dealing with both aquatic and terrestrial ecosystems.

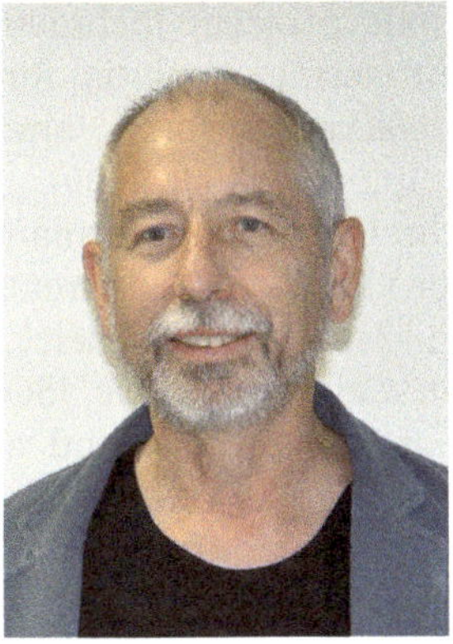

**Norbert JAKUBOWSKI**, BAM - Federal Institute for Materials Research and Testing, Berlin, Germany

(Photo: *Silke Richter*; Copyright: *BAM*; Courtesy of: *BAM*)

Norbert Jakubowski is presently head of the division Inorganic Analysis at the BAM – Federal Institute for Materials Research and Testing, Berlin, Germany. He studied plasma physics at the University of Essen/Duisburg and received his PhD at the University of Stuttgart Hohenheim with summa cum laude in 1991. He started his scientific career at the Institute for Analytical Sciences in Dortmund. In 2013 he received the "European Award for Plasma Spectrochemistry" for his contribution in elemental mass spectrometry. His research interests are related to Analytical Chemistry in general with special interest in development of instruments, methods and problem-orientated procedures based on the use of plasma sources (inductively coupled plasma, glow discharge) for elemental mass spectrometry of solid and liquid samples. Recently, he started a new research direction towards labelling of biomolecules with metals by use of chelating compounds for detection of biomarkers in clinical research. "*Single cells are the most fascinating machineries on earth and it is my challenge for the rest of my career to make life processes visible*"

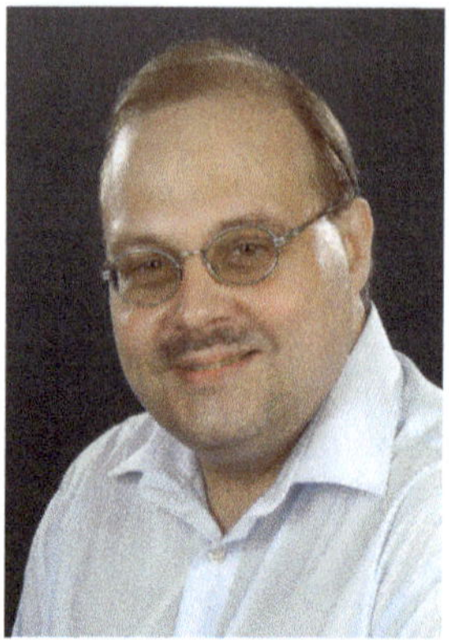

**Harald JUNGNICKEL,** Federal Institute for Risk Assessment, Department of Product Safety, Berlin, Germany

(Photo Copyright: *H. Jungnickel*; Courtesy of: *H. Jungnickel*)

Harald Jungnickel (PhD. Analytical Chemistry and Chemical Biology) was an honorary lecturer for biochemistry at UMIST (now University of Manchester, UK). His research at the University of Manchester focused on the development of ToF-SIMS imaging mass spectrometry for biological samples with focus on metabolomics studies. Additionally, he has worked in the pharmaceutical industry (Research & Development). Previous projects included analytical chemistry, mass-spectrometer development (surface analysis) for biological samples, high-throughput analysis, biomarker research, and biostatistics for mass spectrometry. He is currently employed (since 2009) at the Federal Institute for Risk Assessment (BfR in Berlin Germany, a German governmental organisation) in the department of product safety. Current projects are related to biomarker development, imaging mass spectrometry, metabolomics and nanomaterials. He is a German delegate of the ISO Committee TC201/WG 4 Surface Characterization of Biomaterials and a German delegate of the ISO Committee TC201/SG 1 Nano Materials Characterization. He is a member of the German National Committee for Normation and a Delegate of the DIN National German Committee for Material Testing (NMP) NA 062-08-16 AA "Chemical surface analysis and raster electron microscopy".

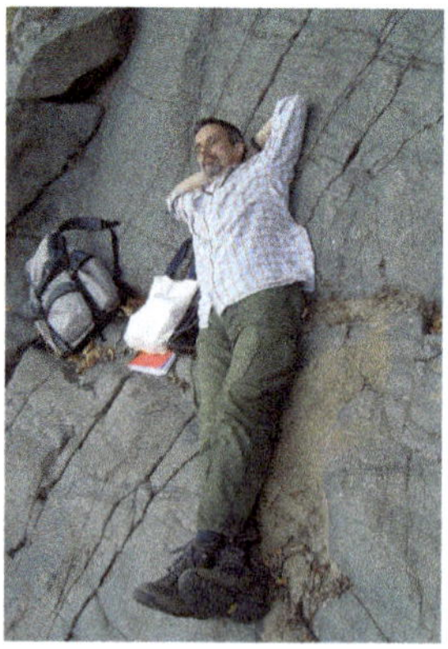

**Urs KLÖTZLI,** University of Vienna, Department of Lithospheric Research, Vienna, Austria

(Photo: *Urs Klötzli*)

Urs Klötzli is professor for geology at the University of Vienna (Austria). He studied Earth Sciences at the University of Berne (Switzerland) and received his PhD in 1991 on the application of negative thermal ionisation mass spectrometry of boron to Earth materials. He then moved to Vienna where he is working with a number of isotope systems of interest to Earth Sciences. His main working fields are the application of the U-Th/Pb, Rb/Sr and Sm/Nd geochronometers to geological problems using IDMS, MC ICP-MS and SIMS with a special focus on analytical improvements of microanalytical techniques.

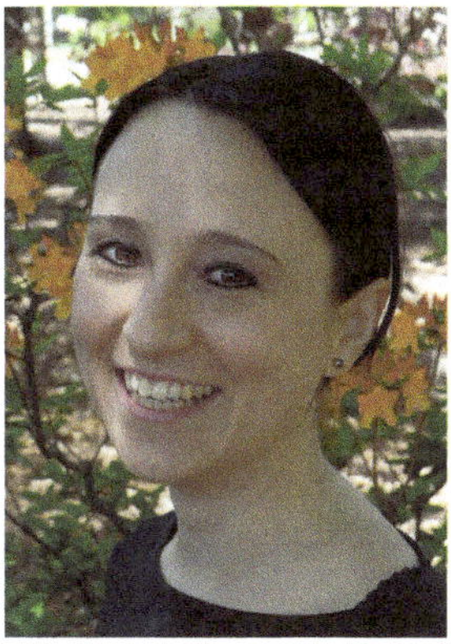

**Stefanie KONEGGER-KAPPEL**, Clemson University, Department of Chemistry, Clemson, South Carolina, USA

(Photo: *Thomas Konegger*; Copyright: *Stefanie Konegger-Kappel*; Courtesy of: *Stefanie Konegger-Kappel*)

Stefanie Konegger-Kappel is a postdoctoral researcher at Clemson University, SC, USA, in the research group of Dr. R. Kenneth Marcus. She graduated from Vienna University of Technology in Technical Chemistry, focusing on Materials Technology and Materials Analytics, in 2009, and in 2012 she obtained a PhD from the University of Natural Resources and Life Sciences (BOKU), Vienna. From 2009 to 2013, she was a consultant for the International Atomic Energy Agency (IAEA), dedicating her research at the IAEA laboratories in Seibersdorf and the VIRIS laboratory (BOKU) mainly to ICP mass spectrometric method development for actinide analysis of single particles. During her time as a PhD student, she was also a visiting researcher at the Swiss Federal Institute of Technology Zurich (ETH Zürich). Her current research activities are directed towards the development of miniaturised, field-deployable plasma sources for environmental and nuclear mishap screenings.

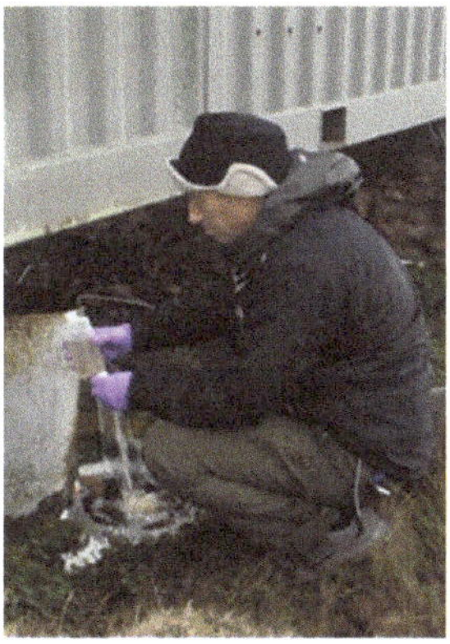

**Mark LAVELLE**, Imperial College, London, UK

(Photo: *Mark Lavelle*)

Mark Lavelle studied geology and geochemistry at Imperial College, the University of Cape Town and the University of Cambridge. He spent seven seasons in Antarctica and Cambridge with the British Antarctic Survey, working on a series of multinational projects exploring the earliest history of regional and continent-wide glaciation. As a Fulbright Fellow at the Oak Ridge National Laboratory, he established a programme applying standard geochemical tools to novel biological problems. He is now a Senior Research Fellow at the Institute for Security Science & Technology, Imperial College.

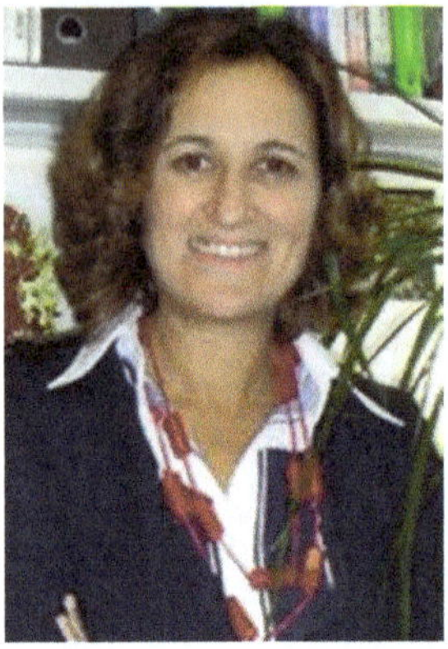

**Cristina MAGUAS,** University of Lisbon, Portugal

(Photo: *Cristina Maria Filipe Máguas da Silva Hanson*; Copyright: *Cristina Maria Filipe Máguas da Silva Hanson*; Courtesy of: *Cristina Maria Filipe Máguas da Silva Hanson*)

Cristina Máguas graduated in Biology (1987) and holds a PhD in Biology – Ecology and Systematics from the University of Lisbon (1997). She is currently an Auxiliary Professor at the Faculty of Sciences, University of Lisbon, where she is responsible for two masters courses: Agronomy and Forests, and Plant Ecology, as well as a Plant Ecophysiology Master course at the University of S. Paulo (Brazil). At the moment she is member of the Executive Committee of Centre for Environmental Biology, Group Coordinator of the Environment Functional Ecology there, and Coordinator of the Stable Isotopes Facility Lab; Coordinator of the Portuguese Committee/WP1 leader of Stable Isotope Biosphere Atmosphere Exchange COST Action, Member of the Executive Committee of Portuguese Ecological Society, Vice-President of the European Ecological Federation and Member of the Expertise Group – Stable Light Isotopes (C, N, O, S, H) by IRMS for Food Traceability of the IAEA. The focuses of her research were ecophysiological and ecological approaches in: Mediterranean Ecosystems, Invasive Alien Plants, Carbon and Water Fluxes and Cryptogrammic Vegetation. Particularly important is the application of stable isotope mass spectrometry to ecosystem and plant functioning as well as geographic origin of plant products. She was the coordinator of more than 18 national and international projects and author or coauthor of 60 international publications and 14 book chapters.

**Rodrigo MAIA,** University of Lisbon, Portugal

(Photo: *Rodrigo Maia*; Copyright: *Rodrigo Maia*)

Rodrigo Maia is head technician at SIIAF – Stable Isotopes and Instrumental Analysis Facility – at the Centro de Biologia Ambiental of the Faculdade de Ciências da Universidade de Lisboa, Portugal, since 1999. He is a Biologist but has dedicated most of his career to Stable Isotopes, especially in applications related to ecology, ecophysiology and environmental sciences. He is especially interested in introducing and developing the application of stable isotope analysis in ecological research, namely assisting in implementing sample preparation methods, training students and helping research activities.

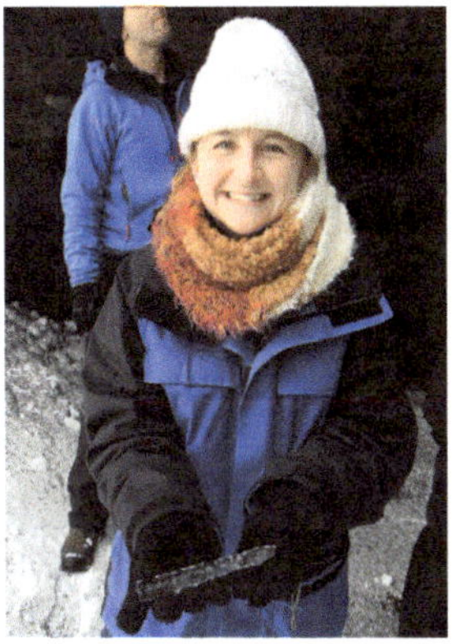

**Laurie NUNES**, Curtin University, Australia
(Photo: *Laurie NUNES*)

Laurie Nunes studied Applied Physics at Curtin University in Perth, Australia and received her PhD with Chancellors' Commendation in 2010. She became a Post-Doctoral Research Fellow in Curtin's Department of Imaging and Applied Physics the same year, running two research laboratories, a TIMS instrumental facility and the Advanced ultraClean Environment facility, the latter of which she also helped set up. Additionally, during her three years in this position she continued the pioneering ice core research of Professor Kevin Rosman, initiated her own research into environmental uranium characterisation and worked with the Chief Scientist of Western Australia to promote greater participation by young women in science. She is currently working as a Petro/Rock Physicist with an oil and gas consulting company in Perth, and maintains a position as Adjunct Research Fellow with Curtin's Applied Physics department, continuing to contribute to TIMS-related research.

**Jorge PISONERO,** University of Oviedo, Department of Physics, Spain
(Photo Courtesy of: *Jorge Pisonero*)

Dr. Jorge Pisonero is a Senior Lecturer in the Department of Physics (University of Oviedo, Spain) and deputy leader of the Laser and Plasma Spectroscopy Research Group (www.unioviedo.es/gelp). He was awarded a Postdoctoral Marie Curie Intra-European Fellowship to work at ETH (Switzerland), and afterwards an excellency "Ramón y Cajal" Research Contract at the University of Oviedo. His current research interests focus on fundamentals, instrumental developments and applications of laser- and glow discharge-based spectroscopies. He has published over 50 articles in high-impact journals, with 700+ citations (h-index = 17). Major distinctions include the International Masao Horiba Award (Kyoto, 2009) and the Spanish Royal Society of Physics Prize for Best Junior Experimental Researcher in Physics (Madrid, 2011).

**Thomas PROHASKA,** University of Natural Resources and Life Sciences Vienna (BOKU), Department of Chemistry, Division of Analytical Chemistry, VIRIS Laboratory for Analytical Eco-geochemistry, Tulln, Austria

(Photo: *Sportograf*; Copyright: *Thomas Prohaska*; Courtesy of: *Thomas Prohaska*)

Thomas Prohaska is professor for analytical chemistry at the BOKU (University of Natural Resources and Life Sciences Vienna, Tulln, Austria). He studied chemistry at the Vienna University of Technology and received his PhD with summa cum laude in 1995. He became a scientific researcher at the BOKU in the same year to build up a laboratory for elemental trace analysis. From 1998 to 2000 he was researcher at the EC joint research center IRMM in Geel, Belgium. He returned to the BOKU with a new focus on isotope research and became associate professor in 2002. In 2004, he received the START award for the setup of a new research laboratory (VIRIS). "*I have dedicated my research to isotope science and metrology keeping the momentum of curiosity driven science*."

**Carla RODRIGUES,** Diverge Grupo Nabeiro Innovation Centre, R&D Projects, Portugal

(Photo: *Carla Rodrigues*; Copyright: *Carla Rodrigues*; Courtesy of: *Carla Rodrigues*)

Carla Rodrigues graduated in Applied Plant Biology in 1996 at the Faculty of Science of the University of Lisbon (FCUL). In the last year of her degree studies, she worked at the Laboratory of Microbiology, Gulbenkian Science Institute, Oeiras Portugal, on the production of bioalcohol by sugar fermentation using yeasts. In 1999, she attended postgraduation studies in Food Technology and Quality at the Faculty of Science and Technology of the New University of Lisbon. In the same year, she participated in a project at the Laboratory of Microbiology, Agronomy Institute (Technical University, Lisbon), developing a microbiological culture medium for detection of fungi causing browning of PDO cheeses that resulted in a patent on behalf of the Institute. In 2001, she began to work as Laboratory Supervisor in the Stable Isotope Laboratory (SIL) of the Institute of Applied Science and Technology (ICAT) in FCUL. In 2007, she began her doctoral studies in Chemistry, specialty of Analytical Chemistry, under the theme *Geographical Origin Discrimination of the Green Coffee Bean and Qualification of Roasting Analytical Profiles*, concluded with honours and distinction in June 2011. She is now a researcher from the Grupo Nabeiro Innovation

Center and a member of the Innovation + R&D team and is involved in different innovation projects as a researcher and product quality advisor.

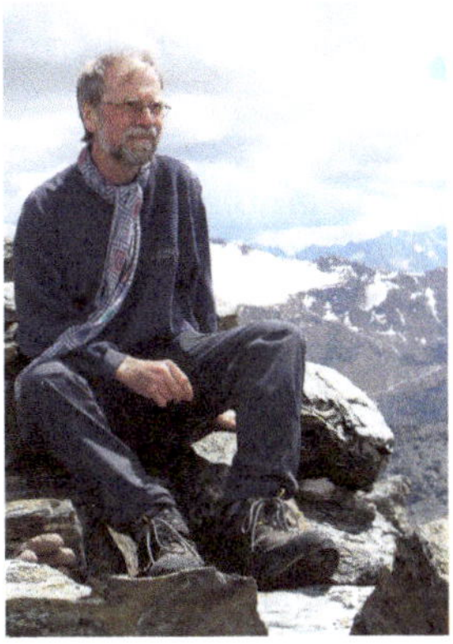

**Peter ROOS,** Ruhr University Bochum, Department of System Biochemistry, Medical Faculty, Germany

(Photo: *Fabian Roos*; Copyright: *Peter H. Roos*)

The scientific "career" of Peter Roos started in early childhood by doing chemical experiments and by zoological studies – both interests triggering his later professional focus on biochemistry. He studied biology at the Ruhr-University in Bochum and received his PhD in 1982 with an investigation on the chloroplast ATP synthase. After a switch from plants to animals and humans, his research at several German universities concentrated on xenobiotic metabolism and on the effects of xenobiotics on the interwoven signal transduction pathways in cells particularly concerning carcinogenesis processes. During the last 12 years he has acted as project leader of the Molecular Toxicology group at the Leibniz Research Centre for Working Environment and Human Factors at the Technical University Dortmund. Investigations were always driven by the notion to recognise and analyse biochemical processes in their complexity. Consequently, there was always a concomitant interest in developing methods for comprehensive analyses of the studied complex processes using chromatography, electrophoresis, UV-VIS-spectrometry and finally LA-ICP-MS. Besides, the aspect of complexity touches his further interests concerning other biological spatial-temporary networks such as phylogenetics, evolution and the ecology of tropical forests. *"As I will retire when writing this CV, I will nevertheless let me inspire and fascinate by the diversity, imagination and beauty of life and nature. So I will keep up to maintain my curiosity and look forward to new insights and developments."*

**Lothar ROTTMANN,** Thermo Fisher Scientific, Bremen, Germany

(Photo Copyright: *Lothar Rottmann*)

Lothar Rottmann is the head of R&D for the Trace Elemental Analysis (TEA) products (ICP-MS, ICP-OES, AAS, GDMS) at Thermo Fisher Scientific in Bremen, Germany. He obtained his PhD in Analytical Chemistry in 1994 at the University of Regensburg under the supervision of Prof. Klaus Heumann. Afterwards he joined the R&D department of Thermo Fisher Scientific where he started as a research scientist. Since 1995 he was as project manager responsible for the development of several high-resolution sector field instruments (*Element2*, *Element XR*, and *Element GD*), the quadrupole-based ICP-MS *iCAP Q* and for some years also for the software of several instruments. From 2006 to 2011 he led the inorganic technical support team in the factory in Bremen. Since 2012 he has had several product management roles within the TEA business of Thermo Fisher Scientific, before he joined the R&D department in his current role in 2014. *"The mission of Thermo Fisher Scientific is to enable our customers to make the world healthier, cleaner and safer. When I look at all the fascinating applications of our global customers, especially done with our sector field instruments, then I am convinced that in sum they make the world healthier, cleaner and safer."*

**Laure SANGELY,** International Atomic Energy Agency, Austria

(Photo: *Mathieu Rosenzweig;* Copyright: *Laure Sangely*)

Laure Sangély graduated in geological engineering at the Ecole Nationale Supérieure de Géologie (Nancy, France) in 2000. In the scope of her PhD thesis under the supervision of Dr. Marc Chaussidon (2004), she investigated the microanalysis of fossil organic matter for stable carbon isotopes using a CAMECA *IMS 1270* at the Centre de Recherches Pétrographiques et Géochimiques (Nancy, France). In 2005, she joined the Direction of Military Applications at the Commissariat à l'Energie Atomique et aux Energies Renouvelables (Bruyères-le-Châtel, France), where she was in charge of CAMECA *IMS 7f* applications in the field of nuclear nonproliferation. Since 2009, she has been with the Safeguards Analytical Laboratory at the International Atomic Energy Agency (Vienna, Austria). She is currently in charge of the CAMECA *IMS 1280* operation for Nuclear Safeguards. What she thinks is the most enthusiastic fact about SIMS is the outstanding variety of instrumental configurations and potential applications, which opens the door to still explore the field method development and optimisation.

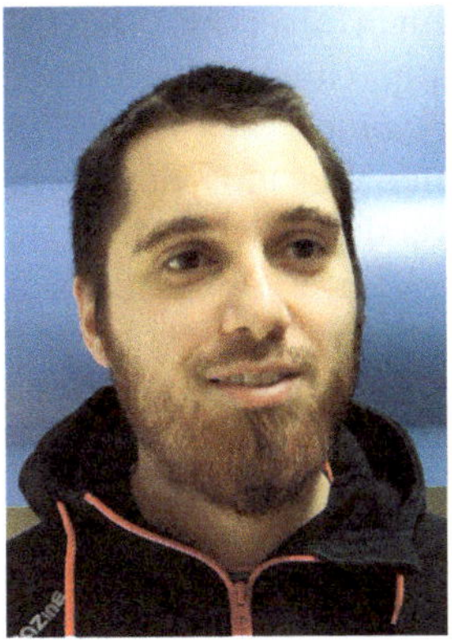

**Jakob SANTNER,** University of Natural Resources and Life Sciences, Vienna, Institute of Soil Research, Tulln, Austria

(Photo: *Andreas Kreuzeder*; Copyright: *Jakob Santner*)

Jakob Santner is a postdoctoral research associate at the Institute of Soil Research of the University of Natural Resources and Life Sciences, Vienna (Austria). He did his PhD on plant–soil interactions in plant root P acquisition, including the development and application of chemical imaging methods for solute mapping in plant rhizospheres. In 2009 and 2010 he worked at Lancaster University (UK) and at the Katholieke Universiteit Leuven (Belgium). His current research is focused on the development and application of passive solute sampling techniques for bulk soil analysis as well as chemical imaging applications for investigating biological hotspot in soils.

**Nathalie VALLE,** Centre de Recherche Public - Gabriel Lippmann, Luxembourg

(Photo Copyright: *Nathalie Valle*)

Nathalie Valle obtained her PhD degree in 2000 in the field of material science from the Institut Polytechnique de Lorraine-Nancy (France). There, she studied the alteration of nuclear waste glasses by using isotopic tracing and the SIMS technique under Etienne Deloule's supervision (research director at the Centre de Recherches Pétrographiques et Géochimiques, Nancy, France). Afterwards, she joined the Science and Analysis of Materials department of the Centre de Recherche Public-Gabriel Lippmann (Luxembourg) under Henri-Noël Migeon's guidance where she started as a researcher specialised on SIMS analysis. Since 2007, she has been a project leader, responsible for research projects involving the characterisation of inorganic materials mainly by SIMS instruments with optimised capabilities (Sc-Ultra and NanoSIMS).

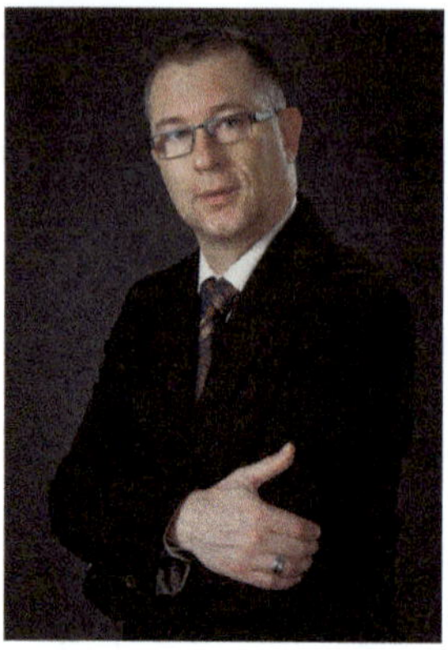

**Frank VANHAECKE**, Ghent University, Department of Analytical Chemistry, Ghent, Belgium

(Photo Copyright: *Frank Vanhaecke*)

Frank Vanhaecke obtained his PhD degree in 1992 from Ghent University (Belgium). He continued carrying out scientific research as a postdoctoral fellow at the same university and also enjoyed a postdoctoral stay at the Johannes Gutenberg University of Mainz (Germany). Currently, he is Senior Full Professor in Analytical Chemistry at Ghent University and head of the "Atomic & Mass Spectrometry A&MS" research group (www.analchem.ugent.be/A&MS), which is specialised in the determination, speciation and isotopic analysis of (trace) elements via ICP-mass spectrometry (ICP-MS). Both fundamentally oriented aspects of the technique and the development of methods for solving challenging scientific problems in an interdisciplinary context are paid attention to. He is (co-)author of >250 journal papers, >10 book chapters and has edited a book (together with Patrick Degryse), entitled "Isotopic analysis – fundamentals and applications using ICP-MS". In 2011, he received a European Award for Plasma Spectrochemistry for his contributions to this research field and he was nominated Fellow of the Society for Applied Spectroscopy – SAS in 2013. "*For me, research is a life-long passion and high-precision isotopic analysis via multicollector ICP-MS a fascinating tool to gain novel insights into a large variety of fields*".

**Cornel VENZAGO,** AQura GmbH, Germany

(Photo: *Christa Venzago*; Copyright: *Christa& Cornel Venzago*)

Cornel Venzago (Director of Inorganic Analytics at AQura GmbH) graduated from an engineering school in Winterthur, Switzerland with a degree in Chemistry in 1986. The same year he joined a producer of high-purity metals for the semiconductor industry, where he was responsible for process and product development as well as for a solid-state mass spectrometry lab equipped with an *SSMS AEI 702R*. In 1989 he joined the Institute of Nuclear Physics at the University of Frankfurt, Germany and worked in a joint industry/university project on the development for spark source mass spectrometry for ultratrace analyses in GaAs on a *CEC SSMS*. In 1991 he joined Degussa AG in Germany to start up a GDMS lab with a *VG 9000*. He is currently director of inorganic analytics at AQura GmbH covering the techniques GDMS, ICPMS, ICPOES, AAS, XRF, and many other techniques. The company AQura is a subsidiary of Evonik Industries (ex Degussa). *"There is always something to improve."*

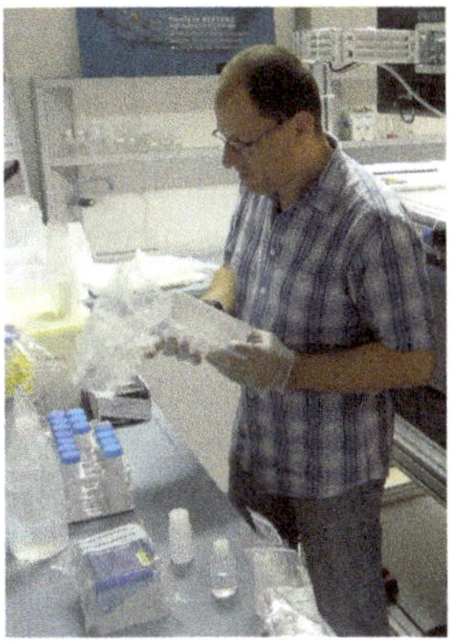

**Jochen VOGL,** BAM Federal Institute for Materials Research and Testing, Berlin, Germany

(Photo: *Olaf Rienitz*; Courtesy of: *Olaf Rienitz*)

Jochen Vogl studied chemistry at the Universities of Regensburg and Mainz in Germany. He finished his diploma thesis on iodine species in 1994 and his doctoral thesis on heavy-metal-humic-acid species in 1997. Since 1993 he worked in the group of Klaus G. Heumann on speciation analysis with online IDMS. Thereafter, he worked for two years on reference measurements by IDMS at the IRMM in Geel, Belgium. Since January 2000 he has been in charge of IDMS and isotope analysis at BAM Federal Institute for Materials Research and Testing, Berlin, Germany. His main working fields are IDMS, isotope analysis, certification of isotope reference materials and "elemental" mass spectrometry in general.

**Michael WIEDENBECK,** Helmholtz-Zentrum Potsdam Deutsches GeoForschungsZentrum GFZ, Potsdam, Germany

(Photo: *Martin Rosner*)

Michael Wiedenbeck has a BSc in geology from the University of Michigan, a diploma in mineralogy from the ETH-Zurich and a PhD in isotope geochemistry from the Australian National University. He held postdoc positions related to SIMS in Nancy, France and Ahmedabad, India and Oak Ridge, USA before becoming manager of the SIMS facility in Albuquerque, USA. Since 1998 he has been head of the SIMS user facility in Potsdam Germany. Michael's work over the past 15 years has involved the widest possible exposure to "novel" analytical needs within the geosciences. One of his key interests is metrology and the characterisation of better reference materials to improve the calibration of microanalytical methods.

**Andreas ZITEK,** University of Natural Resources and Life Sciences Vienna (BOKU), Department of Chemistry, Division of Analytical Chemistry, VIRIS Laboratory for Analytical Ecogeochemistry, Tulln, Austria

(Photo: *Johanna Irrgeher*; Copyright: *Andreas Zitek*; Courtesy of: *Andreas Zitek*)

Andreas Zitek is currently a postdoctoral researcher at the University of Natural Resources and Life Sciences, Vienna (BOKU) and a member of the VIRIS Laboratory for Analytical Ecogeochemistry in Tulln. He received his diploma degree in Landscape Architecture and Landscape Management in 1999 and his PhD degree in 2006, both at the Institute of Hydrobiology and Aquatic Ecosystem Management (IHG) at the University of Natural Resources and Life Sciences Vienna (BOKU) (AT). In 2008 he completed an additional MSc in Geographical Information Science and Systems at the Paris Lodron University of Salzburg (UNIGIS). His current research activities aim at the application of elemental and isotopic analyses to basic questions in the field of aquatic ecology ("aquatic ecogeochemistry"), with a special focus on the chemical analysis of hard parts of freshwater fish. Another focus of his work is the application of GIS methods for chemical imaging from μm (small compartment) to km (landscape, isoscape) scale. *"Elemental and isotopic analysis is able to provide the most basic information about the interrelation between the living and nonliving world, which is why I predict the transdisciplinary research direction of analytical ecogeochemistry a great future!"*

# List of Abbreviations

| | |
|---|---|
| 2D | two-dimensional |
| 3D | three-dimensional |
| *A* | abberation |
| *a* | ionization efficiency |
| AC | alternating current |
| $A_C$ | Achromatic point |
| AFRIMETS | Intra Africa Metrology System |
| AMS | accelerator mass spectrometer |
| *APE* | atom percent excess |
| APM | automated particle monitoring |
| APMP | Asia Pacific Metrology Programme |
| ARL | Applied Research Laboratories |
| *B* | magnetic field strength |
| *b* | beam width at the collector plane |
| B-scan | magnetic scan |
| BABI | basaltic achondrite best initial |
| BAM | The Federal Institute for Materials Research and Testing, Germany |
| BCE | Before the Common era (time reference) |
| BCR | Bureau Communautaire de Références |
| BEC | background equivalent concentration |
| BIPM | International Bureau of Weights and Measures |
| *c* | concentration |
| CBNM | Central Bureau for Nuclear Measurements |
| cc STP | cubic centimeter at standard temperature and pressure |
| CCD | charge coupled device |
| CCQM | comité consultatif pour la quantité de matière |
| CE | capillary zone electrophoresis |

New Developments in Mass Spectrometry No. 3
Sector Field Mass Spectrometry for Elemental and Isotopic Analysis
Edited by Thomas Prohaska, Johanna Irrgeher, Andreas Zitek and Norbert Jakubowski

Published by the Royal Society of Chemistry, www.rsc.org

| | |
|---|---|
| CE | Common Era (time reference) |
| CETAMA | Commission for the Establishment of Analytical Methods |
| CF | continuous flow |
| CHUR | Chondritic uniform reservoir |
| CIPM | International Committee for Weights and Measures |
| CMP | chemical measurement process |
| ConFlo | continuous flow |
| COOMET | Euro-Asian Cooperation of National Metrological Institutions |
| CRM | certified reference material |
| *CSF* | chemical sensitivity factor |
| CVG | chemical vapour generation |
| CYP | cytochromes P450 |
| $D$ | number of radiogenic (daughter) atoms (mole) |
| $D(m/z)$ | mass to charge difference |
| $D^*$ | number of daughter nuclei formed during radioactive decay |
| $D_0$ | initial number of radiogenic (daughter) atoms (mole) |
| DAC | digital to analog converter |
| DC | direct current |
| DC GD | direct current glow discharge |
| DC GDSFMS | direct current glow discharge sector field mass spectrometry |
| DCP | direct current plasma |
| $D_E$ | energy dispersion |
| DI | dual inlet |
| DI-mode | dual inlet mode |
| DIC | dissolved inorganic compound |
| DIHEN | direct injection high efficiency nebulizer |
| $D_M$ | mass dispersion |
| DOC | dissolved organic compound |
| DOTA | 1,4,7,10-tetraazacyclododecane-1,4,7,10-tetraacetic acid |
| DRC | dynamic reaction cell |
| DS TIMS | double spike TIMS |
| DU | depleted uranium |
| $e$ | elemental charge |
| $E$ | electric field strength |
| E-scan | electric scan |
| EA | elemental analyzer |
| EC | European Community |
| EC | European Commission |
| ${}^{e}D_E$ | energy dispersion in an electric field |
| $E_{EA}$ | electron affinity |
| $E_i$ | ionisation potential |
| EM | electron multiplier |
| EMPDA | Educational Materials Production and Distribution Agency |

| | |
|---|---|
| EPMA | electron probe micro analysis |
| EQS | environmental quality standards |
| ERM | European Reference Materials |
| ESA | electrostatic analyzer |
| ESI | Electro Spray Ionization |
| ESI-FTICR-MS | electrospray ionization Fourier transform ion cyclotron |
| ETV | electrothermal vaporization |
| EURAMET | European Association of National Metrology Institutes |
| EW-GDS | European Working Group on Glow Discharge Spectroscopy |
| EXLIE-SIMS | Extreme Low Impact Energy - SIMS |
| $f$ | electron work function of the surface material for electron emission |
| FAB | fast atom bombardement |
| FC | Faraday cup |
| FFF | Field Flow Fractionation |
| FPC | focal plane camera |
| FT | Fourier transformation |
| *FWHM* | full width at half maximum |
| $g_+$ | statistical weights for the ionic state |
| $g_0$ | statistical weights for the atomic state |
| GC | gas chromatography |
| GC-IRMS | gas chromatopgraphy isotope ratio mass spectrometry |
| GD | glow discharge |
| GDMS | glow discharge mass spectrometry |
| GDOES | glow discharge optical emission spectroscopy |
| GDSFMS | glow discharge sector field mass spectrometry |
| GDTOFMS | glow discharge time of flight mass spectrometer |
| GE | guard electrode |
| HPA | high pressure asher |
| HPIC | high performance ion chromatography |
| HPLC | high performance liquid chromatography |
| HR | high resolution |
| HR-RBS | high-resolution Rutherford Backscattering |
| IAEA | International Atomic Energy Agency |
| IBR | ion beam ratio |
| ICP | inductively coupled plasma |
| ICP-CC-MS | inductively coupled plasma collision cell mass spectrometry |
| ICP-DRC-QMS | ICP dynamic reaction cell quadrupole mass spectrometry |
| ICP-MS | inductively coupled plasma mass spectrometry |
| ICP-QMS | ICP quadrupole mass spectrometer |
| ICP-SFMS | inductively coupled plasma sector field mass spectrometry |
| ID | isotope dilution |
| ID ICP-MS | isotope dilution ICP-MS |
| ID TIMS | isotope dilution thermal ionisation mass spectrometry |
| IDMS | isotope dilution mass spectrometry |
| ILAC | International Laboratory Accreditation Cooperation |

| | |
|---|---|
| IMF | instrumental mass fractionation |
| IMS | imaging mass spectrometry |
| IMS | ion mobility separation |
| INAA | instrumental neutron activation analysis |
| irm | isotope ratio monitoring |
| IRMM | Institute for Reference Measurements and Materials |
| IRMS | isotope ratio mass spectrometry |
| ISO | International Organization for Standardization |
| IT | identical treatment |
| ITV | International Target Value |
| IUPAC | International Union of Pure and Applied Chemistry |
| IUPAP | International Union of Pure and Applied Physics |
| JRC | Joint Research Centre |
| $k_B$ | Boltzmann constant |
| $l$ | decay constant (years$^{-1}$) |
| $l''_e$ | image distance in the electric field |
| $l'_m$ | object distance in the magnetic field |
| LA | laser ablation |
| LA-ICP-MS | laser ablation ICP-MS |
| LA-ICP-SFMS | laser ablation ICP-SFMS |
| Lab-Ref gas | laboratory reference gas |
| LC | liquid chromatography |
| LDR | ligase detection reaction |
| LGM | Last Glacial Maximum |
| lhrb | low and high resolution board |
| LI | laser induced |
| LI | laser ionization |
| $l_i'$ | object distance |
| $l_i''$ | image distance |
| LIMS | laser ionization mass spectrometry |
| LMD | laser micro dissection |
| *LOD* | limit of detection/detection limit |
| LR | low resolution |
| $m$ | mass |
| $M$ | magnification |
| MALDI | matrix-assisted laser desorption ionization |
| MC | multi collector |
| MC ICP-MS | multicollector ICP-MS |
| MCP | micro channel plate |
| $^mD_E$ | energy dispersion in a magnetic field |
| MDF | mass dependent fractionation |
| MeCAT | metal-coded affinity tag |
| MEIS | Medium-Energy Ion Scattering |
| MIF | mass independent fractionation |
| MIP | microwave induced plasma |
| MORB | mid ocean ridge basalt |

| | |
|---|---|
| MR | medium resolution |
| MRA | Mutual Recognition Agreement |
| MRP | mass resolving power |
| MS | mass spectrometry |
| MS | mass spectrometer |
| MSI | mass spectrometric imaging |
| MTE | modified total evaporation |
| $N$ | number of atoms |
| $N$ | number of radioactive (mother) atoms (mole) |
| N-TIMS | negative TIMS |
| $n_+$ | flux of ions |
| $n_0$ | flux of atoms |
| $N_0$ | initial number of radioactive (mother) atoms (mole) |
| NAA | neutron activation analysis |
| NASA | National Aeronautics and Space Administration |
| NBL | New Brunswick Laboratory |
| NBS | National Bureau of Standards |
| NEG | normal electron gun |
| NG | noble gas |
| NIST | National Institute for Standards and Technology |
| NRC | National Research Council of Canada |
| NRL | Naval Research Lab |
| OES | optical emission spectrometry |
| OIML | International Organization of Legal Metrology |
| OLS | ordinary least squares |
| $P$ | pressure (GPa) |
| P-TIMS | positive TIMS |
| $P'$ | object point |
| $P''$ | image point |
| PBMF | primary beam magnet filter |
| PCB | printed circuit board |
| PFA | Perfluor-alkoxy-Polymer |
| PHA | pulsed height analyser |
| PL | photoluminescense |
| ppm | parts per million |
| (PP)-TOFMS | Pulsed Profile time of flight mass spectrometry |
| PSE | periodic system of the elements |
| PTFE | Polytetrafluorethylen |
| PVG | photochemical vapour generation |
| $q$ | charge of an ion |
| QMS | quadrupole mass spectrometer |
| qPCR | quantitative polymerase chain reaction |
| $R$ | mass resolution |
| R&D | research and development |
| RAE | resistive anode encoder |
| RBS | Rutherford backscattering spectroscopy |

| | |
|---|---|
| $r_e$ | electric sector radius |
| REE | rare earth elements |
| REIMEP | Regular European Interlaboratory Measurement Evaluation Programme |
| RF | radio frequency |
| RF GD | radio frequency glow discharge |
| RF GDTOFMS | radio frequency glow discharge time of flight mass spectrometer |
| RIMS | resonance ionization mass spectrometry |
| $r_m$ | radius of a magnet/of an ion path through the magnet for the mass m |
| RM | reference material |
| RMO | regional metrology organisations |
| RNAA | radiochemical neutron activation analysis |
| ROI | region of interest |
| RP | resolving power |
| RPQ | retarding potential quadrupole |
| RSD | relative standard deviation |
| RSF | relative sensitivity factor |
| $s'$ | width of the entrance slit |
| $s''$ | width of the exit slit |
| sccm | standard cubic centimeter |
| SEC | Size Exclusion Chromatography |
| SEM | secondary electron multiplier |
| SEM-EDX | scanning electron microscopy with energy dispersive x-ray spectroscopy |
| SDS-PAGE | sodium dodecyl sulfate polyacrylamide gel electrophoresis |
| SF | sector field |
| SFMS | sector field mass spectrometry or spectrometer |
| SHRIMP | sensitive high resolution microprobe |
| SI | système international d'unités |
| SIM | Inter American Metrology System |
| SIMS | secondary ion mass spectrometry |
| SIOMS | surface ionisation organic mass spectrometry |
| SNMS | sputtered neutral mass spectrometry |
| SPME | solid phase microextraction |
| SR-XRF | synchroton radiation x-ray fluorescence |
| SRM | standard reference material, registered trademark NIST |
| SS | spark ion source |
| SSMS | spark source mass spectrometry |
| SSRM | scanning spreading resistance microscopy |
| STP | Standard Temperature and Pressure |
| $t$ | time (years) |
| $T$ | temperature |
| TC | thermal conversion |
| TE | total evaporation |

| | |
|---|---|
| TEM | transmission electron microscopy |
| TIMS | thermal ionisation mass spectrometry |
| TOF | time of flight |
| TOT | time over threshold |
| TSN | thermo spray nebulisation |
| $u$ | mass unit |
| $U$ | expanded uncertainty |
| $U_0$ | acceleration voltage |
| $u_c$ | standard uncertainty |
| UHV | ultra high vacuum |
| USN | ultrasonic nebulizer |
| $V$ | voltage or potential |
| VIM | Vocabulaire International de Métrologie |
| WARP | wide-angle energy retarding filter |
| WFD | water framework directive |
| WW II | Second World War |
| XPS | X-Ray photoemission spectroscopy |
| XRF | x-ray fluorescence |
| $z$ | number of charges of an ion |
| $\alpha$ | angle of divergence |
| $\Delta E$ | energy spread/difference (*e.g.* of ions) |
| $\Delta m$ | mass difference |
| $\varepsilon'$ | magnetic sector entrance angle of main trajectory |
| $\varepsilon''$ | magnetic sector exit angle of main trajectory |
| $\Theta$ | angle |
| µDG | micro droplet generator |
| $v$ | velocity of an ion |
| $\phi_e$ | electric sector angle |
| $\phi_m$ | magnetic sector angle |

CHAPTER 1

# *Introduction*

THOMAS PROHASKA* AND ANDREAS ZITEK*

University of Natural Resources and Life Sciences Vienna (BOKU), Department of Chemistry, Division of Analytical Chemistry, VIRIS Laboratory for Analytical Ecogeochemistry, Tulln, Austria
*Email: thomas.prohaska@boku.ac.at; andreas.zitek@boku.ac.at

## 1.1 Why Another Book on Mass Spectrometry?

Mass spectrometry has been used since its advent as a tool to investigate challenging scientific questions that could not be tackled with conventional analytical methods before in many different scientific disciplines. As a consequence of the steadily growing interest in mass spectrometry, the increasing number of installed instruments and the massive expansion in augmenting numbers of fields of applications, mass spectrometry has developed steadily into a black box for the common user often having little knowledge about what happens between the sample introduction and the final data provided in electronic form (Figure 1.1). However, a basic understanding of the technical and analytical factors that might significantly influence the measurement and the final result obtained is crucial to avoid misleading and even wrong interpretations. It is especially the increasingly transdisciplinary[1] application of mass spectrometry that creates the basic need for a transfer of the related analytical knowledge between scientific disciplines, *e.g.* about the general methodological background and uncertainties related to the final result including a clear picture on the potential and limitations of each method. Therefore, a complex, extensive and far-ranging topic such as mass spectrometry requires to be covered by

New Developments in Mass Spectrometry No. 3
Sector Field Mass Spectrometry for Elemental and Isotopic Analysis
Edited by Thomas Prohaska, Johanna Irrgeher, Andreas Zitek and Norbert Jakubowski

Published by the Royal Society of Chemistry, www.rsc.org

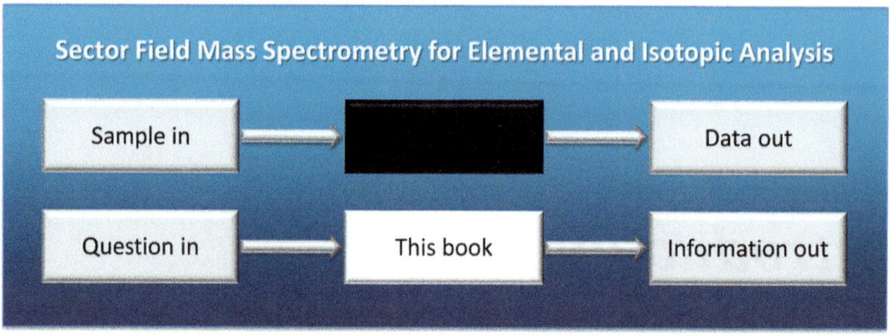


**Figure 1.1** Sector field mass spectrometry has steadily developed into an analytical black box. The aim of this book is to provide fundamental information from different angles and viewpoints to facilitate understanding and provide links to relevant literature.

a variety of educating tools such as textbooks organised along specific main subjects and providing different points of view on necessarily redundant topics.

### 1.1.1 The Success of Magnetic Sector Field Mass Spectrometry

Mass spectrometry has developed from a scientific instrument providing profound knowledge of atoms and isotopes into an affordable and routine analytical tool in many laboratories. Magnetic sector fields as mass separator laid the basis for modern mass spectrometry. From the early days at the beginning of the 20th century until around 1940, mass spectroscopy was mainly applied for the determination of masses and abundances of isotopes.[2] Since then, the applications have spread rapidly from fundamental investigations into elemental and isotopic applications in inorganic mass spectrometry and into the gathering of compositional and structural information of organic molecules in organic mass spectrometry. Historically, the first mass spectrometers at the beginning of the 20th century were based on a sector field device, which were exclusively used for a long period of time. Their importance increased in the second half of the 20th century mainly due to their application for determining organic molecules, where high mass resolution capabilities were required. New designs of mass separators were developed during the second half of 20th century. The major advantages of sector field mass spectrometers have underpinned their importance ever since: the high sensitivity for ultratrace levels of elements, the simultaneous measurement capabilities of multiple isotopes for precise isotope ratio measurements and the high-resolution capabilities to resolve spectral interferences.

### 1.1.2 Magnetic Sector Field Devices for Elemental and Isotopic Analysis

Throughout the last 110 years, the basic principle of all mass spectrometers has remained the same: A sample is ionised in an ion source, the ions are separated *via* a mass separator according to their mass/charge ($m/z$) ratio and finally detected at a detection unit. The devices, which are described in this book, are based on a magnetic sector field as mass separator and applied for quantitative elemental analysis and isotope ratio measurements (of isotopes of one element or of different elements).

Modern mass spectrometers use sector field devices for quantitative low-level elemental analysis due to their high sensitivity and for precise isotope ratio measurements due to their outstanding possibility of separating and letting pass ions of different $m/z$ ratio simultaneously. Another important feature of magnetic sector field devices is the capability to be operated at higher mass resolution (*i.e.* greater than unity) in order to resolve interfering ions from the ions of interest. Most often, the combination of a magnetic sector and an electric sector can be found in a variety of geometries (see Chapter 4) and results in a so-called double-focusing mass spectrometer. Whereas the magnet is used as mass separator and for directional focusing, the electrostatic sector is used for focusing ions of different energies that the energy spread of the ions does not compromise the achievable mass resolution. The electrostatic sector field is not necessarily required if the ionisation source provides ions with little energy spread.

The major difference of the described instruments within this book is based on different ionisation strategies. The goal of this book is to discuss the main features, principles of operation and the resulting applications of these different sector field mass spectrometers. The operation principles are explained in detail based on currently commercially available state-of-the-art instruments.

### 1.1.3 Metrology in Chemistry – The Science of Measurement

We want to make use of this chapter to underpin the necessity of the full understanding of the analytical process, what is usually known under the topic "Science of Measurements" or "Metrology", as defined by the International Bureau of Weights and Measures (BIPM).[3] It is a declared aim of this book to enhance knowledge about the measurement and thus to contribute to a better understanding of the applied methods. Even though instrumental developments have resulted in robust and field-proven devices, all measurement strategies require in any case a sound metrological background – and thus a full understanding of the analytical process from sampling to the final result.

Metrology in chemistry has become of crucial importance in order to avoid the situation that scientific conclusions are drawn from analytical artefacts sometimes provided by highly specialised equipment if not looked at thoroughly. This is even more important as new phenomena can be observed by steadily improving analytical techniques. The recent developments in mass spectrometers used for elemental and isotopic analysis have improved in the analytical performance especially considering sensitivity and measurement precision. Today, both parameters enable the measurement of effects of processes on the elemental and isotopic composition, which have been hidden by the measurement performance for a long time. For example, natural isotopic variations of a number of elements that have – for a long time – been considered as constant, are increasingly exploited. This has a major impact on science such as, *e.g.* transport phenomena in ecosystems or metabolic processes that can be described by isotopic fractionation. Furthermore, this has a consequence on the atomic weight, which is (for some elements) no longer reported as single value but as interval.[4]

This development makes it evident that new scientific questions can be formulated requiring to collaborate across disciplinary boundaries. Therefore, analytical chemistry has been a genuine trigger of transdisciplinary science ever since: New research questions trigger the instrumental development and new developments in the analytical techniques provide access to a large variety of fields of research.

### 1.1.4 A Wide Field for Modern Mass Spectrometers with Unforeseen Development

As most of the magnetic sector field mass spectrometers were applied in fundamental research in the early stages, they soon gained their importance in elemental and isotopic research as well as in organic chemistry and related fields. In elemental and isotopic research, the first applications were aimed towards environmental sciences, geology and radionuclide research, where magnetic sector field mass spectrometers had an important impact from the early beginning (see Chapter 2). Nowadays, the fields of applications are manifold: geology and environmental sciences are still the most prominent fields, followed by nuclear research and technological applications in, *e.g.* the metal and semiconductor industries. More and more, the devices have made an important impact on the life sciences, namely food science and technology, biochemical and biomedical research or medical research. In addition, forensic science and archaeometry have become research fields, where, *e.g.* the measurements of isotope ratios for determining the provenance of forensic evidence or cultural goods have become indispensable tools.

## 1.2 A New Book on Sector Field Mass Spectrometry for Elemental And Isotopic Analysis!

As a consequence of all these considerations, this book was inspired by the idea to provide both a basic and deep insight into currently applied mass

**Table 1.1** Recommended books.

| Topic | References |
|---|---|
| Mass Spectrometry - History | 5–9 |
| Mass Spectrometry - General/Fundamentals | 10–16 |
| Organic Mass Spectrometry | 17 |
| Inorganic Mass Spectrometry | 18–24 |
| Isotopic Mass Spectrometry | 25, 26 |
| ICP-MS | 27–32 |
| GDMS | 6, 8, 20, 22, 24, 33–41 |
| TIMS | 42, 43 |
| SIMS | 44, 45 |
| IRMS | 46–57 |
| AMS | 58, 59 |

spectrometric techniques for elemental and isotopic mass spectrometry using a magnetic mass separator as common principle linking analytical chemistry and the different fields of science. This requires giving a full picture of the technique along with the provision of the spectrum of modern instruments and their potential and actual applications. Some information within the book might be redundant to existing literature but were found to be necessary to guide the reader from theoretical basics *via* an overview of up-to-date instruments to a summary of modern applications within one work. Even though sector field devices have been used in other devices applied for elemental and isotopic analysis (such as AMS, SNMS, SSMS, RIMS or LIMS), this book focuses on the major techniques generally applying sector field devices namely ICP-MS, GDMS, TIMS, SIMS and IRMS (including dynamic and static IRMS).

(***Note:*** for the abbreviations see the List of Abbreviations. Instruments, which are further described in Chapters 12–17, are written in italic without the details on the manufacturer as all these instruments will be explained in detail. All other equipment, which is not listed in these sections, is given with details on the manufacturer.)

The book should encourage reading, encourage the blitheness to gain knowledge and encourage the urge of understanding – especially between the field of mass spectrometry and other scientific disciplines – to force the transdisciplinary application of mass spectrometry. The book shall be comprehensive without being overburdened with details with regard to the technical and analytical background needed to understand the potential and limitations of each method. Many aspects of mass spectrometers are covered in other works, as well, which the interested reader shall be referred to for further reading (Table 1.1).

# References

1. J. Mittelstraß, *Trames: J. Humanit. Soc. Sci.*, 2011, **15**, 329.
2. A. O. Nier, *J. Am. Soc. Mass. Spectrom.*, 1991, **2**, 447.

3. BIPM, http://www.bipm.org/en/convention/wmd/2004/, (accessed 20.03.2014).
4. M. E. Wieser and T. B. Coplen, *Pure Appl. Chem.*, 2011, **83**, 359.
5. H. Budzikiewicz and R. D. Grigsby, *Mass Spectrom. Rev.*, 2006, **25**, 146.
6. I. Cornides, Mass Spectrometric Analysis of Inorganic Solids – The Historical Background, in *Inorganic Mass Spectrometry*, eds. A. Adams, R. Gijbels and R. Van Grieken, John Wiley & Sons Ltd, New York, 1988, ch. 1, pp. 1–15.
7. H. W. Ewald and H. Hintenberger, *Methoden und Anwendungen der Massenspektroskopie*, Verlag Chemie, Weinheim, 1953, 288.
8. K. R. Jennings, ed., *A History of European Mass Spectrometry,* IM Publications LLP, Chichester, 2012, 285.
9. R. W. Kiser, *Introduction to Mass Spectrometry and its Application*, Prentice-Hall, Englewood-Cliffs, NJ, 1965, 356.
10. C. Brunnée and H. Voshage, *Massenspektrometrie,* Karl Thiemig KG, München, 1964, 316.
11. C. Dass, *Fundamentals of Contemporary Mass Spectrometry*, John Wiley & Sons, Inc., Hoboken, NJ, 2007, 608.
12. H. E. Duckworth, R. C. Barber and V. S. Venkatasubramanian, *Mass Spectroscopy*, Cambridge University Press London and New York, 2nd edn., 1990, 337.
13. R. Ekman, J. Silberring, A. M. Westman-Brinkmalm and A. Kraj, eds., *Mass Spectrometry: Instrumentation, Interpretation and Applications*, John Wiley & Sons, Inc., Hoboken, NJ, 2009, 358.
14. J. H. Gross, *Mass Spectrometry - a Textbook*, Springer, Berlin-Heidelberg, 2011, 754.
15. C. A. McDowell, *Mass Spectrometry*, McGraw-Hill, New York, 1963, 639.
16. J. T. Watson and O. D. Sparkman, *Introduction to Mass Spectrometry: Instrumentation, Applications, and Strategies for Data Interpretation*, John Wiley & Sons, Ltd, West Sussex, 4th edn., 2007, 862.
17. E. de Hoffmann and V. Stroobant, *Mass Spectrometry, Principles and Applications*, John Wiley & Sons, Ltd, Chichester a.o., 3rd edn., 2007, 502.
18. A. Adams, R. Gijbels and R. Van Grieken, eds., *Inorganic mass spectrometry*, John Wiley & Sons, Inc., New York, 1988, 404.
19. A. J. Ahearn, *Mass Spectrometric Analysis of Solids*, Elsevier, Amsterdam, 1966, 175.
20. C. M. Barshick, D. C. Duckworth and D. H. Smith, eds., *Inorganic Mass Spectrometry: Fundamentals and Applications*, Dekker, New York, rev. edn., 2000, 512.
21. D. Beauchemin and D. E. Matthews, eds., *The Encyclopedia of Mass Spectrometry: Elemental and Isotope Ratio Mass Spectrometry*, Elsevier, Oxford, 2010, vol. 5, 1088.
22. J. S. Becker, *Inorganic Mass Spectrometry: Principles and Applications*, John Wiley & Sons, Ltd, Chichester, 2007, 514.
23. J. R. de Laeter, *Application of Inorganic Mass Spectrometry*, John Wiley & Sons, Inc., New York a.o., 2001, 474.

24. H. J. Dietze, *Massenspektrometrische Spurenanalyse mit Funken- und Laserionisation*, Akademische Verlagsgesellschaft Geest & Portig K.-G., Leipzig, 1975.
25. J. I. G. Alonso, M. Moldovan and P. R. Gonzalez, *Isotope Dilution Mass Spectrometry*, Royal Society of Chemistry, Cambridge, 2013, 350.
26. F. Vanhaecke and P. Degryse, eds., *Isotopic Analysis: Fundamentals and Applications Using ICP-MS*, Wiley-VCH Weinheim, 2012, 550.
27. Plasma Ionization: Inductively Coupled Plasma, in *The Encyclopedia of Mass Spectrometry: Elemental and Isotope Ratio Mass Spectrometry*, eds. D. Beauchemin and D. E. Matthews, Elsevier, Oxford, 2010, vol. 5, ch. 1, pp. 1–260.
28. K. Jarvis, A. L. Gray and R. S. Houk, *Handbook of Inductively Coupled Plasma Mass Spectrometry*, Blackie, Glasgow, London, 1992, 380.
29. A. Montaser, ed., *Inductively Coupled Plasma Mass Spectrometry*, Wiley-VCH, Hoboken, NJ, 1998, 1004.
30. S. Nelms, ed., *Inductively Coupled Plasma Mass Spectrometry - Handbook*, Blackwell Publishing Ltd., Oxford, 2005, 485.
31. R. Thomas, *Practical Guide to ICP-MS: a Tutorial for Beginners*, CRC Press, Boca Raton, 2nd edn, 2007, 347.
32. F. Vanhaecke and G. Köllensperger, Detection by ICP-Mass Spectrometry, in *Handbook of Elemental Speciation: Techniques and Methodology*, eds. R. Cornelis, H. Crews, J. Caruso and K. Heumann, John Wiley & Sons, Ltd, Chichester, 2003, ch. 5.3, pp. 281–312.
33. Plasma Ionization: Glow Discharge Plasmas and Rf Spark Source, in *The Encyclopedia of Mass Spectrometry: Elemental and Isotope Ratio Mass Spectrometry*, eds. D. Beauchemin and D. E. Matthews, Elsevier, Oxford, 2010, vol. 5, ch. 3, pp. 291–360.
34. C. M. Barshick, Glow Discharge Processes, in *Inorganic Mass Spectrometry: Fundamentals and Applications*, eds. C. M. Barshick, D. C. Duckworth and D. H. Smith, Dekker, New York, 2000, ch. 2, pp. 31–66.
35. B. Chapman, *Glow Discharge Processes: Sputtering and Plasma Etching*, John Wiley & Sons, Inc., New York a.o., 1980, 432.
36. W. W. Harrison, C. Yang and E. Oxley, Mass Spectrometry of Glow Discharges, in *Glow Discharge Plasmas in Analytical Spectroscopy*, eds. R. K. Marcus and J. A. C. Broekaert, John Wiley & Sons, Ltd, Chichester, 2003, ch. 3, pp. 71–96.
37. R. F. Herzog, The Transmission of Ions through Double Focusing Mass Spectrometers, in *Trace Analysis by Mass Spectrometry*, ed. A. J. Ahearn, Academic Press, New York and London, 1972, ch. 3.1, pp. 58–99.
38. N. Jakubowski, A. Bogaerts and V. Hoffmann, Analytical Glow Discharges, in *Atomic Spectroscopy in Elemental Analysis*, ed. M. Cullen, Blackwell Publishing and CRC Press, Oxford and Boca Raton, 2004, ch. 4, pp. 91–156.
39. R. K. Marcus, Handbook of Elemental Speciation: Techniques and Methodology, in *Glow Discharge Plasma as Tunable Sources for Elemental*

*Speciation* eds. R. Cornelis, H. Crews, J. Caruso and K. Heumann, John Wiley & Sons, Ltd, Chichester, 2003, ch. 5.5, pp. 335–355.
40. R. K. Marcus and J. A. C. Broekaert, eds., *Glow Discharge Plasmas in Analytical Spectroscopy*, John Wiley & Sons, Ltd, Chichester, 2003, 498.
41. R. K. Marcus, Glow Discharge Plasma as Tunable Sources for Elemental Speciation in *Handbook of Elemental Speciation: Techniques and Methodology*, eds. R. Cornelis, H. Crews, J. Caruso and K. Heumann, John Wiley & Sons, Ltd, Chichester, 2003, ch. 5.5, pp. 335–355.
42. Thermal Ionization MS, in *The Encyclopedia of Mass Spectrometry: Elemental and Isotope Ratio Mass Spectrometry*, eds. D. Beauchemin and D. E. Matthews, Elsevier, Oxford, 2010, vol. 5, ch. 7, pp. 575–612.
43. I. T. Platzner, *Modern Isotope Ratio Mass Spectrometry*, John Wiley & Sons Ltd, Chichester a. o., 1997, 530.
44. Secondary Ion and Neutral MS, in *The Encyclopedia of Mass Spectrometry: Elemental and Isotope Ratio Mass Spectrometry*, eds. D. Beauchemin and D. E. Matthews, Elsevier, Oxford, 2010, vol. 5, ch. 5, pp. 381–522.
45. A. Benninghoven, F. G. Rudenauer and H. W. Werner, *Secondary Ion Mass Spectrometry: Basic Concepts, Instrumental Aspects, Aplications and Trends*, John Wiley & Sons Ltd, New York a.o., 1987, 1264.
46. Isotope Ratio Mass Spectrometry (IRMS): Measurement of Isotope Ratios of Minerals, Metals and Gases in *The Encyclopedia of Mass Spectrometry: Elemental and Isotope Ratio Mass Spectrometry*, eds. D. Beauchemin and D. E. Matthews, Elsevier, Oxford, 2010, vol. 5, ch. 12, pp. 765–896.
47. W. A. Brand, Mass Spectrometer Hardware for Analyzing Stable Isotope Ratios in *Handbook of Stable Isotope Analytical Techniques*, ed. P. A. de Groot, Elsevier Amsterdam, 2004, vol. 1, ch. 38, pp. 835–856.
48. P. Burnard, ed., *The Noble Gases as Geochemical Tracers*, Springer, Heidelberg a.o., 2013, 391.
49. P. Burnard, L. Zimmermann and Y. Sano, History and Background, in *The Noble Gases as Geochemical Tracers*, ed. P. Burnard, Springer, Heidelberg a.o., 2013, pp. 1–15.
50. P. A. de Groot, ed., *Handbook of Stable Isotope Analytical Techniques*, Amsterdam, Elsevier 2004, vol. 1, 1258.
51. A. P. Dickin, *Radiogenic Isotope Geology*, Cambridge University Press, Cambridge, 2nd edn, 2005.
52. K. Habfast, *Advanced Isotope Ratio Mass Spectrometry I: Magnetic Isotope Ratio Mass Spectrometers*, ed. I. T. Platzner, John Wiley & Sons Ltd, Chichester a.o., 1997, ch. 3, pp. 11–82.
53. H. R. Krouse, Isotope Ratio Mass Spectrometry, in *Encyclopedia of Analytical Chemistry*, John Wiley & Sons, Ltd, 2006.
54. R. Mitchener and K. Lajtha, eds., *Stable Isotopes in Ecology and Environmental Science*, Blackwell Publishing, Malden, MA a.o., 2nd edn, 2007, 594.
55. M. Ozima and F. A. Podosek, *Noble Gas Geochemistry*, Cambridge University Press, Cambridge, 2nd edn, 2002, 300.

56. M. Wendeberg and W. A. Brand, Isotope ratio mass spectrometry (IRMS) of light elements (C, H, O, N, S): The principles and characteristics of the IRMS instrument, in *The Encyclopedia of Mass Spectrometry: Elemental and Isotope Ratio Mass Spectrometry*, eds. D. Beauchemin and D. E. Matthews, Elsevier, Oxford, 2010, vol. 5, ch. 10, pp. 739–748.
57. M. E. Wieser and W. A. Brand, Isotope Ratio Studies using Mass Spectrometry, in *Encyclopedia of Spectroscopy and Spectrometry*, eds. J. C. Lindon, G. E. Tranter and J. L. Holmes, Academic Press, San Diego, CA, 1999, pp. 1072–1086.
58. Accelerator Mass Spectrometry, in *The Encyclopedia of Mass Spectrometry: Elemental and Isotope Ratio Mass Spectrometry*, eds. D. Beauchemin and D. E. Matthews, Elsevier, Oxford, 2010, vol. 5, ch. 8, pp. 613–676.
59. R. Hellborg, ed., *Electrostatic Accelerators, Fundamentals and Applications*, Springer, Berlin – Heidelberg, 2005, 620.

CHAPTER 2

# History

THOMAS PROHASKA

University of Natural Resources and Life Sciences Vienna (BOKU), Department of Chemistry, Division of Analytical Chemistry, VIRIS Laboratory for Analytical Ecogeochemistry, Tulln, Austria
Email: thomas.prohaska@boku.ac.at

## 2.1 Where it All Started

> "In reality, there is only atoms and emptiness" (Figure 2.1)
> (Democrit)

The general principle of the idea of separating our world into the smallest increments possible, atoms, started with Democrit (460–370 BC), the Greek philosopher, a pupil of Leucippus, who asked the origin of all questions: "What is that in truth being?" His answer was simple, but with consequences. Democrit can be named as a major influence on the formulation of an atomic theory of the universe. He is often cited as the "father of modern science". According to the theory of Democrit, matter is composed of impartible (greek = *atomos*) basic modules, which implement already the properties of the matter, which they compose. According to Democrit, these atoms act comparable to more or less hard balls. Atoms and atomic theories went along with the discovery of the elements composing the matter. The definition of elements as major components of our environment had major consequences in science. John Dalton (1766–1844), *e.g.* introduced the fundamental law of multiple proportions as one of the basic laws of

New Developments in Mass Spectrometry No. 3
Sector Field Mass Spectrometry for Elemental and Isotopic Analysis
Edited by Thomas Prohaska, Johanna Irrgeher, Andreas Zitek and Norbert Jakubowski

Published by the Royal Society of Chemistry, www.rsc.org

‚In reality, there are only atoms and emptiness'
Democrit (460 – 370 BC), influential on the formulation of an atomic theorie of the universe, is often cited as father of modern science.


**Figure 2.1** Democrit.
(Source: composed by Johanna Irrgeher; Photo background: Johanna Irrgeher; Photo Democrit: Marc Scheimann, City University of London.)

stoichiometry. The law is based on his atomic hypothesis following the work of Jeremias Benjamin Richter (1762–1807), a German chemist.

Since then, the efforts of scientists have had the goal to visualise these atoms inventing tools to monitor the material world around us into its smallest possible increments. As a result, scientists involved in spectroscopy have invented and established tools to identify the elemental composition of our material world and to measure the weight of atoms by a very special balance: the mass spectrometer.

## 2.2 Cathode Rays and Kanalstrahlen

(Hittdorf, Goldstein, Wien) (Figure 2.2)

The roots of sector field instruments date back to the late 19th century. In 1886, the German physicist Eugen Goldstein (1850–1930), a scholar of Hermann von Helmholtz (1821–1894), coined the term "cathode rays" for the negatively charged electrons, which are extended from a negative electrode (in a glow discharge). The existence of these rays had been discovered by Julius Plücker (1801–1868) and his scholar Johann Wilhelm Hittdorf (1824–1914), who found that these rays could be deflected by a magnetic field. Franz Arthur Friedrich Schuster (1851–1934) determined the mass to charge ratio of the particles of cathode rays by measuring the degree of deflection in a magnetic field in 1890. Goldstein also discovered "*Kanalstrahlen*" (canal rays), positively charged particles formed when electrons are removed from gas particles in a

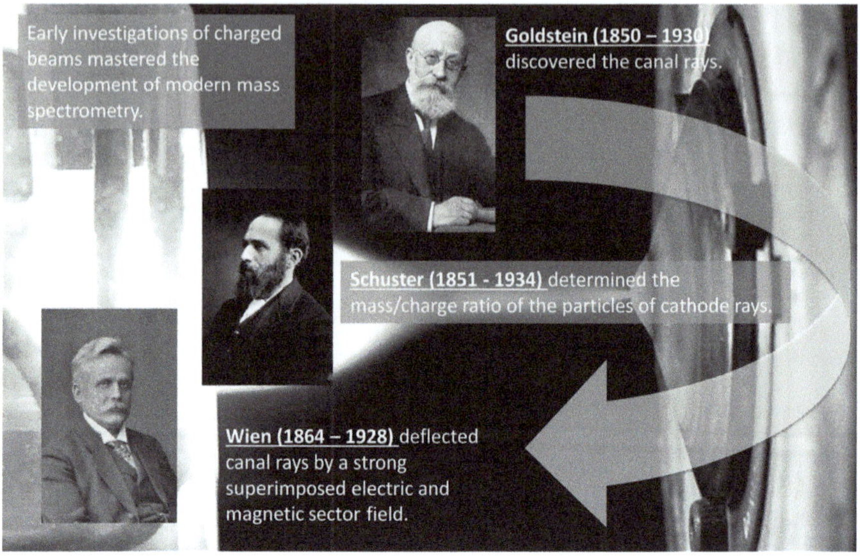


**Figure 2.2** Work on charged rays – Goldstein, Schuster, Wien.
(Source: composed by Johanna Irrgeher; Photo Wien: American Institute of Physics, AIT Publishing; Photo Schuster: The Physical Laboratories of the University of Manchester, At the University Press, 1906, p.15; Photo Goldstein: from http://icp-group4.wikispaces.com/Eugene + Goldstein {{PD-EU-no author disclosure}}; Photo plasma: with courtesy of Clarisse Mariet, CEA Saclay, France.)

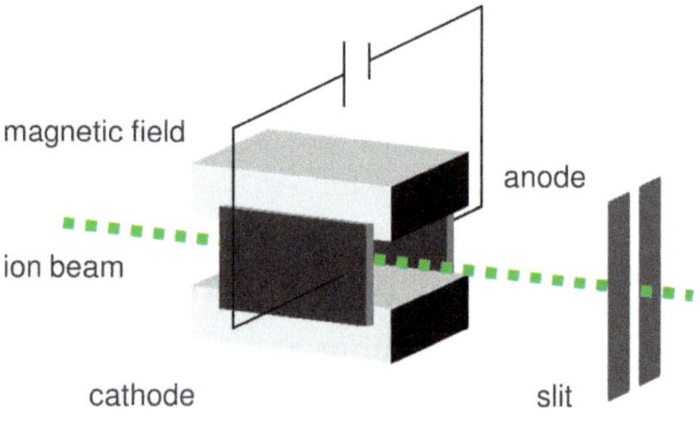


**Figure 2.3** Wien filter.
(Source: Ondrej Hanousek.)

glass tube filled with gas at reduced pressure and equipped with a perforated cathode. In 1898, Wilhelm Wien (1864 – 1928) demonstrated that these canal rays had a positive charge and could be deflected by a strong superimposed electric and magnetic sector field. The resulting velocity filter for charged particles (Wien filter) consisted of perpendicular electric and magnetic fields and could be applied as an energy analyser or mass separator of charged particles (Figure 2.3).

## 2.3 The First Mass Spectrometer and the Determination of Isotopes

(Thomson, Soddy, Aston) (Figure 2.4)

J. J. Thomson (1846–1940),[2] a British physicist, showed that cathode rays consisted of an unknown negatively charged particle – the electron – and determined the *m*/*z* ratio of this particle. He received the Nobel prize in 1906 "in recognition of the great merits of his theoretical and experimental investigations on the conduction of electricity by gases".[3]

"At first there were very few who believed in the existence of these bodies smaller than atoms. I was even told long afterwards by a distinguished physicist who had been present at my [1897] lecture at the Royal Institution that he thought I had been "pulling their legs"."[4]

Thomson had started his studies of the "*Kanalstrahlen*" in 1905 and he improved the Wien filter by reducing the pressure in his apparatus.[5] Thomson was soon assisted by his student Francis William Aston (1877–1945), who started to work at the Cavendish Laboratories in 1909. In 1910, Frederick Soddy (1877–1956) discovered that the element lead

**Figure 2.4** Mass spectrometry and isotopes in the early years – Thomson, Soddy, Aston.
(Source: composed by Johanna Irrgeher; Photo Thomson: Sharma;[1] Soddy: University of Glasgow Archive Services, University Photographic collection, GB0248 UP1/503/2; Aston: Sharma;[1] Aston (background): Cavendish Laboratory, Department of Physics, Photography Collection.)

differed in mass depending whether it had been formed *via* the decay of thorium or the decay of uranium. This was originally considered as a peculiarity of radioactive materials, which he called "isotopes", from the Greek, words "*iso*" (same) and "*topos*" (place) (*ισο* τόπος = at the same place) – as they can be found at the same position within the periodic table of the elements. Even though the term nuclide is the more general expression for an atomic species characterised by the constitution of the nucleus (*i.e.* number of protons and neutrons), the term isotope is still commonly used.[6]

In 1912, Thomson and Aston guided a collimated beam of ions - generated in an electrical discharge in a gas at low pressure (Figure 2.5) – through a magnetic and an electric field based on the positive ray apparatus first employed by Walter Kaufman (1871–1947) in 1901 (Figure 2.6).

The charged particles are deflected in the electric and magnetic field and fall on a parabola in the *y*/*z* plane according to their mass/charge ratio and energy. Thomson's instrument was the first mass spectrometer, called a

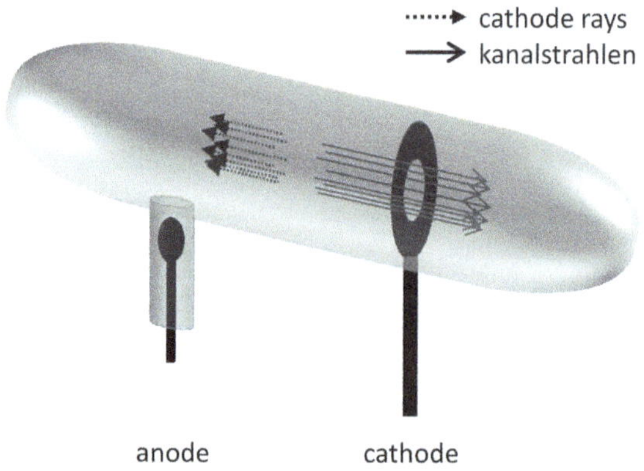


**Figure 2.5** Gas discharge tube by Thomson (1912). (Source: Ondrej Hanousek, after Thomson.[7])

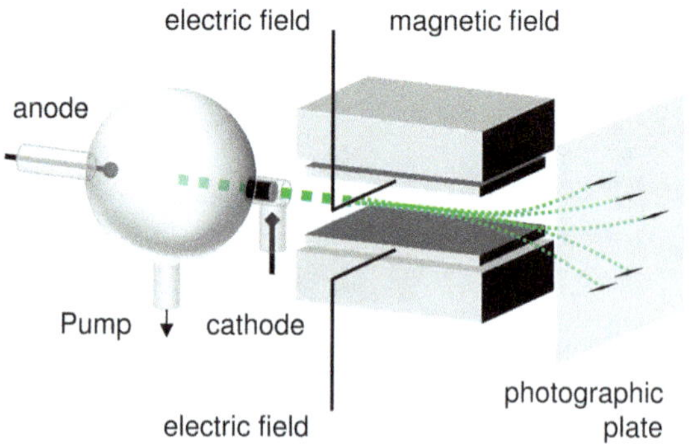


**Figure 2.6** Positive-ray apparatus by Thomson (1912). (Source: Ondrej Hanousek, after Thomson.[8])

**Figure 2.7** Thomson and his first mass spectrograph (1913). (Source: composed by Johanna Irrgeher; Photo Thomson: Science and Society, Picture Library; Photo mass spectrum: from Thomson:[7] Photo mass spectrograph: from Sharma.[1])

parabola spectrograph, named after the parabola formed on the photo plate used as detector. Thomson and Aston observed different patches of light on the photographic plate, which were interpreted according to the *m*/*z* ratio of different elements.[7] Besides He and Ne (at mass 20), they found an additional gas with the mass of 22, which was visible with all specimen of Ne examined (Figure 2.7).

The mind-blowing conclusion was that neon is composed of atoms of two different atomic masses. In the following years, Aston worked on the improvement of the technique and started to separate Ne isotopes by fractional distillation. World War I interrupted his work until 1919, when he finally could design an improved mass separator for the determination of the masses of isotopes by allowing the particles first to pass an electric field before they entered the magnetic field. Therefore, he could focus all ions of the same mass but different speed at the same spot. The position of the spot depended only on the mass and not on the kinetic energy of the particles. The spots were recorded on a straight line giving the indication of the mass of ions. The instrument still lacked in direction focusing (*i.e.* no focusing of particles of the same mass/charge ratio entering the magnet in slightly different directions). Aston was the first to use the term "mass spectrum"[9] (Figure 2.8) (Figure 2.9).

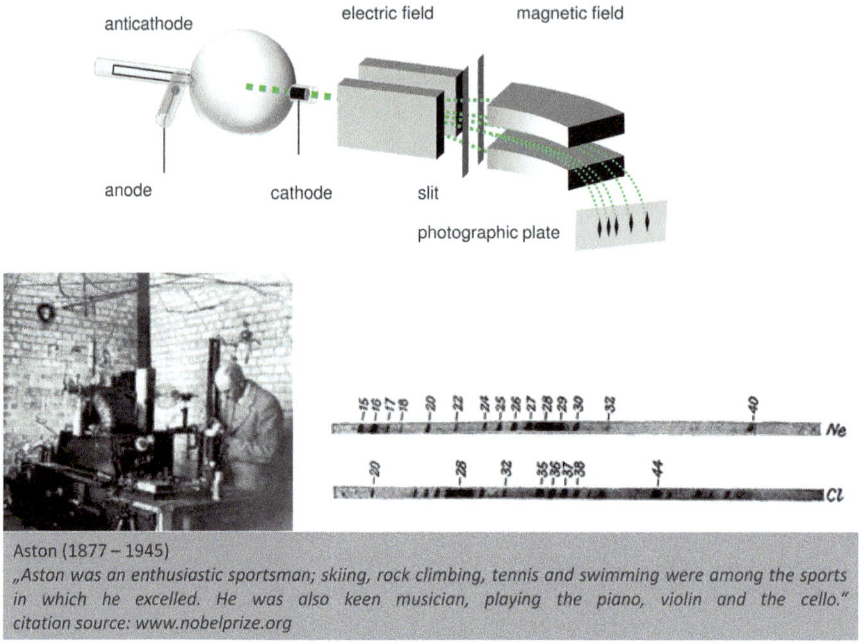


**Figure 2.8** Aston and the schematic of his first spectrograph. (Source: composed by Johanna Irrgeher; Mass spectrometer: Ondrej Hanousek, after http://www-outreach.phy.cam.ac.uk/camphy/massspectrograph/massspectrograph2_1.htm; Photo Aston: see Figure 2.4; Photo Mass spectrum: after Aston.[10])

**Figure 2.9** Aston's first spectrograph. (Source: from Sharma.[1])

In 1921, he created an improved version of his mass separator by applying a curved electric field and a magnet of 1.6 Tesla. Thus, he was able to identify the isotopes of mercury and within a few years he quickly could determine isotopes of more than thirty elements. Aston received the Nobel Prize in Chemistry 1922 for his work on isotope studies carried out with this type of instrument.[10] One important observation by Aston was that masses of particles were always whole numbers. Scientists believed at that time that an atom consisted of a whole number of protons and the same number of electrons. For example, 20 electrons would balance 20 protons for Ne. The model was physically unrealistic until 1932 when James Chadwick (1881–1974) discovered the neutrons, neutral particles of approximately the same mass as protons.

At about the same time as Aston, the American Arthur Jeffrey Dempster (1886–1950) built independently his first mass spectrometer in 1918 with a significantly different design as compared to Aston's apparatus: The positive ions were accelerated through a high potential and entered into a 180° magnetic field where ions were deflected according to their mass/charge ratio. The ions passed through a small slit at the exit and the ion current was measured electrically. Ions of different mass/charge ratios can pass through the slit by varying the potential used for accelerating the ions or the strength of the magnetic field. This apparatus made use of the fact that ions of the same mass/charge ratio and given energy were brought to a focus behind the magnetic field independent of the incident angle. Therefore, directional focusing took place.

The further developments of double-focusing sector field mass spectrometry, documented well in a paper by Nier,[11] soon led to nowadays state-of-the-art machines.

## 2.4 Towards Modern Mass Spectrometers

(Dempster and Bartky, Bainbridge and Jordan,
Mattauch and Herzog, Nier and Johnson)
(Figures 2.10 and 2.11)

During the 1930s a boom in nuclear physics was guided by the interest of scientists to determine the energy in a nucleus and boosted by the development of cyclotrons and high-voltage electrostatic generators.[11] In 1929, at the University of Chicago, based on ion optical calculations performed together with Bartky (1901–1958), Dempster designed and built a spectrometer with an 180° magnetic sector geometry preceded by a 90° electric sector with both velocity and directional focusing – a format adopted commercially and still in use in modern mass spectrometers (Figure 2.10). Dempster also combined the instruments with new ion sources, such as electron impact and spark sources in 1934. In 1935, Dempster developed also the first "scanning" mass spectrometer.[12] In this device, the separation of ions focused into an exit slit was already related directly to the *m*/*z* ratio.

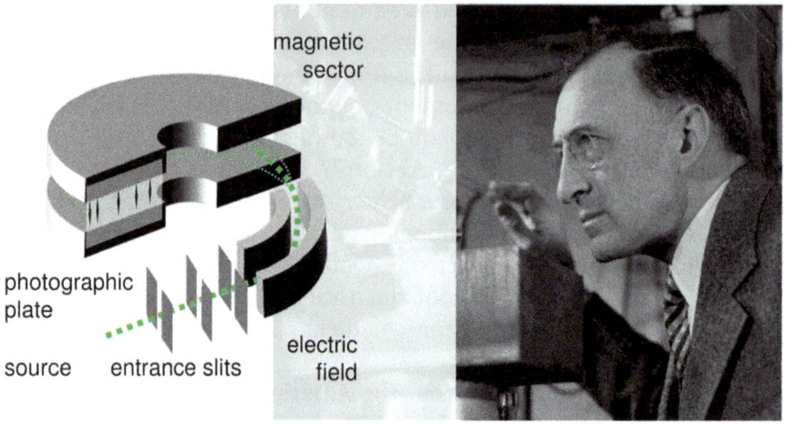


**Figure 2.10** Dempster and the spectrograph using the Dempster geometry. (Source: composed by Ondrej Hanousek; Schematics Dempster geometry: Ondrej Hanousek after Dempster;[12] Photo Dempster: American Institute of Physics, AIP Publishing.)

Scanning was still performed either by changing the magnetic field or the acceleration voltage. He applied a Faraday detector and electronic amplification for detection of the ions. Later, he used his electron impact source for ion generation in molecular mass spectrometry as well, resulting in the first molecular mass spectrum. It is also noteworthy that Dempster's research led to the discovery of the uranium isotope $^{235}U$ in 1935.

In Austria, at the physics department of the University of Vienna, Josef Mattauch (1895–1976) and his student Richard F. K. Herzog (1911–1999) contributed significantly to the development of double-focusing instruments in 1934 using a geometry that became known as Mattauch–Herzog geometry – consisting of a 31.8° ($\pi/4\sqrt{2}$) electric sector field and a 90° magnetic sector field of opposite curvature direction (see Chapter 4).[13] With this geometry, all ions of a mass spectrum could be "double focused" in a plane in the magnetic sector field, and thus could be detected simultaneously on a 25-cm long photo plate. This instrument provided a mass resolution of 6500. Several other efforts were conducted in developing a double-focusing mass spectrometer. For example, Bondy and Popper presented a spectrometer based on considerations by Bartky and Dempster in 1933[14] and Taylor described an instrument based on the Aston design in 1935.[15] In 1936, Bainbridge and Jordan[16] from Harvard presented a paper on an instrument designed for isotopic measurements fulfilling the conditions of high resolving power (Figure 2.11). By obtaining a mass resolution of up to 10 000 and more they were able to proof the existence of isobaric pairs such as $^{113}Cd$–$^{113}In$, $^{115}In$–$^{115}Sn$, and $^{123}Sb$–$^{123}Te$. In 1938, Bleakney and Hipple described an arrangement of a superimposed electric and magnetic field that were at right angles to one another.[17]

In 1937, Aston built his third and final mass spectrograph with externally adjustable slits and an improved magnetic field yielding a 20 times higher mass resolution as compared to the previous instrument. Before his death, 202 nuclides of 71 elements were discovered by Aston.[18] The mass spectrometers are still on display at the Science Museum London and the Museum of Cavendish Laboratory, Cambridge (Figure 2.12). With the development of monoenergetic electron impact ion sources, single magnetic sector instruments became powerful alternatives to the relatively bulky double-focusing devices. For instance in 1940, Alfred O.C. Nier (1911–1994)[19] (Figure 2.13) designed a 60° single magnetic sector field instrument. In 1942, he constructed a 180° sector field instrument, which became commercially available for organic analysis using an electron impact ion source. Nier became substantially influential on modern mass spectrometry in the 1940s and 1950s.[20]

During World War II, the developments came partially to a halt.[11] Nonetheless, the growing interest in separating $^{235}U$ for the production of nuclear weapons led to a significant improvement, *e.g.* in vacuum and electronic techniques. A magnetic sector instrument created by Nier at the department of physics at the University of Minnesota during World War II was designed to do isotopic analysis, with separation of $^{235}U$ from $^{238}U$ obviously of special importance.[21,22] Nier isolated the first sample of plutonium ($10^{-9}$ g) by MS for its first actual characterisation. At the beginning of the 1940s, the US

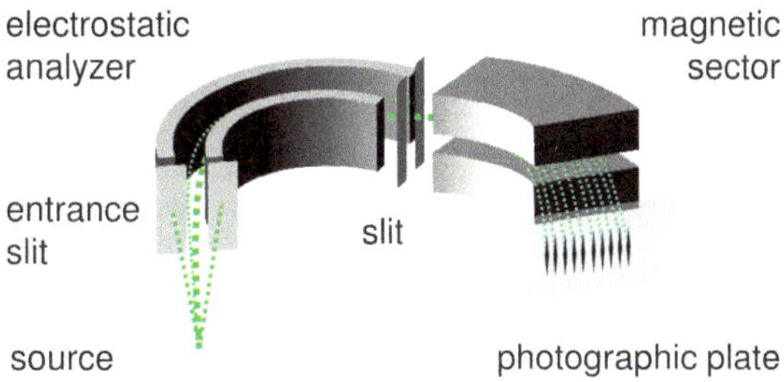


**Figure 2.11** Bainbridge–Jordan geometry.
(Source: Ondrej Hanousek after Bainbridge and Jordan.[16])

**Figure 2.12** Aston's third spectrograph at the Cavendish Laboratory Museum, Cambridge.
(Source: http://www-outreach.phy.cam.ac.uk/camphy/massspectrograph/massspectrograph8_1.htm.)

**Figure 2.13** Alfred O.C. Nier (1911–1994).
(Source: American Institute of Physics, AIP Publishing.)

government started to assemble a large number of scientists and created a vast arrangement of laboratories, pilot plants and manufacturing facilities. The "Manhattan Project" became the code name of the network of research conducted across the country. Enrichment on an industrial scale was first achieved at the Oak Ridge Y-12 plant by using Calutrons, three-story-high cyclotrons invented by Ernest O. Lawrence during the Manhattan project, for enrichment of $^{235}U$ intended for its use for the first atomic bomb (Figure 2.14). Calutron (Cal. U.-tron) is in tribute to the University of California. Magnetic separation was further abandoned by gaseous diffusion, which proved to be more efficient. The gaseous diffusion plant at Oak Ridge supplied uranium-235 for the subsequent explosions of nuclear bombs.

In 1948, Edgar G. Johnson joined Nier as a graduate student looking for a master's thesis.[11] Together, they considered the design of a double-focusing mass spectrometer with improved focusing conditions. Nowadays, most modern sector field devices are based on the theory first discussed by Alfred O.C. Nier and Edgar Johnson in the early 1950s, known today as Nier–Johnson geometry.[23] They reasoned that it would be easier to obtain second-order double focusing at a single point, a slit, rather than across a plane, as compared to the Mattauch–Herzog geometry. In the Nier–Johnson geometry, the 90° electric sector is arranged symmetrically and the 60° magnetic sector asymmetrically in the same curvature direction (see Chapter 4). This geometry accepts a large divergence angle of the ion beam and guarantees second-order directional focusing.

Since the 1950s, sector field instruments have become dominant in organic mass spectrometry where the high-resolution capabilities of the Mattauch–Herzog and Nier–Johnson geometries were required for the identification of organic compounds.

**Figure 2.14** 184-inch cyclotron converted from a calutron to a synchrocyclotron in 1945-46. Ernest O. Lawrence and staff posed with the magnet. (Photo: Roy Kaltschmidt; Source: © 2012 The Regents of the University of California, Lawrence Berkeley National Laboratory.)

## 2.5 The Early Commercialisation of Mass Spectrometry

In the late 1930s, MS began to move from academic to commercial. In 1937 Consolidated Engineering Corporation (CEC) (Pasadena, CA) built a mass spectrometer based on the Dempster design and Westinghouse Electric (Monroeville, PA, US) built a mass spectrometer based on the Nier design, both for the oil industry. John Hipple designed the first portable mass spectrometer in 1941, which was commercialised by Westinghouse Electric (Figure 2.15). The Consolidated Engineering Corporation (CEC) (Pasadena, CA, US) was the first company to achieve market success in 1943 selling the CEC Model 21-101 mass spectrometer. The instrument used the single-focusing design of Dempster's instrument and was applied for the analysis of organic gas mixtures in WW II by the petrochemical industry. Nier directed the instrument development for the Kellex Corporation (New York, NY, US) from 1943–1945. A prototype of a Nier design instrument was replicated by General Electrics (Schenectady, NY, US) and hundreds of instruments were sold to the Manhattan Project. In addition, the major companies manufacturing mass spectrometer in the 1940s were Metropolitan Vickers in

the Mass Spectrometer
A NEW ELECTRONIC
METHOD FOR
Fast, accurate gas analysis
Climaxing 20 Years of Research, Westinghouse now introduces a scientific instrument of tremendous importance to the petroleum, chemical, and synthetic rubber industries.
The Mass Spectrometer, it's called. By means of this electronic device, gas analysis which previously took days can now be completed in a matter of minutes. Purity of the gases being analyzed can be determined within a fraction of one per cent.
For Quick and Accurate Analysis of the purity of gas components, essential to insure product quality in the production of synthetic rubber and high octane gasoline, this new electronic method offers broad possibilities. Production rates may be greatly stepped up, because delays involved in complicated laboratory analysis are eliminated. Waste production may be sharply reduced, because purity of ingredients can be accurately determined, and deviations from fixed standards detected almost immediately.
So Significant is this Westinghouse development to the synthetic rubber, chemical and petroleum industries that several leading representatives of these industries have been working for months, side-by-side with Westinghouse engineers, to perfect its application.
Through this Pooling of Skills, the possibilities of the Mass Spectrometer as a production tool have been amply demonstrated. Today, it stands ready to help speed America's production of vitally needed rubber and aviation fuel. For further information, call the nearest Westinghouse office. Or write
Westinghouse Electric International Company
40 Wall St., New York 5, U. S. A.
TO FACILITATE ITS USE
the Mass Spectrometer has been developed as a wholly self-contained unit. The only outside connections required are for 110-volt a-c power, and water for cooling pumps.
PLANTS IN 25 CITIES . . . OFFICES EVERYWHERE
Westinghouse

**Figure 2.15** Advertisement Westinghouse Electric. (Source: http://masspec.scripps.edu/mshistory/timeline/time_pdf/1943_FirstMassSpecAd.pdf Westinghouse Electric International Company, **1943**; Permission: Marie A. Podvorec, Law Department, CBS Corporation (formerly Westinghouse Electric Corporation), 20 Stanwix Street, Pittsburgh PA 15222.)

England (later Associated Electrical Industries, then VG and Micromass) and the Atlas Werke in Germany (later Atlas MAT (Mess und Analysentechnik)).

## 2.6 The Last 50 Years – Development of Mass Separators and Mastering the Ion Sources

Single- (Dempster design) and double-focusing (Mattauch–Herzog and Nier–Johnson design) magnetic sector field instruments were the dominating instruments in high-performance mass spectrometry until the early 1990s. In addition to sector field mass separators, the cyclotron mass spectrometer, invented by Ernest O. Lawrence in 1931, had dominated the field of mass spectrometry in the early days. Hippler *et al.* presented their ion cyclotron mass spectrometer in 1946.[24] Other mass separators evolved in parallel to the predominant magnetic sector field mass spectrometer. Time of flight was introduced in the 1940s[25] (Stephens, 1946) and Paul developed the quadrupole mass filters and quadrupole ion traps[26] for which he received the Nobel Prize in 1989. The Fourier transform ion cyclotron resonance mass spectrometer was presented by Comisarov and Marshal in 1974 and revolutionised ICR MS.[27] Makarov developed the orbitrap mass spectrometer in 1999.[28]

From the 1960s, Matsuda developed and built double-focusing spectrometers of second order resulting in highest resolution.[29] In 1968 at the Osaka University, they measured the atomic mass difference of $^{40}$Ca and $^{40}$Ar with a resolving power of 700 000 to 1 000 000.[30] In the 1980s Matsuda and his team calculated a spectrometer with an 85° electric sector and a 72.5° magnetic dipole.[31] The Matsuda configuration is used in the Sensitive High Resolution Ion Micro Probe, SHRIMP.[32]

Developments within the second half of the 20th century concerning magnetic sector field instruments applied in elemental and isotope ratio analysis were mainly based on the development, coupling and optimisation of new ion sources. It turned out that double-focusing instruments are perfectly suited for a combination with ion sources generating ions with a broad energy distribution. The first commercially available TIMS instrument was introduced in 1948. Even though ion sputter sources had been used for long time from the 1950s, it took more than 50 years until the first commercially available SIMS instrument with a magnetic sector field was introduced in the mid-1960s. Spark source mass spectrometry (SS-MS) was considered until the mid-1980s as one of the reference techniques for direct solid analysis. The first commercial instrument based on a glow discharge ion source, the VG 9000, was introduced in 1985. The first sector field instrument with an ICP source, the *Plasmatrace*, followed in 1988, after Fassel and Houk had successfully coupled an ICP source to a quadrupole mass spectrometer in 1978. In 1992, the first MC ICP-SFMS was introduced (*Plasma 54*). A more detailed commercial history of sector field instruments can be found in the corresponding chapters.

Modern commercially available magnetic sector field mass spectrometers in elemental and isotopic analysis are mature instruments and mainly based on double-focusing designs (reverse Nier–Johnson, forward Nier–Johnson, Mattauch–Herzog or Matsuda geometry). The major remaining challenges are to improve resolution, sensitivity and isotope ratio precision along with the control of interferences and instrumental mass fractionation. All these topics will be elaborated in detail in the following chapters.

More detailed information on the general history of mass spectrometry were reviewed for example in the books by Ewald and Hintenberger,[33] by Duckworth[34] and by Kiser[35] along with a number of selected articles of the history[36,32,37] and early history of mass spectrometry[1,38,39,40,41] along with numerous web-links, *e.g.* Refs. 42–44.

## References


1. K. S. Sharma, *Int. J. Mass Spectrom.*, 2013, **349–350**, 3–8.
2. J. J. Thomson and G. P. Thomson, *Conduction of Electricity through Gases*, vol. 1, Cambridge, University Press, 1928.
3. *Nobel Lectures: Physics 1901–1921*, Elsevier, Amsterdam, 1967.
4. J. J. Thomson, *Recollect. Reflect.*, G. Bell and Sons, London, 1936. p. 341.
5. J. J. Thomson, *Philos. Mag. Series 6*, 1907, **13**, 561–575.
6. H. Budzikiewicz and R. D. Grigsby, *Mass Spectrom. Rev.*, 2006, **25**, 146–157.
7. J. J. Thomson, *Proc. Roy. Soc. Lond. Series A*, 1913, **89**, 1–20.
8. J. J. Thomson, *London, Edinburgh, and Dublin Philos. Mag. J. Sci.*, 1912, **24**, 209–253.
9. F. W. Aston, *Philos. Mag.*, 1919, **XXXXVIII**, 707.
10. F. W. Aston, *Nature*, 1920, **105**, 617.
11. A. O. Nier, *J. Am. Soc. Mass Spectrom.*, 1991, **2**, 447–452.
12. A. J. Dempster, *Proc. Am. Philos. Soc.*, 1935, **75**, 755–767.
13. R. Herzog, *Z. Phys.*, 1934, **89**, 447–473.
14. H. Bondy and K. Popper, *Ann. Phys.*, 1933, **409**, 425–444.
15. D. D. Taylor, *Phys. Rev.*, 1935, **47**, 666–671.
16. K. T. Bainbridge and E. B. Jordan, *Phys. Rev.*, 1936, **50**, 282–296.
17. W. Bleakney and J. J. A. Hipple, *Phys. Rev.*, 1938, **53**, 521–529.
18. F. W. Aston, *Mass Spectra and Isotopes*, Longmans, Green & Co.; E. Arnold & Co., New York; London, 1942.
19. A. O. Nier, *Rev. Sci. Instrum.*, 1940, **11**, 212–216.
20. M. A. Grayson, *J. Am. Soc. Mass Spectrom.*, 1992, **3**, 685–694.
21. A. O. Nier, *Phys. Rev.*, 1939, **55**, 150–153.
22. A. O. Nier, E. T. Booth, J. R. Dunning and A. V. Grosse, *Phys. Rev.*, 1940, **57**, 546–546.
23. E. G. Johnson and A. O. Nier, *Phys. Rev.*, 1953, **91**, 10–17.
24. J. A. Hipple, H. Sommer and H. A. Thomas, *Phys. Rev.*, 1949, **76**, 1877–1878.
25. W. Stephens, *Bull. Am. Phys. Soc.*, 1946, **21**, 22.

26. W. Paul and H. Steinwedel, *Z. Naturforschg.*, 1953, **8a**, 448.
27. M. B. Comisarow and A. G. Marshall, *Chem. Phys. Lett.*, 1974, **25**, 282–283.
28. A. Makarov, *Analyt. Chem.*, 2000, **72**, 1156–1162.
29. H. Matsuda, Double Focusing Mass Spectrometers of Second Order, in *Atomic Masses and Fundamental Constants 5*, eds. J. H. Sanders and A. H. Wapstra, Springer US, 1976, ch. 26, pp. 185–191.
30. S. Fukumoto, T. Matsuo and H. Matsuda, *J. Phys. Soc. Jpn.*, 1968, **25**, 946.
31. H. Matsuda, *Mass Spectrosc.*, 1985, **33**, 227.
32. G. Münzenberg, *Int. J. Mass Spectrom.*, 2013, **349–350**, 9–18.
33. H. W. Ewald and H. Hintenberger, *Methoden und Anwendungen der Massenspektroskopie*, Verlag Chemie GmbH, Weinheim/Bergstr, 1953, **IV**, 288.
34. H. E. Duckworth, *Mass Spectroscopy*, Cambridge Univ. Press, London and New York, 1958.
35. R. W. Kiser, *Introduction to Mass Spectrometry and its Applications*, Prentice-Hall, 1965.
36. Y. Litvinov and K. Blaum, *Int. J. Mass Spectrom.*, 2013, **349–350**, 1–276.
37. J. Griffiths, *Analyt. Chem.*, 2008, **80**, 5678–5683.
38. K. M. Downard, *J. Mass Spectrom.*, 2012, **47**, 1034–1039.
39. K. M. Downard, *J. Am. Soc. Mass Spectrom.*, 2009, **20**, 1964–1973.
40. K. M. Downard, *Eur. J. Mass Spectrom.*, 2007, **13**, 177–190.
41. K. M. Downard, *Mass Spectrom. Rev.*, 2007, **26**, 713–723.
42. The Cavendish Laboratory, http://www-outreach.phy.cam.ac.uk/camphy/massspectrograph/massspectrograph1_1.htm, (accessed 17.06.2014).
43. Scripps Center for Metabolomics and Mass Spectrometry, http://masspec.scripps.edu/mshistory/mshistory.php, (accessed 17.06.2014).
44. NIH / NIGMS Biomedical Mass Spectrometry Resource, http://msr.dom.wustl.edu/history-mass-spectrometry/, (accessed 17.06.2014).

# Fundamentals

The fundamental principles and technical aspects, which are the same or similar for the types of sector field instruments covered in this book, are described in detail within this section in order to provide understanding of fundamental processes of magnetic sector field devices. More specific information concerning technical aspects of the specific techniques is given in Chapters 11–17.

CHAPTER 3

# *General Overview*

THOMAS PROHASKA

University of Natural Resources and Life Sciences Vienna (BOKU), Department of Chemistry, Division of Analytical Chemistry, VIRIS Laboratory for Analytical Ecogeochemistry, Tulln, Austria
Email: thomas.prohaska@boku.ac.at

The operation principle of a mass spectrometer is (Figure 3.1):

1. to generate ions from a sample;
2. to separate these ions according to their mass/charge ($m/z$) ratio;
3. to detect the ions quantitatively by their respective mass/charge ratio.

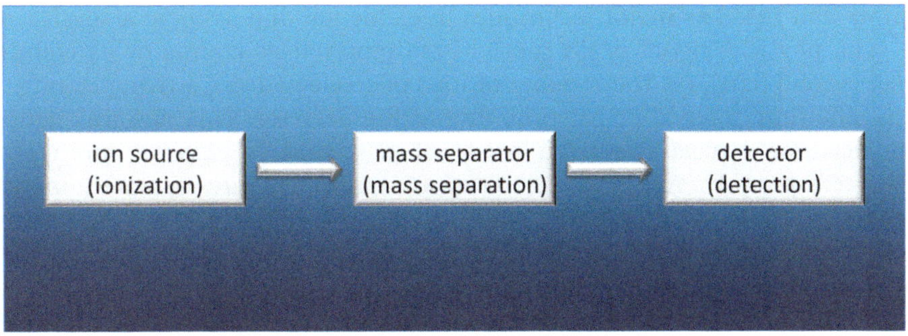


**Figure 3.1** Schematic operation principle of a mass spectrometer (general).

New Developments in Mass Spectrometry No. 3
Sector Field Mass Spectrometry for Elemental and Isotopic Analysis
Edited by Thomas Prohaska, Johanna Irrgeher, Andreas Zitek and Norbert Jakubowski

Published by the Royal Society of Chemistry, www.rsc.org

(***Note***: The $x$-axis of a mass spectrum (Figure 3.3) represents either simply the mass ($m$) or a relationship between the mass of a given ion and the number of its elementary charges. The IUPAC standard is $m/z$, which is the quantity formed by dividing the mass of an ion by the unified atomic mass unit and its charge number (absolute value).[1] On occasions, $m/Q$ or $m/q$ is found. $m$ is the symbol for mass and $Q$ or $q$ the symbol for charge with the units u/$e$ or Da/$e$. On rare occasions, instead of the mass/charge ($m/z$) ratio, the Thomson (Th) is used as a unit of the $m/z$, where negative ions would have negative values.[2])

$$1\text{Th} = 1\frac{u}{e} = 1\frac{\text{Da}}{e} \tag{3.1}$$

Th: Thomson
$u$: unified atomic mass unit
$e$: elementary charge ($1.6022 \times 10^{-19}$ C)[3]
Da: Dalton

Figure 3.2 shows a more detailed schematic sketch of mass spectrometers, which are used for elemental or isotopic analysis and described in this book (ICP-MS, GDMS, TIMS, SIMS, IRMS (gas source mass spectrometer both used in dynamic and static mode)). The mass separator is usually operated under vacuum, whereas the ion source can be placed under ambient pressure (ICP-MS), reduced pressure (GDMS), or vacuum (TIMS, SIMS, IRMS) (see also Chapter 4).

(***Note***: Even though sector field devices have been used in other instruments applied for elemental and isotopic analysis (such as AMS, SNMS, SSMS, RIMS or LIMS) as well, this book focuses on ICP-MS, GDMS, TIMS, SIMS and IRMS as major techniques in elemental and isotopic analysis.

The physical signal is (in most cases) reported as counts or as counts per second (cps), *i.e.* the number of ions of a selected $m/z$ arriving during a specific time interval or per time unit at the detector. This physical signal reflects the typical operation mode of a secondary electron multiplier (see Chapter 4). Alternatively, electron multipliers are operated in an analogue mode, measuring a voltage, which can be converted into cps. In other detectors, *e.g.* in Faraday cups, the signal is usually given in volts but sometimes converted into cps, as well.

(***Note:*** to convert the measured voltage into cps, the following equation can be used:

$$\text{SI} = \frac{U}{R} \times \frac{1}{e} \tag{3.2}$$

SI: signal intensity (cps)
$U$: voltage (V)
$e$: elementary charge ($1.6022 \times 10^{-19}$ C)[3]
$R$: resistance Ω

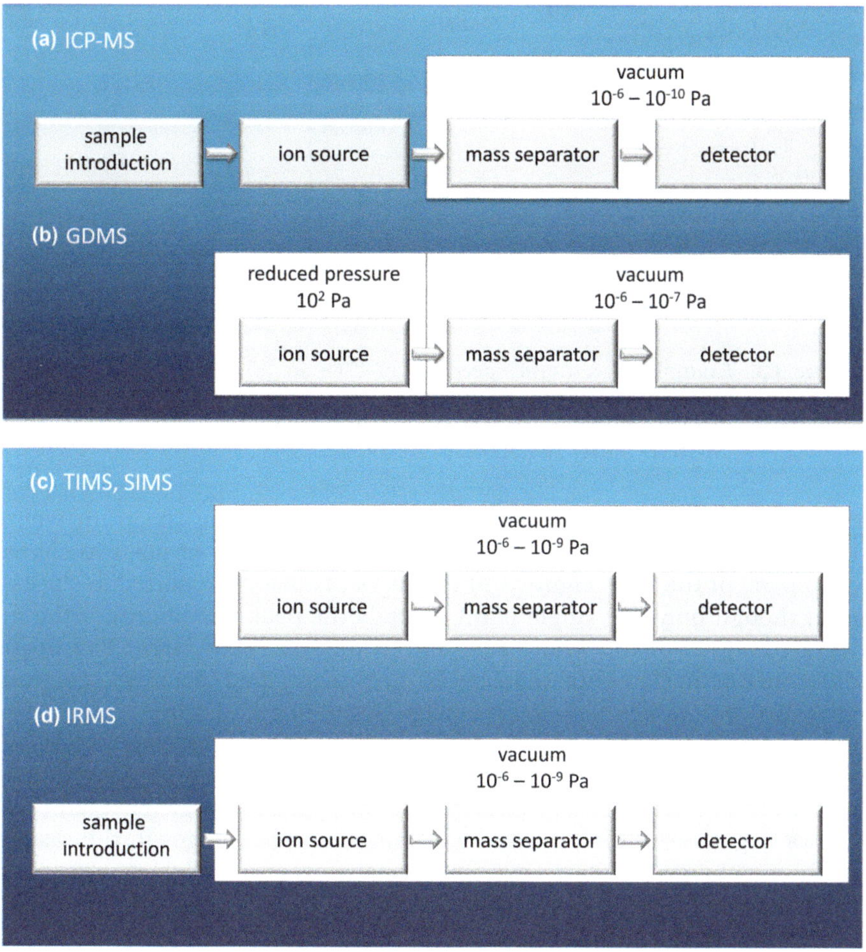


**Figure 3.2** Schematic diagram of different types of mass spectrometer.

a) ICP-MS: Ionisation source is not under vacuum. The generated ions have to be transferred from ambient pressure to high vacuum. The sample is introduced *via* a nebuliser into the ionisation source, but the system can be hyphenated with alternative introduction systems (*e.g.* chromatographic systems, laser ablation).
b) GDMS: The sample is under reduced pressure in, *e.g.* an Ar gas atmosphere. Coupling (*e.g.* of gas chromatography) is rare but possible.
c) TIMS, SIMS: the ionisation source is under vacuum and the sample has to be brought into the vacuum chamber.
d) IRMS: the ionisation source is under vacuum and the sample has to be brought in gaseous form into the vacuum chamber. Direct coupling of an alternative gas inlet system (*e.g.* a gas chromatograph) is possible.

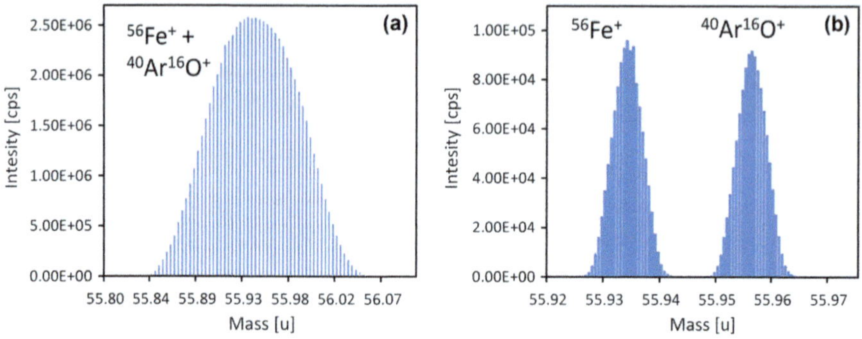


**Figure 3.3** Example of a mass spectrum of $^{56}$Fe in (a) low-resolution and (b) medium-resolution mode using single-collector ICP-SFMS (accomplished using an *Element XR* ICP-MS).
(Source: Ondrej Hanousek.)

The measurement output is typically reported in a mass spectrum, where the measured signal intensity is drawn *versus* the mass or mass to charge ratio ($m/z$) of the ion. (Figure 3.3) The signal is usually reported as "peak" even though only one single point (usually the peak maximum) or the integrated signal of the whole peak is recorded in most cases. The major difference of an elemental mass spectrum as compared to a mass spectrum obtained in organic mass spectrometry is that $m/z$ values are – generally – reported from 0 to about 250 (to cover the mass of Uranium at mass 238 and leave enough flexibility at $m/z > 238$ (*e.g.* U-oxides can be measured as molecular species)). The generated ions are typically single-charged positive or negative atomic or molecular ions (multiple charges in the case of positively charged ions are possible) consisting of usually not more than three atoms.

The success of magnetic sector field devices as mass separator is based on three major advantages:

1. the high sensitivity;
2. the potential of providing high mass resolution;
3. the principle to separate ions of different mass/charge ratios spatially (thus allowing for simultaneous detection of ions of different $m/z$ depending on the geometry of the sector field device).

The major drawbacks can be seen as:

1. the complexity of the systems;
2. the higher costs as compared to other mass separators (*e.g.* quadrupole or time of flight);
3. the limited time resolution between subsequent measurement points (this is especially true if the scanning of isotopes has to be accomplished by switching the magnetic field).

Due to counting statistics (eqn (3.3)) high-sensitivity mass spectrometers need shorter dwell times that can partly compensate the drawback of limited time resolution especially if low signal intensities are measured (*e.g.* in laser-ablation ICP-MS, online speciation ICP-MS and isotope ratio analysis).

$$\sigma_{\mathrm{CS}} = \sqrt{p_{\mathrm{m}}} \tag{3.3}$$

$$p_{\mathrm{m}} = q_{\mathrm{p}} \cdot T_{\mathrm{s}} \tag{3.4}$$

$\sigma_{\mathrm{CS}}$: counting statistics noise
$p_{\mathrm{m}}$: mean total particle (ion) counts
$q_{\mathrm{p}}$: number of particles (ions) per second
$T_{\mathrm{S}}$: total counting time/s

Even though this book focuses on machines using a sector field mass separator, Table 3.1 provides an overview as to whether the described types of instruments have been equipped with alternative mass separators such as quadrupoles, time-of-flight tubes, cyclotrons, ion traps or an accelerator mass separator. In some cases, when these combinations are not commercially available, selected literature references are given for further reading.

The main features of other types of mass separators, which are commonly used in elemental and isotopic mass spectrometry, are only summarised briefly. Figure 3.4 shows the schematics of common mass separators used in elemental mass spectrometry with a short explanation of the basic separation principles and features. More information on the different mass separators can be found in the dedicated literature.

As a consequence, the intended use has led to two major fields of application of mass spectrometers:

## 3.1 (Multi-)element Analysis

The first category of mass spectrometers comprises instruments that are used for multielement analysis (qualitative screening and quantitative measurement of the amount or amount content). The signal intensity (*e.g.* in cps or volts) is directly proportional to the number of generated ions, which are – for their part – proportional to the amount of substance of the element (isotope) in the sample. Thus, the signal can be interpreted as amount of substance (elemental or isotopic content) in the sample. This is usually accomplished *via* calibration of the mass spectrometer. For example, an external calibration makes use of adequate calibration standards by simply comparing the physical signal of a sample to the signal of a standard with known content (*e.g.* concentration or fraction) (see Chapter 8). In general, (linear) dynamic ranges of several orders of

**Table 3.1** Overview of the combination of different ionisation sources used in elemental and isotopic analysis with different selected mass analysers (status August 2013). Noncommercially available research systems or abandoned systems are given as (×) with selected literature reference.

| Ion source | Inductively coupled plasma | Glow discharge | Thermal ionisation | Ion beam | Ion beam | Electron beam | Laser | Spark source |
|---|---|---|---|---|---|---|---|---|
| Type of instrument | ICP-MS, SNMS[a] | GDMS | TIMS | SIMS | AMS | IRMS, SNMS[a] | RIMS, LIMS, SNMS[a] | SSMS |
| Magnetic sector field | × | × | × | × | ×[c] | × | × | (×) |
| Quadrupole | × | (×) | (×)[4] | × | | | × | |
| Time of flight | × | (×)[5,6] | (×)[7] | × | | | × | |
| Fourier transform cyclotron | (×)[8] | (×)[9,10] | | (×)[11,b] | | | | |
| Ion trap | (×)[12,13] | (×)[14] | (×)[15] | (×)[16] | | | | |
| Accelerator | | | | (×)[17,18] | × | | | |

[a]sample is sputtered by an ion beam and sputtered neutrals are ionised by an electron ion beam, a laser or an argon plasma.
[b]not for elemental analysis.
[c]AMS uses magnetic sector separators in combination with an accelerator mass separator.

magnitude are preferred for elemental quantitative analysis. The instruments used for quantitative analysis are usually equipped with one detector (generally an electron multiplier) even though they can be equipped with an additional detector (*e.g.* a Faraday cup) in order to extend the linear dynamic detection range. These instruments are usually operated in scanning mode, measuring the intensity of preselected isotopes sequentially. Thus, either the magnetic field or alternatively, the acceleration voltage is changed in order to sequentially transmit ions of increasing *m*/*z* ratio to the detector. Sometimes, the combination of both approaches is applied. For example, ICP-MS, GDMS and SIMS are found as single collector mass spectrometers for quantitative analysis.

(a) magnetic sector field MS

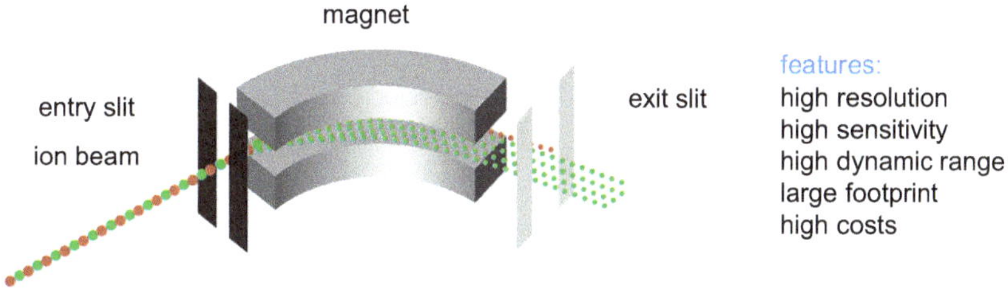


Ions are accelerated to high velocity and pass through a magnetic sector in which a magnetic field is applied perpendicular to the direction of ion motion.

(b) quadrupole MS

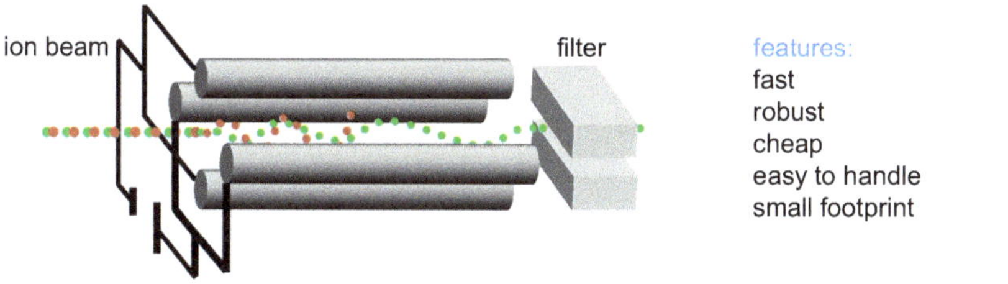


Ions are accelerated in a quadrupole field, which is built up between four rods. The rods are two positive and two negative electrodes of a DC source. Variable radiofrequency AC potentials (180°) out-of-phase are applied to each pair of electrodes. Ions of different *m*/*z* keep a stable trajectory in the x-z plane (high passmass filter) by modulating the AC/DC potentials.

(c) time of flight MS

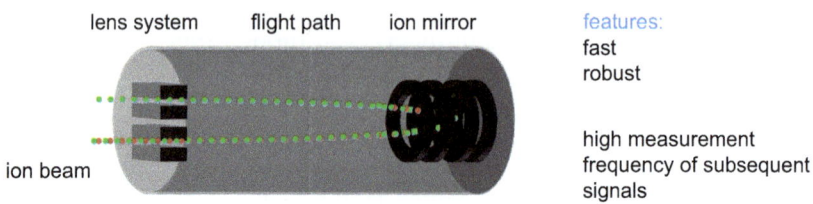

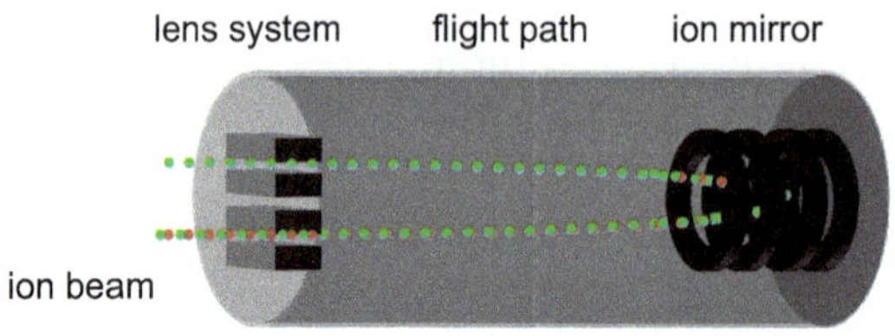


features:
fast
robust

high measurement frequency of subsequent signals

Ions are accelerated with the same kinetic energy into a fiel d-free drift tube. Ions are separated according to their different flight time as they have different velocity according to their mass. Ions are recorded at different times at one detector (time of flight) or at different positions at the same time (position of flight).

(d) Fourier transform cyclotron MS

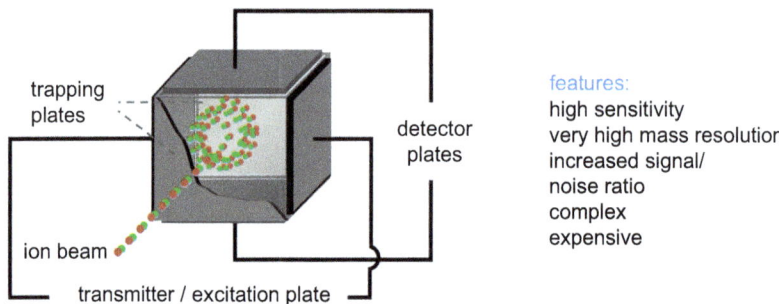

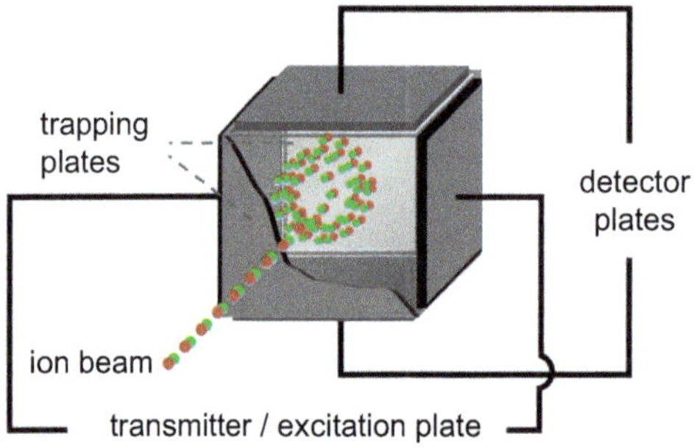


features:
high sensitivity
very high mass resolution
increased signal/
noise ratio
complex
expensive

Ions drift into a spatially uniform static magnetic field leading to a circular motion and are trapped in a Penning trap. The angular frequency is inversely proportional to the m/z value. The ions of a given *m/z* are excited at their resonant cyclotron frequencies to a larger orbital radius applying an oscillating electric field of a few milliseconds across the cell. The ions rotate in phase and induce a charge on a pair of electrodes when the package passes them. The resulting interferogram consists of a superposition of sine waves. The signal is transformed to a mass spectrum performing Fourier transformation.

(e) quadrupole ion trap MS

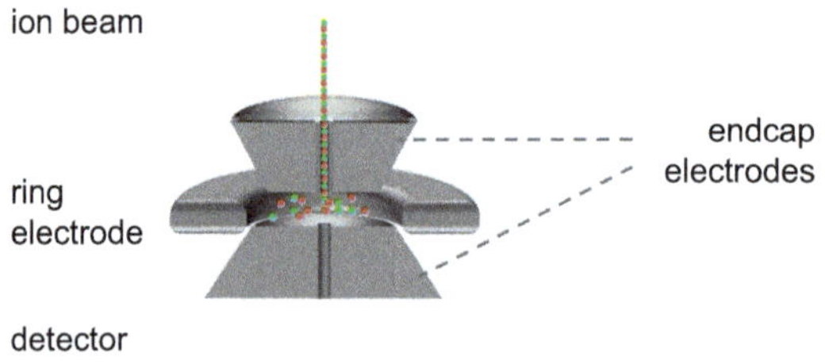


features:
low cost
easy to operate
sensitive

Ions are decelerated by collisional cooling into an ion trap using static direct current (DC) and radio frequency (RF) oscillating electric fields to trap ions. The ion trap can be a linear trap or a 3D trap (Paul trap). The ions of a certain *m/z* ratio are ejected from the ion trap by (multiple) frequency excitation.

(f) accelerator MS

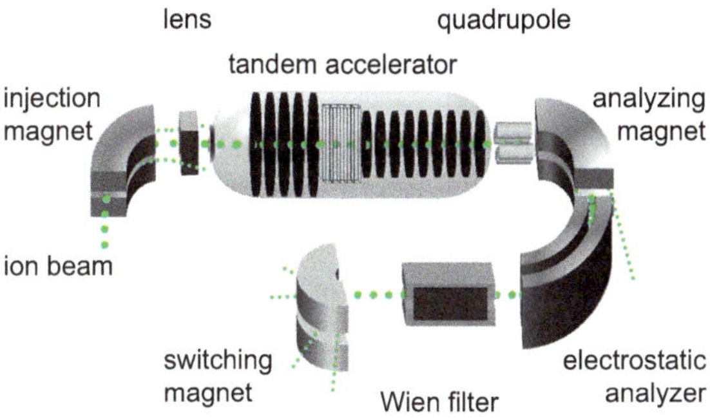


features:
measurement of large abundance ratios
extremely large footprint
expensive
complex

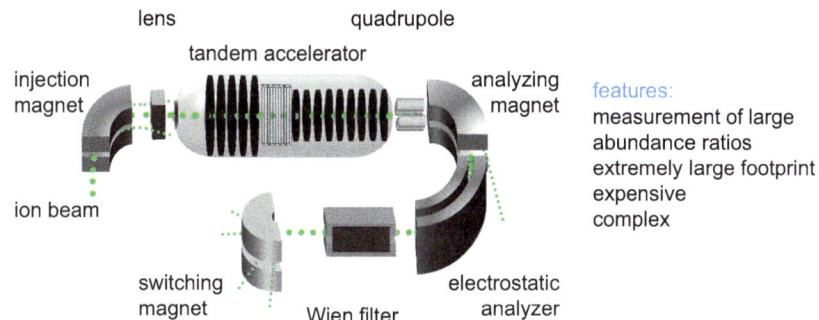

Negative ions are pre-accelerated and pre-separated in a sector field mass spectrometer. They are then accelerated to extremely high kinetic energies by a Tandem Van de Graaff accelerator operating at several million volts with two stages. At the connecting point, the ions pass a thin layer (stripping gas or thin carbon foil) changing their charge from negative to positive and molecules break apart. The positive ions pass an additional mass separator before they are finally detected by single-ion counting (by silicon surface-barrier detectors, ionization chambers, and/or time of flight telescopes).

**Figure 3.4** Schematic drawings of different types of mass analysers commonly applied in elemental and isotopic analysis: (a) magnetic sector field (b) quadrupole (c) time of flight (d) FT cyclotron (e) ion trap and (f) accelerator MS. (Source: Ondrej Hanousek.)

## 3.2 Isotope Ratio Analysis

The second category applies usually to instruments operated in static mode in order to detect multiple ion beams simultaneously on a set of detectors (Faraday cups and/or multiple ion counters or on channel plate-type array detectors) primarily for the purpose of high-precision isotope ratio measurements. Under this operation principle, the magnetic field is kept constant once set for a range of isotopes. These instruments are mostly referred to as multicollector devices. ICP-MS, TIMS, SIMS and IRMS instruments are most commonly used for isotope ratio measurements and equipped with a set of multiple detectors accordingly. Even though isotope ratio measurements can be accomplished using single-collector devices operated in scanning mode, as well, multicollectors are preferred as they are superior with respect to the resulting isotope ratio precision and are – more or less – independent of short and long time signal variations. The resulting signal (cps or volts) is used to compute an isotope ratio. Several correction procedures have to be used to convert a measured ratio into an absolute ratio or ratios are set in relation to a standard of known isotopic composition by calculating so called delta values (reflecting the difference – delta or $\delta$ – of a sample to a standard) (see also Chapter 6 and 8).

The main difference in instruments used for elemental and isotope ratio analysis is based on the type of ion sources such as, *e.g.* an inductively coupled plasma (used in ICP-MS), a glow discharge source (in GDMS), a thermal ionisation source (in TIMS), an electron impact source (in IRMS) or the use of an ion sputter source (in SIMS or SNMS). The ion source determines the type of sample but also the range of elements, which can be analysed. Table 3.2 gives an overview about general features of the techniques used, including information on the possible matrices to be analyzed (gaseous (g), liquid (l) or solid (s)).

Figure 3.5 shows the periodic table of elements giving an overview on which instruments have been used for which elements and whether the isotopic composition or the quantitative amount of an element can be addressed by the respective technique (regardless of whether the capability has been fully exploited, yet). A more detailed description about the already accomplished investigations in elemental and isotopic analysis is given in Chapters 11–17. It is obvious that the described types of devices have the potential to cover almost the complete periodic table of elements for elemental and isotopic analysis.

## 3.3 General Terminology and Abbreviations of Mass Spectrometric Techniques Described in this Book

The general definitions of terms used in mass spectrometry and recommended by the IUPAC are applied throughout this book,[19] involving appropriate terminology in analytical chemistry (The Orange Book)[20] or chemistry (The Gold Book)[1] or for quantities, units and symbols (The Green Book).[21] As for the abbreviations of the instrumentations, the book applies the commonly applied terms as recommended by the IUPAC.

Even though IUPAC recommends the use of hyphens preferably to describe variants of separation techniques (*e.g.* gas-liquid chromatography)[19] and slashes to separate abbreviated terms of combined techniques, the latter are rarely used (*e.g.* MS/MS for tandem mass spectrometers in organic mass spectrometry or GC/MS for the combination of gas chromatography and mass spectrometry). Most commonly, hyphens are used to abbreviate hyphenated techniques, instead. Nonetheless, the use of the hyphen appears to be inconsistent: The combination of an ionisation source with a mass spectrometer can be either seen as combination of two techniques (ionisation and separation) or as one technique (a mass spectrometer using a specific ionisation source). Abbreviations with hyphen (ICP-MS (inductively coupled plasma mass spectrometry)) and without hyphen (GDMS (glow discharge mass spectrometry)) and TIMS (thermal ionisation mass spectrometry)) are used for the combination of an ionisation source

**Table 3.2** Descriptive overview of different types of magnetic sector field instruments (overview covers only common features).

| Device | Ion source | Ions (preferably generated) | Detector single | Multiple | Measurement quantitative | Isotope ratio | Can be hyphenated | Sample matrix |
|---|---|---|---|---|---|---|---|---|
| ICP-MS | inductively coupled plasma | positive | × | × | × | × | × | g,l,s |
| GDMS | glow discharge | positive | × | | × | × | | s |
| TIMS | thermal ionisation filament | positive, negative | | × | | × | | l,s[a] |
| SIMS | ion sputter gun | positive, negative | × | × | × | × | | s |
| IRMS | electron impact source | positive negative | | × | | × | × | g,l,s |

[a]particles.

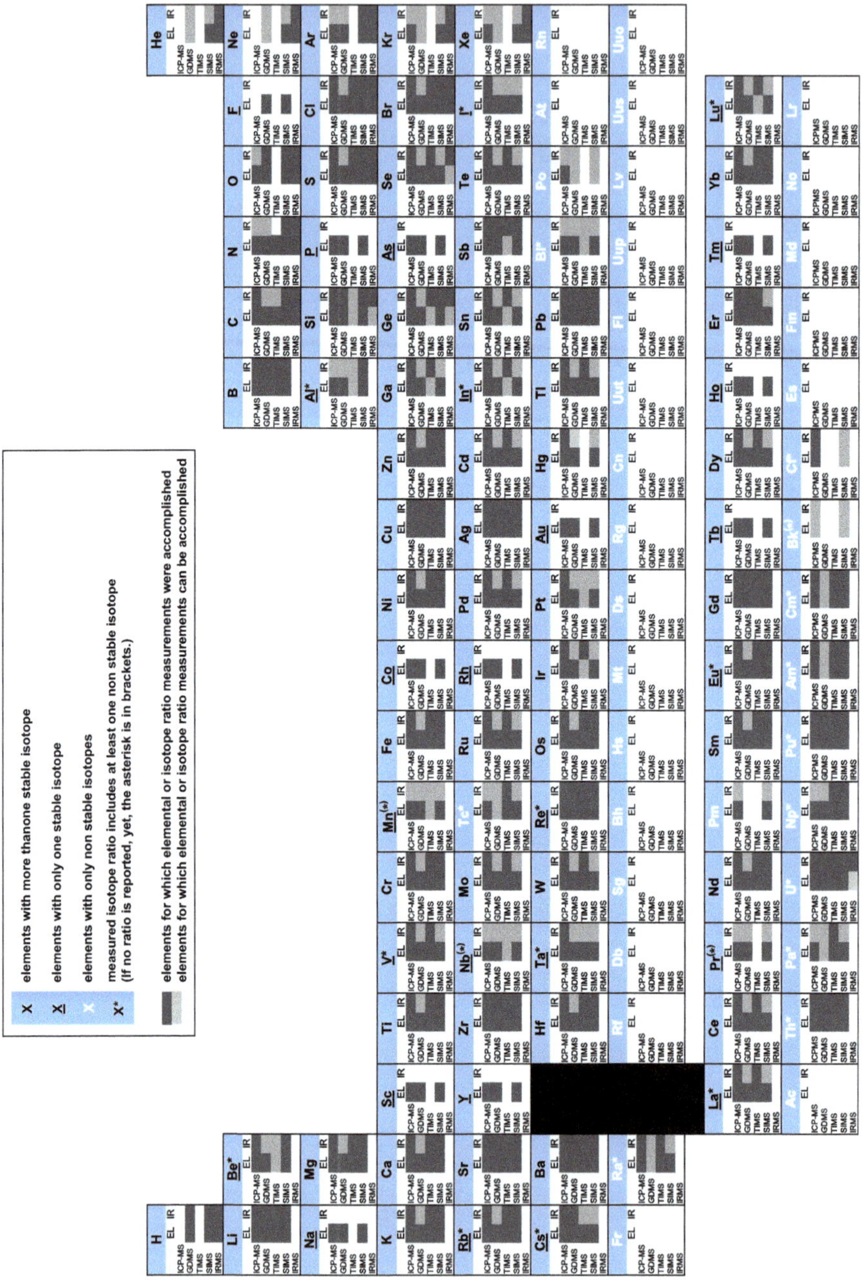
X elements with more thanone stable isotope
X elements with only one stable isotope
X elements with only non stable isotopes
X* measured isotope ratio includes at least one non stable isotope
(If no ratio is reported, yet, the asterisk is in brackets.)
elements for which elemental or isotope ratio measurements were accomplished
elements for which elemental or isotope ratio measurements can be accomplished

and a mass spectrometer. If consistent, ICP-MS could be written as ICPMS. The hyphen – nonetheless – appears within the publication of the inventor of the instrumentation and has therefore been referred to as ICP-MS ever since.[22] In addition, multicollector devices are commonly abbreviated as MC. Nonetheless, they only provide an additional information about the type of instruments and not a hyphenated technique. Therefore, MC should advance the abbreviation before the technique without hyphen. As a consequence, multicollector ICP-MS devices will be described as MC ICP-MS throughout this book.

SIMS (secondary ion mass spectrometry) refers to the use of secondary ions for measurement. The ion source would be an ion gun. The mass spectrometer for measuring isotope ratios of H, C, O, N, S (as primary elements) is usually referred to as IRMS (isotope ratio mass spectrometry) or SIRMS (stable isotope ratio mass spectrometer), not indicating the ionisation source but its intended use. IRMS used for the analysis of light stable isotopes (mainly H, C, N, O, S) is also referred to as gas source or stable isotope mass spectrometry without a commonly used abbreviation whereas IRMS used for noble gas (or light gas) isotope ratio analysis (mainly H, He, Ne, Ar, Kr, Xe) is commonly referred to as noble gas or static gas mass spectrometry (also without a commonly used abbreviation).

(***Note***: IRMS is defined as "Measurement and study of the relative abundances of the different isotopes of an element in a material using a mass spectrometer."[19] Therefore, the term is not necessarily used for the isotopic measurement of light stable isotopes by using a gas-source mass spectrometer, exclusively, but would apply for all mass spectrometers, which allow for isotope ratio measurements, *i.e.* basically all mass spectrometers. Nonetheless, IRMS is usually used when stable light isotopes or noble gas isotopes are measured with the appropriate instruments.)

**Figure 3.5** Overview of the described techniques and their application for quantitative or isotopic measurements with respect to elements in the PSE (reported in dark grey and not reported, yet, in light grey).
(***Note***: under IRMS - both conventional (dynamic) and noble gas (static) MS are reported
Black letter: elements with more than one stable isotope
White letter: elements with only nonstable isotopes
Underlined black letter: element with only one stable isotope
***Note***: All elements with a reported half-life are quoted as nonstable isotopes independent of the magnitude of the half-life. (*e.g.* $^{151}Eu$ has been reported with a half-life of $t_{1/2} > 1.7 \times 10^{18}$ a and $^{209}Bi$ has been reported with a half-life $t_{1/2} = 1.9 \times 10^{19}$ a)
Asterisk: The measured isotope ratio includes at least one nonstable isotope. (If no measured ratio is reported with any type of instrument, yet, the asterisk is in brackets.))

## References


1. K. K. Murray, R. K. Boyd, M. N. Eberlin, G. J. Langley, L. Li and Y. Naito, *Pure Appl. Chem.,*, 2013, **85**, 1515–1609.
2. R. G. Cooks and A. L. Rockwood, *Rapid Commun. in Mass Spec.*, 1991, **5**(2), 93.
3. in *CODATA Value: Elementary Charge*, ed. U. The NIST Reference on Constants, and Uncertainty, US National Institute of Standards and Technology, 2011, pp. Retrieved 2014-2005-2009.
4. A. L. Yergey, N. E. Vieira and J. W. Hansen, *Anal. Chem.*, 1980, **52**, 1811–1814.
5. A. Sanz-Medel, L. Lobo, N. Bordel, R. Pereiro, A. Tempez and P. Chapon, *J. Anal. Atom. Spectrom.*, 2011, **26**, 798–803.
6. C. G. de Vega, L. Lobo, B. Fernandez, N. Bordel, R. Pereiro and A. Sanz-Medel, *J. Anal. Atom. Spectrom.*, 2012, **27**, 318–326.
7. D. M. Wayne, W. Hang, D. K. McDaniel, R. E. Fields, E. Rios and V. Majidi, *Spectrochim. Acta Part B-Atom. Spectrosc.*, 2001, **56**, 1175–1194.
8. K. E. Milgram, F. M. White, K. L. Goodner, C. H. Watson, D. W. Koppenaal, C. J. Barinaga, B. H. Smith, J. D. Winefordner, A. G. Marshall and R. Houk, *Anal. Chem.*, 1997, **69**, 3714–3721.
9. C. H. Watson, J. Wronka, F. H. Laukien, C. M. Barshick and J. R. Eyler, *Anal. Chem.*, 1993, **65**, 2801–2804.
10. C. M. Barshick and J. R. Eyler, *J. Am. Soc. Mass. Spectrom.*, 1993, **4**, 387–392.
11. D. F. Smith, A. Kiss, F. E. Leach, E. W. Robinson, L. Pasa-Tolic and R. M. A. Heeren, *Anal. Bioanalyt. Chem.*, 2013, **405**, 6069–6076.
12. C. J. Barinaga, G. C. Eiden, M. L. Alexander and D. W. Koppenaal, *Fresenius J. Anal. Chem.*, 1996, **355**, 487–493.
13. D. W. Koppenaal, C. J. Barinaga and M. R. Smith, *J. Anal. Atom. Spectrom.*, 1994, **9**, 1053–1058.
14. S. A. McLuckey, G. L. Glish, D. C. Duckworth and R. K. Marcus, *Anal. Chem.*, 1992, **64**, 1606–1609.
15. A. Held, G. J. Rathbone and J. N. Smith, *Aerosol Sci. Technol.*, 2009, **43**, 264–272.
16. J. C. Ingram, A. D. Appelhans and G. S. Groenewold, *Int. J. Mass Spectrom.*, 1998, **175**, 253–262.
17. R. M. Ender, M. Dobeli, M. Suter and H. A. Synal, *Nucl. Instrum. Methods Phys. Res. Sect. B-Beam Interact. Mater. Atoms*, 1997, **123**, 575–578.
18. J. M. Anthony, J. F. Kirchhoff, D. K. Marble, S. N. Renfrow, Y. D. Kim, S. Matteson and F. D. McDaniel, *J. Vac. Sci. Technol. a-Vac. Surf. Films*, 1994, **12**, 1547–1550.
19. K. K. Murray, R. K. Boyd, M. N. Eberlin, G. J. Langley, L. Li and Y. Naito, *Pure Appl. Chem.*, 2013, **85**, 1515–1609.

20. J. Inczedy, T. Lengyel and A. M. Ure, *IUPAC. Compendium of Analytical Nomenclature, 3rd edn. (the "Orange Book")*, Blackwell Science, Oxford, 1998.
21. E. R. Cohen, T. Cvitas, J. G. Frey, B. Holmström, K. Kuchitsu, R. Marquardt, I. Mills, F. Pavese, M. Quack, J. Stohner, H. L. Strauss, M. Takami and A. J. Thor, *IUPAC. Quantities, Units and Symbols in Physical Chemistry, 3rd edn. (the "Green Book")*, Royal Society of Chemistry, Cambridge, UK, as edn., 2007.
22. R. S. Houk, V. A. Fassel, G. D. Flesch, H. J. Svec, A. L. Gray and C. E. Taylor, *Anal. Chem.*, 1980, **52**, 2283–2289.

CHAPTER 4

# *Technical Background*

LOTHAR ROTTMANN,[a] NORBERT JAKUBOWSKI,[b] STEFANIE KONEGGER-KAPPEL,[c] ONDREJ HANOUSEK[d] AND THOMAS PROHASKA*[d]

[a] Thermo Fisher Scientific, Bremen, Germany; [b] BAM - Federal Institute for Materials Research and Testing, Berlin, Germany; [c] Clemson University, Department of Chemistry, Clemson, South Carolina, USA; [d] University of Natural Resources and Life Sciences Vienna (BOKU), Department of Chemistry, Division of Analytical Chemistry, VIRIS Laboratory for Analytical Ecogeochemistry, Tulln, Austria
*Email: thomas.prohaska@boku.ac.at

The major components of sector field instruments can be listed as follows (Figure 4.1) and will be described in more detail in the following chapter. (Specific details for the different types of instruments will be given in the corresponding instrumental sections in Chapters 11–17.) The general principle of all mass spectrometers follows ionisation after sample introduction, separation of the ions according to their *m*/*z* ratio and final detection at the detection unit. The transport and separation of ions follow the principles of ion optics.[1]

Major components of sector field instruments:

- **sample-introduction system** (Section 4.1)
  The sample-introduction system provides the sample in the form to be suitable for the ion source. Some systems can use alternative sample-introduction systems in order to transport the sample into the ion source (*e.g.* into ICP-MS, IRMS).

New Developments in Mass Spectrometry No. 3
Sector Field Mass Spectrometry for Elemental and Isotopic Analysis
Edited by Thomas Prohaska, Johanna Irrgeher, Andreas Zitek and Norbert Jakubowski

Published by the Royal Society of Chemistry, www.rsc.org

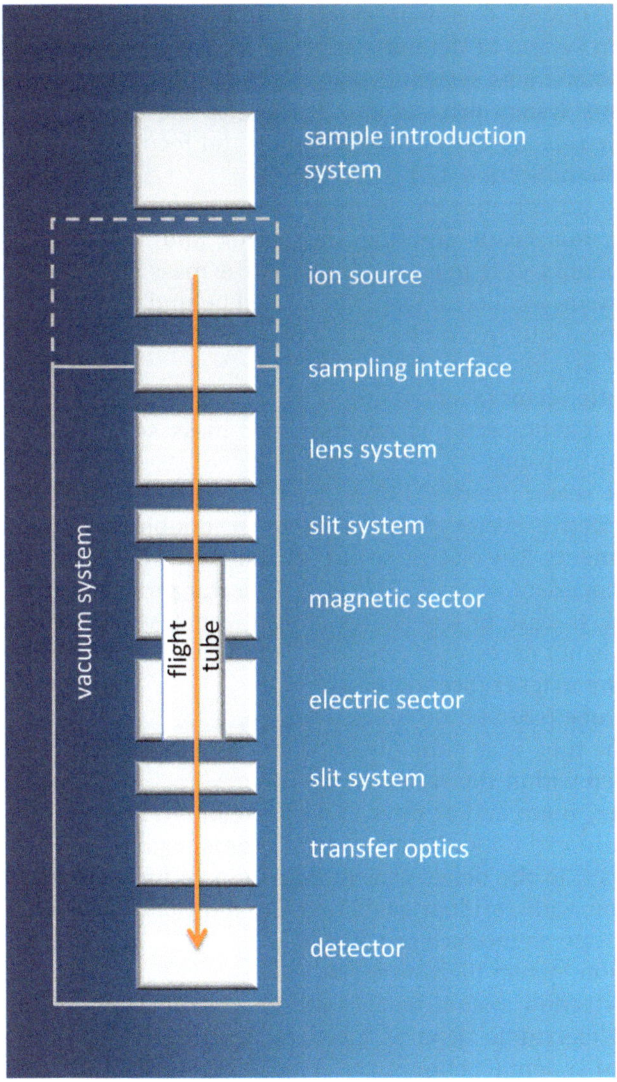


**Figure 4.1** Major components of a magnetic sector field MS used in elemental or isotopic analysis. The red arrow indicates the direction of the ions. (***Note***: The sequence of the magnetic and electrostatic sectors depends on the geometry.)

- **ion source** (Section 4.2)
  The ion source leads to the ionisation of the sample components and generates positively or negatively charged ions.
- **sampling interface** (Section 4.3)
  The sampling interface is mainly required if the ion source and mass analyser are not operated under the same vacuum conditions (ICP-MS,

GDMS, IRMS). The main purpose is to transport the ions into the vacuum system of the mass analyser ideally with minimal loss.

- **electrostatic lens system** (Section 4.4)
  The lens system extracts and accelerates the ions. Moreover, electrostatic lenses are used to shape, guide and focus the ion beam.
- **slit system** (Section 4.5)
  Usually an entrance slit in front of the sector field defines the width of the ion beam entering the sector and thus – amongst other components – the mass resolution of the mass analyser. Depending on the system, a collector (exit) slit is used behind the sector field, as well. Additional slits (*e.g.* alpha slits) can be used to reduce beam aberrations caused by a diverging beam.
- **magnetic sector** (Section 4.6)
  The magnetic sector is necessary for mass separation and provides angular focusing.
- **electric sector** (Section 4.7)
  The electric sector field is used in double-focusing sector field instruments. The electric sector separates ions of different energies and is independent of mass. The electric sector provides angular focusing, as well. In case of low energy spread (*e.g.* TIMS or IRMS), an electric sector is not necessarily required. The electric sector is placed before or after the magnet (Section 4.8).
- **flight tube** (see Section 4.10)
  A flight tube is a prerequisite when the magnet pole pieces are not installed within the vacuum.
- **transfer, zoom and/or filter optics** (Section 4.11)
  An additional optics before the detector is usually applied to guide the ion beam to the detector or to discriminate ions that have undergone collision with residual gas along the ion optical pathway and would deteriorate abundance sensitivity.
- **detection system** (Section 4.12)
  The detection system detects the number of incoming ions. The resulting electronic signal is translated into a digital output.
- **vacuum system** (Section 4.13)
  The vacuum is a prerequisite to minimize collision with residual gases in order to enable ion beams with a distinct geometry to maintain high resolution and to minimise the loss of ions travelling from the ion source to the detector.

## 4.1 Sample Introduction

The sample to be analysed has to be converted into a form that is compatible with the requirements of the ion source. Therefore, the samples (its condition of aggregation and size) are already predefined by the used ion source. In TIMS, the sample is deposited on a filament, which is manually introduced into the vacuum chamber of the ion source. In order to analyse several samples without opening the vacuum chamber, multiple filaments are mounted onto a carousel. In GDMS, the solid sample is directly mounted in the GD source, since it acts as the cathode of the ion source. In SIMS, the sample is mounted as such within the vacuum chamber where it is bombarded by ions generated by an ion gun.

Alternative sample-introduction systems can be usually used if the sample can be continuously transferred into the ion source. This is usually the case if the system operates under ambient pressure (*e.g.* ICP-MS) or if the sample is introduced into the ionisation source as gas (*e.g.* IRMS).

In the case of ICP-MS, the sample is introduced continuously either as a wet or dry aerosol or as a gas. Thus, any system providing a continuous flow of a liquid (usually a nebuliser combined with a spray chamber), a gas or of small particulates can be coupled to an ICP-MS, such as *e.g.* pumped systems, chromatographic systems, or hydride generation systems. But also cold vapour (CV) techniques,[2] electrothermal vaporisation (ETV)[3] and gas exchange devices (GED)[4] have already been coupled to an ICP-MS. Alternatively, solid samples can be introduced into the system by applying laser ablation (LA),[5] which allows for the ablation of small particulates of the solid surface. In all cases, the sample is dispersed in a carrier gas stream (Ar or He) and directly transferred into the inner stream of the ICP. Sample-introduction systems for ICP-MS are explained in detail in dedicated literature.[6,7] IRMS needs a gaseous sample, which is continuously introduced into the ionisation source (electron impact ionisation). Therefore, the system can be coupled, *e.g.* with a GC as a sample-introduction system.

## 4.2 Ion Sources

A prerequisite for all mass spectrometers is the generation of positively or negatively charged particles, molecules or atoms out of the sample. The energy in the ionisation source is needed for different physical processes (in elemental or isotopic mass spectrometry): evaporation, dissociation, atomisation, excitation and finally ionisation. Thus, most ion sources used in elemental mass spectrometry have a significantly different function as compared to organic mass spectrometry, where the aim is to preserve some of the structural information by applying "soft ionisation" techniques. In elemental mass spectrometry, the destruction of the molecular structure leads ideally to the formation of singly charged atomic ions. Nonetheless, multiple-charged atomic ions or molecular ions (which usually consist of a maximum of 3 atoms) are generated, as well. In the case of IRMS, the applied electron impact source aims to ionise the gas (*e.g.* $CO_2$) without breaking the molecular structure of the gas.[8]

The ideal ion source in elemental/isotopic mass spectrometry should fulfil the following basic requirements:

- The ion source should generally create the major ionic species with the same charge (preferably single charged) with high efficiency in order to achieve the lowest possible detection limits or determine the isotope abundance ratios with high precision.
- The ion densities of the individual elements should be directly proportional to the amount of the element, which is introduced into the ion source *via* the sample.
- The ion density of other ionic species should ideally be low in order to not interfere with the analyte ion of the same mass.
- The ions should have a minimum energy spread.
- The ion generation should be free of fractionation effects in the ion source with respect to different isotopes of the same element, between elements or between different elemental species.
- The ion source should be free of memory effects.
- The ionisation should be free of any matrix effect of the concomitant matrix elements.
- The number of formed ions should be stable with time.

All requirements are hardly fulfilled by any of the used ion sources. Adequate measurement and correction strategies have therefore to be applied in the analytical process (*e.g.* blank correction, correction for instrumental isotopic fractionation, nonlinearity of ion sources, matrix effects or spectral interferences).

Ion sources used in elemental mass spectrometry can be generally grouped into plasma ion sources, beam ion sources or field desorption sources. Figure 4.2 gives a schematic overview of the major ion sources used in elemental mass spectrometry. From the figure, it can also be seen that these ion sources are operated at a variety of different pressures: The inductively coupled plasma is operated at atmospheric pressure, the glow discharge at reduced pressure, whereas thermionic, secondary and electron impact ion sources are operated at high- or even ultrahigh-vacuum conditions. Thus, all sources need specific sample-introduction systems, ion source containers and interfaces to couple them to the mass analyser.

### 4.2.1 Plasma Ion Sources

A large group of ion sources in elemental mass spectrometry is based on plasma ion sources, which are very universal and have a long tradition.[9] In the plasma state, ionisation takes place and the resulting ions can be sampled for the further mass separation process. The main commercially available ionisation sources using plasmas are the inductively coupled plasma (ICP, see Chapter 12) and the glow discharge (GD, see Chapter 13). Besides, direct current plasma (DCP),[10] the spark ion source (SS)[11], the

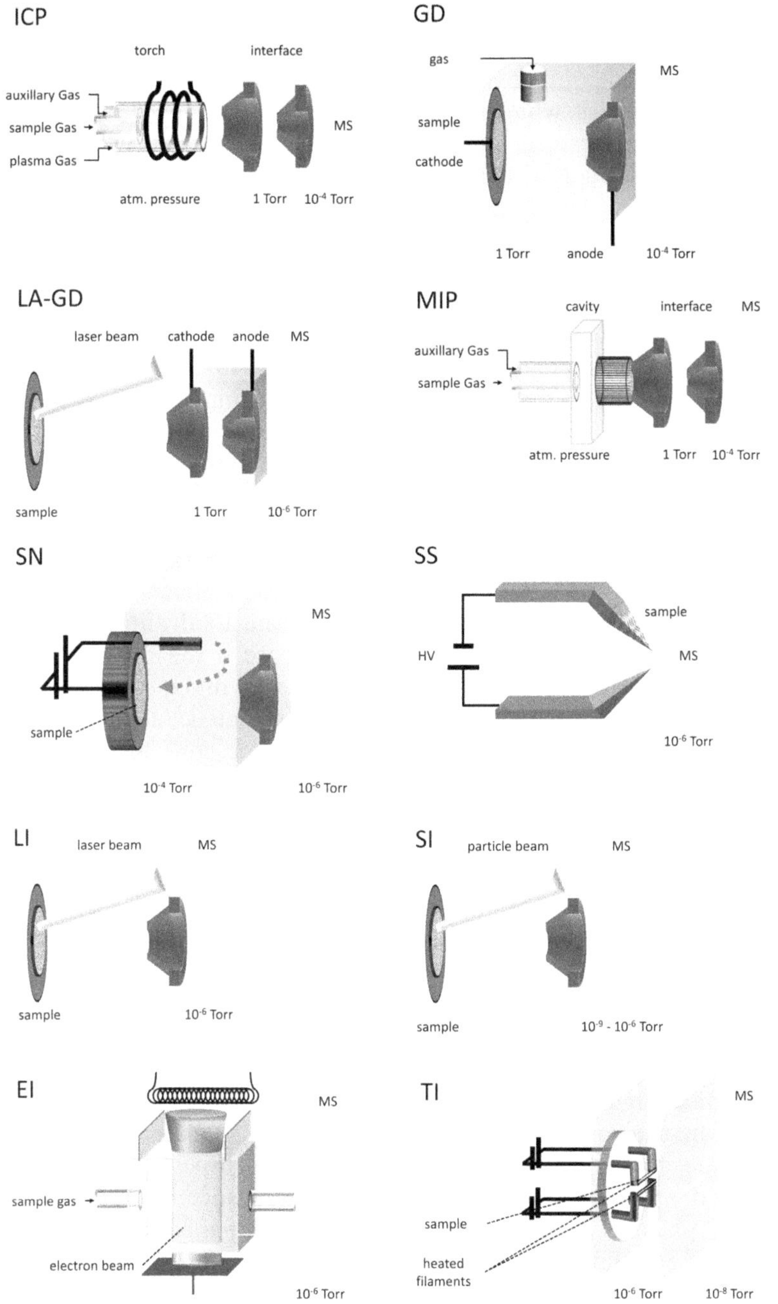


**Figure 4.2** Overview of ion sources used in elemental mass spectrometry: ICP (inductively coupled plasma), GD (glow discharge), LA-GD (laser ablation-glow discharge), MIP (microwave induced plasma), SN (secondary neutral source), SS (spark ion source), LI (pulsed laser introduced), SI (secondary ion source), EI (electron ionization), TI (thermal ionization), partially after Adams and Vertes.[8]

pulsed laser induced (LI)[12,13] or microwave-induced (MIP)[14] plasma sources have been used in combination with elemental mass spectrometers.

From a physical point of view, the plasma is a quasineutral gas of charged and neutral particles, which exhibits collective behaviour.[15] Within a gas of high enough temperature, collision of atoms and electrons will ionise some of the atoms leading to a state of matter where electrons, ions and neutrals coexist. The Debye length defines this collective behaviour. A charged particle in a plasma will attract opposite charges and repel equal charges to the point that its electric field is shielded by the charged particles it has attracted. The Debye shielding length defines the scale length beyond which the electric field is collectively shielded out.[15]

Plasmas used as analytical ion source have to fulfil certain requirements. Depending on the physical state of the sample introduced into the plasma, the material has to be vaporised, dissociated and/or atomised, excited and finally ionised. The energy for all these processes has to be provided by the plasma (and thus the plasma generator). An ionisation source should lead to a large number of mainly singly charged ions of ideally all elements of the periodic table with as uniform sensitivities as possible. The plasma should not show any matrix effects,[16] which means that the sensitivity should not change with the total amount of sample material in plasma. This is most often the case if the density of the vaporised material introduced into the plasma is much smaller than the density of the gas in which the discharge is initiated. Most often a noble gas, *e.g.* argon, is used as discharge gas for ion sources in mass spectrometry.

Plasmas can be generated in different ways, but always external energies are needed for instance from chemical heat-producing processes, by compression of gases, by application of photons in laser-generated plasmas or by bombardment of fast particles, such as electrons (electrical currents), ions or neutrals. Technical plasmas – and this holds also true for plasma ion sources – are often generated by a discharge in a noble gas by application of high voltages. These voltages can be applied to electrodes or can be induced electrode-less by the application of high-frequency electric or magnetic fields. Electrodes are used to initiate corona discharges, sparks, arcs and low and high-pressure glow discharges. In the case of AC discharges radio frequency (RF) and microwave-induced plasmas (MIP) are of important technical interest and have been used as sources for emission and mass spectrometry, too. Most often, RF plasmas are generated electrode-less by means of capacitive or inductively coupling of high voltages in a flowing noble gas atmosphere (see ICP-MS, Chapter 12).

Plasmas are physically characterised by a couple of parameters, among which particle densities (neutrals, electrons, ions) and temperatures play an important role, because they define the physical properties of a plasma already – for instance the ionisation properties. For more details see Chen 1974.[15] Most plasmas are not in a thermal equilibrium, which means that all particles (electrons, ions and neutrals) might have different temperatures. However, plasmas, which are in a thermodynamic equilibrium, can be characterised by a single temperature – the thermodynamic temperature. This temperature describes the energy distribution of all particles, the state of excitation and ionisation, the abundance of chemical species and the spectral energy distribution by the Maxwell-Boltzmann's law and Saha-Eggert's equation, the law of Mass Action and Planck's law. In particular the Saha-Eggert equation (eqn (4.1)) is very useful to calculate the state of ionisation for each element and thus the ionisation efficiency:

$$K_{\mathrm{ION}} = \frac{n_{\mathrm{i}} n_{\mathrm{e}}}{n_{\mathrm{a}}} = \left(\frac{2\pi m_{\mathrm{e}} k T_{\mathrm{ION}}}{h^2}\right)^{3/2} \frac{2Z_{\mathrm{i}}}{Z_{\mathrm{a}}} e^{-\frac{E_{\mathrm{ION}}}{kT_{\mathrm{ION}}}} \qquad (4.1)$$

$K_{\mathrm{ION}}$ : ionisation equilibrium constant
$n_{\mathrm{i}}$, $n_{\mathrm{e}}$, $n_{\mathrm{a}}$ : number of ions, electrons and atoms/volume (densities)
$m_{\mathrm{e}}$ : mass of the electron
$k$ : Boltzmann constant ($1.381 \times 10^{-23}\ \mathrm{J \cdot K^{-1}}$)
$T_{\mathrm{ION}}$ : ionisation temperature
$h$ : Planck's constant ($6.626 \times 10^{-34}\ \mathrm{J \cdot s}$)
$Z_{\mathrm{i}}$, $Z_{\mathrm{a}}$ : partition function of the ionic and atomic state
$E_{\mathrm{ION}}$ : ionisation energy of the element

It is evident from the equation that the ionisation depends strongly on the ionisation energy. The Saha-Eggert equation – with certain assumptions of the temperatures and densities – has been used for instance by Ramendik *et al.*[17] to calculate relative sensitivity factors for ICP-MS and to compare them with those actually measured with an ICP-QMS. However, they found that the ionisation process in inductively coupled plasmas is described only approximately by the Saha-Eggert equation. They had to include some other physical parameters such as the electronegativities of atoms to improve the approximation.

**Inductively Coupled Plasma (ICP)**

An ICP is a high-temperature Ar plasma, which is generated by applying a radio frequency at an RF load coil that is positioned around a torch. Ar gas flows through the three concentric quartz tubes of the torch at different flow rates. After initial generation of Ar ions and free electrons by a spark, the free electrons are accelerated first in one and then in the opposite direction by the oscillating magnetic field generated by the applied RF (usually ~27 MHz or ~40MHz) at the load coil. When the accelerated electrons collide with Ar gas atoms, $Ar^+$ ions are formed by mainly electron impact ionisation. This continues until a steady state is reached (*i.e.* the rate of generation of free electrons and the loss by recombination is equal). Temperatures up to $10^4$ K can be reached in an ICP[18] (see Chapter 12).

**Glow Discharge Plasma**
A glow discharge plasma is formed by applying a high voltage between two electrodes in the presence of a noble gas (mainly Ar) at reduced pressure. In the case of a DC glow discharge ion source, the sample acts as cathode. Therefore, the sample has to be conductive. Particular measures are applied for allowing measurements of nonconductors. Ions (mainly gas ions) that are close to the cathode (cathode dark space) are accelerated towards the surface of the cathode (sample), where they sputter atoms off the sample surface. Inside the plasma, sample ions are generated by charge transfer, penning or electron impact ionisation[19,20] (see Chapter 13). In combination, laser ablation and GDMS can be applied.[21]

**Direct Current Plasma (DCP)**
A direct current plasma is created by an electrical discharge between two or three electrodes usually using argon as plasma support gas. The ionisation of the gas between the anode(s) (graphite electrodes) and the cathode (tungsten electrode) produces a Y-shaped plasma in which the sample can be introduced together with the Ar support gas between the two anodes. Samples can be also deposited on one electrode or make up one electrode. Insulating solid samples are placed near the discharge and ionised gas atoms sputter sample atoms into the gas phase where they are excited (and ionised) (see also DC-GD in Chapter 13). Direct current plasmas produce plasma temperatures between 3000 K and 6000 K. These plasma sources were used as early ion sources for elemental analysis in the 1970s but the plasmas were prone to spectral interferences and were not reliable. Moreover, much of the analyte goes around the plasma. Nonetheless, the DC plasma is very tolerant, *e.g.* for slurry samples and high matrix contents.[10]

**Microwave-Induced Plasma (MIP)**
Microwave-induced[14] plasma sources form a plasma without using electrodes. The microwave energy (usually 100–200 W) is supplied to a gas inside a quartz or glass tube by an excitation cavity. Inside the bulk of the MIP, temperatures between 2000 K and 5000 K are achieved. The MIP sources often lead to matrix effects and are usually less robust, but, *e.g.* nitrogen MIP, may provide a less expensive alternative to ICP.[22–25]

**Spark Source Plasma**
Spark source plasmas have long been used as robust, sensitive and comprehensive ionisation sources for mass spectrometers. The RF spark source has been the most widely used spark source. A vacuum spark discharge is generated between two electrodes of the sample material by a pulsed high-voltage (25–100 kV) AC potential.

Another discharge is formed in the triggered low-voltage DC vacuum arc discharge. The discharge is triggered by a high-voltage pulse as well, but is maintained as an arc discharge by a low voltage DC potential for about 100 μs. Only the ions generated during this phase are finally used for analysis. As the spark source leads to ions with a large energy spread, the ion source was generally used in combination with a double-focusing magnetic sector field instrument.[26,27]

**Laser-Induced Plasma**

Laser-induced (LI) plasmas use a high-energy laser pulse as ionisation source. The laser is focused to the surface and ablates material in the range of nanograms or even picograms. The plasma plume, which is formed by the ablated material, is then used to atomise and ionise the sample.[12,13] The ions generated in the laser plume are further transferred into a mass spectrometer.[28,29]

## 4.2.2 Beam Sources

The second large group of ion sources are based on particle or photon beams. The incoming beam leads – depending on the beam used – to either vaporisation, atomisation and ionisation of the samples or just to ionisation. As ion beam sources in elemental mass spectrometry, an ion beam (in SIMS, see Chapter 15), a beam of electrons (EI, electron impact sources, see Chapter 16) or a photon beam generated by laser interaction (in resonance ionisation mass spectrometry, RIMS or sputtered neutral mass spectrometry (SNMS)) are the most common techniques.

**Ion Beam Sources**

In SIMS, a positive (*e.g.* $Cs^+$) or negative (*e.g.* $O^-$) ion beam is focused onto the surface. The ion beam is used to vaporise the sample and leads to partial ionisation of the vaporised material. ($Cs^+$ ion sources are used for SIMS applied in organic mass spectrometry, as well). Also in accelerator mass spectrometry (AMS), a $Cs^+$ ion source is used to create negative ions for further analysis (of, *e.g.* $^{14}C^-$).[20,30,31]

In SNMS (secondary (or sputtered) neutral mass spectrometry), the sample is sputtered by an incident ion beam of, *e.g.* $Ar^+$, $Kr^+$ or $Ne^+$ with an energy in the range of 0.5–5.0 keV. The neutrals are ionised separately by an additional ion source.[32] This postionisation is accomplished by, *e.g.* a laser beam or by an electron cyclotron wave resonance plasma. Post-ionisation is an important strategy to decouple the sputtering and ionisation process. A laser allows for the selective postionisation making use of resonant ionisation as in RIMS. The advantage is a reduction of matrix effects as they occur usually in SIMS. Alternatively, conventional desorption or sputtering techniques can be coupled to resonance ionisation mass spectrometry.

**Electron Beam Sources**

Electron ionisation sources (or electron impact sources) use a beam of electrons, which passes through a gas and collide with the neutral molecules, which leads to the ionisation of these molecules. Energies of about 50–100 eV are often used. The ionisation technique is used for stable and static isotope ratio mass spectrometry (see Chapter 16) but is mainly found in organic mass spectrometry.[33,34]

**Photon Beam Sources**
Photons generated by a laser beam are often used as postionization sources after sputtering events. Ionisation is accomplished by removing an electron from an atom by excitation by providing enough energy or selectively by applying laser energy at the resonance frequency (RIMS). The method is based on the resonant excitation and ionisation of atoms by means of a pulsed narrow continuous wave laser with a specific wavelength.[35,36] Atoms are removed from the sample by ion sputtering or laser desorption. The secondary atoms are intercepted by multiple pulsed laser beams tuned to excite successive electronic transitions in a preselected element of interest leading to their ionisation. Atoms of other elements are generally transparent to the laser radiation and hardly ionised.[37] Laser sources are also used for ablating material and ionising the volatilised sample in a plasma plume (see Section 4.2.1).

### 4.2.3 Thermal Ionisation Source

A desorption ion source in elemental mass spectrometry is used in thermal ionisation mass spectrometry (TIMS). Controlled desorption/evaporation and ionisation processes of the sample take place either at an electrically heated filament or with the aid of two (or three) filaments (one for evaporation and one for supporting the ionisation)[38] (see Chapter 14).

## 4.3 Sampling Interface

The sampling interface is a crucial and nontrivial part of the mass spectrometer. Within the sample interface, ions have to be extracted from the ion source - which is operated at ambient pressure in the case of ICP, a few hundred Pa in the case of GD or at vacuum (SIMS, TIMS) – to the mass analyser, which is operated under high vacuum. Stationary atmospheric plasmas like the ICP or GD have to be interfaced with one or more stages of a differential pumping system.

Depending on the pressure of the ion source with respect to the mass analyser different mechanisms of the ion transport can be applied:[8]

- When the ion source is operating at a vacuum similar to the mass analyser pressure or during the expansion of the plasma particles have negligible interactions, the ions can be extracted by electric fields.
- When the ion source is operating at atmospheric pressure, the ions are mainly transported by gas dynamics towards the vacuum of the mass analyser. When the ions reach a pressure where the mean free path of

the particles is large enough compared to the dimensions of the ion source, they can be guided by an electric field.
- When the ion source is operating at vacuum in between the two above-mentioned cases (*e.g.* fast flow GD ion source at approx. 100 Pa), the transport mechanism is a mixture of the flow towards a higher vacuum and extraction by an external electric field.

If ions are transported mainly by gas flow dynamics, they usually show an excess and high variation of kinetic energy. Stationary plasma sources also provide a high density of ions. As a consequence, the external electric field accomplishing ion extraction can only penetrate into the plasma following the Debye length. Therefore, only surface ions could be sampled. Moreover, a high charge density will lead to a mutual repulsion of ions, the so-called space-charge effects, resulting in matrix effects.

## 4.4 Electrostatic Lens System

Electrostatic lenses are generally used in sector field mass spectrometers in order to assist the transport of an ion beam. Depending on the kind of the lens, they are used to accelerate or decelerate ions, to shape the ion beam, to confine it or to focus it. Electrostatic lenses consist of electrodes that are usually held at different static potentials. Commonly, electrostatic lenses are made of coaxial apertures or cylinders, multipoles (*e.g.* quadrupole) or slits.

Since there is a lot of analogy of ion optics to light optics, the terminology and also the applied equations are similar. For instance, the law of refraction (Snell's law), which describes the relationship between angles and velocities of waves passing through a boundary between two different isotropic media, has its analogon in ion optics. However, in the latter case, the ions will either be accelerated or decelerated and the trajectories depend on the angle of incidence with respect to the equipotential surfaces of the corresponding electric field generated by the lens.

Thus, instead of the ratio of the refractive indices or velocities, which are equal to the ratio of the sines of the angles of incidence ($\theta_1$) and refraction ($\theta_2$) in light optics, in ion optics the ratio of the sines of these angles with respect to the normal of the equipotential surfaces is equal to the ratio of the velocities. Since it is more practical, this ratio is usually expressed by the ratio of the square root of the two potentials $V_1$ and $V_2$ (eqn (4.2)).

$$\frac{\sin\theta_1}{\sin\theta_2} = \sqrt{\left(\frac{V_1}{V_2}\right)} \tag{4.2}$$

$\theta_{1,2}$ : angle of incidence (1) and refraction (2)
$V_{1,2}$ : potential 1 and 2

It is also possible to define the focal length, focal points and principal planes for any electrostatic lens as for light optical lenses, and thus the entire properties of the optical system. Therefore, all ideal lens equations for light optics also apply for electrostatic lenses.

However, there are a few major differences:

a) In light optics, there is no effect caused by space charge (*i.e.* the Coulomb repulsion of ions of equal charge).
b) In ion optics, there is an infinite number of equipotential surfaces that bend the ion beam at different regions around the lens, whereas there is only one refractive surface in light optics when the light is passing from one media to another.

The simplest ion optical lens is an immersion lens, which consists of two electrodes. The ion beam travels from one cylinder lens at a potential $V_1$ to an adjacent cylinder lens at a different potential $V_2$. Therefore, the ions have a different energy before entering the lens and after leaving the lens (*i.e.* this arrangement is either used when acceleration or deceleration of the ion beam is required). Depending on the ratio of $V_1$ and $V_2$ different focusing properties can be achieved. In Figure 4.3, the trajectories of an ion beam for an immersion lens arrangement can be seen. For the following descriptions, it is assumed that the ions starting from the left and are flying to the right side.

The upper example in Figure 4.3 uses an accelerating arrangement (*i.e.* $|V_2|>|V_1|$) with the ratio of $V_1/V_2$ set in a way that the parallel ion beam (focal length $\infty$) is focused after passing the second lens. The configuration in the bottom picture in Figure 4.3 is still an accelerating arrangement ($|V_2|>|V_1|$) but with a different $V_1/V_2$ ratio. In this configuration, a diverging ion beam is refocused behind the second lens.

Another arrangement with three electrodes that is often used in mass spectrometry is an Einzel lens. This arrangement is mainly applied when a

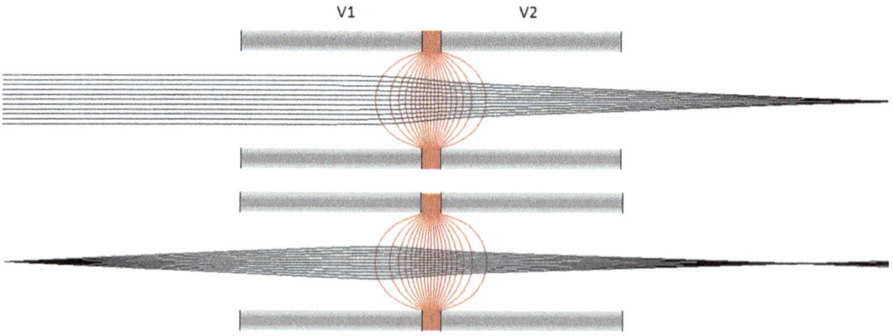


**Figure 4.3** Trajectories of an ion beam for an immersion lens in an accelerating arrangement ($|V_2|>|V_1|$) with two different $V_1/V_2$ ratios (upper and lower picture).

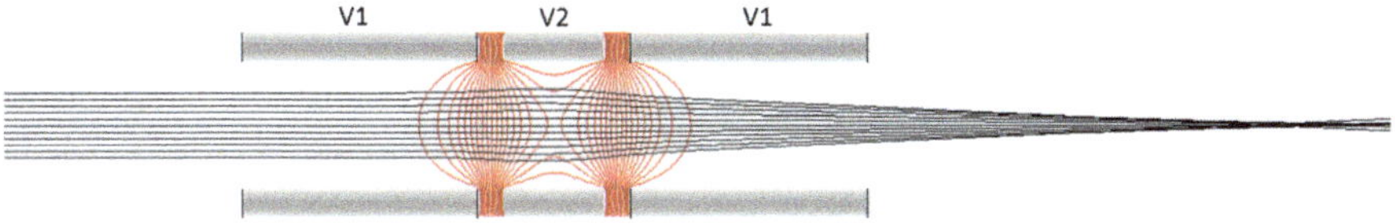


**Figure 4.4** Trajectories of an ion beam in an Einzel lens arrangement with a decelerating lens in the centre (*i.e.* $|V_2|<|V_1|$).

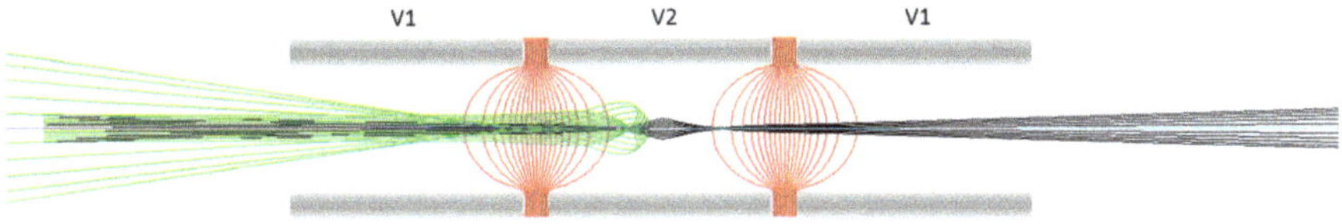


**Figure 4.5** Trajectories of an ion beam (with different energies for the green and black trajectories) in an Einzel lens arrangement consisting of three cylindrical lenses. The voltage at the centre electrode is set to 1000 V and the two outer electrodes are at ground potential.

focus of an ion beam is required, but the final energy of the ions has to be the same as before the lens. In Figure 4.4 the trajectories of an ion beam in an Einzel lens arrangement is shown with a decelerating lens in the centre (*i.e.* $|V_2|<|V_1|$).

This kind of arrangement can also be used to discriminate ions with a "wrong" energy, for instance ions that underwent collisions with residual gas along the pathway before the lens. In a sector field mass spectrometer, these ions are the main reason for the tailing of peaks and can influence abundance sensitivity.

The trajectories of an Einzel lens arrangement consisting of three cylindrical lenses are shown in Figure 4.5. In contrast to the lenses shown in Figure 4.4, a longer centre electrode is used in this arrangement. In this particular example, the voltage at the centre electrode is set to 1000 V and the two outer electrodes are at ground potential.

Figure 4.5 shows two kinds of ions: (1) ions displayed as black trajectories with an energy of 988 eV relative to ground potential and (2) ions displayed as green trajectories with an energy of 985 eV relative to ground potential. In order to use this arrangement as energy filter (*i.e.* the ions with lower energy are rejected, whereas the higher energy ions pass the centre lens), it is essential that the ion beam is confined in the centre.

The functional principle can be better recognised from the display of the potential energy surfaces. In Figure 4.6 the trajectory of the lower-energy ions is displayed, and in Figure 4.7 the trajectories of the higher-energetic ions.

The potential at the centre electrode forms a small cavity in the middle. The higher-energetic ions are able to pass the centre electrode, whereas the energy of the low-energetic ions is too small to pass this potential and are rejected. This is a simplified model of an energy filter where ions of lower

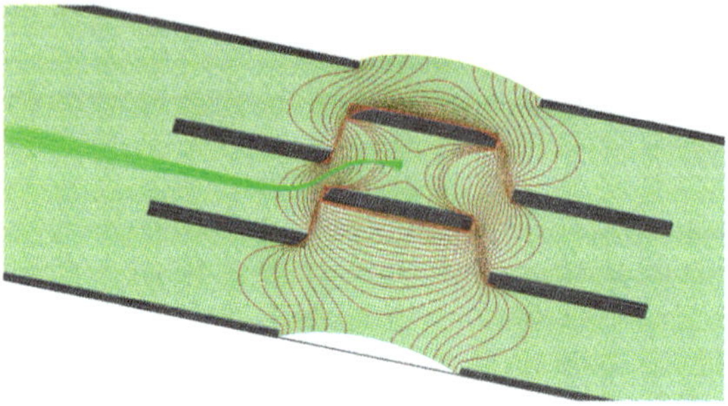

**Figure 4.6** Trajectories of the low-energy ions displayed on a potential-energy surface.

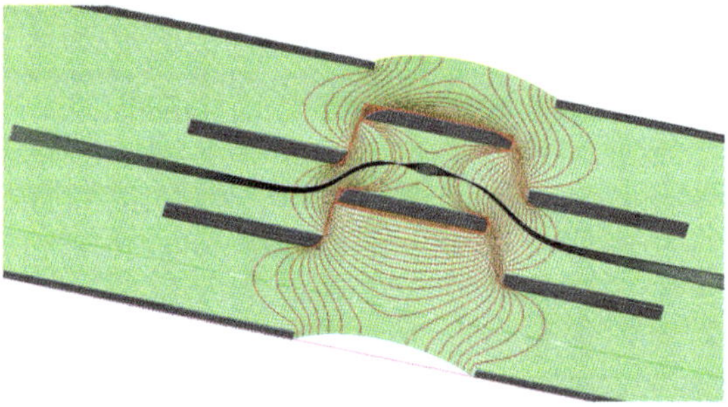

**Figure 4.7** Trajectories of the high-energy ions displayed on a potential-energy surface.

energy can be discriminated. The systems in commercially available systems usually include the confinement of the ion beam in the corresponding lens and are designed slightly differently, although the same principle applies.

Quadrupole lenses are often used in commercially available instruments for focusing the ion beam. The quadrupole lenses discussed here work with just electrostatic DC voltages and are therefore not acting as a mass filter or are RF-only ion guides (usually used in collision cells). The quadrupole lens (Figure 4.8) consists usually of four rods with identical diameter and length. The potential on the rods is the same, but of opposite polarity for neighbouring rods.

Lower voltages are required for quadrupole lenses as compared to other electrostatic lenses (*e.g.* cylindrical lenses) to achieve the same focusing properties. The electric field strength is increasing with the radial distance from the ion optical axis in the centre. Depending on the polarity of the electric fields the electrostatic quadrupole is focusing (defocusing) in the $x$ direction and defocusing (focusing) in the $y$ direction as can be seen in Figures 4.9(a) and (b) for positive ions. In order to focus the ion beam in both directions, either a quadrupole doublet (two quadrupoles with 90 degrees rotated polarity of the second quadrupole) or a quadrupole triplet is used.

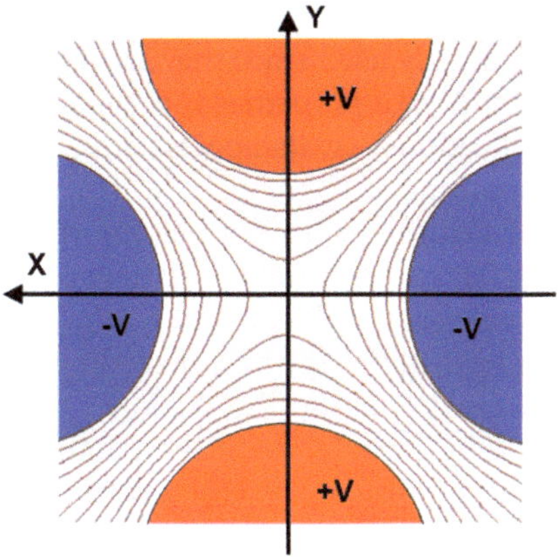


**Figure 4.8** Quadrupole lens field. Voltages ($V$) are the same but of opposite polarity.

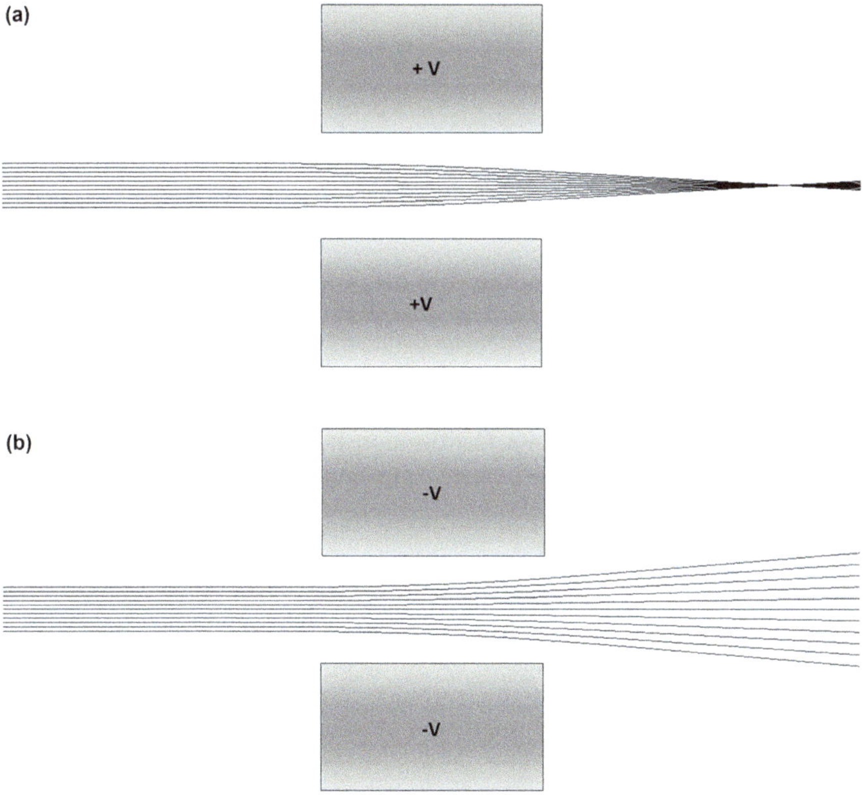


**Figure 4.9** Focusing (a) and defocusing (b) properties of a quadrupole depending on the polarity of the electric fields.

The shape of the rods of an ideal quadrupole is hyperbolic, but in practice a circular cross section is mainly used since they are much easier to manufacture and the deviation from the ideal properties is acceptable. Other rod geometries find use in specific applications, *e.g.* flatapoles with a rectangular cross section.

Beside the above-mentioned electrostatic lenses, other configurations are used in sector field instruments to correct for mechanical tolerances of particular components. In commercial ICP-MS instruments, the lens system behind the skimmer cone is mainly used to guide, focus and accelerate the positively charged ion beam onto the entrance slit and to shape the circular peak profile of the ion beam behind the skimmer to a more rectangular (in reality elliptical) profile in accordance to the geometry of the entrance slit. It usually consists of a number of electrostatic lenses (cylindrical and quadrupole). Therefore, this lens system is more complex in a sector field device than, *e.g.* in quadrupole-based instruments.

A magnetic sector field mass spectrometer has some typical ion optical aberrations, which are – among other – a function of the angular spread of the ion beam. To obtain a high resolving power and to reduce transmission losses, these aberrations must be kept below certain limits. The angular divergence $\alpha$ of the ion beam is inversely proportional to the square root of the accelerating voltage $U_0$ (eqn (4.3))

$$\alpha \propto \frac{1}{\sqrt{U_0}} \tag{4.3}$$

$\alpha$ : angular divergence
$U_0$ : accelerating voltage

Therefore, it is usually preferred to use a high acceleration voltage (*e.g.* around 10 kV). In practice, this is achieved by applying a high voltage at the extraction lens and additional acceleration voltages at the lenses towards the entrance slit. The limitation for the magnitude of these high accelerating voltages is usually driven by practical and economic factors. For instance, the cost of components like a feedthrough increases significantly with increasing voltages as well as the electronic components used at the printed circuit boards (PCB).

## 4.5 Slit System (Entrance Slit and Collector (Exit) Slit)

A slit system is generally required in order to shape the geometry of the ion beam. The resolution depends on the width of the incident ion beam, which is defined by the width of the entrance (source defining) slit (see also Chapter 5). After passing through the entrance slit, the ions are injected perpendicular to the magnetic field and traverse the field on different circular trajectories, according to their $m/z$ ratio. The image of the entrance slit is focused onto a second (exit, collector or imaging) slit. Decreasing the source (and/or collector slit widths) can be used to increase mass resolution.

Full high-resolution spectra are accomplished when decreasing the slit width of both entrance and exit slit simultaneously, whereas multicollector devices are typically operated in edge resolution mode, reducing only the width of the entrance slit (see also Chapter 5).

The slit width of the entrance slit relative to the exit slit also determines the peak shape obtained with sector field instruments. Assuming a magnification of the analyser of 1, flat-top peaks will be obtained when the width of the entrance slit is significantly smaller than the exit slit. This is particularly advantageous for isotope ratio measurements, since the intensity in the flat-top area is not influenced by small variations of the mass position of the peak, caused by small variations in electric or magnetic fields. If the entrance and exit slits are of the same width, theoretically triangular peaks are obtained, whereas practically they are more of a Gaussian shape (see also Chapter 5). Because a higher resolving power requires a smaller slit width, which reduces transmission and thus reduces sensitivity, it is preferred to measure isotopes with the lowest possible resolution.

Some instruments use a continuously variable slit by either applying a stepper motor or using a piezoelectric actuator. This approach allows a high flexibility, but the disadvantages are the speed when changing the resolution and the reproducibility of the positioning. Another faster and more reliable approach in terms of reproducibility is the movement of various predefined slits with different width into the ion optical pathway. This can be accomplished either manually (if no continuous change of resolution is required such as in multicollector devices operated in static mode like the *Nu Plasma HR*) or using automated systems such as, *e.g.* pneumatic devices. (For example, the slit system of the *Element2*, *Element XR* and *Neptune plus* consists of three fixed slits arranged on one single holder. The holder is pneumatically operated under computer control and functions like a pendulum, which can swing to the left- and right-hand side, thus switching between the different, fixed mass resolutions within less than one second.)

## 4.6 Magnetic Sector

The magnetic sector is the core part of a sector field mass spectrometer. Its function can easily be explained with the help of the motion of a charged particle (electron or ion) in a magnetic field. Charged particles with a charge $q$ ($= z \cdot e$; $z =$ number of charges; $e =$ elementary charge ($1.6022 \times 10^{-19}$ C)),[39,40] velocity $v$ and mass $m$ injected into a magnet field are forced by the Lorentz force (perpendicular to the direction of ion motion and magnetic field) onto a circular path with radius $r_{\mathrm{m}}$. The centripetal force is then equal to the Lorentz force, which gives

$$q \cdot v \cdot B = m \cdot \frac{v^2}{r_{\mathrm{m}}} \tag{4.4}$$

$$r_m = \frac{m \cdot v}{q \cdot B} \tag{4.5}$$

$$\frac{m \cdot v}{q} = B \cdot r_m \tag{4.6}$$

$q$ : charge of the ion
$v$ : velocity of the ion
$m$ : mass of the ion
$B$ : magnetic field strength
$r_m$ : radius of the ion path through the magnet

Therefore, the magnetic analyser is dispersive with respect to the mass and energy of the ions (or better the momentum), such that it is a "momentum separator" and can be ideally used for mass separation of ions of different mass when they have the same energy. To achieve this condition to the best possible extent, the ions generated in the ion source are accelerated over a potential difference $U_0$ (the acceleration voltage), thereby minimising the ratio of $\Delta E/E$ ($\Delta E$ = energy spread of the ions; $E$ is the energy of the ion). The kinetic energy of the accelerated ion is:

$$q \cdot U_0 = \frac{1}{2} \cdot m \cdot v^2 \tag{4.7}$$

and therefore:

$$r_m = \frac{1}{B} \cdot \sqrt{\frac{2 \cdot m \cdot U_0}{q}} \tag{4.8}$$

or

$$\frac{m}{q} = \frac{B^2 \cdot r_m^2}{2 \cdot U_0} \tag{4.9}$$

$q$ : charge of the ion
$U_0$ : acceleration voltage
$m$ : mass of the ion
$v$ : velocity of the ion
$r_m$ : radius of the ion path through the magnet
$B$ : magnetic field strength

A sketch of the geometry of a magnetic sector is shown in (Figure 4.10). It consists of an ion source, an entrance (source) slit, the magnetic sector with radius $r_m$, the exit (collector) slit and a collector/detector, for instance a Faraday cup.

There are three basic features of a magnetic sector as an ion optic element, which will be briefly discussed in the following sections (more details are given in dedicated books[41–43]):

- mass dispersion;
- angular focusing;
- energy dispersion.

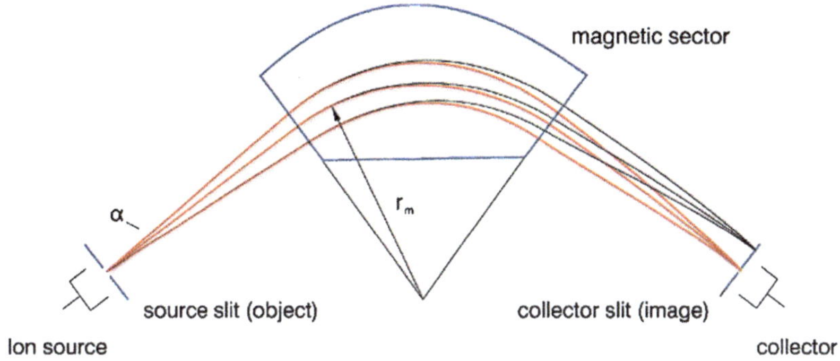


**Figure 4.10** Schematic representation of a magnetic sector.

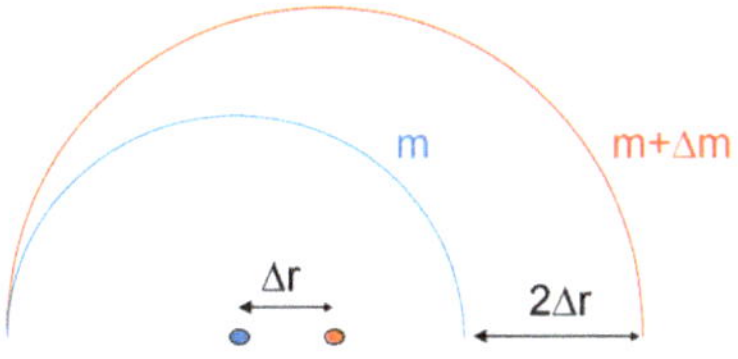


**Figure 4.11** Dispersion at the focal plane at 180°.

**Dispersion**

The term dispersion might be already known from optical spectroscopy, where it describes the power of refraction of a dispersive optical element, such as a prism, where the refraction is typically a function of the wavelength. In mass spectrometry, the magnetic sector is a mass-dispersive element and therefore dispersion $D_m$ can be defined as the physical separation distance $\Delta r$ (*e.g.* in mm) of two masses with a mass difference $\Delta m$ along the focal plane in the region of the collector slit (the collector plane).

A magnetic sector has many physical properties that can be compared with "optical" devices and as mentioned above, it best can be compared with a prism, because it has dispersive and magnification properties, depending on its geometry. For a magnetic sector with a 180° geometry, the dispersion at the focal plane at 180° is $2\Delta r_m$ (Figure 4.11) when two ion trajectories of the same origin show a difference in the deflection radii of $\Delta r_m$.

Rewriting eqn (4.9) results in eqn (4.10) and differentiation results in the relation of $\Delta r$ and $\Delta m$, the dispersion $D_m$ (eqn (4.11)).

$$r_m = \sqrt{\frac{2 \cdot U_0}{q \cdot B^2}} \cdot \sqrt{m} \tag{4.10}$$

$$\Delta r_m = \left.\frac{\partial r_m}{\partial m}\right|_{U_0,B} \cdot \Delta m \Rightarrow \Delta r_m = r_m \cdot \frac{\Delta m}{2 \cdot m} \tag{4.11}$$

$$D_{\mathrm{m}} = 2 \cdot \Delta r_{\mathrm{m}} = \frac{\Delta m}{m} \cdot r_{\mathrm{m}} \tag{4.12}$$

$r_{\mathrm{m}}$ : radius of the ion path through the magnet
$U_0$ : acceleration voltage
$B$ : magnetic field strength
$m$ : mass of the ion
$\Delta m$ : *mass difference*
$D_{\mathrm{m}}$ : mass dispersion

It is clear that for a mass difference of $\Delta m$ at mass $m$, the dispersion is $2\Delta r_{\mathrm{m}}$ for the above-mentioned 180° geometry, or in other words, the dispersion is proportional to the magnet radius and indirectly proportional to the mass. This holds true for other geometries as well.

The dispersion in a general form can be written as

$$\Rightarrow D_{\mathrm{m}} = \frac{\Delta m}{m} \cdot D_0 \tag{4.13}$$

$D_{\mathrm{m}}$ : mass dispersion
$\Delta m$ : mass difference
$m$ : mass of the ion
$D_0$ : dispersion coefficient

The dispersion coefficient $D_0$ only depends on geometrical parameters of the magnet (angle of deflection, entrance angle, exit angle) and the magnitude is usually in the range of the radius of the magnetic sector.

As can be seen from eqn (4.13), the peak width $\Delta m$ is in contrast to, *e.g.* quadrupole-based instruments not constant with mass, *i.e.* heavier masses have a larger peak width and therefore peaks of heavier ions are spaced more closely.

The dispersion is an important parameter of a sector field instrument, because it is directly related to the resolution achievable with the device. During a scan, *e.g.* the ion beams of a mass $m$ and a mass $m + \Delta m$ are moved across the collector slit consecutively. They are 100% resolved, when the ion beam of mass $m$ already passed the slit and the ion beam of mass $m + \Delta m$ is just to enter the slit. The slit widths of the entrance and collector (exit) slit are $s'$ and $s''$, respectively. The distance between the centres of both ion beams is the dispersion $D$ and the ion beam has the width $b$ (Figure 4.12a and b).

Therefore, the dispersion can be expressed by eqn (4.14):

$$\begin{aligned} D_{\mathrm{m}} &= s'' + b = D_0 \cdot \frac{\Delta m}{m} \\ \Rightarrow \frac{m}{\Delta m} &= \frac{D_0}{s'' + b} \end{aligned} \tag{4.14}$$

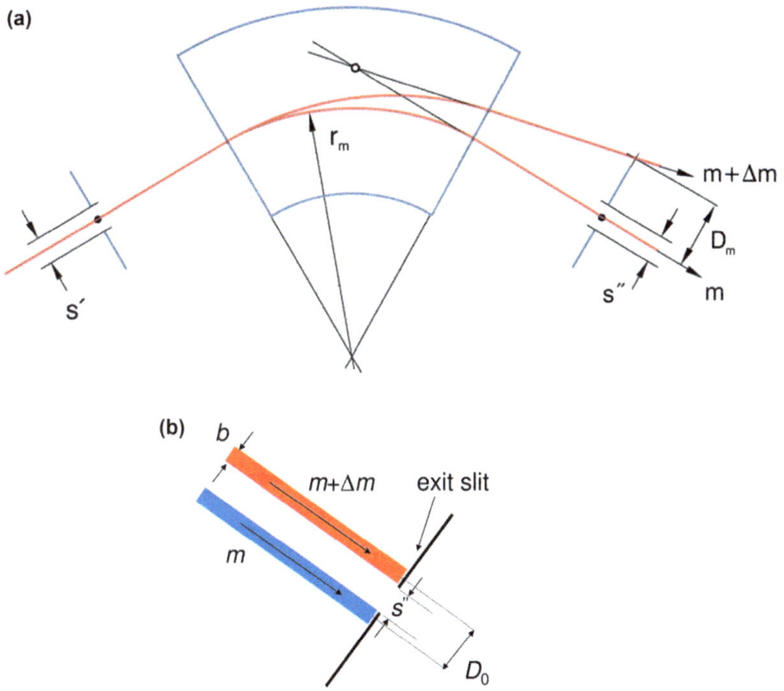


**Figure 4.12** (a) Mass dispersion and (b) mass resolution in a magnetic sector.

$D_m$ : mass dispersion
$s''$ : width of the exit slit
$b$ : beam width at the collector plane
$\Delta m$ : mass difference
$m$ : mass of the ion
$D_0$ : dispersion coefficient

The beam width $b$ at the collector plane is equal to the width of the source slit multiplied by the magnification $M$ of the instrument, a parameter defined by the geometry of the magnet that is usually around 1. The beam width also depends on the sum of aberrations $A$ of the system, so that the resolution can be expressed as follows:

$$R = \frac{m}{\Delta m} = \frac{D_0}{s'' + |M| \cdot s' + \sum A} \tag{4.15}$$

$R$ : mass resolution
$m$ : mass of the ion
$\Delta m$ : mass difference
$D_0$ : dispersion coefficient
$s'$ : width of the entrance slit

$s''$ : width of the exit slit
$b$ : beam width at the collector plane
$M$ : magnification
$A$ : abberation

For instance, in the case of the *Element2* ICP-MS with $M \sim 1$, the dispersion is about 140 mm and the aberrations are negligible compared to the slit width. In order to achieve a resolution of 4000 (for ideally a triangular peak shape), a slit width of the entrance and the exit slit of about 17 μm is required (a 16-μm slit width is currently used in the actual instrument).

## 4.7 Electric Sector

Equation (4.9) shows that the *m/z* value for a given radius is indirectly proportional to the accelerating voltage – or more precisely, to the kinetic energy of the ion. Thus, ions with the same mass, but with different kinetic energies will be focused at different positions. This results in peak broadening and a loss of resolution. The spread of the kinetic energy of the different ions depends on the ion source. For example, in ICP-MS, the ions have a kinetic energy which is directly related to the temperature of the plasma and that can be calculated by the Boltzmann equation. Additionally, ions are generated in the plasma at a potential different from the potential of the interface and additionally, in the interface the ions are accelerated by gas dynamics to the same velocity as the argon gas. This means that the kinetic energy increases proportional to the mass of the ion. All these factors contribute to the final ion kinetic energies and can affect the resolution. This is why a device is needed by which differences in energy can be compensated for. In most instruments, an electric sector is used for this purpose. Sometimes, the electric sector is also referred to as electrostatic analyser (ESA). (When the ion source produces ions with a narrow energy distribution, an electric sector is not necessary, resulting in single (angular) focusing magnetic sector instruments (*e.g.* for TIMS or IRMS)).

In an electric sector, the centrifugal force of motion (left-hand side of eqn (4.16)) must be equal to the centripetal force of the electric field for a particle, with charge $q$, mass $m$ and velocity $v$, injected into an electric field ($E$) produced by a cylindrical condenser at a radius $r_e$.

$$\frac{m \cdot v^2}{r_e} = q \cdot E \tag{4.16}$$

$m$ : mass of the ion
$v$ : velocity of the ion
$r_e$ : radius of the electrostatic field
$q$ : charge of the ion
$E$ : electric field strength

The kinetic energy (right-hand side of eqn (4.17)) of a particle accelerated in an electric field is:

$$q \cdot U_0 = \frac{1}{2} \cdot m \cdot v^2 \tag{4.17}$$

$q$ : charge of the ion
$U_0$ : acceleration voltage
$m$ : mass of the ion
$v$ : velocity of the ion

Thus, the velocity can be calculated by:

$$v = \sqrt{\frac{2 \cdot q \cdot U_0}{m}} \tag{4.18}$$

$q$ : charge of the ion
$U_0$ : acceleration voltage
$m$ : mass of the ion
$v$ : velocity of the ion

and combining with eqn (4.16) results in:

$$\frac{m \cdot 2 \cdot q \cdot U_0}{r_e \cdot m} = q \cdot E \tag{4.19}$$

$m$ : mass of the ion
$q$ : charge of the ion
$U_0$ : acceleration voltage
$r_e$ : radius of the electrostatic field
$q$ : charge of the ion
$E$ : electric field strength

$$\frac{2 \cdot U_0}{r_e} = E \tag{4.20}$$

$$r_e = \frac{2 \cdot U_0}{E} \tag{4.21}$$

$r_e$ : radius of the electrostatic field
$U_0$ : acceleration voltage
$E$ : electric field strength

This equation is similar to eqn (4.8) but does not contain the mass of the ion. Therefore, the electric sector is not dispersive with respect to the mass of the ions. If a slit is placed at a particular position behind the system at the radius $r_e$, the system acts as an energy filter.

## 4.8 Double-Focusing Conditions and Sector Field Geometries

In order to obtain separation of ions of the isotope of interest from interferences at the same *m*/*z* ratio, a sufficient mass resolution has to be obtained. If the energy spread is not compromising the resolution of a mass spectrometer (when the ion source generates ions with a low energy spread (*e.g.* ionisation accomplished in a thermal surface ionisation source combined with a low degree of generation of interfering ions)), single-focusing (*i.e.* angular focusing) instruments can be sufficient. Angular focusing is accomplished by the magnetic sector field (see Section 4.6).

Alternatively, a collision cell can be used to reduce the ion energy spread (*e.g.* of ions generated in an ICP source) and thus a single focusing can be applied in this case. In one type of instrument (*IsoProbe* (Micromass, Manchester, UK), a multicollector ICP-MS, which is no longer commercially available), collisional cooling and focusing was accomplished by collision of the ions with a buffer gas in an RF-mutipole cell.

Summarising so far, we know that the magnetic sector is a dispersive ion optical element with respect to mass and energy (momentum) and the electric sector is a dispersive ion optical element with respect to energy, only. Both systems also provide angular focusing properties. This means that an ion beam with a certain angular divergence is focused to the same position on an image plane. When combining a magnetic and an electric sector in a way that the energy dispersion of the magnet (${}^{m}D_E$) and the electric sector (${}^{e}D_E$) are equal in magnitude, but of opposite direction, then the magnet and electric sector focus both ion angles (first focusing) and ion energies (double-focusing), while being dispersive for *m*/*z* only. Such devices fulfilling the following condition (eqn (4.22)) are termed "double-focusing" instruments:

$$ {}^{e}D_E + {}^{m}D_E = 0 \tag{4.22} $$

${}^{m}D_E$ : energy dispersion in a magnetic field
${}^{e}D_E$ : energy dispersion in an electric field

Only for some well-defined combinations of electric and magnetic sector angles does the angular (directional) focus coincide with the energy focus. The mostly employed designs are the forward and reverse Nier–Johnson geometry, the Matsuda and the Mattauch–Herzog geometry. Other geometries, *e.g.* the Bainbridge–Jordan, Hinterberg–König or Takeshita geometry are less commonly used in mass spectrometers and are described in more detail elsewhere.[1,44–47]

An overview of common geometries in commercially used sector field mass spectrometer is given in Table 11.1.

**Nier–Johnson Geometry**

The Nier–Johnson geometry (Figure 4.13) was developed at the University of Minnesota in the 1950s,[48] and is the dominating design.

In this arrangement, the ion beam is shaped by the ion optics on the entrance slit and further into the radial electrostatic analyser (ESA), in which the beam is deflected by 90°. The ESA is followed by an intermediate slit and a 60° deflection magnetic field.[49] The ESA radius is somewhat larger than that of the magnetic sector. This arrangement is based on the fact that the double-focusing conditions are fulfilled on a curved path to which the collector arrangement is fitted. Due to the ion dispersion in the magnetic field, the instruments in Nier–Johnson geometry are preferably used to determine multiple analytes (isotopes) simultaneously (*i.e.* in multicollector designs). The size determines the mass dispersion, which needs to be selected in a way to have space for the different detectors.

In the reverse Nier–Johnson geometry (Figure 4.14), the electrostatic analyser is positioned after the magnetic field. The deflection angles remain the same (90° in ESA and 60° in the magnetic field).

The reverse Nier–Johnson geometry is mainly used in scanning single-collector sector field mass spectrometers. An advantage of reverse geometry instruments is that due to the finite entry restriction of the ESA and because of its properties, some scattered ions (usually by collisions with residual gas molecules along the ion optical pathway) are not transmitted. Such scattered ions cause the beam size at the collector plane to broaden, resulting in peak tailing, and deteriorating abundance sensitivity.

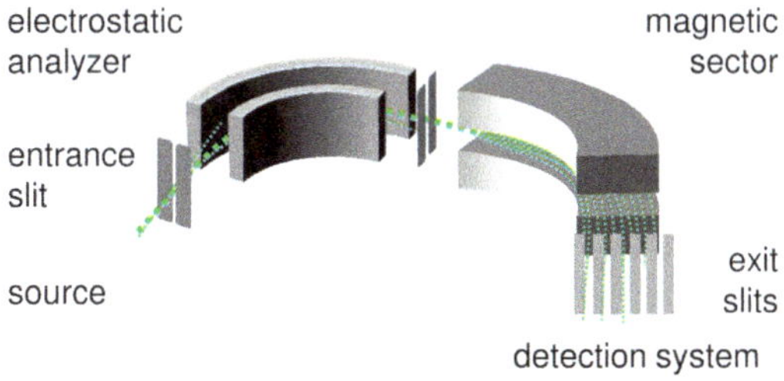


**Figure 4.13** Schematic representation of a Nier–Johnson geometry.

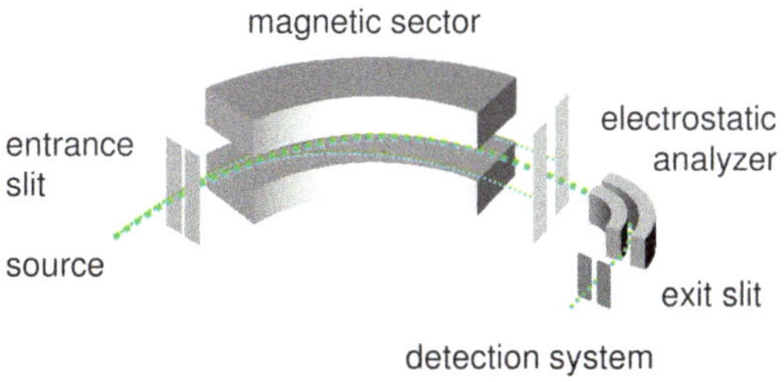


**Figure 4.14** Schematic representation of a reverse Nier–Johnson geometry.

In the following paragraph, eqn (4.22) is discussed with respect to the Nier–Johnson geometry. The parameters and angles are shown in Figure 4.15.

The energy dispersion ${}^{m}D_{E}$ for a homogenous magnetic sector of a total angle $\phi_m$ is given as:

$$ {}^{m}D_{E} = \frac{r_m}{2}\left(1 - \cos\phi_m + \frac{l'_m}{r_m}\cdot\sin\phi_m\right) \tag{4.23} $$

${}^{m}D_{E}$ : energy dispersion in a magnetic field
$r_m$ : radius of the magnetic field
$\phi_m$ : magnetic sector angle
$l'_m$ : object distance in the magnetic field

The energy dispersion for a homogenous cylindrical electric sector of a total angle $\phi_e$ is given as:

$$ {}^{e}D_{E} = \frac{r_e}{2}\cdot\left[\left(1 - \cos\sqrt{2}\cdot\phi_e\right) + \frac{l''_e}{r_e}\cdot\sqrt{2}\sin\sqrt{2}\cdot\phi_e\right] \tag{4.24} $$

${}^{e}D_{E}$ : energy dispersion in an electric field
$r_e$ : radius of the electric field
$\phi_e$ : electric sector angle
$l''_e$ : image distance in the electric field

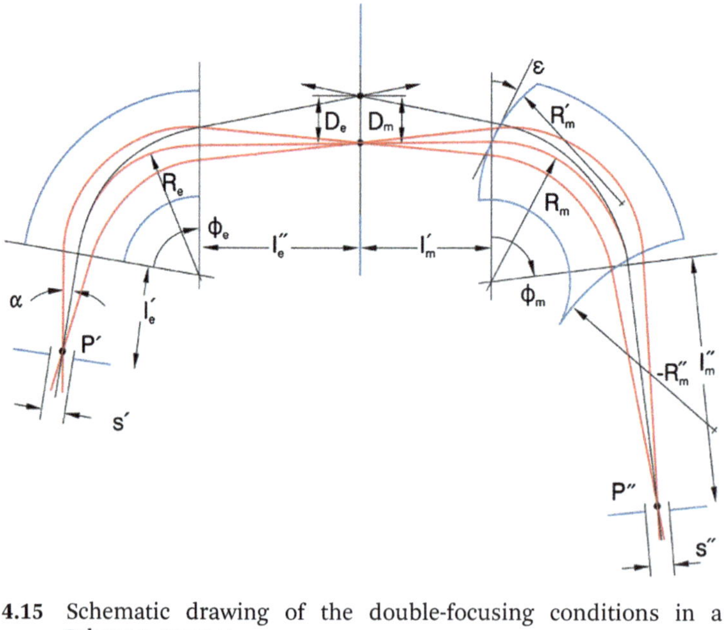


**Figure 4.15** Schematic drawing of the double-focusing conditions in a Nier–Johnson geometry.

Using the double-focusing condition of eqn (4.22), this equation results in:

$$r_e \cdot \left(1 - \cos\sqrt{2} \cdot \phi_e\right) + l''_e \cdot \sqrt{2} \sin\sqrt{2} \cdot \phi_e + r_m(1 - \cos\phi_m) + l'_m \cdot \sin\phi_m = 0 \quad (4.25)$$

$r_e$ : radius of the electric field
$\phi_e$ : electric sector angle
$l''_e$ : image distance in the electric field
$r_m$ : radius of the magnetic field
$\phi_m$ : magnetic sector angle
$l'_m$ : object distance in the magnetic field

Using eqn (4.22) for $r_m$ it can be recognised that the equation is mass dependent and thus the double-focusing condition is valid only for a single mass focused at the exit slit or a small mass range. As a consequence, such a geometry is very well suited for a single-collector arrangement operated in a scanning mode.

**Mattauch–Herzog Geometry**

The Mattauch–Herzog design (Figure 4.16) was developed in 1934 at the University of Vienna for one of the very first double-focusing mass spectrographs.[50] In this arrangement, the ion beam passes through the entrance slit where its width is reduced, to the ESA, and continues through an intermediate (also "energy") slit into the magnetic sector (direction of curvature opposite to the ESA).

The radius of the ESA is approximately twice that of the magnetic sector. Ions are deflected according to their $m/z$ ratio to the inner side of the magnetic analyser, on which a detector plane is positioned. Originally, a photo plate served as a detector. Nowadays, *e.g.* a Direct Charge Detector is used (see Section 4.12.4). The main advantage of this arrangement is the possibility of a real simultaneous measurement of a complete mass spectrum. (***Note:*** A resolution of 10000 and more was achieved with large instruments and photo plates. The resolution depends on a lot of parameters and current instruments with a direct charge detector do not achieve high mass resolution – amongst others – due to the width of the detector.)

Equation (4.25) becomes "simple" in case of the Mattauch–Herzog geometry (Figure 4.17).

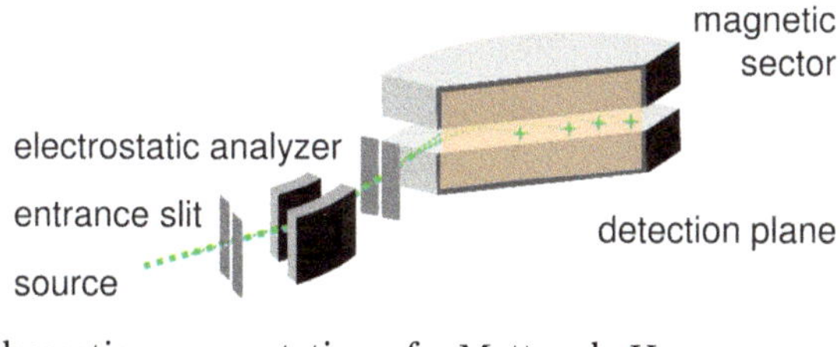


**Figure 4.16** Schematic representation of a Mattauch–Herzog geometry.

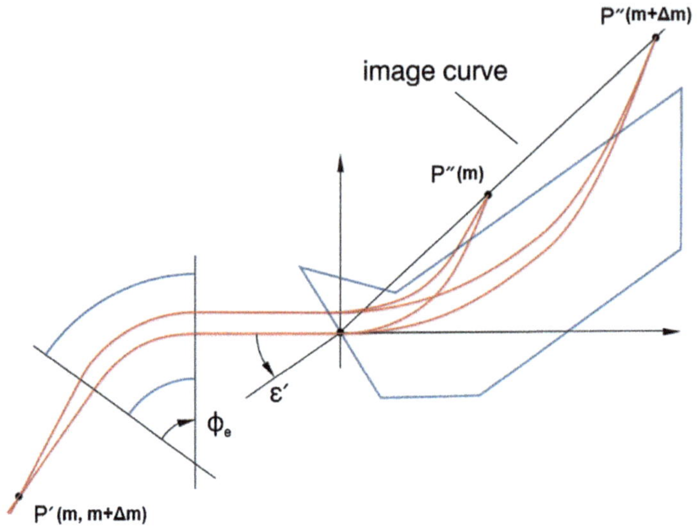


**Figure 4.17** Schematic drawing of the double-focusing conditions in a Mattauch–Herzog geometry.

Equation (4.25) can also be expressed as shown in eqn (4.26), with $\Delta = l''_e + l'_m$ :

$$\frac{r_e}{l''_e} \cdot \left(1 - \cos\sqrt{2} \cdot \phi_e\right) + \sqrt{2}\sin\sqrt{2} \cdot \phi_e = \frac{r_m}{l''_e}(1 - \cos\phi_m) + \left(\frac{\Delta}{l''_e} - 1\right) \cdot \sin\phi_m \quad (4.26)$$

$r_e$ : radius of the electric field
$l''_e$ : image distance in the electric field
$\phi_e$ : electric sector angle
$r_m$ : radius of the magnetic field
$\phi_m$ : magnetic sector angle
$\Delta$ : $l''_e + l'_m$
$l'_m$ : object distance in the magnetic field

The magnetic and electric sector for a Mattauch–Herzog geometry are described by the following parameters:

$$\phi_m = 90^\circ$$

$$\phi_e = \frac{\pi}{4\sqrt{2}} = 31.8^\circ$$

$$\varepsilon' = 0$$

$$\varepsilon'' = -45^\circ$$

$\phi_m$ : magnetic sector angle
$\phi_e$ : electric sector angle
$\varepsilon'$ : magnetic sector entrance angle of main trajectory
$\varepsilon''$ : magnetic sector exit angle of main trajectory

When the source slit is placed in the focal point of the electrostatic sector, then $l_e'' \rightarrow \infty$. This simplifies eqn (4.26), which results in the special double-focusing condition:

$$\sqrt{2}\sin\sqrt{2}\cdot\phi_e = -\sin\phi_m \tag{4.27}$$

$\phi_e$ : electric sector angle
$\phi_m$ : magnetic sector angle

Since eqn 4.27 is – for these conditions – no longer dependent on $r_m$, the *foci* for all masses are located on a plane. Therefore, this geometry is perfectly suited for simultaneous detection and found widespread applications with photoplates for ion detection in SSMS. Nowadays, instruments with such a setup are commercially available under the form of transportable gas analysers with an electronic focal plane camera. Prototype GD- and ICP-MS instruments relying on the Mattauch–Herzog geometry have been developed[51] and recently, Spectro Analytical Instruments GmbH, Kleve, Germany, launched an ICP-MS instrument based on the Mattauch–Herzog geometry to the market.[52]

**Other Geometries**

Variations in deflection angles, dimensions or addition of other elements have led to a number of suitable designs. One of the early designs presents the Bainbridge–Jordan geometry (Figure 4.18) from the 1930s.[53]

In this design, the 127.3° ESA is followed by a 60° magnetic sector with the same direction of curvature. Further development was performed,

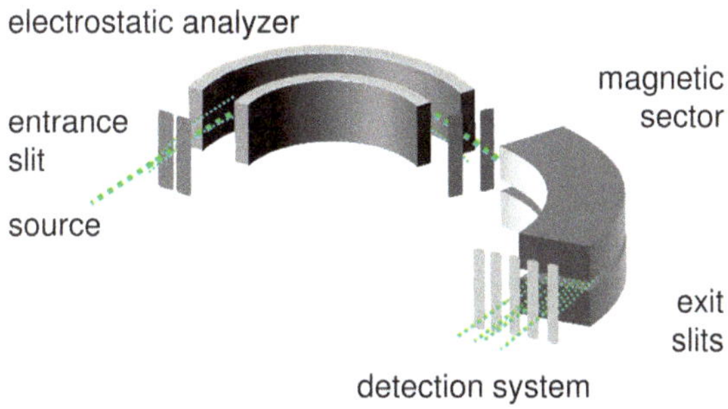


**Figure 4.18** Schematic representation of a Bainbridge–Jordan geometry.

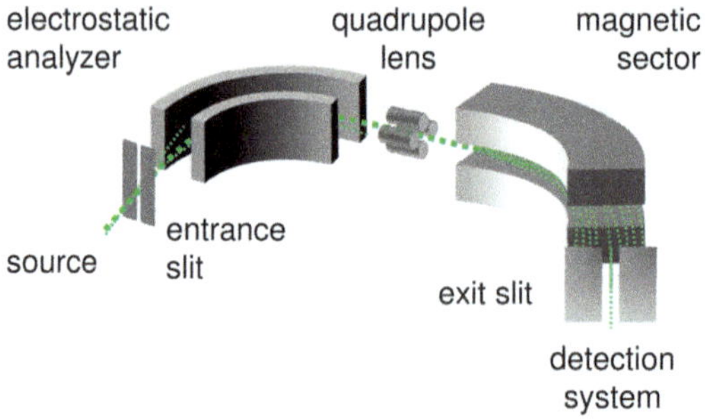


**Figure 4.19** Schematic representation of Matsuda geometry.

*e.g.* by Hinterberger and König, who calculated a large number of double-focusing geometries.[54] In the Takeshita geometry, two Mattauch–Herzog-type electric sectors are used in order to eliminate some second-order aberrations to independently control the kinetic energy spread of the beam divergence.[47]

Matsuda aimed at achieving highest resolving power and designed a double-focusing mass spectrometer that eliminated all second-order aberrations. A quadrupole lens was placed between the magnetic and electric sector to improve focusing in the *y*-direction.[46] In the Matsuda design from the 1980s, the mass spectrometer comprises an 85° electric sector and a 72.5° magnetic dipole[55] (Figure 4.19).

The geometry aims towards high resolution and can be used for both single-, as well as multicollector detection systems. The Matsuda design has been applied, *e.g.* for a large diameter double-focusing SIMS (*SHRIMP*, Australian Scientific Instruments, Canberra, Australia).

A number of other approaches were achieved in designing double-focusing mass spectrometer, *e.g.* using different multipole combinations in addition to the electric and magnetic sectors, a Wien filter in combination with a magnetic sector or a combination of a cylindrically symmetrical electric sector followed by several wedge magnets.[1]

## 4.9 Acquiring a Mass Spectrum

It can already be seen from eqn (4.9), how a mass spectrum can be acquired:

**E-Scan Measurement**

Keeping $B$ and $r_m$ constant and changing the accelerating voltage $U_0$ results in a so-called electric scan (E-Scan). Due to a loss in sensitivity and focusing properties (increase of aberrations), the mass range that can be scanned in this mode is limited to approx. 30–40% of the magnet mass. However, this operation mode can be used for fast scanning, because alteration and stabilisation of the involved voltages can be achieved very quickly.

**B-Scan Measurement**

Keeping $r_m$ and $U_0$ constant but varying the magnetic field $B$ results in a so-called magnetic scan (or B-Scan). In principle, the mass range is unlimited and only dependent on the applied magnetic field, which is technically limited for iron core magnets to about 2 T. To retain sufficient field homogeneity for high mass resolution, a practical limit is about 1.5 T. This still allows using ion energies of some keV. In comparison to other mass spectrometers, such as quadrupoles, this scan mode is slower, because the speed of magnetic field changes is limited by the self-induction of the magnet (eddy currents) and the required power supplies. Therefore, a certain jump time (changing of the magnetic field) and afterwards a stabilisation time (settling time) is needed. This settling time depends on the magnitude of the change in magnetic field (required for "jumping" from one mass to another) in a "peak-hopping mode". However – in practice – for analyte masses within a relatively narrow mass range, the same magnetic mass is used. Scanning across a spectral peak is accomplished *via* E-scanning, followed by a jump of the electric field to the next mass followed by the next E-scan across the next spectral peak of interest. When the scan range limit for the E-scan is reached, the magnetic field is adapted, resulting in a "jump" to a new magnetic mass. This mode is, *e.g.* the preferred operation mode in scanning ICP-SFMS.

**Static Measurement**

The two previous ways to obtain a mass spectrum are mainly used for scanning single-collector instruments. By keeping $U_0$ and $B$ constant, the operation in static mode is used for multicollector instruments. In this case, the detectors are located at different positions along the image plane, corresponding to different $r_m$. This mode allows the simultaneous measurements of the signals of ions of different masses, thus providing high precision for isotope ratio analysis. A switch of the magnet mass can be used to change the mass range (guiding a different set of isotopes to the detectors), which leads to a dynamic option of measurement. The obtained ratios have to be crosscalibrated (*i.e.* at least one ratio should be measured within two different settings) and the method can only be applied if no mechanical movement of the detectors is required between the different settings. As all isotopes are measured simultaneously in case of a Mattauch–Herzog geometry in combination with a detector array, such instruments are only operated in static mode using a permanent magnet.

## 4.10 Flight Tube

Usually, sector field mass spectrometers include a flight tube, which is positioned between the pole faces of an electromagnet. This is necessary when the magnet pole edges are not in vacuum to guide the ions through the curvature of the magnet. It is curved in accordance with the radius of the ion trajectories of the masses of interest within the magnetic field

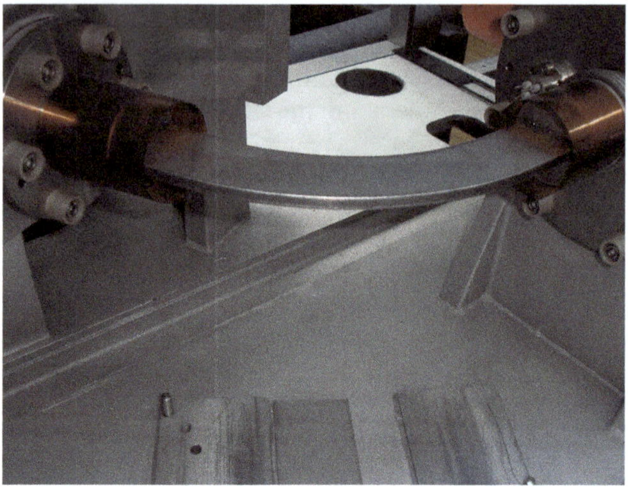

**Figure 4.20** Photo of a flight tube (*Element*, Finnigan MAT, Bremen, Germany). (Photo: Ondrej Hanousek.)

(Figure 4.20). Ideally, the flight tube is made out of a material with a very low magnetic permeability – or in other words, a low degree of magnetisation is obtained in response to the applied magnetic field, as this would otherwise affect the effective magnetic field at the ion beam.

The flight tube is typically created from a thin-walled metal sheet. A thin-walled tube is important as it – together with the electrical conductivity – governs the eddy currents that will be generated when the magnetic field is varied. A thin-walled tube results in a smaller magnet gap for the same internal height, which is very important amongst others for the magnet coil and power supply design. However, there is a practical limitation on the thickness because these walls have to withstand the force of the atmospheric pressure when the flight tube is evacuated down to $10^{-10}$ mbar. It is clear that the walls should neither bend nor collapse. The inner surfaces of the flight tube should be clean in order to ensure a high conductivity when ions of smaller or higher masses than the actual analyte mass hit the surface.

There are also sector field mass spectrometers that do not have a flight tube in a typical way. One example is the stable gas isotope ratio mass spectrometer *Delta V*. The pole faces of the magnet (which is precision mounted into a monolithic analyser), determine the free flight space for the ions, thus eliminating the traditional flight tube. The gain in ion beam height directly translates into increased sensitivity. However, for scanning magnetic sector instruments, the pole faces are usually laminated in order to minimise eddy currents and allow faster and more precise changes in the magnetic field. Because the process used in production of the laminated pieces is not compatible with the high-vacuum requirements, this is a practical limitation for the above-mentioned design.

When the ion source and interface is at ground potential (*e.g.* in specific ICP-MS systems such as *Element2*, *Element XR* and *Neptune* ICP-MS instruments), it is necessary to have the flight tube at a high negative potential. The pole faces of the magnet have thus to be electrically insulated by a thin foil against the flight tube, allowing the magnet to be at ground potential.

# 4.11 Transfer, Zoom and Filter Optics

These additional lenses are applied to focus the ion beams and/or to guide the ion beams towards the centre of the detector. In order to improve abundance sensitivity (see Chapter 5), special energy filters, permitting ions with sufficiently high kinetic energies to pass only, are sometimes applied in sector field devices and located after the magnet before the detector.

**Improvement of Abundance Sensitivity in Sector Field Instruments**
Although the vacuum is better as compared to, *e.g.* quadrupole-based instruments, there are still ions that collide with residual gas molecules when travelling through the mass analyser. During a collision, an ion usually loses some kinetic energy as well as it gets a different direction (different angle compared to the main trajectory).[56] Furthermore, scattering processes at the edges of the beam-defining apertures can have the same effect. Both effects are finally responsible for peak tailing.

(***Note:*** Peak tailing is often expressed as abundance sensitivity, which is defined as the ratio of the maximum ion current recorded at a given mass $m$ to the ion current arising from the same species recorded at an adjacent mass unit ($m \pm 1$ u),[57] and often expressed in ppm.)

The low kinetic energy ions appear in the mass spectrum at a mass slightly shifted to the low mass side which results in a tailing of the peak.[58] This tail will inadvertently contribute a signal to the neighboring masses.[59] In order to reduce this effect, sector field instruments can be equipped with energy filters in front of the detectors, which finally discriminate the ions with the lower kinetic energy or the different direction of motion.[56] Wieser and Schwieters[59] stated values for the abundance sensitivity in the range of 1 ppm to 5 ppm for a state of the art multicollector instrument (*Neptune* MC ICP-MS) without the use of an energy filter. Depending on the instrument manufacturer, such energy filters installed in front of an SEM are referred to as, *e.g.* RPQ (Retarding Potential Quadrupole lenses, Thermo Fisher Scientific), deceleration or retardation filter (Nu Instruments Ltd.) or WARP (wide-angle energy retarding filter, IsotopX Ltd.). In Figure 4.21, a RPQ lens with a SEM (*Neptune* MC ICP-MS) is shown. This RPQ reduces the abundance sensitivity from about 5 ppm to about 500 ppb with an ion optical transmission of more than 90%. However, at 50% ion transmission an abundance sensitivity of 150 ppb can be achieved by adjusting the energy barrier of the RPQ filter.[60]

These energy filters are applied in single-collector instruments, as well, to improve abundance sensitivity. For example, the *Element XR* ICP-MS uses a

**Figure 4.21** Photo of a RPQ in combination with a SEM installed in a *Neptune* MC ICP-MS.
(Source: © Thermo Fisher Scientific.)

Filter Lens in combination with a pair of slits for achieving flat-top peaks with a resolution of approx. 2000. Using this combination, an abundance sensitivity of less than 100 ppb could be achieved for the contribution of $^{232}Th$ at $^{230}Th$ when analysing a solution of 200 ppb Th (*IRMM-036*, IRMM, Geel, Belgium) with a $^{230}Th/^{232}Th$ ratio of 3.11 ppm.

## 4.12 Detection Systems

After the separation of ions according to their mass-to-charge ratios, ions are guided to the detection system, which is located at the back end of the mass spectrometer. Generally, the aim of a detection system is the detection of each ion of interest and the conversion and amplification of ion signals into a physical signal that is suitable for further data processing. Faraday cups and secondary electron multipliers are the most common detection systems in modern SF mass spectrometers. Other detection systems such as the Daly detector, microchannel plate detectors and the direct charge detector are less employed.

### 4.12.1 Electron Multipliers

The function of an electron multiplier is based on secondary electron emission caused by the release of electrons from atoms on the surface by striking charged particles (ions or electrons). The number of secondary electrons depends on the type of incident primary particle, its energy and the characteristics of the surface.

Generally, there are two different types of electron multipliers, which are both based on signal multiplication by secondary electrons emitted as a consequence of impinging ions (or electrons in case a conversion dynode is

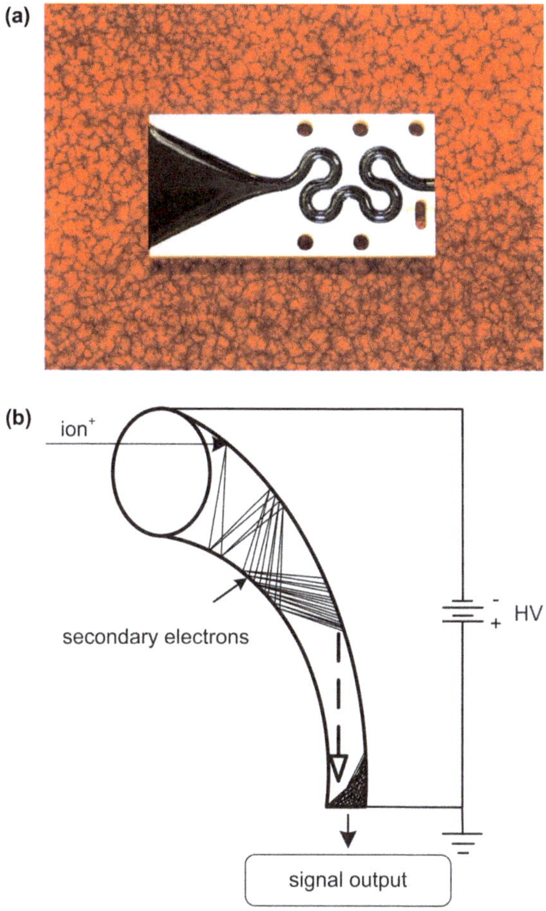


**Figure 4.22** Photo (a) and schematic representation (b) of a CEM. (Source: (a) © Dr. Sjuts Optoelectronics, Göttingen, Germany, (b): Stefanie Konegger-Kappel.)

used before the multiplier): (1) discrete-dynode secondary electron multipliers (SEMs) (Figure 4.23) and (2) continuous dynode or channel electron multipliers (CEMs) (Figure 4.22). Electron multipliers can be used for the measurement of negative or positive ions, depending on the applied voltages at the dynodes and are operated in pulse-counting or analogue mode. In the pulse-counting mode, the output is measured by counting single incoming pulses representing single incident ions. The analogue mode is applied for measuring higher ion beam currents. In this case only the first dynodes are used and the output is measured as a continuously variable current, which is directly proportional to the number of incoming ions.

CEMs were primarily used in early single-collector instruments as they were more stable in air than the SEMs available at that time. However, the development of air-stable dynode materials in SEMs favoured the replacement of CEMs in single-collector instruments by discrete-dynode SEMs during the 1990s, mainly due to their higher dynamic range.[61] Operating an SEM in the so-called dual mode (*i.e.* counting mode and analogue mode; see for example[61]) allows a linear dynamic range of up to nine orders of magnitude to be achieved in modern single-collector instruments.[60] The linear dynamic range can be extended to more than twelve orders of magnitude in *e.g.* the *Element XR* (ICP-MS), the *Element GD* (GDMS), the *AttoM* (ICP-MS) or the *Astrum* (GDMS) by a Faraday cup detector in addition to the SEM. This allows the measurement of trace elements using the SEM and matrix elements using the Faraday cup detector in the same measurement run.

In multicollector instruments, SEMs are typically positioned off-axis to the focal plane of the mass analyser[62] as they are too large in size to be incorporated in the collector block compared to Faraday cups or miniaturised channeltron detectors. Because of the tight space requirements in their multicollector array, Thermo Fisher Scientific has used miniaturised channeltron-type detectors in its *Neptune* MC ICP-MS and *Triton* MC-TIMS instruments, having the same dimensions as the applied Faraday cups (width: 3.5 mm *versus* 20 mm for discrete dynode SEMs). However, miniaturised channeltron-type detectors had a lower dynamic range compared to discrete-dynode SEMs,[63,4] and Thermo Fisher Scientific has switched to compact discrete dynode (CDD) multipliers in the multi-ion counting (MIC) setup of their latest MC ICP-MS (*Neptune Plus*), MC-TIMS (*Triton Plus*), and noble gas mass spectrometers. These CDD detectors yield the same dynamic range (1 cps to 1 400 000 cps) and performance as standard-sized discrete dynode SEMs but they have only about twice the width of standard Faraday cups (6 mm to 7 mm). Hence, they are used as an alternative to the previously applied miniaturised CEMs.[63]

**Channel Electron Multipliers (CEM)**

A CEM consists of a curved glass tube with an internal diameter of about 1 mm.[64] The interior is coated with a lead oxide semiconducting material[65] that has a high inner resistance.[64] Typically, voltages of −2600 V and −3500 V are applied to the multiplier, whereby ions are attracted into the funnel opening.[64] The back of the tube near the collector is held at ground potential.[65] Ions that have entered the glass tube impinge on the surface, which results in an ejection of secondary electrons from the surface. These electrons are accelerated down the tube as a result of the potential gradient across the tube,[64] resulting from the continuous variation of the resistance of the interior coating across the tube.[65] The accelerated electrons are subsequently colliding with the inner surface of the tube. As a consequence, further emissions of secondary electrons result in amplification ($10^7$–$10^8$-fold) of the initial ion signal. At the end, the electron pulses ($\sim$50 to 100 mV), which last about 10 ns, are recorded.[64] A schematic drawing of a CEM is shown in Figure 4.22.

Nowadays, full-size continuous dynode channeltron multipliers are used in multicollector instruments, as well. (For example, in the *Phoenix* MC TIMS,[66] which can fit up to six ion counting detectors alongside their Faraday collector array. The conversion dynode, which is fitted to each multiplier, has the same width as the Faradays and diverts the ion beam to 90°.)

**Discrete Dynode Secondary Electron Multiplier (SEM)**

Today, SEMs are widely employed in sector field instruments, in which they are used in ion-counting mode for detecting low ion signal intensities. SEMs are currently among the most sensitive detectors for measuring extremely small ion currents ($< 10^{-15}$ A),[67] which is a prerequisite for the detection of isotopes that are present in very small concentrations. Typically, the dynamic range goes up to $10^5$–$10^6$ cps in this mode.[67] A more detailed description of the operating principle of SEMs in both counting and analogue mode can be found elsewhere.[61] A SEM typically consists of between 12 and 24 dynodes that are assembled into a dynode array to which a high voltage is applied (Figure 4.23).

The dynodes used in SEMs consist of layered metals (*e.g.* Be on Cu, Mg on Ag) and are constructed in a curved shape in order to focus the electrons onto the subsequent dynode.[68] Ions that enter the detector are accelerated to the first (or conversion) dynode, from which secondary electrons are emitted

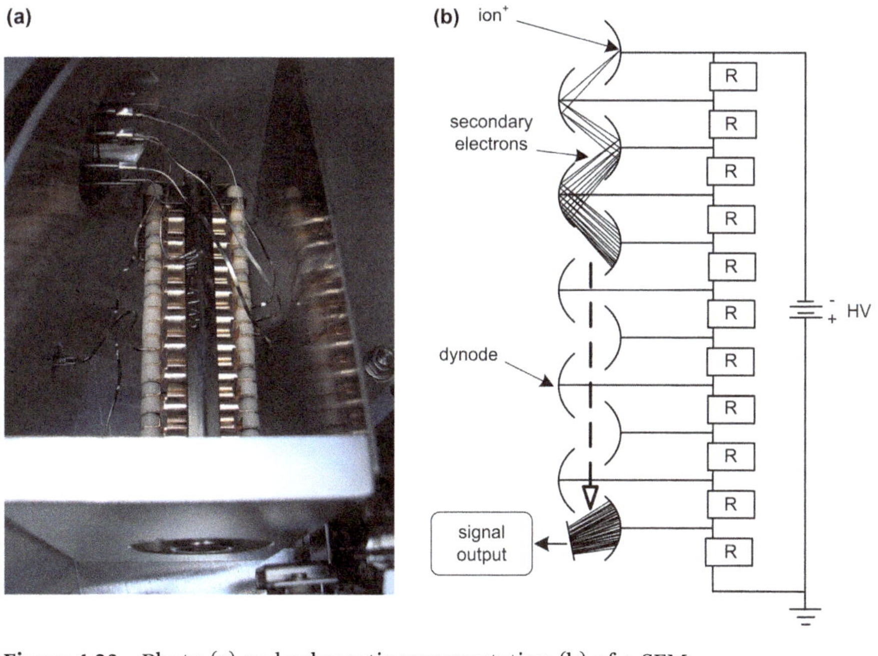


**Figure 4.23** Photo (a) and schematic representation (b) of a SEM. (Source: (a) Photo: Ondrej Hanousek, (b): Stefanie Konegger-Kappel.)

by the impinging ion. The secondary electrons are accelerated by means of the electron optics to the next dynode, where several secondary electrons are generated per incident electron.[61] Thereby, an electron cascade (with an operating gain between $10^4$–$10^8$ electrons)[67] is achieved over several dynodes. The electron pulse that is generated from a single incident ion is finally collected at the detector's output electrode.

The output of the detector is characterised by its gain, which is defined as the average number of electrons collected for each input ion initiating an electron cascade. The gain can be adjusted by means of the voltage applied to the multiplier, which changes the interdynode voltage and as a consequence the electron impact energy.[61] The gain furthermore depends on the mass and the energy of an incident ion as these parameters determine the average number of secondary electrons emitted at the first dynode.[61] Higher gains are achieved by applying higher voltages or using more dynodes.

**High-Voltage Setting and SEM Aging**

The high-voltage setting that yields the maximum gain cannot be considered as the optimum value for a prolonged operation of the multiplier.[69] In Figure 4.24, a typical ion-counting plateau curve is shown.

The curve illustrates the relationship between the high voltage applied to the multiplier and the detected count rate for a given ion signal. The curve is acquired by stepwise increasing the voltage applied to the multiplier while maintaining a stable ion signal to the detector. At the beginning, a low operating voltage is applied so that no signal pulses are measured. In this case the multiplier voltage is set to a value that generates a pulse at the last dynode that is below the amplifier discrimination threshold. This amplifier threshold has to be exceeded in order to assign valid pulses from the multiplier to recorded signals. Below this threshold, real pulses from ions and pulses from other, unwanted sources are not distinguishable.

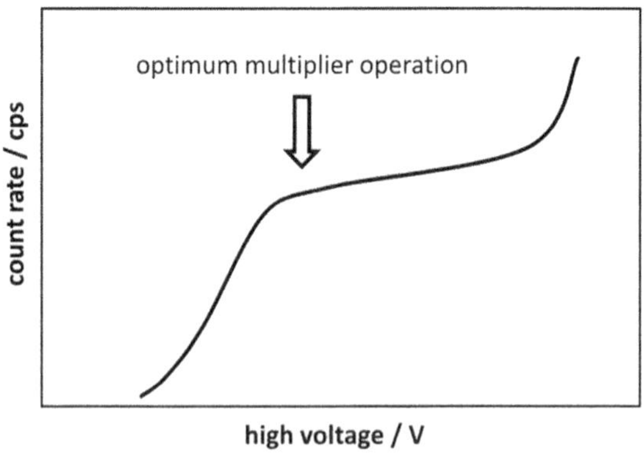


**Figure 4.24** Schematic drawing of an ion-counting plateau of a SEM.

The amplifier thresholds are set up in a way so that the unwanted signals are not recognised, *i.e.* low dark noise and the loss of sensitivity of the detector is minimal. At a certain multiplier voltage the pulse height at the output of the multiplier is above the amplifier threshold and thus a measurable count rate is observed. Further increase of the multiplier voltage results in a steep increase of the count rate. However, at some point – at the plateau region – an increase in the multiplier voltage has only a small effect on the observed count rates. If the multiplier voltage is increased even further, again a sharp increase of the count rate is observed. This can be attributed to 1) ion-feedback in the SEM and 2) electronic double counting of pulses. Ion-feedback occurs as a consequence of the generation of positively charged residual gas molecular ions within the multiplier. If these ions strike a dynode, additional secondary electrons are generated. Hence, a second pulse can be initiated. Electronic double-counting is a result of an impedance mismatch between the signal output electrode and the detection electronics.[61] These so called after-pulsing[70] effects result in an increase of the measured count rates. This means that more counts are detected as one would expect from the ions entering the detector (*i.e.* a gain $>100\%$).

Generally, the optimum voltage should be set just above the knee of the plateau as this is the region of best overall ion detection sensitivity, linearity, gain stability, and operating life. On the knee of the curve already small voltage fluctuations are leading to a pronounced increase or decrease of the detected count rate.[61] The optimum setting of the operation voltage is also important with regard to the SEM lifetime as it becomes shorter with higher voltages. However, the operation voltage has to be gradually increased as the gain decreases with ageing of the detector.[71] A possible cause for this ageing effect is carbon deposition.[71,72] Carbon that is present in the residual gas is bound to the dynodes' surface due to incident secondary electrons, with the amount of deposited carbon being directly related to the total accumulated charge of electrons per unit area on the dynode. Since the number of secondary electrons is highest at the last dynodes, the highest amount of carbon is deposited there. As the accumulated charge is distributed over a larger area in the case of larger surface areas, SEM lifetime can be extended by increasing the surface area of the dynodes. Because of their larger surface area, SEMs (surface area $>1000\ mm^2$ for *ETP Active Film Multipliers*™, SGE Incorporated, Austin, TX, USA) typically have a longer lifetime than standard CEMs (surface area 100–150 $mm^2$).[72]

## Dead Time

Another phenomenon accompanied with a pulse-counting multiplier is the dead time. During this finite time, the system is incapable of determining another subsequent pulse and therefore, the signal of the latter pulse is ignored.[70] Thus, dead time results in count rate losses[65] and has to be accounted for accordingly, especially in high-precision isotope ratio measurements.

Ion counting systems are classified into paralysable or nonparalysable systems.[61,73] In a paralysable system, the dead time can be extended indefinitely as each impinging ion leads to an extension of the dead time, whether or not a pulse is counted. On the other hand, in nonparalysable systems, the dead time is not prolonged by noncounted ions.[61,74] In practice, the SEMs behave in a way that is in between these two.[75] Williamson *et al.*[74] attributed both the detector and the amplifier discriminator to the paralysable system, whereas the electronic components (*i.e.* pulse shaper within the amplifier discriminator and counter) were attributed to the nonparalysable system. Thus, dead time is a feature of both the detector and the ion-counting electronics.[75]

Dead time results in a nonlinear behaviour of the SEM, whereupon nonlinearity and count-rate losses are more pronounced at higher count rates.[65] Hence, measured intensities have to be corrected for dead time effects in order to circumvent the determination of biased isotope ratios. The equation usually applied for correcting for the dead time effect[73,75] is given in eqn (4.28):

$$I_t = \frac{I_0}{(1 - I_0\tau)} \tag{4.28}$$

$I_t$ : dead time corrected signal intensity/cps
$I_0$ : measured (raw) intensity/cps
$\tau$ : dead time/s

Equation (4.28) is only valid for nonparalysable systems. However, at count rates $\ll 1/\tau$ the paralysable and the nonparalysable model are in good approximation.[61,75]

A common approach for determining the dead time of SEMs is based on isotope ratio measurements as a function of the analyte concentration.[76] Nelms *et al.*,[75] for example, compared four different methods of dead time calculations with regard to their ease of application and uncertainty. Three of these methods were based on lead (Pb) isotope ratio measurements. However, another method for determining the dead time is based on pulse-to-pulse timing measurements, in which the time distribution of the pulses from the amplifier is recorded. In this approach, the observed dead time is the time during which no pulse is recorded.[70] More information about dead time and its determination can be found elsewhere.[57,67,75,76]

**SEM Linearity**

Ideally, a linear relationship between the number of ions entering the SEM and the measured count rate is achieved. The slope of this linear relationship represents the gain of the SEM (*i.e.* the ratio of the measured count rate and the actual number of ions entering the SEM), whereas the intercept represents the dark noise. If a linear relationship is present, the gain is independent of count rate. Typically, the gain is – relative to a Faraday cup in a multicollector array – between 90% and 100%, when the high voltage is set to a value just above the knee of the plateau (Figure 4.24). The gain of a Faraday cup is close to 100%.[67,77] However, dead times result in a nonlinear behaviour of a SEM, which is critical for isotope ratio measurements as it leads to inaccurate isotope ratio results.[65] Detailed information on SEM linearity can be found elsewhere.[67,77–80]

### 4.12.2 Daly Detector

The Daly detector is an ion-counting detector that was first described by Daly in 1960.[81] Secondary electrons are produced by ions that are accelerated with several thousand electron volts ($\sim$25 keV)[82] onto a conversion dynode (*e.g.* aluminium surface) which is – in contrast to SEMs – located off-axis to the ion beam (Figure 4.25). The secondary electrons are accelerated in the same field as the ions and are striking a scintillator (*e.g.* phosphor screen), whereby photons are generated. The photons are detected by a photomultiplier that is external to the vacuum system.[81]

### 4.12.3 Faraday Cup

Faraday cups (FC) are – due to their high stability and linearity – regarded as detectors of choice for high-precision isotope ratio measurements.[83] Moreover, their width (*e.g.* 3.5 mm in the case of a Faraday cup used in the *Neptune Plus* MC ICP-MS[63]) allows for easily arranging them in parallel to each other in a multicollector array for the simultaneous detection of separated ions. Faraday cups are also employed in single-collector instruments in addition to SEMs.

Faraday cups are typically applied for the detection of comparatively high ion currents.[62,84] The ion signal is detected in a direct mode where the intensity of the ion beam is measured as a current (charge per unit time) in contrast to pulse-counting systems where individual ions are counted.[62]

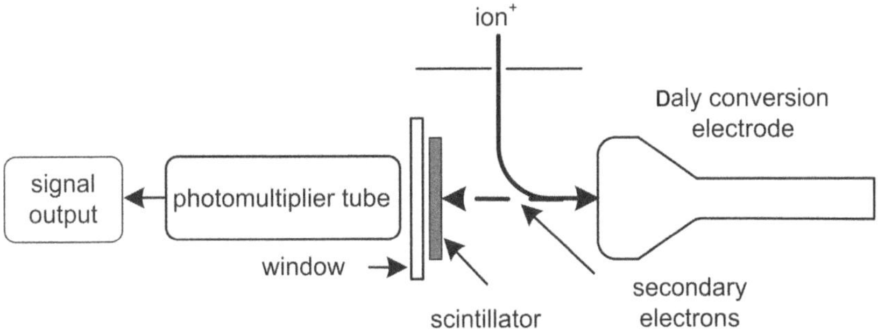


**Figure 4.25** Schematic representation of a Daly detector.

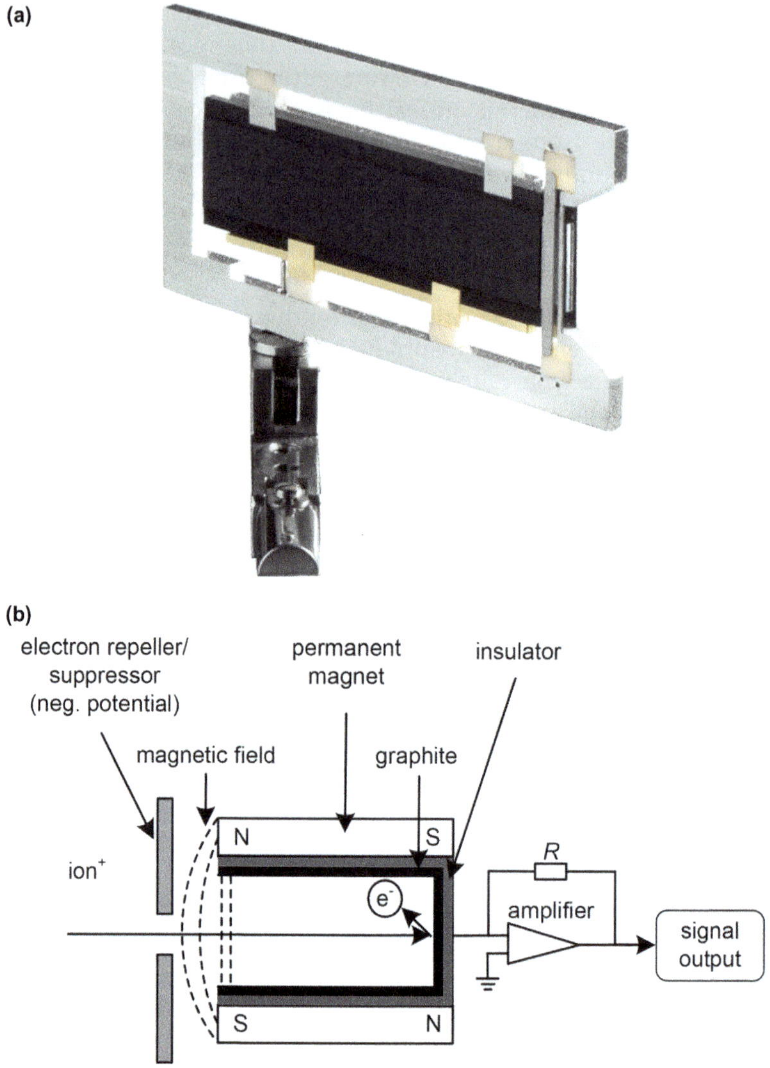


**Figure 4.26** Photo (a) and schematic representation (b) of a Faraday cup. (Source: (a) © Thermo Fisher Scientific.)

Faraday cups are usually designed as a rectangular metal box with one open end into which the separated ions are guided (Figure 4.26).

When ions enter the cup, electrons are needed in order to neutralise the incoming positively charged ions.[68] The electron current flows through a high-value resistor, across which a potential difference is created. This potential difference is recorded. Typically $10^{11}$ Ω high-value resistors are used in MC ICP-MS or MC-TIMS instruments.[85] However, it is also feasible to apply resistors $<10^{11}$ Ω to increase the dynamic range or $>10^{11}$ Ω,[63] to improve the signal-to-noise ratio ($10^{12}$ Ω and $10^{13}$ Ω have been recently applied in the *Neptune plus* instrument). An electron suppressor, which is negatively biased[34] is located near the opening of the Faraday cup.[35] This electron suppressor ensures that secondary electrons are not released from the detector. The escape of secondary electrons is also hindered by small magnets sending surface ejected electrons into a circular path.[35] In addition, Faraday cups are designed in such a way so that the secondary electron escape angle is minimised. This is achieved by increasing the Faraday cup length.[34] The detection efficiency of Faraday cups is close to 100% as virtually all incoming ions are detected as a current.[83]

The lower end of a Faraday cup's dynamic range is limited by the detector noise (*i.e.* Johnson noise[59]) that is associated with the resistance value.[86] The relationship between the random noise and the resistance is given in eqn (4.29):[85]

$$i_{\mathrm{n}} = \sqrt{\frac{4k_{\mathrm{B}}T\Delta f}{R}} \tag{4.29}$$

$I_{\mathrm{n}}$ : random noise/A
$k_{\mathrm{B}}$ : Boltzmann constant/J $K^{-1}$
$T$ : absolute temperature/K
$\Delta f$ : measurement bandwidth/Hz
$R$ : resistance/Ω

Generally, detection systems in multicollector arrays have to be cross- (or gain) calibrated against each other for performing high-precision and accurate isotope ratio measurements. In the case of Faraday cup cross-calibrations, electronic cross-calibration factors are determined, usually by sequentially connecting the amplifier inputs to a stable reference current. The response of the amplifier is measured, whereupon the ratios of the measured signals yield the cross- (or gain) calibration factors.[59] Thermo Fischer Scientific implemented the so-called "Virtual Amplifier" design for the Faraday cup cross-calibration (see Figure 4.27). In comparison to the classical approach in which each Faraday cup has a fixed connection to an amplifier, the "Virtual Amplifier" design uses a relay matrix, whereby the amplifiers can be switched between different Faraday cups. The advantage of this design is that biases caused by uncertainties in the gain calibration of the amplifiers are minimised.[59]

However, electronic cross-calibration does not account for different Faraday cup efficiencies (or cup factors). Faraday cups degrade due to ion beam exposure over time, resulting in a nonuniform behaviour of the Faraday cups in a multicollector array. A measurement strategy for cancelling out such cup factors is discussed in, *e.g.* Wieser and Schwieters.[59]

In order to improve the Faraday cup performance, Thermo Fisher Scientific further employs a larger ion optical magnification (*i.e.* $M = 2$). Since the mass dispersion is directly proportional to the ion optical

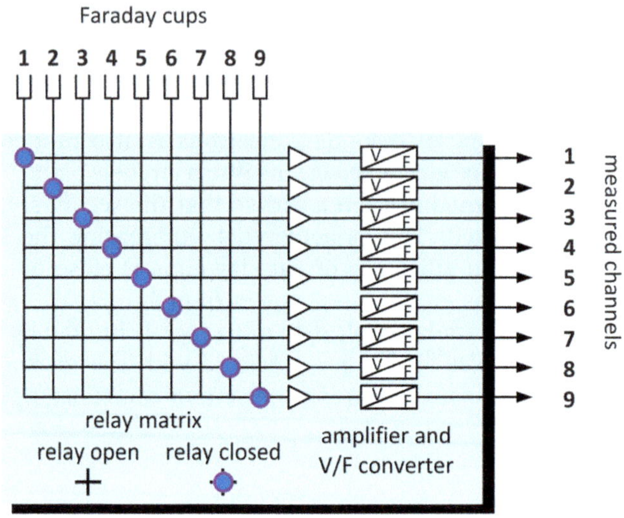


**Figure 4.27** Schematic drawing of the "Virtual Amplifier" design used in MC instruments by Thermo Fisher Scientific.
(Source: © Thermo Fisher Scientific.)

magnification, the Faraday cups can be enlarged (*i.e.* made twice as wide) with $M = 2$ in comparison to an ion optical magnification of $M = 1$. As the angular divergence of the ion beam at the detector is proportionally reduced with the ion optical magnification, the effective depth at which the ions strike the side walls of the Faraday cup is increased. Hence, there is a reduced risk that secondary electrons will escape from the cup.[59]

Faraday cups are the detectors of choice for precise and accurate isotope ratio measurements. However, when it comes to data acquisition of short, fast-changing signals as is the case with transient signals resulting from, for example, laser ablation, Faraday cup amplifier outputs lag behind input signals. Differences in amplifier response times lead to signal intensities that are enhanced or reduced relative to each other.[86] As a consequence, drifts in the determined isotope ratios can be observed during the measurement of a transient signal.[87] A summary of the literature dealing with this issue and with respective data evaluation strategies is given in Kappel *et al.*[88]

## 4.12.4 Focal Plane Detectors

A simultaneous analysis of multiple ions across the periodic table (like in sector field mass spectrometers using Mattauch–Herzog geometry or distance of flight mass spectrometers) or ion imaging (microscope mode) as in SIMS requires focal plane detectors providing high spatial resolution.[89]

**Electro-Optical Imaging Detectors (EOID), Microchannel Plate Detector (MCP)**
Electro-optical imaging detectors (EOIDs) are based on electron multiplier arrays, in which several conversions are used to detect photons that are created

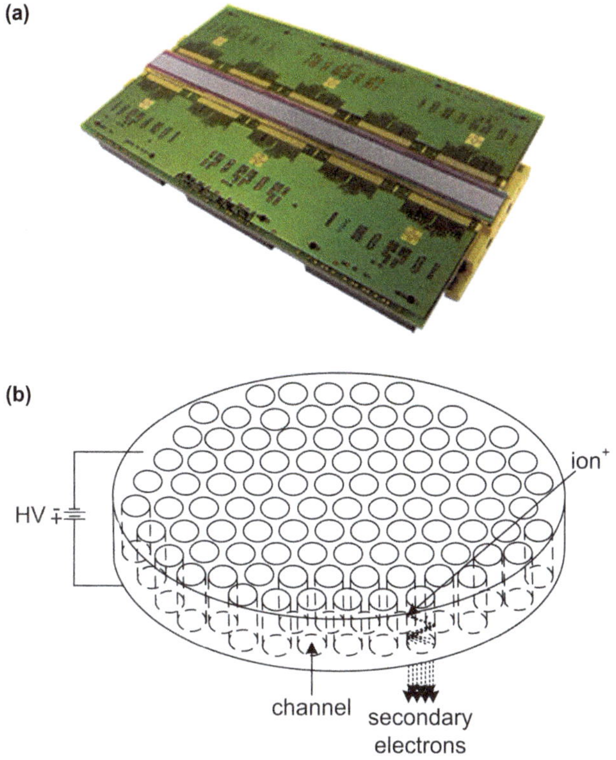


**Figure 4.28** Photo (a) and schematic representation (b) of a MCP. (Source: (a) © AMETEK.)

as a result of "microchannel plate (MPC) electrons" impinging on a scintillator.[90]

Microchannel plates (MCP) typically consist of an array of $10^4$–$10^7$ miniaturised channel electron multipliers[91] (Figure 4.28). The operation principle of these channels is the same as that of standard-sized CEMs (*i.e.* the conversion of incoming ions into secondary electrons and signal amplification by the generation of an electron cascade). The individual channels are orientated in parallel to each other, whereas the channel axes can be normal or at a small angle ($\sim 8°$) to the input surface. Channel diameters range from 10–100 μm and length-to-diameter ratios are between 40 and 100.[91] MCPs usually suffer from a limited dynamic range and cannot be fabricated to cover the entire focal plane at a reasonable expense.[89] In sector field instruments, MCPs are used in SIMS instruments as direct ion image detectors. The MCP detector is either coupled to a fluorescent screen (MCP/FC) or a resistive anode encoder (RAE).

If the MCP detector is coupled to a subsequent fluorescent screen (*e.g.* $^{31}$P screen), secondary electrons exiting the MCP are converted into photons by said screen. The direct ion image can be observed *via* a charge-coupled device (CCD) camera. The spatial resolution of the MCP/FC detector is limited by the channel diameter (~10 μm) and by the so-called electron beam blooming effect (~70 μm), which takes place in the acceleration space between the MCP and the fluorescent screen.[71]

The resistive anode encoder (RAE) system consists of a dual MCP and a resistive film that is deposited on a ceramic plate. Ion-to-electron conversion is accomplished by the MCP, whereas the RAE is used for the determination of the position of the charge. Secondary electrons that leave the second MCP strike the RAE, which results in the generation of a charged pulse. This pulse is shared between the four electrodes of the RAE. Thus, the spatial position of the charged pulse can be determined and an ion image can be derived. However, a dead time of about 3 μs limits the use of this detector with respect to small count rates.[71]

**Direct Charge Detector (DCD); Focal Plane Camera (FPC)**

The basic principle of a direct charge detector (DCD) is similar to that of a Faraday cup. A detector signal is generated due to the neutralisation of incoming positively charged ions by electrons. The efficiency of the DCD is close to 100% as each ion arriving at the detector contributes to the signal. The focal plane camera (Figure 4.29) belongs to a group of detectors developed in a collaboration among the University of Arizona, Pacific Northwest National Laboratory, and Indiana University. The focal plane camera is a linear array of Faraday-strip detectors, each with its own integrating operational amplifier that has a gain inversely proportional to the capacitance in its feedback loop.[92–95]

The fourth-generation focal plane camera features 1696 individual detector elements, each with two independent levels of gain and its own amplifier. The sample-and-hold feature enables nondestructive readout of all channels simultaneously. The current isotope ratio precision, which can be reached with these types of cameras, is 0.015%.[89]

Currently, a DCD is installed in the focal plane of the *SPECTRO MS*, a double-focusing sector field mass spectrometer with Mattauch–Herzog geometry (see Chapter 12). This DCD, the *Ion 120*, which was developed for SPECTRO Analytical Instruments GmbH, is a 12-cm long complementary metal oxide semiconductor (CMOS) array with 4800 separate channels.[96] The size of one channel (pixel) is about 25 μm.[52] This enables the simultaneous measurement of the entire mass range from $^{6}Li$ to $^{238}U$.[96] One isotope peak is typically collected by 5 to 20 channels over its spatial width in the focal plane.[52] As each channel is equipped with separate low- and high-gain detector elements, each channel can handle a wide range of signal intensities separately. The dynamic range of the *Ion 120* DCD is up to eight orders of magnitude in the basic integration cycle, but it can be extended by the readout system in the case of longer measurements. Detector saturation can be overcome by automatically logging the integrated signal, resetting the channel and restarting collection.[96] Different signal acquisition strategies are discussed by Ardelt *et al.*[52]

**Active Pixel Detectors**

The active pixel detector is an imaging detector that has recently been applied for SIMS microscope mode imaging.[97–99] The main advantages of this

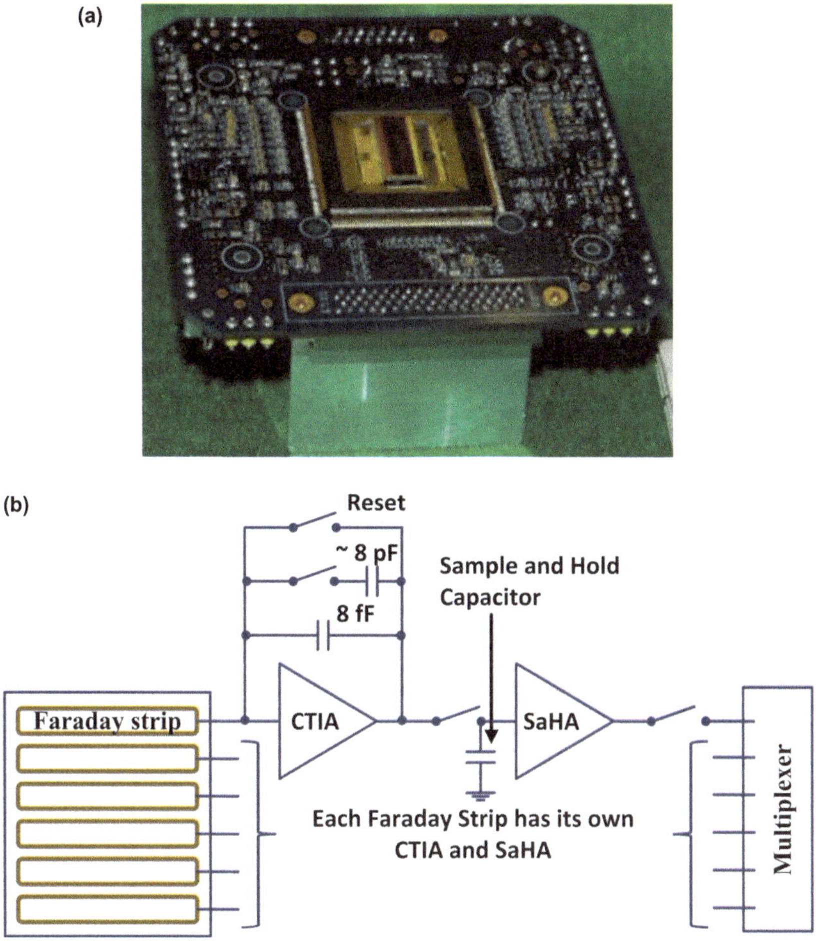


**Figure 4.29** Photo (a) and schematic representation (b) of a FPC. (Source: (a) © Indiana University with courtesy of Gary Hieftje, (b) : Ref. 89.)

detector are both position- and time-resolved imaging of different masses simultaneously, while offering high dynamic range, high throughput, and high spatial resolution. The in-vacuum active pixel detectors belong to the family of *Medipix 2*[100]/*Timepix*[101] fast readout chip detectors. An assembly of MCP/*Timepix* detectors can be operated in particle counting, time-over-threshold (TOT), and time-of-flight (TOF) mode. More details on this detector setup and its potential can be found elsewhere (*e.g.* Kiss *et al.*[97]).

## 4.13 Vacuum System

The vacuum system of a sector field instrument is of crucial importance since the high kinetic energy and the longer flight path (compared to quadrupole-based instruments) would otherwise lead to a significant scatter of ions, loss of transmission and deteriorated abundance sensitivity by collision of ions with residual gas molecules.[102]

The lens system is usually operated in a medium vacuum range ($\sim 10^{-3}$ Pa), whereas the analyser must be operated under higher (compared to a quadrupole-based instrument) vacuum conditions ($\sim 10^{-5}$ Pa). The vacuum required is achieved by using a differential pumping system (vane-type rotary pump in combination with turbomolecular pumps). An even higher vacuum has proven advantageous in the case of multicollector devices. Multicollector instruments use can additional ion getter pumps in order to provide a vacuum down to $10^{-7}$ Pa in the analyser part of the instrument. The analyser part can be separated from the lens system by one or more additional valves.

## References


1. T. W. Burgoyne and G. M. Hieftje, *Mass Spectrom. Rev.*, 1996, **15**, 241–259.
2. S. Christopher, S. Long, M. Rearick and J. Fassett, *Anal. Chem.*, 2001, **73**, 2190–2199.
3. M. Resano, M. Aramendìa and F. Vanhaecke, *J. Anal. Atom. Spectrom.*, 2009, **24**, 484–493.
4. D. Tabersky, K. Nishiguchi, K. Utani, M. Ohata, R. Dietiker, M. B. Fricker, I. M. de Maddalena, J. Koch and D. Günther, *J. Anal. Atom. Spectrom.*, **28**, 831–842.
5. C. Latkoczy and D. Gunther, *J. Anal. Atom. Spectrom.*, 2002, **17**, 1264–1270.
6. B. Neidhart and E. Welter, *Fresenius J. Anal. Chem.*, 1993, **346**, 536–538.
7. A. Montaser, *Inductively Coupled Plasma Mass Spectrometry*, Wiley, 1998.
8. F. Adams and A. Vertes, *Fresenius J. Anal. Chem.*, 1990, **337**, 638–647.
9. N. Jakubowski, *J. Anal. Atom. Spectrom.*, 2008, **23**, 673–684.
10. S. Valente and W. Schrenk, *Appl. Spectrosc.*, 1970, **24**, 197–205.
11. J. S. Becker, *Inorganic Mass Spectrometry - Principles and Applications*, John Wiley & Sons, Chichester, 2007.
12. L. J. Radziemski and D. A. Cremers, *Laser-Induced Plasmas and Applications*, Marcel Dekker, Inc., New York, USA, 1989.
13. K. Song, Y.-I. Lee and J. Sneddon, *Appl. Spectrosc. Rev.*, 1997, **32**, 183–235.
14. J. A. Broekaert and V. Siemens, Spectrochim. Acta Part B, *Atom. Spectrosc*, 2004, **59**, 1823–1839.
15. F. F. Chen, *Introduction to Plasma Physics and Controlled Fusion - Plasma Physics*, Springer Science + Business Media, LLC, New York, NY, USA, 2nd edn., 1974.
16. R. Thomas, *Beginner's Guide to ICP-MS*, CRC Press, Taylor & Francis Group, LLC, 2002, 26–33.
17. G. Ramendik, E. Fatyushina, A. Stepanov and V. Sevast'yanov, *J. Anal. Chem.*, 2001, **56**, 500–506.
18. R. S. Houk, V. A. Fassel, G. D. Flesch, H. J. Svec, A. L. Gray and C. E. Taylor, *Anal. Chem.*, 1980, **52**, 2283–2289.

19. C. D. Quarles, J. Castro and R. K. Marcus, Glow Discharge Mass Spectrometry, in *ONLINE Encyclopedia of Spectroscopy and Spectrometry*, eds. J. C. Lindon, G. E. Tranter and D. Koppenaal, Elsevier Science, 2nd edn., 2010, vol. 3.
20. M. Betti, *Int. J. Mass Spectrom.*, 2005, **242**, 169–182.
21. M. Tarik, G. Lotito, J. A. Whitby, J. Koch, K. Fuhrer, M. Gonin, J. Michler, J.-L. Bolli and D. Günther, *Spectrochim. Acta, Part B*, 2009, **64**, 262–270.
22. W.-L. Shen, T. M. Davidson, J. T. Creed and J. A. Caruso, *Appl. Spectrosc.*, 1990, **44**, 1003–1010.
23. Q. Jin, C. Zhu, M. W. Border and G. M. Hieftje, *Spectrochim. Acta, Part B*, 1991, **46**, 417–430.
24. J. Hubert, M. Moisan and A. Ricard, *Spectrochim. Acta, Part B*, 1979, **34**, 1–10.
25. C. I. M. Beenakker, *Spectrochim. Acta, Part B*, 1976, **31**, 483–486.
26. H. Beske, A. Hurrle and K. Jochum, *Fresenius Z. Anal. Chem.*, 1981, **309**, 258–261.
27. H. Beske, *Int. J. Mass Spectrom. Ion Phys*, 1982, **45**, 173–181.
28. F. Novak, K. Balasanmugam, K. Viswanadham, C. Parker, Z. Wilk, D. Mattern and D. Hercules, *Int. J. Mass Spectrom. Ion Phys*, 1983, **53**, 135–149.
29. Y. Lin, Q. Yu, W. Hang and B. Huang, *Spectrochim. Acta, Part B*, 2010, **65**, 871–883.
30. R. Hellborg, ed., *Electrostatic Accelerators: Fundamentals and Applications (Particle Acceleration and Detection)*, Springer Verlag, Berlin Heidelberg, 2005.
31. M. Claeys and J. Claereboudt, Fast Atom Bombardment Ionization in Mass Spectrometry, in *Encyclopedia of Spectroscopy and Spectrometry*, eds. G. Tranter, J. Holmes and J. Lindon, Academic Press, 2000, vol. 1–3, pp. 505–512.
32. H. Oechsner, *Int. J. Mass Spectrom. Ion Processes*, 1995, **143**, 271–282.
33. C. Dass, *Fundamentals of Contemporary Mass Spectrometry*, John Wiley & Sons, Inc., Hoboken, NJ, USA, 2007.
34. J. H. Gross, Practical Aspects of Electron Ionization, in *Mass Spectrometry*, Springer, Berlin Heidelberg, 2011, ch. 5, pp. 223–248.
35. J. Fassett, L. Moore, J. Travis and J. DeVoe, *Science*, 1985, **230**, 262–267.
36. J. Young, R. Shaw and D. Smith, *Anal. Chem.*, 1989, **61**, 1271A–1279A.
37. J. Levine, M. R. Savina, T. Stephan, N. Dauphas, A. M. Davis, K. B. Knight and M. J. Pellin, *Int. J. Mass Spectrom.*, 2009, **288**, 36–43.
38. D. Beauchemin and D. E. Matthews, eds., *Encyclopedia of Mass Spectrometry - Elemental and Isotope Ratio Mass Spectrometry*, Elsevier, Amsterdam, 2010.
39. L. E. J. Vandevelde and S. van Winckel, *ATB Metall*, 1990, **30**, 99–102.
40. P. J. Mohr and B. N. Taylor, *Rev. Mod. Phys.*, 2005, 77, 1.

41. P. J. Turner, D. J. Mills, E. Schröder, G. Lapitajs, G. Jung, L. A. Iacone, D. A. Haydar and A. Montaser, Basic Concepts of High Resolution ICPMS in *Inductively Coupled Plasma Mass Spectrometry*, ed. A. Montaser, Wiley-VCH, New York, 1998, pp. 446–459.
42. C. Brunnée and H. Voshage, eds., *Massenspektrometrie,* Verlag Karl Thiemig KG, München, 1964.
43. A. Benninghoven, F. G. Rudenauer and H. W. Werner, *Secondary Ion Mass Spectrometry: Basic Concepts, Instrumental Aspects, Applications and Trends*, John Wiley and Sons, New York, 1987.
44. K. T. Bainbridge and E. B. Jordan, *Phys. Rev*, 1936, **50**, 282–296.
45. H. Hintenberger and L. A. König, Mass spectrometers and mass spectrographs corrected for image defects, in *Advances in Mass Spectrometry*, ed. J. D. Waldron, Pergamon Press, New York, 1959, vol. 1, pp. 16–35.
46. H. Matsuda, *Int. J. Mass Spectrom. Ion Phys*, 1974, **14**, 219–233.
47. I. Takeshita, *Rev. Sci. Instrum*, 1967, **38**, 1361–1367.
48. E. G. N. Johnson and O. Alfred, *Phys. Rev.*, 1953, **91**, 10–17.
49. A. D. McNaught and A. Wilkinson, *Compendium of Chemical Terminology, The Gold Book*, 2nd edn., Wiley Blackwell, Oxford, 1997.
50. N. P. Jakubowski, T. Prohaska, L. Rottman and F. Vanhaecke, *J. Anal. Atom. Spectrom.*, 2011, **26**, 693–726.
51. G. D. Schilling, S. J. Ray, A. A. Rubinshtein, J. A. Felton, R. P. Sperline, M. B. Denton, C. J. Barinaga, D. W. Koppenaal and G. M. Hieftje, *Anal. Chem.*, 2009, **81**, 5467–5473.
52. D. Ardelt, A. Polatajko, O. Primm and M. Reijnen, *Anal. Bioanal. Chem.*, 2013, **405**, 2987–2994.
53. K. T. Bainbridge and E. B. Jordan, *Phys. Rev*, 1936, **50**, 282.
54. H. Hintenberger and L. A. Konig, *Mass Spectrometers and Mass Spectrographs Corrected for Image Defects*, Pergamon Press, New York, 1959.
55. H. Matsuda, *Mass Spectrosc*, 1985, **33**, 227.
56. Nu Instruments Ltd., *NP Abundance Sensitivity*, Wrexham, UK.
57. D. C. Baxter, I. Rodushkin and E. Engström, *J. Anal. Atom. Spectrom.*, 2012, **27**, 1355–1381.
58. IUPAC, http://old.iupac.org/goldbook/A00048.pdf, (accessed 01.08.2012).
59. M. E. Wieser and J. B. Schwieters, *Int. J. Mass Spectrom.*, 2005, **242**, 97–115.
60. Thermo Fisher Scientific Inc., *Neptune Plus Hardware Manual, Revision B - 1250710*, Waltham, MA, USA, 2009.
61. K. Hunter and D. Stresau, Ion Detection, in *Inductively Coupled Plasma Mass Spectrometry Handbook*, ed. S. M. Nelms, Blackwell Publishing Ltd, 2005.
62. M. Rehkämper, M. Schönbächler and C. H. Stirling, *Geostandards Newslett*, 2001, **25**, 23–40.
63. D. Tuttas, J. B. Schwieters, C. Bouman and M. Deerberg, *New Compact Discrete Dynode Multipliers Integrated into the Thermo Scientific TRITON*

*Variable Multicollector Array*, AN30192_E 12/09C, Thermo Fisher Scientific Inc., Bremen, Germany, https://http://www.thermo.com/eThermo/CMA/PDFs/Product/productPDF_56696.PDF, 2008.
64. G. O'Connor and E. H. Evans, Fundamental Aspects of Inductively Coupled Plasma - Mass Spectrometry (ICP-MS), in *Inductively Coupled Plasma Spectrometry and its Applications*, ed. S. J. Hill, Blackwell Publishing Ltd., 2007.
65. K. E. Jarvis, A. L. Gray and R. S. Houk, eds., *Handbook of Inductively Coupled Plasma Mass Spectrometry*, Blackie & Son Ltd, 1992.
66. IsotopX, http://www.isotopx.com/docs/PhoenixBrochure.pdf, (accessed 11.06.2013).
67. S. Richter, S. A. Goldberg, P. B. Mason, A. J. Traina and J. B. Schwieters, *Int. J. Mass Spectrom.*, 2001, **206**, 105–127.
68. R. A. Meyers, ed., *Encyclopedia of Analytical Chemistry - Applications, Theory and Instrumentation*, John Wiley & Sons Ltd., 2000.
69. Nu Instruments Ltd., *NP Multiplier Operation*, Wrexham, UK.
70. H. Ramebäck, M. Berglund, D. Vendelbo, R. Wellum and P. D. P. Taylor, *J. Anal. Atom. Spectrom.*, 2001, **16**, 1271–1274.
71. E. de Chambost, *The Cameca Electron and Ion Detector Handbook*, 2007.
72. A. D. Cutter, K. L. Hunter, P. J. K. Paterson, R. W. Stresau, http://www.sge.com/uploads/42/74/427499f9019d7cb429885d70f67383fd/TA-0072-A.pdf, (accessed 14.06.2013).
73. J. W. Müller, *Nucl. Instrum. Methods*, 1973, **112**, 47–57.
74. J. A. Williamson, M. W. Kendall-Tobias, M. Buhl and M. Seibert, *Anal. Chem.*, 1988, **60**, 2198–2203.
75. S. M. Nelms, C. R. Quetel, T. Prohaska, J. Vogl and P. D. P. Taylor, *J. Anal. Atom. Spectrom.*, 2001, **16**, 333–338.
76. F. Vanhaecke, G. de Wannemacker, L. Moens, R. Dams, C. Latkoczy, T. Prohaska and G. Stingeder, *J. Anal. Atom. Spectrom.*, 1998, **13**, 567–571.
77. S. Richter, A. Alonso, Y. Aregbe, R. Eykens, F. Kehoe, H. Kühn, N. Kivel, A. Verbruggen, R. Wellum and P. D. P. Taylor, *Int. J. Mass Spectrom.*, 2009, **281**, 115–125.
78. K. J. R. Rosman, W. Lycke, R. Damen, R. Werz, F. Hendrickx, L. Traas and P. De Bièvre, *Int. J. Mass Spectrom. Ion Process*, 1987, **79**, 61–71.
79. U. Nygren, H. Ramebäck, M. Berglund and D. C. Baxter, *Int. J. Mass Spectrom.*, 2006, **257**, 12–15.
80. D. L. Hoffmann, D. A. Richards, T. R. Elliott, P. L. Smart, C. D. Coath and C. J. Hawkesworth, *Int. J. Mass Spectrom.*, 2005, **244**, 97–108.
81. N. R. Daly, *Rev. Sci. Instrum*, 1960, **31**, 264–267.
82. Z. Palacz, *An assessment of the linearity of the ion-counting Daly detector and Hamamatsu photomultiplier using NBS U500*, Isotopx Technical Note T11010, Middlewich, Cheshire, UK, http://www.isotopx.com/docs/T11010 Performance of the ion-counting Daly on the Phoenix TIMS.pdf.
83. H. P. Longerich and W. Diegor, Introduction to mass spectrometry, in *Principles and Applications of Laser-Ablation ICP-Mass Spectrometry in the*

*Earth Sciences*, ed. P. Sylvester, Mineralogical Association of Canada, Short Course Series Vol. 29, 2001.
84. J. S. Becker, ed., *Inorganic Mass Spectrometry. Principles and Applications*, John Wiley & Sons, Ltd., West Sussex, UK, 2007.
85. Nu Instruments Ltd., *NP Manual* V2, Wrexham, UK.
86. T. Pettke, F. Oberli, A. Audetat, U. Wiechert, C. R. Harris and C. A. Heinrich, *J. Anal. Atom. Spectrom.*, 2011, **26**, 475–492.
87. T. Hirata, Y. Hayano and T. Ohno, *J. Anal. Atom. Spectrom.*, 2003, **18**, 1283–1288.
88. S. Kappel, S. Boulyga, L. Dorta, D. Günther, B. Hattendorf, D. Koffler, G. Laaha, F. Leisch and T. Prohaska, *Anal. Bioanal. Chem.*, 2013, **405**, 2943–2955.
89. J. A. Felton, G. D. Schilling, S. J. Ray, R. P. Sperline, M. B. Denton, C. J. Barinaga, D. W. Koppenaal and G. M. Hieftje, *J. Anal. Atom. Spectrom.*, 2011, **26**, 300–304.
90. J. H. Barnes IV and G. M. Hieftje, *Int. J. Mass Spectrom.*, 2004, **238**, 33–46.
91. J. L. Wiza, *Nucl. Instrum. Methods*, 1979, **162**, 587–601.
92. J. H. Barnes, R. Sperline, M. B. Denton, C. J. Barinaga, D. Koppenaal, E. T. Young and G. M. Hieftje, *Anal. Chem.*, 2002, **74**, 5327–5332.
93. J. H. Barnes, G. D. Schilling, R. Sperline, M. B. Denton, E. T. Young, C. J. Barinaga, D. W. Koppenaal and G. M. Hieftje, *Anal. Chem.*, 2004, **76**, 2531–2536.
94. G. D. Schilling, F. J. Andrade, J. H. Barnes, R. P. Sperline, M. B. Denton, C. J. Barinaga, D. W. Koppenaal and G. M. Hieftje, *Anal. Chem.*, 2006, **78**, 4319–4325.
95. G. D. Schilling, S. J. Ray, A. A. Rubinshtein, J. A. Felton, R. P. Sperline, M. B. Denton, C. J. Barinaga, D. W. Koppenaal and G. M. Hieftje, *Anal. Chem.*, 2009, **81**, 5467–5473.
96. SPECTRO Analytical Instruments GmbH; http://www.labbulletin.com/downloads/20130407_4/download. Accessed 12 June 2013.
97. A. Kiss, J. H. Jungmann, D. F. Smith and R. M. Heeren, *Rev. Sci. Instrum*, 2013, **84**, 013704.
98. J. H. Jungmann, L. MacAleese, R. Buijs, F. Giskes, A. De Snaijer, J. Visser, J. Visschers, M. J. Vrakking and R. Heeren, *J. Am. Soc. Mass Spectrom*, 2010, **21**, 2023–2030.
99. J. H. Jungmann, D. F. Smith, L. MacAleese, I. Klinkert, J. Visser and R. M. Heeren, *J. Am. Soc. Mass Spectrom*, 2012, **23**, 1679–1688.
100. X. Llopart, M. Campbell, R. Dinapoli, D. San Segundo and E. Pernigotti, *Nucl. Sci., IEEE Trans*, 2002, **49**, 2279–2283.
101. X. Llopart, R. Ballabriga, M. Campbell, L. Tlustos and W. Wong, *Nucl. Instrum. Methods Phys. Res. Sect. A: Accelerators, Spectrometers, Detectors Assoc. Equip*, 2007, **581**, 485–494.
102. E. d. Hoffmann and V. Stroobant, *Mass Spectrometry: Principles and Applications*, John Wiley & Sons, Chichester, West Sussex, England, 3rd edn., 2007.

CHAPTER 5

# *Mass Resolution*

ONDREJ HANOUSEK,[a] LOTHAR ROTTMANN[b] AND THOMAS PROHASKA*[a]

[a] University of Natural Resources and Life Sciences Vienna (BOKU), Department of Chemistry, Division of Analytical Chemistry, VIRIS Laboratory for Analytical Ecogeochemistry, Tulln, Austria; [b] Thermo Fisher Scientific, Bremen, Germany
*Email: thomas.prohaska@boku.ac.at

Mass resolution is generally understood as a measure for the separation of pattern (peaks) observed in a mass spectrum. The resolving power is defined by the IUPAC as the measure of the ability of a mass spectrometer to provide a specific value of mass resolution.[1] Mass resolution in general depends on the mass/charge ($m/z$) ratio of the signal and on the difference in mass/charge quotients between two separated ion signals. According to IUPAC, the mass resolution ($R$) in a mass spectrum is defined as the observed $m/z$ value divided by the smallest difference $\Delta(m/z)$ for two ions that can be separated (eqn (5.1)).[1]

$$R = \frac{(m/z)}{\Delta(m/z)} \tag{5.1}$$

$R$: mass resolution
$m$: mass of measured signal (peak)
$z$: absolute charge number
$\Delta m$: mass difference between two peaks at masses $m$ and $m + \Delta\text{m}$.

(***Note:*** The $m/z$ value at which the measurement was made should be reported.

New Developments in Mass Spectrometry No. 3
Sector Field Mass Spectrometry for Elemental and Isotopic Analysis
Edited by Thomas Prohaska, Johanna Irrgeher, Andreas Zitek and Norbert Jakubowski

Published by the Royal Society of Chemistry, www.rsc.org

***Note:*** The definition and method of measurement of $\Delta(m/z)$ should be reported. Commonly this is performed using peak width measured at a specified percentage of peak height.

***Note:*** Alternatively $\Delta(m/z)$ is defined as the separation between two adjacent equal-magnitude peaks such that the valley between them is a specified fraction of the peak height, for example as measured by peak matching.)[1]

For a single peak at mass/charge ratio $m/z$ in the mass spectrum, the symbol $\Delta m/z$ stands for a peak width in a certain part of the peak maximum height. These two different definitions will be discussed in the following paragraphs.

**10% Valley Definition (see Figure 5.1)**

"Value of $(m/z)/\Delta(m/z)$ measured for two peaks of equal height in a mass spectrum at $m/z$ and $m/z \pm \Delta(m/z)$ that are separated by a valley that is – at its lowest point – 10% of the height of either peak. For peaks of similar height separated by a valley, let the height of the valley at its lowest point be 10% of the lower peak. Then the resolution (10% valley definition) is $(m/z)/\Delta(m/z)$ and should be given for a number of values of $m/z$."[1]

**Peak Width Definition (see Figure 5.2)**

The condition concerning equal intensities of two neighbouring peaks is met rather rarely in practice. Therefore, a peak-width definition of the mass resolution is used more often. Here, the $m/z$ is the mass/charge quotient of the signal peak and $\Delta(m/z)$ is defined in most cases as the width of the peak profile at 5% of the peak maximum height.[1] This 5% width definition corresponds well to the 10% valley definition.

"For a single peak corresponding to singly charged ions at mass $m$ in a mass spectrum, the resolution may be expressed as $(m/z)/\Delta(m/z)$, where

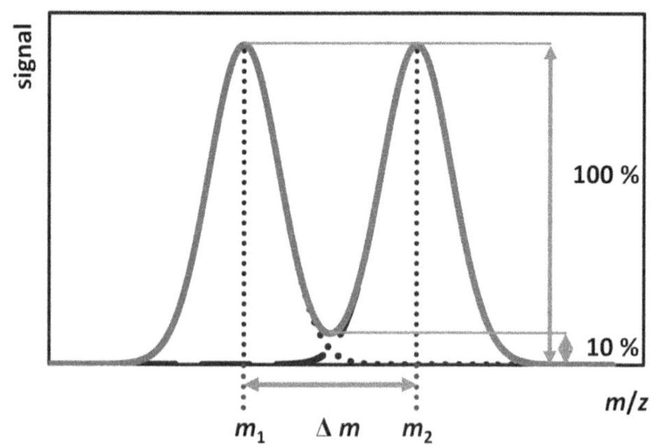


**Figure 5.1** 10% valley definition of mass resolution.

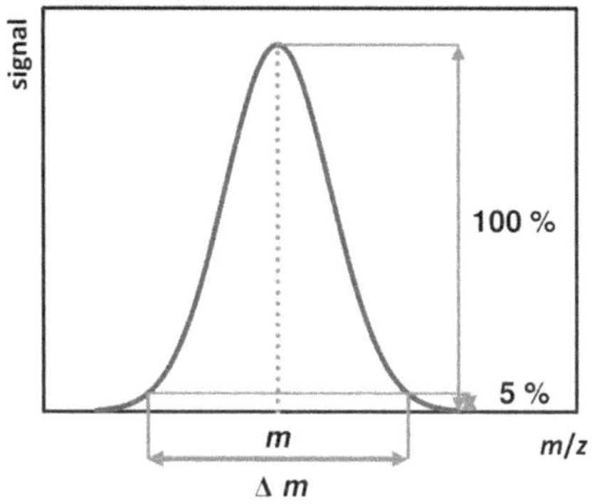


**Figure 5.2** Peak width definition of mass resolution.

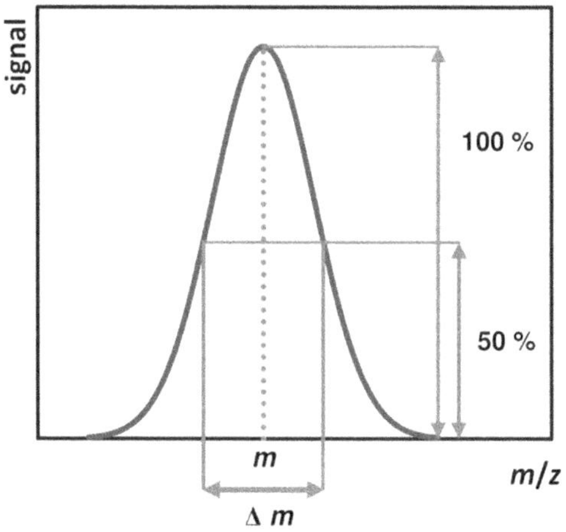


**Figure 5.3** Full width at half-maximum definition of mass resolution.

$\Delta(m/z)$ is the width of the peak at a height that is a specified fraction of the maximum peak height. It is recommended that one of three values 50, 5, or 0.5% should always be used."[1]

**Full Width at Half-Maximum Definition (see Figure 5.3)**

A common approach is the mass-resolution definition based on the use of the peak full width at half-maximum (FWHM) as $\Delta m$. The resulting value calculated with the FWHM definition is generally about twice as high as the value obtained applying the 10% valley definition.

"For an isolated symmetrical peak recorded with a system that is linear in the range between 5% and 10% levels of the peak, the 5% peak width definition is equivalent to the 10% valley definition. A common standard is the definition of resolution based upon $\Delta m$ being full width of the peak at half its maximum (FWHM) height."[1]

## 5.1 Achieving High Mass Resolution with a Magnetic Sector Field

The resolution of a sector field instrument is in the first instance independent of mass. Taking into account eqn (5.1), the peak width ($\Delta(m/z)$) gets wider with increasing mass since $R$ is constant.

(***Note:*** For example, in a quadrupole analyser the peak width is constant, so that according to eqn (5.1), the mass resolution changes with mass.)

The achieved mass resolution in a sector field instrument is governed by

1. the beam width (defined mainly by the entrance slit);
2. the dispersion of the magnet sector (defined by the geometry of the magnet);
3. the lateral magnification of the system (defined by the geometry of the magnet);
4. the energy spread of the ions generated in the ion source;
5. the sum of aberrations of the mass analyser;
6. the width of the exit slit.

To achieve different mass resolutions within one instrument, usually the width of the slits are varied. Decreasing the slit width for the entrance (and exit) slit provides increasing mass resolution. Consequently, this leads usually to lower signal intensities as fewer ions pass through the narrower slit.

## 5.2 Peak Shapes as a Result of Mass Resolution

The combination of the widths of the entrance and the exit slits define the shape of a peak of the obtained mass spectrum. The peak shape can be monitored by scanning the ion beam across the exit slit (*i.e.* scanning the mass spectrometer according to the $m/z$ ratio).

Usually, flat-top peaks are preferred, especially for high-precision isotope ratio measurements. The reason behind is the fact that variations due to small possible drifts in the mass position do not lead to signal variations. Flat-top peaks are achieved if the width of the exit slit is significantly larger than the width of the incoming ion beam (*i.e.* the width of the entrance slit in combination with the lateral magnification). Scanning the ion beam across the exit slit leads in the beginning to a steady increase of the ion beam intensity until the entire ion beam passes through the exit slit. During further scanning, the entire ion beam keeps passing through the exit slit, which appears as a steady signal along increasing $m/z$, until the ion beam moves out of the exit slit, when a signal decrease in the same way as the increase can be observed (Figure 5.4). Flat-top peaks are usually preferred when operating magnetic sector field devices under so-called low-resolution conditions ($R \leq 400$), allowing for the separation of approximately unity $m/z$ at high masses.

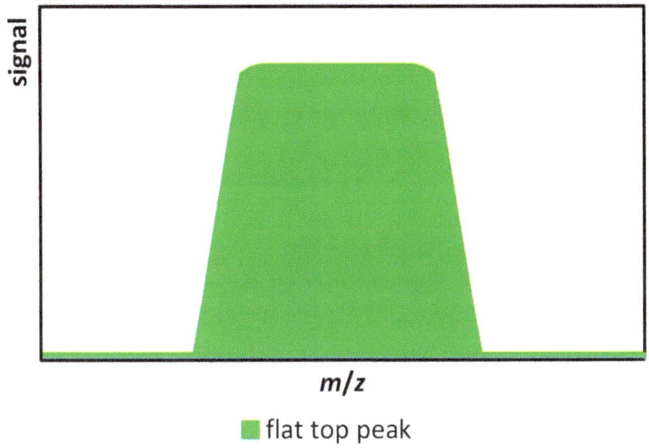


**Figure 5.4** Schematics of a flat-top peak.

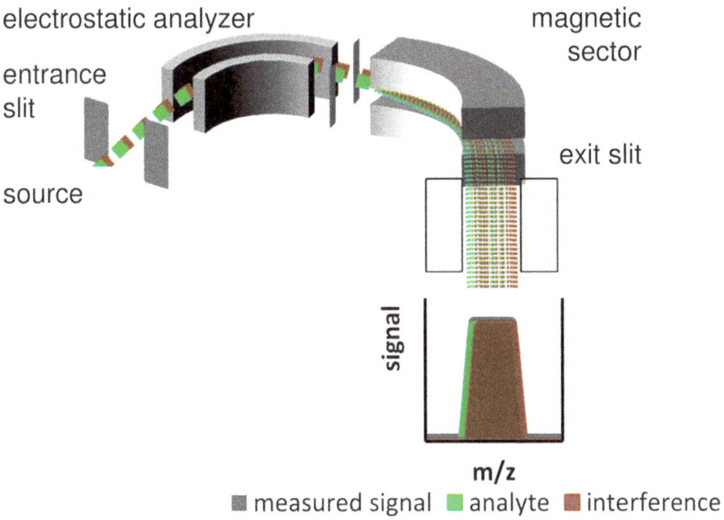


**Figure 5.5** Schematics of a flat-top peak obtained at low mass resolution (in the case of overlapping interferences, these are hidden within one peak).

If the entrance slit width is set to the maximum width for the defined "low resolution" and the ion beam is focused ideally with the aid of ion optical lenses onto the entrance slit, the ion beam enters the analyser in its largest width (Figure 5.5). For example, in the case of the ICP-MS instrument *Element2*, the width of the exit slit is about twice of the width of entrance slit (*e.g.* 600 μm). As a result, a flat-top peak is obtained. The actual mass resolution *R* is in this case is about 300. It is evident that spectral interferences, which would require a higher mass resolution, appear within the same peak. The benefits of using low mass resolution are to achieve the highest sensitivity possible and a flat-top peak profile.

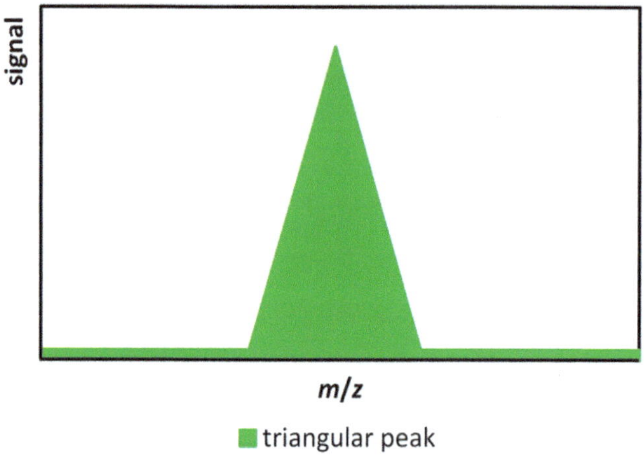


**Figure 5.6** Schematics of a triangular peak obtained at high mass resolution.

Reducing the exit slit width to the width of the ion beam (*i.e.* width of the entrance slit) results in a triangular peak shape (Figure 5.6). Only at the top of the triangle the entire ion beam passes through the exit slit during the scan of the ion beam across the exit slit (Figure 5.7). This is usually observed when instruments are operated under high mass resolution conditions. The width of the exit slit is usually kept similar to the width of the entrance slit in order to achieve the highest mass resolution. Reducing the width of entrance and exit slits means increasing the mass resolution but in parallel decreasing the number of ions passing through the analyser.

Higher mass resolution is usually applied to resolve spectral interferences. This is again shown on the example of the ICP-MS instrument *Element2*. The so-called medium resolution is achieved when the numeric value of the resolution $R$ equals to about 4000 (with entrance and exit slit width are both of the order of 16 μm). In this measurement mode, it is possible to separate typical interferences (*e.g.* $^{15}N^{16}O^+$ from $^{31}P^+$, $^{16}O^{16}O^+$ from $^{32}S^+$ or $^{40}Ar^{16}O^+$ from $^{56}Fe^+$). The high resolution is achieved with this instrument, when $R$ reaches values greater than 10 000 (entrance and exit slit width both are in the order of 5 μm). In this case, *e.g.* signals of rare-earth elements are separated from their interferences, as well as $^{39}K^+$ from $^{38}Ar^1H^+$ or $^{75}As^+$ from $^{35}Cl^{40}Ar^+$. On the other hand, signal intensities are reduced to about 2–3% of those achieved in low resolution and triangular-shaped peaks are obtained (Figure 5.7). Both can lead to a slight decrease of measurement precision.

Of course, also in high resolution a wider exit slit can be combined with a narrower entrance slit in order to obtain flat-top peaks. On this occasion, the resolution is somewhat compromised for obtaining flat-top peaks. For example, this combination of slits (width of the entrance slit of around 25 μm, exit slit width about 50 μm) is used to obtain flat-top peaks with a mass resolution of approx. 2000 with the *Element2* ICP-MS. This is sufficient to separate, *e.g.* $^{16}O^{16}O$ from $^{32}S$, $^{40}Ar^{12}C$ from $^{52}Cr$ or $^{56}Fe$ from $^{40}Ar^{16}O$

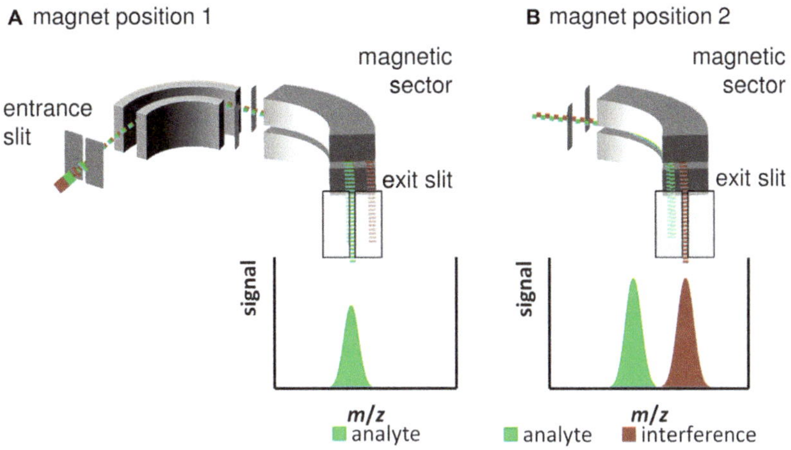


**Figure 5.7** Schematics of the separation of spectral interferences using high-resolution settings (triangular shaped peaks are obtained).

and allows for accurate and precise measurements of isotope ratios due to the flat-top peaks and the interference-free detection of the isotopes of interest.

In the case of multicollector devices, another strategy can be used to resolve spectral interferences by high mass resolution as these instruments are operated in static mode usually. As before, the ion beam width is reduced by the entrance slit. The interfering species are spatially separated from the analyte beam according to their *m*/*z* difference by the magnetic sector. Instead of equipping each single detector with a high-resolution exit slit, the incident ion beam of the interference can be clipped away by simply guarding the beam towards the edge of the detector. Therefore, only the ion beam of the analyte of interest is detected. When scanning the mass spectrometer, the observed peak consists of an interference-free shoulder, a combined signal of the analyte and its interference and a second shoulder representing the signal of the interference only (Figure 5.8). This is generally referred to as ''Pseudo High Resolution'' or ''edge resolution'' mode.

This combination of high resolution entrance slit and a low resolution exit slit (or the width of the Faraday cup, which is wider than the entrance slit) is used mostly by MC instruments to study interfered isotopic systems (*e.g.* Si, S, Fe or Cu in MC ICP-MS). For the final measurement, the magnetic field is set to the corresponding mass of the flat top of the interference-free shoulder of the peak. Thus, the interference is separated effectively and the peak resists better small variations in its position caused by the electric or magnetic field. The numerical value of the mass resolution is in this mode calculated as a quotient between the mass of the analyte flat top shoulder and a $\Delta m$ that is defined as a difference between masses at the 5% and at 95% of the shoulder height.[2] It is worth mentioning, that mass resolution calculated on

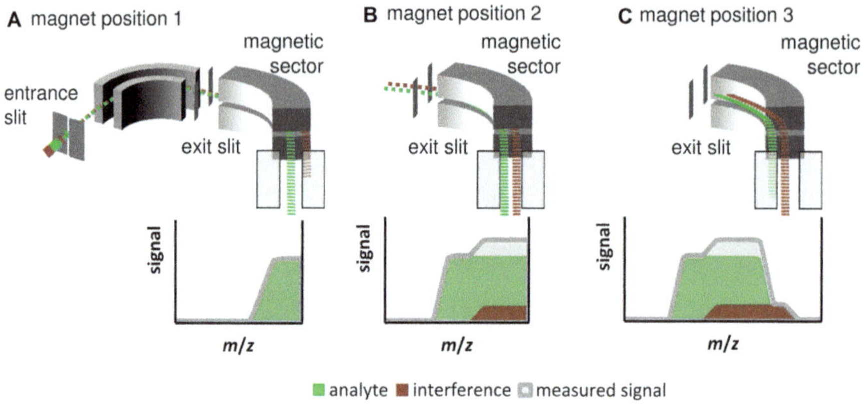


**Figure 5.8** Schematics of the separation of spectral interferences using edge resolution.

this bases differs significantly from the standard 10% valley definition. Resulting *R* values are generally 3–4 times lower in the latter case.

### 5.2.1 Abundance Sensitivity

Another important property of a mass spectrometer is the abundance sensitivity. Since the abundance sensitivity is also dependent on the mass resolution, it should be mentioned here, but is discussed in relation to technical achievements in more detail in Chapter 4.

According to IUPAC, the abundance sensitivity is defined as follows:

"Ratio of the maximum ion current recorded at a specified *m*/*z* value to the maximum ion current arising from the same species recorded at a neighbouring *m*/*z* value.

(***Note:*** The abundance sensitivity is a measure of the contribution of the peak "tail" of a major isotope (with a certain *m*/*z* value) to an adjacent *m*/*z* value, which in some cases might be more than 1 *m*/*z* removed. Its value is dependent on the resolving power of the mass spectrometer.)"[1]

### 5.2.2 Alternative Mass Separators to Achieve High Mass Resolution

Even though high mass resolution is one of the major features of magnetic sector field instruments, it is not a capability exclusive to sector field instruments. Other mass analysers – including quadrupoles – can be operated at higher mass resolution, as well. Conventionally, QMS instruments provide approximately unit mass resolution, but they can also be operated in higher regions of stability and reach for instance in the 2nd region of stability resolutions of up to 9000.[3] Another approach based on quadrupole instruments are tandem devices with a subsequent series of quadrupoles. If both quadrupoles have reflecting grids, ions can be reflected in this trap many times and by multiple passes, resolutions of up to 20 000 were realised.[4] Another trapping concept is provided by ion traps. They have been investigated *e.g.* for their use in ICP-MS by Eiden *et al.*[5] Ion traps consist of a ring electrode in the centre, with two hyperbolic end caps above and below. They are operated with RF voltages and are read out by a high-voltage pulse, which injects the ions into a channeltron multiplier. Depending on the number of cycles the ions spent in the trap, the resolution achieved can be increased from $(m/z)/\Delta(m/z)$ 200 to 2000 (FWHM).

A different type of mass analyser is the time-of-flight (TOF) analyser. For example, an TOFMS was operated with a reflectron to overcome differences in ion energy reaching, *e.g.* a mass resolution of up to 2200 (FWHM) in ICP-MS and GD-MS applications.[6]

In ion cyclotron resonance Fourier transform (FT) analysers, the ions are injected into a strong magnetic field and are gyrating on circles. The gyration frequency is dependent on the mass of the ion and is picked up by a condenser-like detector. By transforming the frequency domain into a mass domain, an approach known as Fourier transformation, mass spectra are obtained. Resolutions in GD- and ICP-MS operation of an $R$ of up to 100 000,[7] and even higher than 1 000 000 (FWHM) have been reported.[8,9] A similar device in which ions are gyrating in a field - but this time in an electrical field - has been invented by Makarov[10] for organic applications and has become known under the trade mark: *Orbitrap*. Again, Fourier transformation results in mass spectra. Koppenaal started with a challenging experiment to use an *Orbitrap* in combination with a linear quadrupole ion trap for an ICP ion source achieving resolutions of $m/\Delta m$ up to 150 000. They also coupled a liquid-sampling atmospheric-pressure glow discharge ionisation source to an *Orbitrap* mass spectrometer for elemental mass spectrometry.[11]

## References

1. K. K. Murray, R. K. Boyd, M. N. Eberlin, G. J. Langley, L. Li and Y. Naito, *Pure Appl. Chem.*, 2013, **85**, 1515–1609.
2. S. Weyer and J. B. Schwieters, *Int. J. Mass Spectrom.*, 2003, **226**, 355–368.
3. Z. Du, D. Douglas and N. Konenkov, *J. Anal. Atom. Spectrom.*, 1999, **14**, 1111–1119.
4. M. W. Soyk, Q. Zhao, R. S. Houk and E. R. Badman, *J. Am. Soc. Mass. Spectrom.*, 2008, **19**, 1821–1831.
5. G. C. Eiden, C. J. Barinaga and D. W. Koppenaal, *J. Am. Soc. Mass. Spectrom.*, 1996, 7, 1161–1171.
6. P. P. Mahoney, S. J. Ray and G. M. Hieftje, *Appl. Spectrosc.*, 1997, **51**, A16–A28.
7. K. E. Milgram, F. M. White, K. L. Goodner, C. H. Watson, D. W. Koppenaal, C. J. Barinaga, B. H. Smith, J. D. Winefordner, A. G. Marshall and R. Houk, *Anal. Chem.*, 1997, **69**, 3714–3721.

8. C. H. Watson, J. Wronka, F. H. Laukien, C. M. Barshick and J. R. Eyler, *Spectrochim. Acta Part B: Atom. Spectrosc.*, 1993, **48**, 1445–1448.
9. C. H. Watson, C. M. Barshick, J. Wronka, F. H. Laukien and J. R. Eyler, *Anal. Chem.*, 1996, **68**, 573–575.
10. A. Makarov, *Anal. Chem.*, 2000, **72**, 1156–1162.
11. R. K. Marcus, C. D. Quarles Jr, C. J. Barinaga, A. J. Carado and D. W. Koppenaal, *Anal. Chem.*, 2011, **83**, 2425–2429.

CHAPTER 6

# Instrumental Isotopic Fractionation

JOHANNA IRRGEHER* AND THOMAS PROHASKA*

University of Natural Resources and Life Sciences Vienna (BOKU), Department of Chemistry, Division of Analytical Chemistry, VIRIS Laboratory for Analytical Ecogeochemistry, Tulln, Austria
*Email: johanna.irrgeher@boku.ac.at; thomas.prohaska@boku.ac.at

In this chapter, the focus is set on the instrumental fractionation between nuclides of the same element ("instrumental isotopic fractionation") resulting in erroneous results of isotope ratios.

"Fractionation" is a commonly used term, which is used inconsistently in mass spectrometry to describe different phenomena of natural as well as instrumental processes.[1] The general term "fractionation" comprises processes leading to separation between two different entities (*e.g.* chemical compounds, elements or isotopes). "Isotopic fractionation" is the recommended term to be used for the effects leading to a difference of isotope ratios of one element (*e.g.* between two substances). The concept is found in the literature also under the term "isotopic enrichment", "discrimination" or "isotopic discrimination".[2]

The isotopic fractionation factor is expressed as ratio of isotope ratios of one element between two substances[3] (eqn 6.1[2])

$$({}^{i}E/{}^{j}E)_P/({}^{i}E/{}^{j}E)_Q \tag{6.1}$$

*P* and *Q*: represent the different substances
${}^{i}E$ and ${}^{j}E$: the heavier ($i$) and lighter ($j$) isotope of an element ($E$)

New Developments in Mass Spectrometry No. 3
Sector Field Mass Spectrometry for Elemental and Isotopic Analysis
Edited by Thomas Prohaska, Johanna Irrgeher, Andreas Zitek and Norbert Jakubowski

Published by the Royal Society of Chemistry, www.rsc.org

In mass spectrometry, the term "instrumental isotopic fractionation" can be recommended to describe the sum of the effects in a mass spectrometer occurring during sample introduction, ion formation, ion extraction, ion transmission, ion separation and ion detection, leading to a difference of the measured isotope ratio from the true isotope ratio in a sample. The term can be used irrespective of whether the observed difference can be related to mass-dependent or mass-independent effects and includes all discriminating effects. In the literature, the term "instrumental isotopic fractionation" is also referred to as "instrumental mass bias", "instrumental mass discrimination" or "instrumental mass fractionation". In many cases, the term "instrumental" is not added even though only instrumental effects are considered.

Instrumental fractionation between elements is usually more pronounced during sample introduction (*e.g.* evaporation, diffusion, transport efficiency) and ionisation (*e.g.* different ionisation efficiency) and matrix dependent and represents a severe source of systematic error in isotope ratio mass spectrometry. Therefore, proper calibration of isotope amount ratios is a prerequisite for accurate isotope ratio determination (see Chapter 8). The term "mass bias correction" is a popular metaphor for this procedure.[4] This correction is of particular importance when absolute isotope ratios are assessed. In case of the use of $\delta$-values (generally used, *e.g.* in IRMS) all factors contributing to a bias of the measured ratio to the true ratio are accounted for irrespective of their origin (*e.g.* instrumental mass fractionation, dead time, detector cross calibration for different detector efficiencies/yields in case of multicollector instruments) given that the standard and the sample behave identically and no cross-contamination has occurred.

The extent of instrumental isotopic fractionation can be reported as "mass bias per mass unit ($MB$)" (eqns (6.2a) and (6.2b))[5] and strongly differs between mass spectrometric techniques. The effect is more pronounced in case of, *e.g.* ICP-MS compared to TIMS. In ICP-MS, the mass bias per mass unit usually ranges from 0.5–1.5% for heavier masses (*e.g.* Pb, U)[6] and up to 25% for light masses (*e.g.* Li, B).[7–9]

$$f = \frac{R_{\text{true}}}{R_{\text{measured}}} \tag{6.2a}$$

$f$: factor to correct for instrumental isotopic fractionation (also termed "mass bias factor" or "fractionation factor")

$R_{\text{true}}$: true ratio

$R_{\text{measured}}$: measured ratio

$$\text{IIF}(\%) = \frac{(f-1) \times 100}{\Delta m} \tag{6.2b}$$

IIF(%): instrumental isotopic fractionation in per cent

$\Delta m$: difference in the mass of the heavier ($i$) and lighter ($j$) isotope of one element ($E$)

In TIMS analysis, the observed instrumental isotopic fractionation is one order of magnitude lower compared to ICP-MS and originates from a preference in evaporation and ionisation of the lighter ion compared to the heavier ion in the thermal ionisation source.[10] In GDMS and SIMS instrumental isotopic fractionation has been discussed explicitly to a lesser extent. In the case of SIMS, matrix-dependent effects dominate the instrumental isotopic fractionation. In IRMS, isotopic fractionation is often used for describing the systematic bias of the determined delta value (see Chapter 8) potentially caused by crosscontamination between samples of different isotopic composition.

The understanding of the factors causing instrumental isotopic fractionation is still limited. There is, however, a conjoint consensus among users of the different MS-instrumentations about the determination of the major effects causing instrumental isotopic fractionation. Generally, instrumental isotopic fractionation is classified into mass-dependent and mass-independent effects. The latter is less studied and understood, so far. Factors such as diffusion repulsion, space-charge effects, absolute isotope mass, supersonic expansion in pressure transition areas, ion density or, ionisation efficiency are only a number of possible sources of instrumental isotopic fractionation. The following considerations review a number of observations for the different techniques described in this book.

## 6.1 Instrumental Isotopic Fractionation in ICP-MS

Instrumental isotopic fractionation can occur potentially within all sections of an ICP-MS instrument. The sample-introduction system and the ionisation source together with the interface are supposed to be the major determining factors regarding instrumental isotopic fractionation. Instrumental isotopic fractionation was shown to depend strongly on the configuration of the sample-introduction system and the ionisation source (*e.g.* gas flow rates, sampling depth, torch design or type of cones). Fontaine *et al.* compared different sample introduction systems (wet and dry plasma) and studied in detail the effects of the matrix composition and varying operating parameters (*i.e.* sampling depth, carrier gas flow, ion lens settings) of a MC ICP-MS instrument (*Nu Plasma HR*) on the isotopic fractionation of Nd. The authors observed a systematic dependence of instrumental isotopic fractionation on all studied parameters, *e.g.* observing reduced instrumental isotopic fractionation when carrier gas flow rates are increased.[11]

The consequence of using different cone geometries was investigated by Newman *et al.*[12] who tested different high-sensitivity skimmer cones (different in internal/external angle and geometry at the tip) provided by *Nu Instruments*. Nonlinear mass dependent isotopic fractionation was observed on Nd, which made accepted mass bias correction laws (see Chapter 8) inapplicable due to a large nonlinear effect. It was proposed that the origin of this phenomenon was caused by isotope-dependent oxide formation, which was favoured when using high-performance skimmer cones. A comparable study on Nd was presented by Newman[13] on a *Thermo Scientific Neptune* MC ICP-MS. Similar observations about preferential oxide formation when using X skimmer cones (*i.e.* high-sensitivity skimmer cones, Thermo Fisher Scientific) were reported.

Following ion-beam formation, the interface – acting as passage between the ion source at atmospheric pressure and the mass spectrometer under vacuum – is tainted with being the region with the major impact on instrumental isotopic fractionation in ICP-MS. Most of the bias most likely occurs in the field-free region between sampler and skimmer cone.[14] Atomic ions of heavier and lighter isotopes coexisting in the generated ion beam are extracted after the ionisation through the interface region to cross the barrier to high vacuum and to be accelerated to pass the mass analyser and reach the detector unit. Due to the different masses of the matrix elements, ions of the heavier masses remain closer to the beam centre and ions of lower masses diffuse more easily into the outer regions of the beam. Barling and Weis showed radial and axial plasma profiles in MC ICP-MS, where a mass-dependent distribution of isotopes in the plasma confirmed previous observations.[15] As a consequence, ions of higher masses get extracted more efficiently by the lens system and are transmitted with a greater yield than ions of lighter masses. Even though the understanding of the ion extraction is still poor, space-charge effects causing more pronounced deflection of lighter isotopes than of heavier isotopes in the interface region are believed to make the greatest contribution to the instrumental isotopic fractionation. Niu and Houk were one of the first to describe in detail the complex fundamentals of ion extraction in ICP-MS, stating that the discrimination was a combination of different factors, mainly supersonic expansion of the neutral plasma into the vacuum between sampler and skimmer cone and space charge effects in the wake of the skimmer cone.[14] Since then, a number of papers have aimed at the visualisation of supersonic expansion in the first vacuum stage of an ICP (*i.e.* the region between sampler and skimmer cone). Preferential transmission and nonuniform, mass-dependent distribution of isotopes of different mass were investigated by deposition experiments, where the sampler as well as skimmer cones were removed after analysis of standards and analysed for the deposited material.

Andrén *et al.* visualised instrumental isotopic fractionation effects by leaching areas of used cones or used extraction lenses and subsequent isotopic analysis of the deposited material. A linear relationship between the levels of lighter isotope depletion and mass ratio was observed on the example of Fe, Zn and Tl isotopes leached from used extraction lenses confirming that the ion beam diverges into a relatively wide solid angle in the field-free region behind the skimmer.[16] Additionally, plasma-operating conditions such as sample gas flow rate and thus sampling depth strongly affected the magnitude of the mass bias for a wide range of elements (Li, B, Fe, Ni, Cu, Sb, Ce, Hf and Re).[16] Mills *et al.* visualised the density distribution of barium atoms and ions (at ground state as well as excited state) between the load coil and sampling cone and observed a nonuniform distribution as the ion beam is compressed into the sample cone orifice.[17] Radicic *et al.* characterised the supersonic expansion in an ICP-MS based on visualisation of neutral Ar atoms by fluorescence excitation spectroscopy, which are consistent with fluid dynamic theory.[18] The effects of collision on the composition of the plasma through the first stage of vacuum of an ICP-MS were investigated in detail by Macedone *et al.*, who observed the most pronounced effect in the first few mm of the expansion where the number of densities and temperatures of the plasma species are highest.[19] Kivel *et al.* investigated isotope fractionation during ion beam formation as well as ion sampling through the interface during MC ICP-MS by implanting metal and graphite targets at the earliest stage after the interface for visualisation of the extracted ion beam with respect to the isotopic distribution along the beam cross section.[20] Based on the results, the authors concluded that about 50% of the entire mass discrimination within the ICP-MS already occurs at this very early stage of ion extraction. To sum up, processes occurring in the expansion chamber, where the cloud of electrons, ions, photons, neutral particles and gas atoms expands quickly and enormously due to a pressure drop (referred to as supersonic expansion) as well as in the *zone of silence* behind the skimmer cone, where an additional pressure drop leads to lateral diffusion of the ion beam combined with repulsion between positively charged ions significantly contribute to instrumental isotopic fractionation in ICP-MS. Once extracted into the mass spectrometer, the main source of mass bias probably derives from ion repulsion due to the high ion density in the ion beam. All these effects strongly depend on absolute mass as well as on the mass difference between the measured isotopes.[21,22]

Various studies showed that the vacuum in the interface has an influence on the instrumental isotopic fractionation (along with interference formation). As a consequence, several instrument manufacturers use interface pumps with increased efficiency to reach a vacuum below 1 mbar.[23,24]

Some additional sources of instrumental isotopic fractionation have been identified for special setups such as collision or reaction-cell mode where the magnitude of instrumental isotopic fractionation varies strongly upon cell conditions, which influence in-cell fractionation effects, space-charge effects and kinetic effects.[25,26]

Other factors influencing instrumental isotopic fractionation in ICP-MS analysis are nonspectral matrix effects like the first ionisation potential of the matrix element.[15,27] The studies of Barling and Weis[27] and Irrgeher *et al.*[28] showed that near-perfect matrix matching and/or analyte separation is a prerequisite when external mass bias correction is done.

In single-collector devices, another phenomenon of instrumental isotopic fractionation can be observed. Since scanning is accomplished by changing the acceleration voltage, different isotopes are analysed at different acceleration voltages causing additional mass-bias effects. This mass fractionation is in favour of the lighter ions, resulting in reverse mass bias compared to the previously described phenomena, and is a consequence of the "Liouville theorem".[29,30]

All of the above-described sources of instrumental isotopic fractionation are mass dependent. However, a number of publications showed deviations from the mass-dependent mass-bias correction models that have been proposed in literature (see Chapter 8) to describe instrumental isotopic fractionation in ICP-MS. These effects are referred to as mass-independent isotopic fractionation (MIF). Yang *et al.* reported significant deviations in observed instrumental isotopic fractionation for Ge and Pb from the Russell's mass bias correction law (see Chapter 8) including an evaluation of possible causes for the effect. However, the actual reason for the observed MIF remains unknown.[31] As mentioned before, Newman *et al.*[12] and Newman[13] observed MIF depending on the use of different cone geometries. Shirai and Humayun described the evidence of an odd–even isotope separation in tungsten isotopic analysis depending on the type of cones used, as well.[32] In the literature, possible causes of mass-independent isotopic fractionation in MC ICP-MS were attributed to the nuclear field shift, the magnetic isotope effect, the oxide formation rate or high first ionisation potentials.[33]

## 6.2 Instrumental Isotopic Fractionation in LA-ICP-MS

Since the advent of LA-ICP-MS both instrumental interelement as well as intraelement nuclide fractionation have been observed. The reasons for both phenomena are similar, mostly caused by differences in volatility, ionisation efficiencies and condensation processes during ablation, differences in transport efficiencies and fractionation within the ICP during vaporisation and ionisation of the ablated particles.[34,35] The easiest way to compensate for laser-induced isotopic fractionation is to use an isotopically characterised matrix reference material for correction.

Hirata *et al.* reported on time-dependent isotopic fractionation of Cu and Fe isotope ratios during time-resolved analysis of metallic reference materials by LA-MC ICP-MS using a 193-nm ArF LA system, under Ar carrier gas atmosphere. The measured $^{65}Cu/^{63}Cu$ ratio was a linear function with the rate of intensity change. Thus, the extent of fractionation was believed to originate from the slow response of the Faraday preamplifier and allowed for the definition of a correction factor.[36]

Jackson and Günther investigated in detail the isotopic fractionation of Cu and found significant influence of pulse energy as well as of the particle-size distribution on laser-induced isotopic fractionation. Higher-energy pulses of the laser led to decreased deviation of the measured isotope ratio from the true isotope ratio in copper metal. Filtering off larger particles of the aerosol yielded isotopic values that were closer to the true value than unfiltered aerosols. The authors recommended ablating at high energy densities, to use laser systems that produce small particle sizes and to filter large particles from the aerosol.[34]

Guillong *et al.* showed a dependence of the laser wavelength on the particle-size distribution, thus likely not only influencing laser-induced elemental fractionation but also isotopic fractionation.[37] In 2007, Horn and von Blanckenburg were the first to report on the differences between nanosecond (ns) and femtosecond (fs) laser-ablation systems, with special emphasis on elemental and isotopic fractionation. Isotopic fractionation was investigated on the example of stable Fe isotope ratios and was insignificant during fs LA-MC ICP-MS analysis. This was explained by the much shorter pulses generated, which allow for a more efficient energy transfer to the sample and thus producing a bias that is beyond the instrumental isotopic fractionation observed in the ICP-MS.[38] Similar observations were shown by Diwakar *et al.* in their paper on elemental fractionation between Cu and Zn during fs LA-ICP-MS of brass samples.[39] Koch and Günther reported on the trends in fs LA-ICP-MS for quantification using nonmatrix-matched calibration and commented on the challenges entailed.[40]

Kuhn *et al.* gave insight into the influence of particle-size distribution during Cu isotope ratio analysis by fs LA-MC ICP-MS ablating pure metallic copper, where a time-dependent preferential vaporisation of the lighter isotopes was observed during an ablation event, most likely because of a change in residence time of the generated droplets in close proximity to the laser plume.[41] In their review on the use of fs LA-ICP-MS Shaheen *et al.* comprehensively summarised the fundamentals of laser–solid interactions and compared in detail fs-systems to ns-systems.[42] This included evaluation of thermal effects on elemental and isotopic fractionation, pulse width, wavelength, pulse frequency, beam profile, power density and fluence, ablation cell configuration and transport system as well as ablation gas used.

## 6.3 Instrumental Isotopic Fractionation in GDMS

Even though instrumental isotopic fractionation has also been observed in GDMS analysis, the extent of the observed bias can be regarded as low compared to ICP-MS. The most comprehensive study on instrumental isotopic fractionation in GDMS was performed by Riciputi *et al.*, who investigated systematically the isotopic bias during the analysis of B, Cu, Sr, Ag, Sb, Re and Pb isotope ratios.[43] No systematic isotopic bias by both the preferential transmission of heavier or lighter isotopes was observed, possibly depending on the type of detector used. However, an absolute difference of the measured values to the true values of less than 1% was reported in most cases (with the exception of $^{88}Sr/^{86}Sr$ where a difference of more than 4% was reported, which was explained by the detector saturation on $^{88}Sr$). In a review on the use of GDMS on the characterisation of nuclear and radioactively contaminated environmental samples, bias in the isotopic composition compared to TIMS data usually ranged from 0.01–1%, being less pronounced with increasing mass.[44] One study by Chartier and Tabarant[45] on Gd and B isotope ratio measurements by GDMS reported on instrumental isotopic fractionation for B isotope ratios of 1.4%, whereas no correction for instrumental isotopic fractionation was necessary for Gd, since regarded as insignificant considering measurement uncertainties. Ba isotope ratio analysis was performed by Betti *et al.*, who determined a bias attributed to fractionation effects in the measured ratios of up to 1.5% from literature values, except for $^{137}$Ba, which was prone to a $^{137}$Cs interference.[46] Isotopic abundances of Li, B, Si, Zr, U an Pu isotopes were assessed by Betti *et al.*[47] Positive and negative biases were reported. Ecker and Pritzkow observed isotope ratio reproducibilities of less than 0.1% ($2\sigma$) for $^{235}$U/$^{238}$U and 0.8% ($2\sigma$) for $^{10}$B/$^{11}$B.[48]

## 6.4 Instrumental Isotopic Fractionation in TIMS

Thermal ionisation mass spectrometry (TIMS) mainly suffers from time-dependent instrumental isotopic fractionation as a result of increasing depletion of the lighter isotope in the sample with the increasing amount of material removed during evaporation. As a consequence, a drift of the isotope ratio can be observed as a result of varying evaporation and ionisation efficiency over time. Details on measurement strategies (*e.g.* total evaporation TIMS) coping with the implication of varying mass fractionation over time are given in Chapter 14.

The major part of literature on mass fractionation during TIMS analysis deals with the evaluation and development of adequate model laws to correct for instrumental isotopic fractionation rather than to describe the fundamental processes behind. A number of empirical fractionation laws have been developed and applied to TIMS data (see Chapter 8). A few fundamental studies on isotope fractionation in TIMS are mentioned here:

In the early 1980s Habfast investigated comprehensively the fractionation in the thermal ionisation source, coming to the major conclusion that the observed fractionation effects are mainly explainable by the concept of molecular evaporation as well as mass- and temperature-dependent dissociation of molecular species. He provided additionally correction models to the TIMS community in order to cope with the observed instrumental isotopic fractionation effects.[49]

Hart and Zindler tested the applicability of coexisting laws for instrumental isotopic fractionation correction in TIMS and showed that isotope fractionation of calcium follows the exponential law of Russell with better fit to the data compared to the linear, power or Rayleigh laws, with maximum deviations of about 0.01% per mass unit. The authors discuss the possible reasons for instrumental isotopic fractionation probably originating from mixing/demixing processes in isotopically heterogeneous sample reservoirs on the filament.[50] Similar observations were found by Jungck *et al.*,[51] Esat *et al.*[52] and Esat[52,53] for other isotopic systems. In 1998, Habfast[54] re-evaluated existing instrumental isotopic fractionation laws applied to TIMS data and presented a theory for the correction of time- and mass-dependent evaporation-related instrumental isotopic fractionation effects based on both the Rayleigh distillation equation and the Langmuir equation describing evaporation. Fractionation based on diffusion controlled emission during TIMS on the example of U isotope ratios on Pt coated Rh filaments was studied by Ramebäck *et al.*[55] Andreasen and Sharma investigated the nature of instrumental isotopic fractionation in a TIMS source on the example of Nd isotope ratios measurements.[56] The existing correction laws for mass-dependent instrumental isotopic fractionation were evaluated and their applicability discussed. Evaporation and mixing from variably depleted domains on the filament indicated both normal as well as inverse fractionation. However, the exponential law was considered fully adequate in correcting for mass-dependent instrumental isotopic fractionation of today's achievable measurement precision in TIMS analysis. In a recent study, Cavazzini provides an interpretation of the exponential law to describe instrumental isotopic fractionation in TIMS, which can – according to the authors – be described as a result of a Rayleigh-type distillation process.[57] The exponential law is most suitable to describe the mass-dependent evaporation processes of multiple isotopes of one element within a mixture of isotopes.

## 6.5 Instrumental Isotopic Fractionation in SIMS

In SIMS analysis, a list of factors contribute to instrumental isotopic fractionation, either related to matrix effects or instrumental conditions, usually favouring the lighter isotopes to be preferentially transmitted through the mass spectrometer finally causing instrumental isotopic fractionation. As a consequence, fractionation is observed at multiple stages of the mass spectrometer, such as the sputtering process, ionisation, extraction, transmission and detection.

During sputtering and the ionisation process isotope ratios underlie fractionation. As pointed out in a work by Shimizu and Hart the easiest way to cope with instrumental isotopic fractionation in SIMS is to generate a correction factor based on the analysis of a certified reference material, which is chemically identical to the sample and that is analysed under the same instrumental conditions as the sample.[58] However, obviously, this scenario cannot be met in many cases due to the lack of suitable (solid) certified (matrix) reference materials. Issues to be addressed include the possible impact of the sample matrix, the secondary ion energy and possible systematic observations over the observed mass range that allow for the application of mathematical models for correction of instrumental isotopic fractionation.

Secondary ion generation by sputtering is often enriched in lighter isotopes than in heavier isotopes. Isotopic fractionation is linearly dependent on the mass difference and the degree of enrichment depends on the spatial location and the energy of the extracted ions.[59–61] Ionisation yields clearly depend on the elemental composition of the target samples.[62] Many of these processes could be well-described by the quantum-mechanical model developed,[63] which is based on the process by which an atom is excited or ionised as it is sputtered from a metal surface.

The effect of matrix components, mixed matrices and the relation between sample size and sputtered sample volume has been the subject of numerous studies, where isotope ratio variation as a function of the chemical composition of the matrix were discussed in detail and where instrumental isotopic fractionation caused by matrix-effects were reported.[64–71] Lyon *et al.*[72] observed an effect of the crystallographic orientation of a sample relative to the secondary ion beam focused on the sample on isotope fractionation. This effect was shown to be significant for magnetite and hematite but not for other samples such as quartz or olivine. A bias of $\delta^{18}O$ values caused by crystal orientation effects has been identified for hematite at similar levels as for magnetite, as well.[73] The phenomenon is a serious source of systematic error and has to be checked by, *e.g.* measurements of the sample at two different orientations. The impact of the matrix composition can be compensated for if suitable matrix reference materials are available for calibration, which in turn is often the limiting factor, since solid standards with complex matrix composition are required (see Chapter 15). Gnaser and Oechsner as well as Weathers *et al.* reported on isotope fractionation in Mo and Ge isotope ratios as a function of emission angle in the sputtering geometry and projectile fluence.[74,75] Beside the influence of matrix effects and sample topography on mass fractionation, instrumental characteristics such as, *e.g.* species, charge and acceleration of the primary beam or detector settings contribute to instrumental isotopic fractionation in SIMS.[70]

## 6.6 Instrumental Isotopic Fractionation in IRMS

Isotopic fractionation is used as a general term in IRMS to describe a systematic bias of the determined delta value (see Chapter 8) and not isotopic fractionation effects directly related to instrumental phenomena. The major sources of the measured bias derive from cross contamination, diffusion effects and flow effects through orifices and capillaries, most of which happening in the sample-introduction and ion-source area. (Cross-contamination does not necessarily cause a fractionation of the original ratio but potentially leads to an alteration of the ratio.)

A substantial influence on isotopic fractionation can derive from the sample inlet system and gas inlet strategy. Sample amounts, leakages in the changeover valves, residual gas in capillaries, valves or the ionisation chamber, inlet changeover equilibration time (also referred to as the "idle-time effect") and inlet pressure have a significant impact on possible memory effects and thus crosscontamination. Halsted and Nier[76] reviewed the important factors in the design of the gas inlet into a mass spectrometer with special emphasis on the "viscous" leak design, including characterisation of three different flow regions in the instrument: flow through the mass spectrometer, flow through constriction areas and flow through tubing capillaries. Brand outlined in detail the key factors to be considered for achieving precise and accurate data in isotope ratio mass spectrometry including the requirements on the ion source design.[77] Verkouteren *et al.* designed an exercise among research groups, all operating a *Finnigan Model 252* isotope ratio mass spectrometer, of two carbon dioxide intercomparison materials prepared at the National Institute of Standards and Technology (NIST).[78,79] The aim of the study was to explore the effects of ion source material substitution, source conductance, inlet pressure, electron emission, acceleration potential and inlet changeover equilibration time, whereby observations made at various laboratories were consistent. Memory effects or crosscontamination were observed for the measurement of isotope ratios of small $CO_2$ samples by dual-inlet mass spectrometry, where two gases with different isotopic compositions were alternately introduced into the mass spectrometer. Increasing sample size, source conductivity and dual-inlet equilibration time reduced the fractionation effects caused by cross-contamination significantly. The authors discussed comprehensively the role of adsorption/desorption, ion implantation, and sputtering in the ionisation chamber on crosscontamination in mass-spectrometry systems.

Elsig and Leuenberger investigated in detail possible sources on isotopic fractionation during the analysis of $^{13}C$ and $^{18}O$ by continuous-flow isotope ratio mass spectrometry. Instrumental isotopic fractionation in open splits and nonlinear effects occurring in the mass spectrometer were detected and three major processes were defined as most influencing: (1) fractionation in the open split, (2) adsorption/desorption effects in the ion source, and (3) amplitude-dependent fractionation processes during ionisation.[80] A method for minimising the observed fractionation processes was presented, where modifications on the position of the capillaries or the flow rates were set in order to reduce the level of ion-induced metal sputtering and redeposition within the ion source and thus avoid re-emission of gas molecules. The authors gave practical recommendations to improve measurement precision during $CO_2$ – IRMS analysis, when a dual-inlet system is used. Summing up, any sources of instrumental isotopic fractionation lead to systematic errors in the data generated and as a consequence proper instrument and method design as well as calibration strategies are of utmost importance.

## References


1. F. Vanhaecke and K. Kyser, The Isotopic Composition of the Elements, in *Isotopic Analysis*, eds. F. Vanhaecke and P. Degryse, Wiley-VCH Verlag GmbH & Co. KGaA, Weinheim, Germany, 2012, ch. 1, pp. 1–29.
2. T. B. Coplen, *Rapid Commun. Mass Spectrom.*, 2011, **25**, 2538–2560.
3. M. Nic, J. Jirat and B. Kosata, International Union of Pure and Applied Chemistry, Durham, North Carolina, 2012.
4. J. Meija, L. Yang, Z. Mester and R. E. Sturgeon, Correction of Instrumental Mass Discrimination for Isotope Ratio Determination with Multi-Collector Inductively Coupled Plasma Mass Spectrometry, in *Isotopic Analysis*, eds. F. Vanhaecke and P. Degryse, Wiley-VCH Verlag GmbH & Co. KGaA, 2012, ch. 5, pp. 113–137.

5. Q. Xie and R. Kerrich, *J. Anal. Atom. Spectrom.*, 2002, **17**, 69–74.
6. S. Kappel, S. F. Boulyga and T. Prohaska, *J. Environ. Radioact.*, 2012, **113**, 8–15.
7. R. Millot, C. Guerrot and N. Vigier, *Geostand. Geoanalyt. Res.*, 2004, **28**, 153–159.
8. J. K. Aggarwal, D. Sheppard, K. Mezger and E. Pernicka, *Chem. Geol.*, 2003, **199**, 331–342.
9. J. Vogl, Calibration strategies and quality assurance, in *Inductively Coupled Plasma Mass Spectrometry Handbook*, ed. S. Nelms, Blackwell Publishing Ltd, Oxford, UK, 1st edn., 2005, pp. 147–181.
10. S. K. Sahoo and A. Masuda, *Chem. Geol.*, 1997, **141**, 117–126.
11. G. H. Fontaine, B. Hattendorf, B. Bourdon and D. Günther, *J. Anal. Atom. Spectrom.*, 2009, **24**, 637–648.
12. K. Newman, P. A. Freedman, J. Williams, N. S. Belshaw and A. N. Halliday, *J. Anal. Atom. Spectrom.*, 2009, **24**, 742–751.
13. K. Newman, *J. Anal. Atom. Spectrom.*, 2012, **27**, 63–70.
14. H. Niu and R. S. Houk, *Spectrochim. Acta Part B: Atom. Spectrosc.*, 1996, **51**, 779–815.
15. J. Barling and D. Weis, *J. Anal. Atom. Spectrom.*, 2012, **27**, 653–662.
16. H. Andrén, I. Rodushkin, A. Stenberg, D. Malinovsky and D. C. Baxter, *J. Anal. Atom. Spectrom.*, 2004, **19**, 1217–1224.
17. A. A. Mills, J. H. Macedone and P. B. Farnsworth, *Spectrochim. Acta Part B: Atom. Spectrosc.*, 2006, **61**, 1039–1049.
18. W. N. Radicic, J. B. Olsen, R. V. Nielson, J. H. Macedone and P. B. Farnsworth, *Spectrochim. Acta Part B: Atom. Spectrosc.*, 2006, **61**, 686–695.
19. J. H. Macedone, D. J. Gammon and P. B. Farnsworth, *Spectrochim. Acta Part B: Atom. Spectrosc.*, 2001, **56**, 1687–1695.
20. N. Kivel, I. Günther-Leopold, F. Vanhaecke and D. Günther, *Spectrochim. Acta Part B: Atom. Spectrosc.*, 2012, **76**, 126–132.
21. C. P. Ingle, B. L. Sharp, M. S. A. Horstwood, R. R. Parrish and D. J. Lewis, *J. Anal. Atom. Spectrom.*, 2003, **18**, 219–229.
22. T. Hirata, *Analyst*, 1996, **121**, 1407–1411.
23. C. Latkoczy and D. Gunther, *J. Anal. Atom. Spectrom.*, 2002, **17**, 1264–1270.
24. J. M. Cottle, A. J. Burrows, A. Kylander-Clark, P. A. Freedman and R. S. Cohen, *J. Anal. Atom. Spectrom.*, 2013, **28**, 1700–1706.
25. S. F. Boulyga and J. S. Becker, *Fresenius J. Anal. Chem.*, 2001, **370**, 618–623.
26. F. Vanhaecke, L. Balcaen, I. Deconinck, I. De Schrijver, C. M. Almeida and L. Moens, *J. Anal. Atom. Spectrom.*, 2003, **18**, 1060–1065.
27. J. Barling and D. Weis, *J. Anal. Atom. Spectrom.*, 2008, **23**, 1017–1025.
28. J. Irrgeher, T. Prohaska, R. E. Sturgeon, Z. Mester and L. Yang, *Anal. Methods*, 2013, **5**, 1687–1694.
29. C. Brunnée, *Int. J. Mass Spectrom. Ion Process.*, 1987, **76**, 125–237.
30. C. R. Quetel, T. Prohaska, M. Hamester, W. Kerl and P. D. P. Taylor, *J. Anal. Atom. Spectrom.*, 2000, **15**, 353–358.

31. L. Yang, Z. Mester, L. Zhou, S. Gao, R. E. Sturgeon and J. Meija, *Anal. Chem.*, 2011, **83**, 8999–9004.
32. N. Shirai and M. Humayun, *J. Anal. Atom. Spectrom.*, 2011, **26**, 1414–1420.
33. A. L. Buchachenko, *Chem. Rev.*, 1995, **95**, 2507–2528.
34. S. E. Jackson and D. Günther, *J. Anal. Atom. Spectrom.*, 2003, **18**, 205–212.
35. T. Hirata, Advances in Laser Ablation–Multi-Collector Inductively Coupled Plasma Mass Spectrometry, in *Isotopic Analysis*, eds. F. Vanhaecke and P. Degryse, Wiley-VCH Verlag GmbH & Co. KGaA, Weinheim, Germany, 2012, ch. 4, pp. 93–112.
36. T. Hirata, Y. Hayano and T. Ohno, *J. Anal. Atom. Spectrom.*, 2003, **18**, 1283–1288.
37. M. Guillong, I. Horn and D. Gunther, *J. Anal. Atom. Spectrom.*, 2003, **18**, 1224–1230.
38. I. Horn and F. von Blanckenburg, *Spectrochim. Acta Part B: Atom. Spectrosc.*, 2007, **62**, 410–422.
39. P. K. Diwakar, S. S. Harilal, N. L. LaHaye, A. Hassanein and P. Kulkarni, *J. Anal. Atom. Spectrom.*, 2013, **28**, 1420–1429.
40. J. Koch and D. Günther, *Anal. Bioanal. Chem.*, 2007, **387**, 149–153.
41. H.-R. Kuhn, N. J. Pearson and S. E. Jackson, *J. Anal. Atom. Spectrom.*, 2007, **22**, 547–552.
42. M. E. Shaheen, J. E. Gagnon and B. J. Fryer, *Chem. Geol.*, 2012, **330–331**, 260–273.
43. L. R. Riciputi, D. C. Duckworth, C. M. Barshick and D. H. Smith, *Int. J. Mass Spectrom. Ion Process.*, 1995, **146–147**, 55–64.
44. M. Betti and L. Aldave de las Heras, *Spectrochim. Acta Part B: Atom. Spectrosc.*, 2004, **59**, 1359–1376.
45. F. Chartier and M. Tabarant, *J. Anal. Atom. Spectrom.*, 1997, **12**, 1187–1193.
46. M. Betti, S. Giannarelli, T. Hiernaut, G. Rasmussen and L. Koch, *Fresenius J. Anal. Chem.*, 1996, **355**, 642–646.
47. M. Betti, G. Rasmussen and L. Koch, *Fresenius J. Anal. Chem.*, 1996, **355**, 808–812.
48. K. H. Ecker and W. Pritzkow, *Fresenius J. Anal. Chem.*, 1994, **349**, 207–208.
49. K. Habfast, *Int. J. Mass Spectrom. Ion Phys.*, 1983, **51**, 165–189.
50. S. R. Hart and A. Zindler, *Int. J. Mass Spectrom. Ion Process.*, 1989, **89**, 287–301.
51. M. Jungck, T. Shimamura and G. Lugmair, *Geochim. Cosmochim. Acta*, 1984, **48**, 2651–2658.
52. T. M. Esat, R. Spear and S. R. Taylor, *Nature*, 1986, **319**, 576–578.
53. T. M. Esat, *Geochim. Cosmochim. Acta*, 1988, **52**, 1409–1424.
54. K. Habfast, *Int. J. Mass Spectrom.*, 1998, **176**, 133–148.
55. H. Ramebäck, M. Berglund, R. Kessel and R. Wellum, *Int. J. Mass Spectrom.*, 2002, **216**, 203–208.
56. R. Andreasen and M. Sharma, *Int. J. Mass Spectrom.*, 2009, **285**, 49–57.

57. G. Cavazzini, *Int. J. Mass Spectrom.*, 2012, **309**, 129–132.
58. N. Shimizu and S. Hart, *Annu. Rev. Earth Planet. Scie.*, 1982, **10**, 483.
59. G. Slodzian, J. Lorin and A. Havette, *J. Phys. Lett.*, 1980, **41**, 555–558.
60. N. Shimizu and S. Hart, *J. Appl. Phys.*, 1982, **53**, 1303–1311.
61. R. L. Hervig, *Rapid Commun. Mass Spectrom.*, 2002, **16**, 1774–1778.
62. G. Slodzian, Dependence of Ionization Yields Upon Elemental Composition; Isotopic Variations, in *Secondary Ion Mass Spectrometry SIMS III*, eds. A. Benninghoven, J. Giber, J. László, M. Riedel and H. W. Werner, Springer, Berlin Heidelberg, 1982, vol. 19, pp. 115–123.
63. J. M. Schroeer, T. Rhodin and R. Bradley, *Surf. Sci.*, 1973, **34**, 571–580.
64. L. R. Riciputi and J. P. Greenwood, *Int. J. Mass Spectrom.*, 1998, **178**, 65–71.
65. L. R. Riciputi, B. A. Paterson and R. L. Ripperdan, *Int. J. Mass Spectrom.*, 1998, **178**, 81–112.
66. M. Betti, *Int. J. Mass Spectrom.*, 2005, **242**, 169–182.
67. V. Deline, W. Katz, C. Evans and P. Williams, *Appl. Phys. Lett.*, 1978, **33**, 832–835.
68. J. M. Eiler, C. Graham and J. W. Valley, *Chem. Geol.*, 1997, **138**, 221–244.
69. D. Vielzeuf, M. Champenois, J. W. Valley, F. Brunet and J. L. Devidal, *Chem. Geol.*, 2005, **223**, 208–226.
70. N. T. Kita, T. Ushikubo, B. Fu and J. W. Valley, *Chem. Geol.*, 2009, **264**, 43–57.
71. M. E. Hartley, T. Thordarson, C. Taylor, J. G. Fitton and Eimf, *Chem. Geol.*, 2012, **334**, 312–323.
72. I. Lyon, J. Saxton and S. Cornah, *Int. J. Mass Spectrom. Ion Process.*, 1998, **172**, 115–122.
73. J. M. Huberty, N. T. Kita, P. R. Heck, R. Kozdon, J. H. Fournelle, H. Xu and J. W. Valley, in *conf.proc. of 5th International Archean Symposium*, Geological Survey of Western Australia.
74. D. L. Weathers, S. J. Spicklemire, T. A. Tombrello, I. D. Hutcheon and H. Gnaser, *Nucl. Instrum. Methods Phys. Res. Sect. B: Beam Interact. Mater. Atoms*, 1993, **73**, 135–150.
75. H. Gnaser and H. Oechsner, *Nucl. Instrum. Methods Phys. Res. Sect. B: Beam Interact. Mater. Atoms*, 1990, **48**, 544–548.
76. R. Halsted and A. O. Nier, *Rev. Sci. Instrum.*, 1950, **21**, 1019–1021.
77. W. A. Brand, Mass Spectrometer Hardware for Analyzing Stable Isotope Ratios, In *Handbook of Stable Isotope Analytical Techniques*, ed. P. A. de Groot, Elsevier Science, 835–858, 2004, vol. 1, ch. 38, pp. 835–856.
78. R. M. Verkouteren, C. E. Allison, S. A. Studley and K. J. Leckrone, *Rapid Commun. Mass Spectrom.*, 2003, **17**, 771–776.
79. R. M. Verkouteren, S. Assonov, D. B. Klinedinst and W. A. Brand, *Rapid Commun. Mass Spectrom.*, 2003, **17**, 777–782.
80. J. Elsig and M. C. Leuenberger, *Rapid Commun. Mass Spectrom.*, 2010, **24**, 1419–1430.

CHAPTER 7

# *Interferences*

THOMAS PROHASKA

University of Natural Resources and Life Sciences Vienna (BOKU), Department of Chemistry, Division of Analytical Chemistry, VIRIS Laboratory for Analytical Ecogeochemistry, Tulln, Austria
Email: thomas.prohaska@boku.ac.at

An interference in spectroscopy can be defined as anything that alters, modifies, or disrupts a measurement signal (see Figure 7.1). Therefore, interferences are usually grouped in (A) spectral interferences (also termed spectroscopic interferences) and (B) nonspectral interferences. The latter refer mainly to matrix effects (also termed matrix interferences). These are specific for the different ionisation sources or sample introduction systems and will not be discussed in detail within this chapter. A third source of interferences (C) would result from the system electronics or stray electrons in the system or the detection unit. For example, scattering electrons can be observed when Faraday detectors are used (see Chapter 4). An electron suppressor or a small magnet ensures that no secondary electrons are released from the detector. In addition, the Faraday cup design minimises the secondary electron escape angle by increasing the Faraday cup length.

Within this section, we will mainly focus on the occurrence of spectral interferences (A):

Spectral interferences are signals determined at a certain nominal *m*/*z*, which do not originate from the analyte ion under investigation. These can mainly result from:

New Developments in Mass Spectrometry No. 3
Sector Field Mass Spectrometry for Elemental and Isotopic Analysis
Edited by Thomas Prohaska, Johanna Irrgeher, Andreas Zitek and Norbert Jakubowski

Published by the Royal Society of Chemistry, www.rsc.org

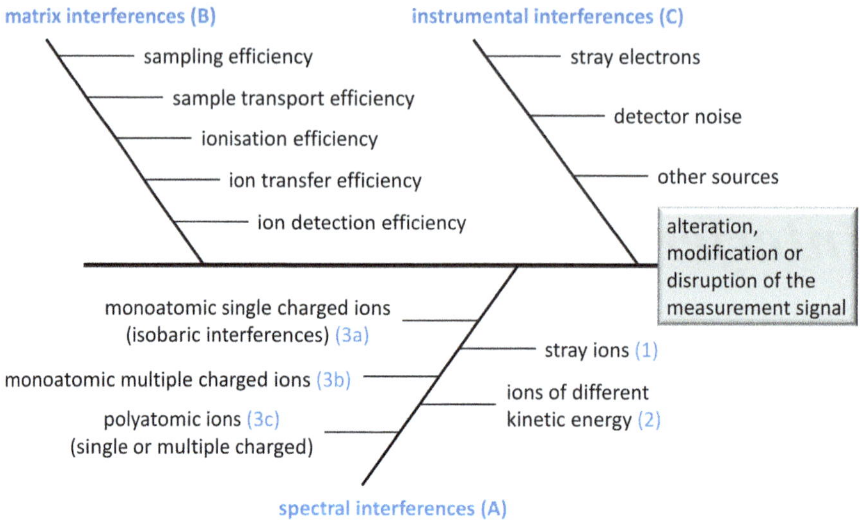


**Figure 7.1** Types of interferences occurring in mass spectrometry.

(1) stray ions resulting in a signal at the time of measurement;
(2) ions of different nominal *m*/*z* with different kinetic energy resulting in a signal at another (mostly adjacent) nominal *m*/*z*;
(3) single- or multiple-charged mono- or polyatomic ions of the same nominal *m*/*z* ratio deriving from other than the analyte ion.

Stray ions (1) are usually removed by applying specific filters such as retarding potential quadrupole lenses, deceleration or retardation filters or wide-angle energy-retarding filters.

One main feature of a double-focusing mass analyser is energy focusing and thus reducing interferences from source (2). This is a prerequisite especially if the ionisation source produces ions with a broad energy scattering (*e.g.* ICP or GD sources – see also Chapter 4). Nonetheless, an influence by peaks of adjacent *m*/*z* can still be observed.

Spectroscopic interferences of source (3) may be subdivided into several groups, as they can be attributed to the presence of isobaric atomic ions, multiply charged ions and polyatomic ions of various origins. One of the major features of a sector field mass spectrometer is the ability to provide high mass resolution in order to separate these interferences from the analyte signal making use of small mass differences of the different incoming ions.

Isobaric (from the Greek words "*ίσος*" – isos, equal and "*βαρύς*" - barýs weight) overlaps (3a) exist when the signals of nuclides of different elements coincide at the same nominal mass (as they have the same number of nucleons). Since the mass difference between isobars is in general very small, resolutions between $10^4$ and $10^8$ are required. The maximum mass resolution setting of most SFMS instruments (in the range of about $10^4$) is often not sufficient to overcome interferences of this kind. Therefore, alternative approaches to cope with this problem are used. In most cases, at least one isotope of each element is free from isobaric overlap – with the only exception of indium – and can be monitored instead (even though in many cases this isotope will not be the most abundant). Isobaric interferences can usually be corrected mathematically by using the natural isotopic abundances.

Multiple charged monoatomic ions (3b) will be monitored in the mass spectrum at a position $m/z$ (*i.e.* for instance doubly charged ions are found at half of their nominal mass). Nonetheless, multiple-charged ions usually have a significantly smaller formation rate as compared to single-charged ions depending on the ionisation potential.

Interferences by unwanted polyatomic ions (3c) cause the most severe problems.

(***Note***: in some cases (*e.g.* IRMS – see Chapters 16 and 17) the polyatomic species are the analytes of interest.)

Polyatomic interferences are less predictable and depend on the sample composition (analytes, matrix and eventual ionisation support gases) and the operational parameters of the system. These interferences are difficult and sometimes impossible to correct for by data reduction using the isotopic abundances or ion formation probabilities. The required mass resolutions usually range from $10^3$ to $10^4$ up to mass 70 and quite often more than $10^4$ at higher masses. Polyatomic interferences may be introduced by the sample itself but may also arise from the discharge gas, contaminants, entrained air or the reagents and solvents used.

A large number of interferences can therefore be resolved by applying higher mass resolution. Besides high mass resolution as a key feature of a sector field mass spectrometers, different approaches have been discussed in the literature and selected examples will be covered only briefly in this chapter.

## 7.1 Mathematical Correction Procedures

Mathematical correction is usually applied in the case of simple corrections (isobaric interferences) or if all other measures fail. In the case of isobaric interferences, the interfering element is measured at another isotope mass along with the analyte of interest and the contribution of the interference is calculated *via* the isotopic abundances. The method applied is generally called peak stripping.

"Peak stripping: Data analysis technique used in elemental mass spectrometry to correct the intensity of an unresolved multiplet peak monitored for a specific element for contributions from a multi-isotopic contaminant. The correction is calculated from the measured intensity of a different isotopic peak of the contaminant and the known natural isotopic abundance ratio".[1]

The situation is more complicated for polyatomic interferences as they are not easily predictable and depend on the operational parameters as well as on the matrix. Therefore, the formation probability has to be estimated using external matrix-matched solutions or the same interference has to be observed on a different mass and used for further correction.

This is demonstrated *via* the following example: In ICP-MS, $^{75}As^+$ suffers from a spectral interference in Cl-containing solutions as a result of the occurrence of the $^{40}Ar^{35}Cl^+$ ion. As a consequence, the following mathematical strategies can be used:

(a) The $^{40}Ar^{35}Cl^+$ interference is measured along with the $^{35}Cl^+$ (or $^{37}Cl^+$) signal in pure matrix solutions free of As, containing only the matrix and Cl of increasing content. The ratio $^{40}Ar^{35}Cl^+/^{35}Cl^+$ is calculated from these solutions and an average ratio is further used as a correction factor. Now, the signal on $m/z=75$ is measured in the solution along with the signal of $^{35}Cl^+$ on $m/z=35$ (or $^{37}Cl^+$ on $m/z=37$). The $^{35}Cl^+$ (or $^{37}Cl^+$) signal is now multiplied by the previously calculated factor, resulting in the probable interference of $^{40}Ar^{35}Cl^+$ on mass 75. The remaining signal is most probably the requested As signal.
(b) Another possibility would be that the $^{40}Ar^{37}Cl^+$ signal on $m/z=77$ is measured along with the signal on $m/z=75$. *Via* the isotopic abundances of Cl, the $^{40}Ar^{35}Cl^+$ signal can be calculated from the $^{40}Ar^{37}Cl^+$ signal. Again, the remaining signal is most probably the requested As signal. In this case, we can observe an additional problem: We find $^{77}Se^+$ on $m/z=77$ along with $^{40}Ar^{37}Cl^+$ interference. Therefore, we have to subtract the signal of $^{77}Se^+$ prior to calculate the $^{40}Ar^{35}Cl^+$ from the $^{40}Ar^{37}Cl^+$ signal. We can correct $^{77}Se^+$ *via* the isotopic abundances using, *e.g.* the $^{82}Se^+$ signal. Nonetheless, we have to consider that we find an isobaric interference of $^{82}Kr^+$ on $m/z=82$. Not to consider any $^{40}Ar_2{}^1H_2{}^+$ interference…

It is obvious that this strategy leads only to satisfying results when a relatively high analyte concentration and low matrix concentration coincide.

As a rule of thumb, it is always wise to monitor more than one isotope (if possible), even if the other isotopes are less abundant.

## 7.2 Sample Preparation and Introduction Systems

Among various strategies discussed over many years, the appropriate selection of a sample preparation or an adequate introduction system for the separation of the matrix and the analyte of interest is usually best suited to overcome many matrix-based spectral interferences (along with nonspectral matrix effects).

Direct separation of the matrix from the analyte of interest is accomplished already as integral part of the analytical process (*e.g.* as offline preparation in TIMS (Chapter 14) or online separation in IRMS or noble gas IRMS (Chapters 16 and 17) or is accomplished as additional feature (*e.g.* online coupling of a chromatographic separation or hydride generation or offline sample/matrix separation in ICP-MS (Chapter 12)). Due to the versatility of coupling a large number of devices to an ICP source (as it is operated at ambient pressure), various sample-introduction systems have been employed for matrix removal (in most cases removal of the solvent or volatile species) such as cooled spray chambers or membrane desolvation units (to remove the water content in the formed aerosol).

It is anyhow evident that all these measures do not lead to satisfying results for solid sampling techniques (SIMS (Chapter 15), GDMS (Chapter 13) or LA-ICP-MS (Chapter 12)) are applied.

## 7.3 Modified Instrumental Parameter

As a number of interfering ions are formed within the ionisation source, depending on the instrumentation used, various strategies are employed such as, *e.g.* temperature programs and specially prepared filaments in TIMS or cool/cold plasma conditions in ICP-MS.

The application of collision and reaction cells in front of a mass analyser had a significant impact especially in ICP-MS, where these devices have been applied in combination with a low-resolution quadrupole analyser in order to overcome spectral interferences. In principle, collision cells are also suitable for ICP-SFMS, as was commercialised in the *IsoProbe* MC ICP-SFMS. In this case, a collision cell was used for energy focusing in a single-focusing magnetic sector field instrument.

## Reference

1. K. K. Murray, R. K. Boyd, M. N. Eberlin, G. J. Langley, L. Li and Y. Naito, *Pure Appl. Chem.*, 2013, **85**, 1515–1609.

CHAPTER 8

# *Measurement Strategies*

JOHANNA IRRGEHER,*[a] JOCHEN VOGL,[b] JAKOB SANTNER[c] AND THOMAS PROHASKA*[a]

[a] University of Natural Resources and Life Sciences Vienna (BOKU), Department of Chemistry, Division of Analytical Chemistry, VIRIS Laboratory for Analytical Ecogeochemistry, Tulln, Austria; [b] BAM Federal Institute for Materials Research and Testing, Berlin, Germany; [c] University of Natural Resources and Life Sciences, Vienna, Institute of Soil Research, Tulln, Austria
*Email: johanna.irrgeher@boku.ac.at; thomas.prohaska@boku.ac.at

Designing an appropriate measurement strategy for a particular analytical question is not always obvious, since a number of factors have to be considered, whereby some of them might be difficult to define. A set of key questions generally precede the experimental design in analytical measurements and help to choose the measurement strategy, which is fit for the intended use – in the particular case of the content of this book – either for quantification, elemental ratio or isotope ratio analyses, accordingly. (***Note***: chemical imaging will be addressed in Chapter 9.)

These questions comprise:

- What is the analyte? What is the measurand?
- What does the matrix constitute of? Are there any possible sources of bias due to interfering substances in the matrix?
- What is the background/blank of my method and of my analytical measurement?

New Developments in Mass Spectrometry No. 3
Sector Field Mass Spectrometry for Elemental and Isotopic Analysis
Edited by Thomas Prohaska, Johanna Irrgeher, Andreas Zitek and Norbert Jakubowski

Published by the Royal Society of Chemistry, www.rsc.org

- What is the required uncertainty of the result (including all input sources to the standard uncertainty)?
- What is my working range?
- How do I calibrate my method?
- Which (certified) reference materials are available?
- How much effort is within one's means (timewise as well as moneywise)?

All considerations contribute to the design of a measurement. Within the following sections, we will focus mainly on considerations with respect to calibration in elemental and isotopic analysis. The used terminology used is in accordance with recent IUPAC recommendations.[1,2]

## 8.1 Calibration

Definition (according to VIM): "Operation that, under specified conditions, in a first step, establishes a relation between the quantity values with measurement uncertainties provided by measurement standards and corresponding indications with associated measurement uncertainties and, in a second step, uses this information to establish a relation for obtaining a measurement result from an indication.

(***Note 1***: A calibration may be expressed by a statement, calibration function, calibration diagram, calibration curve, or calibration table. In some cases, it may consist of an additive or multiplicative correction of the indication with associated measurement uncertainty.

***Note 2***: Calibration should not be confused with adjustment of a measuring system, often mistakenly called "self-calibration", nor with verification of calibration.

***Note 3***: Often, the first step alone in the above definition is perceived as being calibration.)"[3]

The calibration of an analytical instrument is an integral part of establishing the calibration hierarchy being itself an essential prerequisite for proving metrological traceability (see also Chapter 10).

Calibration of a mass spectrometer requires the establishment of the relation between the physical quantity (*e.g.* measured signal in counts per second or volts) and the known measurement parameter (*e.g.* content or isotope ratio) in the calibrant resulting in a function that expresses this comparison (calibration function). The calibration function is further applied for calculating the unknown measurement parameter from a measured signal in a sample (analytical function) (For examples, see eqns (8.1) to (8.4))

**Examples**

a) Example of a calibration for quantification (*e.g.* linear external calibration)

$$\text{calibration function: } I = aw + b \tag{8.1}$$

$$\text{analytical function: } w = \frac{(I - b)}{a} \tag{8.2}$$

$I$: signal intensity
$a$: slope of a linear calibration curve
$w$: mass fraction
$b$: intercept of a linear calibration curve

b) Example of a calibration for isotope ratio measurement (*e.g.* external calibration)

$$\text{calibration function: } R_{\text{standard,measured}} = R_{\text{standard,certified}} \cdot \left(\frac{1}{K}\right) \tag{8.3}$$

$$\text{analytical function: } R_{\text{sample,corrected}} = R_{\text{sample,measured}} \cdot K \tag{8.4}$$

$R_{\text{standard, measured}}$: measured isotope ratio of the standard
$R_{\text{standard, certified}}$: certified isotope ratio of the standard
$K$: external correction factor

**Calibration Standards and Matrix Matching**

Most mass-spectrometric techniques for elemental and isotopic analysis are matrix dependent to various extents. Therefore, the ideal situation deals with identical matrices of standard reference materials and samples and an identical treatment ("IT-principle"). In addition, the requirement of finding different internal normalisation standards to account mainly for (1) long- and short-term signal fluctuations and (2) matrix-induced signal reduction or enhancement is inherent.

If the initial situation deals with different matrices, the removal of all other elements (the "matrix") from the element of interest is a desirable option, since a homogeneous sample, which is free of other components prone to interfere during measurement, is generated. This is accomplished already as integral part of the measurement (*e.g.* in TIMS or IRMS). The sample matrix can lead to spectral interferences (see Chapter 7) or lead to, *e.g.* different ionisation efficiency of the analyte of interest in presence or absence of a matrix. In addition to that, the matrix can lead to a bias already within the preparation step: In the case of the direct nebulisation of a solution (*e.g.* in ICP-MS), the variation of the matrix can lead to different vaporisation in the nebuliser and thus effect the ionisation in the ICP. The situation is more complex when using solid sampling techniques (SIMS, LA-ICP-MS or GDMS), as an analyte/matrix separation is often not achievable. In addition to the previously mentioned interferences, the different matrix can have an effect on sputtering efficiencies (*e.g.* number and size of particles, vaporisation of elements) resulting in different transport and ionisation efficiencies.

However, the lack of matrix-matched standard reference materials especially for the latter techniques often leads to insurmountable limitations since matrix-derived interferences are potential sources of analytical bias in the measurement. Thus, the situation remains unsatisfying with respect to the demand of new CRMs. Matrix matching of reference materials used for data reduction (quantification or isotope ratio analysis) is preferable to the use of nonmatrix-matched CRMs since near-perfect similarity in behaviour of the analyte in the sample and in the standard is favoured. As a consequence, a trend towards the preparation of in-house standards for the analysis of samples to circumvent the nonavailability of commercially available reference materials can be observed.

## 8.2 Quantification

Sound quantification of the elemental content in a sample requires accurate calibration of the mass spectrometer. ICP-MS, GDMS and SIMS are among the preferred methods used for (multi-) elemental quantification, whereby various strategies for calibrating a mass spectrometer are available and the choice strongly depends on, *e.g.* the desired accuracy of the result, the availability of certified reference materials, the matrix of the sample, the presence of interfering components and the expected working range.

In the following, general approaches of calibration for quantification are given, whereby the reader is referred to Chapters 12–17 in this book for technique-specific strategies, if applicable.

The most frequently applied calibration strategies for quantification can be distinguished into (1) external (one- or multipoint calibration with or without matrix matching) and (2) internal calibration (internal standardisation, standard addition and isotope dilution) approaches.

The general procedure of a calibration starts with the preparation of calibration standards (*e.g.* from a certified stock solution or solid starting products) with (preferably) defined composition and known purity as well as known (reported) mass or amount fraction of the analyte.

(***Note:*** As traceability of a result requires a full traceability chain, it is of the utmost importance that the primary material for preparation of the calibrants is traceable to a common standard (preferably the SI).)

The measurement itself is performed using an adequate analytical technique where usually the establishment of the relation (*i.e.* comparison) between input and output parameters of the method (*i.e.* establishment of the calibration function) is followed by the measurement of the sample(s) and the calculation of the unknown amount of substance (*e.g.* the concentration or fraction) in the sample using the corresponding analysis function.

### 8.2.1 External Calibration

External calibration refers to a method of calibration where an external set of one or more calibration standard(s) with known amounts of analyte are used to establish a relation between the content (*e.g.* amount of substance) and instrument response (*i.e.* physical signal such as counts, counts per second or volts) – namely the calibration function.

**One-Point Calibration**

A very fast and simple way of external calibration can be performed by using one calibration standard with a known concentration of the analyte of interest. It is preferable to choose a standard concentration as close as possible to the expected concentration in the unknown sample, since in the case of a one-point calibration, the proof of linearity is not guaranteed over an arbitrary dynamic range of the calibration function, see Figure 8.1.

(***Note:*** an assumption of the linearity of a calibration function in the working range has to be proven.)

In some analytical scenarios, single-point anchoring is the only reasonably possible way of calibration due to the lack of suitable reference standards with different concentrations of the analyte of interest. (This can be the case, *e.g.* for GDMS, SIMS and LA-ICP-MS when matrix-matched solid standards are required.)

**Calibration Using Relative Response Factors**

Using one single standard and establishing response factors or response curves is an alternative approach. This is achieved by measuring a standard containing one or more elements of interest with known quantities and the calculation of element-specific and matrix-specific sensitivity or response factors as a function of the concentration (usually called relative sensitivity factors, RSF, in GDMS and SIMS).[4] More details about these calibration strategies and related approaches are given in Chapter 13 (GDMS) and Chapter 15 (SIMS).

Some instrument manufacturers offer more sophisticated modifications of this calibration strategy. The software routine includes correction for, *e.g.*, interferences and differences in ionisation efficiency when calculating the response curve that finally allows the quantification of the whole range of elements through interpolation over the entire mass range through one analysis.

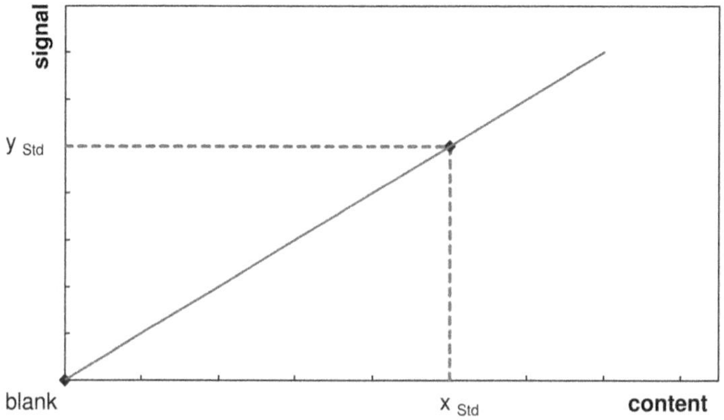


**Figure 8.1** Schematic of a one-point calibration.

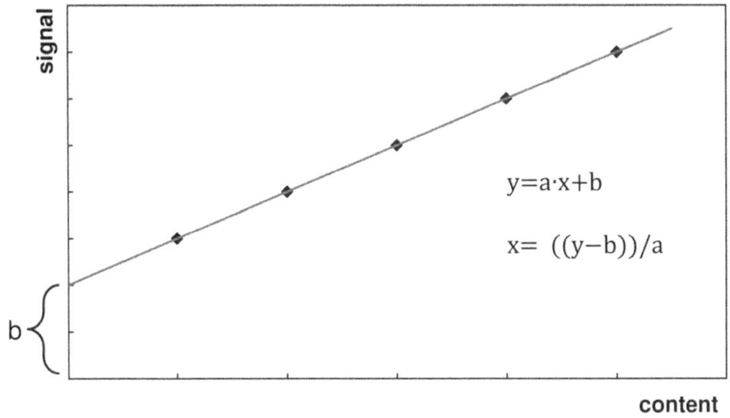


**Figure 8.2** Schematic of a multiple-point calibration.

**Multiple-Point Calibration**

External multiple-point calibration (Figure 8.2) is performed if more than only one external calibration standard with a known quantity of a given property (*e.g.* mass fraction or concentration of interest) is used. External calibration using standards at different concentrations over a desired dynamic range can be performed with or without matrix matching (defined chemical, physical or biological properties), the latter to be preferable in many cases.

By measuring a set of calibration standards, a mathematical function (calibration function) can be established at the beginning of a measurement sequence and subsequently used to calculate the unknown analyte concentration or amount in the sample by comparing the output signal (*e.g.* intensity) of the unknown sample to the calibration function established. The number of calibration points to be selected depends on the purpose of the calibration experiment, the availability of standards at different concentrations, the desired working range, the relevant international standard regulation or other factors. In most cases, equidistant differences in the content of the calibration standards are preferable. External calibration using multiple standards represents a very convenient and thus often applied quantification strategy in solution-based analyses because of the multielement capabilities, but is also used for solid-sampling techniques such as GDMS, SIMS and LA-ICP-MS. To ensure qualified performance of the method, the matrix of sample and standards should be matched at best and the correct mathematical model (which represents the calibration function best) has to be chosen to describe the calibration function (especially when no linear fit is present).

### 8.2.2 Standard Addition

Calibration using standard addition is a promising alternative to external calibration and is especially of interest when significant (but stable) matrix

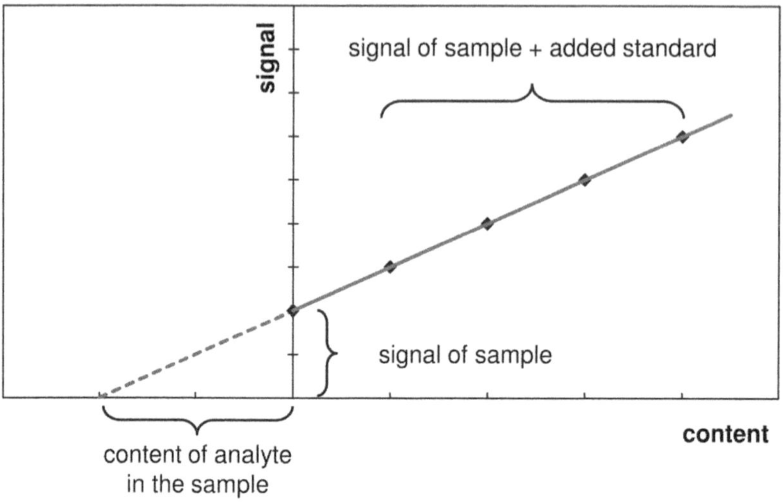


**Figure 8.3** Schematic of a standard addition calibration.

effects are expected that either suppress or enhance the analyte signal and/or matrix-matched reference materials are not available. Standard addition is performed by splitting the sample into several aliquots and adding increasing amounts of an internal calibration standard that contains a known amount of the analyte of interest to the sample aliquots. One aliquot is left without adding an internal standard. Since the standard is added to the sample, matrix effects are mitigated and the need for using matrix-matched reference materials is avoided. The analyte of interest is measured in all aliquots (both the unspiked as well as the spiked aliquots) and a (mostly linear) calibration function can be established. The concentration of the analyte in the unknown sample corresponds to the point on the $x$-axis, where the intercept $y=0$ (Figure 8.3). It is of crucial importance to ensure the sample and the spike reach chemical equilibrium prior to measurement in order to avoid differences in ionisation efficiency and inconsistency along a measurement sequence.

Standard addition is mostly applied to solution-based analyses even though alternative calibration approaches, *e.g.* for standard addition during LA-ICP-MS analyses have been reported.[5] The calibration strategy is a good alternative for certain scenarios, however, an increased number of sample replicates and time-consuming sample preparation have to be taken into account and a linear response of the calibration function is a prerequisite. For details the interested reader is referred to the specialised literature.

### 8.2.3 Internal Normalisation

Internal normalisation is used in order to correct for certain types of noise when, *e.g.* the analysed sample amount may vary or when matrix effects or signal-related sample-to-sample variations in sensitivity may limit the

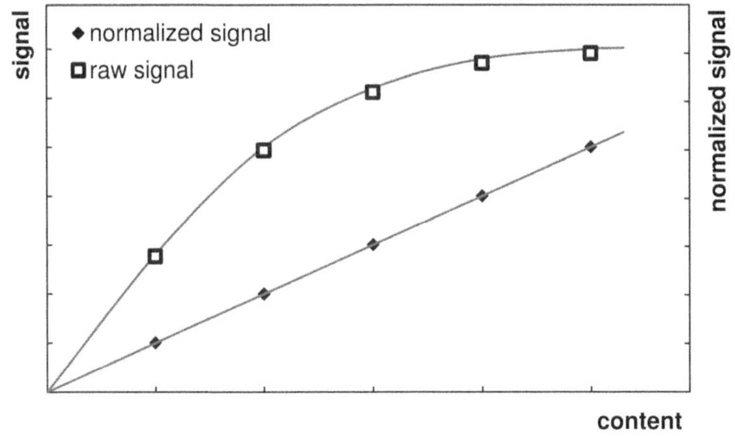


**Figure 8.4** Schematic of a calibration with and without internal normalisation.

method performance. The internal normalisation standard (often a non-analyte isotope) is usually added to the calibration standards and the samples in a defined aliquot (*e.g.* from a single-element standard stock solution) or a homogeneously distributed element in the sample is used. The latter is of particular interest for solid sampling strategies, such as LA-ICP-MS, where isotopes of a homogeneously distributed element at reasonable concentration in the sample are commonly used as internal normalisation standard (*e.g.* $^{13}C$ in organic tissues or other major matrix components in, *e.g.* metals or minerals) (Figure 8.4).

The choice of internal normalisation standard depends on the analyte isotope as well as the sample matrix. It is recommended that the physical and chemical behaviour of the analyte and internal standard are as similar as possible. In the case of different responses throughout the mass range, a mixture of several internal standards can be used.

## 8.2.4 (Linear) Regression Analysis

Quantification in chemical analysis is usually done by establishing a functional (often linear) relationship between a measured physical quantity ($y$; *e.g.* the signal intensity) and the amount of analyte in a sample ($x$; *e.g.* concentration or mass fraction). The linear model for the signal obtained from the instrument is (eqn (8.5)):

$$y_i = a \cdot x_i + b + e_{y_i} \tag{8.5}$$

$a$: slope of the linear calibration line
$b$: intercept of the linear calibration line
$e_y$: random measurement uncertainty of the signal intensity
$x_i$: $x$ - variable
$y_i$: $y$ - variable

As $e_y$ is a random parameter that cannot be accounted for, the calibration function reads

$$y_i = a \cdot x_i + b \tag{8.6}$$

$$x_i = \frac{y_i - b}{a} \tag{8.7}$$

The most common algorithm for estimating $a$ and $b$ is the Gaussian or ordinary least squares estimation method (OLS). This procedure computes the line for which the sum of squares of the $y$ residuals $(y_i - \hat{y}_i)$ is a minimum. (Here, $y_i$ denotes the measured signal intensity and $\hat{y}_i$ the adjusted value on the fitted line). However, the OLS line estimation can only be applied if (1) the uncertainties of the $x_i$ values are negligible compared to the uncertainties of the $y_i$ values and (2) the $e_{yi}$ are uncorrelated and homoscedastic – which means that $e_y$ represents the randomly distributed measurement uncertainty. If these conditions are not met, alternatives to OLS have to be applied. Weighted least squares (WLS) estimation is used if homoscedasticity in the $y_i$ is not given. If significant uncertainty in the $x_i$ exists, the model function for the instrument response changes to[6]

$$y_i = a \cdot x_i + b + e_{y_i} - ae_{x_i} \tag{8.8}$$

$a$: slope of the linear calibration line
$b$: intercept of the linear calibration line
$e_y$: random measurement uncertainty of the signal intensity
$e_x$: random measurement uncertainty of the signal intensity
$x$: $x$ - variable
$y$: $y$ - variable

As the known concentration or amount of analyte in calibration standards is always affected by some uncertainty, Sayago *et al.* suggested to consider uncertainties in $x$ significant if the relative standard uncertainty ($u_{c,\ rel}$) in $x$ is equal to or larger than a tenth of the $u_{c,\ rel}$ in $y$ ($u_{c,\ rel}(x)/u_{c,\ rel}(y)) \geq 0.1$).[7] The additional uncertainty in $x$ has to be taken into account by using a suitable line-estimation algorithm. Danzer and Currie recommend the used of orthogonal least squares (also termed major-axis regression) for cases where uncertainties in both variables occur.[8] For information on this method the reader is referred to Adcock[9] and Kermack and Haldane.[10] However, major-axis regression it is only applicable if the uncertainty ratio $u_{c,\ rel}(x)/u_{c,\ rel}(y)) = 1$ is constant in all calibration points. This restriction is due to the fact that in orthogonal regression the minimised distance is the orthogonal distance between the observed (measured) calibration point and the fitted line. Thus, the inherently attributed influence (weight) of the $u_{c,\ rel}$ (x) and $u_{c,\ rel}$ (y) in the line-fitting procedure is equal, and, as no actual weights are used in this method, it is required that the uncertainty ratio is constant in all points. It is of course unlikely that the uncertainty ratio is unity for all calibration points. Standardised major-axis regression (SMA)[11] introduces constant weighting factors for the $x_i$ and $y_i$, so that $u_{c,\,rel}\,(x)/u_{c,\,rel}\,(y)$ is constant but not necessarily 1. Despite this improvement, nonconstant uncertainty ratios are still not considered. A solution to this problem is presented by York,[12,13] who reported an algorithm that uses the $u_{c,\,rel}\,(x)$ and $u_{c,\,rel}\,(y)$ as weights for each point of the regression and additionally considers a possible correlation in the $x$ and $y$ uncertainties. Generally, York's method yields the same result as the major axis and standardised major axis methods if the corresponding weights are used, *i.e.* $u_{c,\,rel}\,(x_i)/\,u_{c,\,rel}\,(y_i) = 1$ for major axis regression and $u_{c,\,rel}\,(x_i)/u_{c,\,rel}\,(y_i) = \text{constant} \neq 1$.[12] In a later publication[14] it is shown that the least squares and maximum likelihood solution of the problem yield the same set of equations. Moreover, a simple algorithm is described in this publication that allows the implementation of this method for every user with basic programming skills. We omit the reproduction of the whole set of equations here as they are comprehensively presented in York *et al.*[14] A schematic of the different regression models is given in Figure 8.5.

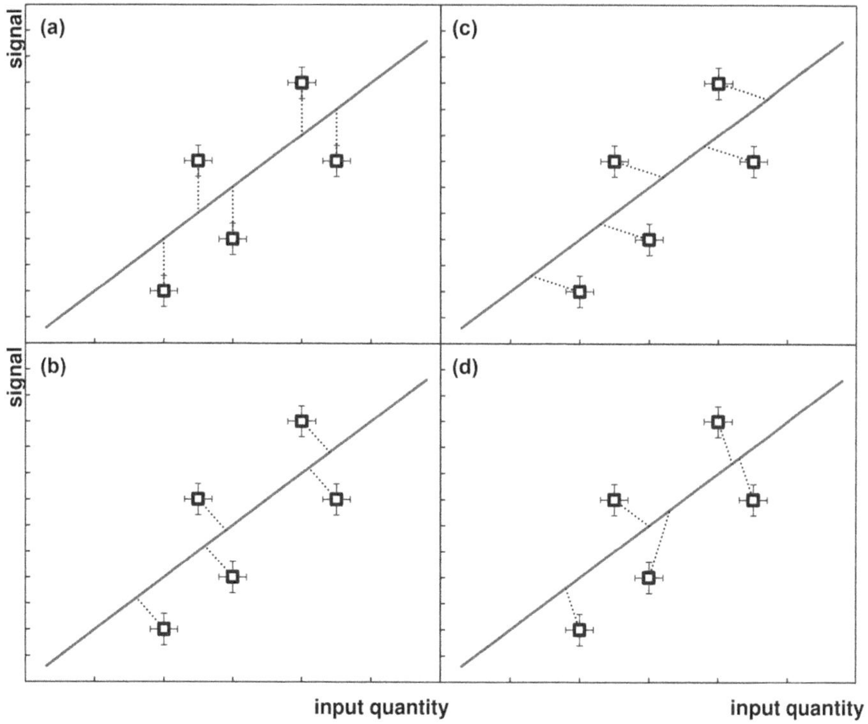


**Figure 8.5** Schematic of different regression models: (a) ordinary least squares, (b) major axis regression, (c) standardised major axis regression and (d) bivariate weighted least squares.

### 8.2.5 Isotope Dilution

The isotope dilution principle is a very accurate approach to calibrate mass spectrometry for quantification based on internal calibration. IDMS represents a method to measure an amount content based on the comparison of amounts of isotopes.

The origin of the isotope dilution principle dates back to the beginnings of isotope research in the early 20th century.[15] With improving mass spectrometry and the availability of enriched stable isotopes the isotope dilution principle was boosted in the 1940s and 1950s and developed to a widely accepted and acknowledged reference method for the quantification of elemental mass fractions.[15] This technique soon became well known under the name isotope dilution mass spectrometry (IDMS). Due to its specific advantages IDMS was accepted by the Consultative Committee for Amount of Substance (CCQM) as a primary method of measurement, which generally offers the highest metrological quality for amount of substance measurements[16] as the values are directly traceable to the SI with a closed uncertainty chain. Meanwhile, IDMS has developed to a routing method providing highest metrological standards.[17,18]

The principle of IDMS is displayed in its simplified form in Figure 8.6. The sample with known isotopic composition of the analyte element but unknown element content is mixed with a known amount of the spike. This spike contains the analyte element in a different isotopic composition – ideally enriched in the less-abundant natural isotope. After complete mixing

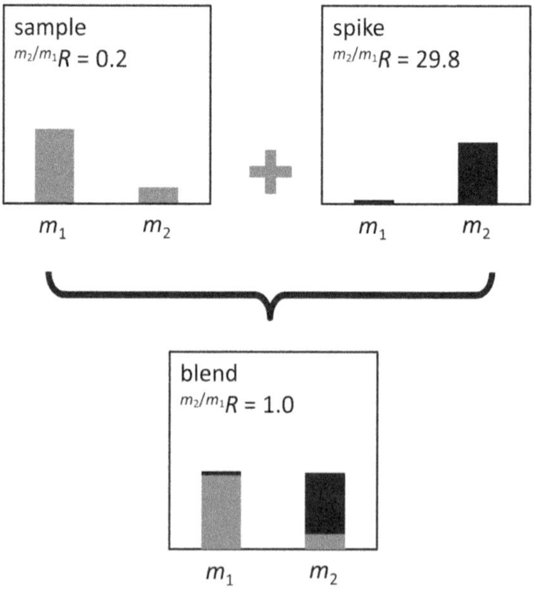


**Figure 8.6** Schematic principle of IDMS.

of sample and spike, the sample–spike blend isotope ratio directly reflects the analyte concentration in the sample.

The final calculation of the amount of analyte is accomplished by the "IDMS equation" in different approaches:

Equation (8.9) gives the functional relationship between the mass fraction in the sample and the blend isotope ratio, as used by BAM.[15]

$$w_x = w_{y,b} \times \frac{M_x \times m_y}{M_b \times m_x \times a_{x,b}} \times \frac{(R_y - R_{xy})}{(R_{xy} - R_x)} \tag{8.9}$$

The most frequently used alternative equations in literature are eqn (8.10) according to Heumann:[19,20]

$$w_x = \frac{10^6}{N_A} \times \frac{M_x}{m'_x} \times N_y \times \frac{a_{y,a} - a_{y,b} \times R_{xy}}{a_{x,b} \times R_{xy} - a_{x,a}} \tag{8.10}$$

Equation (8.10) according to de Bièvre *et al.*:[21,22]

$$c_x = c_y \times \left(\frac{m_y}{m_x}\right) \times \frac{(R_y - R_{xy})}{(R_{xy} - R_x)} \times \left(\frac{\sum_{i=1}^{n} R_{xi}}{\sum_{i=1}^{n} R_{yi}}\right) \tag{8.11}$$

or eqn (8.12) according to Paulsen and Kelly:[23]

$$w_x = \frac{m'_{y,element}}{m_x} \times \frac{M_x}{M_y} \times \frac{a_{y,a} - a_{y,b} \times R_{xy}}{a_{x,b} \times R_{xy} - a_{x,a}} \tag{8.12}$$

$N_a$, $N_b$: number of atoms of isotope $a$ and $b$
$N_x$, $N_y$: number of atoms in sample, in spike
$a_{x,a}$, $a_{y,a}$: abundance of isotope $a$ in sample and in spike
$a_{x,b}$, $a_{y,b}$: abundance of isotope $b$ in sample and in spike
$R_{xy}$: isotope ratio $(a/b)$ in sample-spike blend
$R_x$, $R_y$: isotope ratio $(a/b)$ in the sample and in the spike
$m_x$: mass of sample/kg
$m_y$: mass of spike solution/kg
$m_x'$: mass of sample/mg
$m'_{y,\mathrm{element}}$: mass of spike element/mg
$c_x$, $c_y$: amount content of the element in the sample, in the spike/mol kg$^{-1}$
$M_x$, $M_y$: standard atomic weight of the element in sample and in spike
$M_b$: atomic weight of isotope $b$
$w_x$, $w_y$: mass fraction of the element in the sample and in the spike/mg kg$^{-1}$
$w_{y,b}$: mass fraction of isotope $b$ in the spike/mg kg$^{-1}$

These equations only consist of isotope ratios, masses of sample and spike and tabulated quantities (isotope abundances, atomic weights).[24,25] It has to be noted here that the different forms are not redundant, but reflect directly the measurement strategy. In eqns (8.10) and (8.12) both a measured isotope ratio and tabulated isotope abundances are used. As a consequence, the measured ratio has to be translated into an absolute ratio (*i.e.* mainly corrected for instrumental isotopic fractionation). Therefore, eqns (8.10) and (8.12) are used when only $R_{xy}$ is determined. In eqns (8.9) and (8.11) only isotope ratios are used. In eqn (8.11), the measured isotope ratios can be used as measured. The correction factors for mass fractionation/discrimination will cancel out, if all isotope ratios have been measured within the analytical sequence. In order to avoid matrix-dependent instrumental isotopic fractionation, the matrix of sample, spike and blend has to match as close as possible. (If tabulated values are taken as they are (*e.g.* from certificates), the measured blend ratios have to be corrected to obtain absolute ratios, accordingly).[26]

Therefore, IDMS only requires the measurement of (1) isotope ratios and (2) of the mass. Once the complete isotopic equilibrium established, a possible loss of analyte does not affect the accuracy of the analytical result as the measurand – the blend isotope ratio – is equal in all subsamples of the blend. The accuracy of the result is therefore independent of, *e.g.* the recovery during analyte/matrix separation (if no massive fractionation processes are occurring). As a consequence, the spike may also be regarded as the optimum internal standard, because it has the same physical and chemical properties. Thus, it can compensate for all types of physical and chemical effects, such as losses during sample preparation or transport. Additionally, it also compensates for plasma effects because it has the same ionisation energy and identical mass.

The isotope dilution technique requires the determination of at least two isotopes, ideally completely free from any isobaric or polyatomic interference. In some cases even monoisotopic elements can be determined, if a long-living radionuclide such as $^{129}I$ is available. A prerequisite for the successful application of IDMS is that at least at one stage of the procedure a complete homogenisation of the blend is established (*i.e.* a complete mixture of sample and spike isotopes). Using a digestion step in the elemental analysis most commonly ensures this. Moreover, the measured isotopes should be preferably available in the same form or species to reach an isotopic equilibrium. (When, *e.g.*, Cr(VI) is present in the sample and the spike consists of Cr(III) it has to be guaranteed that Cr is present in the sample and spike in the same form (species) before analyte–matrix separation is accomplished.)

In order to determine the optimum blend ratio, the uncertainty of the ratios (sample, spike, blend) is propagated with the goal to find a blend ratio with a given sample and spike ratio to obtain a minimum uncertainty. This leads to an expression describing the variation of an "error-magnification factor" $\varepsilon$ as a function of the blend ratio. The error-magnification factor $\varepsilon$ is defined as the ratio between the relative precision of $(\sigma(N_x/N_y)/(N_x/N_y))$ to the relative precision of the measurement of an isotope ratio $(\sigma(R)/R)$.

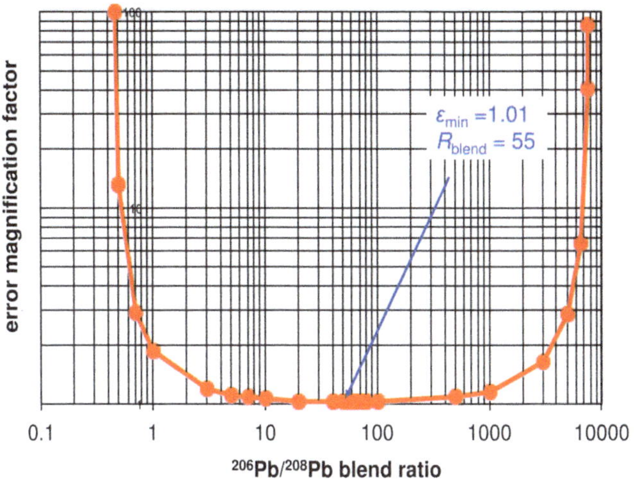


**Figure 8.7** Schematic of the error magnification factor $\varepsilon$ on the example of a $^{206}Pb/^{208}Pb$ ratio.

The magnification factor $\varepsilon$ allows computing the expected uncertainty on a given quantitative determination of an element by IDMS *via* a simplified equation (Figure 8.7):

$$\varepsilon = \frac{(R_{sample} - R_{spike}) \cdot R_{blend}}{(R_{blend} - R_{sample}) \cdot (R_{spike} - R_{blend})} \quad (8.13)$$

$\varepsilon$: error magnification factor
$R$: isotope ratio ($a/b$) in the sample, the spike and the blend

In IDMS, four main calibration strategies are applied: single IDMS, double IDMS, triple IDMS and matching IDMS: (1) Single IDMS is the most simple and straightforward approach, and therefore is the most common calibration approach in routine applications. A certified or calibrated spike is added to the sample, the digestion step is performed (or another equivalent sample-preparation step to ensure isotopic equilibration) and the blend isotope ratio between two isotopes is determined in the blend. The mass fraction can be calculated by using eqns (8.9)–(8.12). In this case, the measurement uncertainty typically is the largest among the four calibration strategies, whereas the effort is smallest. (2) In double IDMS, the spike is characterised by preparing a blend with a high-purity elemental standard with natural composition (named "back-spike"). This blend is now measured for the blend isotope ratio. Introducing the results in the IDMS equation of the sample, the isotopic abundances and atomic weights cancel out, leading to significantly lower uncertainties as compared to single IDMS by at least one order of magnitude. (3) In matching IDMS, the sample – spike blend is iteratively changed until the sample–spike blend ratio is close to the spike–back-spike blend ratio before double IDMS is accomplished. The goal is to avoid uncertainties (*e.g.* dead time) when different ratios are measured. The advantages are clearly adumbrated by the time-consuming preparation. (4) In triple IDMS, the spike ratio does not need to be determined. This can be useful in case of a small isotopic enrichment in the spike (*e.g.* in enriched organic species). The uncertainty is improved only marginally as compared to double IDMS and is significantly more laborious.[27,28] For more details on the calibration strategies and on the blank calculations the reader is referred to the literature.[15,27,29] Independent of the calibration strategy, a blank of the complete analytical procedure has to be measured and usually the blank can be spiked and treated exactly like a sample.

Meanwhile, the isotope dilution technique has become a powerful online method in ICP-MS. The equations have to be modified as the sample and spike masses are replaced by the mass flow. By online spiking of solutions with a multi-isotope spike, routine multielemental analysis is accomplished whereby the spike flow is determined by reverse IDMS.[18] In addition, online IDMS has become popular for the quantification of elemental species *via* the coupling of chromatographic systems to a mass spectrometer. Applications where liquid chromatography (LC) is coupled with ICP-MS as was demonstrated for the first time in 1994[30] and had developed to a gold standard in speciation analysis. The sample flow – usually at the outlet of a chromatographic system – is mixed with a continuous spike flow, which even enables the IDMS analysis of species that are unknown or not synthesisable. Additional advantages and disadvantages occur and higher measurement uncertainties have to be faced. For more details the reader is referred to the current literature.[31,32]

## 8.3 Isotope Ratio Measurements

In isotope amount ratio analyses, calibration of the mass spectrometer is achieved by establishing a relationship between the "true" isotope amount ratio of a certified standard or a stable isotope amount ratio in the sample and the value obtained from the measurement of the same isotope amount ratio. This is necessary since all mass spectrometric techniques are subject to the effect of mass discrimination resulting in a systematic shift from the "true" isotope amount ratio in a sample/or standard to the measured isotope amount ratio in the same. This deviation is referred to as instrumental isotopic fractionation in this book (see Chapter 6). With the objective to calibrate a mass spectrometer for the effect of mass discrimination, a number of different strategies have been proposed over the years, of which a few have developed as frequently applied approaches for isotope amount ratio analysis. The different approaches take also into account whether absolute or relative ratios are of interest.

### 8.3.1 Isotope Ratio Determination – Calibration Strategies

As a basic rule, the calibration for isotope amount ratios can be done by measuring the measurand (isotope ratio of interest in the sample) and the calibrant (reference isotope ratio in a standard or sample) simultaneously or sequentially, also often referred to as internal or external calibration. The calibrant can either be an isotope pair of the same element or another element (close in mass), which will be referred to as "intraelemental" and "interelemental" calibration in the following. The terminology is in accordance with Meija *et al.*[33]

**Internal Calibration of Isotope Amount Ratios (Internal Instrumental Isotopic Fractionation Correction)**

The major requirements to perform internal instrumental isotopic fractionation correction – that is to say to use an invariant isotope ratio of the same element (intraelemental internal instrumental isotopic fractionation correction) or a dopant element (interelemental internal instrumental isotopic fractionation correction) to correct for instrumental isotopic fractionation – are similar fractionation behaviours and chemical properties – which is true for isotopes of the same element but might differ significantly between two elements. In interelemental internal instrumental isotopic fractionation correction, the calibrant element can be part of the sample already or can be added to the sample before measurement. In the latter case, the standards are preferably certified reference materials with certified isotopic composition, which are available for a range of commonly investigated elements. Alternatively, "natural" standards can be used and their isotopic composition taken from the IUPAC tables taking into account the uncertainty that is generally reflected by the natural isotopic variation. The use of isotope pairs of an element close in mass to correct instrumental isotopic fractionation of the isotope ratio of interest has become a common technique, *e.g.* in (MC) ICP-MS. In addition, this approach serves as a tool to correct the bias for elements that do not possess an invariant isotope pair that can be used for correction by internal instrumental isotopic fractionation correction. However, the assumption of having the same fractionation behaviour requires that the fractionation factor $f$ (see Chapter 6, eqn (6.2a)) is invariant for the element of interest and the spike element. Increasingly, it has been shown that fractionation effects are strongly reliant on mass. The correlation of fractionation and mass shows that mass bias per mass unit decreases with increasing atomic number[34] and should be corrected for accordingly.

Intraelemental internal instrumental isotopic fractionation correction has long been applied *e.g.* for Sr isotope ratio measurements by using the $^{88}Sr/^{86}Sr$ isotope ratio in the sample for calculating the fractionation factor $f$ and correcting the $^{87}Sr/^{86}Sr$ isotope ratio (which has long been the only Sr isotope ratio of interest) with the same fractionation factor (Figure 8.8(a)). However, this calibration approach requires that the internal pair of isotopes used for mass fractionation correction is stable. The $^{88}Sr/^{86}Sr$ isotope ratio, however, underlies natural fractionation, therefore it cannot be used for internal mass bias correction of natural samples. Alternatively, Sr isotope ratios can be corrected for mass bias by using another element, which is

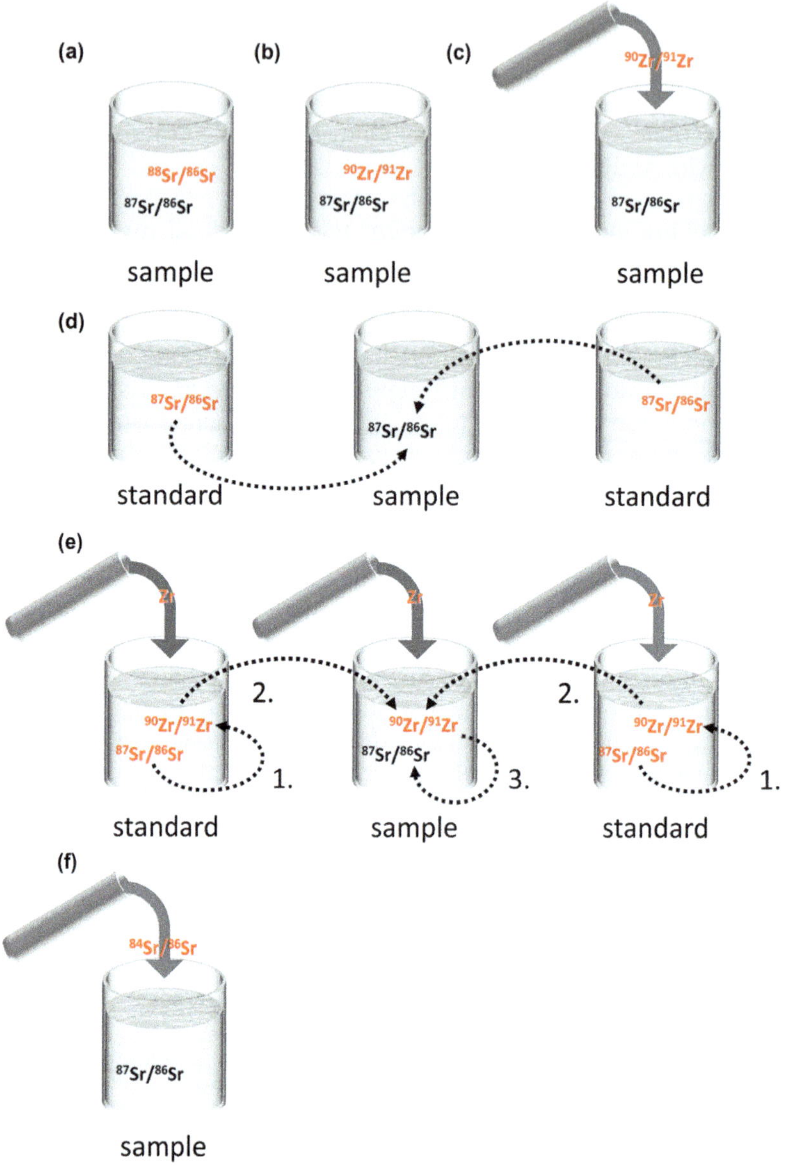


**Figure 8.8** Calibration strategies for isotope ratio determination: (a) internal calibration using the same element, (b) internal calibration using different element, (c) internal calibration with added element, (d) external calibration, (e) combined internal and external calibration and (f) double spike calibration. Red colour represents calibrants, black font stands for analyte.

either already present in the sample solution (Figure 8.8(b)) or added to the solution for internal intraelemental correction (Figure 8.8(c)). This approach has already been applied by adding zirconium and either using $^{92}$Zr/$^{91}$Zr,

$^{92}Zr/^{90}Zr$ or $^{91}Zr/^{90}Zr$.[35] In this approach, the mass difference between Sr and Zr might, however, significantly influence the degree of mass bias. The addition of Tl to sample solutions of Pb also sets an often-applied example for internal interelement calibration, whereby the mass discrimination factor obtained for Tl serves as a correction coefficient for the observed Pb isotope ratio.[34,36,37] Another example is set by Cu and Zn,[38–40] where mass bias is often corrected by means of doping experiments.

**External Calibration of Isotope Amount Ratios**

In this approach, a solution containing an isotopic standard of the target element (with a known isotopic composition) is measured following a bracketing procedure with the sample and the isotopic standard in a consecutive sequence in most cases (*i.e.* intraelemental calibration). A correction factor is calculated on the basis of the observed bias between the measured value and the true value of the isotope ratio of interest and subsequently applied to the sample measured in between (eqns (8.14a) and (8.14b)).

The advantage of external calibration in isotope ratio analysis lies in the simplicity of the strategy. However, it can be inferior in accuracy compared to other strategies such as double-spike calibration.

$$K = \frac{R_{\text{standard, certified}}}{R_{\text{standard, measured}}};\, R_{\text{sample, corrected}} = R_{\text{sample, measured}} \times K \tag{8.14a}$$

$$R_{\text{sample, corrected}} = R_{\text{standard, certified}} \times \frac{R_{\text{sample, measured}}}{\dfrac{R_{\text{standard1, measured}} + R_{\text{standard2, measured}}}{2}} \tag{8.14b}$$

*K*: correction factor
*R*: isotope ratio

In the example of Sr isotope ratio measurements, external calibration (Figure 8.8(d)) is usually done by using the only commercially available isotope certified reference material available for calibration, which is provided by NIST (National Institute of Standards and Technology). The NIST SRM 987 is delivered as $SrCO_3$ powder with a certified natural Sr isotopic composition. Typically, solutions of NIST SRM 987 are measured before and after a sample, which is subsequently corrected for mass bias by taking a mean of the detected isotope ratios of interest in the standard as a measure for instrumental mass fractionation (eqn (8.14b)).

As matrix exerts quite a pronounced influence on the extent of mass discrimination (depending on the mass of the isotopes), it is preferable that the target element is isolated from the matrix in the sample. If this is not possible, matrix and concentration matching between the standard and the sample should be fulfilled, which is usually rather easy to accomplish when solutions are measured. Isotope ratio analysis usually demands the existence of an isotope certified reference material, which is often the central issue. Besides, the isotope ratio of the standard should be as close as possible to the isotopic composition of the sample. When, *e.g.*, the isotopic composition of the sample analyte is very different from that of the available standard reference materials, the mass spectrometer can be calibrated using gravimetrically prepared mixtures of synthetic isotopes (of known purity) from which the isotopic composition can be calculated and a correction factor can be deduced.[41,42] Only recently, a comprehensive overview of the advantages and characteristics of gravimetric blending of isotopes as the primary calibration method for isotope amount ratios measurements was published.[43]

**Combining Internal and External Calibration Strategies**

Combining external and internal mass bias correction can set a promising alternative, combining the advantages of the two strategies and has also been presented for Sr isotope amount ratio analysis. In this case, Zr is added to samples where Sr isotope ratios are to be measured and the Zr isotope ratios are used to correct for the mass bias on Sr, still taking a bracketing standard (in this case NIST SRM 987 containing Zr as well) into consideration.[44–47] The advantage lies in the possibility to correct for short-term drifts in mass bias using the internal (interelement) approach through the addition of Zr and still performing sample-standard-bracketing using the same element (Sr) for mass-bias correction (Figure 8.8(e)).

**Double-Spike Calibration**

The double-spike technique is a well-established technique to correct for instrumental mass discrimination and is applicable to any element that has four or more isotopes. The isotopic composition (and preferably also the concentration) of the added double spike is known, the added element should not interfere with any isotopes of interest (Figure 8.8(f)). The determining factors on the performance of the double-spike technique lie in the choice of isotopic composition to be added as well as in the proportion in which the sample and the double spike are mixed. Recommendations about how to well design a double spike experiment were described amongst others by Rudge *et al.*[48]

When the concentration of the added spike is known as well, both, the isotopic composition as well as the elemental content can be determined in one analysis (by making use of the isotope-dilution equation). Double-spike analysis is applied for many different mass spectrometric techniques and stands out to the mitigation of matrix effects in combination with low uncertainties. However, the analyst has to make sure that the sample and the double spike are in chemical equilibrium before starting the measurement. Again, similar behaviour of the isotope of interest and the added isotopes in the double spike should preferably be met during sample analysis. Details on the use of double spikes in TIMS analysis can be seen in Chapter 14.

**Other Approaches**

Isotope ratio evaluation based on three-isotope plots has been used as evaluation approach to visualise natural fractionation processes between isotope pairs (of the same element), above all in geochemistry, for decades. Examples are given in the literature[49–51] where three-isotope plots were also used for the identification of different sources to the total amount of an element including the determination of contribution of the identified sources.

Besides, three isotope plots are commonly used for evaluation of methodological issues such as the selection of appropriate mass fractionation models[52] by the visualisation of mass-dependent *vs.* mass-independent mass fractionation or the decision on the applicability of, *e.g.*, a linear model compared to an exponential model for mass fractionation correction. Another popular approach to picture mass bias behaviour or also isotopic mixing is the use of *ln-ln space plots.*[53] Three-isotope plots can serve to illustrate the presence of interferences, which has been summarised in detail, as well.[54]

Only recently, a novel approach for the measurement of isotope abundance variations in nature by gravimetric spiking isotope dilution (GS-IDA) was published,[55] based on the example of magnesium. Compared to the classic double-spike technique, which requires the measurement of at least four isotopes, this technique only requires the measurement of three isotopes, since the sample and the spike are mixed gravimetrically and thus the equation for isotope dilution can be applied for data analyses.

A simplified alternative evaluation strategy to double-spike isotope-dilution calculations based on isotope pattern deconvolution (IPD) has amongst others been suggested.[56–59] Details about the basic mathematical considerations behind multiple linear regression models being used for evaluating isotope data can be found elsewhere.[59–61] This mathematical approach stands out due to the potential of deducing the individual contribution of each input source to a blend of the natural isotope source and the introduced isotopically enriched tracer, thus allowing for the deconvolution of interconverting analytes, especially in cases where the amount of tracer finally present in the sample is unknown (*e.g.* in biological tracer studies[58]).

## 8.3.2 Relative Notation *vs.* Absolute Notation of Isotope Amount Ratios

The certification of reference materials for their isotopic composition (isotopic reference standards) or the determination of the absolute abundances of isotopes (e.g. for the accurate determination of the atomic masses) requires accurate absolute isotope ratio measurements with minimum uncertainty. In contrast, most studies dealing with the change of the isotopic composition in a system or the differences of the isotopic composition between samples aim at the relative differences of these isotope amount ratios.

In the latter case, delta values (mostly expressed as per mill difference to an international certified reference standard) have become the gold standard, taking into account that these are linked to the same reference standard.[62]

As a consequence, isotope ratio notations in both absolute as well as relative values coexist. As different laboratories perform different data-correction strategies (*e.g.*, for blank, mass bias) international comparability is generally difficult to establish when using absolute isotope ratios even if certified reference materials are being used. In addition, many existing isotope certified reference materials are provided with high uncertainties assigned to the certified values due to, *e.g.*, heterogeneity of the material or certifications dating back several decades and thus provide out-dated uncertainties. This issue can be bypassed when relative isotope ratios are expressed, which increasingly conquers the scientific community. In this way, measurement results become easily comparable on an international level if analysts use reference CRM as anchor to a common reference.

Relative isotope ratios are expressed in delta ($\delta$) values (eqn (8.15)) as the ratio of an unknown sample to a preferably internationally accepted standard and are expressed in parts per thousand (or per mille, ‰) using the following equation

$$\delta({}^{i}E/{}^{j}E)_{\mathrm{sample/standard}} = \left[\frac{R_{\mathrm{sample}} - R_{\mathrm{standard}}}{R_{\mathrm{standard}}}\right] \tag{8.15}$$

$\delta$: delta value
${}^{i}E$: isotope $i$ of the element $E$ (heavier isotope)
${}^{j}E$: isotope $j$ of the element $E$ (lighter isotope)
$R$: isotope ratio of sample or standard (after blank correction)

Alternatively, a notation in epsilon ($\varepsilon$) values as the ratio of an unknown sample to a preferably internationally reference material in parts per ten thousand has arisen during the last years to express very small variations in the isotopic composition. This notation, however, is recommended to be omitted[1]

When isotope ratios are expressed in delta ($\delta$) values no further model for mass-discrimination correction has to be applied, as all sources of mass discrimination are expected to be similar for the used reference material and the sample under investigation and without exactly knowing the absolute isotopic abundances of an element. This approach has an especially long history in stable isotope ratios measurements using gas source mass spectrometry. The relative isotope ratio notation is usually also applied in secondary ionisation mass spectrometry (SIMS) analysis. In the case of (MC) ICP-MS and TIMS measurements delta ($\delta$) values are increasingly used which facilitates comparability of internationally published data since anchoring values are easily comprehensible, which can be more complex in case of absolute isotope ratios. However, in the case of ICP-MS and TIMS, especially when isotope ratios are calibrated for another pair of isotopes of the same element or a pair of isotopes of another element, more sophisticated mass-discrimination models are needed. Therefore, a number of different mathematical models for correcting mass discrimination have been used frequently over the past decades and thus have become conventional approximations.

In the case of Sr, the use of delta ($\delta$) values has mainly been adopted for reporting relative values of the $^{88}Sr/^{86}Sr$ isotope ratio, where the NIST SRM 987 is used as a scale anchor with a delta ($\delta$) value = 0. Delta $\delta(^{87}Sr/^{86}Sr_{NIST\ SRM\ 987})$ values have only rarely been reported so far, but are increasingly used. The use of delta $\delta(^{87}Sr/^{86}Sr_{NIST\ SRM\ 987})$ will enhance international transparency and comparability, also when considering the large uncertainties of the absolute range of $^{87}Sr/^{86}Sr$ in the NIST SRM 987. So far, many publications – especially those focusing on applications in diverse fields of science – lacked transparency with respect to details on the use as well as results of the CRM used. Some sources of uncertainty entailed with absolute isotope ratio measurements will be eliminated when $\delta$ values finally replace absolute values, as the impact of using different strategies for mass-bias correction as well as the impact of uncertainties of reference materials will be significantly reduced or excluded.

Recently, Brand *et al.*[62] have reported a comprehensive list of reference materials, which should be used as primary reference ($\delta = 0$) along with delta values of other reference materials, which are commonly used as delta reference. The primary reference has per definition a $\delta$ of 0 as well as an uncertainty of 0 (even though the uncertainty as a result of the homogeneity of the CRM has to be taken into account for the calculation of combined standard measurement uncertainties).

It has to be noted that in the case of the use of delta values, they account for all mass-fractionation effects such as

- natural mass-dependent fractionation;
- natural mass-independent fractionation;
- radiogenic formation/depletion;
- differences in instrumental mass fractionation (due to matrix effects *e.g.*);
- time-dependent instrumental mass fractionation (short time fluctuations (Standard–sample–standard bracketing)).

This problem is evident if one seeks to make a difference between, *e.g.*, the radiogenic effect and natural mass-dependent fractionation as occurs in, *e.g.*, Sr isotopes or U isotopes. The effect on the reported $\delta(^{87}Sr/^{86}Sr)$ was recently discussed by Neymark *et al.*[63] The authors conclude that it is not possible to separate the mass-fractionation effect and the radiogenic ingrowth effect for $^{87}Sr/^{86}Sr$ and, therefore, the $\delta(^{87}Sr/^{86}Sr)$ based on the "true" $^{87}Sr/^{86}Sr$ ratio determined using external correction for mass-dependent instrumental isotopic fractionation in double-spike TIMS or MC ICPMS does not bear any useful information about Sr mass fractionation in nature.

### 8.3.3 Mass-Fractionation Correction Laws

Throughout the continuity of isotope ratio mass spectrometry, different correction models for mass bias or mass discrimination have been presented. Most of these models were developed based on TIMS measurements and subsequently adapted for MC ICP-MS, even though the two techniques differ significantly with respect to extent and nature of mass discrimination observed. Both techniques stand out due to the possibility of simultaneous detection of multiple ion beams, since these instrumentations are usually designed with arrays of several detectors. In conrast to TIMS, isotope ratios measured by MC ICP-MS do not show a varying mass bias with time of a single measurement. However, MC ICP-MS suffers from much larger bias over longer time periods and is also strongly dependent on the nuclide mass of the isotope(s) measured.

In this section, a brief overview of commonly used models will be given. The advantages and shortcomings of different models have been extensively discussed in the literature.[33,64]

Three main calibration models that are frequently being used for TIMS and MC ICP-MS measurements are the linear law (eqn (8.16)), the Power (eqn (8.17a)) or exponential law (eqn (8.17b)) and Russell's equation (eqn (8.18)), which are selected for a specific measurement question depending on the assumption about the underlying function of mass discrimination present.

$$R_{\text{standard, certified}}/R_{\text{standard, measured}} = (1 + \varepsilon_{\text{linear}} \cdot \Delta m)\,; \qquad R_{\text{sample, corrected}} = R_{\text{sample, measured}} \cdot (1 + \varepsilon_{\text{linear}} \cdot \Delta m) \tag{8.16}$$

$$R_{\text{standard, certified}}/R_{\text{standard, measured}} = (1 + \varepsilon_{\text{power}})^{\Delta m}\,; \qquad R_{\text{sample, corrected}} = R_{\text{sample, measured}} \cdot (1 + \varepsilon_{\text{power}})^{\Delta m} \tag{8.17a}$$

$$R_{\text{standard, certified}}/R_{\text{standard, measured}} = \mathrm{e}^{\varepsilon \cdot \Delta m}\,; \qquad R_{\text{sample, corrected}} = R_{\text{sample, measured}} \cdot \mathrm{e}^{\varepsilon \cdot \Delta m} \tag{8.17b}$$

$$R_{\text{standard, certified}}/R_{\text{standard, measured}} = (m_1/m_2)^{\beta}\,; \qquad R_{\text{sample, corrected}} = R_{\text{sample, measured}} \cdot (m_1/m_2)^{\beta} \tag{8.18}$$

$R$: measured or certified isotope ratio of sample or standard
$\Delta m$: mass difference
$m_{1,2}$: mass of isotope 1 and 2
$\varepsilon$: correction factor for instrumental isotopic fractionation
$\beta$: correction factor for instrumental isotopic fractionation (Russell's law)

It should be noted that in the literature the power and exponential laws have been used as two separate models, even though based on the same mathematical considerations. Other fractionation laws use different approaches such as the equilibrium fractionation law or the Rayleigh fractionation law (applying different approximations). In the latter, the mass-dependent fractionation depends on the square root of the masses involved.[65–67]

There is an ongoing debate as to which correction method provides the most accurate result. It is evident that the different instrumentations (*i.e.* ion sources) demand different laws. This is an issue of increasing importance as a result of the continuously improving isotope ratio precision attainable.

All these models correct for mass-dependent mass bias but do not consider mass-independent mass fractionation (as discussed in Chapter 6).

The diversity of coexisting mass bias correction approaches shows the inconsistency in opinions in the field as well as the challenge to pick the most accurate description of mass-discrimination effects for a specific analytical scenario. Meija *et al.*[33] gave a comprehensive overview of measurement strategies and mass-bias-correction models when modern methodologies are applied.

# References

1. T. B. Coplen, *Rapid Commun. Mass Spectrom.*, 2011, **25**, 2538–2560.
2. K. Murray, R. K. Boyd, M. N. Eberlin, G. J. Langley, L. Li and Y. Naito, *Pure Appl. Chem.*, 2013, **85**, 1515–1609.
3. Joint Commitee for Guides in Metrology – ISO, *Evaluation of measurement data — Guide to the expression of uncertainty in measurement*, JCGM100:2008, 2012.
4. M. Betti, *Int. J. Mass Spectrom.*, 2005, **242**, 169–182.
5. F. Claverie, J. Malherbe, N. Bier, J. L. Molloy and S. E. Long, *Anal. Chem.*, 2013, **85**, 3584–3591.
6. W. Hyk and Z. Stojek, *Anal. Chem.*, 2013, **85**, 5933–5939.
7. A. Sayago, M. Boccio and A. G. Asuero, *Crit. Rev. Anal. Chem.*, 2004, **34**, 39–50.
8. K. Danzer and L. A. Currie, *Pure Appl. Chem.*, 1998, **70**, 993–1014.
9. R. T. Adcock, *The Analyst*, 1878, **5**, 53–54.
10. K. A. Kermack and J. B. Haldane, *Biometrika*, 1950, **37**, 30–41.
11. A. G. Worthing and J. Geffner, *Treatment of Experimental Data*, Jon Wiley & Sons, Inc., New York, 1943.
12. D. York, *Can. J. Phys.*, 1966, **44**, 1079–1086.
13. D. York, *Earth Planet. Sci. Lett.*, 1968, **5**, 320–324.
14. D. York, N. M. Evensen, M. L. Martínez and J. De Basabe Delgado, *Am. J. Phys*, 2004, **72**, 367–375.
15. J. Vogl, *J. Anal. Atom. Spectrom.*, 2007, **22**, 475–492.
16. J. Vogl and W. Pritzkow, *MAPAN*, 2010, **25**, 135–164.
17. K. G. Heumann, *Anal. Bioanal. Chem.*, 2004, **378**, 318–329.

18. J. I. G. Alonso and P. Rodriguez-Gonzalez, *Isotope Dilution Mass Spectrometry*, Royal Society of Chemistry, 2012, 453.
19. K. Heumann, *Inorganic Mass Spectrometry*, Wiley, New York, 1988, 301.
20. K. G. Heumann, *TRAC Trends Anal. Chem.*, 1982, **1**, 357–361.
21. P. De Bièvre, in conf. proc. of *Analytical Proceedings*, Royal Society of Chemistry, 30, pp. 328–333.
22. P. De Bièvre, *Tech. Instrum. Anal. Chem.*, 1994, **15**, 169–183.
23. P. Paulsen and W. Kelly, *Anal. Chem.*, 1984, **56**, 708–713.
24. J. R. de Laeter, J. K. Böhlke, P. De Bièvre, H. Hidaka, H. Peiser, K. Rosman and P. Taylor, *Pure Appl. Chem.*, 2003, **75**, 683–800.
25. R. Loss, *Pure Appl. Chem.*, 2003, **75**, 1107–1122.
26. J. Diemer, C. Quétel and P. Taylor, *J. Anal. Atom. Spectrom.*, 2002, **17**, 1137–1142.
27. J. Vogl, *Rapid Commun. Mass Spectrom.*, 2012, **26**, 275–281.
28. C. Frank, O. Rienitz, C. Swart and D. Schiel, *Anal. Bioanal. Chem.*, 2013, **405**, 1913–1919.
29. J. Vogl and W. Pritzkow, *J. Anal. Atom. Spectrom.*, 2010, **25**, 923–932.
30. K. G. Heumann, L. Rottmann and J. Vogl, *J. Anal. Atom. Spectrom.*, 1994, **9**, 1351–1355.
31. K. G. Heumann, S. M. Gallus, G. Rädlinger and J. Vogl, *Spectrochim. Acta Part B: Atom. Spectrosc.*, 1998, **53**, 273–287.
32. G. Koellensperger, S. Hann, J. Nurmi, T. Prohaska and G. Stingeder, *J. Anal. Atom. Spectrom.*, 2003, **18**, 1047–1055.
33. J. Meija, L. Yang, Z. Mester and R. E. Sturgeon, Correction of Instrumental Mass Discrimination for Isotope Ratio Determination with Multi-Collector Inductively Coupled Plasma Mass Spectrometry, in *Isotopic Analysis*, Wiley-VCH Verlag GmbH & Co. KGaA, 2012, pp. 113–137.
34. T. Hirata, *Analyst*, 1996, **121**, 1407–1411.
35. T. Ohno and T. Hirata, *Anal. Sci.*, 2007, **23**, 1275–1280.
36. N. S. Belshaw, P. A. Freedman, R. K. O'Nions, M. Frank and Y. Guo, *Int. J. Mass Spectrom.*, 1998, **181**, 51–58.
37. G. D. Kamenov, P. A. Mueller and M. R. Perfit, *J. Anal. Atom. Spectrom.*, 2004, **19**, 1262–1267.
38. C. Archer and D. Vance, *J. Anal. Atom. Spectrom.*, 2004, **19**, 656–665.
39. C. N. Maréchal, P. Télouk and F. Albarède, *Chem. Geol.*, 1999, **156**, 251–273.
40. K. Peel, D. Weiss, J. Chapman, T. Arnold and B. Coles, *J. Anal. Atom. Spectrom.*, 2007, **23**, 103–110.
41. W. Pritzkow, S. Wunderli, J. Vogl and G. Fortunato, *Int. J. Mass Spectrom.*, 2007, **261**, 74–85.
42. E. Ponzevera, C. R. Quetel, M. Berglund, P. D. P. Taylor, P. Evans, R. D. Loss and G. Fortunato, *J. Am. Soc. Mass Spectrom.*, 2006, **17**, 1412–1427.
43. J. Meija, *Anal. Bioanal. Chem.*, 2012, **403**, 2071–2076.
44. J. Irrgeher, T. Prohaska, R. E. Sturgeon, Z. Mester and L. Yang, *Anal. Methods*, 2013, **5**, 1687–1694.

45. H. C. Liu, C. F. You, K. F. Huang and C. H. Chung, *Talanta*, 2012, **88**, 338–344.
46. L. Yang, C. Peter, U. Panne and R. E. Sturgeon, *J. Anal. Atom. Spectrom.*, 2008, **23**, 1269–1274.
47. A. Y. Kramchaninov, I. V. Chernyshev and K. N. Shatagin, *J. Anal. Chem.*, 2012, **67**, 1084–1092.
48. J. F. Rudge, B. C. Reynolds and B. Bourdon, *Chem. Geol.*, 2009, **265**, 420–431.
49. M. F. Miller, *Geochim. Cosmochim. Acta*, 2002, **66**, 1881–1889.
50. D. Rumble, M. F. Miller, I. A. Franchi and R. C. Greenwood, *Geochim. Cosmochim. Acta*, 2007, **71**, 3592–3600.
51. Y. Erel, A. Veron and L. Halicz, *Geochim. Cosmochim. Acta*, 1997, **61**, 4495–4505.
52. F. Albarède and B. Beard, *Rev. Mineral. Geochem.*, 2004, **55**, 113–152.
53. T. F. D. Mason, D. J. Weiss, M. Horstwood, R. R. Parrish, S. S. Russell, E. Mullane and B. J. Coles, *J. Anal. Atom. Spectrom.*, 2004, **19**, 218–226.
54. M. Wieser, J. Schwieters and C. Douthitt, Multi-Collector Inductively Coupled Plasma Mass Spectrometry, in *Isotopic Analysis*, Wiley-VCH Verlag GmbH & Co. KGaA, 2012, pp. 77–91.
55. G. Chew and T. Walczyk, *Anal. Chem.*, 2013, **85**, 3667–3673.
56. H. Hintelmann and R. Evans, *Fresenius' J. Anal. Chem.*, 1997, **358**, 378–385.
57. S. Stürup, C. Chen, J. Jukosky and C. Folt, *Int. J. Mass Spectrom.*, 2005, **242**, 225–231.
58. J. Irrgeher, A. Zitek, M. Cervicek and T. Prohaska, *J. Anal. At. Spectrom.*, 2014, **29**, 193–200.
59. J. Meija, L. Yang, J. A. Caruso and Z. Mester, *J. Anal. Atom. Spectrom.*, 2006, **21**, 1294–1297.
60. J. Meija, *Anal. Bioanal. Chem.*, 2006, **385**, 486–499.
61. J. A. Rodriguez-Castrillón, M. Moldovan, J. Ruiz Encinar and J. I. García Alonso, *J. Anal. Atom. Spectrom.*, 2008, **23**, 318–324.
62. W. A. Brand, T. B. Coplen, J. Vogl, M. Rosner and T. Prohaska, *Pure Appl. Chem.*, 2014, **86**, 425–467.
63. L. A. Neymark, W. R. Premo, N. N. Mel'nikov and P. Emsbo, *J. Anal. Atom. Spectrom.*, 2014, **29**, 65–75.
64. J. Meija, L. Yang, R. Sturgeon and Z. Mester, *Anal. Chem.*, **81**(16), 6774–6778.
65. M. Ranen and S. Jacobsen, in conf. proc. of *Lunar and Planetary Science Conference*, 39, p. 1954.
66. K. Habfast, *Int. J. Mass Spectrom. Ion Phys.*, 1983, **51**, 165–189.
67. G. Cavazzini, *Int. J. Mass Spectrom.*, 2012, **309**, 129–132.

CHAPTER 9

# *Chemical Imaging*

ANDREAS ZITEK,*[a] JÉRÔME ALÉON[b] AND THOMAS PROHASKA*[a]

[a] University of Natural Resources and Life Sciences Vienna (BOKU), Department of Chemistry, Division of Analytical Chemistry, VIRIS Laboratory for Analytical Ecogeochemistry, Tulln, Austria; [b] Centre de Sciences Nucléaires et de Sciences de la Matière (CSNSM), University Paris Sud laboratory in Orsay, France
*Email: andreas.zitek@boku.ac.at; thomas.prohaska@boku.ac.at

Visualisation is probably one of the most powerful approaches for analysing the physical, chemical and biological world around us.[1] Therefore, the creation of chemical images of different physical and biological structures has gained significant importance in a variety of disciplines. In principle, "chemical imaging" refers to the creation of two- or three-dimensional visual representations of spatial distributions of chemical features (*e.g.* elements, isotopes, molecules or other species) in any kind of sample.

Imaging, from a generic viewpoint, can be described as measuring contrasts (*e.g.* reflected light or the varying occurrence and concentration of different chemical compounds) at a given surface. This measuring can be done simultaneously (referred to as "global imaging" represented by microscopy techniques) or by measuring properties of individual points sequentially and by subsequently combining this point information to create an image (referred to as "mapping").[2]

The principle of "chemical imaging" can be applied across multiple scales combining multiple information levels and multiple commodities, *e.g.* by combining the information from single cells to higher hierarchical levels,

New Developments in Mass Spectrometry No. 3
Sector Field Mass Spectrometry for Elemental and Isotopic Analysis
Edited by Thomas Prohaska, Johanna Irrgeher, Andreas Zitek and Norbert Jakubowski

Published by the Royal Society of Chemistry, www.rsc.org

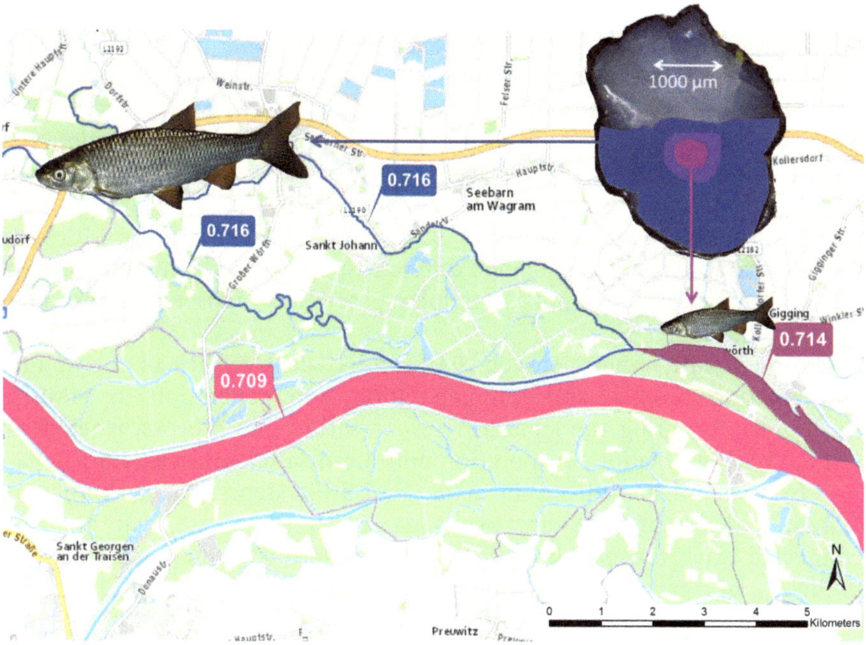


**Figure 9.1** Aquatic isoscape for the $^{87}Sr/^{86}Sr$ isotope ratio composition in a Danube-oxbow-tributary system in Austria matched with time-resolved otolith chemistry data gathered by LA-MC ICP-SFMS describing a habitat shift of an individual fish (map data source: basemap.at, http://www.basemap.at).

like a complete animal,[3] or by linking single compartments or plant and animal species to landscape information, *e.g.* with modelled geographical continuous pictures of the distribution of elements and isotopes ("isoscapes" as geographical images of distributions of elements and isotopes)[4,5] (Figure 9.1).

Spectroscopic (*e.g.* XRF, SEM) or mass spectrometric methods (*e.g.* LA-ICP-MS, SIMS) can be used to achieve spatially resolved images of the distribution of chemical species.[6–10] Secondary ion mass spectrometry (SIMS) or laser ablation inductively coupled plasma sector field mass spectrometry (LA-ICP-SFMS) are sensitive surface analytical techniques capable of performing direct solid analysis at the micrometre down to the nanometre scale.[11] For example, the *NanoSIMS* (CAMECA, Genevilliers, France) is a magnetic sector field multicollector SIMS providing the ultimate resolution down to 50 nm combined with high sensitivity, multielement and molecular-species collection capability. Thus, isotope studies of certain elements can be conducted with 50–100 nm resolution, making the *NanoSIMS* an indispensable tool in many research fields.[12] The small amount of detected signals at the 50–100 nm scale, even with the high sensitivity of the NanoSIMS (useful yield > 10%), precludes high-precision determinations of low-abundance elements (*e.g.* minor isotopes, trace elements). At moderate lateral resolution (1–5 μm) conventional SIMS and MC SIMS (multicollector SIMS) provide both the capability of chemical/isotopic imaging of spatial differences and precise quantification.[13] An alternative approach is to use LA-ICP-MS analysis as a complementary technique to SIMS, where SIMS is used to detect relative spatial differences with high spatial resolution and LA-ICP-MS is used for quantification[1,14,15] with a spatial resolution usually ranging between 2 and 500 μm. Although in some cases, it has been demonstrated that magnetic sector ICP-MS and magnetic sector MC SIMS give comparable results, albeit at a different spatial scale,[16] the sensibility of SIMS to matrix effects implies complex procedural approaches that have their limitations.

The major difference between SIMS and LA-ICP-MS lies in the sputter and ablation process and ion generation. Ion generation in SIMS is accomplished during the sputtering event within one process, whereas in LA-ICP-MS the ablation, evaporation and atomisation processes are separated in time and space from the actual ionisation step.[17] The laser ablation system and the ICP-MS are two separated instruments that are only connected *via* a transport line of the ablated material, which is transported to the ICP by a carrier gas stream (mostly He and/or Ar) (see also Chapter 12 and Chapter 15). Table 9.1 compares some selected parameters of different methods typically used for elemental and isotopic imaging (SIMS, LA-ICP-MS, SEM, XRF and MALDI-MS).

Different denominations for the process of creating chemical images by mass-spectrometric methods can be found in the literature depending on the scientific fields. Chemical imaging has been repeatedly used in SIMS since the early days of SIMS instrumentation, in many cases associated with the need of earth and planetary scientists to study tiny and highly heterogeneous samples.[20–23] In such studies, it is commonly referred to as "ion imaging" or "ion micrography". Depending on the SIMS method employed (see below) it can also be designated as "ion microscopy" or "isotopography".[24,25]

In material and biological sciences, the term "imaging mass spectrometry" has been used especially with respect to the applications of SIMS and MALDI-MS[26–31] for collecting and visualising molecules or relative element distributions in materials science. Becker *et al.*[32] – in reference to the already existing term for chemical imaging by SIMS and MALDI-MS[31] – first used the term "imaging mass spectrometry" for laser ablation ICP-MS, while later the term "mass spectrometric imaging" (MSI) was suggested for SIMS[33] to avoid confusion with the same abbreviation (IMS) used for "ion-mobility separation". Later, MSI was also used for LA-ICP-MS.[34] However, although both terms are still in use, within this book "mass spectrometric imaging" (MSI) is further used. A SCOPUS search showed a significant increase in publications during the last years, with 246 publications containing the terms "mass spectrometric imaging" or "imaging mass spectrometry" in 2013.

**Table 9.1** Comparison of techniques for elemental and isotopic imaging adapted from several authors.[8,17–19]

| Method | Energy source | Particle/ray analysed | Spatial resolution | Depth resolution | Mass analyser | Applications | Quantification | Detection limits | Limitation |
|---|---|---|---|---|---|---|---|---|---|
| SIMS | primary ions | secondary ions | 50 nm–50 μm | >1 nm | quadrupole, **SFMS**, ToF-MS | elements, small molecules, isotope ratios | difficult because of strong matrix effects | ≤ng $g^{-1}$ | strong matrix effects, difficult to quantify |
| LA-ICP-MS | laser | neutrals postionised in ICP | 2–300 μm | >100 nm | quadrupole, **SFMS**, ToF-MS | elements, isotope ratios | possible with matrix-matched standards | μg $g^{-1}$–ng $g^{-1}$ | sample fully ablated, spatial resolution |
| SEM-EDX | electron beam | X-ray | ~5 nm | none | none | elements | difficult | 0.1% (m/m) | low sensitivity |
| (SR)XRF | X-Ray | fluorescent X-ray | ~100 nm | 10 μm | none | elements | difficult | 0.1% (m/m) | quantification, very expensive(in the case of synchrotron radiation) |
| MALDI-MS | photon beam | ions | ~20 μm | >μm | ToF-MS, tandem MS, ion mobility MS, FTICR-MS | biomolecules | difficult | nmol $L^{-1}$–μmol $L^{-1}$ | quantification |

"Chemical imaging" by SIMS and LA-ICP-MS (using sector field mass spectrometers) are explored in more detail in the following sections. The clear advantages of a sector field mass spectrometer are the high sensitivity, the high mass resolution for resolving interferences and the possibility of measuring multiple ions at different *m*/*z* simultaneously. The major disadvantage is the limited time resolution (especially when single-collector instruments are applied and the switching between different *m*/*z* has to be accomplished by changing the magnetic field of the magnet). Multicollector instruments usually allow for the simultaneous measurement of a limited number of different ions with the exception of sector field devices applying a Mattauch–Herzog geometry. With this configuration, almost the whole mass range can be monitored simultaneously by using a focal plane detector (see Section 4.12). Nonetheless, measuring faster with instruments providing less sensitivity is not *a priori* an advantage as counting statistics limits the quality of the measurement especially at low signal intensities (see also Chapter 3). Thus, magnetic sector field MS are still applied where "high-sensitivity mass spectrometry" is required.

## 9.1 Mass Spectrometric Imaging by SIMS

Secondary ion mass spectrometry (SIMS) has been applied for a long time for the analysis of trace elements and isotopes in geological, extraterrestrial rock samples[35–39] and other abiotic materials[40,41] as well as in biological samples.[40,42,43] In particular, its performance characteristics of high sensitivity, good lateral resolution, the ability to detect basically all elements and isotopes and also the possibility for three-dimensional analysis make it an important technique for chemical imaging.[44] NanoSIMS has been designed especially for maximising lateral resolution, sensitivity, and mass resolution.[24] Thus, SIMS can routinely reach spatial resolutions at the submicrometre level. While having been used for more than 40 years in earth, planetary and material sciences, it is now increasingly used in biological studies of the chemical composition at subcellular biological structures.[1] As SIMS is one of the most sensitive surface analytical methods, it is also highly sensitive to surface contamination, notably for volatile and light elements. In that respect, appropriate and careful sample preparation is needed. Before the wide use of laser ablation techniques, it was a chief technique for *in situ* chemical analysis and has the advantage of minimal chemical treatments of the samples. However, the high-vacuum requirements during analysis renders the analysis of fragile biological samples challenging.[42] Therefore, sample preparation has remained a central research area in organic and inorganic SIMS and may differ according to the needs of the study and requiring careful validation.[45–47] The quantification of elements and isotopes by SIMS requires careful calibrations due to strong matrix effects[48–52] (see also Chapter 15). Matrix-matched standards are therefore a prerequisite for accurate quantification. Nonetheless, the lack of the required variety of these materials still presents a major obstacle for accurate quantification by SIMS.

MSI using SIMS is accomplished by sputtering ions from the sample in a spatially resolved manner.[31] SIMS can be operated in static, dynamic and imaging mode (see also Chapter 15). The static and dynamic modes are characterised by different primary doses of ions applied. The static mode is less destructive and used for qualitative imaging. The sample surface is irradiated with very low doses of primary ions retrieving information from a few monolayers of the sample surface yielding near atomic resolution.[10] In the dynamic mode, higher doses of primary ions are used resulting in deeper penetration into the sample surface and actual sputtering of surface layers. This mode is applied for quantitative elemental and isotopic mapping.[13,33,53] Data gained by static SIMS is typically presented as mass spectra, whereas for dynamic SIMS data are provided as mass spectra, depth profiles and images.[10]

### 9.1.1 Key Aspects in the Image Production Process using SIMS

Isotopic and elemental SIMS-images can be created by two processes, which are referred to as (1) "direct ion imaging" or "ion microscopy" (more rarely "isotopography") and (2) "scanning ion imaging" or "ion microprobe imaging" (Figure 9.2 and see also Chapter 15). The first method uses a broad primary ion beam for "global imaging" and the second a scanning focused probe for "mapping" individual points that are later on combined into images.[14,22]

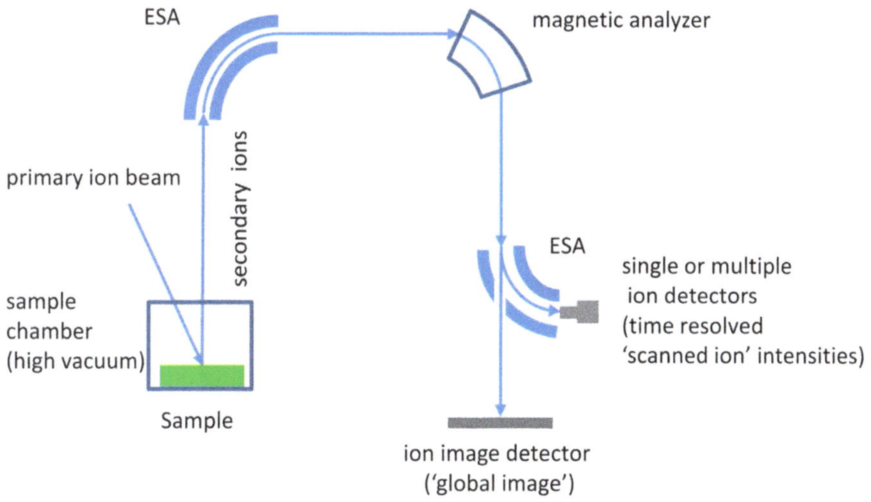


**Figure 9.2** Schematic of the image creation process by secondary ion mass spectrometry (SIMS).

Different commercial instruments are available for SIMS imaging but not all of them are able to perform both "global imaging" and "scanning imaging". Only instruments with an ion optic specifically designed to preserve the sample image within the ion beam along its path in the mass spectrometer (which in fact are the most widely used) can perform "ion microscopy" or "direct imaging". This ion optic is referred to as the transfer optic. The sample image can be recovered using position-sensitive detectors placed inside an image plane. Some instruments are devoid of the transfer optic and can only perform "scanning imaging" (*e.g. NanoSIMS*).

"Global imaging" with SIMS using a broad beam viewing the analysed area simultaneously was described for the first time by Castaing and Slodzian.[54] The shape of ion images acquired in the "ion microscopy" mode depends on the shape of the aperture used to limit the analysed area. This aperture is commonly called the "field aperture" and is placed after the transfer optic at the entrance of the mass spectrometer. In most SIMS instruments, these apertures are discrete round holes of various sizes and the images are typically round shaped (Figure 9.3). Direct images are square in instruments with a motorised tuneable square field aperture (Figure 9.4).

Ion microscopy following the principle of "global imaging" can achieve lateral resolution of 0.5 μm to 1 μm independent of the primary beam width. Another advantage lies in the possibility to reduce measurement times when high sample throughput is needed, *e.g.* during the analysis of degrading (biological) samples.[55] However, to gather statistically valid counts for masses with low count rates, extended counting times per pixel might be

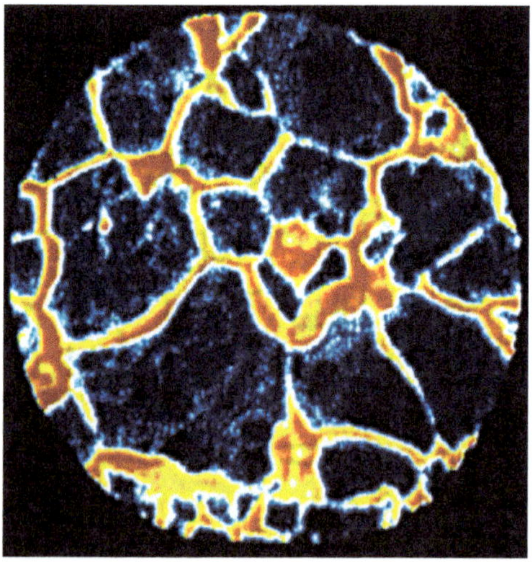

**Figure 9.3** Round image from "ion microscopy", $^{18}O$- mapping in NiO polycrystal, Field of view 150 μm, *Cameca IMS7f.*
Source: © Cameca, by courtesy of Cameca.

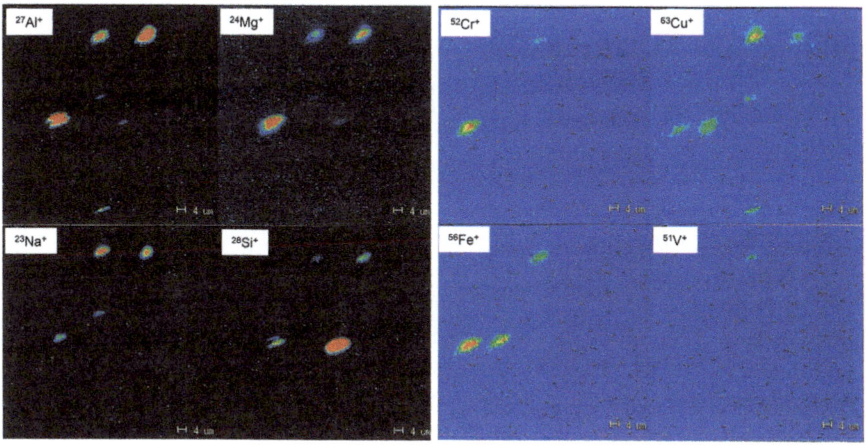


**Figure 9.4** Squared image from "ion microscopy". Elemental mapping of major elements (Al, Mg, Na, Si) and minor elements (Cr, Cu, Fe, V) in airborne desert dust particles collected in Oujda, Morocco and pressed onto a high-purity indium mount for imaging. Positive ions collected in peak-jumping mode by *IMS 1270*, CRPG, Nancy, France, using a large defocused $O^-$ beam.

required.[56] Also for images acquired at the highest spatial resolution, even small amounts of specimen or stage drift can lead to blurring of the image.[56] To improve the imaging process in the microscope mode by higher signal-to-noise ratio and parallel acquisition of arrival time and position, active pixel detectors (see Section 4.12) have been developed and applied.[25,55,57]

During "scanning ion imaging" an ion beam is directed over the sample surface in a defined raster pattern, while software saves secondary ion intensities as a function of beam position from which maps of secondary ion images are generated.[10,13] The image shape follows the raster pattern shape, which is the reason that these images are usually square. Typical areas that are scanned by SIMS range from 5 μm × 5 μm to 500 μm × 500 μm. The creation of continuous images/maps by an integration of multiple images is possible ("mosaic mode rastering") for larger sampling areas.[58] Analytical results are often presented as images of secondary-ion intensities across the sample surface as determination of concentrations is difficult.[10] Quantitative determination can be done if instrumental corrections similar to those used in quantitative spot mode analysis are accomplished.[13,53] Scanning ion imaging achieves a lateral resolution down to 100 nm, mostly depending on the primary beam cross section on the sample surface (see also Chapter 15). Specific primary-beam conditions allow for continuous depth profiling at nm resolution.[10]

In the scanning mode, the sample is scanned in a raster and the acquisition time $t_a$ for a picture depends on the dwell time $t_d$ in each pixel and upon the number of $N \times N$ pixels in the picture. If the time neglected for switching between two pixels:[24]

$$t_a = t_d * N^2 \tag{9.1}$$

$t_a$: acquisition time
$t_d$: dwell time
$N$: number of pixels in $x$- and $y$-direction

The number of secondary ions per pixel is a function of the sputtering yield, the useful yield (ions detected/atom sputtered), the intensity in the probe, the atomic concentration and the dwell time per pixel.[24] The analysed volume per pixel is called a "voxel". It is important to consider the relation between pixel spacing and beam focus for a continuous microchemical representation of the surface. Besides selecting a useful magnification for mapping an object area to a given number of pixels given a certain spot size, the travel time of ions from the sample to the detector needs to be considered to maintain a signal synchronous with the raster scan. This can be achieved by defining dwell times as being longer than travel times.[40,59] The effect of the travel time (also called "time-of-flight") is also corrected using an image-acquisition scheme based on two interlaced images acquired by forward and backward scanning of the ion beam. In this mode the travel time results in shifted images. It is adequately corrected when both images are perfectly superimposed. The amplitude of the shift depends on the mass analysed so that a mass calibration is required. Also, the signal-to-noise ratio is important, especially in relation to the time that needs to be spent on one spot to collect ions for sufficient signal intensities. However, increasing the number of ions can always be done by increasing the dwell time, which consequently increases the total acquisition time and thus integration time per measurement.[24]

Creation of three-dimensional (3D) images can be achieved by two approaches: (i) serial two-dimensional analysis of sample sections later on combined into stacked 3D visualisations, and (ii) depth profiling as a function of sputtering time.[60]

Aligning images acquired by other techniques for the purpose of integrated analyses can be done by the careful location of reference points across the sample.[56] Interestingly, the detection of secondary electrons by the *NanoSIMS* allows acquisition of secondary electron micrographs that are perfectly superimposed with the ion images (and have the same lateral resolution) and can be used for intercomparison with SEM of higher lateral resolution. Multi-isotopic or multielemental acquisitions by SIMS can also be used to localise specific regions of interest where other *in situ* imaging techniques can be applied such as Auger nanoprobe, transmission electron microscopy, synchrotron X-ray absorption near-edge spectroscopy or scanning high-resolution magnetic superconducting quantum interference device.[61–63]

### 9.1.2 Software Tools for Image Production

Software packages for data visualisation are delivered, *e.g.* with the CAMECA instruments (*e.g. WinCurve*, WinImage, APM). Typically data sets of MSI consist of a large number of mass spectra acquired with identical settings. *L'IMAGE* is a specific *PV-WAVE* image processing software developed by Larry R. Nittler at the Carnegie Institution of Washington, for the processing of *NanoSIMS* datasets. To allow the flexible and efficient exchange of MS imaging data between different instruments and data analysis software, the data format imzML was developed.[64] Several imaging MS software tools already support this format (*e.g.* the open source software tools *Datacube Explorer*,[65] *MSiReader*[66] and *MIRION software*[67]).

### 9.1.3 Summary of the Picture-Creation Process with SIMS

The analytical procedure for the image creation and interpretation process using SIMS is described in Table 9.2. Figures 9.5 and 9.6 represent typical SIMS images.

**Table 9.2** Summary of relevant steps in the imaging process using SIMS.

| Step | Explanation |
|---|---|
| Sample preparation | Analysis is performed under vacuum. Flat and conductive sample surface required. |
| Selecting the area on the sample for analysis | As the total area on the sample to be analysed is typically small (*e.g.* 500 × 500 μm), it needs to be carefully selected according to the study aims. |
| Setting of reference points | For direct spatial linking with multiple SIMS pictures from different sessions, or data from other sources gathered from the same sample, the setting of reference points for unambiguous spatial matching and perfect superimposition of the same area is needed. |
| Microscope photographs | For improving data interpretation it is necessary to take microscopy pictures containing the reference points prior to analysis. |
| Selection of imaging mode | Imaging can be either done by using the ion-microscopy mode ("global imaging") or ion-scanning mode ("mapping"). Results are either a "global image" of the simultaneously sputtered region, or time-resolved data of ion intensities linked to specific *x*- and *y*-coordinates. The choice depends on the lateral resolution and sensitivity required. It constrains the choice of the instrument to be used. |
| Setting of instrumental parameters | Appropriate setting of pixel size, magnification, beam width, integration time and other instrumental sampling parameters influencing the creation of secondary ions according to the targeted spatial resolution, mainly depending on the spatial |

**Table 9.2** (*Continued*)

| Step | Explanation |
|---|---|
| | structures to be analysed and analyte concentration in the sample; dwell-time/pixel has to exceed the flight time to gain signals that can be unambiguously linked to spatial positions on the sample. |
| Analysis of the sample | MSI using SIMS is done by sputtering ions from the sample in a spatially resolved manner (see Chapter 15). |
| Analysis of the calibration standard | For the quantification of SIMS data, the additional analysis of calibration standards is needed. However, compared to LA-ICP-MS, quantification is much more complicated because of potential interferences and more pronounced matrix effects. Details are given in Chapter 15. |
| Documentation | Documentation of the instrumental parameters and sampling positions. |
| Data reduction and quantification | Data reduction involves appropriate correction procedures and quantification using, *e.g.* RSF for elemental ratios or IMF for isotopic ratios. |
| Picture creation | Processing of pictures gained in the ion microscopy mode (*e.g.* distribution of ion intensities, quantified data or isotope ratios). Picture creation based on time-resolved intensities that are matched to specific x-y coordinates of the sample. |
| Picture interpretation | Alignment of SIMS-data with photographs or other data gathered from the same sample by other techniques for analysing specific regions of interest. Interpretation of pictures related to regions of interest. |

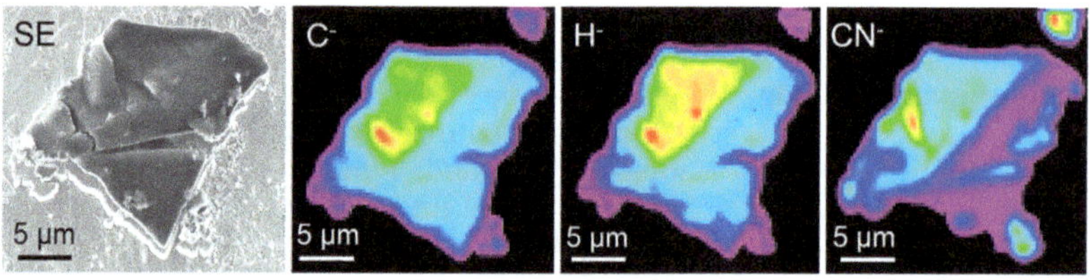


**Figure 9.5** Example of SIMS image showing the elemental distribution of C, H and N (measured as CN at $m/z = 26$) in a type-III kerogen and compared with a secondary electron image. The sample is pressed into a high-purity Au foil. C, H and CN measured as negative ions in scanning imaging mode with a 1.5-μm $Cs^+$ beam with an *IMS 1270*, CRPG, Nancy, France. Note that the C and H images are qualitatively similar, while that of N is slightly different due to the different emission process of molecular ions and thus a different sensitivity to surface conditions.

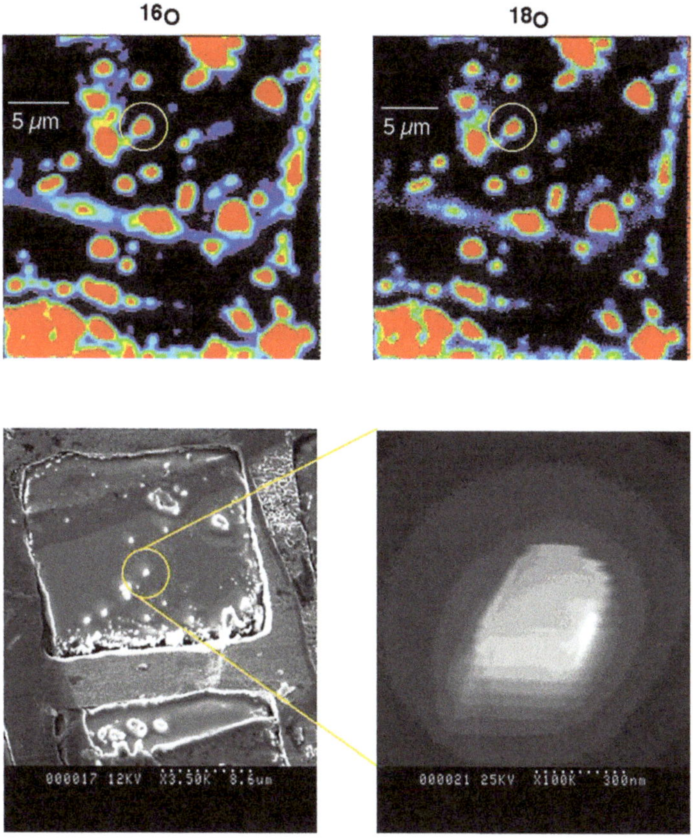


**Figure 9.6** Example of a SIMS image showing the $^{16}O$ and $^{18}O$ isotopic distribution in reference quartz sample powder pressed into a high-purity indium foil. The secondary electron image shown for comparison shows that micrometre to submicrometre grains can be easily relocated after analysis. A TEM Cu grid has previously been pressed in In and removed to provide easy relocation of the analysed area. Blurring in the high-magnification SEM image is due to charging in the absence of conductive coating. O isotope images acquired in scanning imaging mode with a ~1-μm $Cs^+$ beam with a *IMS1270*, CRPG, Nancy, France. At this scale, providing a ratio image would only show a uniform image consistent with the terrestrial value and fairly large uncertainty at the pixel scale. Only images of individual grains larger than 2 μm, with the raster adjusted to fit the grains, would provide sufficient counting and precision to resolve natural variations of a few ‰ before complete sputtering during analysis.

### 9.1.4 Applications

In the following, only a couple of examples will highlight the potential of chemical imaging by SIMS (more applications are given in Chapter 15).

Typical applications of SIMS are the surface 2D and 3D characterisation of different materials in material science.[10,68–70] Blanc *et al.*[71] for example studied the composition of nanoparticles in optical fibres by SIMS. SIMS imaging studies in the earth and planetary science field are numerous. They include elemental and isotopic characterisation of lunar samples.[20] Nittler *et al.*[23] developed an automated imaging procedure to map the isotopic composition of thousands of micrometer-sized grains and recognised those of circumstellar origin with extreme isotopic anomalies from those formed in the solar system. Such presolar grains in the 50 nm–10 μm size range with exotic isotopic compositions inherited from interstellar chemistry or stellar nucleosynthesis have been extensively studied by SIMS and NanoSIMS using scanning and direct imaging.[12,72–74] Isotopic and elemental SIMS imaging have also been used together for chemical characterisation of the carrier phases of isotopic anomalies.[21,75,76] Ion imaging has been used for the study of airborne dust particles and aerosols[77,78] as well as for characterising zoning patterns of trace elements in individual minerals.[79] The distribution of elemental ratios has also been used to shed light on the biomineralisation of extinct and extent corals.[80,81] SIMS-based imaging techniques are also capable of characterising elements, isotopes and molecules in biological samples even at subcellular resolution. Since SIMS has been recognised as a powerful tool for biological research,[42] it has gained an important role in biology and medicine[15,45,60,82,83] ever since. Slodzian *et al.*,[24] *e.g.* used SIMS for elemental mapping in human organs (in the kidney of a person poisoned by Ag). Linton and Goldsmith[43] provide a review on these early applications of elemental and isotopic imaging for biological applications using SIMS in comparison with other techniques. NanoSIMS has been used for tracking stable isotopes in biological systems at subcellular level by dynamic SIMS and static SIMS was used for subcellular imaging with molecular identification.[84] Using dynamic SIMS it was, *e.g.* possible to image single $^{13}C$ labelled bacterial cells in the soil.[85] Ghosal *et al.*[86] studied, *e.g.* the water and ion incorporation in bacillus spores.

Dérue *et al.*[87] mapped boron isotopes in plants, while Moore *et al.*[15] mapped different elements in plant structures with NanoSIMS. NanoSIMS has been used to analyse the subcellular distribution of As and Se in cereal grain[88] and As and Si in rice roots.[44] Smart *et al.*[89] mapped nickel and other elements in leaf tissue of *Alyssum lesbiacum*, a nickel hyperaccumulator plant, at a spatial resolution of about 100 nm using NanoSIMS together with an advanced sample-preparation technique. Si and Ge were mapped at the subcellular level in root and leaf tissues of annual blue grass (*Poa annua*) and orchard grass (*Dactylis glomerata*) by Sparks *et al.*[90] *In situ* mapping of nitrogen uptake in the rhizosphere has also been done.[91] The use of SIMS for the analysis of stable isotope tracers in plant research to study root physiology in relation to the uptake and translocation of mineral nutrients was already proposed by Lazof *et al.*[92] Due to the ability of small-scale isotopic and elemental analysis of samples, SIMS was also applied to the investigation of the physiology of living and ancient microbial communities allowing for the analysis of the specific role of micro-organisms in the biogeochemical cycling of elements.[9]

## 9.2 Imaging by Laser Ablation ICP-MS

ICP-MS has become one of the most widely used methods in trace and ultratrace analysis in a broad variety of applications. The coupling of a laser to an ICP-MS device allows for combining the possibility of a spatially resolved investigation of solid samples with the low detection limits of ICP-MS.[93,94] The usage of a sector field device offers the advantages of higher mass resolution, sensitivity and the possibility for simultaneous detection of elements and isotopes as a prerequisite for the determination of highly precise elemental mass fractions or isotope ratios. Due to its potential for quantification using matrix-similar standards and reduced need for sample preparation (as the analysis is not being performed under vacuum), laser ablation inductively coupled plasma mass spectrometry (LA-ICP-MS) has become a central tool for the spatially distinct quantitative analysis of the distributions of elements and isotopes in wide variety of abiotic[95] and biological[18,96–98] samples. In contrast to SIMS, quantification of ion signals is more easily achievable with LA-ICP-MS by using external calibration standards[99] even though still, a close matrix matching of the calibration standards is preferred. In addition, internal normalisation intensities by applying homogeneously distributed matrix elements are used to correct for differences in the amount of ablated material during the ablation process (see Chapter 8). In minerals or metals, the major matrix element can be used as an internal normalisation standard (under the assumption that the element is distributed homogeneously and the concentration has been estimated by using complementary techniques (*e.g.* XRF)). In other examples, Ca or P was used as internal standard for bone tissues, S was used, *e.g.* for hair and C for the investigation of biological tissues. In particular, the latter is critically discussed within the scientific community as the ablation of carbon leads to $CO_2$ and C-containing particles. For more details the reader is referred to the literature.[100]

Detection limits typically lie in the ng $g^{-1}$ range[8] and the spatial resolution of laser ablation typically ranges from 2–300 μm.[101] Laser ablation coupled to ICP-MS was introduced by Gray,[102] and early in the 1990s, first incrementally growing biological structures (bivalve shells) were analysed in cross-sections.[103–105] The principle of time-resolved analysis of incrementally growing structures by LA-ICP-MS then was extended, *e.g.* to mammalian teeth and the fin rays of fish.[106,107] The first full 2D chemical image of a biological material by LA-ICP-MS was created by Wang *et al.*[108] for the spatially resolved analysis of the distribution of an element (in this case, strontium, Sr) in a fish scale. Outridge *et al.*[109] presented a 2D map of Pb counts on the surface of the tooth of a walrus using a geographic information system (GIS) tool (for more examples see Section 9.3 and Chapter 12).

Although most studies used quadrupole mass analysers coupled to a laser ablation system for chemical imaging of elemental distributions, so far, sector field devices allow higher sensitivity and thus lower detection limits and higher mass resolution for the separation of spectral interferences. In particular, the latter is important as many other measures (*e.g.* sample/matrix separation) cannot be accomplished. In addition, sector field devices in combination with multiple detectors allow for the creation of images of precise isotope ratios (*e.g.* $^{87}Sr/^{86}Sr$).[95]

### 9.2.1 Key Aspects in the Image Production Process using LA-ICP-MS

A focused laser beam achieves the spatially resolved sampling with a given diameter ablating certain amounts of material from a specific site on the sample positioned in a special laser ablation chamber, which are then transported by a carrier gas (He, Ar) into the ICP for ionisation (see Chapter 12). Laser systems typically in use are solid-state Nd:YAG laser systems with a fundamental wavelength of 1064 nm. The frequency is commonly quadrupled or quintupled to generate UV laser light. The most common laser wavelengths are 266, 213 and 193 nm. Whereas 266 and 213 nm Nd:YAG lasers are the most commonly used systems, 193-nm excimer lasers show better ablation properties with respect to equal ablation rates in samples with different absorption behaviour and similar matrix. Moreover, 266-nm lasers produce larger particles compared to 193-nm lasers (in the order: particle size produced by 193-nm lasers < 213-nm lasers < 266-nm lasers).[110] Laser pulse widths vary between 3–5 ns (Nd:YAG-based lasers) and 10–15 ns (excimer lasers).[111,112] About ten years ago, femtosecond lasers were introduced to LA-ICP-MS[113] with the advantage of ultrashort laser pulses reducing the heat being generated enabling the production of very fine aerosol particles.[114]

Different types of commercial laser ablation chambers are in use, with different applications requiring different laser cells.[115] Some biological applications require cooled ablation chambers ("cryo-cells"),[116] while others require specific large ablation cells.[117] An Ar or He carrier gas stream further transports the ablated material to the inductively coupled plasma. Only recently, ablation even in air has been successfully performed, which is of special relevance to samples that are too large to be introduced into the laser chamber or too valuable to cut.[118] After ablation, the air was exchanged to Ar in a gas-exchange device to transport the aerosol to the plasma.

After ionisation in the ICP, the generated ions are further separated in a mass spectrometer according to their mass-to-charge ratios, and finally detected by a specific arrangement of detector(s). Time-resolved scanned ion intensities are produced and saved by software. These time-resolved data need to be assigned to specific $x$–$y$ coordinates on the sample for the creation of a chemical image. Careful data interpolation can be performed to visualise the chemical composition of an entire surface. Figure 9.7 shows a schematic of the image-creation process by LA-ICP-MS.

One major factor determining the spatial resolution of LA-ICP-MS is the diameter of the laser beam in combination with the sensitivity of the detectors[119] and the wash-out of the aerosol from the cell (dispersion). The laser beam width has to be adjusted in order to ablate sufficient material

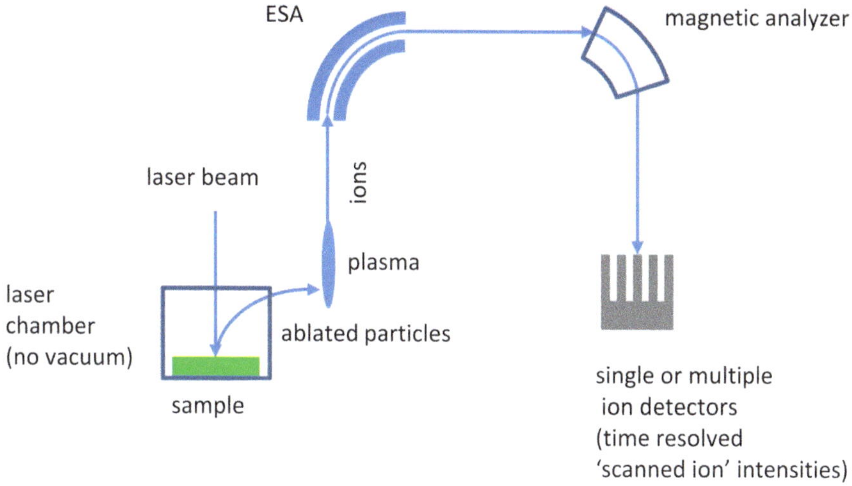


**Figure 9.7** Schematic of the image creation process by laser ablation inductively coupled plasma mass spectrometry (LA-ICP-MS).

to provide signals above detection limits.[120] On commercial instruments most experiments are performed with a laser spot size between 50 and 250 μm, which is mainly determined by sensitivity and the spatial resolution required.[27] However, depending on the analyte concentration and the sensitivity of the detectors, much smaller spot sizes down to 2 μm can be achieved with commercially available systems. In addition, it was reported that it is possible to extend LA-ICP-MS to the nanometre scales with near-field techniques, *e.g.* by using the tip of an silver needle to enhance the laser beam energy.[6,19,121,122] This technique was discussed critically in the scientific community and has not been exploited further since. Also, a laser microdissection (LMD) apparatus was used as sampling device attached to an ICP-MS for nano imaging of elements in life science, biology and medicine (analyses of single cells, cell organelles or biological structures in the nanometre range).[122]

Finally, the potential spatial resolution of the analysis and therefore the element image quality is determined by a complex interplay between different measurement conditions (laser scan mode, beam diameter, laser fluence, repetition rate, scanning speed, gas flow rate, dwell time, instrument settling time (of the ICP-MS) and the resulting total acquisition time),[123] which will be discussed in more detail in the following paragraphs.

**Line Scan *vs.* Spot Scan**

In LA-ICP-MS, typically two modes of ablation – single-point rastering and line-scanning mode – are used. During a raster scan with single-point ablation, the laser stays on a defined position generating sample material from which the signal is acquired by the MS for a certain amount of time before it moves to another position. The signals can be directly linked to a specific ablation site, as for each time-resolved signal a distinct spot on the sample with clearly defined *x*- and *y*-coordinates exists. During line scans, the laser is moved across the sample in a continuous manner at a given speed, with many overlapping spots being created. The spatial difference between these spots is determined by the scan speed in combination with the repetition rate of the laser.[124] Often, line scans are preferred as a "mapping" method for the creation of chemical images, as by line scanning equivalent or better information about the distribution of elements and isotopes can be gathered than by discrete spot analysis at much reduced time (Figure 9.8).

However, one of the main limitations of LA-ICP-MS is the time-dependent change of elemental and isotopic ratios during the ablation affecting both the accuracy and the precision of measurements.[125] A major influence is given by the fact, that some time is needed before equilibrium is reached in the plasma until a continuous flow of ablated materials is established. In particular, at the start of the ablation process, a dramatic change in the fractionation can be observed until equilibrium is achieved. Thus, the starting signal of each spot and the starting signal of each line is prone to aberrant results and has to be taken into account. Liu *et al.*[126] demonstrated that using, *e.g.* water or ethanol to moisten the ablated aerosols after the ablation chamber leads to enhanced and stable laser ablation signals.

In general, it is not possible to quantify the information from structures in the sample smaller than the spot size since the resulting signal represents a "mixture" of the elemental composition of the structure with material from surrounding areas. Nonetheless, these structures can be detected qualitatively by line scans when the laser moves stepwise across the investigated structure. During this process, the signal may change, when the feature of interest is reached. During the line scan, spots are usually overlapping leading to the effect that new material is ablated mainly from small (sickle-shaped) areas when the laser beam moves forward. Therefore, line scans are the preferred strategy when features smaller than the spot size are to be detected.[94,124] However – theoretically – deconvolution techniques can be applied for improving the quantitative interpretation of results.[127]

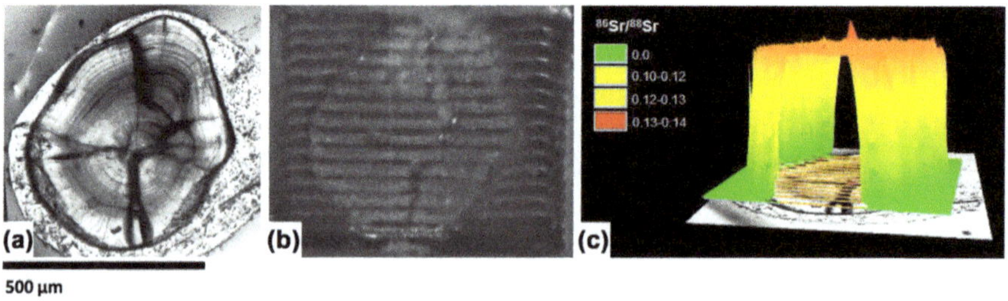


**Figure 9.8** $^{86}Sr/^{88}Sr$ ratio of a carp (*Cyprinus carpio* L.) otolith (a) mapped by multiple line scans (b) to detect a maternally transmitted $^{86}Sr$ spike in the core of the otolith (c)[128]; 35 µm spot size, 2 µm/s scan speed and 20 Hz repetition rate using a *NWR 193* (ESI, Portland, OR, US) coupled to a *NexION 300D* ICP-MS (Perkin Elmer, Waltham, MC, USA).

**Depth Profiling and Three-Dimensional Mapping**

Spatially resolved chemical profiles using laser ablation can also be produced by continuous ablation at one single sampling spot creating a depth profile. Depth profiling can provide a better spatial resolution in the direction of vertical ablation, than it would be possible by lateral scanning of the same features.[129] The depth resolution is hereby dependent on the rate of ablation per laser pulse, and depends on the material analysed in combination with the laser energy, pulse length and repetition rate.[130] Vertical profiling has been successfully applied to abiotic materials[130] and biological tissues.[131] The two main approaches currently in use for the creation of three-dimensional elemental images are (i) the consecutive 2D ablation of sample sections followed by stacking of multiple 2D maps[132] or (ii) laser drilling on a grid, with subsequent combination of data maps gained from different depths.[133,134]

**Data Processing and Creation of Spatial Coordinates for the Laser Ablation Data**

The final results of the ablation process are intensities of ions at the detectors stored per time (integration time) in data files. These data need to be further processed to create the desired spatially resolved images of elemental concentrations and/or isotopic information. As a major step towards chemical imaging, processing of LA-ICP-MS data always requires the attachment of spatial coordinates to the data gathered by laser scanning.

As LA-ICP-MS involves sample ablation, particle, molecule and atom transport, diffusion, ionisation and ion transport, the attribution of a signal to a specific location on the sample can be complex, especially in the line-scan mode.[135] A time lag between laser firing and signal detection occurs during which the laser might move forward, which needs to be considered when linking line-scan signals to specific locations on a sample.[136] Plotnikov *et al.*[137] developed a mathematical model for establishing the correlation between spatial and temporal coordinates of the analytical signal from line scans correctly. The model is based on the estimation of the signal delay caused by material washout as an effect of the size of the laser ablation chamber. The delayed washout of the ablated material may also lead to mixing of ablated material from one site with newly ablated material from the next site during forward movement of the laser before it enters the mass spectrometer.[95] Sanborn and Telmer[124] developed an experimental approach for the direct determination of the effect of washout times on the spatial resolution of line scans. They found that the spatial resolution is controlled by the magnitude of concentration gradients, the direction of concentration shifts and laser parameter (cell size and type and scan speed). Spatial resolution is reduced for transitions from high to low concentrations by a factor of two.

The problem of delayed washout has been tackled by improving the design of ablation cells, *e.g.* by the use of so-called "jet cells" that introduce the carrier gas onto the ablation site as a high-velocity jet reducing mixing and fractionation in the ablation cell.[138] Another additional unwanted effect affecting the spatial resolution lies in potential resampling of material that was deposited on the sample alongside the ablation craters, which can be avoided, *e.g.* by leaving a certain distance between subsequent line scans (*e.g.* one spot size).[95] Both effects, mixing of ablated material from different sampled spots and resampling may lead to a blurring of the signal and needs special consideration in situations where high spatial resolution is required.[95]

When all these effects are under control, $x$ and $y$-coordinates representing distinct positions of the sample can be calculated for each data point based on the raster pattern, dimensions of the ablated area and laser speed. However, it needs to be considered that each data point and coordinate reflects an ablated area of a specific length and width. The lateral spatial resolution of each data point in the $y$-direction (the width) is determined by the spot diameter (given the laser is appropriately focused). In the line-scanning mode, the spatial dimension of the data point in the $x$-direction can be determined less accurately. The laser beam moves across the sample surface at a set scan speed during the time when the signal is integrated. Therefore, the $x$-dimension of the acquired signal represents the spot size plus the travelled distance during the integration time. (For example, a spot size of 20 μm, an integration time of 0.5 s, and a scan speed of 20 μm $s^{-1}$ would yield 20 μm in the y-direction and 30 μm as the dimension of the analysed spot in the $x$-direction.) An experimental approach for determining the spatial resolution that can be achieved by a given instrumental setup seems to be most appropriate.[124]

The repetition rate can be seen as a limiting factor for the scan speed in order to obtain a certain resolution as usually repetition rates in the Hz region are applied. (For instance, a repetition rate of 10 Hz and a scan speed of 100 μm $s^{-1}$ means that the spatial distance between 2 laser pulses would be 10 μm.) Newer systems allow for kHz repetition rates and thus much faster scan speeds.

It has also to be mentioned that sector field mass spectrometers are limited in the number of elements that can be measured within a short time period, especially in the case of single-collector devices. As a consequence, the acquisition time of the mass spectrometer is a critical parameter, which determines the possible resolution. The time that is needed to analyse a number of $n$ isotopes within a sequential measurement sequence is given by the acquisition time $t_a$ that determines the time resolution between two measurement points by the mass spectrometer with $t_s$ being the time needed for a sequence of isotopes and $t_d$ being the dwell time:

$$t_a = \sum_{i=1}^{n} (t_{s,i} + t_{d,i}) \tag{9.2}$$

As a consequence, the time shift of the measured signal of a random isotope $k$ in the measurement sequence from the starting point of the measurement is given in eqn (9.3).

$$\Delta t_a(k) = \sum_{i=1}^{k} (t_{s,i} + t_{d,i}) \tag{9.3}$$

This time shift might have to be taken into account when long dwell times are combined with fast scan rates of the laser for allocating an isotopic signal to a specific sample surface position. As simultaneous measurements are accomplished in the case of multi collector devices in static mode, the acquisition time $t_a$ is determined by the dwell time (or integration time) $t_d$ (eqn (9.4)). The minimum dwell time (integration time) is still the limiting factor for the time resolution using sector field mass spectrometers.

$$t_a = \sum_{i=1}^{n} t_{d,i} \tag{9.4}$$

### 9.2.2 Software Tools for Data Processing and Image Production

Different approaches have been developed for simplifying the data reduction process of time-resolved data gathered by LA-ICP-MS and subsequent image creation. Typical approaches build upon the import and processing of the raw data in data tables using either an in house developed spreadsheet or existing spreadsheet packages like *SILLS* (ETH Zürich, Zürich, CH) relying on *MATLAB* (The MathWorks, Natick, MA, US).[139] Others are standalone applications like the Java-based *AMS* software.[140] More sophisticated software tools typically support the import of raw data, possibilities to inspect, manipulate and process the time-resolved data and also the initial creation of two-dimensional representations of the data. The *IMAGENA* software (Ostbayrische Technische Hochschule Regensburg, Regensburg, Germany) provided one of the first approaches combining data reduction of laser ablation data with the possibility of creating images.[141] The chemical composition in freely drawn regions of interest can be extracted and analysed by using *PMOD* software (*PMOD* Technologies Ltd., Zürich, CH). However, *IMGAGENA* has been applied mainly inhouse so far.[142,143] Other examples of freely available software tools for data reduction and visualisation of two-dimensional element maps of LA-ICP-MS data are *IOLITE* (University of Melbourne, Melbourne, AUS)[144] and the add-on package "LAICPMS" for the R language for statistical computing.[145] *IOLITE* is a self-contained system that runs within the Igor pro framework (WaveMetrics Inc., Lake Oswego, Oregon, US). A comparison of the functions provided by the R add-on package *LAICPMS* and *IOLITE* is presented by Rittner and Müller.[145]

Probably the most specific integrated tool for data reduction and semiautomated creation of spatially registered laser ablation images is *CellSpace*,[136] a module to be used within the *IOLITE* software package. *CellSpace* is based on the synchronisation of laser log files created *e.g.* by the *GeoStar* program used by the RESOlution laser ablation system (Resonetics, Nashua, New Hampshire, USA and Australian Scientific Instruments, Fyshwick ACT, Australia) with mass spectrometry data. The final result is a plot of the mass-spectrometric data in the cell coordinate space. There also exists the possibility to combine the mass-spectrometric data with referenced pictures from other sources.

However, in many publications, no information is given on how coordinates are created and matching of pictures and mass-spectrometric data is achieved. Some publications describe that matching of data from laser ablation and pictures or data from other sources was done by aligning chemical information to pictures purely by visual interpretation of concentration levels with regard to structures visible in the pictures. A more structured approach is needed for accurate spatial matching of chemical data with pictures or data gathered from other analysis. Latkoczy *et al.*[119] for example describe the creation of clearly visible spatial marks to assure that the data recorded at the same regions by different methodologies can be matched. Another approach providing advanced possibilities for the creation of chemical images and spatially distinct matching of mass spectrometry data with pictures or data from other sources is the use of geographical information software (GIS-software).[146–148] After data reduction and the determination of *x*- and *y*-coordinates, data are loaded into GIS software and a chemical image can be directly displayed. When using clearly visible reference points, GIS software offers the capability to deal with the spatial alignment of data from different sources in the most efficient way, while also offering possibilities for advanced spatial and statistical analysis, *e.g.* in regions of interest, across the different input data, and the application of a variety of advanced fully documented interpolation methods.

### 9.2.3 Summary of the Picture Creation Process with LA-ICP-MS

The analytical procedure for the image creation and interpretation process using LA-ICP-MS is described in Table 9.3. Figure 9.9 represents a typical LA-ICP-MS image.

**Table 9.3** Summary of relevant steps in the imaging process using LA-ICP-MS.

| Step | Explanation |
|---|---|
| Sample preparation | Sample preparation for LA-ICP-MS typically involves embedding, cutting, grinding and cleaning, with potential effects of the sample-preparation process to be carefully considered. In some cases the sample can be taken as it is. |
| Selecting the area on the sample for analysis | Typically, total areas ranging from $\mu m^2$ to $cm^2$ range are assessed. |
| Setting of reference points | Reference points set by single laser shots allow for the matching of data gathered by different methods. |
| Microscope photographs | Microscopy pictures containing the reference points should be made for improving data interpretation. |

**Table 9.3** (*Continued*)

| Step | Explanation |
|---|---|
| Selection of imaging mode | Determination of the scan mode (line or point raster) in spatial relation to the reference points for later matching of data with photographs. The main factors that might influence the choice are the area to be ablated, spatial differences in concentrations to be detected, considerations on sampling time. Furthermore potential resampling should be avoided by keeping adequate distances between sampling points. The direction of the scan in relation to the gas flow should be considered to avoid the contamination of nonablated sites during the transport of the ablated material by the carrier gas. |
| Setting of instrumental parameters | Setting of laser spot diameter, number of masses scanned, integration time, dwell time, repetition rate, scan speed, laser energy as these parameters need to be chosen according to the targeted spatial resolution and mainly depend on the spatial structures to be analysed and analyte content in the sample. |
| Analysis of the sample | The laser ablation process involves the ablation of the sample material, the transport of the ablated material from the laser ablation chamber to the ICP, the ion generation and the transport through the MS, the separation by the MS and the collection of ions per ablated sample region at the collectors. Here the following points should be considered: washout time – as this might lead to a delay in the signal detected by the collectors and thus potential sample mixing within the ablation chamber or resampling of deposited material stemming from prior ablations leading to mixed signals for one ablated site hampering the potential spatial resolution achievable. |
| Analysis of calibration standards | Matrix-matched standards (with a matrix as close as possible to the sample) need to be analysed together with the sample. Preferably the standards can be put in the ablation chamber together with the sample to maintain comparable conditions. |
| Documentation | Documentation of laser ablation parameters, laser positions and ablations performed. |
| Data reduction and quantification | Data reduction involves correction for blank, instrumental isotopic fractionation and interferences. Finally, isotope ratio determination or quantification based on matrix-matched standards can be performed. Several software tools are available to support this step. |

**Table 9.3** (*Continued*)

| Step | Explanation |
|---|---|
| Assignment of *x*- and *y*-coordinates to time-resolved data | Data reduction for the creation of chemical images from LA-ICP-MS data involves the assignment of *x*- and *y*-coordinates to time-resolved data. The relation between dwell time and scan speed needs consideration for determining the resolution in the *x*-direction. |
| Picture creation | Picture creation based on time-resolved data that are matched to specific x-y coordinates of the sample and spatial interpolation of data. |
| Picture interpretation | Alignment of LA-ICP-MS data with photographs or data gathered from the same sample by other techniques for investigating specific regions of interest.<br>Interpretation of chemical pictures related to regions of interest. |

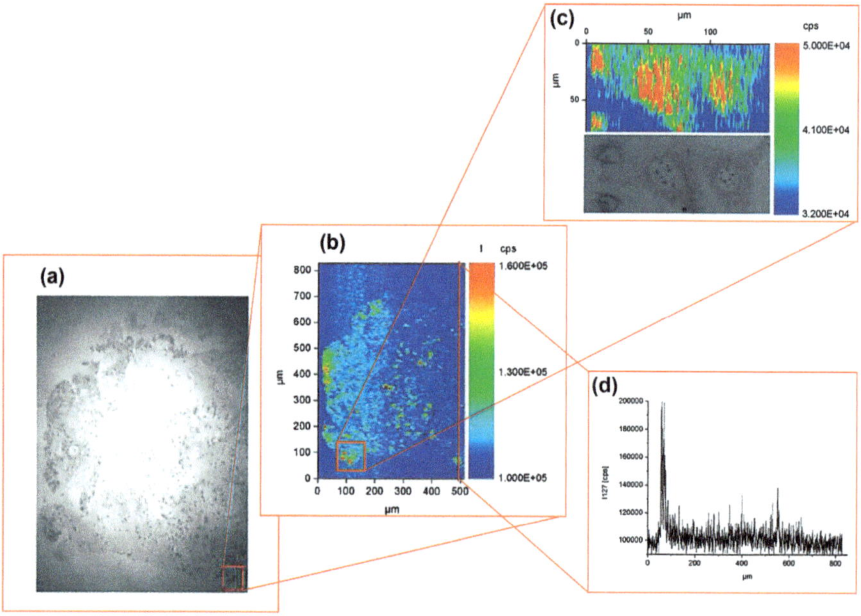


**Figure 9.9** Example of a LA-ICP-MS image showing how the principle of chemical imaging can be applied across different scales, from tissue section level down to the cellular level (a, b, c); (d) shows the signal intensity/cps for $^{127}$I of a line scan along the right side of the sample b.[149]

## 9.2.4 Applications

In the following, only a very limited number of examples will highlight the potential of chemical imaging by LA-ICP-SFMS (more applications are given in Chapter 12).

Biological samples have been widely investigated by LA-ICP-MS to create chemical images. For example, Becker *et al.*,[93] Zoriy and Becker[150] and Dobrowolska *et al.*[151] performed two-dimensional imaging of copper, zinc, and other elements in thin sections of human brain samples. Becker *et al.*[152] performed quantitative imaging of selenium, copper, and zinc in thin sections of slugs of the *genus Arion*. Stadlbauer *et al.*[153] performed time-resolved monitoring of heavy-metal (Hg and Pt) intoxication in a single hair by laser ablation ICP-QMS in DRC mode. A similar approach was applied by Sela *et al.*[154] for the spatially resolved analysis of Zn, Fe and Cu and toxic elements like Cr, Pb and U in single hair strands. Zoriy *et al.*[155] performed imaging of copper, zinc, and platinum in thin sections of a kidney from a mouse treated with cis-platin. Jones and Chen[120] applied line rastering to determine trace element distribution in fish otoliths and Woodhead *et al.*[95] measured two-dimensional spatial variation of the $^{87}Sr/^{86}Sr$ ratio in fish otoliths.

Becker *et al.*[156] utilised imaging mass spectrometry using LA-ICP-SFMS to study the lead and uranium isotope composition of the age dating of single Precambrian zircon grains from the Baltic Shield. The same authors utilised imaging mass spectrometry using LA-ICP-SFMS to study the isotope composition of uranium oxide single particles with a diameter of 1 μm or less in a spatial manner.

# References


1. E. J. Lanni, S. S. Rubakhin and J. V. Sweedler, *J. Proteom.*, 2012, **75**, 5036.
2. M. J. Pelletier and C. C. Pelletier, Spectroscopic Theory for Chemical Imaging, in *Raman, Infrared, and Near-Infrared Chemical Imaging*, eds. S. Šašić and Y. Ozaki, John Wiley & Sons, Inc., Hoboken, New Jersey, 2010, pp. 1–20.
3. P. Chaurand, D. S. Cornett, P. M. Angel and R. M. Caprioli, *Mol. Cell. Proteom.*, 2011, **10**, O110.004259.
4. K. W. McMahon, L. L. Hamady and S. R. Thorrold, *Limnol. Oceanogr.*, 2013, **58**, 697.
5. J. B. West, G. J. Bowen, T. E. Dawson and K. P. Tu, eds., *Isoscapes: Understanding Movement, Pattern, and Process on Earth Through Isotope Mapping*, Springer, Dordrecht, 2010, 487.
6. B. Wu and J. S. Becker, *Metallomics*, 2012, **4**, 403.
7. R. Lobinski, C. Moulin and R. Ortega, *Biochimie*, 2006, **88**, 1591.
8. J. S. Becker and D. Salber, *TrAC, Trends Anal. Chem.*, 2010, **29**, 966.
9. V. J. Orphan and C. H. House, *Geobiology*, 2009, 7, 360.
10. D. S. McPhail, *J. Mater. Sci.*, 2006, **41**, 873.
11. B. Fernández, J. Costa, R. Pereiro and A. Sanz-Medel, *Anal. Bioanal. Chem.*, 2010, **396**, 15.
12. P. Hoppe, S. Cohen and A. Meibom, *Geostand. Geoanal. Res.*, 2013, **37**, 111.

13. J. Aléon, M. Chaussidon, M. Champenois and D. Mangin, *Geostand. Newsletter*, 2001, **25**, 417.
14. J. M. Chabala, K. K. Soni, J. Li, K. L. Gavrilov and R. Levi-Setti, *Int. J. Mass Spectrom. Ion Processes*, 1995, **143**, 191.
15. K. Moore, E. Lombi, F.-J. Zhao and C. M. Grovenor, *Anal. Bioanal. Chem.*, 2012, **402**, 3263.
16. T.-H. Luu, M. Chaussidon and J.-L. Birck, *C. R. Geosci.*, 2014, **346**, 75.
17. J. Pisonero, B. Fernandez and D. Gunther, *J. Anal. Atom. Spectrom.*, 2009, **24**, 1145.
18. J. S. Becker, M. Zoriy, A. Matusch, B. Wu, D. Salber, C. Palm and J. S. Becker, *Mass Spectrom. Rev.*, 2010, **29**, 156.
19. B. Wu and J. S. Becker, *Int. J. Mass Spectrom.*, 2011, **307**, 112.
20. C. A. Andersen and J. R. Hinthorne, *Earth. Planet. Sci. Lett.*, 1972, **14**, 195.
21. K. D. McKeegan, R. M. Walker and E. Zinner, *Geochim. Cosmochim. Acta*, 1985, **49**, 1971.
22. S. Elphick, C. Graham, F. Walker and M. Holness, *Mineral. Mag.*, 1991, **55**, 347.
23. L. R. Nittler, C. M. O'D Alexander, X. Gao, R. M. Walker and E. K. Zinner, *Nature*, 1994, **370**, 443.
24. G. Slodzian, B. Daigne, F. Girard, F. Boust and F. Hillion, *Biol. Cell*, 1992, **74**, 43.
25. H. Yurimoto, K. Nagashima and T. Kunihiro, *Appl. Surf. Sci.*, 2003, **203–204**, 793.
26. L. A. McDonnell and R. M. A. Heeren, *Mass Spectrom. Rev.*, 2007, **26**, 606.
27. R. M. A. Heeren, D. F. Smith, J. Stauber, B. Kükrer-Kaletas and L. MacAleese, *J. Am. Soc. Mass. Spectrom.*, 2009, **20**, 1006.
28. M. Pacholski and N. Winograd, *Chem. Rev.*, 1999, **99**, 2977.
29. M. Stoeckli, P. Chaurand, D. E. Hallahan and R. M. Caprioli, *Nature Med.*, 2001, 7, 493.
30. P. Chaurand, S. A. Schwartz and R. M. Caprioli, *Curr. Opin. Chem. Biol.*, 2002, **6**, 676.
31. S. S. Rubakhin, J. C. Jurchen, E. B. Monroe and J. V. Sweedler, *Drug Discov. Today*, 2005, **10**, 823.
32. J. S. Becker, M. V. Zoriy, J. Dobrowolska and A. Matusch, *Eur. J. Mass Spectrom.*, 2007, **13**, 1.
33. E. R. Amstalden van Hove, D. F. Smith and R. M. A. Heeren, *J. Chromatogr. A*, 2010, **1217**, 3946.
34. J. S. Becker, A. Matusch, J. S. Becker, B. Wu, C. Palm, A. J. Becker and D. Salber, *Int. J. Mass Spectrom.*, 2011, **307**, 3.
35. J. F. Lovering, *N.B.S. Spec. Pub.*, 1975, **427**, 135.
36. N. Shimizu and S. Hart, *Annu. Rev. Earth Planet. Sci.*, 1982, **10**, 483.
37. S. Reed, *Mineral. Mag.*, 1989, **53**, 3.
38. R. Hinton, Ion microprobe analysis in geology, in *Microprobe Techniques in the Earth Sciences*, eds. P. J. Potts, J. F. W. Bowles, S. J. B. Reed

and M. R. Cave, Chapman and Hall, London a.o., 1995, vol. 6, ch. 6, pp. 235–289.
39. T. R. Ireland, Ion microprobe mass spectrometry: Techniques and applications in cosmochemistry, geochemistry, and geochronology, in *Advances in Analytical Geochemistry*, eds. M. Hyman and M. Rowe, JAI Press Ltd., Middlesex, 1995, vol. 2, pp. 1–118.
40. R. Levi-Setti, G. Crow and Y. L. Wang, Imaging SIMS at 20 nm Lateral Resolution: Exploratory Research Applications, in *Secondary Ion Mass Spectrometry SIMS V*, eds. A. Benninghoven, R. Colton, D. Simons and H. Werner, Springer, Berlin Heidelberg, 1986, vol. 44, ch. 31, pp. 132–138.
41. R. W. Linton, in conf. proc. of *51st Annual Meeting Microscopy Society of America*, pp. 498–499.
42. M. S. Burns, *J. Microsc.*, 1982, **127**, 237.
43. R. W. Linton and J. G. Goldsmith, *Biol. Cell*, 1992, **74**, 147.
44. K. L. Moore, M. Schröder, Z. Wu, B. G. H. Martin, C. R. Hawes, S. P. McGrath, M. J. Hawkesford, J. Feng Ma, F.-J. Zhao and C. R. M. Grovenor, *Plant Physiol.*, 2011, **156**, 913.
45. S. Chandra, *Appl. Surf. Sci.*, 2008, **255**, 1273.
46. M. Chaussidon, F. Robert, D. Mangin, P. Hanon and E. F. Rose, *Geostand. Newsletter*, 1997, **21**, 7.
47. E. Deloule, M. Chaussidon and P. Allé, *Chem. Geol.: Isotope Geosci. Sect.*, 1992, **101**, 187.
48. E. Deloule, C. France-Lanord and F. Albarède, D/H analysis of minerals by ion probe. In stable isotope geochemistry: a tribute to Samuel Epstein, eds. H. Taylor, J. O'Neil and I. Kaplan, *The Geochemical Society Special Publication*, 1991, vol. 3, pp. 53–62.
49. É. Deloule, O. Paillat, M. Pichavant and B. Scaillet, *Chem. Geol.*, 1995, **125**, 19.
50. C. Rollion-Bard and J. Marin-Carbonne, *J. Anal. Atom. Spectrom.*, 2011, **26**, 1285.
51. J. M. Eiler, C. Graham and J. W. Valley, *Chem. Geol.*, 1997, **138**, 221.
52. R. I. Gabitov, A. C. Gagnon, Y. Guan, J. M. Eiler and J. F. Adkins, *Chem. Geol.*, 2013, **356**, 94.
53. A. Thomen, F. Robert and L. Remusat, *J. Anal. Atom. Spectrom.*, 2014, 512.
54. R. Castaing and G. Slodzian, *J. de Microsc.*, 1962, **1**, 395.
55. A. Kiss, J. H. Jungmann, D. F. Smith and R. M. A. Heeren, *Rev. Sci. Instrum.*, 2013, **84**.
56. G. McMahon, B. J. Glassner and C. P. Lechene, *Appl. Surf. Sci.*, 2006, **252**, 6895.
57. K. Nagashima, T. Kunihiro, I. Takayanagi, J. Nakamura, K. Kosaka and H. Yurimoto, *Surf. Interface Anal.*, 2001, **31**, 131.
58. A. Broersen, R. van Liere, A. F. M. Altelaar, R. M. A. Heeren and L. A. McDonnell, *J. Am. Soc. Mass. Spectrom.*, 2008, **19**, 823.
59. R. Levi-Setti, Y. Wang and G. Crow, *Le J. de Phys. Colloq.*, 1984, **45**, C9.

60. E. H. Seeley and R. M. Caprioli, *Anal. Chem.*, 2012, **84**, 2105.
61. C. Floss, F. J. Stadermann, A. F. Mertz and T. J. Bernatowicz, *Meteorit. Planet. Sci.*, 2010, **45**, 1889.
62. W. W. Fischer, D. A. Fike, J. E. Johnson, T. D. Raub, Y. Guan, J. L. Kirschvink and J. M. Eiler, *Proc. Natl. Acad. Sci.*, 2014, **111**, 5468.
63. C. Le Guillou, S. Bernard, A. J. Brearley and L. Remusat, *Geochim. Cosmochim. Acta*, 2014, **131**, 368.
64. A. Römpp, T. Schramm, A. Hester, I. Klinkert, J.-P. Both, R. A. Heeren, M. Stöckli and B. Spengler, imzML: Imaging Mass Spectrometry Markup Language: A Common Data Format for Mass Spectrometry Imaging, in *Data Mining in Proteomics: From Standards to Applications*, eds. M. Hamacher, M. Eisenacher and C. Stephan, Humana Press, New York a.o., 2011, vol. 696, ch. 12, pp. 205–224.
65. I. Klinkert, K. Chughtai, S. R. Ellis and R. M. A. Heeren, *Int. J. Mass Spectrom.*, 2014, **362**, 40.
66. G. Robichaud, K. Garrard, J. Barry and D. Muddiman, *J. Am. Soc. Mass. Spectrom.*, 2013, **24**, 718.
67. C. Paschke, A. Leisner, A. Hester, K. Maass, S. Guenther, W. Bouschen and B. Spengler, *J. Am. Soc. Mass. Spectrom.*, 2013, **24**, 1296.
68. M. Grasserbauer, H. J. Dudek and M. F. Ebel, *Angewandte Oberflächenanalyse mit SIMS Sekundär-Ionen-Massenspektrometrie, AES Auger-Elektronen-Spektrometrie, XPS Röntgen-Photoelektronen-Spektrometrie*, Springer, Berlin a.o., 1986, 300.
69. M. Rosner, C. Eisenmenger-Sittner and H. Hutter, *J. Trace Microprobe Tech.*, 2001, **19**, 91.
70. L. C. Feldman and J. W. Mayer, *Fundamentals of Surface and Thin Film Analysis*, North-Holland, New York, Amsterdam, London, 1986, 352.
71. W. Blanc, C. Guillermier and B. Dussardier, *Opt. Mater. Exp.*, 2012, **2**, 1504.
72. S. Messenger, *Nature*, 2000, **404**, 968.
73. A. N. Nguyen and E. Zinner, *Science*, 2004, **303**, 1496.
74. K. Nagashima, A. N. Krot and H. Yurimoto, *Nature*, 2004, **428**, 921.
75. J. Aléon, C. Engrand, F. Robert and M. Chaussidon, *Geochim. Cosmochim. Acta*, 2001, **65**, 4399.
76. J. Aléon, F. Robert, M. Chaussidon and B. Marty, *Geochim. Cosmochim. Acta*, 2003, **67**, 3773.
77. J. Aléon, M. Chaussidon, B. Marty, L. Schütz and R. Jaenicke, *Geochim. Cosmochim. Acta*, 2002, **66**, 3351.
78. B. W. Sinha, P. Hoppe, J. Huth, S. Foley and M. O. Andreae, *Atmos. Chem. Phys.*, 2008, **8**, 7217.
79. N. MacRae, *Can. Mineral.*, 1995, **33**, 219.
80. A. Meibom, S. Mostefaoui, J.-P. Cuif, Y. Dauphin, F. Houlbreque, R. Dunbar and B. Constantz, *Geophys. Res. Lett.*, 2007, **34**, L02601.
81. J. Stolarski, A. Meibom, R. Przeniosło and M. Mazur, *Science*, 2007, **318**, 92.

82. G. Gillen, A. Fahey, M. Wagner and C. Mahoney, *Appl. Surf. Sci.*, 2006, **252**, 6537.
83. J.-L. Guerquin-Kern, T.-D. Wu, C. Quintana and A. Croisy, *Biochim. Biophys. Acta. (BBA) - Gen. Subj.*, 2005, **1724**, 228.
84. M. R. Kilburn and P. L. Clode, *Methods Mol Biol.*, 2014, **1117**, 733.
85. S. Chandra, G. Pumphrey, J. M. Abraham and E. L. Madsen, *Appl. Surf. Sci.*, 2008, **255**, 847.
86. S. Ghosal, T. J. Leighton, K. E. Wheeler, I. D. Hutcheon and P. K. Weber, *Appl. Environ. Microbiol.*, 2010, **76**, 3275.
87. C. Dérue, D. Gibouin, M.-C. Verdus, F. Lefebvre, M. Demarty, C. Ripoll and M. Thellier, *Microsc. Res. Tech.*, 2002, **58**, 104.
88. K. L. Moore, M. Schröder, E. Lombi, F.-J. Zhao, S. P. McGrath, M. J. Hawkesford, P. R. Shewry and C. R. M. Grovenor, *New Phytol.*, 2010, **185**, 434.
89. K. E. Smart, J. A. C. Smith, M. R. Kilburn, B. G. H. Martin, C. Hawes and C. R. M. Grovenor, *The Plant J.*, 2010, **63**, 870.
90. J. Sparks, S. Chandra, L. Derry, M. Parthasarathy, C. Daugherty and R. Griffin, *Biogeochemistry*, 2011, **104**, 237.
91. P. L. Clode, M. R. Kilburn, D. L. Jones, E. A. Stockdale, J. B. Cliff, A. M. Herrmann and D. V. Murphy, *Plant Physiol.*, 2009, **151**, 1751.
92. D. Lazof, R. W. Linton, R. J. Volk and T. W. Rufty, *Biol. Cell*, 1992, **74**, 127.
93. J. S. Becker, M. V. Zoriy, C. Pickhardt, N. Palomero-Gallagher and K. Zilles, *Anal. Chem.*, 2005, **77**, 3208.
94. T. Ulrich, B. S. Kamber, P. J. Jugo and D. K. Tinkham, *Can. Mineral.*, 2009, **47**, 1001.
95. J. D. Woodhead, J. Hellstrom, J. M. Hergt, A. Greig and R. Maas, *Geostand. Geoanal. Res.*, 2007, **31**, 331.
96. A. B. Moradi, S. Swoboda, B. Robinson, T. Prohaska, A. Kaestner, S. E. Oswald, W. W. Wenzel and R. Schulin, *Environ. Exp. Bot.*, 2010, **69**, 24.
97. T. Prohaska, C. Latkoczy, G. Schultheis, M. Teschler-Nicola and G. Stingeder, *J. Anal. Atom. Spectrom.*, 2002, **17**, 887.
98. T. Prohaska, C. Stadlbauer, R. Wimmer, G. Stingeder, C. Latkoczy, E. Hoffmann and H. Stephanowitz, *Sci. Total Environ.*, 1998, **219**, 29.
99. D. Hare, C. Austin and P. Doble, *Analyst*, 2012, **137**, 1527.
100. D. A. Frick and D. Gunther, *J. Anal. Atom. Spectrom.*, 2012, **27**, 1294.
101. M. Resano, E. Garcia-Ruiz and F. Vanhaecke, *Mass Spectrom. Rev.*, 2010, **29**, 55.
102. A. L. Gray, *Analyst*, 1985, **110**, 551.
103. W. T. Perkins, R. Fuge and N. J. G. Pearce, *J. Anal. Atom. Spectrom.*, 1991, **6**, 445.
104. R. Fuge, T. J. Palmer, N. J. G. Pearce and W. T. Perkins, *Appl. Geochem.*, 1993, **8**(Supplement 2), 111.
105. N. Imai, *Anal. Chim. Acta*, 1992, **269**, 263.

106. R. D. Evans, P. M. Outridge and P. Richner, *J. Anal. Atom. Spectrom.*, 1994, **9**, 985.
107. R. D. Evans, P. Richner and P. M. Outridge, *Arch. Environ. Contam. Toxicol.*, 1995, **28**, 55.
108. S. Wang, R. Brown and D. J. Gray, *Appl. Spectrosc.*, 1994, **48**, 1321.
109. P. Outridge, G. Veinott and R. Evans, *Environ. Rev.*, 1995, **3**, 160.
110. M. Guillong, H.-R. Kuhn and D. Günther, *Spectrochim. Acta Part B: Atom. Spectrosc*, 2003, **58**, 211.
111. C. Vogt and C. Latkoczy, Chapter 6: Laser Ablation ICP-MS, in *ICP Mass Spectrometry Handbook*, ed. S. M. Nelms, Blackwell Publishing and CRC Press, Oxford UK and Boca Raton, USA, 2005, pp. 228–258.
112. F. Vanhaecke, Single-collector inductively coupled plasma mass spectrometry, in *Isotopic Analysis – Fundamentals and Applications using ICP-MS*, eds. F. Vanhaecke and P. Degryse, Wiley-VCH, Weinheim, 2012, ch. 2, pp. 31–75.
113. R. E. Russo, X. Mao, J. J. Gonzalez and S. S. Mao, *J. Anal. Atom. Spectrom.*, 2002, **17**, 1072.
114. J. Pisonero and D. Günther, *Mass Spectrom. Rev.*, 2008, **27**, 609.
115. M. B. Fricker, *Design of ablation cells for LA-ICP-MS: from modeling to high spatial resolution analysis applications*, ETH Zürich, Zürich, CH, 2012, 173.
116. M. V. Zoriy, M. Kayser, A. Izmer, C. Pickhardt and J. S. Becker, *Int. J. Mass Spectrom.*, 2005, **242**, 297.
117. M. B. Fricker, D. Kutscher, B. Aeschlimann, J. Frommer, R. Dietiker, J. Bettmer and D. Günther, *Int. J. Mass Spectrom.*, 2011, **307**, 39.
118. L. Dorta, R. Kovacs, J. Koch, K. Nishiguchi, K. Utani and D. Gunther, *J. Anal. Atom. Spectrom.*, 2013, **28**, 1513.
119. C. Latkoczy, Y. Müller, P. Schmutz and D. Günther, *Appl. Surf. Sci.*, 2005, **252**, 127.
120. C. M. Jones and Z. Chen, in conf. proc. of *The Big Fish Bang. The 26th Annual Larval Fish Conference*, Institute of Marine Research, Bergen, Norway, 2003, pp. 431–443.
121. M. Zoriy, D. Mayer and J. S. Becker, *J. Am. Soc. Mass. Spectrom.*, 2009, **20**, 883.
122. J. S. Becker, A. Gorbunoff, M. Zoriy, A. Izmer and M. Kayser, *J. Anal. Atom. Spectrom.*, 2006, **21**, 19.
123. J. Triglav, J. T. van Elteren and V. S. Šelih, *Anal. Chem.*, 2010, **82**, 8153.
124. M. Sanborn and K. Telmer, *J. Anal. Atom. Spectrom.*, 2003, **18**, 1231.
125. M. Guillong and D. Gunther, *J. Anal. Atom. Spectrom.*, 2002, **17**, 831.
126. S. Liu, Z. Hu, D. Gunther, Y. Ye, Y. Liu, S. Gao and S. Hu, *J. Anal. Atom. Spectrom.*, 2014, **29**, 536.
127. G. M. Fuchs, T. Prohaska, G. Friedbacher, H. Hutter and M. Grasserbauer, *Fresenius J. Anal. Chem.*, 1995, **351**, 143.
128. A. Zitek, J. Irrgeher, M. Cervicek, M. Horsky, M. Kletzl, T. Weismann and T. Prohaska, *Marine Freshwater Res.*, http://dx.doi.org/10.1071/MF13235.

129. A. Plotnikov, C. Vogt, V. Hoffmann, C. Taschner and K. Wetzig, *J. Anal. Atom. Spectrom.*, 2001, **16**, 1290.
130. P. R. D. Mason and A. J. G. Mank, *J. Anal. Atom. Spectrom.*, 2001, **16**, 1381.
131. M. Holá, J. Kalvoda, H. Nováková, R. Škoda and V. Kanický, *Appl. Surf. Sci.*, 2011, **257**, 1932.
132. D. J. Hare, J. K. Lee, A. D. Beavis, A. van Gramberg, J. George, P. A. Adlard, D. I. Finkelstein and P. A. Doble, *Anal. Chem.*, 2012, **84**, 3990.
133. J. T. van Elteren, A. Izmer, M. Sala, E. F. Orsega, V. S. Selih, S. Panighello and F. Vanhaecke, *J. Anal. Atom. Spectrom.*, 2013, **28**, 994.
134. S. Peng, Q. Hu, R. P. Ewing, C. Liu and J. M. Zachara, *Environ. Sci. Technol.*, 2012, **46**, 2025.
135. J. Kaiser, M. Galiová, K. Novotný, R. Červenka, L. Reale, J. Novotný, M. Liška, O. Samek, V. Kanický, A. Hrdlička, K. Stejskal, V. Adam and R. Kizek, *Spectrochim. Acta Part B: Atom. Spectrosc.*, 2009, **64**, 67.
136. B. Paul, C. Paton, A. Norris, J. Woodhead, J. Hellstrom, J. Hergt and A. Greig, *J. Anal. Atom. Spectrom.*, 2012, **27**, 700.
137. A. Plotnikov, C. Vogt and K. Wetzig, *J. Anal. Atom. Spectrom.*, 2002, **17**, 1114.
138. S. E. Jackson, N. J. Pearson, W. L. Griffin and E. A. Belousova, *Chem. Geol.*, 2004, **211**, 47.
139. A. Murray, *LA-ICP-MS Analysis with Sills V 1.0.4*, (2008) Eidgenössische Technische Hochschule Zürich, CH and the University of Leeds, UK.
140. S. R. Mutchler, L. Fedele and R. J. Bodnar, Analysis Management System (AMS) for reduction of laser ablation ICPMS data, in *Laser-Ablation-ICPMS in the Earth Sciences: Current Practices and Outstanding Issues*, ed. P. Sylvester, Mineralogical Association of Canada, Vancouver, BC, 2008, vol. Short Course Series, pp. 318–327.
141. T. Osterholt, D. Salber, A. Matusch, J. S. Becker and C. Palm, *Int. J. Mass Spectrom.*, 2011, **307**, 232.
142. B. Wu, F. Andersch, W. Weschke, H. Weber and J. S. Becker, *Metallomics*, 2013, **5**, 1276.
143. M.-M. Pornwilard, R. Weiskirchen, N. Gassler, A. K. Bosserhoff and J. S. Becker, *PLoS One*, 2013, **8**, e58702.
144. C. Paton, J. Hellstrom, B. Paul, J. Woodhead and J. Hergt, *J. Anal. Atom. Spectrom.*, 2011, **26**, 2508.
145. M. Rittner and W. Müller, *Comput. Geosci.*, 2012, **42**, 152.
146. L. T. Humphrey, M. C. Dean, T. E. Jeffries and M. Penn, *Proc. Natl. Acad. Sci.*, 2008, **105**, 6834.
147. B. Jackson, S. Harper, L. Smith and J. Flinn, *Anal. Bioanal. Chem.*, 2006, **384**, 951.
148. D. Kang, D. Amarasiriwardena and A. Goodman, *Anal. Bioanal. Chem.*, 2004, **378**, 1608.
149. C. Giesen, L. Waentig, T. Mairinger, D. Drescher, J. Kneipp, P. H. Roos, U. Panne and N. Jakubowski, *J. Anal. Atom. Spectrom.*, 2011, **26**, 2160.

150. M. V. Zoriy and J. S. Becker, *Int. J. Mass Spectrom.*, 2007, **264**, 175.
151. J. Dobrowolska, M. Dehnhardt, A. Matusch, M. Zoriy, N. Palomero-Gallagher, P. Koscielniak, K. Zilles and J. S. Becker, *Talanta*, 2008, **74**, 717.
152. J. S. Becker, A. Matusch, C. Depboylu, J. Dobrowolska and M. V. Zoriy, *Anal. Chem.*, 2007, **79**, 6074.
153. C. Stadlbauer, T. Prohaska, C. Reiter, A. Knaus and G. Stingeder, *Anal. Bioanal. Chem.*, 2005, **383**, 500.
154. H. Sela, Z. Karpas, M. Zoriy, C. Pickhardt and J. S. Becker, *Int. J. Mass Spectrom.*, 2007, **261**, 199.
155. M. Zoriy, A. Matusch, T. Spruss and J. S. Becker, *Int. J. Mass Spectrom.*, 2007, **260**, 102.
156. J. S. Becker, H. Sela, J. Dobrowolska, M. Zoriy and J. S. Becker, *Int. J. Mass Spectrom.*, 2008, **270**, 1.

CHAPTER 10

# Metrology

JOHANNA IRRGEHER* AND THOMAS PROHASKA*

University of Natural Resources and Life Sciences Vienna (BOKU), Department of Chemistry, Division of Analytical Chemistry, VIRIS Laboratory for Analytical Ecogeochemistry, Tulln, Austria
*Email: johanna.irrgeher@boku.ac.at; thomas.prohaska@boku.ac.at

In the "Vocabulaire International de Métrologie" (VIM) as published by the "Bureau International des Poids et Mésures" (BIPM) the term "metrology" is defined as "the science of measurement and its application" and covers all theoretical and practical aspects of any measurement, including uncertainty and field of application.[1]

A number of institutions dedicated to/with a special focus on metrology regularly publish recommendations and guidelines concerning metrology, method validation and terminology and thus form a global metrological organisational structure. The most relevant institutions are interlinked *via* the MRA (Mutual Recognition Arrangement) of 1999, aiming at demonstrating international comparability and harmonisation of metrological aspects on a global basis. The Comité Consultatif de Quantité de Matière (CCQM) represents the overall commission of national metrology institutes. Figure 10.1 shows a schematic of the interlink between the most relevant institutions in the field and whose activities mainly concern primary methods for measuring amount of substance, international comparisons, establishment of international equivalence between national laboratories and advice to the CIPM (International Committee for Weights and Measures) on matters concerned with metrology in chemistry. CCQM operates worldwide and is subdivided into regional subgroups, namely EURAMET (Europe), COOMET (North Asia), APMP (South Asia and Australia), AFRIMETS (Africa)

New Developments in Mass Spectrometry No. 3
Sector Field Mass Spectrometry for Elemental and Isotopic Analysis
Edited by Thomas Prohaska, Johanna Irrgeher, Andreas Zitek and Norbert Jakubowski

Published by the Royal Society of Chemistry, www.rsc.org

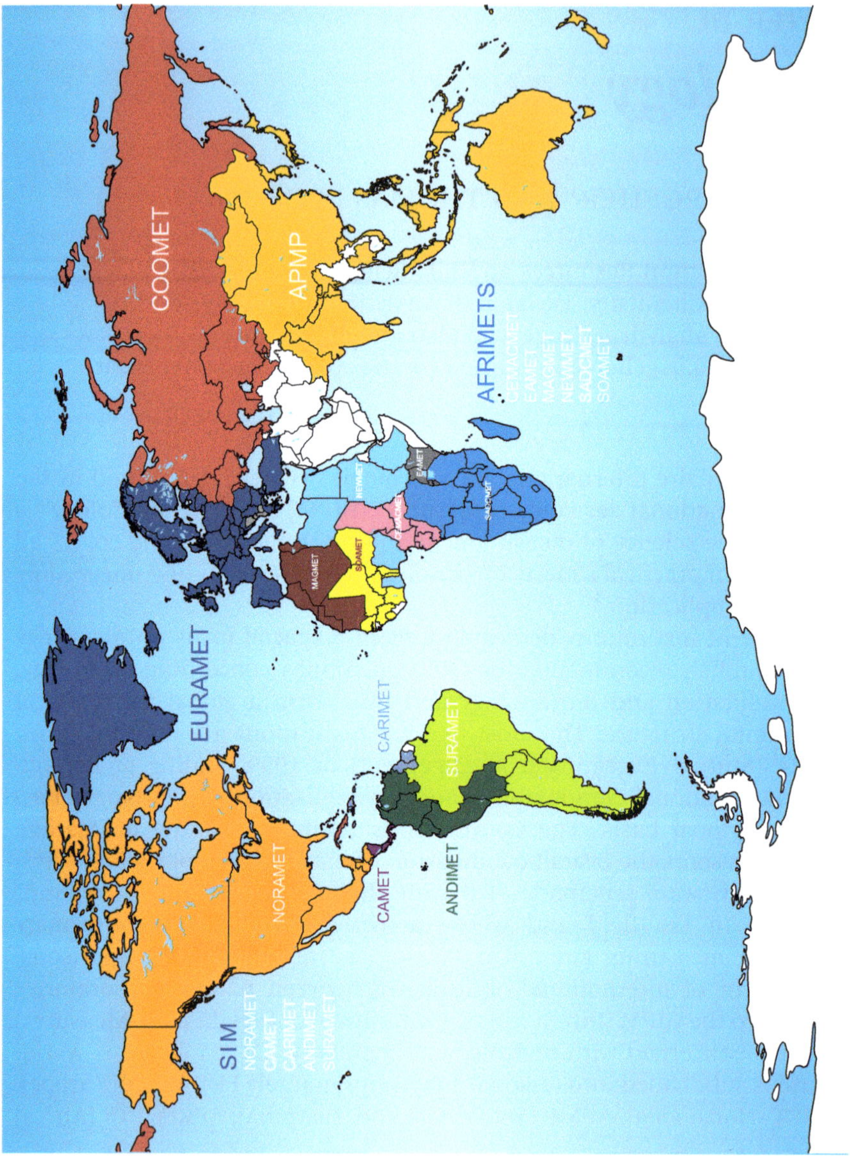


**Figure 10.1** Map of regional metrology organisations around the world. (Source: © EURAMET.[2])

and SIM (America). Details on the different institutions can be found in the respective EURAMET documents.[2]

Additional metrological institutions are, *e.g.* the ILAC (International Laboratory Accreditation Cooperation) and the OIML (International Organization of Legal Metrology), which are dedicated to the development of international laboratory accreditation practices and procedures as well as the release of metrological guidelines for the elaboration of national and regional requirements concerning legal metrology applications, respectively.

IUPAP (International Union of Pure and Applied Physics) and IUPAC (International Union of Pure and Applied Chemistry) are international, nongovernmental organisations, aiming at the exchange of information within the international scientific community and thus are concerned with aspects such as nomenclature, terminology, standardised methods of measurements, atomic weights, *etc.*

In this section, the basic terminology and concepts of metrology according to ISO/VIM and IUPAC with aspects such as traceability, reference materials and uncertainty calculations will be briefly discussed.

## 10.1 General Terminology and Explanatory Notes

**Validation**

Definition (according to VIM):

> "Verification, where the specified requirements are adequate for an intended use."

The validation of a method verifies that the method is suitable for its intended use, meeting the requirements of an analytical procedure. Validation is established by a set of validation specifications, which should be regarded as a guidance checklist during method development. Full validation of an analytical method does not always require the validation of all validation parameters, however those essential for the qualification of the method for the intended use (including the definition of the requirement for each parameter) have to be checked. Method validation serves to better understand the potential and the limits of a method. Method validation is not only to be performed when new methods are implemented or developed, but also on a regular basis in case, *e.g.* the method performance shows significant changes, slight modifications in the method are made or existing methods are implemented in new laboratories or with new instrumentation. Before starting method validation, the scope of validation has to be set according to the requirements of the selected analytical procedure (with respect to the equipment available, the personnel available, the (certified) reference materials available, *etc.*).

A validation report with detailed results is the usual outcome of a full method validation and indicates whether a method is fit-for-purpose. Proper documentation is a prerequisite for quality control of a laboratory and often legally required. A statement about the fit for intended use of a method should complete method validation. Of course, the extent of validation has to be adapted to the performance requirements of a method and is within the analyst's responsibility.

Even though a set of different national and international standards are being used worldwide, it should be pointed out here that method validation according to these standards are often legal requirements for laboratories, *e.g.* for accreditation or legal approval. ISO/IEC 17025, representing one important international standard regulative on general requirements for the competence of laboratories, requires that laboratories demonstrate their competence based on three pillars, which are (1) validation of the measurement procedure, (2) traceability of the measurement result and (3) estimation of uncertainty of measurement results.

A measurement result has to be given in a value (result from a validated method) assigned with a unit (to prove traceability) and an uncertainty (including all sources of uncertainty according to GUM).

The following terms are the basis of the validation and will be used throughout this book, accordingly:

### Analyte and Measurand

There is a clear difference between the "analyte" (*e.g.* selenium) and "measurand" (*e.g.* mass fraction of selenium as selenite in milk) and thus, the terms should be used appropriately by the analyst.[3]

Definition according to VIM:

"The measurand is the quantity intended to be measured.
(***Note 1***: The specification of a measurand requires knowledge of the kind of quantity, description of the state of the phenomenon, body, or substance carrying the quantity, including any relevant component, and the chemical entities involved.
***Note 2***: In the second edition of the VIM and in IEC 60050-300:2001, the measurand is defined as the "particular quantity subject to measurement".
***Note 3***: The measurement, including the measuring system and the conditions under which the measurement is carried out, might change the phenomenon, body, or substance such that the quantity being measured may differ from the measurand as defined. In this case, adequate correction is necessary.)"[1]

### Calibration Function

According to the IUPAC recommendation the "calibration function" is defined as "the functional (not statistical) relationship for the chemical measurement process, relating the expected value of the observed (gross) signal or response variable to the analyte amount".[4]

**Analytical Function**

According to the IUPAC recommendation the "analytical function" relates the measured value to the instrument reading. The analytical function is taken as equal to the inverse of the calibration function.[4]

**Selectivity**

Definition (according to VIM): "...property of a measuring system, used with a specified measurement procedure, whereby it provides measured quantity values for one or more measurands such that the values of each measurand are independent of other measurands or other quantities in the phenomenon, body, or substance being investigated."[1]

Definition (according to IUPAC): "Selectivity refers to the extent to which a method can determine particular analytes in mixtures or matrices without interferences from other components of similar behaviour."[5]

Selectivity refers to the property of the method that indicates its applicability to measure the analyte of interest selectively in presence of possible interferences, *e.g.*, deriving from the matrix or the measurement medium. It is generally recommended to check for the most likely occurring interferences in a measurement sample in order to establish a method that is suitable for the intended use and able to cope with the "worst-case" scenario.

The terms "selectivity" and "specificity" have long been used interchangeably. Specificity, however, has been described by the IUPAC as the "ultimate of selectivity" and thus cannot be graded. The IUPAC finally released recommendations about how to use the terms specificity and selectivity correctly and promotes the use of the term selectivity to describe the capability of an analytical method to deliver "true results"[5]

**Sensitivity**

Definition (according to VIM): "...quotient of the change in an indication of a measuring system and the corresponding change in a value of a quantity being measured."[1]

Sensitivity is a measure of the instrument response and often represented by the slope of a (linear) calibration curve in quantitative analyses. When the calibration curve is not linear, the sensitivity is a function of the analyte concentration or amount.[4]

The greater the sensitivity, the higher the resolution of the method to distinguish between small changes in the input quantity (*e.g.* concentration of the analyte).

**Working Range**

(also termed: measuring interval/working interval)

Definition (according to VIM): "...set of values of quantities of the same kind that can be measured by a given measuring instrument or measuring system with specified instrumental measurement uncertainty, under defined conditions."[1]

The working interval of a method - often represented by a set of calibration standards with defined (increasing) concentration of analyte - is usually chosen according to the expected concentration in the samples. This is often based on the experience of the user or, *e.g.* in case of unknown materials, has to be determined experimentally. In most cases, the working interval is determined by the lowest and highest working standard and should be chosen accordingly to fit the expected concentration range.

**Limit of Detection/Detection Limit (LOD)**

Definition (according to VIM): "...measured quantity value, obtained by a given measurement procedure, for which the probability of falsely claiming the absence of a component in a material is $\beta$, given a probability $\alpha$ of falsely claiming its presence."[1]

Definition (according to IUPAC): "The limit of detection, expressed as a concentration, or a quantity, is derived from the smallest measure that can be detected with reasonable certainty for a given analytical procedure."

**Limit of Quantification/Quantification Limit (LOQ)**

Even though there is no official definition for the limit of quantification by VIM, it is commonly regarded as the lowest quantity of a given kind that can be stated with a given uncertainty.

According to a IUPAC recommendation, quantification limits are performance characteristics that mark the ability of a chemical measurement process (CMP) to adequately "quantify" an analyte.[4]

**Blank Indication**

Definition (according to VIM): "...indication obtained from a phenomenon, body, or substance similar to the one under investigation, but for which a quantity of interest is supposed not to be present, or is not contributing to the indication."[1]

The blank is one of the most crucial quantities in trace analyses and thus care must be taken in establishing a realistic blank for a chemical measurement procedure.

A number of different concepts for the determination of LOD and LOQ have been presented in the literature and are common in practice, whereby three main approaches are usually applied:

(1) blank method (2) signal-to-noise - method and (3) calibration line method. Details about these approaches have been discussed extensively in the literature.[6–11]

**Method Blank**

The method blank is an analyte-free matrix to which all reagents are added in the same volumes or proportions as used in sample processing. The method blank should be carried through the complete sample-preparation and analytical procedure. The method blank is used to document contamination resulting from the analytical process.

**Robustness and Ruggedness**

The terms "robustness" and "ruggedness" have been discussed in detail only recently by the IUPAC, since divergent use of the terms and thus incomparability has been common practice in the analytical community.[12]

Both terms have been used to describe the resistance of a method to changes in the result when slight deviations are made from the experimental design (*e.g.* operating conditions such as ambient temperature or humidity). The definitions for the two different concepts given by the IUPAC in a recent publication are as follows and differ in terms of describing the resistance of the method to changes "within an individual laboratory" or "between various laboratories":

"The relative robustness of an analytical method is defined as the ratio of the ideal signal for an uninfluenced method compared to the signal for a method subject to known and unknown operational parameters as studied in an intralaboratory experiment.

The relative ruggedness of an analytical method is defined as the ratio of the ideal signal for an uninfluenced method compared to the signal for a method subject to known and unknown operational parameters as studied in an interlaboratory experiment."[12]

**Recovery**

Definition (according to VIM): "...measure of the trueness of an analytical procedure."[1]

Recovery is usually checked by the analysis of a common standard, *i.e.* standard reference material that is processed according to the same protocol of the developed method. The portion between the observed (or measured) value to the reference (or known) value, can be expressed as recovery. This portion should give a value of 1 in the case of 100% recovery and significant deviations indicate a systematic analytical bias due to, *e.g.* loss of the analyte (when recovery $<100\%$) or the presence of interfering components (when recovery $>100\%$). Recovery checks should preferably be performed using matrix-matched certified reference materials, since the presence of other components may have a significant impact on the method performance.

The IUPAC recommends to use the term "apparent recovery" instead of "recovery", the latter rather to describe "the yield of a preconcentration or extraction stage of an analytical process for an analyte divided by amount of analyte in the original sample" than the ratio of the observed value, derived from an analytical procedure by means of a calibration graph divided by the reference value.[13]

**Measurement Trueness**

Definition according to VIM: "...closeness of agreement between the average of an infinite number of replicate measured quantity values and a reference quantity value".[1]

(***Note 1***: "Measurement trueness" is not a quantity and thus cannot be expressed numerically, but measures for closeness of agreement are given in ISO 5725.

***Note 2***: "Measurement trueness" is inversely related to systematic measurement error, but is not related to random measurement error.

***Note 3***: "Measurement accuracy" should not be used for "measurement trueness".)"[1]

**Measurement Accuracy**

Definition (according to VIM): "...closeness of agreement between a measured quantity value and a true quantity value of a measurand.

(***Note 1***: The concept "measurement accuracy" is not a quantity and is not given a numerical quantity value. A measurement is said to be more accurate when it offers a smaller measurement error.

***Note 2***: The term "measurement accuracy" should not be used for measurement trueness and the term "measurement precision" should not be used for "measurement accuracy", which, however, is related to both these concepts.

***Note 3***: "Measurement accuracy" is sometimes understood as closeness of agreement between measured quantity values that are being attributed to the measurand)."[1]

It is important to understand, that measurement "accuracy" cannot be reported as a quantity value, since it includes information on both concepts, the uncertainty of a measurement as well as the trueness of a measurement result. Thus, "measurement accuracy" is often referred to as "closeness of agreement between measured quantity values that are being attributed to the measurand" and therefore it is an indicator for the quality of a measurement.

**Measurement Precision**

Definition (according to VIM): "...closeness of agreement between indications or measured quantity values obtained by replicate measurements on the same or similar objects under specified conditions.

(***Note 1***: "Measurement precision" is usually expressed numerically by measures of imprecision, such as standard deviation, variance, or coefficient of variation under the specified conditions of measurement.

***Note 2***: The "specified conditions" can be, for example, repeatability conditions of measurement, intermediate precision conditions of measurement, or reproducibility conditions of measurement (see ISO 5725-1:1994).

***Note 3***: "Measurement precision" is used to define measurement repeatability, intermediate measurement precision, and measurement reproducibility.

***Note 4***: Sometimes "measurement precision" is erroneously used to mean "measurement accuracy")."[1]

"Measurement precision" gives information on random errors and is expressed as a quantity value such as standard deviation of replicate results of a measurement. The "specified conditions" lead to different levels of measurement precision, which are conventionally referred to as "repeatability" (or within-laboratory precision), "intermediate precision" and "reproducibility" (or between-laboratory precision). Surprisingly, measurement results sometimes still lack the indication of an error given as imprecision of the measurement. When measurement results are reported including the measurement imprecision, the statement of the "specified conditions" is not common practice, which makes the evaluation of reported measurement results impossible and (between-laboratory) comparability difficult. Therefore, it is emphasised that the level of measurement precision should be stated with awareness in order to provide solid analytical results.

**Repeatability**
Definition (according to VIM): "...measurement precision under a set of repeatability conditions of measurement, which are defined as condition of measurement, out of a set of conditions that includes the same measurement procedure, same operators, same measuring system, same operating conditions and same location, and replicate measurements on the same or similar objects over a short period of time."[1]

**Intermediate Precision**
Definition (according to VIM): "...measurement precision under a set of intermediate precision condition of measurement, which are defined as condition of measurement, out of a set of conditions that includes the same measurement procedure, same location, and replicate measurements on the same or similar objects over an extended period of time, but may include other conditions involving changes."[1]

**Reproducibility**
Definition (according to VIM): "...measurement precision under a set of reproducibility condition of measurement, which are defined as condition of measurement, out of a set of conditions that includes different locations, operators, measuring systems, and replicate measurements on the same or similar objects."[1]

## 10.2 Uncertainty

The calculation of combined (measurement) uncertainties represents an integral part during method development or at the latest in method validation. Here, the impact of all relevant input parameters that contribute to the total uncertainty of a measurement have to be taken into account and their uncertainties have to be propagated accordingly. Different concepts of uncertainty calculation coexist and the choice of which concept is being used for a particular uncertainty calculation seems to be rather based on the force of habit and sympathy towards one approach over the other than on conceptual mathematical reasons. Amongst others the "Joint Committee for

Guides in Metrology" regularly publishes guidelines for the calculation of uncertainty in measurements in order to provide users with standard procedures for uncertainty assessment and to counteract the lack of international consensus on the expression of uncertainty in measurement.

**Standard (Measurement) Uncertainty**
Definition according to VIM: "...uncertainty of the result of a measurement expressed as a standard deviation."[1]

**Relative Standard (Measurement) Uncertainty**
Definition (according to VIM): "...standard uncertainty divided by the absolute value of the measured quantity value."[1]

**Type-A Evaluation of (Measurement) Uncertainty**
Definition (according to VIM): "Type-A evaluation of a component of measurement uncertainty by a statistical analysis of measured quantity values obtained under defined measurement conditions

***Note 1***: For various types of measurement conditions, see repeatability condition of measurement, intermediate precision condition of measurement, and reproducibility condition of measurement.

***Note 2***: For information about statistical analysis, see, *e.g.* the GUM:1995.

***Note 3***: See also GUM:1995, 2.3.2, ISO 5725, ISO 13528, ISO/TS 21748, ISO 21749."[1]

**Type-B Evaluation of (Measurement) Uncertainty**
Definition (according to VIM): "Type-B evaluation of a component of measurement uncertainty determined by means other than a Type-A evaluation of measurement uncertainty

EXAMPLES: Evaluation based on information

- associated with authoritative published quantity values;
- associated with the quantity value of a certified reference material;
- obtained from a calibration certificate;
- about drift;
- obtained from the accuracy class of a verified measuring instrument;
- obtained from limits deduced through personal experience.

***Note***: See also GUM:1995, 2.3.3."[1]

**Combined Standard (Measurement) Uncertainty ($u_c$)**
Definition (according to VIM): "...standard measurement uncertainty that is obtained using the individual standard measurement uncertainties associated with the input quantities in a measurement model.

***Note***: In case of correlations of input quantities in a measurement model, covariances must also be taken into account when calculating the combined standard measurement uncertainty; see also GUM:1995, 2.3.4."[1]

**Expanded Uncertainty *U***

Definition (according to VIM): "Product of a combined standard measurement uncertainty and a factor larger than the number one."

***Note 1***: The factor depends upon the type of probability distribution of the output quantity in a measurement model and on the selected coverage probability.

***Note 2***: The term "factor" in this definition refers to a coverage factor."[1]

**Coverage Interval**

Definition (according to VIM): "...interval containing the set of true quantity values of a measurand with a stated probability, based on the information available."

***Note 1***: A coverage interval does not need to be centred on the chosen measured quantity value (see JCGM 101:2008).

***Note 2***: A coverage interval should not be termed "confidence interval" to avoid confusion with the statistical concept (see GUM:1995, 6.2.2).

***Note 3***: A coverage interval can be derived from an expanded measurement uncertainty (see GUM:1995, 2.3.5)."[1]

**Uncertainty Budget**

Definition (according to VIM): "...statement of a measurement uncertainty, of the components of that measurement uncertainty, and of their calculation and combination.

***Note***: An uncertainty budget should include the measurement model, estimates, and measurement uncertainties associated with the quantities in the measurement model, covariances, type of applied probability density functions, degrees of freedom, type of evaluation of measurement uncertainty, and any coverage factor."[1]

The following steps have to be undertaken for calculating an uncertainty budget:

1) Specify the measurand (for definition see Section 10.1).
2) Specify the measurement procedure and model function (*i.e.* equation with which the final result is computed by taking all input variables into account, which need to be determined in order to calculate a final result).
3) Define all input parameters to your model.
4) Determine/Quantify the uncertainties to all input parameters
   a) Determine the type of uncertainty (A or B) and the distribution of the error Types of errors (for definitions see also Section 10.1)
   b) Convert the error from expanded uncertainty into standard deviation equivalent.
5) Calculate the combined standard uncertainty by partial derivatives, Kragten spreadsheet approach or Monte Carlo simulation.
6) Calculate the expanded uncertainty.
7) Examine the uncertainty budget (error contributions).

Three major approaches have developed as leading concepts for uncertainty calculation, yet all of them aim to calculate the total uncertainty of a measurement result and to a better understanding of the impact of a single input parameter to the final result. (1) In the recommended approach by the GUM[1] the propagation of uncertainties is accomplished by partial derivatives either manually or by using dedicated software packages, which direct the user through a graphic user interface without the necessity to understand the calculations being made.[14] (2) "Kragten's (spread sheet) method" represents a very reliable, transparent and straightforward way to calculate uncertainties in standard calculation programmes.[15] The condition for both approaches is that the variables are not correlated. (3) In contrast to that, Monte-Carlo simulation is another approach, which is increasingly been used and also encouraged by the Joint Committee for Guide in Metrology (JCGM) of the Bureau International des Poids et Mesures (BIPM) as a valuable alternative to the GUM approach.[16] The simulations rely on the propagation of distributions, instead of propagation of uncertainties like the GUM, and thus are not subjected to its approximations and correlations can be taken into account.[17]

## 10.3 Traceability

**Metrological traceability** requires the establishment of a calibration hierarchy, providing that all input parameters into a model equation are traceable to a common reference – if possible, the SI – unit.

Definition (according to VIM): "...property of a measurement result whereby the result can be related to a reference through a documented unbroken chain of calibrations, each contributing to the measurement uncertainty.

***Note 1***: For this definition, a "reference" can be a definition of a measurement unit through its practical realisation, or a measurement procedure including the measurement unit for a nonordinal quantity, or a measurement standard.

***Note 2***: Metrological traceability requires an established calibration hierarchy.

***Note 3***: Specification of the reference must include the time at which this reference was used in establishing the calibration hierarchy, along with any other relevant metrological information about the reference, such as when the first calibration in the calibration hierarchy was performed.

***Note 4***: For measurements with more than one input quantity in the measurement model, each of the input quantity values should itself be metrologically traceable and the calibration hierarchy involved may form a branched structure or a network. The effort involved in establishing metrological traceability for each input quantity value should be commensurate with its relative contribution to the measurement result."[1]

Establishing metrological traceability requires some effort of the analyst to define and ascertain traceability of all input quantities into the measurement model. This includes considerations about the choice of an adequate reference material (either provided with a certificate or calibrated against a certified standard) for instrument calibration or/and method validation, calibration of balances and pipettes used, *etc.* in order to assure a metrologically valid calibration hierarchy. A recent IUPAC recommendation gives a very comprehensive perspective about the concept of metrological traceability of a measurement result in chemistry as well as its realisation and implementation in practice.[18]

## 10.4 Reference Materials

Reference materials have become of the utmost importance to provide the highest metrological quality of measurements for method validation, uncertainty determination and establishing traceability of measurement results. Certified reference materials are certified for quantity/amount of one or more substances – matrix matched or nonmatrix matched or for their isotopic composition (isotope reference materials – either with natural or enriched isotopic composition). "Matrix reference materials" realise the measurand in a specific matrix - often produced from a natural source – and are consequently preferred for the analysis of complex natural samples. An increased demand in the preparation and provision of reference materials can be observed. Since the production of certified reference materials is complex, laborious and expensive the increasing demand of a large variety of reference materials cannot be satisfied.

Currently, a number of metrological authorities provide certified reference materials, such as the National Institute of Standards and Technology (NIST) in the USA, the Federal Institute for Materials Research and Testing (BAM) in Germany, the National Research Council (NRC) in Canada, the International Atomic Energy Agency (IAEA) or the Institute for Reference Materials and Measurement (IRMM) in Belgium. In order to disseminate information on available certified reference materials on a global scale, the Federal Institute for Materials Research and Testing (BAM) placed a free-of-charge database (www.comar.bam.de/en) listing thousands of certified reference materials produced worldwide by about 220 producers in 25 countries at everyone's disposal.

**Reference Measurement Standard**
Definition (according to VIM): "...measurement standard designated for the calibration of other measurement standards for quantities of a given kind in a given organisation or at a given location."

**Certified Reference Materials**
Definition (according to VIM): "...reference material, accompanied by documentation issued by an authoritative body and providing one or more specified property values with associated uncertainties and traceabilities, using valid procedures."

At present, only a few institutions are dedicated (and authorised) to produce, certify and release reference materials and since much effort has to be put into the entire procedure, placing of a new reference material on the market takes much handling time. The production and certification of new isotope reference materials cannot keep up with the constantly increasing demand by the scientific community. A comprehensive discussion about this issue together with an overview of existing isotope reference materials and its proper use were published during the last years and can be found in the respective documents.[19–21]

## References

1. Joint Commitee for Guides in Metrology – ISO, *Evaluation of measurement data – Guide to the expression of uncertainty in measurement*, JCGM100:2008, 2012.
2. P. Howarth, F. Redgrave, P. AGermany, S. Madsen and S. Grafisk, *Metrology in Short*, 3rd edn, ed. EURAMET, 2008.
3. P. de Bièvre, *Accredit. Quality Assur.*, 2013, **18**, 71–72.
4. L. A. Currie, *Pure Appl. Chem.*, 1995, **67**, 1699–1723.
5. J. Vessman, R. I. Stefan, J. F. Van Staden, K. Danzer, W. Lindner, D. T. Burns, A. Fajgelj and H. Müller, *Pure Appl. Chem.*, 2001, **73**, 1381–1386.
6. T. Meisel, Quality Control in Isotope Ratio Applications, in *Isotopic Analysis*, Wiley-VCH Verlag GmbH & Co. KGaA, 2012, pp. 165–187.
7. L. A. Currie, *Anal. Chim. Acta*, 1999, **391**, 127–134.
8. L. A. Currie, *Appl. Radiat. Isotopes*, 2004, **61**, 145–149.
9. J.-M. Mermet, *Spectrochim. Acta Part B: Atom. Spectrosc.*, 2008, **63**, 166–182.
10. L. V. Rajakovic, D. D. Markovic, V. N. Rajakovic Ognjanovic and D. Z. Antanasijevic, *Talanta*, 2012, **102**, 79–87.
11. J. M. Mermet, G. Granier and P. Fichet, *Spectrochim. Acta Part B: Atom. Spectrosc.*, 2012, **76**, 221–225.
12. D. T. Burns, K. Danzer and A. Townshend, *J. Assoc. Public Analysts (Online)*, 2009, **37**, 40–60.
13. D. T. Burns, K. Danzer and A. Townshend, *Pure Appl. Chem.*, 2002, **74**, 2201–2205.
14. W. C. Losinger, *Am. Statist.*, 2004, **58**, 165–167.
15. J. Kragten, *Analyst*, 1994, **119**, 2161–2165.
16. P. R. G. Couto, J. C. Damasceno and S. P. d. Oliveira, *Monte Carlo Simulations Applied to Uncertainty in Measurement*, 2013, INTECH Open Science, DOI: 10.5772/53014.
17. G. Chew and T. Walczyk, *Anal. Bioanal. Chem.*, 2012, **402**, 2463–2469.
18. P. de Bièvre, R. Dybkaer, A. Fajgelj and D. B. Hibbert, *Pure Appl. Chem.*, 2011, **83**, 1873–1935.
19. J. Vogl and W. Pritzkow, Reference Materials in Isotopic Analysis, in *Isotopic Analysis*, Wiley-VCH Verlag GmbH & Co. KGaA, 2012, pp. 139–163.
20. J. Vogl and W. Pritzkow, *J. Anal. Atom. Spectrom.*, 2010, **25**, 923–932.
21. J. Vogl, M. Rosner and W. Pritzkow, *Anal. Bioanal. Chem.*, 2012, 1–8.

# Magnetic Sector Field Instruments

In the following chapters, the currently most important techniques of elemental and isotopic mass spectrometers using a magnetic sector field are described in detail.

CHAPTER 11

# *Magnetic Sector Field Instruments*

THOMAS PROHASKA

University of Natural Resources and Life Sciences Vienna (BOKU),
Department of Chemistry, Division of Analytical Chemistry,
VIRIS Laboratory for Analytical Ecogeochemistry, Tulln, Austria
Email: thomas.prohaska@boku.ac.at

In the following chapters, the currently most important techniques of elemental and isotopic mass spectrometers using a magnetic sector field are described in more detail. Where more specific technical explanations are required in addition to the fundamental knowledge given in Chapters 3–10, these are given in the respective chapters.

Chapter 12 is addressed to inductively coupled plasma mass spectrometry (ICP-MS) using both single collector or multicollector configuration.
Chapter 13 is addressed to glow discharge mass spectrometer using a magnetic sector field for mass separation (GDMS).
Chapter 14 is addressed to thermal ionisation mass spectrometry (TIMS).
Chapter 15 is addressed to secondary ionisation mass spectrometry (SIMS).
Chapter 16 is addressed to gas source mass spectrometry (IRMS) covering the analysis of light stable isotopes.
Chapter 17 is addressed to static IRMS for the analysis of noble gas isotopes.

New Developments in Mass Spectrometry No. 3
Sector Field Mass Spectrometry for Elemental and Isotopic Analysis
Edited by Thomas Prohaska, Johanna Irrgeher, Andreas Zitek and Norbert Jakubowski

Published by the Royal Society of Chemistry, www.rsc.org

In each chapter, the currently available instruments on the market are briefly described as they provide specific features and interesting details excelling the general instrumental descriptions. You will not find more detailed and comprehensive information on today's instruments than in this book. Table 11.1 provides an overview of the features of selected currently available instruments. (***Note***: in most cases the instrumental details were provided by the manufacturers. In some cases the manufacturers were not able to provide the requested instrumental specifications).

The following instrumental manufacturers are leading the field of sector field mass spectrometers described in this book:

**Australian Scientific Instruments Pty Ltd., Fyshwick, Australia**
(http://www.asi-pl.com.au)
Australian Scientific Instruments Pty Ltd (ASI) was established in 1993 to commercialise a Secondary High Resolution Ion Micro Probe (*SHRIMP*) analytical instrument developed at the Australian National University's (ANU's) Research School of Earth Sciences. In 1998, ASI became a proprietary company with the intent and ability to commercialise products not only from the ANU but also from other Australian scientific organisations. ASI is a wholly owned subsidiary of ANU Enterprise Pty Ltd, which itself is wholly owned by ANU. Even though the major contribution have been SIMS machines, thermochronology and laser ablation are other core activities of the company.

**CAMECA, Gennevilliers Cedex, France** (http://cameca.com)
CAMECA was founded in 1929 as a subsidiary of Compagnie générale de la télégraphie sans fil (CSF). First called "Radio Cinema", its mission was to design and manufacture movie projectors for large cinema screening rooms at the time of the emergence of talking pictures. After World War II, the company also manufactured scientific instruments developed in French university laboratories: spark spectrometers in the early 1950s, Castaing microprobes from 1958, secondary ion analysers from 1968. The company settled in its factory of Courbevoie soon after 1950 where it remained for more than fifty years. It is now headquartered in Gennevilliers, also near Paris.

In 1954, the company was renamed CAMECA, standing for Compagnie des Applications Mécaniques et Electroniques au Cinéma et à l'Atomistique. The business of movie projectors was stopped in the early 1960s. From 1977, the year of the launching of the *IMS 3f*, CAMECA had a virtual monopoly in the field of magnetic SIMS, but shared the market of Castaing microprobes with Japanese competitors. The semiconductor industry became an important outlet for magnetic SIMS, and the company diversified offering in-fab metrology solutions for semiconductor manufacturers. At the end of the 20th century, CAMECA got a foothold in a third analytical technique: "atom probe tomography". CAMECA left the Thomson-CSF group in 1987 and eventually joined the US headquartered group Ametek. Imago Scientific Instruments, a US start-up specialised in atom probes, was acquired in 2010, adding a new production facility in Madison, Wisconsin, USA. Since 1975, the number of employees has been around 200. Offices opened in the US, Japan, Korea, Taiwan, Germany, and more recently in China, India and Brazil. These outlets are committed to sales and maintenance activities and employ a few dozen people.

In 2014, CAMECA evolves in two different markets: scientific instruments dedicated to research activities and metrology equipment for the semiconductor industry. CAMECA's metrology solutions address semiconductor fabrication cleanrooms with a dedicated version of the Castaing electron probe based on the LEXES technique (low-energy electron-induced X-Ray emission spectrometry) developed at the beginning of the 21st century. The CAMECA instruments are widely recognised by the international academic community in such fields as geochemistry, planetary science and microbiology, and have been cited dozens of times in scientific journals such as Nature and Science.

**Compact Science Systems, Newcastle-Under-Lyme, United Kingdom** (http://www.compactsciencesystems.com)
Compact Science Systems is a British-based company focusing on the production of compact, rugged and easy to operate IRMS instruments. Following this concept, the company designed and built an IRMS system for the Beagle II Mars Lander for the European Space Agency.

**IsoPrime-elementar, Cheadle, United Kingdom** (http://www.isoprime.co.uk/)
The company Vacuum Generators (VG) in the UK was started in the 1960s for the manufacture of high-vacuum components. In 1969 Vacuum Generators got involved in mass spectrometry as VG-Micromass. In 1973, VG-Micromass split its businesses into several companies, including the IRMS business, which was operated as VG-Isotopes. In the 1980s VG-Isotopes was split into VG Isotopes and VG Isogas dedicated to IRMS. In 1989, these two companies were again merged to create VG Isotech. VG Instruments was sold to BAT, and in the late 1980s, BAT sold VG Instruments to Fisons, where it was included in the newly constituted Fisons Instruments. Thermo Instruments acquired Fisons Instruments from Fisons in 1994, but was required to divest, *inter alia*, the Isotope ratio mass spectrometry products of Fisons Instruments. These were acquired by a small group of investors and incorporated as Micromass. Micromass was later sold to Waters. Waters shortly divested itself of the isotope technologies, which were taken into private ownership by a small group of investors as GVI, and they took the rights to the Isoprime developed in 1997. 10 years later, GVI was acquired by Thermo Electron Corporation. Once again, the British government forced divestiture of several of the isotope technologies. The stable isotope technology, including the Isoprime, was sold to Elementar Analysensysteme GmbH, a manufacturer of elemental analysers in Hanau, Germany, who set it up as Isoprime Ltd in the UK. Isoprime Ltd. also provides a suite of inlet peripheral systems for all stable isotopic analysis.

**Table 11.1** Instrumental specifications of selected currently available sector field instruments.

| Technique | Instrument | Company | Single/ multicollector | Electron multiplier: Counting mode | Electron multiplier: Analogue mode | Faraday cups | Instruments with extended dynamic range (electron multiplier and Faraday cup) |
|---|---|---|---|---|---|---|---|
| ICP-MS | *Element 2* | Thermo Fisher Scientific Inc. | single-collector | x | x | | |
| ICP-MS | *Element XR* | Thermo Fisher Scientific Inc. | single-collector | x | x | x | x |
| ICP-MS | *AttoM* | Nu Instruments Limited | single-collector | x | | x | x |
| ICP-MS | *NEPTUNE (Plus)* | Thermo Fisher Scientific Inc. | multicollector | x | | x | |
| ICP-MS | *Nu Plasma II* | Nu Instruments Limited | multicollector | x | | x | |
| ICP-MS | *Nu Plasma 1700* | Nu Instruments Limited | multicollector | x | | x | |
| ICP-MS | *SPECTRO MS* | SPECTRO Analytical Instruments GmbH | multicollector | | | | |
| GDMS | *Element GD* | Thermo Fisher Scientific Inc. | single-collector | x | x | x | x |
| GDMS | *Astrum* | Nu Instruments Limited | single-collector | x | | x | x |
| GDMS | *Autoconcept GD90* | Mass Spectrometry Instruments Ltd. | single-collector | x | x | x | |
| TIMS | *TRITON Plus* | Thermo Fisher Scientific Inc. | multicollector | x | | x | |
| TIMS | *Phoenix/ Phoenix62* | Isotopx Limited | multicollector | optional | | x | |
| TIMS | *Nu TIMS* | Nu Instruments Limited | multicollector | x | | x | |
| IRMS | *MAT 253* | Thermo Fisher Scientific Inc. | multicollector | | | x | |
| IRMS | *MAT 253 ultra* | Thermo Fisher Scientific Inc. | multicollector | x | | x | |
| IRMS | *Delta V Advantage/ Delta V Plus* | Thermo Fisher Scientific Inc. | multicollector | | | x | |
| IRMS | *Horizon* | Nu Instruments Limited | multicollector | | | x | |
| IRMS | *Perspective* | Nu Instruments Limited | multicollector | | | x | |

| Daly PAD | CMOS AP DCD | Microchannel plate | RAE | RF | Achievable resolution (or resolving power RP) | Geometry | Focusing | Electric sector radius/ angle | Magnetic sector radius/angle |
|---|---|---|---|---|---|---|---|---|---|
| | | | | | >10 000 | reverse Nier Johnson | double | 105 mm/ 90 deg | 160 mm/ 60 deg |
| | | | | | >10 000 | reverse Nier Johnson | double | 105 mm/ 90 deg | 160 mm/ 60 deg |
| | | | | | 10 000 | Nier Johnson | double | 320 mm/ 120 deg | 250 mm/ 60 deg |
| | | | | | 10 000 | Nier Johnson | double | 220 mm/ 90 deg | 230 mm/ 90 deg |
| | | | | | 3000 (1000 RP) | Nier Johnson | double | 350 mm/ 70 deg | 250 mm/ 70 deg |
| | | | | | 5000 | Nier Johnson | double | 943 mm/ 70 deg | 750 mm/ 70 deg |
| | x | | | | 500 | Mattauch–Herzog | double | 160.3 mm/ 31.82 deg | 90 deg |
| | | | | | >10 000 | reverse Nier Johnson | double | 105 mm/ 90 deg | 160 mm/ 60 deg |
| | | | | | >10 000 | Nier Johnson | double | 320 mm/ 120 deg | 250 mm/ 60 deg |
| x | | | | x | 80 000 | Nier Johnson | double | 90 deg | 90 deg |
| | | | | | 500 | – | single | – | 230 mm/ 90 deg |
| x | | | | | 450 | – | single | – | 270mm/ 90 deg |
| x | | | | | 450 | – | single | – | 300 mm/ 70 deg |
| | | | | | 200 | – | single | – | 230 mm/ 90 deg |
| | | | | | >20 000 RP | Nier Johnson | double | 220 mm/ 90 deg | 230 mm/ 90 deg |
| | | | | | 110 | – | single | – | 90 mm/ 90 deg |
| | | | | | 110 | – | single | – | 120 mm/ 90 deg |
| | | | | | 200 | – | single | – | 240 mm/ 75 deg |

**Table 11.1** (*Continued*)

| Technique | Instrument | Company | Single/ multicollector | Electron multiplier: Counting mode | Electron multiplier: Analogue mode | Faraday cups | Instruments with extended dynamic range (electron multiplier and Faraday cup) |
|---|---|---|---|---|---|---|---|
| IRMS | *Panorama* | Nu Instruments Limited | multicollector | | | x | |
| IRMS | *SerCon 20-22 IRMS* | Sercon Limited | multicollector | | | x | |
| IRMS | *SerCon MIRA* | Sercon Limited | multicollector | | | x | |
| IRMS | *IsoPrime100* | IsoPrime Limited | multicollector | | | x | |
| IRMS | *IDmicro* | Compact Science Systems Limited | multicollector | | | x | |
| IRMS | *Isologger* | Compact Science Systems Limited | multicollector | | | x | |
| noble gas IRMS | *ARGUS VI* | Thermo Fisher Scientific Inc. | multicollector | x | | x | |
| noble gas IRMS | *HELIX MC Plus* | Thermo Fisher Scientific Inc. | multicollector | x | | x | |
| noble gas IRMS | *HELIX SFT* | Thermo Fisher Scientific Inc. | multicollector | x | | x | |
| noble gas IRMS | *Noblesse* | Nu Instruments Limited | multicollector | x | | x | |
| noble gas IRMS | *NGX* | Isotopx Limited | multicollector | x | | x | |
| SIMS | *SHRIMP II* | Australian Scientific Instruments Pty Ltd. | single or multicollector | x | | x | x |
| SIMS | *SHRIMP IIe* | Australian Scientific Instruments Pty Ltd. | single or multicollector | x | | x | x |
| SIMS | *SHRIMP RG* | Australian Scientific Instruments Pty Ltd. | single-collector | x | | x | x |
| SIMS | *SHRIMP SI* | Australian Scientific Instruments Pty Ltd. | multicollector | x | | x | x |
| SIMS | *IMS 7f/6fR/ WF SC Ultra* | CAMECA | single-collector | x | | x | x |
| SIMS | *IMS 1280* | CAMECA | multicollector | x | | x | x |
| SIMS | *NanoSIMS 50L* | CAMECA | multicollector | x | | x | x |

| Daly PAD | CMOS AP DCD | Microchannel plate | RAE | RF | Achievable resolution (or resolving power RP) | Geometry | Focusing | Electric sector radius/ angle | Magnetic sector radius/angle |
|---|---|---|---|---|---|---|---|---|---|
| x | | | | | undefined | Matsuda | double | 1020 mm/85 deg | 800 mm/ 72 deg |
| | | | | | 90 | – | single | – | 110mm/ 120 deg |
| | | | | | 250 | – | single | – | 250 mm/ 120 deg |
| | | | | | 100 | – | single | – | 90 mm/ 90 deg |
| | | | | | 75 | – | single | – | 140 mm/ 90 deg |
| | | | | | 75 | – | single | – | 140 mm/ 90 deg |
| | | | | | >200 | – | single | – | 130 mm/ 90 deg |
| | | | | | >5000 RP | – | single | – | 350 mm/ 120 deg |
| | | | | | >2000 RP | – | single | – | 350 mm/ 120 deg |
| | | | | | 800 (3000 RP) | – | single | – | 240 mm/ 75 deg |
| | | | | | >600 | – | single | – | 270 mm/ 90 deg |
| | | | | | 13 000 | Matsuda | double | 1270 mm/85 deg | 1000 mm/ 72.5 deg |
| | | | | | 13 000 | Matsuda | double | 751 mm/ 89 deg | 1000 mm/ 46 deg |
| | | | | | 35 000 | reverse Matsuda | double | 751 mm/ 89 deg | 1000 mm/ 46 deg |
| | | | | | 13 000 | Matsuda | double | 1270 mm/85 deg | 1000 mm/ 72.5 deg |
| | | x | optional | | 20 000 | Nier Johnson | double | 85 mm/ 90 deg | 117 mm/ 90 deg |
| | | x | optional | | 40 000 | Nier Johnson | double | 585 mm/ 90 deg | 585 mm/ 90 deg |
| | | | | | 20 000 | Mattauch–Herzog | double | 100 mm/ 90 deg | 145–680 mm/90 deg |

**IsotopX Ltd, Middlewich, United Kingdom** (http://isotopx.com)
Isotopx is a British company specialising in the design and manufacture of thermal ionisation mass spectrometers and was founded in 2008. The company traces its heritage to the very first commercially developed TIMS instrument launched by VG Micromass in 1973, the *VG MM30*. Recently, the company announced a new multicollector noble-gas mass spectrometer.

**Mass Spectrometry Instruments Ltd., Dewsbury, United Kingdom** (http://www.massint.co.uk)
Mass Spectrometry Instruments Ltd (MSI) in Dewsbury, West Yorkshire, UK was established in 1994 with the blessing of Kratos Analytical Instruments, when its owner Shimadzu Corporation decided to exit the magnetic sector instruments. MSI had the responsibility of supporting all the installed organic Instruments of Kratos. The initial aim of the company was to improve the after-sales service. A new development program was established to manufacture a new series of high-performance instruments. In early 1998, the first fully automated *Autoconcept Environmental Instrument* was launched. The success of the *Autoconcept Instrument* enabled MSI to jointly work with a petroleum company to develop a fully automated All Glass Heated Inlet System. On the demise of the *VG9000*, the need for its replacement created an opportunity to redevelop the *Concept GD*, which was launched by Kratos.

**Nu Instruments, Wrexham, United Kingdom** (http://www.nu-ins.com/)
Nu Instruments is a British-based company founded in 1995 by Phil Freedman, a former VG R&E scientist. Its initial focus was to develop and market MC ICP-MS technology. Nowadays, Nu instruments provides sector field devices for ICP-MS, IRMS ad TIMS.

**SerCon, Crewe, United Kingdom** (http://www.sercongroup.com/)
Sercon continues the pioneering efforts of Europa Scientific, which was founded in 1984 on the development of continuous-flow isotope ratio mass spectrometry techniques by Preston and Owens. Europa Scientific produced the Tracermass, the first benchtop CF-IRMS. In 1993 the 20-20 was released. During 2010, the 20-22 was released onto the market followed by a number of developments in mass spectrometry. 2012 saw the launch of the *Sercon MIRA* (multiple isotopologue ratio analyser), a new generation of large-radius high-sensitivity machine targeting clumped isotopes and other species such as organohalogens and mineral isotopes.

Sercon are the UK distributor for the complete range of isotopically labelled compounds as well, and supply products of a full range of enrichments and of a large variety of label positions in any quantity required.

**Spectro Analytical Instruments GmbH, Kleve, Germany** (www.spectro.com)
SPECTRO is a member of the AMETEK Materials Analysis Division, which is a global manufacturer of electronic instruments and electromechanical products. AMETEK consists of two operative units: Electronic instruments

and Electromechanical. The Electronic Instruments Group manufactures advanced instruments for the process, aerospace, power and other industries. The Electromechanical Group is a differentiated supplier of electrical interconnects, precision motion-control solutions, specialty metals, thermal management systems, and floor care and specialty motors.

SPECTRO was founded in 1979 and has become a leading supplier of analytical instruments for optical emission and X-ray fluorescence spectrometry and has recently launched a multicollector ICP-MS based on a Mattauch–Herzog geometry.

**Thermo Fisher Scientific, Waltham, Massachusetts, USA**
(http://www.thermoscientific.com)
In 1947, Dr. Ludolf Jenckel, a young physicist at the Atlas-Werke in Bremen, Germany started to build in his spare time a sector field mass spectrometer. After finishing the first prototype, a small division Atlas MAT (Mess- und Analysentechnik) was set up. In 1950 an improved version – the *CH3* – came out. The main application was the quantitative analysis of hydrocarbon mixtures. In the early 1950s MAT started its isotope activities. The first instruments were based on the *CH3* and *CH4*. The first dedicated isotope ratio sector field mass spectrometer was introduced in 1962, namely the *M86*. In these days, IRMS manufacturing was widely distributed in the 1970s (USA, UK, Germany, Russia, Japan) but ended up being centred finally in MAT in Bremen and VG in the UK. In 1967 Varian acquired MAT and later on in 1981, Varian sold MAT to Finnigan Corporation (US) and renamed the company in Bremen Finnigan MAT, to recognise the contributions of both factories. In the early 1990s Finnigan MAT was acquired by Thermo Instruments, a part of Thermo Electron and - after several name changes (Thermo Finnigan, and ThermoQuest Corporation) – is now operated as part of the Chromatography and Mass Spectrometry division of Thermo Fisher Scientific. There has been complete continuity in the production of isotope ratio mass spectrometers from the 1970s to the present day and an addition sector field mass spectrometer for analysis of trace elements in 1993.

CHAPTER 12

# *Inductively Coupled Plasma Mass Spectrometry*

NORBERT JAKUBOWSKI,*[a] MONIKA HORSKY,[b] PETER H. ROOS,[c] FRANK VANHAECKE[d] AND THOMAS PROHASKA*[b]

[a] BAM – Federal Institute for Materials Research and Testing, Berlin, Germany; [b] University of Natural Resources and Life Sciences Vienna (BOKU), Department of Chemistry, Division of Analytical Chemistry, VIRIS Laboratory for Analytical Ecogeochemistry, Tulln, Austria; [c] Ruhr University Bochum, Department of System Biochemistry, Medical Faculty, Germany; [d] Ghent University, Department of Analytical Chemistry, Ghent, Belgium
*Email: norbert.jakubowski@bam.de; thomas.prohaska@boku.ac.at

ICP-MS is based on the formation of (preferentially monovalent positively) charged atomic ions in an inductively coupled Ar plasma at almost 10 000 K. The ions formed are transferred from the plasma source at ambient pressure into a mass separator operated at high vacuum *via* a set of cones. The ions are separated according to their mass/charge ratio in the mass separator (quadrupole, magnetic sector field or time-of-flight mass separator). In most cases, the ions are detected using a secondary electron multiplier; in some set-ups (also) a Faraday cup can be used. Single-collector (scanning mass spectrometer usually used for quantitative elemental analysis) or multicollector (static operation of mass spectrometer for precise isotope ratio analysis) configurations can be found (Figures 12.1 and 12.2).

New Developments in Mass Spectrometry No. 3
Sector Field Mass Spectrometry for Elemental and Isotopic Analysis
Edited by Thomas Prohaska, Johanna Irrgeher, Andreas Zitek and Norbert Jakubowski

Published by the Royal Society of Chemistry, www.rsc.org

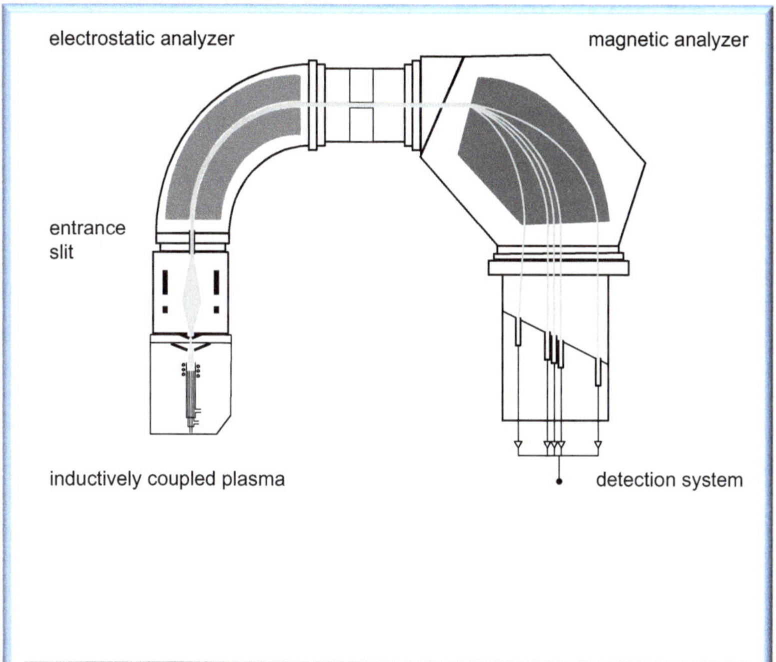


| | |
|---|---|
| Type of instrument | ICP-MS |
| Type of analysis | quantification, isotope ratios |
| Sample types | liquid, solid |
| Ionisation source | inductively coupled plasma |
| Type of detectors | SEM, FC, channeltron, Daly, MCP, ion CCD |
| First commercially available SF-instrument (year) | ICP-SFMS: Plasmatrace (1988)<br>MC ICP-SFMS:<br>Plasma 54 MC ICP-SFMS (1992) |

**Figure 12.1** Features and general schematics of a sector field ICP-MS. (Source: Ondrej Hanousek.)

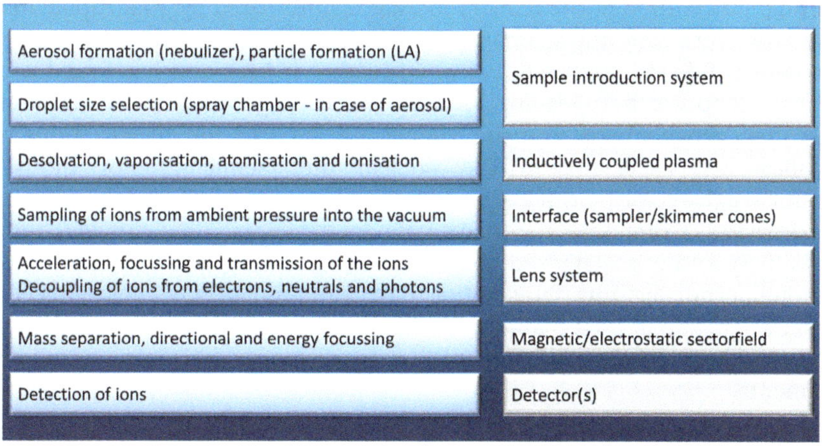

Aerosol formation (nebulizer), particle formation (LA)
Droplet size selection (spray chamber - in case of aerosol)
Sample introduction system
Desolvation, vaporisation, atomisation and ionisation
Inductively coupled plasma
Sampling of ions from ambient pressure into the vacuum
Interface (sampler/skimmer cones)
Acceleration, focussing and transmission of the ions
Decoupling of ions from electrons, neutrals and photons
Lens system
Mass separation, directional and energy focussing
Magnetic/electrostatic sectorfield
Detection of ions
Detector(s)


**Figure 12.2** Operational flow chart of an ICP-MS.

**Name and Abbreviation:** The IUPAC-conform (IUPAC: International Union of Pure and Applied Chemistry) abbreviation of the technique is ICP-MS, coined by the inventors of the technique.[1] ICP-SFMS (ICP sector field MS) is commonly used in the literature for indicating that a sector field is used as mass separator. Sometimes, the instrument is referred to as HR-ICP-MS (high-resolution ICP-MS). This term would be correct if the ICP mass spectrometer was used at high mass resolution. Since other ICP-MS devices can be operated at high resolution, as well, the correct abbreviation would be HR-ICP-SFMS, even though (according to IUPAC), the hyphen would indicate the combination of two techniques, which is not the case. Thus, the latter two abbreviations should be omitted for these reasons. Multicollector (MC) devices are usually referred to as MC-ICP-MS, even though the hyphen would, again, indicate the combination of two techniques – which is not the case. Thus, MC ICP-MS would seem to be more systematic, although MC-ICP-MS is commonly used. As all commercially available MC instruments are sector field-based machines, the abbreviation MC-ICP-SFMS is rarely used. ICP-MS and MC ICP-MS are used as abbreviations throughout this book. ICP-SFMS is used when a particular difference to other mass separators used for ICP-MS needs to be made.

## 12.1 General

30 years after the famous publication of Houk *et al.*, ICP-MS has matured to become one of the most important methods in atomic spectroscopy.[1] The success of ICP-MS is based on the fact that nearly all elements of the periodic table can be determined with their isotopic composition, spectra are easy to interpret and detection limits at pg $g^{-1}$ levels and below can be achieved with a dynamic range of more than 9 orders of magnitude. Additional advantages are a powerful ion source operated under atmospheric pressure, thus allowing straightforward sample introduction and higher sample throughput. Although the ICP-MS market is dominated by quadrupole mass spectrometers with approximately unit mass resolution, the current generation of ICP-SFMS instruments has a broad user base and has found widespread adoption by metrological labs, research groups in environmental research, geochemistry and life sciences, as well as for diverse industrial applications, especially in the nuclear and semiconductor industry. Even though sector field ICP-MS instruments are up to a factor of five more expensive as compared to quadrupole instruments, they provide higher sensitivity and especially the possibility of measuring at higher mass resolution, permitting interference-free determination of nuclides subject to spectral overlap at unit mass resolution. Moreover, they also show the capability of high-precision isotope ratio measurements for a large number of elements.

The main components of commercially available sector field instruments are very similar, although the design of instruments varies significantly between manufacturers. This holds true for both single- and multicollector devices. In this section, we discuss the fundamentals and the peculiarities of both types of ICP-MS instruments: (1) Single-collector instruments, which are equipped with one detector (or more detectors at the same position used to enhance the linear dynamic range). The ions are detected sequentially, and the instruments are operated under dynamic (= scanning) conditions. (2) Multicollector instruments have a set of parallel detectors, allowing simultaneous detection of ions at different mass-to-charge ratios. These instruments are usually operated under static (= fixed magnet field) conditions.

### 12.1.1 Single-Collector ICP-MS

The intended use of a single-collector ICP-MS is the accurate determination of elements at (ultra-) trace levels. After its advent, it has been a tremendous success and has become one of the most important, if not the most important, analytical methods for trace and ultratrace analysis. By fast sequential scanning, isotope ratios can be measured with single-collector devices although the precision achievable is significantly decreased than that achievable *via* multicollector instruments. Nonetheless, single-collector ICP-MS is also a valuable tool for IDMS (isotope dilution mass spectrometry – see Chapter 8 and a new RSC book of Alonso and González)[2] applications or in tracer experiments. In general, two companies drove the commercialisation of ICP-MS: VG Instruments in the UK and Sciex in Canada. The first instruments were based on quadrupole mass separators and there was limited reason to come up with a more complex and more expensive sector field instrument. Only VG pursued the development of ICP sector field instruments based on their existing sector field organic mass spectrometers. The basic instrumental design originated from the *VG series 70*, a high-resolution mass spectrometer with Nier–Johnson geometry used for analysis of organic compounds. The first sector field instrument with an ICP source, the *Plasmatrace*, was produced in 1988. (Figure 12.3) JEOL introduced the *JMS-Plasmax* ICP-SFMS in 1993 (followed by *JMS-Plasmax 2* in 1995), which - like VG's *Plasmatrace* – was based on a previously existing organic mass spectrometer. The JEOL instruments were not successful outside of Japan and the series was discontinued.

The very high purchase price and the complexity of operation were the main issues why the first double-focusing single-collector instruments were not widely adopted. This fact served as an incentive for Finnigan MAT to develop a novel "plasma monitor" based on a sector field instrument. The main design criterion for this instrument was a smaller magnetic sector with a radius of 16 cm in a reverse Nier–Johnson geometry, resulting in reduced production cost and thus lower price, without compromising analytical figures of merit. The instrument was designed to achieve a mass resolution of 3000 in the medium and 7500 in the high mass resolution mode to resolve most of the commonly encountered interferences. The first hardware of this

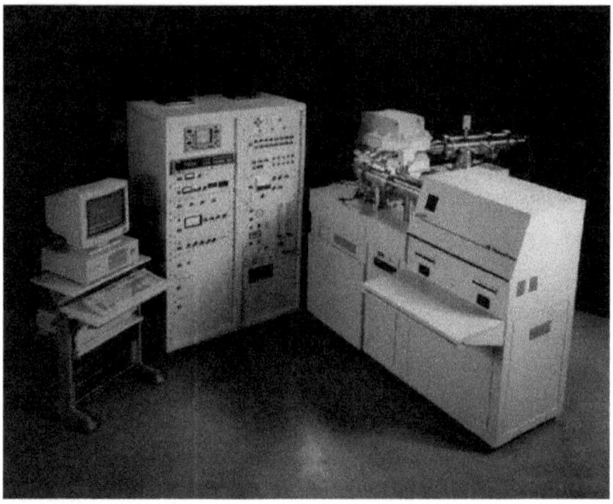

**Figure 12.3** *Plasmatrace*, the first commercial sector field ICP-MS (1988). (Source: courtesy of Thermo Fisher Scientific.)

new instrument was already assembled at the end of 1991, but it took some further time to commercialise the instrument. The resulting instrument, the *Element*, was presented in 1993 and was introduced to the market in 1994.[3–6] Improved versions (the *Element 2* and the *Element XR*) were launched in 1998 and 2004, respectively.

Other instruments launched onto the market are no longer commercially available. After it had acquired the assets of the former VG Elemental subsequent to Thermo's acquisition of Fisons Instruments, Thermo Optek introduced a new sector field ICP-MS, the *Axiom*, in 1998. The *Axiom* was discontinued in 2002, leaving the *Element 2* and *Element XR* as the only commercially available sector field instruments for some time. In 2004, Nu Instruments introduced a new high-resolution double-focusing sector field instrument with Nier–Johnson geometry and vertical orientation, the *Nu AttoM*. The instrument has – like the *Plasmatrace* and *Axiom* – a continuously variable slit system, which can cover resolution settings between 300 and more than 10 000. More instrumental details of currently commercially available instruments will be given in Section 12.3. Currently, sector field instruments have become significantly cheaper in price, more robust and easier to operate, making them a valuable option for ultratrace elemental analysis.

### 12.1.2 Multicollector ICP-MS

The intended use of MC ICP-MS is the precise and accurate determination of isotope ratios of basically all isotopic systems that can be assessed by ICP-MS. The limitation of scanning single-collector instruments for isotope ratio measurements is the timely staggered detection, as a result of which instabilities of the plasma (flicker noise) or of the sample introduction system or drift in the signal intensity deteriorate the isotope ratio precision substantially, especially when short transient signals (*e.g.* in combination with chromatography, laser ablation or in the context of nanoparticle analysis) are assessed. A considerably higher precision can be attained with a multi-collector array.[7]

MC ICP-MS instruments have developed into dedicated tools for isotopic analysis and nowadays, they display isotope ratio precisions similar to those attainable using thermal ionisation mass spectrometry (TIMS), *i.e.* 0.001% RSD at low mass resolution and 0.005% RSD at higher mass resolution, under optimum conditions.[8] ICP-MS also permits atoms of elements with an ionisation energy > 7.5 eV to be efficiently converted into $M^+$ ions, thus overcoming one of the most important drawbacks of TIMS.[9] Although MC ICP-MS is used increasingly for "traditional" TIMS applications, both instruments have their (often complementary) qualification in isotope ratio analysis.

Since the introduction of the first MC ICP-SFMS system in 1992 (the *Plasma 54* from VG Elemental) (Figure 12.4), nearly 250 MC ICP-SFMS units have been installed in more than 200 different laboratories worldwide. The *Plasma 54* was a hybrid of the VG *Plasmatrace* ICP-SFMS and the multi-collector system of the *Sector 54* TIMS instrument. This instrument with an s-type-geometry consists of two electric sectors before and after a magnetic sector with a radius of 54 cm, giving the instrument its name *Plasma 54*. It was equipped with an array of adjustable Faraday collectors and a Daly detector. In this design, the ICP and interface was floating at high potential, while the rest of the instrument was operated at ground potential. The first isotope ratio measurements with this instrument were published by Walder and Freedman in 1992[10] and already one year later an application involving

**Figure 12.4** *Plasma 54*, the first commercial sector field MC ICP-MS (1992). (Source: courtesy of Thermo Fisher Scientific.)

direct analysis of solid samples by using laser ablation as a means of sample introduction was reported.[11]

Just at the beginning, the acceptance of the newly launched instrument was diffident and only a few units were purchased. Nevertheless, the potential of such an instrument was realised to be high and thus, Micromass (later GV instruments), a management buyout company from former VG, introduced the *IsoProbe-P*, which was based to a certain extent on the technology of the *Plasmatrace 2* (ICP and front end design) and the magnet and collector array from the *Sector 54* TIMS instrument. This instrument is a single magnetic sector field device only. Energy focusing is achieved by using a pressurised hexapole collision cell in front of the magnetic sector.

Phil Freedman took the experience gained with the *Plasma 54* and started to design a new MC ICP-SFMS instrument with the newly founded company Nu Instruments in 1995. The *Nu Plasma* was introduced in 1997.[12] The instrument is based on a Nier–Johnson double-focusing geometry and uses a fixed detector array in combination with zoom optics. Later, the instrument was completed with a high-resolution slit system (*Nu Plasma HR*). In 2010, improvements with respect to the detector array, electronics and pumping system led to the launch of the *Nu Plasma II*. A larger geometry instrument, the *Nu Plasma 1700*, was introduced in 2002. The large geometry and large dispersion provide mass resolution of up to 5000, while maintaining flat-top peaks.

Thermo Optek added a multicollector array to the *Axiom,* thus realising the *Axiom MC* in 1998, which was later distributed by Thermo Elemental. This instrument is no longer commercially available. Thermo Quest (now subsumed into Thermo Fisher Scientific) introduced the *Neptune* in 1999. It shares the ICP interface with the *Element 2* and the multicollector technology with the *Triton* TIMS instrument.

Even though the number of detectors is increasing steadily, the total number of isotopes that can be analysed simultaneously is still limited, reducing the multi-isotope capabilities, especially for short transient signals (*e.g.* with LA-ICP-MS). In principle, it should nowadays be possible to design small cups by micromachining, so that all isotopes can be measured continuously as they are separated by a magnet and imaged simultaneously at the same focal plane.[13] This idea is not new and was realised by using a double-focusing sector field device with a Mattauch–Herzog geometry and a focal plane camera (*SPECTRO MS*, Spectro Analytical Instruments, Kleve, Germany).[14]

## 12.2 Technical Background

Most parts in front of a sector field device are very similar to those of other ICP-MS devices, such as quadrupole instruments, and the "front side" consists of the sample introduction system, an inductively coupled plasma ion source and a sampling interface. After extraction, the ions are accelerated in a sector field device over a potential difference of up to

10 keV by using an acceleration lens. The central piece of the sector field instrument is the mass separator (including the slit systems, the widths of which determine the achievable mass resolution of the instrument): entrance slit, magnetic sector, electric sector and exit slit. Transfer or filter optics are used after the exit slit. Finally, the ions are detected in a detection system, most often discrete dynode electron multiplier(s) or Faraday cup(s).

### 12.2.1 Sample Introduction

The robustness of the ion source and the fact that the ion source is operated under ambient pressure enables a large variety of sample introduction systems. In principle, all sample introduction systems, which are used with quadrupole-based ICP-MS instruments, can be used in combination with sector field instruments as well. Many "alternative" sample introduction systems have been investigated with the aim of improving the sensitivity, reducing spectral interferences or handling small volumes of sample. Other sample introduction systems used in the context of speciation (*e.g.* chromatographic systems) or of direct analysis of solids (*e.g.* laser ablation) generate transient signals. Most of these investigations are focused on a specific application to achieve specific improvements and thus will be discussed in the application section (Section 12.5).

**Pneumatic Nebulisers**

The most commonly used nebulisers for analytical plasma source spectrometry are pneumatic nebulisers (for a more detailed description and additional references; the authors refer to the reviews of Sharp,[15,16] as well as to basic literature on ICP-MS.[17–19] Three basic types of pneumatic nebulisers have proved to be useful: concentric, crossflow and Babington-type nebulisers. A pneumatic nebuliser consists of a nozzle to accelerate the transport (or nebuliser) gas and a tubing for introducing the liquid sample into the gas stream. Pneumatic nebulisers working at extremely low sample consumption rates are very common nowadays, but during the last decade, the coupling of low flow rate separation systems with ICP-MS instruments often proved to be challenging. Separation systems operating at low mobile phase flow rates (capillary zone electrophoresis, nano-HPLC) request specific low flow rate nebulisers, such as the microconcentric nebuliser (MCN).[20] For CE-ICP-MS coupling, Schaumlöffel and Prange developed a new interface based on a *modified* microconcentric nebuliser, which permits a flow rate of about 6 μL $min^{-1}$ in the free aspiration mode.[21] This interface construction provided an electrical connection for stable electrophoretic separations and adapted the flow rate of the electro-osmotic flow inside the CE capillary to the flow rate of the nebuliser for efficient transport into the plasma.

Pneumatic nebulisers are typically combined with one of the different types of spray chambers, enabling the removal of larger droplets and resulting in a homogeneous aerosol. Additionally, cooled spray chambers or desolvation systems reduce the solvent load of the plasma, reduce water- and acid-based interferences (*e.g.* H-, O-, N- and Cl-based interfering ions) and improve signal intensities significantly.

Nebulisers and spray chambers manufactured from inert polymers (such as PTFE or PFA) can be used when aggressive chemicals, such as HF, are applied. Another reason for preferring such nebulisers are those applications in which contact of the sample solution with glass needs to be avoided. In this context, these nebulisers are most often combined with a torch injector tube manufactured from sapphire or platinum cones.

**High Efficiency Nebulisers**

ICP-SFMS can be combined with high efficiency nebulisers (*e.g.* ultrasonic nebulisation (USN), thermospray nebulisation (TSN) or direct injection high efficiency nebulisation (DIHEN) to further improve the sensitivity. Depending on the element, the signal can be enhanced between 3- to 20-fold with the latter system compared to a conventional nebuliser-spray chamber arrangement.[22] A microdroplet generator (μDG) sample introduction system with sample uptake rates of less than 20 nL $min^{-1}$ was coupled to a sector field ICP-MS instrument to investigate the analytical figures of merit with respect to single-cell analysis. In this context, the sector field instrument was operated in a fast scanning mode (E-scan) with the shortest time resolution of 100 μs demonstrating a droplet recovery of 100%, which is usually not achieved with pneumatic nebulisation.[23]

**Hyphenation of ICP-SFMS with Chromatographic Systems**

ICP-MS has become a powerful and important element- and isotope-specific detector in chromatography (mainly HPLC, GC).[24] An HPLC system can be coupled to an ICP-MS unit *via* a conventional nebuliser as the HPLC unit provides a liquid sample with a flow rate that is in the range of the sample uptake rate of conventional nebulisers. However, more and more micro- and nanobore columns are applied so that for these low-flow systems also low-flow nebulisers are required. For other separation techniques requiring low-flow conditions, such as CE, special considerations have to be taken into account, as a continuous electric field has to be applied. Coupling a GC is even more challenging, as the gaseous phase is in most cases heated and has to be transferred into the plasma without cooling in order to prevent condensation. Thus, heated interfaces have become common and meanwhile commercially available.[25,26]

**Direct Solid Sample Introduction**

Devices for direct solid sample introduction, such as laser ablation, have gained significant interest, even though the availability of adequate calibration standards is still a prerequisite for obtaining accurate analytical data.[27] In LA-ICP-MS, an ablation cell allowing for rapid transfer of the aerosol (particles and gases) obtained upon ablation to the plasma combined with short washout times and a homogeneous sampling throughout the chamber is required. Laser ablation systems can be combined with liquid nebulisation systems to allow for calibration by using liquid standards.[28]

Electrothermal vaporisation (ETV) as a means of sample introduction allows the direct analysis of solid samples as well. Resano *et al.* describe the coupling of a graphite furnace ETV-unit to an ICP-SFMS instrument and illustrate how the high mass resolution capabilities of the sector field mass spectrometer and the capability of stepwise heating of the furnace with the aim of separating the vaporisation of different compounds in time complement one another.[29] While higher mass resolution allows for overcoming C-based interferences (C originating from the graphite furnace), programmed heating even allows signals of isobaric nuclides to be separated from one another (in time).[30] These isobaric signals would otherwise require a resolving power that is higher than that offered by present-day ICP-SFMS instruments.

**Sample Introduction by Gas-Phase Generation**

Generation of volatile species is a very elegant and often employed methodology for determination of trace and ultratrace levels of selected elements, for matrix/analyte separation or for speciation of elements.[31] Thus, it can be used to improve the analytical figures of merit significantly by overcoming the low aerosol generation efficiency of pneumatic nebulisers (typically below 5%) and to separate the trace elements of interest from complex matrices. The vapours can be directly introduced into the plasma, trapped on cryogenic traps, or – in the case of metallo-organic compounds – can be sampled from the headspace on fibres.

The most popular of these techniques deal with the elements forming volatile hydrides or other volatile compounds, such as atomic mercury or osmium tetraoxide. Usually different derivatisation strategies are implemented through chemical, electrochemical, photochemical or sonochemical gas-phase generation.[32] Aqueous-phase chemical generation of volatile hydrides (chemical vapour generation, CVG) by derivatisation with tetrahydroborate salts (mainly $NaBH_4$ and $KBH_4$) and other borane complexes has been widely employed for addressing elements such as Ge, Sn, Pb, As, Sb, Bi, Se, Te, Hg, Cd and, more recently, several transition and noble metals.[33,34] Recently, photochemical vapour generation (PVG) has been extensively discussed by Sturgeon and coworkers (for more details see Nobrega *et al.*[31]). This approach offers several advantages compared to classical CVG, including ease of implementation, elimination of the need for fresh chemical reductor, simplified flow injection manifolds, a less violent reduction process, which is more compatible with phase separators, reduced generation of $H_2$ (better compatibility with low-power plasma sources), potentially lower blanks and improved limits of detection, as well as being amenable to transition metals, nonmetals and metalloids.

### 12.2.2 Ion Source

The inductively coupled plasma is one of the most successful analytical plasma sources in emission, as well as in mass spectrometry. This plasma is generated electrodeless in an atmospheric-pressure gas (argon), by application of high-frequency voltages to an induction coil located around the top of a torch, consisting of three concentric quartz tubes. Usually three different argon flows are used. A relatively low gas flow rate (inner gas, nebuliser gas, carrier gas) of about 1 L $min^{-1}$, is used to transport the aerosol to and inject it into the hot core of the plasma plume. A much higher flow rate (10–15 L $min^{-1}$) (outer gas, plasma gas or coolant gas), is introduced tangentially. This flow of Ar thermally isolates the outer quartz tube from the plasma and thus prevents melting, while transporting sufficient Ar gas to maintain a stable plasma. The third flow (support gas, intermediate gas or auxiliary gas) can be used for tuning the position of the plasma within the torch (Figure 12.5).

The outer quartz tube is surrounded by a water- or Ar-cooled induction coil (usually Cu tubing in two or three turns), connected to an RF generator power supply with a frequency of either 27.1 MHz or 40.6 MHz. An oscillating magnetic field is developed around the coil, which causes the electrons present in the plasma (after external ignition by a spark) to oscillate at high frequency. The accelerated electrons collide with the Ar atoms, thus exciting and ionising them. The plasma gas is heated to high temperatures (gas-kinetic temperature of 5000–8000 K, ionisation temperature *ca.* 7500 K, excitation temperature 6500–7000 K, electron temperature of about 10 000 K). The electron density is in the range of $10^{15}$ $cm^{-3}$. The ICP is thus maintained by inductive heating and the electrical power applied ranges from 1000 to 1500 Watts, which is much higher than the highest power of typically 100 W for a glow discharge plasma source. Ranges from 500–800 W are usually applied when cold plasma conditions are used. The "cold plasma" technique has been used to improve detection limits in ICP-SFMS significantly by applying reduced plasma power and increased nebuliser gas flow rates. Especially for Ar-based interfering ions, such as $Ar^+$, $ArC^+$, $ArO^+$, $ArCl^+$, $Ar_2{}^+$, the signal intensity is reduced substantially, thus leading to improved LODs for elements such as Ca, K, Cr or Fe. However, these advantages come at

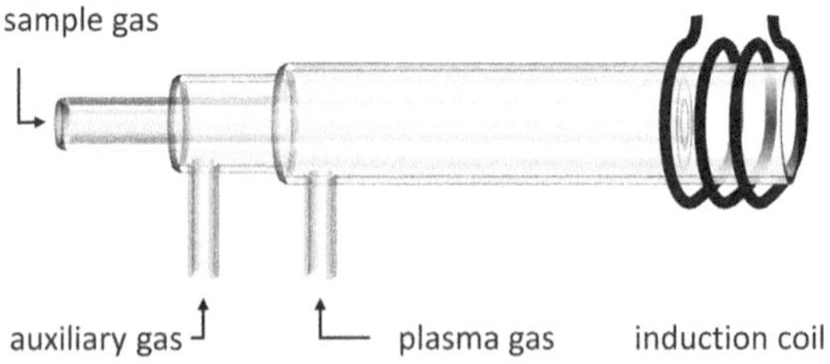


**Figure 12.5** Schematic drawing of a plasma torch. (Source: Ondrej Hanousek.)

the price of reduced signal intensities for elements with a high ionisation energy (*e.g.* Cl, Se), additional spectral interferences resulting from water clusters and increased oxide formation, and thus cold plasma conditions are preferably combined with the use of a membrane desolvation unit.

The injected aerosol (or particles from a laser-ablation system) travels through a narrow axial central channel into the plasma, surrounded by the high-temperature plasma core. The temperature in this channel is still sufficiently high to give rise to efficient volatilisation, atomisation, excitation and possibly ionisation of the sample components. Predominantly singly charged positive ions are generated for most elements in the periodic table. The ionisation efficiencies (%) as calculated from the Saha–Eggert equation (see Chapter 4) using the electron densities and temperatures typical for this discharge show that all elements with first ionisation potentials below 8 eV are completely ionised (>99%) and even elements with first ionisation potentials between 8 and 12 eV are ionised to more than 10%.[35] Thus, the mass spectrum measured consists mainly of signals from singly charged positive atomic ions and the elements can be recognised based on their "isotopic pattern". It has to be noted that even though positive ions are generated predominantly in the ICP, negative ions are formed in the plasma as well, which are not measured in most commercial systems.

Higher sensitivity can also be achieved by applying a "Shielded Torch" or "Guard Electrode" (GE), which is basically a grounded metal foil between the torch and load coil. This grounded metal foil capacitively decouples the coil and the plasma and thus, avoids a potential difference between the plasma and the interface (sampling cone). As a result, the ion energy spread is reduced and the transmission efficiency is improved. The GE is in particular required in some instruments (*e.g.* the *Element 2*), when applying cold plasma conditions to prevent secondary discharges between the ICP and the grounded interface.

### 12.2.3 Sampling Interface

The interface of an ICP ion source is usually water cooled and made of materials with good thermal conductivity. It generally consists of two cones (even though more than two cones can be found), the sampler and skimmer cone, which are in general hat-shaped metal plates with a small central orifice (0.4–1.2 mm) to sample the centre part of the plasma. (Figure 12.6) The diameter of the orifice has to be large enough to allow maximum transmission and reduce clogging, whereas the vacuum still has to be maintained. The skimmer cone is typically sharper and has a smaller orifice than the sampler cone. The cones are placed coaxially at a short distance (several mm) from one another, allowing a sequential pressure decrease. Shape, material and orifice diameter can have an impact on the instrumental performance in terms of, *e.g.*, transmission efficiency or formation of interfering polyatomic ions. To reach higher sensitivity, high-performance cones have been developed, which usually have sharper edges but are more

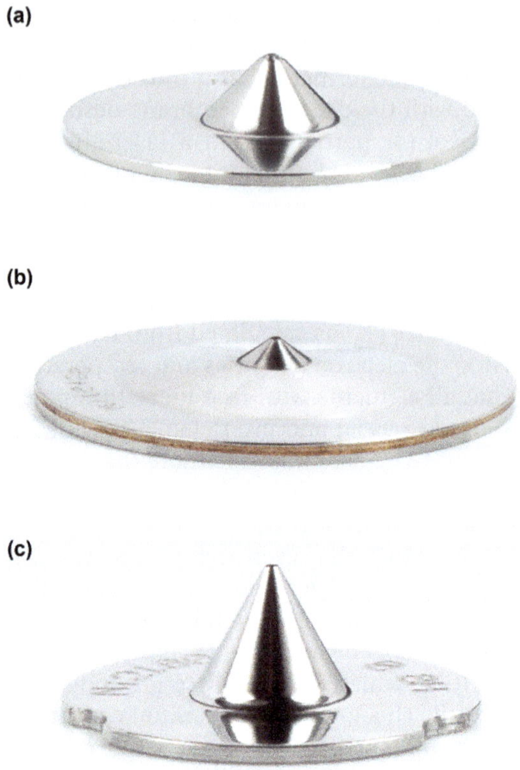


**Figure 12.6** Photo of a set of cones (a) Jet cone; (b) Sampler cone; (c) Skimmer cone. (Source: © Thermo Fisher Scientific.)

prone to erosion. However, changes in mass discrimination are observed when such cones are applied in the context of isotope ratio measurements using MC ICP-MS (see Chapter 6).[36] Cones are usually made of Ni, although Pt-covered cones are preferred for low Ni measurements or for analysing corrosive solutions. Cones manufactured from Al are used as well, especially in combination with laser ablation. In order to reduce effects of electrostatic coupling between the load coil and the plasma discharge – resulting in an electrical discharge between plasma and sampler cone – the load coil is usually grounded at the side facing the sampler cone.

It has been proven that reducing the vacuum at the interface has positive effects on the measurement performance.[37] For example, the Jet interface of the *Element 2*/*Element XR* consists of a modified sampler (Jet-sampler) and skimmer (X-skimmer) cone geometry and a high-capacity dry interface pump. This Jet interface enhanced the sensitivity of the ICP-SFMS operated in dry plasma conditions by one order of magnitude and an additional factor of five was obtained by using the *APEX* sample introduction system (ESI, Omaha, NE) or the *ARIDUS II* membrane desolvator (Cetac

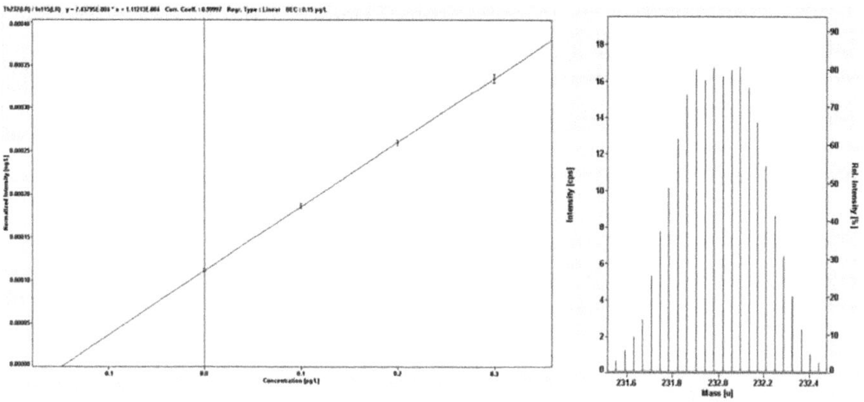


**Figure 12.7** Standard addition calibration curve with additions of (left) 0.1, 0.2 and 0.3 pg $L^{-1}$ $^{232}Th$ and (right) peak shape for 0.15 pg $L^{-1}$ $^{232}Th$. (Reproduced from Jakubowski *et al.*[39])

Technologies, Omaha, NE).[38] With this setup, it was possible to determine the $^{235}U/^{238}U$ ratio in a solution of 1 ng $L^{-1}$ CRM U010 Uranium Isotopic Standard (New Brunswick Laboratories, Argonne, IL, USA) with an external precision of 0.07% RSD. Figure 12.7 shows a standard addition calibration curve for $^{232}Th$ with spike additions of 0.1, 0.2 and 0.3 pg $L^{-1}$ and the peak shape of the initial concentration of 0.15 pg $L^{-1}$ Th, which clearly demonstrates that this is a real peak and not baseline noise. The detection limit (3 SD of 10 blanks) for Th has been calculated to be 5 fg $L^{-1}$. The overall transmission for uranium was approximately 3%.[39]

## 12.2.4 Lens System

The ions coming from the ICP source are accelerated after passing the interface region. An acceleration voltage ranging from 4000 to 10000 V – depending on the instrument – is applied behind the interface. There are two possibilities: a) the ICP and interface are at high potential and the analyser at ground potential or b) the ICP and the interface are at ground potential and the mass analyser is floating at high potential.

An extraction lens is usually placed directly behind the skimmer cone to attract and accelerate ions leaving the extraction chamber *via* the skimmer cone aperture. In comparison to quadrupole instruments, up to three orders of magnitude higher acceleration voltages are used. This is one of the reasons why sector field instruments show a higher transmission of low-mass elements and a more uniform response throughout the mass range. Sector field instruments provide the possibility of either magnetic scanning or B-scanning (by adapting the magnet field strength) or electric mass scanning or E-scanning (by adapting the acceleration voltage) (see Section 12.4.2).

Due to the curved shape of sector field devices, no additional means (*e.g.* photon or shadow stop or ion mirror) are required to hinder neutrals and photons generated in the plasma from reaching the detector. Whereas a neutral "cloud" of ions and electrons can be observed within the interface region, the electrons diffuse much quicker into the high-vacuum region after the skimmer cone and are removed by the vacuum pumps. Together with a first lens displaying a potential more negative than that of the ICP, this results in a positively charged ion beam, which is further guided and shaped by a series of electrostatic lenses. (Even though in ICP-MS generally positive ions are detected, the use of negative-ion detection as an alternative approach has been evaluated for the analysis of nonmetals such as F, that are likely to form negatively charged species in the ICP or in the extraction process.[40])

### 12.2.5 Sector Field Mass Separator

Currently available single-collector MS setups use the Nier–Johnson (*AttoM*, Nu Instruments) or the reverse Nier–Johnson geometry (*Element 2*, *Element XR*, Thermo Finnigan). Modern commercial double-focusing MC ICP-MS systems use the Nier–Johnson geometry (*Nu Plasma II*, Nu Instruments; *Neptune*, Thermo Finnigan) or the Mattauch–Herzog geometry (*Spectro MS*, Spectro Analytical Instruments). A previous setup (*Isoprobe*) used a collision cell prior to the magnetic sector field for energy focusing. The latter setup is no longer commercially available.

**Magnetic Sector**

Single-collector ICP-SFMS instruments require an electromagnet with minimised hysteresis because of the need to scan the magnet. Therefore, laminated magnets in different shapes or geometries (see description of the instruments) are used. Moreover, (water) cooling of the magnet coils is a prerequisite for high mass calibration stability. When using MC-devices with Nier–Johnson geometry for isotope ratio measurements, the magnet is typically operated at constant field strength (static mode). Nevertheless, these devices also require a laminated magnet with minimum hysteresis, (i) because the magnet settings need to be changed when defining another mass range and (ii) for allowing dynamic operation (wherein a change of the magnetic field is required during the isotope ratio measurements). The Mattauch–Herzog geometry does not require a change in the magnetic field as it is operated in full static mode and all ions are focused on one focal plane without scanning. Therefore, this instrument does not require an electromagnet and thus can be equipped with a (cheaper) permanent magnet of constant field strength instead.

**Electric Sector**

In the electric sector, differences in ion energies are compensated for, as otherwise the spread in the ion energy limits the mass resolution attainable. The energy spread of ions originating from the ICP can be up to 20 electron volts (eV), thus requiring an electric sector in order to achieve high-resolution capabilities. The energy spread can already be reduced significantly down to a few eV by using a shielded torch. It should be mentioned that this energy spread is more pronounced for ions coming from an ICP ion source than for those produced, *e.g.* in a GD plasma source.

The combination of radius and angle of the magnetic and electric sectors needs to fulfil certain conditions to guarantee double-focusing conditions, either for a single mass (slit for single-collector devices) or on a well-defined curve (on which the Faraday cups are located in a multicollector device). Small magnets typically require small electric sectors as well, the smallest being 105 mm only (in the *Element 2*). Cylindrical or toroidal geometries of the electric sector are typically chosen (see also Chapter 4).

## 12.2.6 Slit System

The slit system is required to shape the ion beam and defines the width of the incident ion beam and thus, the mass resolution achievable, which is defined by the width of the entrance (alpha or source defining) slit. (For more details see Chapters 4 and 5.) Due to the high density of incoming ions (an immense number of $Ar^+$ ions is generated by the plasma source), it has to be considered that the continuous bombardment of the entrance slit with ions of keV energies leads to a continuous sputtering of material. Even though the sputtered material is not detected during operation (the generated ions do not show sufficient energy to pass through the mass analyser), it leads to a continuous degradation of the slit geometry and results in a deterioration of the beam profile. Additionally, the consequences hereof are loss of resolution and increased abundance sensitivity. Therefore, slit systems have to be exchanged regularly.

The use of a second (imaging) slit in MC ICP-SFMS is not always useful. In the case of moveable detectors (*Neptune*), each detector has to be equipped with a high-resolution imaging slit. This would result in a change of detectors when switching from low to full high resolution. (***Note***: Edge resolution can be accomplished if no high resolution imaging slit is present.) For fixed detectors (*Nu Plasma II*), a grid of slits can be moved mechanically in front of the detector entrances, resulting in the gradual reduction of the entrance width and thus defining an imaging slit width. Nonetheless, high mass resolution results in the loss of flat-top peak shapes and thus deteriorates the precision of isotope ratio measurements significantly. As a consequence, MC ICP-SFMS instruments are more typically operated in edge-resolution mode (sometimes referred to by Nu Instruments as "pseudohigh-resolution mode") when high resolution capabilities are required in order to separate the analyte ion of interest from an interfering ion (see Chapter 5).[41]

## 12.2.7 Transfer and Filter Optics

Abundance sensitivity becomes critical when a minor nuclide has to be determined in the presence of a highly abundant neighbouring nuclide.

This effect becomes even more detrimental when isotope ratio measurements are concerned. Additional lens filters in front of detectors help to eliminate the impact of tailing ions (see Chapter 4).

### 12.2.8 Detection System

All commercially available scanning instruments are equipped with an electron multiplier (discrete dynode detector), which can be operated in analogue or counting mode, thus allowing a large variation in the number of incident ions per unit of time to be handled. In some cases, an additional Faraday cup is available for measuring higher intensities, expanding the linear dynamic range up to twelve orders of magnitude (instrumental details see Section 12.3). This count range corresponds to a concentration range from sub pg $L^{-1}$ to over 1000 mg $L^{-1}$. This has been proven to be of crucial importance when major and trace components of samples have to be analysed simultaneously. (Otherwise, samples needed to be either analysed in two batches (undiluted/diluted), minor isotopes have to be selected and/or the measurement parameters have to be readjusted to decrease the intensity of the ion beam arriving at the detector). Measurement of noninterfered ions can also be carried out at high mass resolution settings with the single aim of reducing signal intensities. This is particularly critical for direct solid analysis by LA-ICP-MS where a major matrix component is often used as internal normalisation standard to correct for the amount of material ablated or sputtered. An extended linear dynamic range is also of importance for large isotope ratios, when adequate measurement of the major isotope requires a Faraday detector, whereas for the minor isotope an electron multiplier needs to be used. This allows the measurement of higher concentrations and leads to better precision. Nonetheless, proper cross-calibration between the electron multiplier and the Faraday detector is a prerequisite for accurate results. A key goal during the development of a combined detector and its control electronics was to allow switching times of less than 1 ms between analogue and Faraday detection modes.

The main feature of multicollector devices is the capability for simultaneous monitoring of isotopes by using discrete detectors. Over the years, the number of Faraday cups used in MC ICP-MS devices has increased significantly and more and more, secondary electron multipliers were (also) fitted into the instruments. This can be seen as an answer to the fact that Faraday cups limit the concentration range to the higher ng $g^{-1}$ levels. This drawback becomes even more evident when laser ablation is applied for direct sample introduction. Lower concentrations of the target element result in low signal intensities and poor isotope ratio precisions. In addition, the introduction of high-resolution systems (slit systems) again reduces the sensitivity of these instruments. The commercial systems use either fixed detector arrays or moveable cup configurations.

The Faraday cup detectors used in the *Nu Plasma II* (Nu Instruments) multicollector array, for example, consist of a ceramic body. The outer walls are gold coated and the interior ones are coated with graphite. The rear of the detector interior consists of a graphite block. The gold coating is grounded when the Faraday cup is mounted in the detector block in order to mitigate charging when ions pass along the outer walls of the detector. An electron suppressor is located near the opening of the Faraday cup.[42] This electron suppressor has a potential of −50 V to −100 V and ensures that secondary electrons are not released from the detector. Moreover, small magnets that are located above and below the collector block send surface-ejected electrons into a circular path, which additionally hinders the escape of electrons.[42] In addition, the Faraday cups are designed in such a way that the possible electron escape angle is minimised, which is achieved by increasing the Faraday cup length. The Faraday cups that are used in the *Nu Plasma II* have a dynamic range of up to 55 V using $10^{11}$ Ω resistors.[43]

The Faraday cups used in the *Neptune Plus* are (like the cups of the *Triton* TIMS) made of pure graphite. They are machined from a solid block of epitaxially grown graphite crystals. The *Neptune Plus* MC ICP-MS is available with Faraday cups with a dynamic range of up to 50 V using $10^{11}$ Ω resistors.[44]

Nowadays, it is possible to manufacture cups by micromachining that are so small that all nuclides can be measured all the time, for instance in a Mattauch–Herzog geometry (see also Chapter 4). In a collaboration of the groups of M. Bonner Denton, Gary Hieftje and Dave Koppenaal an array of micromachined electrodes were developed, which were used to cover the whole focal plane area (see Chapter 4). This is why this detector was given the name "focal plane camera".[45,46] Special low-capacitance integrating amplifiers are applied for signal amplification and the time resolution realised so far is already in the range of 1 ms per image.[45] Gary Hieftje's group had pioneered the technology for many years and they demonstrated that such a concept works for various types of plasma sources including GD[47] and ICP ion sources.[48]

## 12.2.9 Vacuum System

As an ICP is operated at ambient pressure, differential pumping systems are used to decrease the pressure (plasma: about $10^5$ Pa – interface: $10^1$–$10^2$ Pa – analyser: $10^{-6}$–$10^{-7}$ Pa) and thus, ensure maximum transmission efficiency of the ions. The interface region is pumped by a regular rotary pump. As the sampler cone provides a permanent aperture between the ambient atmosphere and the high-vacuum part, the mass analyser can be separated from the interface region *via* a slide valve, which is usually located directly behind the skimmer cone. This valve is only opened when the plasma is operating in a stable way and is closed before plasma shutdown. Therefore, the pumping of the interface is only required if the plasma is in operation. As a consequence, the interface pump is operating separately from the high-vacuum system. A higher amount of corrosive gases can be found in the interface region, which is pumped by the interface pump. Therefore, the oil of the interface pump requires regular changing. The vacuum, which is obtained by the interface pump, has an impact on the instrumental performance features, such as transmission efficiency or abundance sensitivity. As a consequence, stronger vacuum pumps are becoming popular in ICP-SFMS systems (see also Section 12.2.3).

## 12.3 Instrumentation

Over the years, the significant reduction in cost has led to a noticeable increase in the number of sector field instruments, which now account for almost 10% of all ICP-MS instruments worldwide. Table 12.1 gives an overview of instruments, which have been launched onto the market, so far.

In the following section, we provide more detailed information on sector field instruments, which are currently commercially available with the aim of highlighting the peculiarities of these instruments to provide further understanding of the method. Nonetheless, older instruments that are no longer produced are still in use. This holds true for the single-collector

**Table 12.1** Instruments launched to the market.

| Name | Manufacturer | Year of production | Type of Instrument | Geometry |
|---|---|---|---|---|
| Plasmatrace | VG Elemental | 1988–1998 | ICP-SFMS | reverse Nier–Johnson |
| Plasma 54 | VG Elemental | 1992–1998 | MC ICP-SFMS | Nier–Johnson |
| Element | Finnigan MAT | 1993–1998 | ICP-SFMS | reverse Nier–Johnson |
| JMS-Plasmax | JEOL | 1993–1995 | ICP-SFMS | reverse Nier–Johnson |
| JMS-Plasmax 2 | JEOL | 1995–? | ICP-SFMS | reverse Nier–Johnson |
| Nu Plasma HR | Nu Intruments | 1997–ongoing | MC ICP-SFMS | Nier–Johnson |
| Axiom | VG Elemental | 1998–2002 | ICP-SFMS | Nier–Johnson |
| Axiom MC | VG Elemental | 1998–2002 | MC ICP-SFMS | Nier–Johnson |
| Element 2 | Thermo Finnigan | 1998–ongoing | ICP-SFMS | reverse Nier–Johnson |
| Nu Plasma 1700 | Nu Intruments | 1999–ongoing | MC ICP-SFMS | Nier–Johnson |
| Neptune | Thermo Finnigan | 2000–ongoing | MC ICP-SFMS | Nier–Johnson |
| IsoProbe-P | GV Instruments | 2004–? | MC ICP-SFMS | magnetic sector field with hexapole collision cell |
| Element XR | Thermo Finnigan | 2004–ongoing | ICP-SFMS | reverse Nier–Johnson |
| AttoM | Nu Intruments | 2004–ongoing | ICP-SFMS | Nier–Johnson |
| SPECTRO MS | SPECTRO Analytical Instruments GmbH | 2010–ongoing | MC ICP-SFMS | Mattauch–Herzog |
| Nu Plasma II | Nu Intruments | 2010–ongoing | MC ICP-SFMS | Nier–Johnson |

ICP-SFMS (including the *Plasmax* and *Plasma 2* (JEOL), the *Axiom SC* (Thermo Optek), the *PlasmaTrace 2* (Micromass) and the *Element* (Thermo Fisher Scientific)), as well as for multicollector instruments, including the *Plasma 54* (VG Elemental), the *Axiom MC* (Thermo Optek) and the *IsoProbe* (Micromass, then GV Instruments).

## 12.3.1 Single-Collector ICP-MS

**_Element2_ and _XR_ (Thermo Fisher Scientific)**

Both systems evolved from the *Element* and thus share a number of hardware components (along with the *Element GD* (Thermo Fisher Scientific) GDMS instrument – see Chapter 13). They have a small magnetic sector with a radius of 160 mm only, followed by a 105-mm radius toroidal electrostatic sector in a reverse Nier–Johnson geometry. The entire mass analyser, including the lens system, the flight tube and the electric sector is operated at a high potential of about 8 kV and thus, the interface is at ground potential, allowing the interface pumps to be at ground potential as well, and thus providing easy coupling of different inlet devices.

The slit system of the *Element2* and *Element XR* (Figure 12.8), consists of three fixed slits arranged into one single holder (Figure 12.9). The slit width for a resolution of 10 000 is about 5 μm and that for a resolution of 4000 is 16 μm. In the low-resolution mode ($R=300$), flat-top peaks can be achieved. The slit holder is operated pneumatically under computer control and functions like a pendulum, which can swing to the left- and right-hand side, thus switching reliably between low (LR), medium (MR) and high (HR) mass resolution within less than one second. Recently, the precision for isotopic analysis of chromium (RSD for $^{53}Cr/^{52}Cr$: 0.005% at 1 ng mL$^{-1}$) and

**Figure 12.8** *Element 2* (Thermo Fisher Scientific).
(Source: © Thermo Fisher Scientific.)

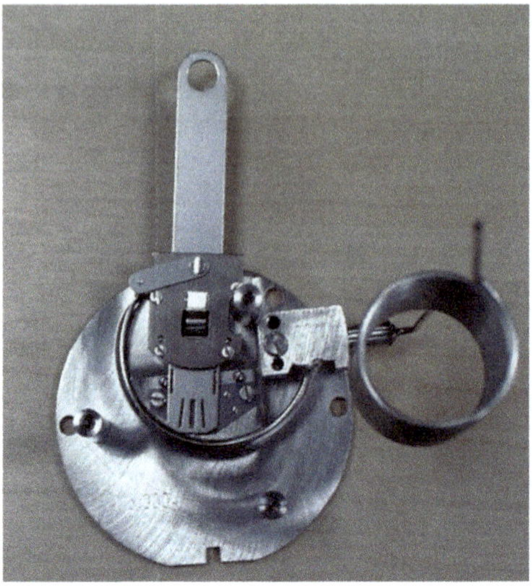

**Figure 12.9** Three fixed slits arranged on one single holder used in the *Element 2* and *Element XR*.
(Source: © Thermo Fisher Scientific.)

sulfur (RSD for $^{34}S/^{32}S$: 0.01% for 100 ng $mL^{-1}$ IRMM-643) could be improved with the *Element 2* and *Element XR* by using flat-top peaks at a resolution of about 2500, as obtained with a built-for-purpose slit (see also Chapter 5).[49]

A single secondary electron multiplier can be operated in a combination of "analogue" and "counting" modes to provide a linear dynamic range of nine orders of magnitude. Both detection modes are "crosscalibrated" automatically. The dark noise measured with the instrument is below 0.2 cps and the sensitivity achieved with a standard inlet system is more than $1\times10^{9}$ cps (mg $mL^{-1})^{-1}$ (measured at a mass resolution of 300 for a solution containing indium at a concentration of 1 µg $mL^{-1}$). With the new introduced "Jet Interface" option, the sensitivity for indium is even more than $2\times10^{10}$ cps (mg $mL^{-1})^{-1}$ under the same conditions. In the *Element GD* (see Chapter 13) and *Element XR*, an extended dynamic range of more than 12 orders of magnitude is provided by addition of a Faraday cup with a fast amplifier, which can be operated with integration times down to a millisecond. To improve abundance sensitivity, the *Element XR* has a retardation filter lens.

### *AttoM* (Nu Instruments)

The *AttoM* (Figure 12.10) is a double-focusing, high-resolution ICP-MS instrument with a 320-mm radius electrostatic analyser and a 250-mm radius magnet in Nier–Johnson geometry. The interface is floating at a potential of 6 kV, whereas the analyser and pumping system is at ground potential. The sample introduction system of the *AttoM* is mounted outside the torch-box for easy access.

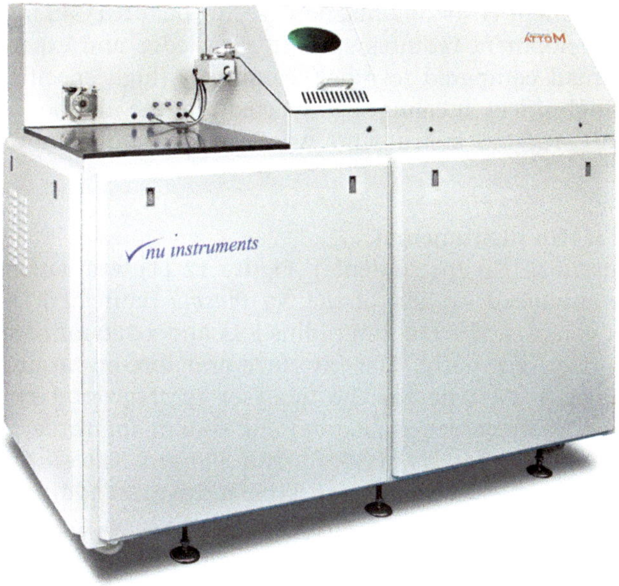


**Figure 12.10** *AttoM* (Nu Instruments).
(Source: © Nu Instruments.)

The *AttoM* uses computer-controlled multiple-slit assemblies to provide resolutions from $R = 300$ to $R > 10000$. A single secondary electron multiplier provides a linear dynamic range of 9 orders of magnitude. In contrast to the *Element 2*, this is not achieved by switching between counting and analogue modes, but by ion-beam attenuation in the lens system. Beam attenuation is performed by electrostatically deflecting the beam through a grid to physically reduce the ion-beam intensity. An extended dynamic range of up to 12 orders of magnitude is provided by an additional Faraday cup with a fast amplifier.

In order to avoid mass-discrimination effects introduced by a change of the acceleration voltage, the *AttoM* incorporates a deflector ion optics system, which allows alteration of the ion trajectories within the magnet. This offers the advantage of beam deflection at constant acceleration energy and ESA voltages and therefore avoiding additional mass-discrimination effects (see Chapter 6). By applying different voltages to the lens assembly, the entrance angle of the ion beam is altered, such that its trajectory is changed. A beam of ions of a certain mass can therefore be deflected either to a lower or higher "apparent" mass on the microsecond time scale. A mass range of up to 40% of a given mass can be covered in this way. This feature is also used in combination with fast magnetic scanning to increase the integration time for selected nuclides during elemental analysis along the whole mass range.

### 12.3.2 Multicollector ICP-MS

The instruments using Nier–Johnson geometry are currently well established and make up 99% of the market of MC ICP-MS instruments at the moment.

As the development costs of a magnetic sector field ICP-MS instrument are high, the development requires specific knowledge and expertise and the market is small compared to other techniques (high specificity of measurements; high price), it cannot be expected that the number of companies providing MC ICP-MS instruments will increase significantly in the near future.

***Nu Plasma II* (Nu Instruments)**

The *Nu Plasma II* (Nu Instruments) (Figure 12.11) was launched in 2010 and is the enhanced version of the *Nu Plasma* (launched in 1997). The instrument consists of a 350-mm radius ESA and a 250-mm radius magnet in Nier–Johnson geometry. The interface and fore-pump are floating at a potential of 6 kV, whereas the analyser is at ground potential. The instrument has an entrance (source) slit system for three preselectable resolutions with slit widths of 30, 50 and 300 μm. The slits are mechanically moved into the ion beam using a linear drive. An additional alpha-slit is placed between the source slit and the ESA entrance and is fully adjustable from 0 to 7 mm by computer control. The alpha slit reduces beam aberrations caused by a diverging beam entering the magnet, further enhancing the high-resolution performance. Where full resolution spectra are required, three sets of moveable slits are located in front of the detector array to reduce the widths of the entrance slits from 1 to 0 mm. This is only possible since the detectors are in a fixed position. Mass resolutions of $R = 300$ to $R > 3000$ can be achieved (10% valley definition) whilst maintaining flat-top peaks. For edge resolution, mass-resolving powers of up to 10 000 can be obtained.

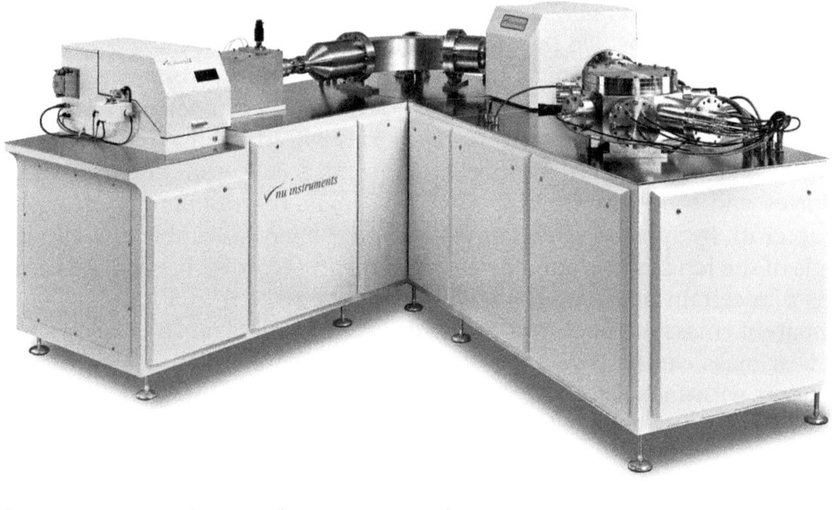


**Figure 12.11** *Nu Plasma II* (Nu Instruments).
(Source: © Nu Instruments.)

The *Nu Plasma II* detector array consists of 16 Faraday cup detectors at fixed positions. The steering of the ions into the cups is accomplished *via* a patented variable zoom lens optics system. The *Nu Plasma II* has a dispersion of about 500 mm, leading to a separation of, *e.g.* the Nd isotopes by about 2.5 mm, which corresponds to the fixed separation of the detectors in the Faraday array. For heavier elements, magnification of the ion beam image is required for alignment into adjacent detectors. In contrast, for the lighter elements, a larger separation of the adjacent isotopes is obtained and therefore, the ion beam image has to be demagnified. However, the range of dispersion without too much distortion is limited. Therefore, isotopes of significantly lighter masses cannot be aligned into adjacent collectors, and every other collector or every third collector may have to be used instead. Within the Faraday collector array, up to six full-size discrete dynode electron multipliers can be interspersed. Small deflector assemblies are located behind the Faraday detector array to deflect the ion beams into the off-axis secondary electron multipliers. Utilising the deflector assemblies, the multipliers can be independently protected against intense signals by deflection of the ion beam away from the first dynode of the multiplier.

In addition, a deceleration lens filter can be placed in front of each secondary electron multiplier. The application of the ion deceleration lens system in the *Nu Plasma II* MC ICP-MS instrument allows, *e.g.* a reduction of the peak tailing from $^{238}U$ ions at $m/z = 236$ down to $3 \times 10^{-9}$, whereas the absolute sensitivity for uranium is reduced by about 10%. Thus, abundance sensitivity is improved by almost two orders of magnitude and the minimum determinable $^{236}U/^{238}U$ ratio was improved by more than one order of magnitude compared with conventional sector field ICP-MS or TIMS.[50]

### *Nu Plasma 1700* (Nu Instruments)

The large geometry of the *Nu Plasma 1700* (Nu Instruments) is used for high-resolution isotope ratio measurements (Figure 12.12). The instrument is

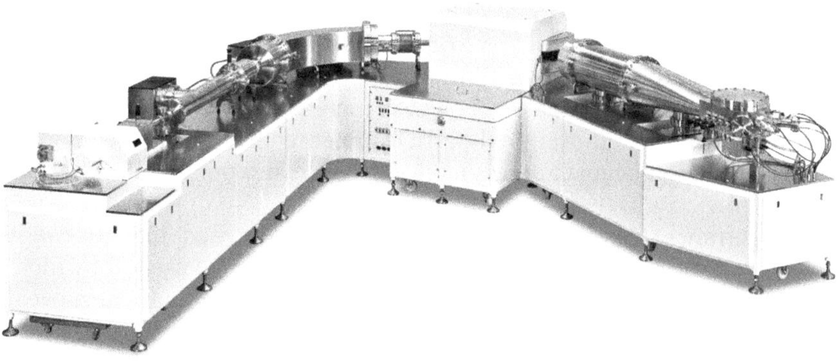

**Figure 12.12** *Nu Plasma 1700* (Nu Instruments). (Source: © Nu Instruments.)

characterised by a high dispersion and large geometry and uses a single 750-mm radius, 70° magnet, combined with a 943-mm radius, 70° ESA. This assembly results in a mass dispersion of about 1760 mm (reflected in the name of the instrument). Curved pole pieces and four multipole elements rotate the focal plane for right-angle intersection with the ion beams and to ensure stigmatic focusing.

As opposed to other MC ICP-SFMS instruments operating in pseudohigh-resolution mode, the *Nu Plasma 1700* is capable of simultaneously rejecting interfering masses at both the low and high mass sides of the beams of interest. High-resolution tuning is achieved by means of a continuously adjustable source-defining slit in combination with computer-controlled adjustable collector slits at all detectors. Each collector slit is independently adjustable in width and central position. Using these collector slits, the resolution capabilities are independently obtainable on the chosen collectors by the analyst. Aberration is minimised by selectable alpha restrictors and by higher-order functions applied to the four multipoles. The instrument achieves a mass-resolving power up to 40 000 (5% peak height) and maintains flat peak tops up to a mass resolution of 5000 (10% valley definition), which is unique in MC ICP-MS.

The *Nu Plasma 1700* has a total of sixteen Faraday detectors, originally ten of which were incorporated in a fixed central array but later instruments consist of thirteen within the fixed array. The other Faraday detectors are located on the low- and high-mass side of the fixed array and are adjustable *via* mechanical drives. The steering of the ions into the detectors is accomplished *via* two quadrupole zoom lenses. Five discrete dynode electron multipliers, one of them equipped with a deceleration lens filter, are interspersed behind the fixed collector array.

### ***Neptune*** **and** ***Neptune Plus*** **(Thermo Fisher Scientific)**

The *Neptune* (Figure 12.13) shares the ICP interface of the *Element 2* (Thermo Fisher Scientific) and the multicollector technology of the *Triton* TIMS instrument (Thermo Fisher Scientific) and combines high mass resolution, variable multicollection, zoom optics and multiple ion counting. The mass dispersion of 812 mm is achieved by an ion optical magnification of two. With increasing ion optical magnification, the divergence angles of the ion beams are reduced and the dispersion is increased. As a result, cups can be wider and deeper and scattered particles released at the cup's side wall by the incoming ion beams are less likely to escape and do not alter the "true" ion current.

The entrance slit system of the *Neptune* is designed like that of the *Element2*. If required, special high-resolution cups can be used, for which an exit slit is directly placed in front of the specific detector. Eight moveable collector supports and one fixed central channel are installed on the optical bench of the high-precision variable multicollector module. The centre channel is equipped with a Faraday cup and optionally an ion counter, with or without a retardation lens (RPQ) to improve the abundance

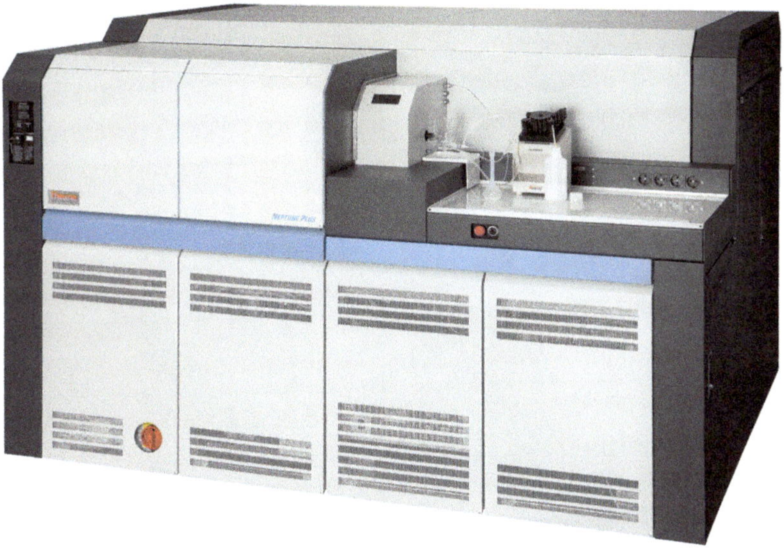

**Figure 12.13** *Neptune* (Thermo Fisher Scientific).
(Source: © Thermo Fisher Scientific.)

sensitivity (see Chapter 4). The eight detector supports where up to 17 collectors (9 Faraday cups and 8 miniaturised secondary electron multipliers) can be mounted, are motor driven and can be precisely positioned along the focal plane. A position sensor is located beneath each variable detector platform inside the vacuum chamber to ensure reliable positioning. The maximum distance between the outermost channels corresponds to a relative mass range of 17%, which is sufficient for the simultaneous measurement of $^{6}Li$ and $^{7}Li$ or for the range from $^{202}Hg$ to $^{238}U$. The amplifier signal is digitised by a high-linearity voltage-to-frequency converter with an equivalent digital resolution of 22 bits to ensure subparts per million digital resolution of all measured signals independent of the intensity.

An enhanced version of the *Neptune*, the *Neptune plus,* was introduced in 2009. Compact discrete dynode multipliers (CDD) with a similar performance as standard-sized discrete dynode multipliers enable simultaneous measurement of U, Th and Pb isotopes in a multicollector setup. The combination with a "Jet Interface" also improved sensitivity. The "Dual RPQ" option improves abundance sensitivity.

***Spectro MS* (Spectro Analytical Instruments)**

The main feature of multicollector devices is the capability for simultaneous detection of all of the isotopes of interest by use of discrete detectors, as mentioned before. Over the years, the number of Faraday cups in the collector array has been increased significantly and nowadays, it is possible to manufacture electrodes by micromachining that are so small that all nuclides can be measured all the time for instance in a Mattauch–Herzog

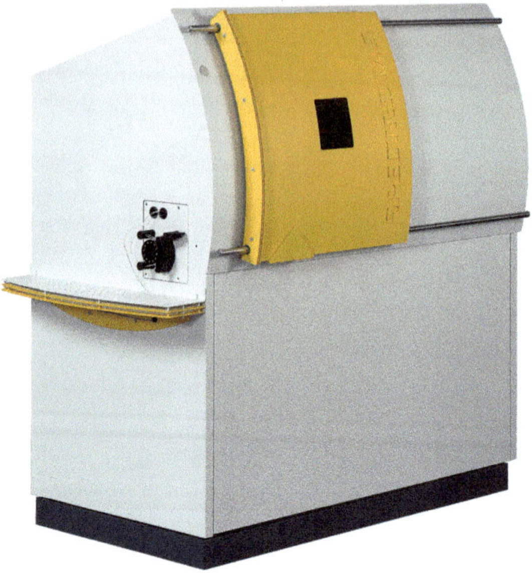

**Figure 12.14** *SPECTRO MS* (SPECTRO Analytical Instruments GmbH). (Source: © SPECTRO Analytical Instruments GmbH.)

geometry (see also Chapter 4). A commercial Mattauch–Herzog instrument with an ICP ion source and a focal plane camera was introduced in 2010 as the *Spectro MS* (Figure 12.14).[14] The instrument's core is a small Mattauch–Herzog mass spectrograph with a 120-mm focal plane permanent magnet, onto which a 120-mm direct charge semiconductor detector with a total of 4800 channels is mounted. With this detector, 210 nuclides can be monitored simultaneously. Every channel is made up with a low- and high-amplification pixel to cover 6 orders of magnitude of working range. Especially when using laser ablation as a means of sample introduction, this full spectrum simultaneous monitoring can be of great interest.

# 12.4 Measurement Considerations

## 12.4.1 Spectral Interferences

Over the years, the applications in the field of ICP-MS changed from rather pure to extremely complex samples and soon, the Achilles' heel of ICP-MS, which had been described in the first paper of Houk already, was discovered again: the limitation by spectral interferences.[51] As a consequence, high mass resolution was discovered as the most general principle to overcome this limitation (see also Chapter 7).

As discussed previously, polyatomic interferences are most problematic as they are less predictable and depend on the sample composition and the operational parameters of the ICP-MS system. They might become extremely complex in organic matrices. The formation of cluster ions from the dominant species in the plasma (Ar, H, O, C, N) is a major source of polyatomic interferences in ICP-MS and they occur preferentially in cooler plasma boundaries (possibly at the walls of the skimmer)[52] or in the sampling and expansion areas of the interface and even in the neighbourhood of vaporising droplets, where the gas temperature is somewhat lower. Nonose and Kubota investigated the optical characteristics of microplasmas behind the sampler and skimmer cone and observed an influence of the acceleration voltage on the formation of polyatomic interferences, as well.[53]

Many approaches to cope with the problem of spectral interferences have been discussed since the early days of ICP-MS.[51] The most straightforward method to overcome limitations from spectroscopic interferences is the application of high mass resolution. The high mass resolution capability of ICP-SFMS allows the identification of interferences and ions of interest on the basis of their exact mass easily and from such observations, we can already define major groups of possible spectroscopic interferences and some of their features:

*Ar-containing ions*

- Ar is introduced for plasma generation at flow rates between 15 and 20 L $min^{-1}$.
- Under standard plasma conditions, $Ar^+$ and $Ar_2^+$ are always observed as very intense signals, thus affecting determination of $K^+$ and $Ca^+$ and $Se^+$.
- $Ar^+$ can form polyatomic species with elements from the solvent, ambient air and/or the matrix of the sample and thus argon-related interferences are the most severe problem of possible spectroscopic interferences.

*Analyte M attached to oxide and hydroxide ions*

- $m/z + 16$ ($MO^+$) and $m/z + 17$ ($MOH^+$) are typical interferences.
- The ratio of $MO^+/M^+$ depends on the M–O bond strength.
- Usually it is observed that $MO^+/M^+ > MOH^+/M^+$
- By optimisation of the instrumental settings, the formation of (hydr)oxide ions usually can be reduced to values of $MO(H)^+/M^+ < 5\%$.
- The interference is a major problem if $m/z(M_1O^+) = m/z(M_2)$ and $c(M_1) > c(M_2)$.
- (Hydr)oxide interferences are formed in the ICP or "survive" the ICP in regions with low temperatures (*e.g.* in the neighbourhood of vaporising droplets, particles).

*Doubly charged ions*

- The degree to which $M^{++}$ ions are formed depends on the difference between the 2nd and the 1st ionisation potential of the element.
- Ions of this type always have a strong effect once they are present as major components of the sample, solvents or the discharge gas, for instance.

- For example, $^{138}Ba^{++}$ can hamper the detection of gallium.
- The formation of $M^{++}$ and $MO^{+}$ signals in ICP-MS are strongly depending on instrumental parameters (for more details see the investigation of Vanhaecke *et al.*[54] but only doubly charged ions can be reduced by application of a grounded shield between the plasma and the induction coil to reduce capacitive coupling of the high frequency voltages.

*Solvent and matrix-based polyatomic ions*
The second most severe spectral interferences observed in many applications belong to this group.

- They are formed by constituents of the solvents used or by components of the matrix (salts) such as: $MNa^{+}$, $MCa^{+}$, $MK^{+}$.

The formation of interferences is governed by thermodynamics (exothermic and endothermic processes) and is possible if the energy, which is needed for the process, is provided by the plasma or the excess reaction heat. The product molecule usually has a binding energy that is high enough to survive the short passage through the plasma and the interface as exposure time of particles in the plasma is only about 2 ms.[55] This is also why polyatomic species with low binding energy such as $ArH_2O^{+}$ and $ArH_3O^{+}$ are not detected. If molecules of this kind are measured in the mass spectrum, this would give us a hint that they are formed in an environment with much lower temperatures. This idea was used by Sears *et al.* to study the origin of spectroscopic interferences.[56]

Of course, high mass resolution is not a panacea against all spectral interferences. As mentioned already, isobaric interferences are far beyond the scope of most commercially available devices. $MO^{+}$ ions can be resolved in the mass range below 100 u and above 140 u with a resolution of less than 10 000. In the mass range between 100 u and 140 u, a resolution of up to 1 000 000 would be necessary. The resolution required exceeds by far the resolution attainable with commercially available instrumentation. This is a crucial point, *e.g.* for low-level rare earth element (REE) analysis if formation of oxides is not prevented by other means.

### 12.4.2 Scan Conditions (Single-Collector ICP-SFMS)

The main characteristic of single-collector ICP-MS is not the number of detectors (because some of the sector field instruments for which this terminology is used may have more than a single detector) but the fact that they are operated in a scanning mode, whereas multicollector instruments are most often operated in static mode. For a long time, sector field instruments have been assumed to be too slow for the detection of transient signals. This does not hold true for the new generation of instruments equipped with laminated magnets. Nonetheless, even with peak-hopping techniques, double-focusing instruments are slightly slower than quadrupole instruments, because a certain settling time is required for each stepwise change of the magnetic field. Therefore, the number of isotopes that can be measured is limited when transient signals are recorded. This is the case with sample introduction techniques such as electrothermal vaporisation (ETV), laser ablation (LA), liquid chromatography (LC) and gas chromatography (GC) (see Section 12.2.1).

Two different scan modes are routinely applied in single-collector sector field devices (see also chapter 4.9). Either the magnetic field strength is varied in a so-called B-scan or, alternatively, the acceleration voltage is varied (or the ion beam is deflected using electric lenses) in a so-called E-scan. Due to self-induction in the magnet coils, the scan speed in the B-scan mode is limited. Owing to special electronics, the cycling from $^{7}Li$ to $^{238}U$ and back to $^{7}Li$ can be achieved with a magnet settling time of 90 ms only. This means that contemporary instruments such as the *Element 2* and *AttoM* can scan from the low end of the mass spectrum to the high end and back again in several hundred milliseconds. E-scan is an even faster scan mode with a settling time of one millisecond per mass unit only. In this scan mode, the electric field ($E$) of the electric sector and the accelerating voltages ($U_0$) are scanned simultaneously, to maintain the ratio $E^2/U_0$ at a constant value, equal to the value providing transfer of the main beam of ions through the electric sector. Alternatively, the acceleration voltage may be kept constant and electro-optics can be used to scan the ion beam, as it is the case of the *AttoM*. In either case, the magnetic field is set at a fixed value, such that main-beam ions of a predetermined $m/z$ are transmitted by the magnetic field. At a first glance, the E-scan may be advantageous, but due to a loss of sensitivity at larger mass ranges, it is never applied to cover the whole mass range and therefore restricted to a partial mass range of about 30–40% around the magnet mass only. In the low-resolution mode, the preferred scanning mode of the instrument is peak hopping, as is the case for quadrupole mass analysers too. In this mode, the whole peak is not scanned but only the central channel(s) of the peak is monitored. With the *Element 2/XR*, this is accomplished by setting the magnet to a particular mass and scanning a small range at the corresponding peak top *via* an E- scan, then jumping to the next mass of interest by changing the accelerating voltage, followed by scanning the next peak, again using an E-scan. This procedure is applied until the maximum E-scan range for the particular magnet mass is reached. Subsequently, the magnet setting is changed to address the next mass range of interest. In the high mass resolution mode, the peak shape usually changes from a flat-topped into a triangular shape.[57] As a consequence, it is no longer the peak top only that is scanned, but the entire peak (or a fraction thereof) of the nuclide of interest. Otherwise, the same scan strategy as for the low-resolution mode is applied.

There are additional scan modes like the SynchroScan at the *Element 2/XR*, wherein a combination of B- and E-scanning is used. In this scan mode, the magnet field is increased continuously and the acceleration voltage is changed in a sawtooth manner. Therefore, the acquisition is carried out at the centre of a peak with nearly optimum acceleration voltage before jumping to the next peak and thus, the duty cycle is high since there is no loss due to magnet settling times. A very fast scan mode of a sector field instrument (*Element XR*) was discussed by Shigeta *et al.* for measurement of single droplets injected into the ICP-MS by use of a piezoelectric μ-droplet generator; more details are given in the application section (Section 12.5.1).[23]

*FastScan* ion optics are used in the *AttoM* to alter ion trajectories within the magnet in combination with fast magnetic scanning to increase the integration time on selected isotopes during elemental analysis along the whole mass range. This method was developed to avoid any potential mass-discrimination issues that could be caused by varying the acceleration voltage. For the fast magnetic scanning, the Hall-probe control of the magnet is switched off. A high voltage is switched onto the magnet for 100 ms, causing the magnet field strength to increase linearly, thus permitting the whole mass range from 6 to 250 u to be scanned. During this period, the ion counter captures data at 10-μs intervals. After 100 ms, the high voltage is turned off and the magnet de-energises. In this way, the mass range from 250 to 6 u is rescanned. During both (up and down) periods, two datasets are acquired for each full magnet cycle. The next cycle starts after a 20-ms wait time. Furthermore, the *AttoM FastScan* ion optics can be used in conjunction with the fast magnetic scanning mentioned above. In this mode, the *FastScan* optics is used to allow a greater dwell time at each mass position during the sweep of the magnet, much as with the *SynchroScan* systems described earlier.

### 12.4.3 Mass Calibrations (Single-Collector ICP-SFMS)

Due to the fact that sector field devices are always operated at fixed resolution, the peak width changes over the mass range, with smaller width at lower masses. Compared to the peak width of typically about 0.7 u for quadrupole-based instruments, the peak width with a double-focusing magnetic sector instrument for, *e.g.* $^{7}Li$ at a mass resolution of 10 000 is only 0.0007 u (10% valley definition). Consequently, the requirements concerning accuracy and stability of the mass calibration are significantly higher for sector field instruments than for quadrupole-based instruments.

Therefore, the magnetic field is usually controlled by a feedback loop with the determination of the magnetic field strength at a representative location at the magnet, accomplished *via* either a Hall probe or another type of magnetic sensor. The generation of the magnetic field is accomplished by applying a particular current to the coils of the magnet. The operator just enters the isotope of interest into the steering software and – *via* the mass calibration - this input is converted by means of a DAC (digital-to-analogue converter) into a specific value that is provided to the control electronics, which then applies the required current to the magnetic coils. To perform a mass calibration, a solution containing defined elements is introduced into the ICP and subsequently, a scan at a mass range around the expected position of the nuclides or interferences is performed. The DAC value corresponding to the centroid of the acquired peak together with the accurate mass of the isotope is then entered into the mass calibration table. Depending on the sensor used, the electronics applied and the fact that the mass is proportional to the square of the magnetic field (see Chapter 4), the relationship between mass and the digital control may not be linear. Due to the fact that the peak width at low masses is smaller, it is preferred to have a steeper calibration curve, *i.e.* the change in mass per DAC step is smaller. Fit functions are used to calibrate the entire mass range on the basis of experimentally calibrated masses, distributed over the entire mass range of the mass spectrometer.

In order to reliably operate the instrument at higher resolution while scanning only a small mass range around the nuclide of interest, the mass calibration has to be very stable. This is usually achieved by very stable electronics and by using, *e.g.* the lock mass technique. Here, an ion signal that is always present in the spectrum, *e.g.* that from $^{40}Ar^{40}Ar^{+}$, is used to dynamically and automatically adjust the mass calibration between analyses.

## 12.4.4 Isotope Ratio Analysis

Single-collector ICP-MS instruments have not only been used for trace-element determination, but also for isotope ratio measurements. Obviously, the precision and accuracy of TIMS or MC ICP-SFMS cannot be reached. Nonetheless, ICP-SFMS instruments can achieve more precise isotope ratio measurements under optimum conditions than quadrupole ICP-MS instruments and even the early works reported 0.04% RSD, *e.g.* for both magnesium and lead isotope ratios.[58] The reason for this lies in the flat-top peak shapes (see Chapter 5) observed at low resolution settings and fast sequential measurement capabilities when the magnet is kept at constant field strength (mass) and the isotopes of interest are scanned *via* the E-scan. Flat-top peaks have the important advantage that small shifts of the position of the peak with respect to the mass scale do not significantly affect the result when a sufficiently narrow section of the peak is integrated (*i.e.* about 5–10% of the peak width). Nowadays, instrumentation excels by the superior sensitivity and the capability to avoid spectral interference by operating a mass spectrometer at a higher mass resolution and these are undeniable advantages for isotopic analysis as well.[59] Various applications of isotope ratio measurements have been described in the literature (see Section 12.5.2).[60]

At medium resolution, the peak shape deteriorates from flat top to rather triangular or rounded. This results in a deterioration of the isotope ratio precision. Nevertheless under optimum data acquisition conditions, a precision of $\geq 0.1\%$ RSD can still be achieved.[61] The recent introduction of a new slit system for the *Element 2* generates flat-top peaks at a mass resolution of 2000 and therefore significantly improves the precision for isotope ratios (see also Chapter 5).

In multicollector instruments high mass resolution is usually achieved applying edge resolution settings. This allows isotope ratio measurements with the same precision as compared to low-resolution settings.

Even though multicollector devices are capable of measuring isotopes simultaneously, a short time gap between the detector (between Faraday cups and/or between Faraday cups and secondary electron multipliers) electronics handling the signal can result in a deterioration of isotope ratio results.[62,63] This effect is particularly pronounced when short transient signals are monitored in the case of, for example, high-performance liquid chromatography,[64] gas chromatography,[65] and laser ablation.[66] Krupp and Donard, for example, attributed drifts observed in Pb and Hg isotope ratio measurements by GC-MC ICPMS, to a Faraday preamplifier response lagging.[65] Recently, a method for isotope ratio drift correction by internal amplifier signal synchronisation was presented for the measurement of transient signals by MC ICP-MS.[67] A bias in the responses between Faraday cups and secondary electron multipliers is depicted in Figure 12.15, showing a transient signal acquired by LA-MC ICP-MS of single particles.

### 12.4.5 Speciation Analysis

In the early days of ICP-MS, most analytical problems relating to biological systems or environmental studies were addressed by measuring the total concentrations of elements, in particular "heavy metals", even though there has been increasing awareness of the importance of the chemical form in which an element is present in the environment or in biological systems, *e.g.* the oxidation state, the nature of the ligands or even the molecular structure.[68] In particular, in life-science applications, information on the speciation of an element is required to understand processes related to toxicity, transport and bioavailability of metals or biological processes, in which metals are involved. Although the majority of species studies use quadrupole-based ICP-MS instruments, ICP-SFMS has been used increasingly. The definition, philosophy, methods and novel aspects of speciation have recently been addressed in a few review articles.[69–71]

Speciation has been defined by the International Union of Pure and Applied Chemistry (IUPAC) as: "Speciation analysis is the analytical activity of identifying and/or measuring the quantities of one or more individual chemical species in a sample; the chemical species are specific forms of an element defined as to isotopic composition, electronic or oxidation state, and/or complex or molecular structure; the speciation of an element is the distribution of an element amongst defined chemical species in a system."[72]

The most important tool for speciation studies is the use of a separation and/or a fractionation technique. Most often the ICP-MS instrument is only used as a detector, which has to provide time-resolved measurement capabilities, a feature that is nowadays available on the sector field devices

**Fast transient signals**
**Bias in detector response**

| Time / sec. | $^{238}U$ / V | $^{236}U$ / cps | $^{235}U$ / V | $^{234}U$ / cps |
|---|---|---|---|---|
| 0 | 1.13E-03 | 4.01E+01 | -1.31E-03 | 1.10E+02 |
| 0.1 | 1.18E-03 | 7.01E+01 | -1.40E-03 | 8.71E+02 |
| 0.2 | 1.68E-01 | 4.01E+01 | 3.20E-04 | 7.01E+01 |
| 0.3 | 4.53E-03 | 3.00E+01 | -1.14E-03 | 7.01E+01 |

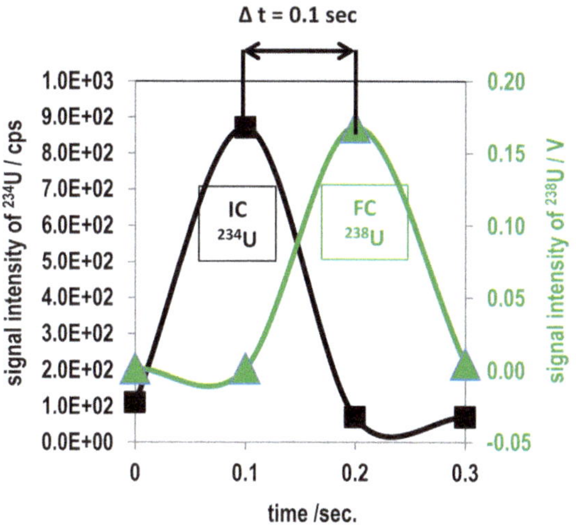


Data are not background corrected

**Figure 12.15** Transient signal of U isotope measurements of single particles by using LA-MC ICP-MS.
(Source: Stefanie Konegger-Kappel, unpublished data.)

from all manufacturers. Some problems might occur for sector field devices due to a limited scanning speed if very short transient signals have to be measured.

Various hyphenated techniques (by online coupling of HPLC, GC or CE with ICP-MS) allow speciation analysis aiming to solve analytical problems in the life sciences and in environmental research. Instrumental details and optimised experimental conditions of hyphenated mass-spectrometric analysis, experimental arrangements, the progress and trends in speciation using ICP-MS with different online separation techniques or enrichment techniques are discussed elsewhere.[73,74] The most frequently applied techniques include LC, HPLC, GC, CE, SEC, FFF and SPME. For example, SEC-ICP-MS permits the size-based separation of macromolecules and has been found to be very useful in the analysis of large biomolecules in particular proteins, and the approach supplies information that is crucial for the elucidation of reaction mechanisms and for the evaluation of the binding capacities of different elements with biological entities.

Considerable work has already been devoted to the performance enhancement of ICP-MS for speciation analysis, with respect to the detection power and also to ease of operation. Detection limits could be improved by preconcentration or by application of new sample introduction techniques, such as electrothermal vaporisation (ETV), thermospray nebulisation (TSN), direct injection nebulisation (DIN), ultrasonic nebulisation (USN), hydraulic high-pressure nebulisation (HHPN) or direct coupling of gas chromatography to ICP-MS.

Moreover, the analysis time can be reduced considerably by application of online techniques. In particular, online high-performance liquid chromatography (HPLC) separation, which combines high chromatographic resolution with a short analysis time, is favourable. In many cases organic eluents are used and so detection *via* plasma source mass spectrometry is generally impeded by interferences owing to the presence of carbon (*e.g.* $^{40}Ar^{12}C^{+}$ on $^{52}Cr^{+}$).[75] Clogging of the sampling orifice is a further problem that arises as a consequence of the high carbon content. This problem is discussed in more detail by Jakubowski *et al.*, who had applied ion pairing chromatography using 25% methanol in the eluent.[76] In this example, a desolvation system was applied to reduce the organic load into the plasma. Addition of oxygen to the aerosol gas could be considered as a means of reducing any problems due to interferences from carbides and clogging. An elegant way to avoid the use of organic eluents was presented by Barnowski *et al.*[77] They used anion-exchange chromatography (Ion Pac-AG5, Dionex) to separate Cr(III) from Cr(VI) using nitric acid as the eluent to separate both species.[78,79] Specific problems in speciation studies will be discussed in more detail in Section 12.5.1.

## 12.4.6 Beam Alignment (Multicollector ICP-SFMS)

Operation principles as described in the following section are mainly based on two currently commercially available instruments, the *Neptune* and the *Nu Plasma II* (see for more details the book by Vanhaecke and Degryse[60]).

In MC ICP-SFMS devices, the electric sector has to be placed before the magnet and thus both the *Neptune Plus* as well as the *Nu Plasma II* are using Nier–Johnson geometry. The magnet is set to constant field strength in order to direct (in most cases) the central mass into the central detector. The other ions have to be directed into the adjacent detectors, whereby the separation between adjacent detectors depends on the mass dispersion of the magnet. The dispersion allows the simultaneous collection across a mass range of up to 17% for the *Neptune plus* and about 15% for the *Nu Plasma II*. The steering of the ions into the detectors is accomplished by two different methods:

The *Neptune Plus* uses movable cups, which are moved mechanically to the position of the ion beam by means of motorised detector carriers, which can be equipped either with Faraday cups, channeltron ion counters or discrete dynode multipliers. An additional dynamic zoom lens system is positioned in-between the magnet and the detectors. It supports analysis at high mass resolution and enhances multidynamic measurements by applying variable zooming for different measurement sequences.

The *Nu Plasma II* has a variable dispersion zoom lens system behind the magnetic sector. As the location of the ion beam is adjusted by changing the dispersion of the zoom lens only, no moveable cups are required. As a consequence, the instrument can be operated in a dynamic mode. By changing the magnetic field along with the lens settings of the zoom lens system, a different set of isotopes can be analysed. As no cup movement is involved, the switching delay between subsequent sequences is only some seconds.

It is evident that small mass drifts have no significant effect on the final ratio when low mass resolution is applied as long as flat-top peaks are maintained. This is different in edge-resolution mode, as small drifts can already lead to a significant change of the isotope ratio precision. Therefore, control of temperature, amongst some other parameters, is of crucial importance.

## 12.5 Applications of Sector Field ICP-MS

The following section is intended to provide insight into the various fields and specific studies in which sector field ICP-MS has been applied until now, but it cannot give a complete overview, due to the high number and diversity of all the papers published. A special focus is set on papers covering the last decade, mainly as the early days have been covered well in review articles.[80,81] A comprehensive bibliography of publications of sector field ICP-MS has been compiled by Douthitt.[82]

The number of papers reporting the use of ICP-SFMS already tripled from 1991 to 1994. In 1997, after the introduction of the *Element* ICP-SFMS unit onto the market, about 50 papers on ICP-SFMS were published. About 1/3 of the papers were dedicated to environmental applications at that time. This fraction changed with time and meanwhile, a major part of the publications (almost 40%) is dedicated to geological applications. The major reason is the increased use of MC ICP-SFMS instruments. Whereas 20–30% of the papers using ICP-MS deal with isotope ratio measurements, this number shifts to more than 50% of the papers when the field of sector field ICP-MS instruments is examined.

Concerning trends, the combination of laser ablation coupled to ICP-MS has turned into a useful tool for geoscientists. Additionally, an increasing awareness for the need of sound metrology in analytical chemistry can be noticed.[83,84] Another trend, where ICP-SFMS has become a powerful detector is the area of speciation analysis, a field in which an increased interest in combining elemental and molecular data can be observed. This fact is reflected by the large number of such papers in the area of life-science applications, even though this number is still relatively low when compared to the total number of ICP-MS systems used in this field.

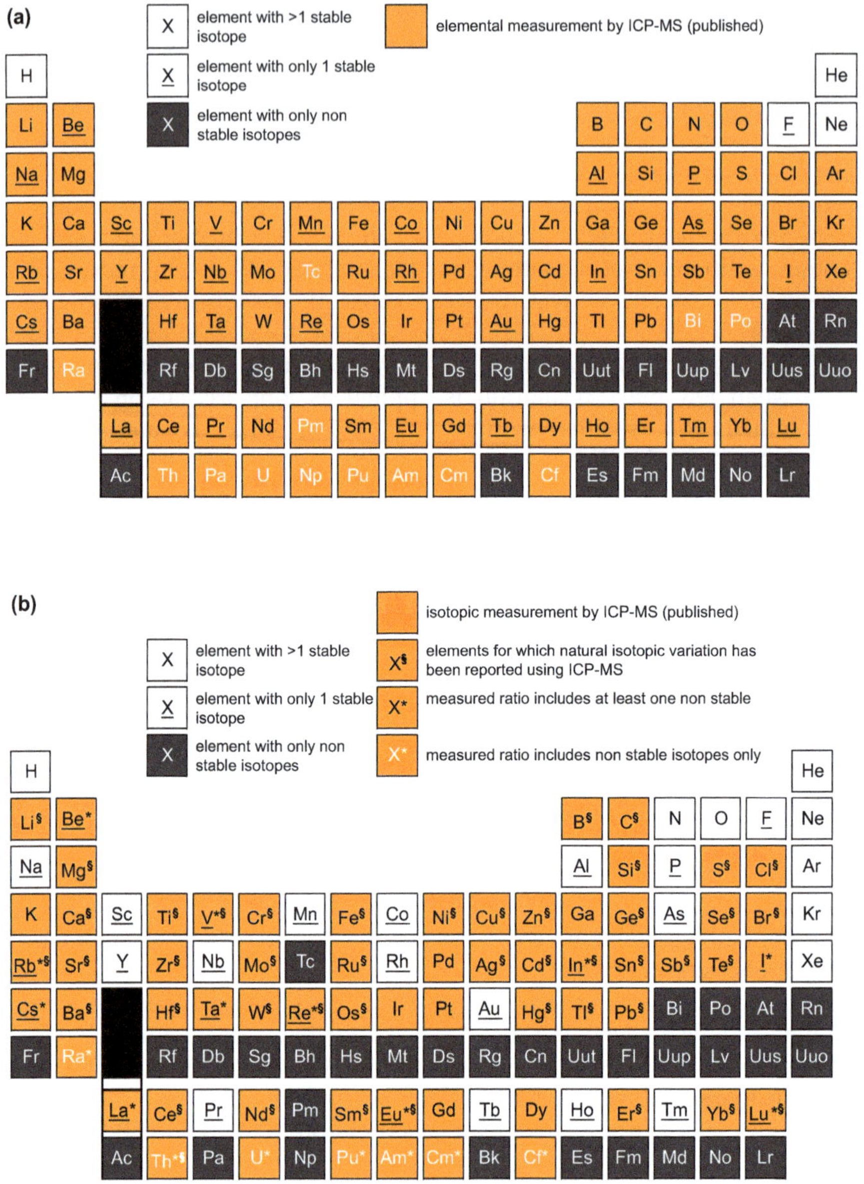


**Figure 12.16** (a) Periodic table of the elements: elements for which elemental data acquired by ICP-MS has been published (highlighted in mandarine orange) – also included if used, *e.g.*, for interference corrections or internal normalisation. (b) Periodic table of the elements: elements for which isotope ratio data acquired by ICP-MS has been published (highlighted in mandarin orange) – also included if measured as inter elemental normalisation standard or in IDMS analysis.

## 12.5.1 Applications of Single-Collector ICP-SFMS

Since more than 3000 articles related to the use of sector field instruments for elemental analysis have been published up to now, it is only possible to discuss some selected examples; these examples focus mainly on the use of the capability of sector field devices to measure at higher mass resolution and higher sensitivity. Most nuclides in the mid-mass range require a mass resolution of less than 4000 to overcome the major interferences.[85] Since single-collector ICP-SFMS has become a mature technique, multielement applications are now carried out in routine laboratories. A significant number of publications in various fields underline the importance of applying high mass resolution in samples with very low analyte concentrations and a complex matrix type (*i.e.* high salt concentration, high reagent or acid concentration in digested or extracted samples or high solvent loads, *e.g.* after chromatographic separation).

### *12.5.1.1 Applications in Industry and Materials Research*

**Solvents and Ultrapure Chemicals**

The major field of applying high mass resolution in industrial and materials research can be allocated to the semiconductor industry and the characterisation of pure and ultrapure materials.[86,87] Even though the number of papers might be limited, sector field instruments have a high impact on routine analysis in this context.

Due to the high price of the first ICP-SFMS instruments, the early customers mainly came from the semiconductor industry, where the quality control of high-purity reagents and solvents is a most important issue. Both the high-resolution capability and the low detection limits of single-collector ICP-SFMS proved to be important features. For instance, for the determination of 26 elements, present at concentration levels below 10 ng $mL^{-1}$ in concentrated $H_2SO_4$ 3 : 1 mixture with $H_2O_2$ for etching Si wafers were used, and ICP-SFMS analysis of ten-fold diluted $H_2SO_4$ allowed the sample throughput to be significantly increased.[88] Indeed, the method replaced single-element graphite furnace atomic absorption spectrometry (GFAAS) and quadrupole-based ICP-MS, for which an evaporation step to remove components otherwise leading to polyatomic interferences prior to the determination of Li, Ti, V, Cr, Mn, Co, Ni and Zn was necessary. The risk of contamination was greatly reduced and results were found to be precise and accurate.

Requirements for the measurement of impurities in ultrapure water used in the semiconductor industries have become considerably more stringent with the development of very large scale integrated circuits. Ultrapure water is of critical importance for the cleaning and etching of the Si substrate. In this context, colloidal silicon is one of the most important contaminants and its maximum level is now below 1 ng $mL^{-1}$, which requires an analytical method capable of determining Si at levels below 0.1 ng $mL^{-1}$.

Takaku *et al.* demonstrated that ICP-SFMS eliminates the spectral interference at mass 28, mainly caused by $CO^+$ and $N_2^+$ (and accounting for a BEC of several 100 ng $mL^{-1}$ Si).[89] Nonetheless, the instrumental background signal, mainly originating from the plasma torch, hampered the determination. Use of appropriate materials and/or preconcentration by evaporation made it possible to determine Si at the ng $mL^{-1}$ level.

Elemental analysis of process chemicals, such as photoresists in the semiconductor industry, had been a challenge for a long time, but could be performed successfully by single-collector ICP-SFMS.[90] Multielement analysis at ultratrace levels (down to ng $L^{-1}$ and pg $L^{-1}$ levels) is currently routinely performed for the analysis of semiconductor grade reagents (especially for critical elements like K, Ca and Fe).[91]

**Semiconductors, Superconductors and Optoelectronics**

Semiconductor products are often investigated by vapour-phase decomposition or drop etching and the sample solution thus obtained is subsequently analysed for its trace-element content. Polyatomic interferences from matrix elements (*e.g.* Si) lead to significant interferences for trace elements like Ca, Sc, Ti, V, Ni, Cu, Zn, Ga or Ge[92] that require medium mass resolution for interference-free determination. Ferrero and Posey[93] improved their detection limits for vapour-phase decomposition-ICP-MS significantly by changing from quadrupole to ICP-SFMS and quantified 42 elements in their samples successfully. A mass resolution of 4000 proved to be sufficient to resolve matrix-based interfering ions occurring in Si- and F-containing samples, such as $SiH^+$, $CF^+$, $NO^+$, $NOH^+$ or $^{28}SiH_3^+$ (affecting the determination of P), $^{28}Si^{28}Si^+$ and $ArO^+$ (affecting $^{56}Fe^+$), $^{28}Si^{16}O_2^+$and $^{28}Si^{30}Si^+$ (affecting $Ni^+$ isotopes), $SiOF^+$ (affecting $Cu^+$) and $SiF_2^+$and $SiOHF^+$ (affecting $Zn^+$) from the analyte ions at the same mass-to-charge ratio (Figure 12.17).

Boulyga *et al.* compared three different ICP-MS instruments (sector field and quadrupole with and without a collision cell) for the determination of the stoichiometry and trace impurities in thin barium strontium titanate perovskite layers, which are of importance for different applications in microelectronics.[94] The maximum sensitivity, lowest detection limits and best precision were achieved with ICP-SFMS. 23 elements were determined in high purity gallium by Xie *et al.*[95] Ryu *et al.* analysed Si wafers coated by a spin-on-glass approach for Al, Cr, Fe, Na, Co and Cu by laser ablation with ICP-SFMS.[96] Vaculovic *et al.* used LA-ICP-SFMS to study corrosion processes as a result of molten fluoride salt treatment of structural materials of a nuclear reactor cooling circuit.[97] LA with ICP-SFMS was used by Latkoczy and Ghislain to analyse the elemental distribution of both major and trace elements in an industrial multiphase magnesium-based alloy.[98]

High mass resolution was further applied to assess the purity of TlBr single crystals, which are used as detectors in space technology.[99] The ICP-MS analysis of $CaF_2$, glass, quartz, ceramics, and other multicomponent materials is hampered due to spectral interferences, which can be overcome by high mass resolution.[100–103]

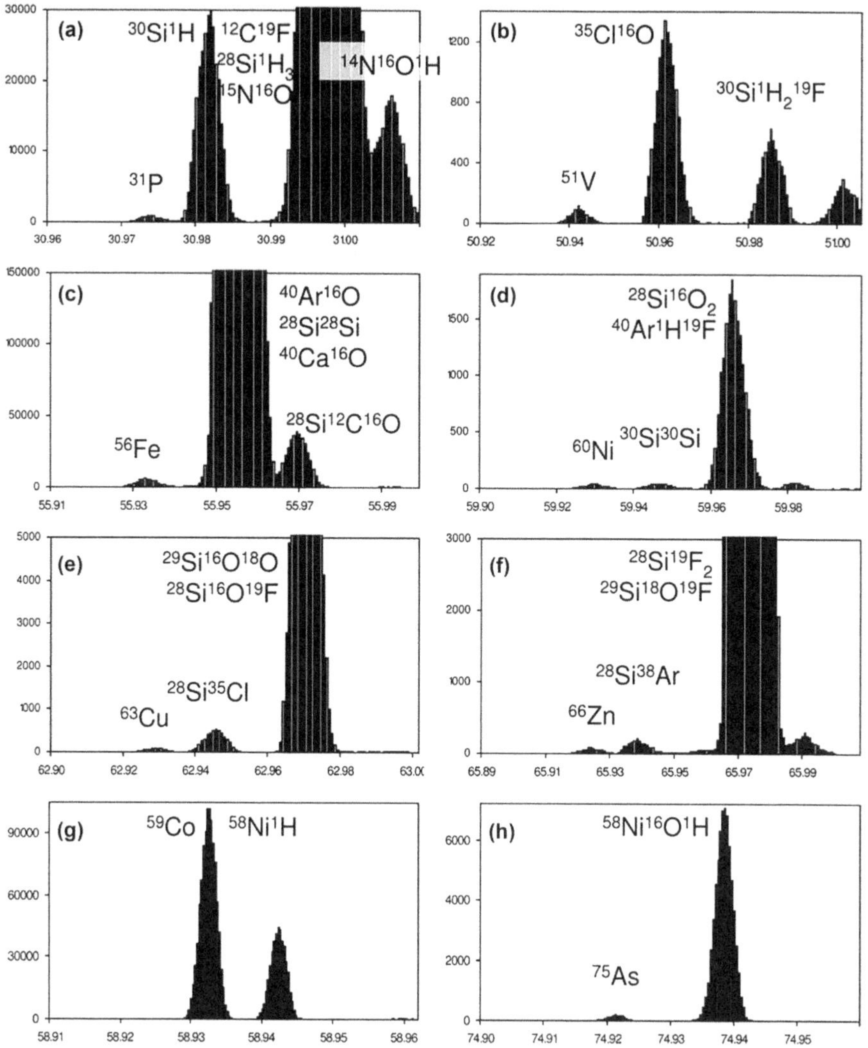


**Figure 12.17** Removal of interferences (a)–(f) in a 1000 mg kg$^{-1}$ Si/6% (*v*/*v*) HF solution at 4000 mass resolution and (g)–(h) in a 1000 mg kg$^{-1}$ Ni solution at mass resolution $R = 10\,000$.
(Adapted from Jakubowski *et al.*[39])

High-purity lanthanide compounds are used in the production of semiconductors, high-temperature superconductors and optoelectronics. In order to determine lanthanide impurities in high-purity $Y_2O_3$ and $Gd_2O_3$, Takaku *et al.* used a sector field instrument, operated at a low resolution setting of 400.[104] Detection limits of the order of 10 pg g$^{-1}$ in the solid material were found for most lanthanides. However, in $Gd_2O_3$, interferences by $GdO^+$, $GdH^+$ and $GdOH^+$ hampered the determination of Tb, Yb, Tm and Lu. To eliminate these interferences, a resolution setting of up to $R > 20\,000$

would be required. Since the instrument used was limited to a resolution $R = 10\,000$, the authors chose to measure the signal intensity of the doubly charged ions of the analytes at the low-resolution setting. Detection limits were thus deteriorated by approximately a factor of 100, but were still in the ng $g^{-1}$ range. Actual concentrations measured in 99.99% pure $Y_2O_3$ and $Gd_2O_3$ ranged between some ng $g^{-1}$ and several thousand ng $g^{-1}$.

**Ceramics, Glass and Paints**

The analysis of ceramic materials is difficult for methods requiring sample dissolution since highly concentrated strong acids and long reaction times are often necessary for digestion.[105] An example of such a matrix is SiC. A mixture of $HNO_3$, $H_2SO_4$ and HF was applied for digestion. As a result, the determination of several transition elements was seriously hampered by the formation of polyatomic sulfur- and silicon-containing species.[106] For example, the determination of the mono-isotopic Sc would not be possible because the contribution of the interfering component cannot be corrected for by a matrix-matched solution. By application of high mass resolution, separation of the isotope of interest from the abundant polyatomic ions originating from the Si matrix is possible. Jakubowski *et al.* compared the results obtained by ICP-QMS in combination with online HPLC for matrix-trace separation and preconcentration of the trace elements of interest with application of ICP-SFMS in the context of analysis of very pure $Al_2O_3$ powders for interfered as well as for noninterfered elements.[5] With ICP-SFMS, the limits of detection are one to two orders of magnitude lower than with ICP-QMS without preconcentration for noninterfered elements – at least on condition that blank contamination is not the limiting factor – and similar values obtained by ICP-QMS with preconcentration. Various new strategies in the analysis of $ZrO_2$, SiC and $Al_2O_3$ by the use of new sample preparation, processing and introduction techniques for AAS, ICP-OES and ICP-MS with low and high mass resolution are discussed by Kohl *et al.*[106] As a final conclusion of this investigation, the authors claim that ICP-SFMS operated in a high mass resolution mode is an important analytical tool to control the completeness of matrix-removal techniques and to overcome limitations due to spectral interferences.

Multielement analysis of materials and objects in the context of forensic studies or provenance determination has been described in literature as well.[107–110] The use of LA-ICP-SFMS for the forensic study of glass was described comprehensively by Latkoczy *et al.*[111] Determination of Fe in glass for forensic purposes by LA-ICP-SFMS was performed by Castro *et al.*[112] Polyatomic interferences such as $^{40}Ca^{16}O^+$ and $^{40}Ar^{16}O^+$, affecting the determination of $^{56}Fe^+$, and $^{40}Ca^{16}O^1H^+$, $^{40}Ar^{16}O^1H^+$ and $^{41}K^{16}O^+$, affecting the determination of $^{57}Fe^+$, are resolved at a mass resolution of 4000. Ancient glass, for instance, was characterised using LA-ICP-SFMS (single collector) in order to determine the provenance of glass from an excavation in Ephesos/Turkey and to investigate the secret production processes of Art Nouveau glass.[113]

ICP-SFMS instrumentation achieves lower method detection limits and its data acquisition is in most cases sufficiently fast for multielemental analysis during laser ablation. Fast transient multielement analysis is performed at a fixed resolution mode setting. Deconinck *et al.*[114] used laser ablation with quadrupole-based ICP-MS and ICP-SFMS for element analysis of car paint fragments found at crime scenes for forensic purposes. It was observed that the complex matrix composition of some of the layers – organic components, kaolin, talc, barite and/or Fe-containing pigments – resulted in spectral interferences, significantly affecting the signal profiles, especially for the transition metals when using the quadrupole-based instrument. ICP-SFMS operated at a mass resolution of 4000 enabled not only interference-free monitoring of the target elements, but also identification of most of the interfering ions, primarily on the basis of their exact mass-to-charge ratio.

**Metals**

ICP-SFMS instruments prove to be most useful in the metal industry, for instance for the analysis of pure and alloyed steel and of coatings on these materials. Trace impurities may seriously deteriorate the properties of the material. Quality control and monitoring during production are therefore necessary. Multielement analysis of metals is still a challenging task.[115–117] Important elements nowadays include Sb, Bi, Pb, Sn and P, which can influence the physicochemical properties even at $\mu g\ g^{-1}$ levels. Phosphorus is difficult to determine at such low levels with ICP-QMS instruments, but with ICP-SFMS instruments its determination at the $\mu g\ g^{-1}$ level can be accomplished without major difficulties. In the work of Finkeldei and Staats, it was shown that P had been overestimated previously in many steel reference standard materials.[118] (The same holds true for the determination of P and 13 other trace elements in cement.[119])

In the work of Traub *et al.*[120] different calibration strategies for the determination of trace elements in pure copper metal by nanosecond laser-ablation ICP-MS were investigated. In addition to certified reference materials (CRMs), pellets of doped copper powder were used for calibration. The microhomogeneity of the CRMs as well as the solution-doped pellets was sufficient to use them as calibration samples in combination with a laser spot size of 200 μm. In contrast, pellets doped with analytes in solid form showed a significant heterogeneity. Thermal fractionation effects during the ablation of the solution-doped pellets were suspected to explain the observed fractionation effects.

**Nanoparticles**

A new research direction in materials science is the production and production control of nanoparticles. Nanoparticles are particles with a diameter of less than 100 nm. Due to their large ratio of active surface to volume, they show very interesting physical, chemical and mechanical properties, which are increasingly exploited in medicine, electronic industry, consumer-care products and even in textiles. The analysis of these particles is challenging, but highly required, as they are already found in the environment. Depending on size, nanoparticles can even penetrate through cell membranes, but little is known about their toxicity. Very sensitive analytical methods urgently need to be developed for characterisation and quantification of nanoparticles in the environment, in foodstuffs and drinking water (for more details see a recent review paper from Krystek *et al.*[121]).

Although most often, analysis of nanoparticles is performed after digestion, more and more applications are reported in which the composition and the size distribution of the particles are measured. Generally, although the application of conventional pneumatic nebulisers (very briefly discussed in a recent paper from Olesik and Gray[122]) is most common, recently a new - approach was presented by Gschwind *et al.* where a microdroplet generator (μDG) as a novel sample introduction system was applied.[123] They coupled the μDG to an ICP-SFMS for the analysis of nanoparticles with particles sizes below 100 nm and achieved a 100% transport efficiency into the ICP.

**Organic Materials and Chemicals**

Organic liquids, such as solvents, or industrial products from petrochemistry are in general difficult matrices because of the high carbon load to the plasma that can cause severe spectral and nonspectral interferences. Additionally, organic solvents are sometimes needed to reduce the viscosity or to digest the samples. For instance, Pohl *et al.*[124] determined 18 trace elements in oil samples by ICP-SFMS after their dilution with xylene. Interferences hampering the determination of Mg, Al, Ca, Ti, V, Cr, Fe, Ni, Cu and Zn were resolved at a mass resolution of 4000. Detection limits were at the low pg $g^{-1}$ level and were improved by a factor of 5–50 in comparison to quadrupole instruments, while the reliability of the analysis was increased, especially for Cr, Fe, Mg, Ni, Ti and V, which suffer from spectral interferences at unit mass resolution. Federov *et al.*[125] quantified 59 trace, rare-earth and other elements in crude oil and Xie *et al.*[126] quantified 20 trace elements in residual oil by ICP-SFMS after microwave digestion.

Yang used an ICP-SFMS instrument to quantify Rh, Pd, Ir, Pt, Au and Ag in coals by ICP-SFMS.[127] A method for the direct multielement determination of Cl, S, Hg, Pb, Cd, U, Br, Cr, Cu, Fe, and Zn in powdered coal samples was developed by Boulyga *et al.*[128] They applied an ICP-SFMS instrument for isotope dilution analysis (IDMS) and used laser ablation for sample introduction into the plasma. A sector field ICP-MS with a mass resolution of 4000 and a laser system with a high ablation rate provided significantly better sensitivity and detection limits compared to a conventional laser-ablation system coupled with a quadrupole ICP-MS unit. Detection limits varied from 450 ng $g^{-1}$ for chlorine and 18 ng $g^{-1}$ for sulfur to 9.5 pg $g^{-1}$ for mercury and 0.3 pg $g^{-1}$ for uranium.

Later, an ICP-MS isotope dilution method for direct determination of sulfur in petroleum products by laser-assisted sample introduction was developed by Boulyga *et al.*[129] The detection limit for sulfur by isotope dilution analysis using LA-ICP-MS was 0.04 μg $g^{-1}$ and the analysis time was only about

10 min with little sample preparation, which therefore qualifies this method for accurate determinations of low sulfur contents in petroleum products on a routine level.

Edler *et al.* have applied silylation of the alcohols to tag an organic compound by a heteroelement (Si in this case), thus enabling their determination by ICP-SFMS (*Element 2*).[130] For this purpose, GC-ICP-SFMS has been used for quantification of n-alkanols (butanol, pentanol, hexanol and heptanol) after silylation using N-methyl-N-trimethylsilyltrifluoracetamide (MSTFA) in pyridine. External and internal calibrations have been performed, providing identical sensitivities for the compounds tested and therefore facilitating a substance-independent quantification. The first results revealing an absolute and a relative detection limit of 100 fmol (1 μL injection volume) and 100 nmol $L^{-1}$, respectively, indicate the capability of this method to become an add-on tool for quantification of organic compounds such as alcohols, amines, acids or ketones, which are routinely analysed with gas-chromatographic techniques.

HPLC-ICP-SFMS has been used at $R = 3000$ to resolve spectral interferences caused by $NOH^+$ and $NO^+$ on $^{31}P^+$ and by $N_2{}^+$ and $CO^+$ on $^{28}Si^+$, thereby facilitating the speciation of these elements.[131] As an example, polydimethylsiloxanes and their silanol breakdown products were investigated by size exclusion and reverse-phase chromatography (RP-HPLC) using high mass resolution ICP-SFMS as an element-specific detector for Si. In a second example, the authors demonstrated the capability of RP-HPLC-ICP-SFMS for separation of P-containing compounds and their detection, enabling the development of a method for quantitative and reproducible determination of common organophosphorus pesticides.

### 12.5.1.2 Applications in Geological, Cosmochemical and Radionuclide Research

Already from the early days, high-resolution ICP-SFMS has been widely applied for studying trace-element concentrations in geological samples and soils.[132,133]

**Rocks and Ores**

The simultaneous multielement capabilities in combination with low detection limits have been of great interest for the analysis of minerals, rocks and ores since the early days of ICP-SFMS. For example, Ir, Ru, Pt and Re levels in geological materials were determined using high-resolution ICP-SFMS with detection limits of 0.9 pg $g^{-1}$ for Re, 2 pg $g^{-1}$ for Ru, 3 pg $g^{-1}$ for Ir and 10 pg $g^{-1}$ for Pt.[134] In total, 39 geologically relevant trace elements were accurately quantified by ICP-SFMS in digests of granitoid reference materials G-2 and GSP-2 by Pretorius *et al.*[135] Several of these elements had to be determined at a mass resolution of 4000 or 10 000 to remove spectral interferences.

In particular, the use of laser ablation as a means of direct introduction of solid material in combination with the capabilities of ICP-SFMS has gained increasing interest in geological sciences for the elemental analysis of a variety of geological materials.[136] We observe, *e.g.* an increasing number of papers using the combination of laser ablation and ICP-SFMS for U-Pb dating of zircons.[137–144] The high sensitivity of ICP-SFMS allows precise and accurate routine analysis of zircons with a laser beam diameter of 20 to 30 μm and an ablation time of 30 s, resulting in an ablation crater depth of 15–20 μm (approximately 35 to 65 ng of zircon). This is significantly lower than the diameter of 120 to >200 μm commonly applied in LA-ICP-QMS. The better lateral and indepth resolution reduces the risk of analysing different growth zones of the zircon laterally.[137] As an example of the multielement capability of ICP-SFMS, Axelsson and coworkers determined about 50 elements in ferromanganese concretions by LA-ICP-SFMS and applied medium mass resolution ($R = 4000$) for spectrally interfered nuclides.[145]

Laser-ablation ICP-SFMS recently proved its capability to remove phosphor-, sulfur-, halide-and argon-based interferences for the measurements of U-Pb ages of uraninite and davidite.[146] Furthermore, sulfur was determined in fluid inclusions by laser ablation with an agreement between ICP-QMS and ICP-SFMS data within about 5%.[147] Trace elements in clinopyroxene crystals from juvenile scoria clasts, lava flows and hypoabyssal magmatic ejecta were determined by laser ablation with ICP-SFMS.[148]

**Radionuclide Analysis**

Sector field ICP-MS has gained more and more importance in the nuclear industries (especially MC ICP-MS instruments used for isotope ratio measurements). As a result of nuclear weapons testing, nuclear fuel reprocessing and the Chernobyl accident, radioactive pollutants have been released into the environment. Careful monitoring of their presence in different environmental compartments is therefore necessary. The improved detection limits of ICP-SFMS turned out to be very useful for the determination of low concentrations of long-lived radionuclides ($^{99}$Tc, $^{232}$Th, $^{237}$Np, $^{232,233,235,238}$U) in different materials (*e.g.* in geological samples, high-purity graphite and nonconducting concrete matrix).[149] A large number of publications covers the field of radionuclide determination *via* sector field devices, making use of the high sensitivity, the isotope ratio capabilities (even at large ratios) and the high mass resolution.[150–153] A review of Ketterer and Szechenyi shows how ICP-SFMS has become one of the most popular methods for the routine and high-throughput determination of plutonium and other transuranic elements because of its speed of analysis, simplicity of sample preparation, straightforward operation, sensitivity, low background and ability to resolve interfering polyatomic ions.[154]

Larivière *et al.* used the high sensitivity of ICP-SFMS for the rapid and automated determination of ultratrace long-lived actinides in air filters.[155] Detection limits of 0.0006 ($^{238}$U), 0.0063 ($^{239}$Pu), 0.0041 ($^{240}$Pu) and

0.062 ($^{241}$Am) μBq m$^{-3}$ were obtained for air filters after sampling of 3000 m$^3$ of air. Ultratraces of U and Pu were determined by nanovolume flow injection ICP-SFMS by Schaumlöffel *et al.*[156] The absolute limits of detection were $9.1\times10^{-17}$ g (~230 000 atoms) for $^{238}$U and $1.5\times10^{-17}$ g (~38 000 atoms) for $^{242}$Pu. ICP-SFMS was also used to quantify $^{226}$Ra in mineral, well and sediment pore waters by taking advantage of the high sensitivity of this instrumentation. Zoriy *et al.* used a method consisting of a preconcentration on a $MnO_2$ filter and a separation on a Sr-specific resin and obtained a limit of detection of 0.02 pg L$^{-1}$ for Ra.[157] Leermakers *et al.* obtained similar limits of detection using ICP-SFMS in combination with diffusive gradients in a thin-film (DGT) sampling technique.[158] The DGT technique is based on a device that accumulates solutes on a binding agent (a resin immobilised in a thin layer of hydrogel) during passage of the sample. ICP-SFMS was used to quantify $^{239}$Pu in faeces[159] and $^{239}$Pu and $^{240}$Pu in urine.[160] A method comparison showed that ICP-SFMS achieved better limits of detection for $^{239}$Pu (0.21 mBq L$^{-1}$) and for $^{240}$Pu (0.19 mBq L$^{-1}$) than ICP-QMS or alpha spectrometry.[161] Fission products in nuclear samples were determined using capillary electrophoresis coupled to ICP-SFMS by Pitois *et al.*[162]

A sensitive analytical method for determining the artificial radionuclides $^{90}$Sr, $^{239}$Pu and $^{240}$Pu at ultratrace levels in groundwater samples from the Semipalatinsk Test Site area in Kazakhstan by ICP-SFMS was developed by Zoriy *et al.*[163] The measurements were performed at medium mass resolution under cold plasma conditions in order to avoid possible spectral interferences at $m/z = 90$ for the determination of $^{90}$Sr (*e.g.* $^{90}Zr^+$, $^{40}Ar^{50}Cr^+$, $^{36}Ar^{54}Fe^+$, $^{58}Ni^{16}O^{2+}$, $^{180}Hf^{2+}$, *etc.*). Analysis at medium mass resolution displayed interfering peaks in the mass region of Pu, which originated most likely from lanthanide-containing polyatomic ions.[164]

As a novelty, isotope dilution was applied for laser-ablation (LINA-Spark-Atomiser) ICP–MS of soil samples.[165] A mass resolution of 4000 was needed to reduce or separate interferences by $UH^+$ and $PbO_2{}^+$ ions and by tailing of the $^{238}U^+$ peak in soil samples for detection of $^{239+240}$Pu. With an ultrasonic nebuliser, detection limits in the order of 1–10 fg mL$^{-1}$ were obtained, corresponding to activities ranging between 0.2 mBq mL$^{-1}$ ($^{226}$Ra) and 0.08 nBq mL$^{-1}$ ($^{232}$Th). The radioactivity levels of these isotopes in the environment are of the order of 1 mBq g$^{-1}$ or less, and with radiometric methods, samples need to be counted for at least one day to collect 100 counts per gram of sample. Application of ICP-SFMS is therefore very advantageous, since measurement times of only 2.5 min are required to reach the detection limits stated above. Yamamoto *et al.* reported the determination of $^{99}$Tc in environmental samples.[166] After leaching, the leachate was purified *via* solvent extraction and ion-exchange chromatography to remove the isobaric interference due to $^{99}$Ru and to reduce the concentration of dissolved matrix elements in the final solution. The resulting method is characterised by an absolute detection limit of 0.25 pg of $^{99}$Tc, corresponding to 0.16 mBq. In sediments from the Esk Estuary in the Irish Sea,

$^{99}Tc$ activities between 24 mBq $g^{-1}$ and 109 mBq $g^{-1}$ were measured. The depth profile of the measured activities reflects the trend in the variation of the discharge from the nearby nuclear fuel reprocessing plant at Sellafield (UK).

Varga demonstrated that U isotopic analysis of $UO_2$ particles down to 10 μm can be accomplished *via* LA single-collector ICP-SFMS with a precision of a few per cent RSD.[167] This precision suffices to gain insight into the intended use of the material, *i.e.* energy production in a fission reactor or the production of nuclear weapons. Especially for the low-abundant isotopes ($^{234}U$ and $^{236}U$), the use of a higher mass resolution ($R = 4000$) was required for avoiding spectral interference. Varga and Surányi also demonstrated that both solution-based single-collector ICP-SFMS and LA-ICP-SFMS can be used for the determination of the production date of $UO_2$ material, based on the $^{230}Th/^{234}U$ ratio.[168] In the case of solution analysis, spectral interferences were avoided by prior chromatographic separation. In the case of LA-ICP-SFMS this was accomplished by measuring at higher mass resolution. Although the precision offered by LA-ICP-SFMS was inferior to that offered by solution ICP-SFMS, it was sufficient in the context of nuclear forensics. Moreover, both ICP-SFMS approaches show a higher sample throughput compared to gamma or alpha spectrometry.

As a result of the accident at the Chernobyl nuclear power plant in 1986, the environment was contaminated with spent nuclear fuel. Boulyga and Becker developed rapid and sensitive analytical procedures for uranium isotopic ratio measurement in environmental samples based on ICP-SFMS with a *MicroMist* nebuliser, a direct-injection high-efficiency nebuliser (*DIHEN*)[169] and a low-flow microconcentric nebuliser with a membrane desolvation system.[170] Boulyga *et al.* further investigated the use of the $^{236}U$ isotope to monitor the spent uranium from nuclear fallout using ICP-MS in soil samples collected in the vicinity of the Chernobyl reactor.[151] Sector field ICP-MS and quadrupole ICP-MS without and with hexapole collision cell (ICP-CC-MS) were used for U isotope analysis. In addition, MC ICP-MS was used for high-precision isotope ratio measurements. An absolute sensitivity of 3600 counts $fg^{-1}$ was achieved for U by using a direct-injection high-efficiency nebuliser for introduction of solutions into ICP-SFMS with a detection limit for $^{236}U$ in the fg $g^{-1}$ range. Boulyga and Heumann developed an ICP-SFMS method that allowed the measurement of $^{236}U$ at concentration ranges down to $3 \times 10^{-14}$ g $g^{-1}$ and extremely low $^{236}U/^{238}U$ isotope ratios of $10^{-7}$ in soil samples.[171] By using the high-efficiency solution-introduction system APEX in connection with ICP-SFMS, a sensitivity of $> 5000$ cps $fg^{-1}$ uranium was achieved. The use of an aerosol desolvation unit reduced the formation rate of uranium hydride ions $UH^+/U^+$ down to a level of $10^{-6}$. An abundance sensitivity of $3 \times 10^{-7}$ was observed for $^{236}U/^{238}U$ isotope ratio measurements at a mass resolution of 4000. The detection limit for $^{236}U$ and the lowest detectable $^{236}U/^{238}U$ isotope ratio were improved by more than two orders of magnitude compared with corresponding values by alpha spectrometry.

A number of papers report the use of U isotopic analysis with the intention to evaluate the presence of depleted uranium (DU). DU is a waste product from the isotopic enrichment of U used as fuel in nuclear fission reactors. Because of its very high density, it is used for manufacturing ammunition capable of penetrating armoured steel and of balance weights in airplanes. Trešl *et al.*, *e.g.* have measured the $^{235}U/^{238}U$ isotope ratio in urine samples of rescue workers potentially exposed to DU at the scene of a plane crash.[172] The isotope ratio precision of single-collector sector field ICP-MS self-evidently is sufficient to distinguish between natural U (0.72% $^{235}U$) and DU (~0.2% $^{235}U$), but preconcentration was required to obtain a sufficiently intense $^{235}U^{+}$ signal.

### 12.5.1.3 Applications in Environmental Research

A review by Krachler highlights the features that make ICP-SFMS a very powerful tool in environmental analysis.[173] The analysis of waters, soils and plant material using direct introduction systems or after digestion has been in the focus of environmental researchers for a long time. Incrementally grown matter such as tree rings, sediments, ores, corals or otoliths (incrementally growing earstones of fish, which record environmental information over its life history) are important archives and deposits of elemental compositions taken up in the past. Laser ablation coupled to an ICP-SFMS has been used increasingly to investigate time-dependent information chronologically stored in incrementally grown matter.[174–177]

**Surface, Ground Water and Aquatic Systems**

The low detection limits of ICP-SFMS, down to the fg $mL^{-1}$ range have proven to be an important feature especially in the analysis of natural waters.

New legislative water directives require new certified reference materials and for a number of elements (*e.g.* As, Cr, Fe, Mn, Ni and Zn), the use of high mass resolution with an ICP-SFMS instrument is the method of choice to obtain accurate results without pretreatment (preconcentration) of the samples.[178] In centrifuged and noncentrifuged surface waters, As, Cd, Cr, Cu, Ni, Pb and Zn were quantified by Popp *et al.*[179] using ICP-SFMS *via* slurry-type nebulisation. According to their distribution, the elements were classified into three groups: (i) Ni, Cr and Cu showing considerable particle-bound fractions; (ii) Pb and Zn with a high tendency towards particle binding; and (iii) As and Se being predominantly present in the liquid phase. Medium mass resolution was necessary for interference-free measurements of $^{52}Cr$, $^{60}Ni$, $^{65}Cu$ and $^{66}Zn$, and high mass resolution ($R = 10\,000$) for $^{75}As$ and $^{77}Se$. Method detection limits were below the current environmental quality standards (EQS) of the European Water Framework Directive (WFD) for inland surface water for all elements.[180,181]

Tracer experiments with stable isotopes can be used to investigate elemental fluxes in, *e.g.* the environment or metabolic processes. Balcaen *et al.*, *e.g.* relied on a tracer experiment using two isotopic tracers to determine the relative importance of the two pathways according to which *Daphnia magna* – a water flea often used as a model species in ecotoxicology – can take up Zn, *i.e.* *via* the respiratory system (gills) or *via* the alimentary channel.[182] The use of medium mass resolution rendered the $Zn^+$ signals free from spectral interference from S-containing polyatomic ions, such as $SO_2H^+$. Arsenic speciation in freshwater snails was studied by Lai *et al.* by means of HPLC-ICP-SFMS.[183] Methanol/water extracts were prepared from the freeze-dried soft tissue for analysis. The ICP-SFMS instrument was operated in the low-resolution mode ($R = 300$) for monitoring $^{75}As$. The most common arsenic compounds found in the snails were tetramethylarsonium, as well as oxo- and thio-arsenosugars.

**Snow and Ice**

The Greenland and Antarctic snow and ice caps are well preserved and detailed archives, consisting of successive datable layers and hence, permit past and recent variations in the composition of the earth's atmosphere to be studied. The investigation of elements at ultratrace levels in order to monitor the fate of heavy metals in the environment still presents a significant challenge.[184–188] As a result of the extremely low concentration levels involved (low and sub-pg $g^{-1}$ levels), an extremely sensitive technique is required and extreme precautions have to be taken during sample collection, storage, treatment and analysis to avoid contamination. Rare-earth elements (REE) were quantified by Gabrielli *et al.* at the sub-pg $g^{-1}$ level in Antarctic ice by ICP-SFMS with a sample introduction system containing a desolvation unit.[189] An interesting application of ICP-SFMS is the determination of Ir and Pt in a Greenland ice core for the calculation of cosmic fallout to the earth.[190] Procedural detection limits of 0.02 and 0.08 fg $g^{-1}$ for Ir and Pt were reported.

Krachler *et al.* determined low Sb and Sc concentrations in arctic ice and snow samples. Detection limits of 0.005 pg $g^{-1}$ for Sc and 0.03 pg $g^{-1}$ for Sb were obtained with ICP-SFMS using an Apex sample introduction system.[191] Pb isotope ratios have been widely applied to distinguish between natural, geogenic Pb (crustal signature) and anthropogenic Pb (ore signature) and calculate the relative contribution of the sources, even with single-collector instrumentation. Döring and coworkers were one of the first groups to deploy single-collector sector field ICP-SFMS in this context and determined Pb isotope ratios and heavy metals (Cr, Cu, Zn, Co, Ni, Mo, Rh, Pd, Ag, Cd, Sb, Pt, Au, and U) in high alpine snow.[192,193] The results were situated on a mixing line between the isotopic signature of the background (crustal) Pb and of the Pb found in leaded fuel (still sold in Europe). Though the Pb emission by automobiles has drastically decreased in the past decade, the authors were able to conclude from these data that cars still contribute as much to the Pb pollution in high alpine areas as other anthropogenic sources (waste incineration, industry) do. This research is remarkable because the Pb concentrations in the snow samples were very low, ranging from 0.02 to 9.7 ng $g^{-1}$, and determinations did not require more than 0.8 ml of sample (microconcentric nebulisation) while yielding an average standard error on the ratios of 0.14% only. Krachler *et al.,* determined Pb isotope ratios at low picogram per gram levels in ice samples by ICP-SFMS with an accuracy and precision that are similar to those obtained by thermal ionisation mass spectrometry (TIMS) at such low Pb concentration levels.[194]

**Seawater and Marine Studies**

Seawater analysis represents again the combination of low elemental concentrations and a complex matrix due to the high salt concentration.[195–198] A large number of matrix-dependent interferences can be observed for a number of elements, thus making ICP-SFMS the analytical tool of choice. Whereas with ICP-QMS a variety of separation/preconcentration methods have already been used, Rodushkin and Ruth succeeded in the analysis of saline water samples using an ICP-SFMS instrument after simple four- to fivefold dilution of the samples.[199]

Field *et al.* combined the *SC-FAST* (ESI, Omaha, NE) sample introduction system with an ICP-SFMS unit to reduce sample uptake and wash times and thus reduce cone deposition and increase sample throughput.[200] This sample introduction system consists of an inert loop on a six-port flow injection valve placed very close to the nebuliser. Samples are loaded onto the loop with the aid of a vacuum pump and pushed to the nebuliser by a pressurised carrier solution. Therefore, any contact of the sample with the peristaltic tubing is avoided.

Incrementally grown matter in the sea has been increasingly applied for studying time-resolved environmental changes. Thorrold *et al.* have used an *Element 2* to identify natural geochemical signatures in otoliths to investigate the ability of trace element and isotopic signatures in otoliths to record the nursery areas of juvenile (young-of-the-year) weakfish (*Cynoscion regalis*) from the east coast of the USA.[201] Spawning site fidelity ranged from 60 to 81%, similar to estimates of natal homing in birds and anadromous fishes. These data were in contrast to genetic analysis of population structure in weakfish. Thorrold's findings highlight the need for consideration of spatial processes in fisheries models and have implications for the design of marine reserves in coastal regions. The same sample materials can reflect the environmental status of the water they (have) reside(d) in *via* their element/Ca ratios.

Sea-water temperatures can be calculated from Mg/Ca ratios in foraminifera since the uptake of Mg into foraminiferal $CaCO_3$ is temperature-dependent.[202–204] Katz *et al.* measured Mg/Ca ratios of benthic foraminifera by ICP-SFMS to calculate seafloor water temperatures and to reconstruct the stepwise transition from the Eocene greenhouse to the Oligocene icehouse.[203] Oppo *et al.* used Mg/Ca ratios of foraminifera for 2000-year-long sea-water temperature reconstructions.[204] Coral skeleton P/Ca ratios were determined by ICP-SFMS to reconstruct seawater phosphate concentrations.[205]

**Soils and Sediments**

The applicability of ICP-SFMS for routine multielement analysis, using both the high- and low-resolution mode of the instrument, is demonstrated by the multielement analysis of soil after digestion or using leachable extracts. The latter is important to understand heavy metal binding and mobility in soils of different composition and from different climate zones; this work required a reliable and fast multielement method. High mass resolution was used for identifying a number of interfered elements in soils, *e.g.* K, Mn, Ni, Cr, V, Fe, Cu and in particular As.[206–208] Chipley and coworkers coupled an online leaching system of soils to an ICP-SFMS unit in order to obtain real-time leaching data.[209] Prohaska *et al.* discussed the application of high mass resolution for the accurate determination of REE.[210]

Townsend *et al.* profited from the high sensitivity offered by single-collector sector field ICP-SFMS to determine Pb isotope ratios in Antarctic sediments, despite the low concentration in samples not affected by anthropogenic contamination.[211] The goal of their research was detecting the presence of anthropogenic contamination and – to the extent possible – deciphering the corresponding source.

**Plant Materials**

Plant materials play an important role in nutrition, but this aspect will be discussed lateron. Additionally they are successfully used as biomonitors of environmental contamination or are used to study the bioavailability of heavy metals. Elements such as Ca, Ti, V, Cr and Fe are spectrally interfered at unit mass resolution and accurate analysis was accomplished by relying on the high mass resolution capabilities of sector field devices.[212–215] Nonetheless, ICP-QMS using a dynamic reaction cell and ICP-SFMS reveals similar results for metal analyses of plant material.[216]

A special interest has grown in the analysis of REE in plant materials, also for using the REE pattern for provenance studies of agricultural products. Most oxide interferences of REE require a mass resolution between 7500 and 10 000, but for all elements finally at least one nuclide is applicable which can be used for analysis. Palmer *et al.* determined background levels of REE and other trace elements in the gammaridean amphipod *Paramoera walkeri* (a sensitive bioaccumulating organism) by ICP-SFMS in low mass resolution mode.[217] Appropriate digestion of the samples resulted in complex matrices, especially in the case of silicate-containing samples. The elemental loss in silicate residues of plant material was found to be up to 30% and therefore digestion methods relying on the use of HF were required. The high concentration of matrix elements led to spectral interferences, which were investigated by measuring the elements with different mass resolutions. The capability of high mass resolution was demonstrated in the direct determination of Sc, Y and REE in reference materials: basalts (BCR-1, BHVO-1, BIR-1, DNC-1), andesite (AGV-1) and ultramafics (UB-N, PCC-1 and DTS-1).[218] Time-consuming ion-exchange separation or preconcentration were found to be unnecessary when using high mass resolution. The mass spectral peak of $^{45}Sc^+$ was completely resolved from $^{29}Si^{16}O^+$ and $^{28}Si^{16}O^1H^+$ using medium resolution. Becker *et al.* determined trace elements including Pt in tree bark by ICP-SFMS. Detection limits in the tree bark were better for sector field instrumentation (0.03 ng $g^{-1}$) than for quadrupole ICP-MS (0.2 ng $g^{-1}$).[219]

Accumulation of 25 elements was compared for seeds of transgenic and of nontransgenic soybeans.[220] Medium-resolution ($R=4000$) ICP-SFMS was used for quantification of nuclides from $^{7}Li^{+}$ to $^{209}Bi$, including some for which important interferences are known, such as $^{75}As$ and $^{31}P$. For most elements, the concentrations were similar in seeds of transgenic and non-transgenic beans, except for Co, Cu, Fe and Sr, which were increased in "transgenic samples".

ICP-SFMS was used by Kempter *et al.* to study the accumulation of atmospheric Pb and Ti in *Sphagnum* mosses.[221] Results from two different sampling sites in Germany were compared and attention was paid to both intrasite and species-dependent variation. Important for drawing meaningful ecological considerations is that the accumulation was seen to largely depend on the annual dry-matter production rate of the mosses.

Airborne contamination by metals has also been determined by accumulation in lichens by Pino *et al.*, who made use of ICP-SFMS in this context.[212] In a subsequent study, the authors determined the iridium, platinum and rhodium content of lichens from different sites in Argentina as a measure of possible airborne contamination by traffic or other anthropogenic activities.[222] Application of medium resolution or mathematical correction solved the problems of spectral interference for nuclides such as $^{193}Ir$ ($^{177}Hf^{16}O$); $^{195}Pt$ ($^{179}Hf^{16}O$, $^{177}Hf^{18}O$ and $^{178}Hf^{17}O$) and $^{103}Rh$ ($^{87}Sr^{16}O$, $^{63}Cu^{40}Ar$, $^{65}Cu^{38}Ar$, $^{67}Zn^{36}Ar$, $^{66}Zn^{37}Cl$, $^{68}Zn^{35}Cl$ and $^{206}Pb^{2+}$). As a final result, the authors excluded emission of Pt, Ir and Rh at the study sites.

Recently, laser ablation of dried green leaves was used by Cizdziel *et al.* for sample introduction into ICP-SFMS. Nine elements were quantified in the low-resolution mode, namely Mg, Ca, Mn, Cu, Sr, Cd, Ba, Hg and Pb.[223]

The bioaccumulation of platinum group elements (PGEs) in hydroponically grown grass samples was studied by Lesniewska *et al.* to answer the question if and how PGEs can enter the food chain.[224] Pt, Pd and Rh were added to the nutrition solution and after extraction, the metal-containing species from roots and leaves were separated from one another by size-exclusion chromatography. The PGEs and some essential metals (Cu, Sr, Mo, Mn) were monitored in the eluates. The elements C, S and Ca served as marker elements for specific classes of plant biomolecules and had to be measured in medium-resolution mode simultaneously with the metals of interest. $^{13}C$ was used as a natural tag representing organic fractions. The PGEs were detected in leaves already 12 h after administration and were bound to more than 10 different molecular fractions.

Mihucz *et al.* studied the uptake and speciation of arsenite and arsenate by cultured cucumber plants by means of HPLC coupled to sector field ICP-SFMS.[225] Medium-resolution condition ($R=4000$) was used, which did not overcome the interference of $^{40}Ar^{35}Cl^{+}$ with $^{75}As^{+}$. Consequently, anion-exchange chromatography was used for separation of chloride prior to analysis. As a biological result, the authors found that the concentration order of the arsenic species in the xylem sap was As(III) > As(IV) > dimethylarsinic acid, independent of the plant treatment by arsenate or arsenite.

Zheng *et al.* also studied the arsenic content and its speciation in plants.[226] For this purpose, various freshwater and terrestrial plant species were collected from an arsenic-contaminated former gold-mining site. Methanol–water (9 + 1) extracts of the collected plant material were analysed by high-resolution ICP-SFMS ($R = 10\,000$). A problem discussed is the low extraction efficiency by $CH_3OH/H_2O$ for As-species from the plant material, which roughly amounts to less than 20% only. Nevertheless, the authors could determine numerous arsenic species in the water-soluble fractions, including arsenate, arsenite, methylarsonic acid, dimethylarsinic acid, trimethylarsine oxide and tetramethyl-arsonium. Total arsenic content and speciation patterns varied between plant species and different parts of the plants. However, the inorganic arsenic species were dominant in all samples, usually making up more than 90% of the totals.

A method based on application of double-spike isotope dilution for the accurate determination of Cr(III), Cr(VI) and total Cr in yeast was developed by Mester and coworkers.[227]

**Environmental Aerosols**

Trace-metal analysis in aerosols has become of increasing interest, especially for the platinum group elements (PGEs). The latter are in the focus of research as they are set free from catalytic converters of present-day cars. Their transport into the environment and possible health effects are still a hot topic in many countries.[228–230] ICP-SFMS was applied for the accurate quantification of Rh and Re in trace levels in size-classified urban aerosol samples and of Rh, Pd and Pt in silica-containing matrices.[231] The two major interferences, $^{40}Ar^{63}Cu^{+}$ and $^{206}Pb^{2+}$, affecting the monoisotopic $^{103}Rh$ could be separated by applying high and medium mass resolution, respectively. Moreover, the authors presented a so far unidentified interference, which shows a peak height of 50% of the $^{103}Rh^{+}$ signal and that was resolved in medium mass resolution. Other interferences (*e.g.* $^{68}Zn^{35}Cl^{+}$) require a higher mass resolution ($R = 9000$) or even above $R = 10\,000$ (*e.g.* $^{86}Sr^{17}O^{+}$). Therefore, other strategies are required, such as the application of membrane desolvation in order to reduce oxide-containing interferences. ICP-SFMS was further applied to analyse airborne particles for their Ir content[232] and to investigate Cr species.[233]

### *12.5.1.4 Applications in Life-Science Research*

Under the umbrella "life sciences", many different and independent disciplines are compiled, covering food and nutrition, as well as biology, biochemistry and medicine. The emerging field of metallomics refers to the entirety of research activities aiming at the understanding of the molecular mechanisms of metal-dependent life processes.[234] Speciation analysis has become a widely applied technique in life sciences. Reviews of applications involving high mass resolution or sector field devices in speciation analysis are given by Houk in the "Handbook of Elemental Speciation",[235] by Krupp *et al.*,[236] Moldovan *et al.*[237] and Montes-Bayón *et al.*[238]

Besides metals and metalloids, in particular the nonmetals phosphorus and sulfur are receiving increasing interest in the context of life-science applications. Unfortunately, phosphorus and sulfur are known to be elements that are difficult to determine *via* plasma-based spectrochemistry, since they are characterised by a high first ionisation potential, low ionisation efficiencies and suffer from overlap with polyatomic ions.[239] This situation gets even worse in the case of HPLC coupling with organic solvents used as the liquid phase.

**Food and Nutrition**

Most published applications in this area can be related to the multielement analysis of beverages, vegetables and fish. For example, Bibak *et al.* fully exploited the possibility of ICP-SFMS for the analysis of onions[240] and peas[241] by using different resolution settings between $R = 400$ and $R = 10\,000$ to determine the concentration of up to 63 elements, including Eu ($BaO^+$ interference at mass 153), for which a resolution of $R = 10\,000$ was employed. Other examples of multielement quantification are the analysis of infant food,[242] or raw nuts and seeds.[243] In the latter study, 70 elements were investigated in more than 40 products.

A comprehensive determination of metals in human milk has been performed by Krachler *et al.* by means of ICP-SFMS, operated in the medium-resolution mode ($R = 4000$) to overcome spectral interferences.[244] It is shown that, *e.g.* $^{52}Cr$ is clearly separated from the otherwise interfering polyatomic ions $^{40}Ar^{12}C^+$, $^{36}Ar^{16}O^+$ and $^{38}Ar^{14}N^+$ under the conditions used. Concentrations of 14 elements, Co, Cr, Cu, Fe, Mn, Ni, Se, V, Al, Ag, As, Au, Pt, Sc and Ti, could be determined simultaneously and reliably by ICP-SFMS.

Toxic elements (in particular As, Cd, Hg and Pb) were determined in the muscle tissue of river fish samples,[245] honey[246] and offal.[247] One major group of elements, which requires high sensitivity and that is gaining increased interest in food science as well, are the rare-earth elements. They can, *e.g.* provide a unique fingerprint as an indication of provenance. The major drawback is their low abundance in food, since transfer factors (soil/plant ratio) can be as low as 0.04 because soil properties like pH, organic matter or cationic exchange capacity have an influence on the exchangeable fraction of REE.[248] Thus, a highly sensitive method with multielement measurement capabilities is required.

Alcoholic beverages are in general a difficult matrix because of the high carbon load to the plasma that can cause severe spectral and nonspectral interferences.[249,250] Castiñeira Gómez *et al.* applied ICP-SFMS for the classification and provenance testing of German white wines based on data for 13 elements at trace and ultratrace levels.[251] $^{25}Mg$ (interfered by $^{12}C_2{}^1H^+$), $^{44}Ca$ (interfered by $^{12}C^{16}O_2{}^+$ and $^{88}Sr^{2+}$), $^{55}Mn$ (interfered by $^{40}Ar^{14}N^1H^+$ and $^{39}K^{16}O^+$), $^{56}Fe$ (interfered by $^{40}Ar^{16}O^+$ and $^{40}Ca^{16}O^+$) and $^{59}Co$ (interfered by

$^{43}Ca^{16}O^{+}$) were measured at medium mass resolution to overcome spectral interferences from the wine matrix, which was simply diluted 1:20. The classification of unknown German wines could be accomplished with a success rate higher than 75%. Smith determined the country of origin of garlic using trace-metal profiling by determining 12 elements with ICP-SFMS.[252] 19 trace elements were determined by Shibuya *et al.* in marijuana samples by ICP-SFMS to classify the samples according to their geographical origins.[253] Ga, Cu, Zn, Fe and Mn were measured at a mass resolution of 3000 because of spectral interferences. Provenance testing has been discussed for fish as well. For example, scales of salmon were analysed directly by LA-ICP-SFMS for stock identification.[254]

As discussed previously, Pd and Pt are gaining interest, also in the context of the analysis of food samples and thus were determined in vegetables and flour products and foodstuffs of animal origin.[255] Evans *et al.* quantified Pu, Am and Np in food samples, ranging from cabbage to milk and meat by a combination of ion chromatography, ultrasonic nebulisation and ICP-SFMS.[256] Because of the high sensitivity of sector field instrumentation, detection limits as low as 0.02 pg $g^{-1}$ ($4.6\times10^{-2}$ Bq $kg^{-1}$) for $^{239}Pu$, 0.011 pg $g^{-1}$ ($3\times10^{-4}$ Bq $kg^{-1}$) for $^{237}Np$ and 0.033 pg $g^{-1}$ ($2.45\times10^{-1}$ Bq $kg^{-1}$) for $^{243}Am$ were obtained.

Sector field devices are also applied for speciation studies by HPLC-coupling in food analysis for interference-free analysis of As or Fe in, *e.g.* vegetables or meat.[225,257]

**Applications in Biology, Biochemistry and Medicine**

Analysis of complex biological matrices, most often rich in carbon and salts, requires analytical instrumentation capable of coping with spectral interferences for determination of target elements at steadily decreasing concentration levels.[210,258–260] High mass resolution has become the method of choice when facing low-level multielement analysis in biological and clinical samples.[173] The capability of high mass resolution devices for the speciation of biomolecules was soon recognised in the scientific community as well and the combination of molecular and elemental mass spectrometry represents a powerful synergy in modern life-science applications.[261–263] This is especially true for the characterisation of metal-containing or metal-binding proteins in biological systems and for protein modifications involving "heteroelements", such as P in protein phosphorylation studies.[264,265]

Analysis of biological materials can be based on their natural element composition or on specifically introduced selective elemental tags, combined with or without spatial resolution. Application of spatially resolved analysis on unfractionated biological material, such as complete tissues or cells, is called bioimaging or immuno-bioimaging when combined with element-labelled antibodies for detection. Clearly, the demands differ for the various applications and strongly depend on the required analytical figures of merit, such as chemical background, sensitivity, interferences and spatial resolution. The latter is a particular challenge for single-cell analyses and the examination of the intracellular distribution of multiple target elements. For larger tissue sections, scanning speed is a further decisive parameter when precise and comprehensive analyses are required.

**Elemental Analyses of Body Tissues and Body Fluids**

The determination of physiologically and toxicologically important trace and ultratrace elements in biological materials such as body fluids and tissues is difficult for ICP-QMS because, in addition to interfering species such as oxides and argides, many polyatomic interferences originate from the matrix elements C, Na, P, S, Cl, K and Ca. Riondato *et al.* were the first to use ICP-SFMS to overcome these limitations.[266] Steuerwald *et al.* have shown that gadolinium may interfere as a double-charged ion or in the form of polyatomic species with several selenium or platinum isotopes, respectively, in urine samples.[267] Gadolinium is usually not present in urine, but occurs at higher concentrations after administration of Gd-containing contrast agents to patients. The seemingly high Se and Pt values in urine determined by ICP-QMS could be explained by Gd-based interferences that could be resolved using ICP-SFMS operated at higher mass resolution.

Size-exclusion chromatography (SEC) coupled to an ICP-SFMS instrument operated at LR, MR and HR has been applied by Wang *et al.* to identify mineral elements in proteins of human serum and of DNA fragments, demonstrating that ICP-SFMS can be a valuable tool to study the interaction of essential and toxic metals with biological macromolecules.[268] Zn, Cu, Se, Cd, Pb, Th and U have been studied as binding partners in biomolecules at LR, whereas for Cr, Mn and Fe, a resolution of $R = 3000$ was required. In both cases, convincing detection limits were achieved. Th and U were determined in a human serum reference material (National Institute of Standards and Technology, USA) at ambient levels in the low pg $mL^{-1}$ range in several fractions of molecular masses in the interval 10–630 kDa.

Methods for rapid determination of 50 to 60 elements in digested blood were described by Rodushkin and coworkers.[269,270] The majority of elements were found at concentrations above the method detection limit. In a subsequent work, they quantified ultratrace levels of multiple low-abundance elements in urine and serum.[271] D'Ilio *et al.* quantified 17 elements and compared ICP-SFMS in medium- ($R = 4000$) and high-resolution ($R = 10\,000$) mode and dynamic reaction cell ICP-QMS.[272] They claimed that both techniques are equally suited for the quantification of Al, As, Ba, Cd, Co, Cr, Li, Mn, Mo, Ni, Pb, Sb, Se, Sn, Sr, V and Zr in blood samples. Sarmiento-González *et al.* quantified Ti, V, Cr, Co, Ni and Mo potentially released from dental implants and prostheses in human body fluids by ICP-SFMS overcoming the difficulties of the occurrence of polyatomic interferences such as with $^{31}P^{16}O^{+}$ and $^{35}Cl^{16}O^{+}$.[273,274] They also determined titanium levels in the organs and blood of rats with a titanium implant *via* ICP-SFMS.[275] The low detection limit of 0.07 μg $L^{-1}$ enabled reliable determination of Ti in organ tissues and blood even at basal levels and independently of the chemical form of Ti.

In a recent study, 18 trace elements were determined in blood, urine and sweat of 20 individuals, either healthy or with health problems.[276] An important finding of the study was that excretion of certain toxic elements occurs preferentially *via* sweat. A set of 18 elements was analysed in blood spots from new-born children.[277] A suitable extraction method was developed, allowing for the subsequent analysis by ICP-SFMS. Co, Cr, Ti, Mo and Mn were quantified in body fluids of patients with total hip arthroplasty by Nuevo Ordóñez *et al.* applying ICP-SFMS operated at a resolution of $R=3000$.[278] The methodology proved to be sensitive enough to determine these five elements accurately in two relevant reference materials (Seronorm Trace Elements Urine Level 1 and Seronorm Trace Elements Whole Blood Level 1). In a subsequent study, the authors relied on ICP-SFMS to demonstrate that titanium present in the serum of patients carrying Ti nails can bind to the iron-transporting protein transferrin.[279] Analysis was done by cation-exchange chromatography hyphenated to ICP-SFMS and measuring $^{47}Ti$, $^{49}Ti$, $^{56}Fe$, $^{57}Fe$, $^{34}S$ and $^{32}S$ isotopes. The authors conclude that titanium can enter cells *via* the transferrin route.

Krachler *et al.* validated the determination of the ultratrace elements Co, Cr, Mo and Ni in whole blood, serum and urine using ICP-SFMS.[280] Vanadium was quantified in biological fluids by Yang *et al.*[281] At a mass resolution of 4000, the $^{35}Cl^{16}O^+$ interference was resolved from the analyte signal. Featherstone and coworkers developed methods for the determination of Se in human plasma using the high-resolution capability of ICP-SFMS, including the identification of some discrepancies in assigned Se levels in Seronorm Trace Elements Serum Level 2.[282,283]

Bocca and coworkers determined Al, Co, Cr, Mn, Ni and V in medium mass resolution mode in biological fluids of patients with Parkinson's disease to investigate a correlation between the disease and the level of these elements.[284,285] They found imbalances in metal concentrations in blood, serum, urine, cerebrospinal fluid and hair of these patients. Praamsma *et al.* determined manganese concentrations in blood and urine and compared different ICP-MS methods (including sector field and quadrupole ICP-MS in standard and dynamic reaction cell mode).[286] The ICP-SFMS run in medium-resolution mode ($R=4000$) clearly resolved the potential interferences $^{54}Fe^{1}H^+$, $^{39}K^{16}O^+$, $^{40}Ar^{15}N^+$ and $^{40}Ar^{14}N^{1}H^+$ affecting $^{55}Mn^+$ at low mass resolution. It was concluded that both ICP-SFMS and ICP-DRC-QMS are suitable for Mn quantitation in blood and urine. However, detection limits are lower for ICP-SFMS, *i.e.* 1.0 and 0.5 $\mu g\ L^{-1}$ for blood and urine, respectively, compared to 6.4 and 0.6 $\mu g\ L^{-1}$ achieved by ICP-DRC-QMS.

Exposure to hexavalent chromium species can have severe health effects in humans and in animals, while trivalent chromium is known as an essential trace element but may exhibit toxicity at higher concentrations.[287–290] Thus, analysis of chromium species in tissues and also in environmental samples is of interest. Spectral interferences can be overcome with high mass resolution.[291,292] Van Lierde *et al.* deployed an experimental setup based on the hyphenation of CE with ICP-SFMS to study the fate of Cr(III), Cr(VI) and other Cr species extracted from chromium-tanned leather in simulated sweat.[293] It could be shown that the amino acid methionine is responsible for the reduction of Cr(VI) into Cr(III), which, in turn, forms a complex with lactic acid. In a subsequent study, the setup was also used in *in vitro* permeation studies with porcine and human skin to study the factors affecting permeation.[294]

Bloom *et al.* quantified Hg, Cd and Pb in the follicular fluid surrounding the oocyte and in blood.[295] While they found a correlation for Hg and Cd concentrations in the two fluids, this was not the case for Pb. Interestingly, the Cd concentration correlated positively with fertilisation and even stronger with pregnancy. This "positive" physiological effect may be attributed to the estrogenic activity of cadmium.[296] A comprehensive analysis of trace elements in follicular fluid by ICP-SFMS was performed by Kruger *et al.*[297]

Alimonti *et al.* compared sector field and quadrupole ICP-MS for their utility to quantify Ba, Cs, Sb and W in urine samples.[298] The ICP-SFMS unit was run in low-resolution mode ($R=300$) and the authors claim that there were no significant interferences although $^{122}Sn^{16}O^{+}$ and $^{177}Hf^{16}O^{+}$ may potentially perturb the signals for $^{138}Ba$ and the internal standard $^{193}Ir$, respectively. A sector field and a quadrupole instrument revealed similar LODs, but SFMS was more sensitive when taking into account all measured elements.

Noble metals have long been considered as harmless, because they are chemically inert. Today, however, it is known that Pd, Pt and their compounds, for instance, are among the most potent sensitisers with a high incidence of various allergic reactions. The proliferation in the environment of Pd, Pt and Rh from catalytic converters, the release of Au, Pd and other noble metals from dental alloys, the potential leaching of noble-metal catalyst residues from polymers and the pollution by anticancer and antirheumatic Pt-containing and Au-containing drugs *via* the sewage system, therefore call for watchfulness. Spezia *et al.* compared quadrupole and sector field ICP-MS for Pt quantification in human urine samples.[299] Both appear suitable for this task but ICP-SFMS in the low-resolution mode ($R=300$) proved to be more sensitive than ICP-QMS. Óvári *et al.* examined urine samples of tram drivers.[228] Palladium used in dental applications may be responsible for observed sensitisation in patients. Based on this assumption, Caimi *et al.* determined palladium in urine, saliva and blood.[300] Three Pd isotopes, $^{105}Pd$, $^{106}Pd$ and $^{108}Pd$, were quantified by ICP-SFMS in medium resolution mode ($R=3000$). The authors present a long list of possible interfering ions for the various isotopes, whereby $^{108}Cd$ is of major concern for the determination of $^{108}Pd$. An ICP-SFMS-based method for the simultaneous determination of the platinum group elements Ir, Pd, Pt and Rh in biological materials, *i.e.* hair, blood and urine, was developed by Petrucci *et al.*[301] The nuclides $^{103}Rh$, $^{105}Pd$, $^{106}Pd$, $^{193}Ir$ and $^{195}Pt$ were selected, based on their abundance and on taking into account potential spectral interferences. It was claimed that for quantification of iridium and platinum in biological materials, interferences can be neglected.

The urinary excretion of gallium following anticancer treatment by *tris*(8-quinolinolato)gallium(III) was studied by Filatova *et al.*[302] In their investigation they have identified a couple of polyatomic or doubly charged ions interfering with both isotopes of Ga ($^{69}Ga^+$ and $^{71}Ga^+$) such as $^{36}Ar^{35}Cl^+$, $^{40}Ar^{31}P^+$ and $^{138}Ba^{2+}$, but they could be separated by ICP-SFMS.

Clerico *et al.* showed that ICP-SFMS is also suitable for pharmacokinetic studies.[303] They determined the area under the curve of plasma platinum concentration *vs.* time of pediatric tumour patients treated with carboplatin. Control of the plasma concentration of the drug is highly important due to its toxic side effects and an optimal balance between therapeutic efficiency and possible adverse effects is aimed at. Bocca *et al.* quantified 30 elements in colorectal biopsies by ICP-SFMS.[304] For analysis, the samples were digested under acidic conditions at 80 °C. Many of these elements had to be determined at a mass resolution of 4000 or 10 000 to remove spectral interferences.

Within the framework of the "German Environmental Survey 1998", gold and platinum levels were determined in urine samples of 1080 persons by ICP-SFMS.[305] The limit of quantitation for both noble metals was stated as 0.1 ng $L^{-1}$. Based on the large study group, it was possible to delineate the main sources of the two metals in the body. The most important exposure is by teeth with metal dental alloy restorations. Urinary concentrations of both Pt and Au increase with the number of restored teeth and are further enhanced by the use of chewing gum.

An important issue in environmental and occupational medicine is the quantification of radioisotopes within body fluids. ICP-SFMS in low-resolution mode was a sensitive and suitable method for the detection of $^{235}U$ and $^{238}U$ in human urine samples.[306–309] Base levels of thorium in urine were quantified by Roth *et al.* in occupationally nonexposed persons.[310] The daily excretion amounted to an average of 1.8 ng of thorium. Ratios of $^{234}U/^{238}U$, $^{235}U/^{238}U$ and $^{236}U/^{238}U$ were determined by ICP-SFMS after solid-phase extraction.[311] With a sample volume of 8 mL, reliable measurements with similar precision for the different isotope ratios are obtained at uranium concentrations down to 1 ng $L^{-1}$.

Plutonium, strontium and its radioactive isotope $^{90}Sr$ taken up by the body are preferentially deposited in the bones. Froidevaux *et al.* studied the content of these elements/nuclides by ICP-SFMS and radiometric methods, respectively, with the aim to determine their biological half-lives.[312] For this, vertebrae samples from people who died between 1960 and 2004 were analysed. It is noted that the analysis of samples containing $^{239}Pu$ and additionally also $^{238}U$ may constitute a problem due to the overlap of the $^{239}Pu^+$ and $^{238}U^1H^+$ signals, which cannot be resolved by ICP-SFMS. The $^{239}Pu/^{240}Pu$ ratio was also determined by ICP-SFMS in different tissues of persons living close to the Semipalatinsk nuclear test site.[313] Pu concentrations in lung, liver and bone were found not to be elevated compared to values from other areas.

ICP-SFMS was deployed for the detection of trace elements in sections of human hairs in order to provide time-resolved hair profiles.[314] Rodushkin and Axelsson developed and validated a method for the quantification of 71 elements in hairs and nails by ICP-SFMS.[315] A further development in elemental hair analysis was the hyphenation of laser ablation with ICP-SFMS.[316] This combination makes it possible to study elemental distribution along the length of hairs. Dependent on the target elements, the MS was operated in the low- or medium-resolution mode for detection of 55 elements in total. Among the elements determined at medium resolution were Si, P, S, Cl and Br.

Penetration of different arsenic compounds into human skin has been analysed in an *in vitro* test setup using microsampling laser-ablation ICP-MS.[317] Detailed parameters for the laser ablation are given that were employed to analyse the lateral and indepth spatial distribution of As within the skin samples. It could be shown that arsenite and dimethylarsinic acid penetrated the skin much faster than arsenate or arsenosugars, an important fact with respect to risk assessment.

Tracer studies are deployed for studying human mineral metabolism as well. Opting for stable isotopic tracers rather than for radionuclides avoids health concerns especially in studies involving "vulnerable populations", such as, *e.g.* pregnant women or infants. Stable isotopes can be administered at levels corresponding to normal daily intake of the element and therefore pose no health risk. Moreover, elements with at least two stable isotopes for which no radionuclide with a suited half-life exists, are now also amenable to tracer studies with a main focus on essential elements. For example, both Stürup and Field *et al.* measured $^{44}Ca/^{42}Ca$ ratios with a precision of 0.05% to 0.06% RSD by ICP-SFMS.[318,319]

Al-Saad *et al.* studied the distribution of various selenium compounds in the organs of rats.[320] Each compound contained a particular selenium isotope for selective identification and quantification by ICP-MS. After feeding rats with excess $^{80}Se$ followed by 10 selenium-free days, only negligible levels of the other natural selenium isotopes were present in the rats. After this, the rats orally received a mixture of $^{82}Se$-selenite, $^{77}Se$-selenate, $^{76}Se$-methylseleninic acid and $^{78}Se$-methylselenonic acid and the isotopes were determined in several organs of the rats by sector field and quadrupole ICP-MS. By this method, the distribution of several selenium-containing compounds in the body can be simultaneously and selectively analysed. There are numerous possible interferences with the Se isotopes, which obviously can be resolved by ICP-SFMS in high-resolution mode ($R = 10\,000$) and by collision/reaction cell ICP-QMS. An exception is the $^{40}Ar_2{}^+$ interference with $^{80}Se^+$ that is not successfully resolved by ICP-SFMS.

The distribution of tungsten in the body of mice 24 h after oral uptake of sodium tungstate was studied by Guandalini *et al.*[321] They used high-resolution ICP-SFMS to quantify the metal in several organs such as liver, kidney, colon, spleen, brain and bone, which were digested using $HNO_3$ and $H_2O_2$ prior to analysis. It could be shown that tungstate predominantly accumulates in bone and spleen. Pallavicini *et al.* present an ICP-SFMS-based method for osmium quantification in biological material.[322] For reliable Os quantification at environmentally relevant concentrations in small-sized biological samples, preparation and handling of the samples is of utmost importance. The isotopes $^{189}Os$, $^{190}Os$ and $^{191}Os$ were determined at low resolution in several organs of vole (rodents), such as liver, kidney, lung and muscle.

The isotopic composition of Pb incorporated in animal and human tissues can also give an indication of its source. On the basis of comparison of Pb isotope ratios measured by single-collector ICP-SFMS in the blood of pre-release Californian Condors (bred in captivity), in the diet of these birds, in the blood of free-flying Condors and in ammunition, found in carcasses of or gut piles from animals killed by hunters, Church *et al.* could demonstrate that bullet Pb is the major source of Pb in these Condors.[323] This may cause a raise in Pb level that is lethal for these endangered birds.

**Element-Specific Detection of Small Molecules in Biological Samples**

Small molecules, such as metabolites, drugs and toxicants can be analysed by ICP-MS if they contain a suitable heteroelement such as sulfur, bromine, chlorine or phosphorus, being present in many drugs, pesticides or flame retardants. For example, De Wolf *et al.* studied the metabolism including conjugation with glutathione of the chlorine-containing antipsychotic drug clozapine.[324] They compared dynamic reaction cell ICP-MS and ICP-SFMS as detectors coupled to RP-HPLC for monitoring the Cl and S signals originating from the drug and from glutathione, respectively. While with the sector field instrument, the signals from $^{16}O^{16}O^+$ and $^{16}O^{18}O^+$ were resolved from those of $^{32}S$ and $^{34}S$, the dynamic reaction cell instrument did not yield optimal results because of the pronounced $ArC^+$ formation at masses 48 and 50, interfering with the $SO^+$ species used for S determination. For detection of common organophosphorus pesticides ICP-SFMS has been used at $R = 3000$ to resolve spectral interferences caused by $NOH^+$ and $NO^+$ on $^{31}P^+$.[131] ICP-SFMS is also a suitable tool for the determination of sulfur-containing amino acids and/or their reaction products. Gas chromatography (GC) with sulfur-specific detection by ICP-SFMS was applied for the determination of homocysteine,[325] which is currently discussed as a risk factor in cardiovascular disease.[326]

**Proteins and Peptides**

Proteins cannot be directly and selectively detected by ICP-MS. However,the naturally occurring heteroelements S, Se or P, as well as bound metals in the case of metalloproteins can be used as "indicator elements". Sulfur is present in two amino acids, methionine and cysteine, and in inorganic form in proteins containing iron–sulfur clusters. Selenium occurs as selenocysteine in a small number of proteins and phosphorus is introduced by kinase-dependent phosphorylation, predominantly of serine, threonine and tyrosine residues of proteins. Finally, many proteins can firmly bind metals and may require them for their function. Furthermore, additional elements can be artificially attached by tagging for detection. However, determination of biomolecules in complex biological and medical samples *via* detection of elements essentially requires application of separation techniques, of selective and specific element labeling or of both prior to ICP-MS analysis.

Szpunar has already comprehensively reviewed the application of hyphenated techniques coupled with atomic detection for bioinorganic speciation.[327] Various online separation techniques have already been used for separation of proteins and polypeptides, such as size-exclusion chromatography, reversed-phase chromatography and capillary zone electrophoresis.[24] Although very powerful, the separation techniques mentioned cannot compete with separations based on slab gel electrophoresis, which is the benchmark for protein separation, and one- or two-dimensional sodium dodecylsulfate polyacrylamide gel electrophoresis (SDS-PAGE) in particular is the state-of-the-art procedure for protein separations nowadays. SDS-PAGE is most often used in biochemistry and in combination with isoelectric focusing the highest separation power can be achieved so that thousands of proteins can be separated from one another in two dimensions.

The most straightforward way to detect electrophoretically separated heteroatom-containing proteins by LA-ICP-MS is the direct ablation of the electrophoretic spots in the gel.[328] This approach has been used for the first time by Marshall *et al.*[329] and Wind *et al.*[330] for detection of phosphorylated proteins. The phosphorylation state of proteins and its analysis is an important topic in biochemistry and fundamental medical research because post-translational modifications of proteins, such as phosphorylation, have a high impact on their activity, subcellular localisation and life span and, thus, on the dynamics and activity of the cells' proteome. The significance is underlined by the fact that a new term and research field, namely phosphoproteomics, has been established.[331–333] Thus, quantitative methods in phosphoproteomics are needed in order to identify characteristics of different physiological and pathological states. There are several ways for LA-ICP-MS-based detection and quantification of electrophoretically separated phosphoproteins: (1) direct measurement of P, (2) detection of a phospho-group binding metal label or (3) use of element-tagged antibodies recognising the phosphorylated site(s) of a protein. For these analyses, electrophoretic blotting of proteins from gels onto membranes offers several advantages or is even necessary for subsequent analyses by LA-ICP-MS as discussed elsewhere.[334] Reduction of matrix effects, enrichment of the protein concentration in a flat surface layer and the possibility to label the proteins or to detect proteins by use of element-tagged antibodies are some of the advantages. Consequently, a special laser-ablation cell suited for detection of proteins after electroblotting onto membranes was developed[335] and its analytical capabilities were demonstrated for detection of phosphorylated proteins by Venkatachalam *et al.*[334]

Becker and Przybylski for the first time used a sector field device to study the phosphorylation of the tau protein, which is the target protein in

Alzheimer's disease (AD).[336] The combination of LA-ICP-SFMS and ESI-FTICR-MS was used to detect the human neurofibrillary tau protein and 17 phosphorylation sites were found. Even the structural identification of the phosphorylated peptides was made possible by the use of the tryptic digest in combination with the ESI-FTICR-MS instrument. Becker *et al.* also applied laser ablation for direct detection of metals, sulfur and phosphorus in human brain proteins using ICP-SFMS.[337] Phosphorus to sulfur ratios were used to calculate the degree of protein phosphorylation directly in the gel. Wind *et al.* have also applied LA-ICP-MS for detection of phosphorylated proteins separated by SDS-PAGE but they used electroblotting of proteins onto a nitrcocellulose membrane.[330] Limits of detection of about 5 pmol were estimated from the signal-to-noise ratio. This technology has then been used to determine the phosphorylation state, measured at medium mass resolution, of a cytoplasmatic proteome of selected bacterial and of eukaryotic cells.[338]

The phosphorylation degree of polypeptides was investigated by coupling capillary liquid chromatography to an ICP-SFMS unit as element-specific detector for P.[339] An acetonitrile/trifluoracetic acid gradient was used for elution, which causes an additional interference originating from the carbon of the mobile phase, which varies proportional to the gradient. Again, the interferences can be resolved at a mass resolution $R = 4000$. By using sulfur, present in the amino acids methionine and cysteine of the peptides, as an internal standard and measured together with phosphorus in medium resolution ($R = 4000$) an easy and straightforward determination of the phosphorylation state was possible.[340] A comparative analysis to assess the general protein phosphorylation status in different tissues of the plant *Arabidopsis thaliana* was performed by Krüger *et al.*[341] An important measure and prerequisite to obtain reliable data is the removal of distracting phosphorus-containing compounds such as oligonucleotides, phospholipids and metabolites. This is achieved by efficient extraction steps. Furthermore, low molecular weight compounds are separated by SDS-PAGE, which is used as a prefractionation step. Gel slices are treated with trypsin and the resultant peptide mixtures are subjected to capillary RPLC. The eluting fractions were analysed for $^{31}P$ and $^{34}S$ by ICP-SFMS as measures for phosphorylation degree and protein content. As a final result, the authors note the overall low protein phosphorylation degree in the different plant samples compared to mouse tissues. Furthermore, there are significant differences between plant tissues such as leaves and roots.

As cysteine and methionine occur in most proteins, sulfur can be used as an indicator element and for normalisation in combination with other elements such as metals or phosphorus as already discussed. For absolute quantification of proteins separated by capillary liquid chromatography, Zinn *et al.* determined the sulfur content of eluted fractions by means of ICP-SFMS.[342] Based on the known amino acid composition of the analysed proteins and thus, their cysteine and methionine content, it was possible to accurately quantify apo-lipoprotein A-1 *via* isotope dilution analysis using a postcolumn $^{34}SO_4^{2-}$-spike. Prerequisites for a reliable quantification are (1) base-line separation of the analysed protein species from other sulfur-containing compounds and (2) quantitative recovery of the protein(s) under consideration. To achieve sufficient resolution of complex protein mixtures, the authors recommend multidimensional chromatography. A natural peptide with unusually high cysteine content is hepcidin, which plays a major regulatory role in iron homeostasis. Of the 25 amino acids, 8 are cysteines. This property can be exploited for quantification of hepcidin in urine by means of ICP-MS. For this, samples were subjected to capillary liquid chromatography coupled to ICP-MS.[343] Compared to quadrupole ICP-MS, double-focusing ICP-SFMS revealed a better limit of detection, *i.e.* 7 ng $mL^{-1}$, as well as an acceptable precision and accuracy.

Selenium is an essential trace element present in amino acids and their derivatives and, thus, comprises a further heteroatom occuring in proteins that can be exploited by ICP-MS. As genetically coded amino acid no. 21, selenocysteine is found in some essential enzymes, such as glutathion peroxidase and thioredoxin reductases;[344] for more details see the review article of Polatajko *et al.*[345] Furthermore, selenomethionine is present to a minor extent in all proteins.[346] Speciation analysis may contribute to the evaluation of its physiological and biochemical roles. The seleno-derivatives of cysteine, cysteamine, methionine and ethionine can be separated by RP-HPLC and subsequently quantified by ICP-SFMS. Amounts of $<2$ pg can be detected in a 200-μl injection volume of the sample by this method.[347] For the detection and quantification of selenoproteins, ICP-SFMS appears to be the method of choice and has been used in combination with size-exclusion chromatography and laser ablation of electrophoretically separated and blotted proteins. By the latter method eleven different selenium-containing proteins were detected in catfish extracts of which six were subsequently identified by ESI-Fourier transform ion cyclotron resonance (FTICR)-MS.[348] A database search revealed several proteins that are located in muscle tissue. However, the very low Se concentration prevents the detection of selenium-containing amino acids in the tryptic peptides by electrospray-MS/MS.[349]

For elements with high first ionisation potential like Se and As, it is well known that improved sensitivities are observed in the presence of C-containing matrices. In this context, the effect of a gas added *via* the nebuliser or directly into the spray chamber can be exploited and has been increasingly used in ICP-MS. For example, this principle is now used analytically by supplying a constant flow of $CH_4$ into the spray chamber or the nebuliser gas flow. Subsequently, sensitivity is enhanced, the levels of oxygen- and argon-based interferences are reduced and matrix effects from C-containing sample matrices are equalised as the C matrix is kept more or less constant. The advantage of high resolution with ICP-SFMS is clearly demonstrated as C-based interferences (*e.g.* on Cr or Ca) are resolved at a mass resolution of $R = 4000$. Hence, the multielemental capabilities of ICP-SFMS are not compromised by methane addition and the method is applied for routine applications.[350]

Metalloproteins or metal-containing proteins are predestined for analysis by ICP-MS. They include (1) proteins that specifically bind certain metals for their transport or that are involved in metal detoxification processes by the biosystem, (2) proteins that need tightly bound specific metal ions for their enzymatic or binding activity and (3) proteins to which metals or metalloids bind more or less unspecifically and that may subsequently be affected in their activities. Metallothioneines are small cysteine-rich proteins (20 Cys residues) of about 6 to 7 kDa size optimised for binding of divalent cations such as $Zn^{2+}$, $Cd^{2+}$ or $Cu^{2+}$. These proteins have widespread functions in homoeostasis of both essential and toxic metals including their transport, storage and detoxification. The separation and analysis of metallothioneines was accomplished by coupling of various separation techniques to high mass resolution devices.[351,352] For example, Prange *et al.* separated cytosolic fractions from brain cells by capillary electrophoresis coupled *via* a nebuliser to ICP-SFMS.[353] Three metallothionein isoforms were identified and the following metal isotopes were monitored: $^{63}Cu$, $^{65}Cu$, $^{64}Zn$, $^{68}Zn$, $^{114}Cd$ and $^{116}Cd$. In addition, $^{32}S$ and $^{34}S$ were used for normalisation. A reverse isotope dilution principle was used for metallothionein quantification: The sheath liquid supplied to the CE nebuliser interface was spiked with isotopically enriched metals and isotopically enriched sulfur ($^{32}S$ or $^{34}S$).[354] Due to the spectral interferences affecting all sulfur isotopes, the instrument was operated at medium mass resolution ($R = 3000$), also for detection of the metals. In this way, possible interferences from ${}^{16}O_2{}^+$ or ${}^{16}O^{18}O^+$, for example, were eliminated. A variety of metallothionein species was detected, since their elution (with natural isotopic abundances) introduced changes in the isotope ratios. In this way, both the amounts and the degree of saturation of individual metallothioneins with specified metals were accessible parameters. Electrophoretic separation techniques often show an extremely high separation power and are most often applied if only low amounts of substances are available. Thus, high instrumental sensitivity is needed (for more details see a recent review[355]) and therefore, ICP-SFMS is the instrument of choice for these applications. Van Lierde *et al.* developed a method based on the combination of CE and ICP-SFMS to decipher the stoichiometric composition (metal/S ratio) of metalloproteins and used *Aeromonas hydrophila* metallo beta-lactamase as a model for validation purposes.[356] For both target nuclides ($^{32}S$ and $^{64}Zn$), spectral overlap (with ${}^{16}O_2{}^+$ and ${}^{32}S^{16}O_2{}^+$ ions, respectively) was avoided by deploying a mass resolution of 3000. Zn/protein ratios of 1 and 2, before and after saturation of the enzyme with Zn, respectively, were obtained using external calibration *versus* albumin as a S standard and $ZnCl_2$ as a Zn standard.

Iron status was analysed in human serum by hyphenating FPLC (fast protein liquid chromatography) and ICP-SFMS.[357] The measurements concentrated on transferrin and its several sialylated forms with respect to iron load and iron saturation. Based on the different degree of sialylation, the glycoprotein was fractionated on the strong ion exchanger Mono-Q and subsequently, the fractions were analysed by ICP-SFMS. By means of iron saturation with isotopically enriched $^{57}Fe$, the amount of naturally present iron and supplemented iron – a measure of the unsaturated iron-binding capacity – could be calculated by pattern deconvolution for each individual sialo-form of transferrin, but only if the measured isotopes of iron are free of interferences, which is guaranteed with medium mass resolution.

Metal-containing proteins present in human-body fluids were analysed by ICP-SFMS as well. For example, Gellein *et al.* separated proteins of cerebrospinal fluid by size-exclusion chromatography and determined their metal content, including Cd, Mn, Fe, Pb, Cu and Zn.[358] The binding of aluminium and vanadium to transferrin was studied by Nagaoka *et al.*[359,360] They applied a mass resolution of 4000 to avoid overlap of the signals of $^{51}V$ with those of $^{35}Cl^{16}O^+$ and $^{38}Ar^{13}C^+$, for example.

Heavy-metal ions taken up by plants can lead to their hyper-accumulation as shown for cadmium-exposed *Arabidopsis halleri*.[361] After uptake, the metal ions are bound by several proteins. Hyperaccumulation was also observed for spinach plants grown in a hydroponic solution and Polatajko *et al.* analysed protein extracts of leaves, stems and roots after uptake of cadmium.[362] They identified three major cadmium-containing protein fractions, which were separated from one another by native PAGE and blotted onto nitrocellulose membranes prior to analysis by LA-ICP-SFMS. Furthermore, they showed that the metal-binding protein with the highest molecular weight contained zinc in addition to cadmium and that the former can be replaced by cadmium at higher concentrations. In a subsequent paper by Polatajko *et al.*, ribulose-1,5-bisphosphate carboxylase/oxygenase (RuBisCO) was identified by LA-ICP-MS combined with ESI-FTICR-MS as the main cadmium-binding protein in $Cd^{2+}$-exposed spinach plants.[363] By means of commercially available purified RuBisCO-protein, the effect of cadmium on bound divalent metal ions such as $Zn^{2+}$, $Mn^{2+}$, $Fe^{2+}$ and $Cu^{2+}$ was studied by size-exclusion chromatography hyphenated to ICP-SFMS.

Hann *et al.* investigated the sulfur to metal content in 5 commercially available proteins (myoglobin, haemoglobin, cytochrome C, arginase and Mn superoxide dismutase) and two in-house produced proteins after heterologous expression in a host organism.[364] They clearly demonstrated the necessity of separation of the metalloproteins prior to sulfur/metal molar ratio determination. Their comparison of ICP-DRC-QMS and ICP-SFMS clearly showed that both methods led to accurate results with similar uncertainties.

Quantitative analysis of the metal content of four different proteins, cytochrome C, haemoglobin, ferritin and transferrin, has been performed by coupling gel electrophoresis directly to ICP-SFMS.[365] The critical points for the analysis are the conditions during electrophoresis. Denaturing separation in the presence of SDS is shown to result in loss of iron ions from proteins, which even disappear completely in the case of transferrin. The ratio of $^{56}Fe^+$ to $^{32}S^+$ proved a good measure for the determination of the iron load of the proteins.

Zn- and Fe-containing proteins were detected in yeast mitochondria membrane protein complexes separated by two-dimensional PAGE and analysed by LA-ICP-MS.[366] Metals as well as P and S were detected in a short analysis time using LA-ICP-MS from more than 60 mitochondrial protein spots in 2D gels. By combination of LA-ICP-MS and MALDI-FTICR-MS on SDS-PAGE gels, structural analysis of metal-containing subunits of membrane protein complexes and quantification of metals could be combined.

Several arsenic species are classified as nongenotoxic carcinogens[367] and are responsible for a number of different health effects.[368] As the element is also widely distributed in the environment and can accumulate in the food chain,[369] it is one of the most studied elements in speciation analysis and arsenic analysis became even more popular, once the capability of high mass resolution for the interference-free determination of the mono-isotopic element had been recognised.[370] It is still a "hot topic" in speciation analysis, even though high mass resolution is mostly not required since Cl-containing products (leading to an $ArCl^+$ interference at $m/z = 75$) are chromatographically separated from the As species in most LC separations. Thus, As can often be measured in low-resolution mode, leading to an increased sensitivity and to low detection limits between 1.2 and 2.4 pg $mL^{-1}$ for different arsenic species.[371] In a recent application, complexation of arsenite with phytochelatins in *Arabidopsis thaliana* was studied by HPLC coupled to ICP-SFMS and to high-resolution ESI-MS (LTQ Orbitrap).[372]

Platinum binding to proteins and structures in kidney tissue was investigated in mice, treated with the anticancer agent cisplatin.[373] For this purpose, tissue slices were subjected to LA-ICP-SFMS. Platinum, copper and zinc were concomitantly measured in low mass resolution mode. Platinum was also detected by ICP-SFMS in kidney and liver tissue of human autopsy tissue samples and could be accurately determined by isotope dilution analysis.[374]

There are several strategies and numerous methods to tag proteomes, proteins or even specific proteins with heteroelements for subsequent detection and quantification by ICP-MS. We can distinguish between direct and indirect tagging, meaning that either the target proteins are modified or that a protein-specific binding partner such as an antibody is labelled with a heteroelement and subsequently bound to the protein for detection, respectively. In both cases, tagging exploits the chemical properties of the various amino acid side chains of proteins/ antibodies, which can be modified by chemical reagents. For example, iodination of tyrosine residues with the nonradioactive isotope $^{127}I$ has been used for direct and indirect tagging for subsequent quantification by ICP-SFMS.[375] For this, a fast and mild method for iodine tagging was specifically designed and optimised for LA-ICP-MS based proteomics. Single proteins, whole proteomes and antibodies have been efficiently iodinated by means of potassium triiodide with minimal losses of antigen properties and antibody binding to iodinated proteins. The same approach had been extended by Giesen *et al.* and iodine was employed as an elemental dye for single fibroblast cells and for thin tissue sections.[376]

To the thiol and amino groups of cysteine and lysine, respectively, chelating reagents can be attached, which can firmly bind different metal ions such as lanthanides. For example, antibodies could be tagged specifically at lysine residues by means of lanthanide-containing isothiocyanato benzyl-(1,4,7,10-tetraazacyclododecane-1,4,7,10-tetraacetic acid)(DOTA).[377] Based on the multiplicity of suitable metal ions, a respectable number of specific tags can be created.[378,379] For instance, Ahrends *et al.*[380,381] used metal-coded affinity tags (MeCAT) for direct tagging of standard proteins and eye-lens proteins for quantitative ICP-SFMS.[380] As MeCAT reagents, they chose derivatives of the lanthanide-chelating DOTA, to which a cysteine-reactive maleimide moiety was attached *via* a spacer. By means of holmium-tagged BSA and lactalbumin, detection limits of 110 attomole and 670 attomole for the respective proteins present in a single SDS-PAGE band could be achieved. A corresponding calibration curve revealed linearity over a range of 4 orders of magnitude. As exemplarily shown for alpha-crystallin A, quantification of proteins in single spots obtained by 2D-gel electrophoresis is feasible.

Lanthanide tagging strategies (see the review of Bettmer *et al.*[382]) for antibodies have been discussed recently in a paper by Waentig *et al.*, who comparatively studied labelling with different reagents, namely MeCAT, MAXPAR and SCN-DOTA and presented the analytical figures of merit using an *Element XR* to hyphenate conventional biochemical work flows like Western blot immunoassays with detection by laser-ablation inductively coupled plasma mass spectrometry (LA-ICP-MS) analysis.[383] A special focus was set to apply the multielement capabilities of ICP-MS for the design of multiparametric immunoassays. In particular, the lanthanide tagging of antibodies with metal-coded affinity tags (MeCAT) has been investigated and optimised for application in a Western blot analysed by LA-ICP-MS for which a particular laser-ablation cell was designed.[335] Sub-fmol LODs have been achieved in a dot blot experiment for the pure antibodies. Dual tagging of a protein *via* its thiol and amino-groups is also possible.[384] More details about the history of immunoassays and detection by ICP-MS are given in a recent review.[385]

The detection of electrophoretically separated and electroblotted cytochromes P450 by element-tagged monoclonal antibodies *via* laser-ablation ICP-SFMS (*Element2*) was described by Roos *et al.*[386] Cytochromes P450 (CYP) are iron-containing enzymes that play an important role in the metabolism of toxicants and drugs. Two different cytochromes P450 with a difference in their molecular weight of 2.8 kDa could be detected simultaneously in one SDS-PAGE separation *via* CYP-specific monoclonal antibodies, differentially tagged with DOTA-$Eu^{3+}$ (CYP1A1) or iodine (CYP2E1). The limit of detection achieved in this study was similar to that obtained *via* chemiluminescence detection. About 160 fmol CYP2E1 present in a SDS-PAGE protein band could be detected by this method.

Based on an optimised tagging protocol, differential expression of 5 cytochromes P450 in liver microsomes of rats treated with 4 different

CYP-inducing compounds was studied by means of antibodies tagged with specific DOTA-lanthanide complexes in a multiparametric Western blot combined with LA-ICP-SFMS.[387] The inducer-dependent modulations of the CYP isoenzyme profile detected by LA-ICP-SFMS were in accordance with published data on expression and inducibility of CYP enzymes obtained by other analytical methods.[388]

To trace the biological fate of a peptide that is devoid of suitable heteroelements, Kretschy *et al.* used an indium-containing DOTA complex for tagging.[389] They relied on ICP-SFMS to show that the tagged peptide was taken up by human umbilical-cord epithelial cells. A similar experiment with the fluorescence tagged peptide revealed higher sensitivity for the ICP-MS compared to the fluorescence detection. It has to be noted that impaired tag-dependent cellular uptake can also be the reason for a quasilower sensitivity.

For protein quantitation by ICP-MS, Esteban-Fernandez *et al.* used direct labelling of the proteins with lanthanide-containing MeCAT reagents.[390] For this, the isotope dilution method was applied with ytterbium enriched in $^{171}$Yb and by measurement of the isotopes $^{170}$Yb, $^{171}$Yb, $^{172}$Yb, $^{173}$Yb, $^{174}$Yb and $^{176}$Yb. Serum albumin and transferrin could be quantified in human blood serum by the method *via* calculation of the $^{172}$Yb/$^{171}$Yb-ratios.

As a promising tool to fully exploit the multiparameter and multiplex potential of ICP-SFMS for protein analyses and proteomics microarrays are the platforms of choice also with respect to downscaling. The combination of high-throughput protein microarrays with element-tagged antibodies and LA-ICP-MS makes it possible to detect and quantify many proteins or biomarkers in multiple samples simultaneously, thus surmounting the capabilities and capacities of the methods discussed before. Consequently, a new multiparametric protein microarray embracing the multianalyte capabilities of LA-ICP-SFMS was developed.[391] A proof-of-concept experiment was performed for the analysis of cytochrome P450 enzymes. With the aid of the LA-ICP-MS based multiparametric protein microarray, it was possible to determine 7 cytochromes P450 and cytochrome $b_5$ in 14 different proteomes in one run. The methodology of which a work flow is presented in Figure 12.18 shows excellent detection limits in the lower attomole range and a very good linearity of $R^2 \geq 0.9996$, which is a prerequisite for development of further quantification strategies.

**Nucleic Acids and Nucleotides**

DNA bears in its backbone structure the heteroelement phosphorus as natural label, which can be used for detection and quantification by ICP-MS. It is known that phosphorus determination in biomolecules is difficult by ICP-MS due to isobaric interferences from $^{14}N^{16}O^{1}H^{+}$ and $^{15}N^{16}O^{+}$ at $m/z = 31$. This limitation can be easily overcome by the use of ICP-SFMS operated at medium mass resolution ($R = 4000$). In the case of phosphorus detection, another limitation causes even more problems and this is related to blank problems, resulting from many phosphorus-containing buffers that are usually applied in biochemical work flows. Thus, such buffers need to be

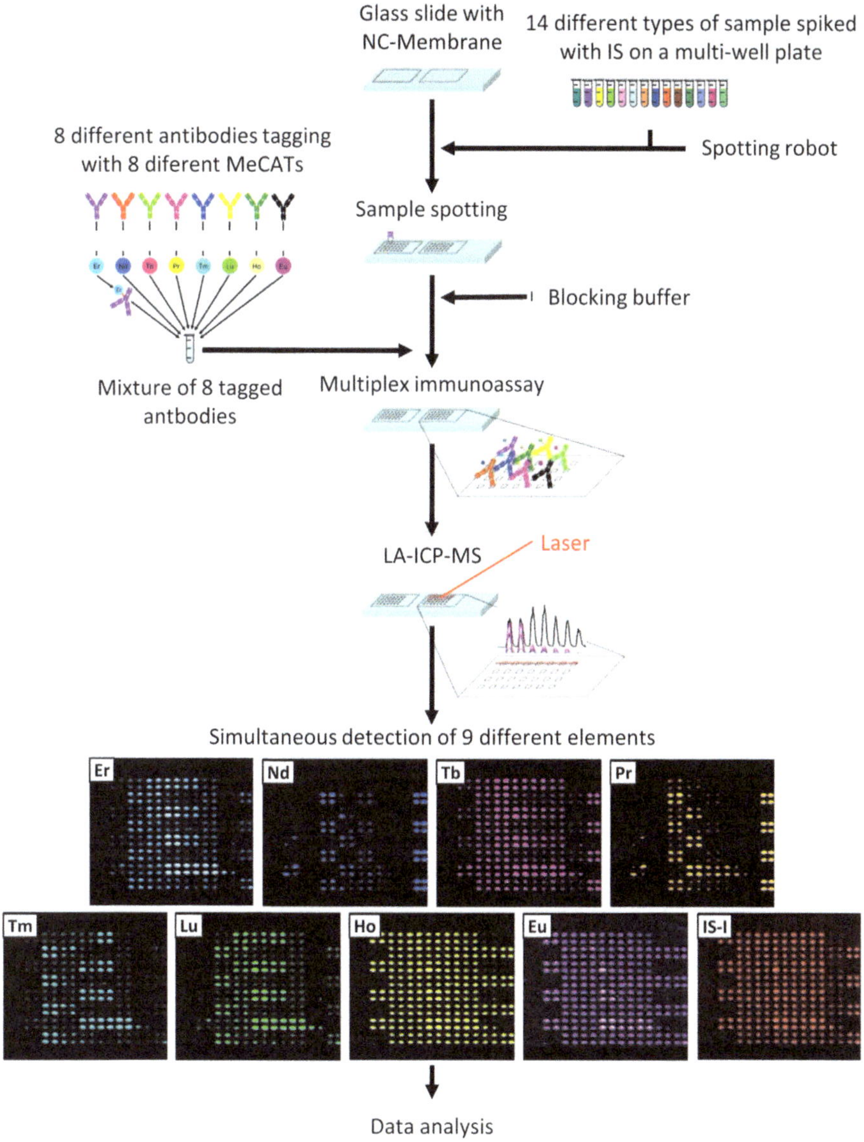


**Figure 12.18** Experimental approach of a multiparametric reverse-phase protein microarray. Protein extracts including an internal standard were spotted onto a NC slide and exposed to a pool of eight lanthanide tagged antibodies. The array was analysed *via* LA-ICP-MS by ablating the whole NC slide in line scans.
(Reproduced from Waentig *et al.*[391])

strictly avoided and to a certain extent, new sample-preparation procedures have to be developed if one wants to fully exploit the capabilities of ICP-MS for sensitive P determination.

Numerous chemicals and drugs are known to directly interact with DNA. Among them are also platinum-containing drugs, which are widely used in cancer chemotherapeutics. More than 15 years ago Siethoff *et al.* studied DNA-adduct formation with cisplatin by use of LC-ESI-MS, but they were not able to quantify the adducts due to differences in the ionisation efficiency.[392] Therefore, Pt was used for quantification of each adduct and was then used for calibration of the LC-ESI-MS device. In this work it was also found that phosphorus measured in MR, which is an essential element of the ribose-deoxyribose phosphate backbone in the RNA and DNA chains, respectively, was also very suited for the quantification of the metal-DNA adducts by using simple inorganic phosphorus standards and this has paved the way for using ICP-MS to even investigate substances, such as DNA or proteins, which do not contain metals at all.

Recently, Leclerc *et al.* determined $^{31}P$ by ICP-SFMS for the quantification of DNA from *Legionella pneumophila*.[393] The authors did overcome interferences from $^{15}N^{16}O^+$ and $^{14}N^{16}O^1H^+$ and achieved a detection limit for phosphorus in DNA of 0.1 ng $g^{-1}$. They further claimed that the accuracy of the method is much better than that of quantitative polymerase chain reaction (qPCR). The method appears to be suitable for the characterisation of DNA reference material.

Methods for quantification of specific DNA-sequences by ICP-MS combined with ligase detection reaction (LDR) technology were developed by Brückner *et al.*[394] Amplified DNA sections were hybridised with two adjacent sequence-specific probes, one labelled with a lanthanide DOTA-complex for detection by ICP-MS and one attached to biotin necessary for subsequent separation from nonhybridised oligonucleotides *via* streptavidin binding. The high multplexing capacity of the method compared to fluorescence-based techniques is stressed.

The interaction of cisplatin and guanosine monophosphate was also studied by Hann *et al.*[395] They found mono- and biadducts, which were separated from one another by HPIC and analysed for of $^{31}P^+$ and $^{195}Pt^+$ by ICP-SFMS. A high mass resolution of 4500 was used to avoid interferences from $^{15}N^{16}O^+$, $^{14}N^{16}O^1H^+$, $^{12}C^{18}O^1H^+$ and $^{13}C^{17}O^1H^+$. To study the kinetics of interactions and the formation of intermediates between cisplatin and specifically designed guanosine-containing nucleotides. Brüchert *et al.* coupled continuous elution gel electrophoresis to ICP-SFMS.[396] The artificial oligonucleotides contained either adjacent guanosines (GG) or not (TG). Mono- and biadducts were detected and the respective binding constants could be determined. A much lower binding level was observed for the TG compared to the GG-containing oligonucleotides. A mass resolution of 4000 was applied to overcome spectral interference of $^{15}N^{16}O^+$ and $^{14}N^{16}O^1H^+$ with $^{31}P^+$ because P is the analyte of interest and the others are "disturbing". The interaction of oxaliplatin with nucleotides was also studied by medium-resolution ICP-SFMS and for structural information simultaneously (flow-splitting) by ESI-MS.[397] Additionally, the authors developed a highly sensitive ICP-SFMS method for the detection of oxaliplatin intrastrand

GG-adducts in DNA. They achieved a detection limit of 0.22 Pt adducts/ $10^6$ nucleotides.

Besides adduct formation, oxidation of DNA is another kind of genetic material damage that is relevant with respect to carcinogenetic processes. Fernández *et al.* used plasmid DNA to analyse oxidative DNA damage induced by the Fenton reaction.[398] The resulting linearised and fragmented circular plasmid DNA strands were separated by continuous elution gel electrophoresis and quantified by ICP-SFMS using $^{31}$P as the natural DNA label. Interference-free determination was achieved in the medium-resolution mode ($R = 4000$). The method is suitable to screen substances for their DNA damaging potential.

**Single-Cell Analysis**

Element analysis of single cells was first performed by Haraguchi, who coined the term metallomics.[399] His team were the first to investigate the more or less complete ionome of single or multiple salmon egg cells. 78 stable isotopes and elements were investigated and 74 were determined or detected in 3 salmon egg cells by ICP-OES and ICP-SFMS after microwave digestion.[400] They could demonstrate that many essential trace elements showed extremely high bioaccumulation factors (compared to sea water) ranging from more than 10 000 for Cu, Zn, Co, Mn, P, Se and Hg to 500 000 for Fe.

A direct analysis of single cells using an ICP-SFMS has been first reported by Houk's group.[401] They measured the signal of U incorporated intrinsically in *Bacillus subtilis* using a microconcentric nebuliser and operated the ICP-SFMS with an integration time of 4 ms to investigate the behaviour of single cells and observed $U^+$ spikes for intact bacteria. They also showed that by sonification, the bacteria increased the $U^+$ response by 30%, which was interpreted as a hint of severe matrix effects, due to the fact that cells behave more as a microparticle than as a droplet. Due to the very high number of droplets produced by the pneumatic nebuliser, no correlation between a single droplet or a single cell embedded in droplets could be made. In a recent study Giesen *et al.* presented for the first time a spatial resolution of a liver biopsy tissue at the single-cell level using LA-ICP-SFMS, as mentioned before.[376] For such an example, the highest sensitivity of ICP-MS is required. Iodination was used as an elemental stain and iodine served as an internal standard for paraffin-embedded thin cuts of the tissue samples.

More recently, a second example where single-cell resolution capabilities of an LA-ICP-SFMS instrument were used was presented by Drescher *et al.* who investigated the distribution of silver and gold nanoparticles in individual fibroblast cells upon different incubation experiments by spatially resolved bioimaging.[402] Single-cell spatial resolution was achieved by optimisation of scan speed, ablation frequency and laser energy. Nanoparticles are visualised with respect to cellular substructures and are found to accumulate in the perinuclear region with increasing incubation time. Based on matrix-matched calibration, the authors developed a method for quantification of the number of metal nanoparticles at the single-cell level. The results provide insight into the kinetics of nanoparticle/cell interactions, which is important information to assess the toxicology of nanoparticles.

To overcome the limitations of long integration times, a very fast scan mode of a sector field instrument (*Element XR*) was discussed by Shigeta *et al.* for measurement of single droplets injected into the ICP-MS by use of a piezoelectric μ-droplet generator (see Section 12.4.2).[23]

Usually, a sample solution is continuously introduced into the plasma by pneumatic nebulisation, with typical integration times for the sector field instrument in the range of a few ms or higher, which is much too long to measure the fast transient signals generated by injection of a single droplet ($< 1$ ms). The *Element XR* allows integration times as low as 100 μs if operated in the fast E-scan mode. This can be done in the low-resolution mode with a flat-top peak but for a single isotope only.

This μDG has now been applied by Shigeta *et al.* as a sample introduction for injecting single selenised yeast cells into the plasma of a sector field ICP-MS instrument (*Element XR*), which was operated in the fast-scanning mode described above.[403] Selenised yeast cells have been used as a model system for first investigations. The single cells to be measured were embedded into droplets. A fixed droplet generation rate of 50 Hz produced equidistant signals in time of each droplet event and was advantageous to separate the contributions from the background and blank from the analytical signal. Multielement analysis of washed yeast cells was performed after open-vessel digestion and absolute amounts per single cell were determined for Na (0.91 fg), Mg (9.4 fg), Fe (5.9 fg), Cu (0.54 fg), Zn (1.2 fg) and Se (72 fg). Signal intensities from single cells have been measured for the elements Cu, Zn and Se and histograms were calculated for about 1000 cell events. For all Se isotopes, a very broad intensity distribution with some structures was obtained. For all isotopes, the background and blank signal levels are low and clearly separated from the cell events. No spectral interference can be identified for $^{77}Se^{+}$ in the background distribution, which means that a contribution from $^{40}Ar^{36}Ar^{1}H^{+}$ can be neglected for this isotope due to the fact that a much lower contribution from spectral interferences is generated as a result of the very low water load of the plasma by the μDG.

### 12.5.2 Applications of Multicollector ICP-MS

It is clearly neither possible to cover all isotopic systems, which have been under investigation, nor all types of studies carried out by today. Therefore, it is rather the aim of this section to demonstrate the versatility of the instrumentation and to highlight publications reporting scientific progress in various disciplines facilitated by data generated using MC ICP-MS instruments. We hope to spark the interest in the diverse applications of the method and refer the reader who wants to learn more to the bibliography given and in particular to several general reviews dealing exclusively with or including MC ICP-MS.[39,59,404,405]

In general, MC ICP-MS can be used to monitor natural variation in the isotopic composition of practically all elements with $\geq 2$ isotopes, be it due to radioactive decay processes, mass-dependent or the more recently discovered mass-independent fractionation. Furthermore, artificially induced changes by use of an enriched tracer can also be investigated. Quantification of elements by means of isotope dilution mass spectrometry have also been carried out using MC ICP-MS instruments, even though the isotope ratio precision of a single-collector instrument is usually sufficient. Balcean *et al.* reviewed the determination of isotope ratios of metals and metalloids by means of ICP-MS for provenance purposes,[406] coming to a final conclusion that MC ICP-SFMS is a very promising tool, complementing TIMS in particular for those elements that are hardly detectable by TIMS (see also the book by Vanhaecke and Degryse[60]).

The first research areas, in which MC ICP-MS instruments were applied, were geo- and cosmochemistry. Publications from these fields still make up the largest fraction of MC ICP-MS application papers. The first applications were limited to a selected number of isotopic systems, but more and more, the "white areas" of the periodic table were investigated. Prior to the introduction of MC ICP-MS instrumentation, isotopic analysis of metals and metalloids had been primarily carried out using TIMS. Today, MC ICP-MS is recognised as a complementary technique, allowing for more elements to be analysed and in some cases MC ICP-MS proved to be a useful alternative to TIMS, *e.g.* due the more limited efforts required in sample preparation, the capability of direct solid sample introduction by using a laser-ablation system and the assessment of isotopic systems, which could not be analysed by TIMS satisfyingly, due to their high ionisation potential (*i.e.* $>7.5$ eV).

With the expansion towards new fields – from environmental applications to authenticity and provenance determination in archaeometry and forensics and, rather recently, towards life sciences – the range of investigated isotopic systems has also been extended. Even isotope ratios of nonmetals like sulfur and carbon – usually analysed by means of gas-source isotope ratio mass spectrometry (IRMS; see Chapter 16) – have been tackled using MC ICP-MS by now.

#### 12.5.2.1 Applications in Geological, Cosmochemical and Radionuclide Research

In 1998, De Laeter described in a review, how significant the development of mass spectrometry was for the advance of geochronology.[407] Although MC ICP-MS was quite a new development at that time, he discussed its potential as a complementary method to TIMS, particularly combined with direct laser ablation of solids as a sample introduction system. In the same year, Halliday *et al.* also published a review solely discussing how MC ICP-MS instruments can be applied to address challenging research questions from the fields of geochemistry, cosmochemistry and paleoceanography – demonstrating clearly that scientists from these research areas were the first to implement MC ICP-MS instruments in their application field.[408] Some more recent reviews on isotopic analysis in the earth sciences, which discuss novel developments, in terms of both instrumentation and analytical protocols, are now available.[409–411]

In the context of geological investigations, on the one hand, MC ICP-MS is used as an alternative to the traditional benchmark technique TIMS, but on the other hand, it also extends the application range to isotopic systems which have been difficult to tackle using TIMS. One such example is the hafnium–tungsten chronometry, that is based on the radioactive decay of the by now extinct mother nuclide $^{182}$Hf ($T_{1/2}$ 9 Ma) to $^{182}$W and the fact, that both highly refractory elements behave differently during planet differentiation (iron core segregation): Hf is a lithophile element, while W exhibits a more siderophile character, thus causing physical separation of mother and daughter nuclides. Consequently, the W isotope ratio in the Earth's crust is influenced by the time of core formation relative to the extinction of $^{182}$Hf.[412] After establishing a method that enabled the precise determination of W isotope ratios using MC ICP-MS,[413] Lee and Halliday could shed more light onto the timing of the formation of the Earth's core[414] and the age and origin of the moon.[415]

Similarly, the availability of the lutetium–hafnium isotope system was enhanced due to simplified measurement using MC ICP-MS, both in solution[416] and spatially resolved by means of direct laser ablation of zircons.[417]

Luo *et al.* demonstrated improved precisions for $^{230}$Th/$^{232}$Th and $^{234}$U/$^{238}$U isotope ratio measurements, opening the door to the uranium–thorium dating method.[418] As an example, Robinson *et al.* applied U-Th dating to a $\delta^{18}$O stratigraphy to constrain the timing of the penultimate interglacial and found that results from MC ICP-MS showed significantly better precision than the TIMS data.[419]

Another popular geochronometer is the rubidium–strontium system. Waight *et al.* developed an improved method for Rb quantification using ID MC ICP-MS[420] and Nebel *et al.* report tenfold improved precisions in Rb isotope ratio measurements using MC ICP-MS compared to TIMS.[421] A LA-based method was presented by Willigers *et al.* for the Rb–Sr dating of biotites and phlogopites.[422]

Further traditional geochronological methods use the isotopic systems of U, Th-Pb and Pb–Pb, which are based on the radioactive series of uranium and thorium, which have been tackled using MC ICP-MS as an (often superior) alternative to using TIMS, particularly when hyphenating a laser ablation system to the MC ICP-MS, as demonstrated, *e.g.* for Pb–Pb dating of apatite, monazite and sphene.[423] Apart from dating applications, lead isotope ratios can also be used, for example, to constrain alterations in ocean circulation caused by climate change over millions of years by analysing ferromanganese crusts using laser-ablation MC ICP-MS.[424]

In addition to radiogenic isotopic variations in geological material, the investigation of isotopic fractionation of metals and metalloids caused by physical and chemical processes gained the interest of the community. Improved measurement precisions even facilitated the observation of uranium isotope fractionation effects, *e.g.* during reduction processes in the ocean[425] or granite weathering.[426] Numerous further elements have been under investigation and this section cannot give but a few examples: Variations in the $^{88}Sr/^{86}Sr$ ratio, which had previously been assumed constant, were first observed by Fietzke and Eisenhauer[427] and later reported, *e.g.* for a glaciated granitic watershed by de Souza *et al.*[428] Bizzarro *et al.* provided estimates for the magnesium isotopic composition of the bulk silicate earth based on MC ICP-MS measurements.[429] A similar conclusion was obtained for zinc by Chen *et al.*, who evaluated the potential of this isotopic system as a proxy for planet differentiation.[430]

Isotopic fractionation of many elements is influenced by ambient conditions during geochemical processes, such as temperature, pH or redox potential. Thus, it is particularly interesting to study isotope ratio variations in natural chronological archives, like sediments, speleothems, corals or foraminifera, to use them as paleoproxies of certain conditions. Boron, for instance, occurs in the ocean as boric acid and borate and the equilibrium between these two species is controlled by pH. Isotopic fractionation occurs during conversion between the two species, but not during incorporation of borate into carbonates of corals and foraminifera. For this reason, the latter were investigated by Foster and provided information on the variation of seawater pH as a function of time.[431] Another exemplary element is antimony, which reflects seawater redox conditions in its fractionation behaviour and can thus be used as a paleoredox tracer.[432] Nakagawa *et al.* developed an MC ICP-MS based methodology for the measurement of molybdenum isotope ratios and applied it to seawater samples.[433] In a later study, Voegelin *et al.* suggested that Mo isotopic analysis in marine carbonates presents a tool for obtaining information on the Mo isotopic composition of the paleo-ocean, as it reflects the ambient water and the redox conditions present therein and is only altered minimally by minor isotope fractionation during its incorporation in biominerals.[434]

MC ICP-MS also proved beneficial for the determination of the contents of siderophile elements such as the platinum-group elements, in meteorites and terrestrial samples using isotope dilution.[435] Rehkämper *et al.* have highlighted the potential of their method for establishing a subduction history based on results showing variably fractionated patterns in the Tanzanian subcontinental lithosphere.[436] Similar methods were developed by Yi *et al.* and applied to the rare volatile siderophile elements cadmium, indium, tellurium and tin in oceanic basalts by means of IDMS.[437] Resano *et al.* demonstrated the applicability of a new Mattauch–Herzog instrument coupled to a laser-ablation system for the direct simultaneous determination of platinum-group elements in NiS buttons prepared from ores.[438]

Apart from the apparent application of isotopic analysis of radioactive and radiogenic nuclides for geochronology, isotope ratios involving natural and artificial long-lived radionuclide have been used as a tool in numerous other areas, ranging from environmental monitoring and nuclear safeguards to the nuclear industry and nuclear waste control. The related analytical tasks are often tackled using mass-spectrometric techniques, as Becker described in a comprehensive review on the topic.[439] Another review focusing on ICP-MS was compiled by Larivière *et al.* and discusses the analytical challenges related to radionuclide determinations and the instrumental configurations allowing the challenges to be overcome in detail.[440]

A few application examples shall be detailed in the following: Günther-Leopold *et al.* demonstrated the versatility of MC ICP-MS for the characterisation of nuclear fuels, which is of high relevance for safeguards considerations.[441] They successfully determined burn-up (the number of fissions per heavy metal atom) *via* monitoring of $^{148}$Nd and the U and plutonium contents by coupling an HPLC system to an MC ICP-MS unit. Furthermore, they used this hyphenated system for fission product analysis and carried out U and Pu isotope ratio measurements along the cross section of an irradiated $UO_2$ fuel sample by means of LA-MC ICP-MS. Pu isotope ratios can indicate fall-out from nuclear weapons tests or discharge from atomic power plants. Nonetheless, high-precision analysis is required due to the small shifts often encountered, but the precision necessary could indeed be achieved, even for fg amounts of Pu by using multiple ion counting devices.[442] Boulyga and Prohaska presented a method for direct analysis of radioactive particles from the environment in the proximity of the reactor accident site in Chernobyl, Ukraine, regarding (artificial) actinide isotopes and fission product isotope ratios (Nd, Ru) at very low levels using LA-MC ICP-MS with multiple ion counters.[443] In a more recent work of this group, submicrometre particles of mixed isotopic composition were investigated and evaluated by applying a newly developed "finite mixture model".[444] Sharabi *et al.* presented a method for the determination of $^{228}$Ra and the radium isotope ratio $^{226}$Ra/$^{228}$Ra in natural waters using MC ICP-MS.[445] Comparison to previous results from $\alpha$- and $\gamma$-counting revealed good agreement and improved ratio precision with MC ICP-MS. MC ICP-MS has also been applied as a tool to determine the half-life of certain long-lived radioisotopes by dividing the measured activity by the amount of atoms. The latter can be measured in a highly accurate and precise way by means of isotope dilution MC ICP-MS, as discussed for the example of $^{60}$Fe by Kivel *et al.*[446] Similarly, Chmeleff *et al.* established the half-life of $^{10}$Be in agreement with previous estimates, but with lower uncertainty.[447]

### 12.5.2.2 Applications in Environmental Sciences

Lead is the element that has been the most extensively studied for its isotopic composition in environmental sciences.[448] Due to the fact, that three of the four stable Pb isotopes ($^{206}$Pb, $^{207}$Pb and $^{208}$Pb) are produced from radioactive decay of uranium and thorium isotopes, crustal Pb and ore Pb show considerable differences in their isotopic compositions, caused by different ages and the different chemical composition of the original material. Consequently, it is quite straightforward to discriminate between "natural" geogenic Pb (crustal signature) and anthropogenic, mined Pb (ore signature) in different environmental compartments and to deduce the relative contributions of the sources. Furthermore, the isotopic composition of crustal lead varies between different geographical locations based on different ages of the bedrock. A comprehensive review on the issue of Pb isotopes in environmental sciences was written by Komarek *et al.*, who summarises studies investigating aerosols, peat deposits, tree rings, sediments and soils.[448] As a result of the fairly pronounced natural variation in the isotopic composition of Pb, most of the studies used single-collector ICP-QMS or ICP-SFMS instruments. One of the exceptions is a paper by Weiss *et al.* who presented a methodology for Pb isotope ratio measurements in diverse environmental samples using MC ICP-MS and included two application examples:[449] First, a peat-bog profile from the Faroe islands allowed changing sources of lead over a time span of 2500 years to be deciphered due to differences in Pb isotopic composition, as further discussed by Shotyk *et al.*[450] The second case study investigated Pb pollution in the proximity of a Cu smelter in a heavily polluted area in Russia and isotopic data for Pb in lichens and vegetables suggested three sources of Pb, two of which were smelter dust and natural background. Purvis *et al.* combined these data with quantitative multielement results for comprehensive biomonitoring.[451] A few years later, Cocherie and Robert developed a method for analysing environmental samples with very low Pb levels, down to 100 pg $mL^{-1}$, using a multiple ion counting system and achieved adequate precisions for the purpose of environmental monitoring.[452] A study that combined TIMS and MC ICP-MS data and found a good agreement between the two methods was published by Vallelonga *et al.*, who investigated an Antarctic ice core that serves as a chronological archive that recorded dust and volcanic Pb over a period of 220 000 years.[453]

A case study, where the isotopic analysis of Pb was not successful for the source appointment of heavy metal pollution caused by a Pb-Zn refinery plant, was carried out by Cloquet *et al.*[454] The reason behind the failure was the fact that throughout its history, the smelter had handled different types of Pb ores from several origins and consequently showing different isotopic signatures. Therefore, an alternative approach was needed and cadmium isotope ratios were found to be a useful tracer as the smelting process causes isotope fraction due to enrichment of the lighter isotopes in the gaseous phase. In this way, soil Cd could be mainly attributed to industrial sources and only to a smaller extent to the use of fertiliser.[454] Shiel *et al.* further investigated whether isotope fractionation occurs during smelting and refining and came to the same conclusion for Cd, while no significant changes were found for zinc.[455] Another study dealt with urban atmospheric metal deposition in France, monitored using lead and zinc isotopic analysis of lichens, ambient particulate matter and bus air-filter aerosols and found decoupled sources for the two elements.[456] Furthermore, Zn isotope ratios in soils were used as tracers for anthropogenic pollution from industrial activity, while also fractionation processes in soils were observed and have to be considered for correct interpretation of results.[457] A possible mechanism for fractionation in soils is adsorption to organic matter, as shown in laboratory experiments by Jouvin *et al.*[458] In addition to soil processes, plants can also discriminate between Zn isotopes during uptake (preferential absorption of heavy isotopes) and transport within the plant (isotopic depletion from root to shoot). This was concluded from greenhouse experiments with rice, lettuce and tomato plants.[459] In the course of efforts to investigate Zn isotope geochemistry in the soil-plant context, a method improvement was presented by Arnold *et al.* who introduced a double-spike MC ICP-MS methodology and applied it to the determination of the isotopic composition of the bioavailable Zn fraction in soils.[460]

In their review about the controlling processes of the stable isotopic composition of lithium, boron, magnesium and calcium in nutrient cycles Schmitt *et al.* discussed measurement strategies including MC ICP-MS, and consequences of the results for the understanding of soil–water–plant interaction.[461] The first two mentioned isotopic systems can give information related to clay mineral formation, while Mg and Ca isotope ratios depend on vegetation growth cycles.

Strontium isotopes can serve as useful tracers for Ca sources in ecosystems, but the corresponding studies mainly relied on TIMS or single-collector ICP-SFMS measurements. Traditionally, the ratio $^{87}Sr/^{86}Sr$ was investigated and internally corrected for instrumental mass discrimination (and theoretically for mass-dependent fractionation in nature, although this was considered insignificant) by means of the "constant" $^{88}Sr/^{86}Sr$ ratio. MC ICP-MS measurements using a sample–standard–sample-bracketing approach revealed, however, that measurable mass-dependent fractionation occurs in the environment.[427] Based on the analysis of speleothem, soil and loess samples, Halicz *et al.*, for instance, suggest that $\delta^{88}Sr/^{86}Sr$ may be useful as a geochemical tracer of, *e.g.* soil pedogenesis.[462]

The use of iron isotopes to improve the understanding of geochemical and biological cycles was broadly discussed by Beard *et al.*, who reported the beneficial application of an MC ICP-MS unit equipped with a collision cell pressurised with $H_2$ as collision gas.[463] MC ICP-MS is favourable to TIMS for Fe isotope ratio measurements, due to the low ionisation efficiency using the latter. Another method for the measurement of high-precision Fe isotope ratios using multicollector ICP-MS has been presented by Arnold *et al.* in 2004 and benefits from the high-resolution capabilities for avoiding polyatomic Ar-based interferences.[464] Another group investigated the Fe isotopic composition of aerosols collected in a parking garage, which contained Cu and Fe at elevated levels because trace-element analyses alone could not solve the question of source appointment.[465] MC ICP-MS measurements of Fe isotopes in two particulate matter size fractions and source material such as brake pads, waste oil and tyre tread revealed that the major contributor to total Fe was most likely dust from brake wear.

Mercury as a toxic pollutant featuring bioaccumulation, *e.g.* in aquatic food webs, is an intensively studied isotopic system. It is interesting to note that isotopic analysis of Hg is not possible using TIMS as a result of its volatility and high ionisation energy. Hg also shows interesting fundamental characteristics. Hg has seven stable isotopes, two of which have odd mass numbers. In addition to mass-dependent fractionation (MDF), Hg exhibits anomalous, *i.e.* mass-independent fractionation (MIF) in the course of certain chemical reactions. MIF affects the odd-numbered isotopes due to the nuclear field shift and the magnetic isotope effect. The advance of MC ICP-MS made the discovery and investigation of these effects possible.[466] It is usually expressed as $\Delta^n$Hg ($n = 199$ and 201, respectively), describing the deviation from the $^n$Hg/$^{198}$Hg ratio expected taking into account MDF only (the $\Delta$ is used for representing the deviation between the experimentally determined value and the ratio expected based on MDF only and not the deviation from a standard value, which would be noted as $\delta$). MIF of Hg has been quantified for the first time by Bergquist and Blum and was associated with the photoreduction of methylmercury (MeHg).[467] The authors could demonstrate the potential of investigating the Hg isotopic composition for understanding environmental processes and reaction pathways, both with data from an experimental setup and from natural samples. They established the correlation between $\Delta^{199}$Hg and $\Delta^{201}$Hg as a useful tool to shed light on underlying reaction mechanisms. Further insights into the Hg cycle in aquatic systems have been gained by Gantner *et al.*, who investigated the Hg distribution and isotopic composition in Arctic foodwebs.[468] Hg isotopic signatures of zooplankton, chironomids and fish were compared to those of sediments. The latter exhibited low MDF, while the largest MIF was found in zooplankton. The authors could show that the extent of MIF in biota allows tracing back the Hg pathways along the food chain. Furthermore, geographical differences between several lakes in Northern Canada in terms of Hg isotopic systematics were found. In a very recent study, Yin *et al.* used the Hg isotopic composition in Chinese rice plants from a Hg-mining area to draw conclusions on sources of Hg, its transportation within the plants and transformation between root, stem, foliage and seed.[469] At a larger scale, mercury isotope ratio measurements could also be successfully applied to trace Hg contamination from a Slovenian mining region to the Gulf of Trieste by analysing sediment samples.[470] A further advance was presented by Epov *et al.*, who developed a method for species-specific Hg isotopic analysis using gas chromatography hyphenated to MC ICP-MS, which constitutes an important step forward in the understanding of Hg cycling as the Hg species strongly influences mobility, bioaccumulation and toxicity.[471] The application of an MC ICP-MS instrument for quantification of mercury in natural waters using an isotope dilution approach was recently validated (and used as a reference to evaluate the performance of quadrupole ICP-MS) with the aim to accurately quantify Hg at ultratrace levels.[472]

In a recent study, Savage and coauthors characterised the silicon isotopic composition of the Upper Continental Crust based on a set of loess and shale samples and found it to be identical within error to the bulk silicate earth.[473] They employed a previously developed sample-preparation protocol,[474] which had also been applied in the characterisation of Alpine river water for its Si isotopic composition using the Nu Plasma 1700, allowing for interference-free measurement of the three silicon isotopes. Insights into the complexity of the system, consisting of weathering and mixing processes, fractionation by biological processes and seasonal variations were gained.[475] Multicollector ICP-MS was also applied to study the silicon isotopic composition of plants: Opfergelt *et al.* published two papers on Si isotopic fractionation in banana, from source to root and root to shoot[476] and during biomineralisation of silica, resulting in deposition of phytoliths, which represent a Si source in soils after plant decay.[477] Their findings contributed to the understanding of the role of plants in the Si cycle. In a later study, Engström *et al.* applied a similar method to investigate different plant species and humus from a Swedish boreal forest to assess the homogeneity of Si isotopic signatures in the biomass.[478] They conclude from their data, that the signatures of plants are determined by the Si isotopic composition of bioavailable dissolved $H_4SiO_4$, fractionation during uptake processes and Si-containing exogenous surface contamination.

Another pollutant of high relevance, *e.g.* in groundwater, is chromium, which is toxic in the oxidation state +VI, but can be removed from the solution by reduction to Cr(III), which is insoluble and adsorbs to surfaces. This process is accompanied by isotope fractionation, as the lighter isotopes are preferentially reduced. This has been shown, *e.g.* by Zink *et al.* who carried out fundamental laboratory experiments to enhance the understanding of the mechanisms behind this process and relied on a MC ICP-MS method for the measurement of Cr isotope ratios.[479]

As mentioned earlier, ice cores present valuable archives of, *e.g.* past climate, atmosphere composition and the presence of dust. In this context, Aciego *et al.* presented a method allowing for isotope ratio determination of uranium, thorium, palladium, radium, strontium, neodymium and hafnium in ice cores with very low concentration levels, after isolation of soluble and dust material, respectively, and chromatographic isolation of the different elements. U, Th, Pa, Ra and Hf isotopic compositions were then measured using MC ICP-MS, while TIMS was applied for Sr and Nd isotope ratio measurements.[480]

### *12.5.2.3 Archaeometry and Forensics*

Similarly to some of the source-appointment questions solved in environmental science, the determination of provenance using isotope ratios of so-called "heavy" elements is also of high relevance in archaeological migration surveys, food authentication studies and forensic investigations as reviewed by Balcaen *et al.*[406] It is based on (small) differences in the isotopic composition, which characterises a material providing a unique "fingerprint" related to its source. In particular, the radiogenic isotopes of Sr, Nd and Pb are widely used for this purpose. Since the advent of MC ICP-MS, elements that show isotopic variations due to mass-dependent fractionation have been applied.[481] This variation can be distinct for a specific source. However, further fractionation, *e.g.* during weathering or along the food chain, has to be considered.

One very popular isotopic system for the purpose of origin determination is strontium. It has four stable isotopes, one of which – $^{87}$Sr – is radiogenic and consequently varies in its abundance in different geological bedrocks, based on the initial relative amount of its mother nuclide $^{87}$Rb, the $^{87}$Sr/$^{86}$Sr ratio at the time of rock formation and the age of the rock. Due to similar ionic radii, Sr can substitute for Ca and consequently ends up in diverse Ca-containing matrices, such as biogenic carbonates or apatites. Examples of the latter are bones and teeth, which store nutrition information from different life spans. Tooth enamel, for example, is formed during childhood and is not subject to metabolic changes later in life. Consequently, Sr isotope ratios in skeletal remains are a unique indicator for migration of an individual during his/her life, and Sr isotopic analysis is extensively applied to archaeological findings.[482,483] Predominantly, these investigations are carried out using thermal ionisation mass spectrometry, a few studies benefit, however, from the possibility to study the isotopic composition with spatial resolution using LA-MC ICP-MS.[484–486] Some limitations of this method prevail, however, due to matrix-derived interferences, hindering accurate isotope ratio measurement.[487] Sr isotopic analysis has been carried out with a modern forensics background, as well: From an attempt to apply Sr isotope ratio investigation of human remains from unidentified individuals found in Belgium as an auxiliary tool to gain information about their area of provenance by simple relating the data to a geological map,[488] it is evident that this is not a straightforward task, due to *e.g.* globalised food production and supply, and consequently, data have to be interpreted with great care. For any origin determination of plants or animals, isotopic reference values of the fraction of Sr that is bioavailable to the respective type of organism, are required. Maurer *et al.* compared different approaches to generate such a Sr isoscape for investigating past migration and applied both TIMS and MC ICP-MS instrumentation.[489] They found water and vegetation the most reliable and era-independent proxies of bioavailable Sr, whereas, *e.g.* a study from the field of large herbivore ecology characterised bioavailable Sr of habitats using LA-MC ICP-MS analysis of rodent teeth, as these small animals "sample" and average Sr from defined territories *via* feeding.[490] Both MC ICP-MS and TIMS were applied in a comprehensive study characterising 650 bottled European mineral waters for their $^{87}$Sr/$^{86}$Sr for future application in food traceability.[491]

As detailed before, lead exhibits comparably large radiogenic variation in its isotopic composition, which is not only useful in source appointment of environmental pollution, but can also be utilised in archaeometry for the determination of provenance of artefacts, as ores from different mines vary in their isotopic composition. In their review on metal provenancing using Pb isotope ratios Stos-Gale and Gale state the superiority of TIMS compared to MC ICP-MS and present the extensive database of Pb isotope ratios of ores and artefacts from Europe and beyond, generated at the University of

Oxford.[492] Ponting *et al.* applied laser-ablation MC ICP-MS to determine Pb isotopic fingerprints of coins from different Roman mints and report a precision similar to that of TIMS results.[493] Similarly, Baker *et al.* found good agreement between solution nebulisation MC ICP-MS, LA-MC ICP-MS and TIMS for archaeological silver and copper samples with high Pb content ($>500$ μg $g^{-1}$), while LA-MC ICP-MS produced inaccurate results for Cu ingots from a shipwreck with Pb contents $<100$ μg $g^{-1}$.[494] Desaulty *et al.* investigated the silver, copper and lead isotopic composition in coins from several centuries and different continents and found distinct differences for Ag between several coin provenances, relating back to the ores, while the original ore Pb and Cu isotopic compositions may be masked during metal processing.[495] Due to their elemental composition, bronzes would suggest Cu and Sn isotopic analysis for their fingerprinting. A method for this purpose was thus presented by Balliana *et al.* recently.[496] It has only been applied to a set of archaeological bronzes from northern Spain, though, which did not allow the distinction of groups.

Fortunato *et al.* analysed lead white pigments from both Flemish-Dutch and Italian Renaissance paintings and found that the respective paints were produced from Pb from different ores. In case of the Flemish-Dutch painters, comparison to data from different mines revealed, that the lead originated from either Germany or England.[497] Pb isotopic analysis has also been applied for modern forensic purposes, *e.g.* in order to link a fired projectile to a weapon – as comprehensively assessed by Zeichner *et al.* who carried out firearm cleaning and shooting experiments and found out that a Pb memory effect in the barrel complicates the association.[498] In specific cases, however, Pb isotopic analysis can provide evidence to assign particular gunshot entries to the respective guns given that significantly different Pb isotopic compositions are present. Another sample type frequently analysed in forensic laboratories, is glass. In a study of 52 representative samples of glass, Sjåstad *et al.* identified Pb isotopic analysis by MC ICP-MS as a method with high discriminative power for different types of glass.[499]

A study investigating Pb isotope ratios in skeletal remains from a 19th century mental-asylum population in Colorado, U.S. against the background of contemporary Pb production and consequential smelter sources applied both TIMS and MC ICP-MS in a complementary way.[500] In an archaeo-anthropological context, De Muynck *et al.* could disclose Pb prenatal intoxication of infants in the Roman era by comparing the isotopic composition of Pb of infant bone tissue excavated in the Netherlands with that of the soil of the corresponding graves and of Pb objects from the same era and found at the same location or nearby.[501]

In present-day forensics, copper isotope ratio measurement using fs-LA MC ICP-MS has been assessed as a method to distinguish between authentic coloured andesines and stones treated by Cu diffusion to obtain similar characteristics.[502] The latter fraudulent practice results in distinct Cu isotopic patterns within a sample and in generally lower Cu isotope ratios compared to the only definitely naturally coloured gemstone from Oregon investigated in the study. Results from Ar isotope ratio analysis supported the findings of heat treatment. It was, however, not possible to assign mineral samples to different provenances, *e.g.* within Asia.

Neodymium isotope ratios ($^{143}Nd/^{144}Nd$) vary in nature due to the radioactive decay of $^{147}Sm$ to $^{143}Nd$, but the range of variation is very small due to the chemical similarity of the two rare-earth elements, which causes limited ranges of Sm/Nd ratios in different rocks. Nonetheless, differences in Nd isotope ratios can be resolved using MC ICP-MS. The method can be applied to distinguish between archaeological glass findings, for the production of which different silica sources have been used. For this purpose, Brems *et al.* analysed sands from different locations around the Mediterranean Sea.[503] Previously, both Nd isotopes and Sr isotopes – the latter reflecting the source of limestone used in glass production – were applied to trace the provenance of raw materials for glass production in the Roman Empire in the 1st to 3rd century A.D. in order to confirm a statement of Pliny the Elder, that resources from Italy, Gaul and Spain had been employed.[504] Devulder *et al.* recently presented a method for B isotopic analysis of (pre-) Roman glass, a means of obtaining information on the origin of the natron used as flux during the manufacturing of the glass,[505] while Lobo *et al.* determined the isotopic composition of the Sb present in glass (added as an opacifier or as a colorant) in an effort to unravel the origin of the corresponding stibnite ore.[506]

As an alternative to traditional gas-source mass spectrometry, sulfur isotope ratio measurements can also be carried out using MC ICP-MS and can prove useful in a forensics context: Clough *et al.* could distinguish between authentic and counterfeit Viagra pills, the active ingredient of which is the S-containing compound sildenafil, using LA-MC ICP-MS analysis of S isotope ratios.[507] Similarly, Santamaria-Fernandez *et al.* could distinguish between authentic samples and suspect counterfeits of an antiviral drug by determining S isotope ratios using MC ICP-MS, as well as carbon isotope ratios by means of IRMS.[508] Mg isotope ratios were also investigated, but did not allow for distinction. In another study by the same author, the spatially resolved direct sample introduction provided by LA-MC ICP-MS was beneficial for documenting variations in the S isotope ratio along a single hair strand. These changes could be related to short and long-distance journeys of the corresponding individual.[509] The method presented may thus be of use in future forensic or archaeological studies.

### 12.5.2.4 *Applications in Life Sciences*

The number of multicollector ICP-MS application papers from the field of life sciences is still limited as compared to the vast collection of papers reporting on the use of single-collector ICP-MS in this context, but an increasing trend is obvious. Only recently, a review has been published about isotopic analysis in clinical samples such as diverse body fluids, hair or teeth using ICP-MS.[510] While the majority of studies mentioned therein apply quadrupole-based systems, some benefit from the improved precision of MC ICP-MS, particularly studying natural variations in the composition of elements with (a) radiogenic isotope(s) or observing natural fractionation effects.

To a large extent, isotopic analysis in the life sciences is carried out in the context of stable isotope tracer studies, as reviewed by Stürup,[511] who only mentions one application of MC ICP-MS for Zn isotopic analysis in feces.[512] Since the date of the review, a few more papers on this topic have been published, however. For example, a recent pilot study demonstrated that the precision of MC ICP-MS allows resolving differences in the Cu metabolism in patients with Parkinson's disease in comparison to healthy subjects by analysing blood samples taken at different time points after the ingestion of a $^{65}Cu$ isotopic tracer.[513] Furthermore, Urgast *et al.* investigated the spatial distribution of two Zn isotopic tracers in rat brain thin sections by coupling a laser-ablation system to a MC ICP-MS instrument.[514] Due to administration of the two tracers at different time points, insights into the kinetics of Zn metabolism in different brain compartments were gained, which is of particular interest as Zn homeostasis plays a role in cognitive functions and neurological disorders like Alzheimer's disease and epilepsy. A critical discussion about possibilities, limitations, and perspectives of isotope ratio measurements in biological tissues using LA-MC ICP-MS is presented in a recent review.[515]

Mostly, however, multicollector ICP-MS instruments are deployed in the investigation of natural variations in the isotopic composition of selected elements for exploring questions in a medical or biological context. Among the first researchers, who investigated human trace-metal metabolism *via* isotope ratio measurements using MC ICP-MS, were Walczyk and von Blanckenburg. They published several papers on iron isotopic fractionation. From analysing blood and dietary Fe sources, they revealed i) a difference between the male and female cohort of a reference population and ii) depletion of human blood in heavier Fe isotopes compared to dietary Fe, providing evidence for Fe isotopic fractionation during intestinal absorption, consequently causing depletion in heavier isotopes along the food chain.[516] In a follow-up paper, they reported improved insights into the mechanisms of Fe uptake, embedded in a detailed discussion of isotope fractionation mechanisms, Fe metabolism and implications to biomedical research.[517] One such application is the use of Fe isotopes as a diagnostic tool for hereditary hemochromatosis, a disease resulting in badly controlled intestinal Fe absorption and consequent overload of tissues, and for the evaluation of treatment success.[518] One possible treatment is phlebotomy, resulting in the release of stored Fe, which causes a shift in blood Fe isotopic composition as described by Hotz *et al.*[519] Isotopic patterns of iron and zinc were analysed in healthy controls and in patients with hemochromatosis.[520]

Multicollector ICP-SFMS was operated in the high-resolution mode for determination of $^{54}$Fe, $^{56}$Fe and $^{57}$Fe and in medium-resolution mode for $^{64}$Zn, $^{66}$Zn, $^{67}$Zn and $^{68}$Zn.

Ohno *et al.* were the first to present a method for zinc isotopic analysis in human samples (red blood cells and hair) using MC ICP-MS.[521] While they could not see significant differences in the Zn isotope ratios between red blood cells of a small population ($n = 6$), a measurable shift in the Zn isotope ratio between red blood cells and hair (which was comparably depleted in heavier Zn isotopes) of one individual was found. The natural variation in zinc, iron and copper isotopic compositions between total blood, red blood cells and serum was investigated by Albarède *et al.*[522] They reported that $\delta^{65}$Cu and $\delta^{56}$Fe carry information of the red blood cell production process reflected in differences in isotopic compositions between erythrocytes and serum. In a medical forensics and anthropology context, Jaouen *et al.* identified an age dependence of blood Cu and Zn isotope ratios in an isolated Yakut population (Vilyuysk, Russia) with a uniform diet, while no age effect was observed for $\delta^{56}$Fe.[523] In order to improve the fundamental understanding of Zn metabolism, the Zn isotopic composition of different organs and body fluids of sheep[524] and mice[525] were investigated revealing large differences between body compartments. Van Heghe *et al.* conducted a study with twelve healthy individuals and detected an influence of diet habits (vegetarian *vs.* omnivore) on Zn isotopic composition and of sex on the Fe and Cu isotopic composition of whole blood.[526] Costas-Rodriguez *et al.* have shown recently that the difference in the isotopic composition of whole-blood Zn between omnivores and vegetarians is linked with the isotopic composition of food products of plant and animal origin, respectively.[527] Van Heghe *et al.* demonstrated that menstruation and the associated loss of mineral elements are at the basis of the gender-based difference in whole blood Fe and Cu isotopic compositions.[528] Postmenopausal women show isotopic signatures that correspond with those of the male cohort of the population.

Calcium is an important nutrient to any organism and particularly important in vertebrates as bone-forming mineral. The isotopic composition of bone Ca has been compared between modern and archaeological sheep and archaeological humans from both an era with and one without dairy consumption.[529] While in sheep, a difference between males and females was found, variation in humans was generally high and interpretation nontrivial. Previously, Chu *et al.* had suggested the use of Ca isotopes as a proxy for dairy consumption due to the distinct isotopic signature of milk with enrichment of the lighter isotopes.[530] Ca isotopes are fractionated during metabolic uptake from dietary sources, as shown by Hirata *et al.* using MC ICP-MS analysis of mouse bones.[531] They could demonstrate using this tool, that lactating females show a significantly different isotopic composition in their bones. From a medical point of view, Ca isotopes can serve as a tool to assess changes in the bone-mineral balance, which is affected by diseases like osteoporosis. In this context, Morgan *et al.* have developed an analytical MC ICP-MS based method to determine Ca isotope ratios in urine.[532]

While lead is one of the most widely investigated isotopic systems in earth and environmental sciences, it has also been used for investigating human health issues using TIMS or single-collector ICP-MS instruments, as summarised in a review.[533] Methods for Pb isotope ratio measurements in blood using MC ICP-MS have, however, also been published.[534,535]

Another toxic element often investigated in a biomedical context is mercury. Humans are exposed to mercury from inhalation of elemental $Hg^0$ (*e.g.* from dental amalgams) and intake of methylmercury (MeHg) *via* consumption of fish. Due to the unique properties of Hg stable isotopes they can be deployed to gain insight into the sources of Hg in human hair and urine as concluded by Sherman *et al.* from their study with dental professionals.[536]

The natural strontium isotope ratio variation due to the radiogenic $^{87}$Sr is mostly used in geochronology or the context of archaeological migration research, but has also been applied successfully to biological systems. For example, Woodhead *et al.* presented a method to study the isotopic composition with spatial resolution in fish otoliths – incrementally growing earstones of fish, which record environmental information over its life history – using LA-MC ICP-MS.[537] Zitek *et al.* applied a similar method to monitor migration patterns and provenance of freshwater fish.[538] In the context of fish ecology and management, stable isotope tracers have also been used to batch-mark the progenies of one specific mother fish. This transgenerational marking results in an alteration of the isotopic composition of the selected element in the core of the otoliths of all offspring, which can be revealed by using laser-ablation sampling. For example, a stable Sr spike has been used for this purpose.[539,540] Alternatively, a barium dual-isotope spiking approach was developed, which allows for individual specific signatures for progenies of differently marked mother fish.[541]

The issue of food authenticity, *e.g.* for goods from Protected Geographical Origins, is of growing concern to protect consumers from fraud. For a broad overview of food authentication using isotope ratio (and multielement) analysis the reader is referred to Kelly *et al.*[542] and concerning the particular use of Sr isotopes to Fortunato *et al.*[543] A few example studies of pricy goods, in which MC ICP-MS instruments were used, are the determination of Sr isotope ratios in, *e.g.* asparagus,[544] vendace caviar,[545] brown rice,[546] wine[547] and cider.[548] Many studies for source determination of food benefit from the combination of different approaches, *e.g.* different isotopic systems that provide information on different aspects of growth conditions. For instance, light stable isotopes like $\delta^{18}$O reflect meteorological conditions, while Sr isotopic composition is determined primarily by geology. By combining the $\delta^{18}$O, $\delta^{34}$S, $^{87}$Sr/$^{86}$Sr and multielement information including some rare-earth elements, Rodrigues *et al.* could discriminate between coffee beans from different regions of Hawaii.[549]

## References


1. R. S. Houk, V. A. Fassel, G. D. Flesch, H. J. Svec, A. L. Gray and C. E. Taylor, *Anal. Chem.*, 1980, **52**, 2283–2289.
2. J. Alonso and P. González, *Isotope Dilution Mass Spectrometry*, RSC Cambridge, 2013.
3. W. Tittes, N. Jakubowski, D. Stuewer, G. Toelg and J. A. C. Broekaert, *J. Anal. Atom. Spectrom.*, 1994, **9**, 1015–1020.
4. I. Feldmann, W. Tittes, N. Jakubowski, D. Stuewer and U. Giessmann, *J. Anal. Atom. Spectrom.*, 1994, **9**, 1007–1014.
5. N. Jakubowski, W. Tittes, D. Pollmann, D. Stuewer and J. A. C. Broekaert, *J. Anal. Atom. Spectrom.*, 1996, **11**, 797–803.
6. U. Giessmann and U. Greb, *Fresenius J. Anal.Anal. Chem.*, 1994, **350**, 186–193.
7. C. Douthitt, *Anal. Bioanal. Chem.*, 2008, **390**, 437–440.
8. J. S. Becker, *Inorganic Mass Spectrometry – Principles and Applications*, John Wiley & Sons, Chichester, 2007.
9. D. Smith, Inorganic Mass Spectrometry: Fundamentals and Applications, in *Inorganic Mass Spectrometry: Fundamentals and Applications*, eds. C. Barshick, D. Duckworth and D. Smith, Marcel Dekker, New York, 2000, ch. 1, pp. 1–30.
10. A. J. Walder and P. A. Freedman, *J. Anal. Atom. Spectrom.*, 1992, **7**, 571–575.
11. A. J. Walder, I. D. Abell, I. Platzner and P. A. Freedman, *Spectrochim. Acta Part B-Atom. Spectrosc.*, 1993, **48**, 397–402.
12. N. S. Belshaw, P. A. Freedman, R. K. O'Nions, M. Frank and Y. Guo, *Int. J. Mass Spectrom.*, 1998, **181**, 51–58.
13. D. A. Solyom, T. W. Burgoyne and G. M. Hieftje, *J. Anal. Atom. Spectrom.*, 1999, **14**, 1101–1110.
14. D. Ardelt, A. Polatajko, O. Primm and M. Reijnen, *Anal. Bioanal. Chem.*, 2013, **405**, 2987–2994.
15. B. L. Sharp, *J. Anal. Atom. Spectrom.*, 1988, **3**, 613–652.
16. B. L. Sharp, *J. Anal. Atom. Spectrom.*, 1988, **3**, 939–963.
17. M. Akbar, *Inductively Coupled Plasma Mass Spectrometry*, Wiley-VCH, New York, 1998.
18. R. Thomas, *Practical Guide to ICP-MS*, Marcel Dekker, Inc., New York, 2004, 336.
19. S. M. Nelms, *Inductively Coupled Plasma Mass Spectrometry Handbook*, Blackwell Publishing Ltd., Oxford, United Kingdom, 2005, 485.
20. A. Woller, H. Garraud, J. Boisson, A. M. Dorthe, P. Fodor and O. F. X. Donard, *J. Anal. Atom. Spectrom.*, 1998, **13**, 141–149.
21. D. Schaumlöffel and A. Prange, *Fresenius J. Anal. Chem.*, 1999, **364**, 452–456.
22. J. A. McLean, J. S. Becker, S. F. Boulyga, H. J. Dietze and A. Montaser, *Int. J. Mass Spectrom.*, 2001, **208**, 193–204.
23. K. Shigeta, H. Traub, U. Panne, A. Okino, L. Rottmann and N. Jakubowski, *J. Anal. Atom. Spectrom.*, 2013, **28**, 646–656.

24. J. Szpunar, R. Lobinski and A. Prange, *Appl. Spectrosc.*, 2003, **57**, 102A–112A.
25. J. Meija, M. Montes-Bayón, D. L. Le Duc, N. Terry and J. A. Caruso, *Anal. Chem.*, 2002, **74**, 5837–5844.
26. D. Pröfrock, P. Leonhard, S. Wilbur and A. Prange, *J. Anal. Atom. Spectrom.*, 2004, **19**, 623–631.
27. M. Ødegård, Ø. Skår, H. Schiellerup and N. J. Pearson, *Geostand. Geoanal. Res.*, 2005, **29**, 197–209.
28. J. Pisonero, B. Fernández and D. Günther, *J. Anal. Atom. Spectrom.*, 2009, **24**, 1145–1160.
29. M. Resano, M. Aramendia and F. Vanhaecke, *J. Anal. Atom. Spectrom.*, 2009, **24**, 484–493.
30. P. P. Mahoney, S. J. Ray, G. Q. Li and G. M. Hieftje, *Anal. Chem.*, 1999, **71**, 1378–1383.
31. J. A. Nobrega, R. E. Sturgeon, P. Grinberg, G. J. Gardner, C. S. Brophy and E. E. Garcia, *J. Anal. Atom. Spectrom.*, 2011, **26**, 2519–2523.
32. P. Wu, L. He, C. Zheng, X. Hou and R. E. Sturgeon, *J. Anal. Atom. Spectrom.*, 2010, **25**, 1217–1246.
33. C. Bendicho, F. Pena, M. Costas, S. Gil and I. Lavilla, *TrAC – Trends Anal. Chem.*, 2010, **29**, 681–691.
34. X. M. Guo, R. E. Sturgeon, Z. Mester and G. J. Gardner, *Anal. Chem.*, 2004, **76**, 2401–2405.
35. F. F. Chen, *Introduction to Plasma Physics*, Plenum Press, New York, 1974.
36. J. M. Cottle, A. J. Burrows, A. Kylander-Clark, P. A. Freedman and R. S. Cohen, *J. Anal. Atom. Spectrom.*, 2013, **28**, 1700–1706.
37. D. Günther, H. P. Longerich and S. E. Jackson, *Can. J. Appl. Spectrosc.*, 1995, **40**, 111–116.
38. T. Lindemann, J. Hinrichs, T. Oki, S. McSheehy, J. Wills and M. Hamester, *"Enhancing Sensitivity of Sector-Field ICP-MS" application note Thermo Fisher Scientific*, 2010.
39. N. Jakubowski, T. Prohaska, F. Vanhaecke, P. H. Roos and T. Lindemann, *J. Anal. Atom. Spectrom.*, 2011, **26**, 727–757.
40. G. H. Vickers, D. A. Wilson and G. M. Hieftje, *Anal. Chem.*, 1988, **60**, 1808–1812.
41. S. Weyer and J. B. Schwieters, *Int. J. Mass Spectrom.*, 2003, **226**, 355–368.
42. Nu Instruments Ltd., *NP Faraday detectors*, Wrexham, UK, 2007.
43. Nu Instruments Ltd., *Nu Plasma II Multi-Collector ICP-MS Brochure*, Wrexham, UK, http://www.nu-ins.com/pdf/Plasma-II-brochure.pdf. Accessed 31 July 2012, 2013.
44. L. Ball, K. W. W. Sims and J. Schwieters, *J. Anal. Atom. Spectrom.*, 2008, **23**, 173–180.
45. J. H. Barnes, R. Sperline, M. B. Denton, C. J. Barinaga, D. Koppenaal, E. T. Young and G. M. Hieftje, *Anal. Chem.*, 2002, **74**, 5327–5332.
46. G. D. Schilling, F. J. Andrade, J. H. Barnes IV, R. P. Sperline, M. B. Denton, C. J. Barinaga, D. W. Koppenaal and G. M. Hieftje, *Anal. Chem.*, 2007, **79**, 7662–7668.

47. A. A. Rubinshtein, G. D. Schilling, S. J. Ray, R. P. Sperline, M. B. Denton, C. J. Barinaga, D. W. Koppenaal and G. M. Hieftje, *J. Anal. Atom. Spectrom.*, 2010, **25**, 735–738.
48. G. D. Schilling, S. J. Ray, A. A. Rubinshtein, J. A. Felton, R. P. Sperline, M. B. Denton, C. J. Barinaga, D. W. Koppenaal and G. M. Hieftje, *Anal. Chem.*, 2009, **81**, 5467–5473.
49. T. Lindemann, M. Hamester, J. Hinrichs, L. Rottmann and J. D. Will, http://www.thermo.com/eThermo/CMA/PDFs/Various/File_53667.pdf, (accessed 17.06.2014).
50. S. F. Boulyga, U. Kloetzli and T. Prohaska, *J. Anal. Atom. Spectrom.*, 2006, **21**, 1427–1430.
51. E. H. Evans and J. J. Giglio, *J. Anal. Atom. Spectrom.*, 1993, **8**, 1–18.
52. N. F. Zahran, A. I. Helal, M. A. Amr, A. Abdel-Hafiez and H. T. Mohsen, *Int. J. Mass Spectrom.*, 2003, **226**, 271–278.
53. N. Nonose and M. Kubota, *J. Anal. Atom. Spectrom.*, 2001, **16**, 551–559.
54. F. Vanhaecke, C. Vandecasteele, H. Vanhoe and R. Dams, *Mikrochim. Acta*, 1992, **108**, 41–51.
55. D. B. Aeschliman, S. J. Bajic, D. P. Baldwin and R. Houk, *J. Anal. Atom. Spectrom.*, 2003, **18**, 1008–1014.
56. K. C. Sears, J. W. Ferguson, T. J. Dudley, R. S. Houk and M. S. Gordon, *J. Phys. Chem. A*, 2008, **112**, 2610–2617.
57. F. Vanhaecke, L. Moens and R. Dams, *J. Anal. Atom. Spectrom.*, 1998, **13**, 1189–1192.
58. F. Vanhaecke, L. Moens, R. Dams and P. Taylor, *Anal. Chem.*, 1996, **68**, 567–569.
59. F. Vanhaecke, L. Balcaen and D. Malinovsky, *J. Anal. Atom. Spectrom.*, 2009, **24**, 863–886.
60. F. Vanhaecke and P. Degryse, *Isotopic Analysis: Fundamentals and Applications Using ICP-MS*, John Wiley & Sons, Weinheim, 2012.
61. F. Vanhaecke, L. Moens, R. Dams, I. Papadakis and P. Taylor, *Anal. Chem.*, 1997, **69**, 268–273.
62. J. M. Cottle, M. S. A. Horstwood and R. R. Parrish, *J. Anal. Atom. Spectrom.*, 2009, **24**, 1355–1363.
63. T. Pettke, F. Oberli, A. Audétat, U. Wiechert, C. R. Harris and C. A. Heinrich, *J. Anal. Atom. Spectrom.*, 2011, **26**, 475–492.
64. I. Günther-Leopold, B. Wernli, Z. Kopajtic and D. Günther, *Anal. Bioanal. Chem.*, 2004, **378**, 241–249.
65. E. M. Krupp and O. F. X. Donard, *Int. J. Mass Spectrom.*, 2005, **242**, 233–242.
66. T. Hirata, Y. Hayano and T. Ohno, *J. Anal. Atom. Spectrom.*, 2003, **18**, 1283–1288.
67. A. Gourgiotis, S. Berail, P. Louvat, H. Isnard, J. Moureau, A. Nonell, G. Manhes, J.-L. Birck, J. Gaillardet, C. Pecheyran, F. Chartier and O. F. X. Donard, *J. Anal. Atom. Spectrom.*, 2014, **29**, 1607–1617.
68. N. Jakubowski, R. Lobinski and L. Moens, *J. Anal. Atom. Spectrom.*, 2004, **19**, 1–4.

69. C. F. Harrington, R. Clough, H. R. Hansen, S. J. Hill, S. A. Pergantis and J. F. Tyson, *J. Anal. Atom. Spectrom.*, 2009, **24**, 999–1025.
70. C. F. Harrington, R. Clough, H. R. Hansen, S. J. Hill and J. F. Tyson, *J. Anal. Atom. Spectrom.*, 2010, **25**, 1185–1216.
71. C. F. Harrington, R. Clough, L. R. Drennan-Harris, S. J. Hill and J. F. Tyson, *J. Anal. Atom. Spectrom.*, 2011, **26**, 1561–1595.
72. D. M. Templeton, F. Ariese, R. Cornelis, L. G. Danielsson, H. Muntau, H. P. Van Leeuwen and R. Lobinski, *Pure Appl. Chem.*, 2000, **72**, 1453–1470.
73. R. Cornelis, J. Caruso, H. Crews and K. G. Heumann, eds., *Handbook of Elemental Speciation, Series I*, Wiley, New York, 2003.
74. R. Cornelis, J. Caruso, H. Crews and K. G. Heumann, eds., *Handbook of Elemental Speciation, Series II*, Wiley, New York, 2005.
75. C. M. Andrle, N. Jakubowski and J. A. C. Broekaert, *Spectrochim. Acta Part B-Atom. Spectrosc.*, 1997, **52**, 189–200.
76. N. Jakubowski, B. Jepkens, D. Stuewer and H. Berndt, *J. Anal. Atom. Spectrom.*, 1994, **9**, 193–198.
77. C. Barnowski, N. Jakubowski, D. Stuewer and J. A. C. Broekaert, *J. Anal. Atom. Spectrom.*, 1997, **12**, 1155–1161.
78. C. Barnowski, *Entwicklung und Untersuchung von Methoden zur Speziation von Chrom in Umgebungsaerosolen*, PhD Thesis, University of Dortmund, Germany, Dortmund, Germany, 2001.
79. D. Metze, H. Herzog, B. Gosciniak, D. Gladtke and N. Jakubowski, *Anal. Bioanal. Chem.*, 2004, **378**, 123–128.
80. L. Moens and N. Jakubowski, *Anal. Chem.*, 1998, **70**, 251a–256a.
81. N. Jakubowski, L. Moens and F. Vanhaecke, *Spectrochim. Acta, Part B*, 1998, **53**, 1739–1763.
82. C. Douthitt, pers. comm.: Bibliography about ICP-SFMS 2013.
83. P. J. Sylvester, *Geostand. Geoanal. Res.*, 2008, **32**, 469–488.
84. W. C. Davis, S. J. Christopher and G. C. Turk, *Anal. Chem.*, 2005, **77**, 6389–6395.
85. A. T. Townsend, *J. Anal. Atom. Spectrom.*, 2000, **15**, 307–314.
86. H. Wildner and R. Hearn, *Fresenius J. Anal. Chem.*, 1998, **360**, 800–803.
87. V. Balaram, *Bull. Mater. Sci.*, 2005, **28**, 345–348.
88. *Fisons Instruments application note, High-resolution ICP-MS analysis of ultrapure* $H_2SO_4$*, Document No. VGE/SM/TM/175*, 1993.
89. Y. Takaku, K. Masuda, T. Takahashi and T. Shimamura, *J. Anal. Atom. Spectrom.*, 1994, **9**, 1385–1387.
90. C.-C. Wan, S.-J. Jiang, M.-T. You and A. C. Sahayam, *J. Anal. Atom. Spectrom.*, 2005, **20**, 1290–1292.
91. J. S. Becker and H.-J. Dietze, *Int. J. Mass Spectrom.*, 2003, **228**, 127–150.
92. M. B. Shabani, Y. Shiina, F. G. Kirscht and Y. Shimanuki, *Mater. Sci. Eng. B*, 2003, **102**, 238–246.
93. E. J. Ferrero and D. Posey, *J. Anal. Atom. Spectrom.*, 2002, **17**, 1194–1201.
94. S. F. Boulyga, H.-J. Dietze and J. S. Becker, *J. Anal. Atom. Spectrom.*, 2001, **16**, 598–602.

95. H. Xie, X. Nie and Y. Tang, *Chin. J. Anal. Chem.*, 2006, **34**, 1570–1574.
96. W. K. Ryu, J. S. Kim, J. S. Lee, H. B. Lim and P. K. Jun, *J. Anal. Atom. Spectrom.*, 2007, **22**, 623–629.
97. T. Vaculovic, P. Sulovsky, J. Machat, V. Otruba, O. Matal, T. Simo, C. Latkoczy, D. Günther and V. Kanicky, *J. Anal. Atom. Spectrom.*, 2009, **24**, 649–654.
98. C. Latkoczy and T. Ghislain, *J. Anal. Atom. Spectrom.*, 2006, **21**, 1152–1160.
99. V. Kozlov, M. Leskelä, T. Prohaska, G. Schultheis, G. Stingeder and H. Sipilä, *Nucl. Instrum. Methods Phys. Res., Sect. A: Accelerators, Spectrometers, Detectors Assoc. Equip.*, 2004, **531**, 165–173.
100. B. Flem, R. B. Larsen, A. Grimstvedt and J. Mansfeld, *Chem. Geol.*, 2002, **182**, 237–247.
101. J. S. Becker, C. Pickhardt, N. Hoffmann, H. Höcker and J. S. Becker, *Atom. Spectrosc.*, 2002, **23**, 1–6.
102. H. R. Kuhn and D. Guenther, *J. Anal. Atom. Spectrom.*, 2004, **19**, 1158–1164.
103. J. Koch, I. Feldmann, B. Hattendorf, D. Günther, U. Engel, N. Jakubowski, M. Bolshov, K. Niemax and R. Hergenröder, *Spectrochim. Acta, Part B*, 2002, **57**, 1057–1070.
104. Y. Takaku, K. Masuda, T. Takahashi and T. Shimamura, *J. Anal. Atom. Spectrom.*, 1993, **8**, 687–690.
105. T. Graule, A. Von Bohlen, J. A. C. Broekaert, E. Grallath, R. Klockenkaemper, P. Tschoepel and G. Toelg, *Fresenius Z. Anal. Chem.*, 1989, **335**, 637–642.
106. F. Kohl, N. Jakubowski, R. Brandt, C. Pilger and J. A. C. Broekaert, *Fresenius J. Anal. Chem.*, 1997, **359**, 317–325.
107. J. E. S. Sarkis, O. N. Neto, S. Viebig and S. F. Durrant, *Foren. Sci. Int.*, 2007, **172**, 63–66.
108. A. M. Dobney, W. Wiarda, P. de Joode and G. J. Q. van der Peijl, *J. Anal. Atom. Spectrom.*, 2002, **17**, 478–484.
109. G. De Wannemacker, F. Vanhaecke, L. Moens, A. Van Mele and H. Thoen, *J. Anal. Atom. Spectrom.*, 2000, **15**, 323–327.
110. S. A. Junk, *Nucl. Instrum. Methods Phys. Res., Sect. B: Beam Interact. Mater. Atoms*, 2001, **181**, 723–727.
111. C. Latkoczy, S. Becker, M. Dücking, D. Günther, J. A. Hoogewerff, J. R. Almirall, J. Buscaglia, A. Dobney, R. D. Koons, S. Montero, G. J. Q. Van Der Peijl, W. R. S. Stoecklein, T. Trejos, J. R. Watling and V. S. Zdanowicz, *J. Foren. Sci.*, 2005, **50**, 1327–1341.
112. W. Castro, T. Trejos, B. Naes and J. R. Almirall, *Anal. Bioanal. Chem.*, 2008, **392**, 663–672.
113. G. Schultheis, T. Prohaska, G. Stingeder, K. Dietrich, D. Jembrih-Simburger and M. Schreiner, *J. Anal. Atom. Spectrom.*, 2004, **19**, 838–843.
114. I. Deconinck, C. Latkoczy, D. Günther, F. Govaert and F. Vanhaecke, *J. Anal. Atom. Spectrom.*, 2006, **21**, 279–287.

115. A. C. Sahayam, S.-J. Jiang and C.-C. Wan, *Anal. Chim. Acta*, 2007, **598**, 214–218.
116. A. V. Izmer, M. V. Zoriy, C. Pickhardt, W. Quadakkers, V. Shemet, L. Singheiser and J. S. Becker, *J. Anal. Atom. Spectrom.*, 2005, **20**, 918–923.
117. H. Xie and X. Nie, *Anal. Sci.*, 2006, **22**, 1371–1374.
118. S. Finkeldei and G. Staats, *Fresenius J. Anal. Chem.*, 1997, **359**, 357–360.
119. W. Devos and C. Moor, *J. Anal. Atom. Spectrom.*, 2002, **17**, 138–141.
120. H. Traub, M. Czerwensky, R. Matschat, H. Kipphardt and U. Panne, *J. Anal. Atom. Spectrom.*, 2010, **25**, 690–696.
121. P. Krystek, A. Ulrich, C. C. Garcia, S. Manohar and R. Ritsema, *J. Anal. Atom. Spectrom.*, 2011, **26**, 1701–1721.
122. J. W. Olesik and P. J. Gray, *J. Anal. Atom. Spectrom.*, 2012, **27**, 1143–1155.
123. S. Gschwind, L. Flamigni, J. Koch, O. Borovinskaya, S. Groh, K. Niemax and D. Guenther, *J. Anal. Atom. Spectrom.*, 2011, **26**, 1166–1174.
124. P. Pohl, N. Vorapalawut, B. Bouyssiere, H. Carrier and R. Lobinski, *J. Anal. Atom. Spectrom.*, 2010, **25**, 704–709.
125. Y. N. Fedorov, K. S. Ivanov, Y. V. Erokhin and Y. L. Ronkin, *Doklady Earth Sci.*, 2007, **414**, 634–637.
126. H. Xie, K. Huang, J. Liu, X. Nie and L. Fu, *Anal. Bioanal. Chem.*, 2009, **393**, 2075–2080.
127. J. Yang, *Fuel*, 2006, **85**, 1679–1684.
128. S. F. Boulyga, J. Heilmann, T. Prohaska and K. G. Heumann, *Anal. Bioanal. Chem.*, 2007, **389**, 697–706.
129. S. F. Boulyga, J. Heilmann and K. G. Heumann, *Anal. Bioanal. Chem.*, 2005, **382**, 1808–1814.
130. M. Edler, D. Metze, N. Jakubowski and M. Linscheid, *J. Anal. Atom. Spectrom.*, 2002, **17**, 1209–1212.
131. J. Carter, L. Ebdon and E. H. Evans, *Microchem. J.*, 2004, **76**, 35–41.
132. J. T. Cullen, M. P. Field and R. M. Sherrell, *J. Anal. Atom. Spectrom.*, 2001, **16**, 1307–1312.
133. Z. Yu, P. Robinson and P. McGoldrick, *Geostand. Newslett.*, 2001, **25**, 199–217.
134. W. Pretorius, D. Chipley, K. Kyser and H. Helmstaedt, *J. Anal. Atom. Spectrom.*, 2003, **18**, 302–309.
135. W. Pretorius, D. Weis, G. Williams, D. Hanano, B. Kieffer and J. Scoates, *Geostand. Geoanal. Res.*, 2006, **30**, 39–54.
136. R. R. Barefoot, *Anal. Chim. Acta*, 2004, **509**, 119–125.
137. D. Frei and A. Gerdes, *Chem. Geol.*, 2009, **261**, 261–270.
138. Z. Chang, J. D. Vervoort, W. C. McClelland and C. Knaack, *Geochem., Geophys., Geosyst.*, 2006, 7, Q08002.
139. J. L. Paquette and M. Tiepolo, *Chem. Geol.*, 2007, **240**, 222–237.
140. G. Meinhold and D. Frei, *Geol. Mag.*, 2008, **145**, 886–891.
141. F. Kalsbeek, D. Frei and P. Affation, *Sediment. Geol.*, 2008, **212**, 86–95.
142. A. Gerdes and A. Zeh, *Earth Planet. Sci. Lett.*, 2006, **249**, 47–61.

143. A. Zeh, A. Gerdes, R. Klemd and J. M. J. Barton, *J. Petrol.*, 2007, **48**, 1605–1639.
144. V. Janousek, A. Gerdes, S. Vrana, F. Finger, V. Erban, G. Friedl and C. J. R. Braithwaite, *J. Petrol.*, 2006, **47**, 705–744.
145. M. D. Axelsson, I. Rodushkin, D. C. Baxter, J. Ingri and B. Öhlander, *Geochem. Trans.*, 2002, **3**, 40–47.
146. D. Chipley, P. A. Polito and T. Kurtis Kyser, *Am. Mineral.*, 2007, **92**, 1925–1935.
147. M. Guillong, C. Latkoczy, J. H. Seo, D. Günther and C. A. Heinrich, *J. Anal. Atom. Spectrom.*, 2008, **23**, 1581–1589.
148. M. Gaeta, C. Freda, J. N. Christensen, L. Dallai, F. Marra, D. B. Karner and P. Scarlato, *Lithos*, 2006, **86**, 330–346.
149. J. S. Becker, C. Pickhardt and H.-J. Dietze, *Int. J. Mass Spectrom.*, 2000, **202**, 283–297.
150. Y. Muramatsu, S. Yoshida and A. Tanaka, *J. Radioanal. Nucl. Chem.*, 2003, **255**, 477–480.
151. S. F. Boulyga, J. L. Matusevich, V. P. Mironov, V. P. Kudrjashov, L. Halicz, I. Segal, J. A. McLean, A. Montaser and J. S. Becker, *J. Anal. Atom. Spectrom.*, 2002, **17**, 958–964.
152. C. K. Kim, C. S. Kim, B. H. Rho and J. I. Lee, *J. Radioanal. Nucl. Chem.*, 2002, **252**, 421–427.
153. E. J. Wyse, S. H. Lee, J. La Rosa, P. Povinec and S. J. de Mora, *J. Anal. Atom. Spectrom.*, 2001, **16**, 1107–1111.
154. M. E. Ketterer and S. C. Szechenyi, *Spectrochim. Acta, Part B*, 2008, **63**, 719–737.
155. D. Larivière, K. Benkhedda, S. Kiser, S. Johnson and R. J. Cornett, *Anal. Methods*, 2010, **2**, 259–267.
156. D. Schaumlöffel, P. Giusti, M. V. Zoriy, C. Pickhardt, J. Szpunar, R. Lobinski and J. S. Becker, *J. Anal. Atom. Spectrom.*, 2005, **20**, 17–21.
157. M. V. Zoriy, Z. Varga, C. Pickhardt, P. Ostapczuk, R. Hille, L. Halicz, I. Segal and J. S. Becker, *J. Environ. Monitor.*, 2005, 7, 514–518.
158. M. Leermakers, Y. Gao, J. Navez, A. Poffijn, K. Croes and W. Baeyens, *J. Anal. Atom. Spectrom.*, 2009, **24**, 1115–1117.
159. C. Li, K. Benkhedda, Z. Varve, V. Kochemin, B. Sadi, E. Lai, G. Kramer and J. Cornett, *J. Anal. Atom. Spectrom.*, 2009, **24**, 1429–1433.
160. D. Larivière, T. A. Cumming, S. Kiser, C. Li and R. J. Cornett, *J. Anal. Atom. Spectrom.*, 2008, **23**, 352–360.
161. C. Li, D. Larivière, S. Kiser, G. Moodie, R. Falcomer, N. Elliot, L. Burchart, L. Paterson, V. Epov, D. Evans, S. Pappas, J. Smith and J. Cornett, *J. Anal. Atom. Spectrom.*, 2008, **23**, 521–526.
162. A. Pitois, L. A. de Las Heras and M. Betti, *Int. J. Mass Spectrom.*, 2008, **270**, 118–126.
163. M. V. Zoriy, P. Ostapczuk, L. Halicz, R. Hille and J. S. Becker, *Int. J. Mass Spectrom.*, 2005, **242**, 203–209.
164. U. Nygren, I. Rodushkin, C. Nilsson and D. C. Baxter, *J. Anal. Atom. Spectrom.*, 2003, **18**, 1426–1434.

165. S. F. Boulyga, M. Tibi and K. G. Heumann, *Anal. Bioanal. Chem.*, 2004, **378**, 342–347.
166. M. Yamamoto, Syarbaini, K. Kofuji, A. Tsumura, K. Komura, K. Ueno and D. J. Assinder, *J. Radioanal. Nucl. Chem.*, 1995, **197**, 185–194.
167. Z. Varga, *Anal. Chim. Acta*, 2008, **625**, 1–7.
168. Z. Varga and G. Suranyi, *Anal. Chim. Acta*, 2007, **599**, 16–23.
169. S. F. Boulyga and J. S. Becker, *Fresenius J. Anal. Chem.*, 2001, **370**, 612–617.
170. S. F. Boulyga and J. S. Becker, *J. Anal. Atom. Spectrom.*, 2002, **17**, 1143–1147.
171. S. F. Boulyga and K. G. Heumann, *J. Environ. Radioact.*, 2006, **88**, 1–10.
172. I. Trešl, G. De Wannemacker, C. R. Quétel, I. Petrov, F. Vanhaecke, L. Moens and P. D. P. Taylor, *Environ. Sci. Technol.*, 2004, **38**, 581–586.
173. M. Krachler, *J. Environ. Monitor.*, 2007, **9**, 790–804.
174. K. Kyser, D. Chipley, A. Bukata, P. Polito, A. Fitzpatrick and P. Alexandre, *Can. J. Anal. Sci. Spectrosc.*, 2003, **48**, 258–268.
175. M. LaVigne, M. P. Field, E. Anagnostou, A. G. Grottoli, G. M. Wellington and R. M. Sherrell, *Geophys. Res. Lett.*, 2008, **35**.
176. S. R. Thorrold, G. P. Jones, S. Planes and J. A. Hare, *Can. J. Fish. Aquat. Sci.*, 2006, **63**, 1193–1197.
177. F. J. Fodrie and L. A. Levin, *Limnol. Oceanogr.*, 2008, **53**, 799–812.
178. M. Segura, C. Cámara, Y. Madrid, C. Rebollo, J. Azcárate, G. N. Kramer, B. M. Gawlik, A. Lamberty and P. Quevauviller, *TrAC – Trends Anal. Chem.*, 2004, **23**, 194–202.
179. M. Popp, G. Koellensperger, G. Stingeder and S. Hann, *J. Anal. Atom. Spectrom.*, 2008, **23**, 111–118.
180. Official Journal of the European Communities L 327/1, *Directive 2000/60/EC of the European Parliament and the Council of 23 October 2000 establishing a framework for Community action in the field of water policy*, 2000.
181. Official Journal of the European Communities L 348/84, *Directive 2008/105/EC of the European Parliament and the Council of 16 December 2008 on environmental quality standards in the field of water policy, amending and subsequently repealing Council Directives 82/176/EEC, 83/513/EEC, 84/156/EEC, 84/491/EEC, 86/280/EEC and amending Directive 2000/60/EC of the European Parliament and of the Council*, 2008.
182. L. I. L. Balcaen, K. A. C. De Schamphelaere, C. R. Janssen, L. Moens and F. Vanhaecke, *Anal. Bioanal. Chem.*, 2008, **390**, 555–569.
183. V. W.-M. Lai, K. Kanaki, S. A. Pergantis, W. R. Cullen and K. J. Reimer, *J. Environ. Monitor.*, 2012, **14**, 743–751.
184. C. Barbante, C. Boutron, C. Morel, C. Ferrari, J. L. Jaffrezo, G. Cozzi, V. Gaspari and P. Cescon, *J. Environ. Monitor.*, 2003, **5**, 328–335.
185. S. Caimi, O. Senofonte and S. Caroli, Chapter 10 – Certified reference materials in Antarctic matrices: Development and use, in *Environmental Contamination in Antarctica*, eds. S. Caroli, P. Cescon and D. W. H. Walton, Elsevier Science, Amsterdam, 2001, pp. 275–292.

186. M. Krachler, J. Zheng, D. Fisher and W. Shotyk, *J. Anal. Atom. Spectrom.*, 2004, **19**, 1017–1019.
187. M. Krachler, J. Zheng, D. Fisher and W. Shotyk, *Anal. Chim. Acta*, 2005, **530**, 291–298.
188. M. P. Field and R. M. Sherrell, *J. Anal. Atom. Spectrom.*, 2003, **18**, 254–259.
189. P. Gabrielli, C. Barbante, C. Turetta, A. Marteel, C. Boutron, G. Cozzi, W. Cairns, C. Ferrari and P. Cescon, *Anal. Chem.*, 2006, **78**, 1883–1889.
190. P. Gabrielli, C. Barbante, J. M. C. Plane, A. Varga, S. Hong, G. Cozzi, V. Gaspari, F. A. M. Planchon, W. Cairns, C. Ferrari, P. Crutzen, P. Cescon and C. F. Boutron, *Nature*, 2004, **432**, 1011–1014.
191. M. Krachler, J. Zheng, R. Koerner, C. Zdanowicz, D. Fisher and W. Shotyk, *J. Environ. Monitor.*, 2005, 7, 1169–1176.
192. T. Döring, M. Schwikowski and H. W. Gäggeler, *Fresenius J. Anal. Chem.*, 1997, **359**, 382–384.
193. C. Barbante, M. Schwikowski, T. Döring, H. W. Gäggeler, U. Schotterer, L. Tobler, K. Van De Velde, C. Ferrari, G. Cozzi, A. Turetta, K. Rosman, M. Bolshov, G. Capodaglio, P. Cescon and C. Boutron, *Environ. Sci. Technol.*, 2004, **38**, 4085–4090.
194. M. Krachler, J. Zheng, D. Fisher and W. Shotyk, *Anal. Chem.*, 2004, **76**, 5510–5517.
195. I. Rodushkin, T. Ruth and D. Klockare, *J. Anal. Atom. Spectrom.*, 1998, **13**, 159–166.
196. J. L. Barriada, A. D. Tappin, E. H. Evans and E. P. Achterberg, *TrAC - Trends Anal. Chem.*, 2007, **26**, 809–817.
197. L. Yang, Z. Mester, L. Abranko and R. E. Sturgeon, *Anal. Chem.*, 2004, **76**, 3510–3516.
198. A. J. Beck, J. K. Cochran and S. A. Sañudo-Wilhelmy, *Estuaries Coasts*, 2009, **32**, 535–550.
199. I. Rodushkin and T. Ruth, *J. Anal. Atom. Spectrom.*, 1997, **12**, 1181–1185.
200. M. P. Field, M. Lavigne, K. R. Murphy, G. M. Ruiz and R. M. Sherrell, *J. Anal. Atom. Spectrom.*, 2007, **22**, 1145–1151.
201. S. R. Thorrold, C. Latkoczy, P. K. Swart and C. M. Jones, *Science*, 2001, **291**, 297–299.
202. H. L. Filipsson, J. M. Bernhard, S. A. Lincoln and D. C. McCorkle, *Biogeosciences*, 2010, 7, 1335–1347.
203. M. E. Katz, K. G. Miller, J. D. Wright, B. S. Wade, J. V. Browning, B. S. Cramer and Y. Rosenthal, *Nature Geosci.*, 2008, **1**, 329–334.
204. D. W. Oppo, Y. Rosenthal and B. K. Linsley, *Nature*, 2009, **460**, 1113–1116.
205. M. LaVigne, K. A. Matthews, A. G. Grottoli, K. M. Cobb, E. Anagnostou, G. Cabioch and R. M. Sherrell, *Geochim. Cosmochim. Acta*, 2010, **74**, 1282–1293.
206. C. Latkoczy, T. Prohaska, G. Stingeder and W. W. Wenzel, *Fresenius J. Anal. Chem.*, 2000, **368**, 256–262.
207. W. J. Bounds and K. H. Johannesson, *Water, Air, Soil Pollut.*, 2007, **185**, 195–207.

208. E. Engström, A. Stenberg, D. C. Baxter, D. Malinovsky, I. Maekinen, S. Poenni and I. Rodushkin, *J. Anal. Atom. Spectrom.*, 2004, **19**, 858–866.
209. D. Chipley, T. K. Kyser, D. Beauchemin and B. MacFarlane, *Can. J. Anal. Sci. Spectrosc.*, 2003, **48**, 270–276.
210. T. Prohaska, S. Hann, C. Latkoczy and G. Stingeder, *J. Anal. Atom. Spectrom.*, 1999, **14**, 1–8.
211. A. T. Townsend and I. Snape, *Sci. Total Environ.*, 2008, **389**, 466–474.
212. A. Pino, A. Alimonti, F. Botre, C. Minoia, B. Bocca and M. E. Conti, *Rapid Commun. Mass Spectrom.*, 2007, **21**, 1900–1906.
213. B. Bocca, M. E. Conti, A. Pino, D. Mattei, G. Forte and A. Alimonti, *Int. J. Environ. Anal. Chem.*, 2007, **87**, 1111–1123.
214. E. Beccaloni, A. M. Coccia, L. Musmeci, E. Stacul and G. Ziemacki, *Microchem. J.*, 2005, **79**, 271–289.
215. N. Rausch, L. Ukonmaanaho, T. M. Nieminen, M. Krachler, G. L. Roux and W. Shotyk, *Anal. Chim. Acta*, 2006, **558**, 201–210.
216. D. Pick, M. Leiterer and J. W. Einax, *Microchem. J.*, 2010, **95**, 315–319.
217. A. S. Palmer, I. Snape, J. S. Stark, G. J. Johnstone and A. T. Townsend, *Marine Pollut. Bull.*, 2006, **52**, 1441–1449.
218. P. Robinson, A. T. Townsend, Z. Yu and C. Münker, *Geostand. Newslett*, 1999, **23**, 31–46.
219. J. S. Becker, D. Bellis, I. Staton, C. W. McLeod, J. Dombovari and J. S. Becker, *Fresenius J. Anal. Chem.*, 2000, **368**, 490–495.
220. L. R. V. Mataveli, P. Pohl, S. Mounicou, M. A. Z. Arruda and J. Szpunar, *Metallomics*, 2010, **2**, 800–805.
221. H. Kempter, M. Krachler and W. Shotyk, *Environ. Sci. Technol.*, 2010, **44**, 5509–5515.
222. A. Pino, A. Alimonti, M. E. Conti and B. Bocca, *J. Environ. Monitor.*, 2010, **12**, 1857–1863.
223. J. Cizdziel, K. Bu and P. Nowinski, *Anal. Methods*, 2012, **4**, 564–569.
224. B. A. Lesniewska, J. Messerschmidt, N. Jakubowski and A. Hulanicki, *Sci. Total Environ.*, 2004, **322**, 95–108.
225. V. G. Mihucz, E. Tatár, I. Virág, E. Cseh, F. Fodor and G. Záray, *Anal. Bioanal. Chem.*, 2005, **383**, 461–466.
226. J. Zheng, H. Hintelmann, B. Dimock and M. S. Dzurko, *Anal. Bioanal. Chem.*, 2003, **377**, 14–24.
227. Z. Mester, S. Willie, L. Yang, R. Sturgeon, J. A. Caruso, M. L. Fernandez, P. Fodor, R. J. Goldschmidt, H. Goenaga-Infante, R. Lobinski, P. Maxwell, S. McSheehy, A. Polatajko, B. B. M. Sadi, A. Sanz-Medel, C. Scriver, J. Szpunar, R. Wahlen and W. Wolf, *Anal. Bioanal. Chem.*, 2006, **385**, 168–180.
228. M. Óvári, G. Muránszky, M. Zeiner, I. Virág, I. Steffan, V. G. Mihucz, E. Tatár, S. Caroli and G. Záray, *Microchem. J.*, 2007, **87**, 159–162.
229. C. P. R. Morcelli, A. M. G. Figueiredo, J. E. S. Sarkis, J. Enzweiler, M. Kakazu and J. B. Sigolo, *Sci. Total Environ.*, 2005, **345**, 81–91.
230. B. Bocca, A. Alimonti, A. Cristaudo, E. Cristallini, F. Petrucci and S. Caroli, *Anal. Chim. Acta*, 2004, **512**, 19–25.

231. K. Kanitsar, G. Koellensperger, S. Hann, A. Limbeck, H. Puxbaum and G. Stingeder, *J. Anal. Atom. Spectrom.*, 2003, **18**, 239–246.
232. I. Iavicoli, G. Carelli, B. Bocca, S. Caimi, L. Fontana and A. Alimonti, *Chemosphere*, 2008, **71**, 568–573.
233. P. Krystek and R. Ritsema, *Int. J. Mass Spectrom.*, 2007, **265**, 23–29.
234. S. Mounicou, J. Szpunar and R. Lobinski, *Chem. Soc. Rev.*, 2009, **38**, 1119–1138.
235. R. S. Houk, Elemental Speciation by ICP-MS with High Resolution Instruments, in *Handbook of Elemental Speciation: Techniques and Methodology*, eds. R. Cornelis, J. Caruso, H. Crews and K. Heumann, John Wiley & Sons Ltd., West Sussex, England, 2003, pp. 378–416.
236. E. Krupp, F. Seby, R. R. Martín-Doimeadios, A. Holliday, M. Moldován, G. Köllensperger, S. Hann and O. F. X. Donard, Trace Metal Speciation with ICP-MS Detection, in *Inductively Coupled Plasma Mass Spectrometry Handbook*, ed. S. M. Nelms, Blackwell Publishing Ltd, Oxford, UK, 2005, pp. 259–335.
237. M. Moldovan, E. M. Krupp, A. E. Holliday and O. F. X. Donard, *J. Anal. Atom. Spectrom.*, 2004, **19**, 815–822.
238. M. Montes-Bayón, K. DeNicola and J. A. Caruso, *J. Chromatogr. A*, 2003, **1000**, 457–476.
239. J. S. Becker, S. F. Boulyga, C. Pickhardt, J. Becker, S. Buddrus and M. Przybylski, *Anal. Bioanal. Chem.*, 2003, **375**, 561–566.
240. A. Bibak, A. Behrens, S. Stürup, L. Knudsen and V. Gundersen, *J. Agricul. Food Chem.*, 1998, **46**, 3139–3145.
241. A. Bibak, A. Behrens, S. Stürup, L. Knudsen and V. Gundersen, *J. Agricul. Food Chem.*, 1998, **46**, 3146–3149.
242. R. Melø, K. Gellein, L. Evje and T. Syversen, *Food Chem. Toxicol.*, 2008, **46**, 3339–3342.
243. I. Rodushkin, E. Engström, D. Sörlin and D. Baxter, *Sci. Total Environ.*, 2008, **392**, 290–304.
244. M. Krachler, T. Prohaska, G. Koellensperger, E. Rossipal and G. Stingeder, *Biol. Trace Element Res.*, 2000, **76**, 97–112.
245. L. Mancini, S. Caimi, S. Ciardullo, M. Zeiner, P. Bottoni, L. Tancioni, S. Cautadella and S. Caroli, *Microchem. J.*, 2005, **79**, 171–175.
246. C. Frazzoli, S. D'Ilio and B. Bocca, *Anal. Lett.*, 2007, **40**, 1992–2004.
247. G. Forte and B. Bocca, *Food Chem.*, 2007, **105**, 1591–1598.
248. G. Tyler, *Plant Soil*, 2005, **267**, 191–206.
249. I. Rodushkin, F. Ödman and P. K. Appelblad, *J. Food Composit. Anal.*, 1999, **12**, 243–257.
250. B. Wyrzykowska, K. Szymczyk, H. Ichichashi, J. Falandysz, B. Skwarzec and S. I. Yamasaki, *J. Agricul. Food Chem.*, 2001, **49**, 3425–3431.
251. M. D. M. Castiñeira Gómez, I. Feldmann, N. Jakubowski and J. T. Andersson, *J. Agricul. Food Chem.*, 2004, **52**, 2962–2974.
252. R. G. Smith, *J. Agricul. Food Chem.*, 2005, **53**, 4041–4045.
253. E. K. Shibuya, J. E. S. Sarkis, O. Negrini-Neto and J. P. H. B. Ometto, *J. Braz. Chem. Soc.*, 2007, **18**, 205–214.

254. B. Flem, V. Moen and A. Grimstvedt, *Appl. Spectrosc.*, 2005, **59**, 245–251.
255. C. Frazzoli, R. Cammarone and S. Caroli, *Food Add. Contamin.*, 2007, **24**, 546–552.
256. P. Evans, S. Elahi, K. Lee and B. Fairman, *J. Environ. Monitor*, 2003, **5**, 175–179.
257. C. F. Harrington, S. Elahi, A. Merson and P. Ponnampalavanar, *Anal. Chem.*, 2001, **73**, 4422–4427.
258. J. M. Marchante-Gayón, C. Sariego-Muñiz, J. Ignacio-Alonso and A. Sanz-Medel, Analysis of biological material by double focusing-inductively coupled plasma-mass spectrometry (DF-ICP-MS), in *Advances in Atomic Spectroscopy*, ed. J. Sneddon, Elsevier Science B.V., Amsterdam, The Netherlands, 2002, vol. 7, p. 117.
259. I. Rodushkin and F. Ödman, *J. Trace Elements Med. Biol.*, 2001, **14**, 241–247.
260. C. Sariego Muñiz, J. M. Marchante Gayón, J. Ignacio García Alonso and A. Sanz-Medel, *J. Anal. Atom. Spectrom.*, 1999, **14**, 1505–1510.
261. N. Jakubowski, C. Thomas, D. Klueppel and D. Stuewer, *Analusis*, 1998, **26**, M37–M43.
262. M. Wind and W. D. Lehmann, *J. Anal. Atom. Spectrom.*, 2004, **19**, 20–25.
263. J. S. Becker and N. Jakubowski, *Chem. Soc. Rev.*, 2009, **38**, 1969–1983.
264. J. L. Gómez-Ariza, T. García-Barrera, F. Lorenzo, V. Bernal, M. J. Villegas and V. Oliveira, *Anal. Chim. Acta*, 2004, **524**, 15–22.
265. A. Prange and D. Pröfrock, *J. Anal. Atom. Spectrom.*, 2008, **23**, 432–459.
266. J. Riondato, F. Vanhaecke, L. Moens and R. Dams, *J. Anal. Atom. Spectrom.*, 1997, **12**, 933–937.
267. A. J. Steuerwald, P. J. Parsons, J. G. Arnason, Z. Chen, C. M. Peterson and G. M. Buck Louis, *J. Anal. Atom. Spectrom.*, 2013, **28**, 821–830.
268. J. Wang, R. S. Houk, D. Dreessen and D. R. Wiederin, *J. Am. Chem. Soc.*, 1998, **120**, 5793–5799.
269. I. Rodushkin, F. Ödman and S. Branth, *Fresenius J. Anal. Chem.*, 1999, **364**, 338–346.
270. I. Rodushkin, O. F. Ödman, R. Olofsson and M. D. Axelsson, *J. Anal. Atom. Spectrom.*, 2000, **15**, 937–944.
271. I. Rodushkin, E. Engström, A. Stenberg and D. C. Baxter, *Anal. Bioanal. Chem.*, 2004, **380**, 247–257.
272. S. D'Ilio, N. Violante, M. Di Gregorio, O. Senofonte and F. Petrucci, *Anal. Chim. Acta*, 2006, **579**, 202–208.
273. A. Sarmiento-González, J. M. Marchante-Gayón, J. M. Tejerina-Lobo, J. Paz-Jiménez and A. Sanz-Medel, *Anal. Bioanal. Chem.*, 2005, **382**, 1001–1009.
274. A. Sarmiento-González, J. M. Marchante-Gayón, J. M. Tejerina-Lobo, J. Paz-Jiménez and A. Sanz- Medel, *Anal. Bioanal. Chem.*, 2008, **391**, 2583–2589.
275. A. Sarmiento-González, J. R. Encinar, J. M. Marchante-Gayón and A. Sanz-Medel, *Anal. Bioanal. Chem.*, 2009, **393**, 335–343.

276. S. J. Genuis, D. Birkholz, I. Rodushkin and S. Beesoon, *Arch. Environ. Contam. Toxicol.*, 2011, **61**, 344–357.
277. E. K. Langer, K. J. Johnson, M. M. Shafer, P. Gorski, J. Overdier, J. Musselman and J. A. Ross, *J. Expos. Sci. Environ. Epidemiol.*, 2011, **21**, 355–364.
278. Y. Nuevo Ordóñez, M. Montes-Bayón, E. Blanco-González, J. Paz-Jiménez, J. M. Tejerina-Lobo, J. M. Peña-López and A. Sanz-Medel, *J. Anal. Atom. Spectrom.*, 2009, **24**, 1037–1043.
279. Y. Nuevo-Ordoñez, M. Montes-Bayón, E. Blanco González and A. Sanz-Medel, *Metallomics*, 2011, **3**, 1297–1303.
280. M. Krachler, C. Heisel and J. P. Kretzer, *J. Anal. Atom. Spectrom.*, 2009, **24**, 605–610.
281. L. Yang, R. E. Sturgeon, D. Prince and S. Gabos, *J. Anal. Atom. Spectrom.*, 2002, **17**, 1300–1303.
282. A. M. Featherstone, A. T. Townsend, G. A. Jacobson and G. M. Peterson, *Anal. Chim. Acta*, 2004, **512**, 319–327.
283. A. Townsend, A. Featherstone, C. C. Chéry, F. Vanhaecke, J. Kirby, F. Krikowa, B. Maher, G. Jacobson and G. Peterson, *Clin. Chem.*, 2004, **50**, 1481–1482.
284. B. Bocca, A. Alimonti, F. Petrucci, N. Violante, G. Sancesario, G. Forte and O. Senofonte, *Spectrochim. Acta, Part B*, 2004, **59**, 559–566.
285. B. Bocca, A. Alimonti, O. Senofonte, A. Pino, N. Violante, F. Petrucci, G. Sancesario and G. Forte, *J Neurol Sci.*, 2006, **248**, 23–30.
286. M. L. Praamsma, J. G. Arnason and P. J. Parsons, *J. Anal. Atom. Spectrom.*, 2011, **26**, 1224–1232.
287. A. L. Holmes, S. S. Wise and J. P. Wise Sr, *Ind. J. Med. Res.*, 2008, **128**, 353–372.
288. D. A. Eastmond, J. T. MacGregor and R. S. Slesinski, *Crit. Rev. Toxicol.*, 2008, **38**, 173–190.
289. M. Costa and C. B. Klein, *Crit. Rev. Toxicol.*, 2006, **36**, 155–163.
290. K. L. Witt, M. D. Stout, R. A. Herbert, G. S. Travlos, G. E. Kissling, B. J. Collins and M. J. Hooth, *Toxicol. Pathol.*, 2013, **41**, 326–342.
291. F. Séby, M. Gagean, H. Garraud, A. Castetbon and O. F. X. Donard, *Anal. Bioanal. Chem.*, 2003, **377**, 685–694.
292. F. Vanhaecke, S. Saverwyns, G. D. Wannemacker, L. Moens and R. Dams, *Anal. Chim. Acta*, 2000, **419**, 55–64.
293. V. Van Lierde, C. C. Chéry, L. Moens and F. Vanhaecke, *Electrophoresis*, 2005, **26**, 1703–1711.
294. V. Van Lierde, C. C. Chéry, N. Roche, S. Monstrey, L. Moens and F. Vanhaecke, *Anal. Bioanal. Chem.*, 2006, **384**, 378–384.
295. M. S. Bloom, K. Kim, P. C. Kruger, P. J. Parsons, J. G. Arnason, A. J. Steuerwald and V. Y. Fujimoto, *J. Assist. Reprod. Genet.*, 2012, **29**, 1369–1379.
296. F. M. Kluxen, P. Diel, N. Höfer, E. Becker and G. H. Degen, *Arch. Toxicol.*, 2013, **87**, 633–643.

297. P. C. Kruger, M. S. Bloom, J. G. Arnason, C. D. Palmer, V. Y. Fujimoto and P. J. Parsons, *J. Anal. Atom. Spectrom.*, 2012, **27**, 1245–1253.
298. A. Alimonti, G. Forte, S. Spezia, A. Gatti, G. Mincione, A. Ronchi, P. Bavazzano, B. Bocca and C. Minoia, *Rapid Commun. Mass Spectrom.*, 2005, **19**, 3131–3138.
299. S. Spezia, B. Bocca, G. Forte, A. Gatti, G. Mincione, A. Ronchi, P. Bavazzano, A. Alimonti and C. Minoia, *Rapid Commun. Mass Spectrom.*, 2005, **19**, 1551–1556.
300. S. Caimi, F. Petrucci, A. Cristaudo and V. Bordignon, *Open Chem. Biomed. Methods J.*, 2009, **2**, 48–54.
301. F. Petrucci, N. Violante, O. Senofonte, M. De Gregorio, A. Alimonti, S. Caroli, G. Forte and A. Cristaudo, *Microchem. J.*, 2004, **76**, 131–140.
302. D. G. Filatova, I. F. Seregina, L. S. Foteeva, V. V. Pukhov, A. R. Timerbaev and M. A. Bolshov, *Anal. Bioanal. Chem.*, 2011, **400**, 709–714.
303. A. Clerico, C. Cappelli, G. Ragni, S. Caroli, M. A. De Ioris, A. Sordi, F. Petrucci, B. Bocca and A. Alimonti, *Annali dell'Istituto Superiore di Sanita*, 2006, **42**, 461–468.
304. B. Bocca, A. Lamazza, A. Pino, E. De Masi, M. Iacomino, D. Mattei, S. Rahimi, E. Fiori, A. Schillaci, A. Alimonti and G. Forte, *Rapid Commun. Mass Spectrom.*, 2007, **21**, 1776–1782.
305. J. Benemann, N. Lehmann, K. Bromen, A. Marr, M. Seiwert, C. Schulz and K. H. Jöckel, *Int. J. Hyg. Environ. Health*, 2005, **208**, 499–508.
306. D. Bland, R. J. Rona, D. Coggon, J. Anderson, N. Greenberg, L. Hull and S. Wessely, *Occupat. Environ. Med.*, 2007, **64**, 834–838.
307. K. Benkhedda, V. N. Epov and R. D. Evans, *Anal. Bioanal. Chem.*, 2005, **381**, 1596–1603.
308. J. S. Becker, M. Burow, M. V. Zoriy, C. Pickhardt, P. Ostapczuk and R. Hille, *Atom. Spectrosc.*, 2004, **25**, 197–202.
309. P. Schramel, *J. Chromatogr. B*, 2002, **778**, 275–278.
310. P. Roth, V. Höllriegl, W. B. Li, U. Oeh and P. Schramel, *Health Phys.*, 2005, **88**, 223–228.
311. J. G. Arnason, C. N. Pellegri and P. J. Parsons, *J. Anal. Atom. Spectrom.*, 2013, **28**, 1410–1419.
312. P. Froidevaux, F. Bochud and M. Haldimann, *Chemosphere*, 2010, **80**, 519–524.
313. M. Yamamoto, S. Oikawa, A. Sakaguchi, J. Tomita, M. Hoshi and K. N. Apsalikov, *Health Phys.*, 2008, **95**, 291–299.
314. K. Gellein, S. Lierhagen, P. S. Brevik, M. Teigen, P. Kaur, T. Singh, T. P. Flaten and T. Syversen, *Biol. Trace Element Res.*, 2008, **123**, 250–260.
315. I. Rodushkin and M. D. Axelsson, *Sci. Total Environ.*, 2000, **250**, 83–100.
316. I. Rodushkin and M. D. Axelsson, *Sci. Total Environ.*, 2003, **305**, 23–39.
317. S. Ouypornkochagorn and J. Feldmann, *Environ. Sci. Technol.*, 2010, **44**, 3972–3978.
318. S. Stürup, *J. Anal. Atom. Spectrom.*, 2002, **17**, 1–7.
319. M. P. Field, S. Shapses, M. Cifuentes and R. M. Sherrell, *J. Anal. Atom. Spectrom.*, 2003, **18**, 727–733.

320. K. A. Al-Saad, M. A. Amr and A. I. Helal, *Biol. Trace Element Res.*, 2011, **140**, 103–113.
321. G. S. Guandalini, L. Zhang, E. Fornero, J. A. Centeno, V. P. Mokashi, P. A. Ortiz, M. D. Stockelman, A. R. Osterburg and G. G. Chapman, *Chem. Res. Toxicol.*, 2011, **24**, 488–493.
322. N. Pallavicini, F. Ecke, E. Engström, D. C. Baxter and I. Rodushkin, *J. Anal. Atom. Spectrom.*, 2013, **28**, 1591–1599.
323. M. E. Church, R. Gwiazda, R. W. Risebrough, K. Sorenson, C. P. Chamberlain, S. Farry, W. Heinrich, B. A. Rideout and D. R. Smith, *Environ. Sci. Technol.*, 2006, **40**, 6143–6150.
324. K. De Wolf, L. Balcaen, E. Van De Walle, F. Cuyckens and F. Vanhaecke, *J. Anal. Atom. Spectrom.*, 2010, **25**, 419–425.
325. R. R. De La Flor St. Rèmy, M. Montes-Bayón and A. Sanz-Medel, *Anal. Bioanal. Chem.*, 2003, **377**, 299–305.
326. T. Lin, J.-C. Liu, L.-Y. Chang and C.-W. Shen, *Atherosclerosis*, 2010, **212**, 501–506.
327. J. Szpunar, *Analyst*, 2005, **130**, 442–465.
328. R. L. Ma, C. W. McLeod, K. Tomlinson and R. K. Poole, *Electrophoresis*, 2004, **25**, 2469–2477.
329. P. Marshall, O. Heudi, S. Bains, H. N. Freeman, F. Abou-Shakra and K. Reardon, *Analyst*, 2002, **127**, 459–461.
330. M. Wind, I. Feldmann, N. Jakubowski and W. D. Lehmann, *Electrophoresis*, 2003, **24**, 1276–1280.
331. M. R. Larsen, G. L. Sørensen, S. J. Fey, P. M. Larsen and P. Roepstorff, *Proteomics*, 2001, **1**, 223–238.
332. Y. Oda, T. Nagasu and B. T. Chait, *Nature Biotechnol.*, 2001, **19**, 379–382.
333. E. Salih, *Mass Spectrom. Rev.*, 2005, **24**, 828–846.
334. A. Venkatachalam, C. U. Koehler, I. Feldmann, P. Lampen, A. Manz, P. H. Roos and N. Jakubowski, *J. Anal. Atom. Spectrom.*, 2007, **22**, 1023–1032.
335. I. Feldmann, C. U. Koehler, P. H. Roos and N. Jakubowski, *J. Anal. Atom. Spectrom.*, 2006, **21**, 1006–1015.
336. J. S. Becker and M. Przybylski, *J. Anal. Atom. Spectrom.*, 2007, **22**, 761–765.
337. J. S. Becker, M. Zoriy, J. S. Becker, C. Pickhardt and M. Przybylski, *J. Anal. Atom. Spectrom.*, 2004, **19**, 149–152.
338. R. Krüger, D. Kübler, R. Pallissé, A. Burkovski and W. D. Lehmann, *Anal. Chem.*, 2006, **78**, 1987–1994.
339. M. Wind, M. Edler, N. Jakubowski, M. Linscheid, H. Wesch and W. D. Lehmann, *Anal. Chem.*, 2001, **73**, 29–35.
340. M. Wind, H. Wesch and W. D. Lehmann, *Anal. Chem.*, 2001, **73**, 3006–3010.
341. R. Krüger, F. Wolschin, W. Weckwerth, J. Bettmer and W. D. Lehmann, *Biochem. Biophys. Res. Commun.*, 2007, **355**, 89–96.
342. N. Zinn, R. Krüger, P. Leonhard and J. Bettmer, *Anal. Bioanal. Chem.*, 2008, **391**, 537–543.

343. T. Konz, M. Montes-Bayón, J. Bettmer and A. Sanz-Medel, *J. Anal. Atom. Spectrom.*, 2011, **26**, 334–340.
344. W. C. Hawkes and Z. Alkan, *Biol. Trace Element Res.*, 2010, **134**, 235–251.
345. A. Polatajko, N. Jakubowski and J. Szpunar, *J. Anal. Atom. Spectrom.*, 2006, **21**, 639–654.
346. M. Siwek, B. Galunsky and B. Niemeyer, *Anal. Bioanal. Chem.*, 2005, **381**, 737–741.
347. I. Feldmann, N. Jakubowski, D. Stuewer and C. Thomas, *J. Anal. Atom. Spectrom.*, 2000, **15**, 371–376.
348. Z. Pedrero, Y. Madrid, C. Camara, E. Schram, J. B. Luten, I. Feldmann, L. Waentig, H. Hayen and N. Jakubowski, *J. Anal. Atom. Spectrom.*, 2009, **24**, 775–784.
349. Z. Pedrero, S. Murillo, C. Camara, E. Schram, J. B. Luten, I. Feldmann, N. Jakubowski and Y. Madrid, *J. Anal. Atom. Spectrom.*, 2011, **26**, 116–125.
350. I. Rodushkin, P. Nordlund, E. Engström and D. C. Baxter, *J. Anal. Atom. Spectrom.*, 2005, **20**, 1250–1255.
351. C. Wolf, D. Schaumlöffel, A. N. Richarz, A. Prange and P. Brätter, *Analyst*, 2003, **128**, 576–580.
352. A. Prange and D. Schaumlöffel, *Anal. Bioanal. Chem.*, 2002, **373**, 441–453.
353. A. Prange, D. Schaumlöffel, P. Brätter, A. N. Richarz and C. Wolf, *Anal. Bioanal. Chem.*, 2001, **371**, 764–774.
354. D. Schaumlöffel, A. Prange, G. Marx, K. G. Heumann and P. Brätter, *Anal. Bioanal. Chem.*, 2002, **372**, 155–163.
355. J. E. Sonke and V. J. M. Salters, *J. Chromatogr. A*, 2007, **1159**, 63–74.
356. V. Van Lierde, C. C. Chéry, K. Strijckmans, M. Galleni, B. Devreese, J. Van Beeumen, L. Moens and F. Vanhaecke, *J. Anal. Atom. Spectrom.*, 2004, **19**, 888–893.
357. M. E. Del Castillo Busto, M. Montes-Bayón, J. Bettmer and A. Sanz-Medel, *Analyst*, 2008, **133**, 379–384.
358. K. Gellein, P. M. Roos, L. Evje, O. Vesterberg, T. P. Flaten, M. Nordberg and T. Syversen, *Brain Res.*, 2007, **1174**, 136–142.
359. M. H. Nagaoka, H. Akiyama and T. Maitani, *Analyst*, 2004, **129**, 51–54.
360. M. H. Nagaoka and T. Maitani, *J. Inorg. Biochem.*, 2005, **99**, 1887–1894.
361. S. Farinati, G. DalCorso, E. Bona, M. Corbella, S. Lampis, D. Cecconi, R. Polati, G. Berta, G. Vallini and A. Furini, *Proteomics*, 2009, **9**, 4837–4850.
362. A. Polatajko, M. Azzolini, I. Feldmann, T. Stuezel and N. Jakubowski, *J. Anal. Atom. Spectrom.*, 2007, **22**, 878–887.
363. A. Polatajko, I. Feldmann, H. Hayen and N. Jakubowski, *Metallomics*, 2011, **3**, 1001–1008.
364. S. Hann, G. Koellensperger, C. Obinger, P. G. Furtmüller and G. Stingeder, *J. Anal. Atom. Spectrom.*, 2004, **19**, 74–79.
365. M. Garijo Añorbe, J. Messerschmidt, I. Feldmann and N. Jakubowski, *J. Anal. Atom. Spectrom.*, 2007, **22**, 917–924.

366. J. S. Becker, M. Zoriy, U. Krause-Buchholz, J. S. Becker, C. Pickhardt, M. Przybylski, W. Pompe and G. Rodel, *J. Anal. Atom. Spectrom.*, 2004, **19**, 1236–1243.
367. L. G. Hernandez, H. van Steeg, M. Luijten and J. van Benthem, *Mutat. Res.-Rev. Mutat. Res.*, 2009, **682**, 94–109.
368. M. M. Rahman, J. C. Ng and R. Naidu, *Environ. Geochem. Health*, 2009, **31**, 189–200.
369. F. J. Zhao, S. P. McGrath and A. A. Meharg, *Annu. Rev. Plant Biol.*, 2010, **61**, 535–559.
370. T. Prohaska, C. Latkoczy, G. Stingeder and W. W. Wenzel, Determination of arsenic in soil extracts using high resolution inductively coupled plasma mass spectrometry, in *Plasma Source Mass Spectrometry Developments and Applications*, ed. G. a. T. Holland, S. D., Spec. Publ. - R. Soc. Chem., 1997, vol. 202, pp. 291–297.
371. J. Zheng and H. Hintelmann, *J. Anal. Atom. Spectrom.*, 2004, **19**, 191–195.
372. W. J. Liu, B. A. Wood, A. Raab, S. P. McGrath, F. J. Zhao and J. Feldmann, *Plant Physiol.*, 2010, **152**, 2211–2221.
373. E. Moreno-Gordaliza, C. Giesen, A. Lázaro, D. Esteban-Fernández, B. Humanes, B. Cañas, U. Panne, A. Tejedor, N. Jakubowski and M. M. Gómez-Gómez, *Anal. Chem.*, 2011, **83**, 7933–7940.
374. E. Rudolph, S. Hann, G. Stingeder and C. Reiter, *Anal. Bioanal. Chem.*, 2005, **382**, 1500–1506.
375. L. Waentig, N. Jakubowski, H. Hayen and P. H. Roos, *J. Anal. Atom. Spectrom.*, 2011, **26**, 9.
376. C. Giesen, L. Waentig, T. Mairinger, D. Drescher, J. Kneipp, P. H. Roos, U. Panne and N. Jakubowski, *J. Anal. Atom. Spectrom.*, 2011, **26**, 2160–2165.
377. N. Jakubowski, L. Waentig, H. Hayen, A. Venkatachalam, A. von Bohlen, P. H. Roos and A. Manz, *J. Anal. Atom. Spectrom.*, 2008, **23**, 1497–1507.
378. S. C. Bendall, E. F. Simonds, P. Qiu, E.-a. D. Amir, P. O. Krutzik, R. Finck, R. V. Bruggner, R. Melamed, A. Trejo, O. I. Ornatsky, R. S. Balderas, S. K. Plevritis, K. Sachs, D. Pe'er, S. D. Tanner and G. P. Nolan, *Science*, 2011, **332**, 687–696.
379. M. Wang, W.-Y. Feng, Y.-L. Zhao and Z.-F. Chai, *Mass Spectrom. Rev.*, 2010, **29**, 326–348.
380. R. Ahrends, S. Pieper, A. Kühn, H. Weisshoff, M. Hamester, T. Lindemann, C. Scheler, K. Lehmann, K. Taubner and M. W. Linscheid, *Molec. Cell. Proteom.*, 2007, **6**, 1907–1916.
381. R. Ahrends, S. Pieper, B. Neumann, C. Scheler and M. W. Linscheid, *Anal. Chem.*, 2009, **81**, 2176–2184.
382. J. Bettmer, N. Jakubowski and A. Prange, *Anal. Bioanal. Chem.*, 2006, **386**, 7–11.
383. L. Waentig, N. Jakubowski, S. Hardt, C. Scheler, P. H. Roos and M. W. Linscheid, *J. Anal. Atom. Spectrom.*, 2012, **27**, 1311–1320.

384. A. H. El-Khatib, D. Esteban-Fernandez and M. W. Linscheid, *Anal. Bioanal. Chem.*, 2012, **403**, 2255–2267.
385. C. Giesen, L. Waentig, U. Panne and N. Jakubowski, *Spectrochim. Acta Part B-Atom. Spectrosc.*, 2012, **76**, 27–39.
386. P. H. Roos, A. Venkatachalam, A. Manz, L. Waentig, C. U. Koehler and N. Jakubowski, *Anal. Bioanal. Chem.*, 2008, **392**, 1135–1147.
387. L. Waentig, N. Jakubowski and P. H. Roos, *J. Anal. Atom. Spectrom.*, 2011, **26**, 310–319.
388. P. H. Roos and A. Mahnke, *Biochem. Pharmacol.*, 1996, **52**, 73–84.
389. D. Kretschy, M. Groger, D. Zinkl, P. Petzelbauer, G. Koellensperger and S. Hann, *Int. J. Mass Spectrom.*, 2011, **307**, 105–111.
390. D. Esteban-Fernandez, F. S. Bierkandt and M. W. Linscheid, *J. Anal. Atom. Spectrom.*, 2012, **27**, 1701–1708.
391. L. Waentig, S. Techritz, N. Jakubowski and P. H. Roos, *Analyst*, 2013, **138**, 6309–6315.
392. C. Siethoff, I. Feldmann, N. Jakubowski and M. Linscheid, *J. Mass Spectrom.*, 1999, **34**, 421–426.
393. O. Leclerc, P. O. Fraisse, G. Labarraque, C. Oster, J. P. Pichaut, M. Baume, S. Jarraud, P. Fisicaro and S. Vaslin-Reimann, *Anal. Biochem.*, 2013, **435**, 153–158.
394. K. Brückner, K. Schwarz, S. Beck and M. W. Linscheid, *Anal. Chem.*, 2014, **86**, 585–591.
395. S. Hann, A. Zenker, M. Galanski, T. L. Bereuter, G. Stingeder and B. K. Keppler, *Fresenius J. Anal. Chem.*, 2001, **370**, 581–586.
396. W. Brüchert, R. Krüger, A. Tholey, M. Montes-Bayón and J. Bettmer, *Electrophoresis*, 2008, **29**, 1451–1459.
397. A. Zayed, G. D. D. Jones, H. J. Reid, T. Shoeib, S. E. Taylor, A. L. Thomas, J. P. Wood and B. L. Sharp, *Metallomics*, 2011, **3**, 991–1000.
398. L. L. Fernández, M. Montes-Bayón, E. B. González, L. M. Sierra, A. Sanz-Medel and J. Bettmer, *J. Anal. Atom. Spectrom.*, 2011, **26**, 195–200.
399. H. Haraguchi, *J. Anal. Atom. Spectrom.*, 2004, **19**, 5–14.
400. H. Haraguchi, A. Ishii, T. Hasegawa, H. Matsuura and T. Umemura, *Pure Appl. Chem.*, 2008, **80**, 2595–2608.
401. F. Li, D. W. Armstrong and R. S. Houk, *Anal. Chem.*, 2005, **77**, 1407–1413.
402. D. Drescher, C. Giesen, H. Traub, U. Panne, J. Kneipp and N. Jakubowski, *Anal. Chem.*, 2012, **84**, 9684–9688.
403. K. Shigeta, G. Koellensperger, E. Rampler, H. Traub, L. Rottmann, U. Panne, A. Okino and N. Jakubowski, *J. Anal. Atom. Spectrom.*, 2013, **28**, 637–645.
404. M. Rehkämper, M. Schönbächler and C. H. Stirling, *Geostand. Newslett*, 2001, **25**, 23–40.
405. L. Yang, *Mass Spectrom. Rev.*, 2009, **28**, 990–1011.
406. L. Balcaen, L. Moens and F. Vanhaecke, *Spectrochim. Acta Part B-Atom. Spectrosc*, 2010, **65**, 769–786.
407. J. R. de Laeter, *Mass Spectrom. Rev.*, 1998, **17**, 97–125.

408. A. N. Halliday, D.-C. Lee, J. N. Christensen, M. Rehkämper, W. Yi, X. Luo, C. M. Hall, C. J. Ballentine, T. Pettke and C. Stirling, *Geochim. Cosmochim. Acta*, 1998, **62**, 919–940.
409. J. D. Woodhead, *Geostand. Geoanal. Res.*, 2008, **32**, 495–507.
410. J. D. Woodhead, *Geostand. Geoanal. Res.*, 2010, **34**, 395–406.
411. M. Wiedenbeck, R. Bugoi, M. J. M. Duke, T. Dunai, J. Enzweiler, M. Horan, K. P. Jochum, K. Linge, J. Košler, S. Merchel, L. F. G. Morales, L. Nasdala, R. Stalder, P. Sylvester, U. Weis and A. Zoubir, *Geostand. Geoanal. Res.*, 2012, **36**, 337–398.
412. S. B. Jacobsen, *Annu. Rev. Earth Planet. Sci.*, 2005, **33**, 531–570.
413. D. C. Lee and A. N. Halliday, *Int. J. Mass Spectrom. Ion Process.*, 1995, **146–147**, 35–46.
414. D. C. Lee and A. N. Halliday, *Nature*, 1995, **378**, 771–774.
415. D. C. Lee, A. N. Halliday, G. A. Snyder and L. A. Taylor, *Science*, 1997, **278**, 1098–1103.
416. J. Blichert-Toft, C. Chauvel and F. Albarède, *Contrib. Mineral. Petrol.*, 1997, **127**, 248–260.
417. M. F. Thirlwall and A. J. Walder, *Chem. Geol.*, 1995, **122**, 241–247.
418. X. Luo, M. Rehkämper, D. C. Lee and A. N. Halliday, *Int. J. Mass Spectrom. Ion Process.*, 1997, **171**, 105–117.
419. L. F. Robinson, G. M. Henderson and N. C. Slowey, *Earth Planet. Sci. Lett.*, 2002, **196**, 175–187.
420. T. Waight, J. Baker and B. Willigers, *Chem. Geol.*, 2002, **186**, 99–116.
421. O. Nebel, K. Mezger, E. E. Scherer and C. Münker, *Int. J. Mass Spectrom.*, 2005, **246**, 10–18.
422. B. J. A. Willigers, K. Mezger and J. A. Baker, *Chem. Geol.*, 2004, **213**, 339–358.
423. B. J. A. Willigers, J. A. Baker, E. J. Krogstad and D. W. Peate, *Geochim. Cosmochim. Acta*, 2002, **66**, 1051–1066.
424. J. N. Christensen, A. N. Halliday, L. V. Godfrey, J. R. Hein and D. K. Rea, *Science*, 1997, **277**, 913–918.
425. S. Weyer, A. D. Anbar, A. Gerdes, G. W. Gordon, T. J. Algeo and E. A. Boyle, *Geochim. Cosmochim. Acta*, 2008, **72**, 345–359.
426. M. B. Andersen, Y. Erel and B. Bourdon, *Geochim. Cosmochim. Acta*, 2009, **73**, 4124–4141.
427. J. Fietzke and A. Eisenhauer, *Geochem. Geophys. Geosyst.*, 2006, 7, 6.
428. G. F. de Souza, B. C. Reynolds, M. Kiczka and B. Bourdon, *Geochim. Cosmochim. Acta*, 2010, **74**, 2596–2614.
429. M. Bizzarro, C. Paton, K. Larsen, M. Schiller, A. Trinquier and D. Ulfbeck, *J. Anal. Atom. Spectrom.*, 2011, **26**, 565–577.
430. H. Chen, P. S. Savage, F.-Z. Teng, R. T. Helz and F. Moynier, *Earth Planet. Sci. Lett.*, 2013, **369–370**, 34–42.
431. G. L. Foster, *Earth Planet. Sci. Lett.*, 2008, **271**, 254–266.
432. O. Rouxel, J. Ludden and Y. Fouquet, *Chem. Geol.*, 2003, **200**, 25–40.
433. Y. Nakagawa, M. L. Firdaus, K. Norisuye, Y. Sohrin, K. Irisawa and T. Hirata, *Anal. Chem.*, 2008, **80**, 9213–9219.

434. A. R. Voegelin, T. F. Nägler, E. Samankassou and I. M. Villa, *Chem. Geol.*, 2009, **265**, 488–498.
435. M. Rehkämper and A. N. Halliday, *Talanta*, 1997, **44**, 663–672.
436. M. Rehkämper, A. N. Halliday, D. Barfod, J. G. Fitton and J. B. Dawson, *Science*, 1997, **278**, 1595–1598.
437. W. Yi, A. N. Halliday, J. C. Alt, D. C. Lee, M. Rehkämper, M. O. Garcia and Y. Su, *J. Geophys. Res. B: Solid Earth*, 2000, **105**, 18927–18948.
438. M. Resano, K. S. McIntosh and F. Vanhaecke, *J. Anal. Atom. Spectrom.*, 2012, **27**, 165–173.
439. J. S. Becker, *Spectrochim. Acta Part B-Atom. Spectrosc.*, 2003, **58**, 1757–1784.
440. D. Larivière, V. F. Taylor, R. D. Evans and R. J. Cornett, *Spectrochim. Acta Part B-Atom. Spectrosc.*, 2006, **61**, 877–904.
441. I. Günther-Leopold, N. Kivel, J. Kobler Waldis and B. Wernli, *Anal. Bioanal. Chem.*, 2008, **390**, 503–510.
442. R. N. Taylor, T. Warneke, J. A. Milton, I. W. Croudace, P. E. Warwick and R. W. Nesbitt, *J. Anal. Atom. Spectrom.*, 2003, **18**, 480–484.
443. S. F. Boulyga and T. Prohaska, *Anal. Bioanal. Chem.*, 2008, **390**, 531–539.
444. S. Kappel, S. F. Boulyga, L. Dorta, D. Guenther, B. Hattendorf, D. Koffler, G. Laaha, F. Leisch and T. Prohaska, *Anal. Bioanal. Chem.*, 2013, **405**, 2943–2955.
445. G. Sharabi, B. Lazar, Y. Kolodny, N. Teplyakov and L. Halicz, *Int. J. Mass Spectrom.*, 2010, **294**, 112–115.
446. N. Kivel, D. Schumann and I. Guenther-Leopold, *Anal. Bioanal. Chem.*, 2013, **405**, 2965–2972.
447. J. Chmeleff, F. von Blanckenburg, K. Kossert and D. Jakob, *Nucl. Instrum. Methods Phys. Res., Sect. B: Beam Interact. Mater. Atoms*, 2010, **268**, 192–199.
448. M. Komarek, V. Ettler, V. Chrastny and M. Mihaljevic, *Environ. Int.*, 2008, **34**, 562–577.
449. D. J. Weiss, B. Kober, A. Dolgopolova, K. Gallagher, B. Spiro, G. Le Roux, T. F. D. Mason, M. Kylander and B. J. Coles, *Int. J. Mass Spectrom.*, 2004, **232**, 205–215.
450. W. Shotyk, M. E. Goodsite, F. Roos-Barraclough, N. Givelet, G. Le Roux, D. Weiss, A. K. Cheburkin, K. Knudsen, J. Heinemeier, W. O. van Der Knaap, S. A. Norton and C. Lohse, *Geochim. Cosmochim. Acta*, 2005, **69**, 1–17.
451. O. W. Purvis, P. J. Chimonides, G. C. Jones, I. N. Mikhailova, B. Spiro, D. J. Weiss and B. J. Williamson, *Proc. Royal Soc. B: Biol. Sci.*, 2004, **271**, 221–226.
452. A. Cocherie and M. Robert, *Chem. Geol.*, 2007, **243**, 90–104.
453. P. Vallelonga, P. Gabrielli, E. Balliana, A. Wegner, B. Delmonte, C. Turetta, G. Burton, F. Vanhaecke, K. J. R. Rosman, S. Hong, C. F. Boutron, P. Cescon and C. Barbante, *Quat. Sci. Rev.*, 2010, **29**, 247–255.
454. C. Cloquet, J. Carignan, G. Libourel, T. Sterckeman and E. Perdrix, *Environ. Sci. Technol.*, 2006, **40**, 2525–2530.
455. A. E. Shiel, D. Weis and K. J. Orians, *Sci. Total Environ.*, 2010, **408**, 2357–2368.

456. C. Cloquet, J. Carignan and G. Libourel, *Environ. Sci. Technol.*, 2006, **40**, 6594–6600.
457. Y. Sivry, J. Riotte, J. E. Sonke, S. Audry, J. Schäfer, J. Viers, G. Blanc, R. Freydier and B. Dupré, *Chem. Geol.*, 2008, **255**, 295–304.
458. D. Jouvin, P. Louvat, F. Juillot, C. N. Maréchal Iii and M. F. Benedetti, *Environ. Sci. Technol.*, 2009, **43**, 5747–5754.
459. D. J. Weiss, T. F. D. Mason, F. J. Zhao, G. J. D. Kirk, B. J. Coles and M. S. A. Horstwood, *New Phytol.*, 2005, **165**, 703–710.
460. T. Arnold, M. Schönbächler, M. Rehkämper, S. Dong, F. J. Zhao, G. J. D. Kirk, B. J. Coles and D. J. Weiss, *Anal. Bioanal. Chem.*, 2010, **398**, 3115–3125.
461. A. D. Schmitt, N. Vigier, D. Lemarchand, R. Millot, P. Stille and F. Chabaux, *Compt. Rendus - Geosc.*, 2012, **344**, 704–722.
462. L. Halicz, I. Segal, N. Fruchter, M. Stein and B. Lazar, *Earth Planet. Sci. Lett.*, 2008, **272**, 406–411.
463. B. L. Beard, C. M. Johnson, J. L. Skulan, K. H. Nealson, L. Cox and H. Sun, *Chem. Geol.*, 2003, **195**, 87–117.
464. G. L. Arnold, S. Weyer and A. D. Anbar, *Anal. Chem.*, 2004, **76**, 322–327.
465. B. J. Majestic, A. D. Anbar and P. Herckes, *Sci. Total Environ.*, 2009, **407**, 5104–5109.
466. D. Malinovsky and F. Vanhaecke, *Anal. Bioanal. Chem.*, 2011, **400**, 1619–1624.
467. B. A. Bergquist and J. D. Blum, *Science*, 2007, **318**, 417–420.
468. N. Gantner, H. Hintelmann, W. Zheng and D. C. Muir, *Environ. Sci. Technol.*, 2009, **43**, 9148–9154.
469. R. Yin, X. Feng and B. Meng, *Environ. Sci. Technol.*, 2013, **47**, 2238–2245.
470. D. Foucher, N. Ogrinc and H. Hintelmann, *Environ. Sci. Technol.*, 2009, **43**, 33–39.
471. V. N. Epov, P. Rodriguez-Gonzalez, J. E. Sonke, E. Tessier, D. Amouroux, L. M. Bourgoin and O. F. X. Donard, *Anal. Chem.*, 2008, **80**, 3530–3538.
472. L. Fischer, M. Brunner, T. Prohaska and S. Hann, J. Anal. Atom. Spectrom., 2012.
473. P. S. Savage, R. B. Georg, H. M. Williams and A. N. Halliday, *Geochim. Cosmochim. Acta*, 2013, **109**, 384–399.
474. R. B. Georg, B. C. Reynolds, M. Frank and A. N. Halliday, *Chem. Geol.*, 2006, **235**, 95–104.
475. R. B. Georg, B. C. Reynolds, M. Frank and A. N. Halliday, *Earth Planet. Sci. Lett.*, 2006, **249**, 290–306.
476. S. Opfergelt, D. Cardinal, C. Henriet, X. Draye, L. André and B. Delvaux, *Plant Soil*, 2006, **285**, 333–345.
477. S. Opfergelt, D. Cardinal, C. Henriet, L. André and B. Delvaux, *J. Geochem. Explor.*, 2006, **88**, 224–227.
478. E. Engström, I. Rodushkin, B. Öhlander, J. Ingri and D. C. Baxter, *Chem. Geol.*, 2008, **257**, 247–256.
479. S. Zink, R. Schoenberg and M. Staubwasser, *Geochim. Cosmochim. Acta*, 2010, **74**, 5729–5745.
480. S. M. Aciego, B. Bourdon, M. Lupker and J. Rickli, *Chem. Geol.*, 2009, **266**, 203–213.

481. J. Aggarwal, J. Habicht-Mauche and C. Juarez, *Appl. Geochem.*, 2008, **23**, 2658–2666.
482. R. A. Bentley, *J. Archaeol. Method Theory*, 2006, **13**, 135–187.
483. J. Montgomery, *Annal. Human Biol.*, 2010, **37**, 325–346.
484. S. R. Copeland, M. Sponheimer, P. J. Le Roux, V. Grimes, J. A. Lee-Thorp, D. J. De Ruiter and M. P. Richards, *Rapid Commun. Mass Spectrom.*, 2008, **22**, 3187–3194.
485. J. Montgomery, J. A. Evans and M. S. A. Horstwood, *Environ. Archaeol.*, 2010, **15**, 32–42.
486. M. S. A. Horstwood, J. A. Evans and J. Montgomery, *Geochim. Cosmochim. Acta*, 2008, **72**, 5659–5674.
487. P. Z. Vroon, B. Van Der Wagt, J. M. Koornneef and G. R. Davies, *Anal. Bioanal. Chem.*, 2008, **390**, 465–476.
488. P. Degryse, D. De Muynck, S. Delporte, S. Boyen, L. Jadoul, J. De Winne, T. Ivaneanu and F. Vanhaecke, *Anal. Methods*, 2012, **4**, 2674–2679.
489. A. F. Maurer, S. J. G. Galer, C. Knipper, L. Beierlein, E. V. Nunn, D. Peters, T. Tütken, K. W. Alt and B. R. Schöne, *Sci. Total Environ.*, 2012, **433**, 216–229.
490. F. G. T. Radloff, L. Mucina, W. J. Bond and P. J. le Roux, *Oecologia*, 2010, **164**, 567–578.
491. S. Voerkelius, G. D. Lorenz, S. Rummel, C. R. Quetel, G. Heiss, M. Baxter, C. Brach-Papa, P. Deters-Itzelsberger, S. Hoelzl, J. Hoogewerff, E. Ponzevera, M. Van Bocxstaele and H. Ueckermann, *Food Chem.*, 2010, **118**, 933–940.
492. Z. A. Stos-Gale and N. H. Gale, *Archaeol. Anthropol. Sci.*, 2009, **1**, 195–213.
493. M. Ponting, J. A. Evans and V. Pashley, *Archaeometry*, 2003, **45**, 591–597.
494. J. Baker, S. Stos and T. Waight, *Archaeometry*, 2006, **48**, 45–56.
495. A. M. Desaulty, P. Telouk, E. Albalat and F. Albarède, *Proc. Natl. Acad. Sci. USA*, 2011, **108**, 9002–9007.
496. E. Balliana, M. Aramendia, M. Resano, C. Barbante and F. Vanhaecke, *Anal. Bioanal. Chem.*, 2013, **405**, 2973–2986.
497. G. Fortunato, A. Ritter and D. Fabian, *Analyst*, 2005, **130**, 898–906.
498. A. Zeichner, S. Ehrlich, E. Shoshani and L. Halicz, *Foren. Sci. Int.*, 2006, **158**, 52–64.
499. K.-E. Sjåstad, S. L. Simonsen and T. Andersen, *J. Anal. Atom. Spectrom.*, 2011, **26**, 325–333.
500. N. W. Bower, S. A. McCants, J. M. Custodio, M. E. Ketterer, S. R. Getty and J. M. Hoffman, *Sci. Total Environ.*, 2007, **372**, 463–473.
501. D. Muynck, C. Cloquet, E. Smits, F. Wolff, G. Quitté, L. Moens and F. Vanhaecke, *Anal. Bioanal. Chem.*, 2008, **390**, 477–486.
502. G. H. Fontaine, K. Hametner, A. Peretti and D. Günther, *Anal. Bioanal. Chem.*, 2010, **398**, 2915–2928.
503. D. Brems, M. Ganio, K. Latruwe, L. Balcaen, M. Carremans, D. Gimeno, A. Silvestri, F. Vanhaecke, P. Muchez and P. Degryse, *Archaeometry*, 2013, **55**, 449–464.
504. P. Degryse and J. Schneider, *J. Archaeol. Sci.*, 2008, **35**, 1993–2000.

505. V. Devulder, P. Degryse and F. Vanhaecke, *Anal.Anal. Chem.*, 2013, **85**, 12077–12084.
506. L. Lobo, P. Degryse, A. Shortland and F. Vanhaecke, *J. Anal. Atom. Spectrom.*, 2013, **28**, 1213–1219.
507. R. Clough, P. Evans, T. Catterick and E. H. Evans, *Anal. Chem.*, 2006, **78**, 6126–6132.
508. R. Santamaria-Fernandez, R. Hearn and J. C. Wolff, *Sci. Justice*, 2009, **49**, 102–106.
509. R. Santamaria-Fernandez, J. Giner Martínez-Sierra, J. M. Marchante-Gayón, J. I. García-Alonso and R. Hearn, *Anal. Bioanal. Chem.*, 2009, **394**, 225–233.
510. I. Rodushkin, E. Engstrom and D. C. Baxter, *Anal. Bioanal. Chem.*, 2013, **405**, 2785–2797.
511. S. Stürup, *Anal. Bioanal. Chem.*, 2004, **378**, 273–282.
512. C. Ingle, N. Langford, L. Harvey, J. R. Dainty, C. Armah, S. Fairweather-Tait, B. Sharp, M. Rose, H. Crews and J. Lewis, *J. Anal. Atom. Spectrom.*, 2002, **17**, 1502–1505.
513. F. Larner, B. Sampson, M. Rehkaemper, D. J. Weiss, J. R. Dainty, S. O'Riordan, T. Panetta and P. G. Bain, *Metallomics*, 2013, **5**, 125–132.
514. D. S. Urgast, S. Hill, I.-S. Kwun, J. H. Beattie, H. Goenaga-Infante and J. Feldmann, *Metallomics*, 2012, **4**, 1057–1063.
515. D. S. Urgast and J. Feldmann, *J. Anal. Atom. Spectrom.*, 2013, **28**, 1367–1371.
516. T. Walczyk and F. von Blanckenburg, *Science*, 2002, **295**, 2065–2066.
517. T. Walczyk and F. Von Blanckenburg, *Int. J. Mass Spectrom.*, 2005, **242**, 117–134.
518. P. A. Krayenbuehl, T. Walczyk, R. Schoenberg, F. Von Blanckenburg and G. Schulthess, *Blood*, 2005, **105**, 3812–3816.
519. K. Hotz, P. A. Krayenbuehl and T. Walczyk, *J. Biol. Inorg. Chem.*, 2012, **17**, 301–309.
520. A. Stenberg, D. Malinovsky, B. Oehlander, H. Andren, W. Forsling, L.-M. Engstroem, A. Wahlin, E. Engstroem, I. Rodushkin and D. C. Baxter, *J. Trace Elements Med. Biol.*, 2005, **19**, 55–60.
521. T. Ohno, A. Shinohara, M. Chiba and T. Hirata, *Anal. Sci.*, 2005, **21**, 425–428.
522. F. Albarède, P. Telouk, A. Lamboux, K. Jaouen and V. Balter, *Metallomics*, 2011, **3**, 926–933.
523. K. Jaouen, M. Gibert, A. Lamboux, P. Telouk, F. Fourel, F. Albarede, A. N. Alekseev, E. Crubezy and V. Balter, *Metallomics*, 2013, **5**, 1016–1024.
524. V. Balter, A. Zazzo, A. P. Moloney, F. Moynier, O. Schmidt, F. J. Monahan and F. Albarède, *Rapid Commun. Mass Spectrom.*, 2010, **24**, 605–612.
525. F. Moynier, T. Fujii, A. S. Shaw and M. Le Borgne, *Metallomics*, 2013, **5**, 693–699.
526. L. Van Heghe, E. Engstrom, I. Rodushkin, C. Cloquet and F. Vanhaecke, *J. Anal. Atom. Spectrom.*, 2012, **27**, 1327–1334.

527. M. Costas-Rodriguez, L. Van Heghe and F. Vanhaecke, *Metallomics*, 2014, **6**, 139–146.
528. L. Van Heghe, O. Deltombe, J. Delanghe, H. Depypere and F. Vanhaecke, *J. Anal. Atom. Spectrom.*, 2014, **29**, 478–482.
529. L. M. Reynard, G. M. Henderson and R. E. M. Hedges, *Geochim. Cosmochim. Acta*, 2010, **74**, 3735–3750.
530. N.-C. Chu, G. M. Henderson, N. S. Belshaw and R. E. M. Hedges, *Appl. Geochem.*, 2006, **21**, 1656–1667.
531. T. Hirata, M. Tanoshima, A. Suga, Y. K. Tanaka, Y. Nagata, A. Shinohara and M. Chiba, *Anal. Sci.*, 2008, **24**, 1501–1507.
532. J. L. L. Morgan, G. W. Gordon, R. C. Arrua, J. L. Skulan, A. D. Anbar and T. D. Bullen, *Anal. Chem.*, 2011, **83**, 6956–6962.
533. B. Gulson, *Sci. Total Environ.*, 2008, **400**, 75–92.
534. M. Takagi, J. Yoshinaga, A. Tanaka and H. Seyama, *Anal. Sci.*, 2011, **27**, 29–35.
535. S. N. Chillrud, N. G. Hemming, J. M. Ross, S. Wallace and N. LoIacono, *Appl. Geochem.*, 2005, **20**, 807–813.
536. L. S. Sherman, J. D. Blum, A. Franzblau and N. Basu, *Environ. Sci. Technol.*, 2013, **47**, 3403–3409.
537. J. Woodhead, S. Swearer, J. Hergt and R. Maas, *J. Anal. Atom. Spectrom.*, 2005, **20**, 22–27.
538. A. Zitek, M. Sturm, H. Waidbacher and T. Prohaska, *Fish. Manag. Ecol.*, 2010, **17**, 435–445.
539. A. Zitek, J. Irrgeher, M. Kletzl, T. Weismann and T. Prohaska, *Fish. Manag. Ecol.*, 2013, **20**, 354–361.
540. J. Irrgeher, A. Zitek, M. Cervicek and T. Prohaska, *J. Anal. Atom. Spectrom.*, 2014, **29**, 193–200.
541. G. Huelga-Suarez, M. Moldovan, A. Garcia-Valiente, E. Garcia-Vazquez and J. I. G. Alonso, *Anal. Chem.*, 2012, **84**, 127–133.
542. S. Kelly, K. Heaton and J. Hoogewerff, *Trends Food Sci. Technol.*, 2005, **16**, 555–567.
543. G. Fortunato, K. Mumic, S. Wunderli, L. Pillonel, J. O. Bosset and G. Gremaud, *J. Anal. Atom. Spectrom.*, 2004, **19**, 227–234.
544. S. Swoboda, M. Brunner, S. F. Boulyga, P. Galler, M. Horacek and T. Prohaska, *Anal. Bioanal. Chem.*, 2008, **390**, 487–494.
545. I. Rodushkin, T. Bergman, G. Douglas, E. Engström, D. Sörlin and D. C. Baxter, *Anal. Chim. Acta*, 2007, **583**, 310–318.
546. A. Kawasaki, H. Oda and T. Hirata, *Soil Sci. Plant Nutrit.*, 2002, **48**, 635–640.
547. M. Barbaste, K. Robinson, S. Guilfoyle, B. Medina and R. Lobinski, *J. Anal. Atom. Spectrom.*, 2002, **17**, 135–137.
548. S. García-Ruiz, M. Moldovan, G. Fortunato, S. Wunderli and J. I. García Alonso, *Anal. Chim. Acta*, 2007, **590**, 55–66.
549. C. Rodrigues, M. Brunner, S. Steiman, G. J. Bowen, J. M. F. Nogueira, L. Gautz, T. Prohaska and C. Máguas, *J. Agricul. Food Chem.*, 2011, **59**, 10239–10246.

CHAPTER 13

# *Glow Discharge Mass Spectrometry*

CORNEL VENZAGO*[a] AND JORGE PISONERO[b]

[a] AQura GmbH, Germany; [b] University of Oviedo, Department of Physics, Spain
*Email: cornel.venzago@aqura.de

GDMS is based on a glow discharge ion source, which generates ions through sputtering (cathodic sputtering) of a solid surface and ionisation (electron impact or Penning ionisation) by a glow discharge plasma. A glow discharge plasma is formed by the passage of an electric current through a low-pressure gas. The ions are separated in a mass spectrometer and finally recorded at a detector. GDMS is a method for the direct determination of trace elements in solid states providing trace-element information of solid samples. Beside the direct current (DC) mode, analytical glow discharges can also be operated in a radio-frequency (RF) mode, most often applied for analysis of nonconducting samples (Figures 13.1 and 13.2).

New Developments in Mass Spectrometry No. 3
Sector Field Mass Spectrometry for Elemental and Isotopic Analysis
Edited by Thomas Prohaska, Johanna Irrgeher, Andreas Zitek and Norbert Jakubowski

Published by the Royal Society of Chemistry, www.rsc.org

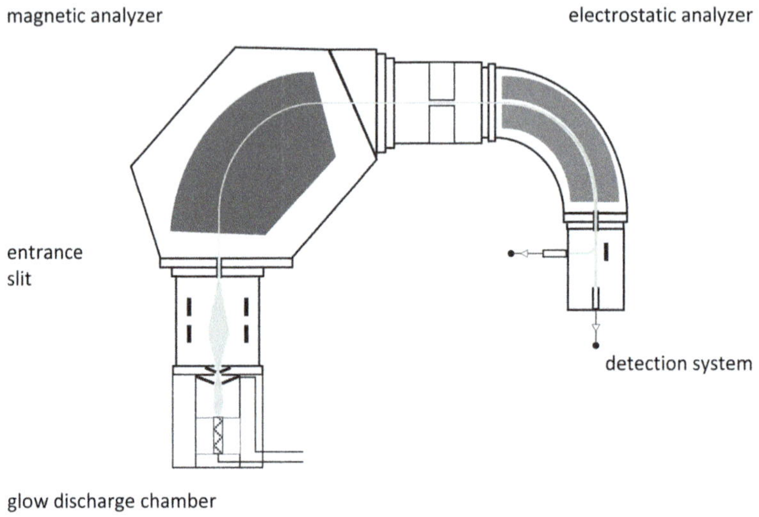


| | |
|---|---|
| Type of instrument | GDMS |
| Type of analysis | quantification, isotope ratios |
| Sample types | solid |
| Ionisation source | glow discharge plasma |
| Type of detectors | SEM, FC, Daly |
| First commercially available instrument (year) | VG 9000 (1985) |

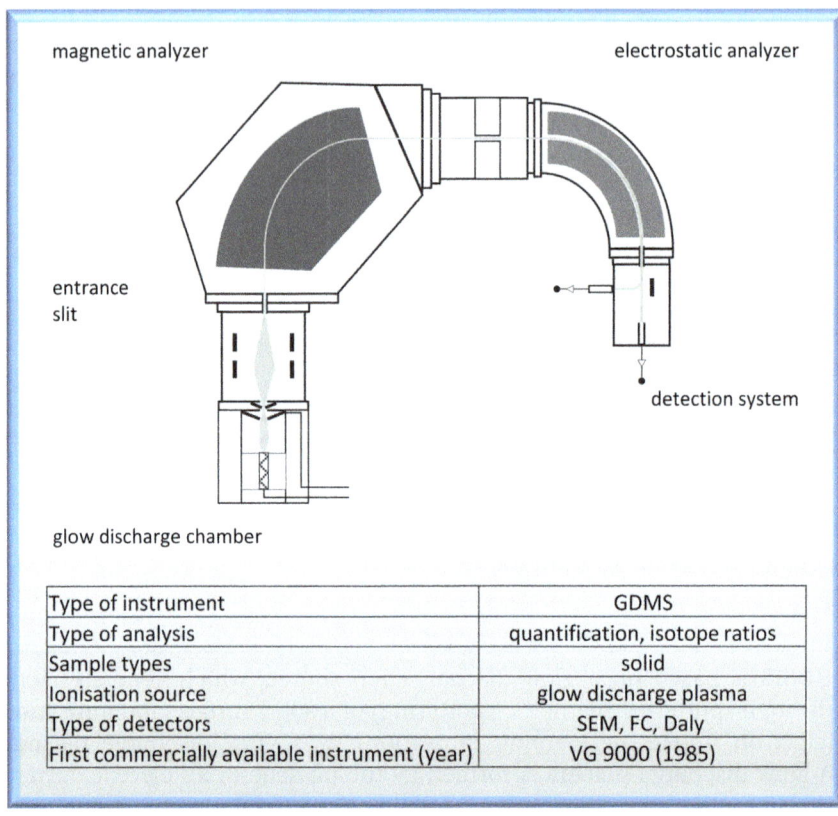

**Figure 13.1** Features and general schematics of a GDMS. (Source: Ondrej Hanousek.)

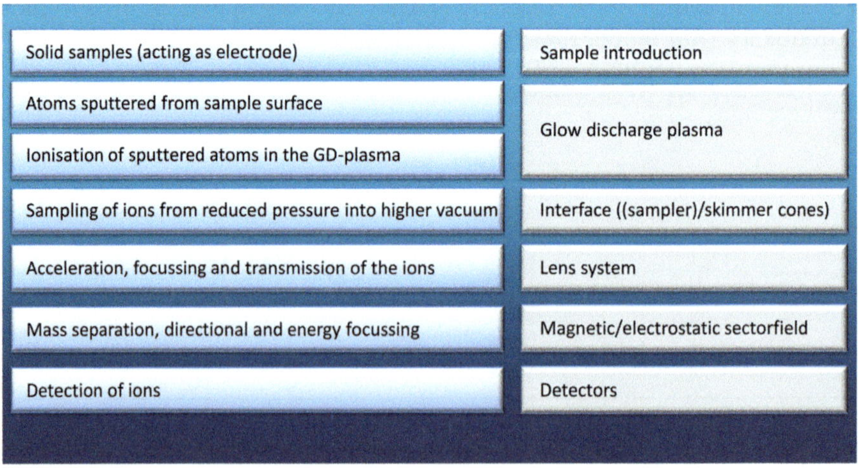


**Figure 13.2** Operational flow chart of a GDMS. (Source: Thomas Prohaska.)

**Name and Abbreviation:** Glow discharge mass spectrometry is usually referred to as GDMS. Neither a hyphen nor a more specific description of the used mass analyser (quadrupole or sector field) is commonly used.

## 13.1 General

Originally, glow discharge (GD) plasmas were used as ion sources in the early days of mass spectrometry, in particular at the beginning of the 20th century on the pioneering studies carried out by J. J. Thomson, A. J. Dempster, and F. W. Aston.[1–3] Nevertheless, later, other ionisation sources such as electron impact and spark sources were developed for their use in combination with mass spectrometers, moving the interest on GD sources into the background. For instance, spark source mass spectrometry (SSMS) was considered until the mid-1980s as one of the reference techniques for direct solid analysis.[4]

SSMS provides multielemental analysis capabilities with high sensitivity. However, its major drawback is its poor precision (>10%). Due to this important limitation other ion sources were further investigated. In this sense, a revival of the GD ion source was initiated by Coburn *et al.*, who used a planar diode sputtering arrangement and introduced the use of radiofrequency (RF) discharges for the direct analysis of conducting and nonconducting materials.[5,6]

The GD ion source, in its simplest form, consists of two electrodes containing a discharge gas, being one of the electrodes the sample to be analysed. Early GD ion sources combined with sector field mass spectrometers (GDSFMS) were mainly operated using a pin-shaped sample as it allowed enhanced ion extraction to the mass analyser. Moreover, it should be stressed that former GDSFMS instruments were based on the use of low discharge power ($\sim$3–6 W), resulting in relatively low sputtering rates and long times (hours) to reach stable signals under equilibrium conditions. Therefore, the sample throughput was a major drawback for these original instruments. In order to overcome this problem, Grimm-type GD sources,[7] which were originally developed for GDOES, and modified Grimm-type GD sources (without suction between anode and cathode), were coupled to different mass spectrometers.[8–10] In particular, these GD sources, which are flat cathode sources where the sample itself serves as the vacuum sealing part, allow fast sample changing and easy source cleaning. In addition to that, higher powers ($\sim$tens of W) could be applied due to more effective cooling, and more stable and reproducible signals could be achieved due to a restricted discharge area on the sample surface, itself defined by the anode diameter. As a result, superior analytical performance was not only achieved for bulk analysis but also for depth-profile analysis.

Nowadays, direct current (DC) GD ion sources combined with SFMS are the most popular analytical techniques in industry for the direct solid analysis of high-purity metals and for other special applications (*e.g.* photovoltaic materials). In particular, during the last 30 years, 6 commercially available GDSFMS instruments of different design were introduced to the market (Table 13.2), being the *VG 9000* (VG Elemental, Winsford, UK) in

**Table 13.1** Lists a number of reviews relevant for the fundamental understanding of GD processes. Additional books on general basics of GDMS are listed in Table 1.1.

| Authors | Title | Reference |
|---|---|---|
| W.W. Harrison, K.R. Hess, R.K. Marcus, F.L. King | Glow discharge mass spectrometry | 11 |
| W.W. Harrison | Glow discharge mass spectrometry: a current assessment | 12 |
| D.J. Hall, P.K. Robinson | Glow discharge MS: A powerful technique for multielement analysis | 13 |
| F. Adams, A. Vertes | Inorganic mass spectrometry of solid samples | 14 |
| R. Gijbels | Elemental analysis of high-purity solids by mass spectrometry | 15 |
| F.L. King, W.W. Harrison | Glow discharge mass spectrometry: An introduction to the technique and its utility | 16 |
| D. Stuewer | Glow discharge mass spectrometry: A versatile tool for elemental analysis | 17 |
| W.W. Harrison, C.M. Barshick, La.A. Klingler, P.H. Ratliff, Y. Mei | Glow discharge techniques in analytical chemistry | 18 |
| W.W. Harrison | Glow discharge: Considerations as a versatile analytical source | 19 |
| A. Raith, W. Vieth, J.C. Huneke, R.C. Hutton | Quadrupole *versus* magnetic sector field glow discharge mass spectrometry: Comparison of quantitative analytical capabilities | 20 |
| M. Grasserbauer | Solid state mass spectrometry for materials science | 21 |
| R. Gijbels, A. Bogaerts | Recent trends in solid mass spectrometry: GDMS and other methods | 22 |
| A. Bogaerts, R. Gijbels | Review: Fundamental aspects and applications of glow discharge spectrometric techniques | 23 |
| A. Bogaerts, R. Gijbels | New developments and applications in GDMS | 24 |
| A. Bogaerts | The glow discharge: an exciting plasma | 25 |
| S. Baude, J.A.C. Broekaert, D. Delfosse, N. Jakubowski, L. Feuchtjohann, N. G. Orellana-Velado, R. Pereiro, A. Sanz-Medel | Glow discharge atomic spectrometry for the analysis of environmental samples – a review | 26 |
| J.S. Becker, H.J. Dietze | State-of-the-art in inorganic mass spectrometry for the analysis of high-purity materials | 27 |
| M. Betti, L. Aldave de las Heras | Review: Glow discharge mass spectrometry for the characterisation of nuclear and radioactively contaminated environmental samples | 28 |

**Table 13.1** (*Continued*)

| Authors | Title | Reference |
|---|---|---|
| V. Hoffmann, M. Kasik, P.K. Robinson, C. Venzago | Review: Glow discharge mass spectrometry | 29 |
| C.B. Douhitt | Commercial development of HR-ICPMS, MC-ICPMS and HR-GDMS | 30 |
| J. Pisonero, B. Fernandez, D. Guenther | Critical review of GD-MS, LA-ICP-MS and SIMS as inorganic mass spectrometric techniques for direct solid analysis | 31 |
| Ph. Belenguer, M. Ganciu, Ph. Guillot, Th. Nelis | Review: pulsed glow discharges for analytical applications | 32 |
| A.A. Ganeev, A.R. Gubal, K.N. Uskov, S.V. Potapov | Analytical glow discharge mass spectrometry | 33 |
| N.H. Bings, A. Bogaerts, J.A.C. Broekaert | Atomic spectrometry | 34 |
| J. Pisonero, N. Bordel, C. Gonzalez de Vega, B. Fernández, R. Pereiro, A. Sanz-Medel | Critical evaluation of the potential of radio-frequency pulsed glow discharge time-of-flight mass spectrometry for depth-profile analysis of innovative materials | 35 |

**Table 13.2** List of commercial SFMS instruments, including the corresponding manufacturer, year of introduction into the market, current status, and geometry of SFMS.

| Instrument | Manufacturer | Year of production | Geometry |
|---|---|---|---|
| *VG 9000* | VG Elemental | 1985–2005 | reverse Nier–Johnson |
| *Concept GD* | Kratos | 1992–1995 | Nier–Johnson |
| *Element 2* ICP-MS with alternative GD source | Finnigan MAT | 1996 | reverse Nier–Johnson |
| *Element GD* | Thermo Electron | 2005–ongoing | reverse Nier–Johnson |
| Autoconcept GD90[a] | Mass Spectrometry Instruments | 2008–ongoing | Nier–Johnson |
| *AstruM* | Nu Instruments | 2010–ongoing | Nier–Johnson |

[a]A previous design of ConceptGD was initially launched in 1992 by Kratos on the platform of the organic MS Concept.

1985 the first commercially available double-focusing GD instrument used for elemental analysis.[36] The first commercial spark source instrument, however, was the *MS-7* from AEI (UK) introduced in 1959.

An alternative approach for GDSFMS uses the Mattauch–Herzog geometry, although no commercially available instrument has been introduced to the market so far. In this case, simultaneous multielemental detection is achieved using an electronic array detector, also called focal plane camera.[37]

Since GDSFMS is mainly present in industry and in contract laboratories, only few research studies have been published during the past five years.

The only remarkable innovation of this technique was the introduction of a microsecond pulsed glow discharge supply in order to improve depth-profiling capabilities and the analysis of thermally sensitive materials.[38] For instance, the development of pulsed-DC GDSFMS plays a major role on the analysis of thin-film photovoltaic layers on glasses,[39] and on other application areas.

On the other hand, most of the recent studies on GDMS focused on the development of a glow discharge time-of-flight-mass spectrometer (GDTOFMS). In particular, the use of pulsed-RF GDTOFMS was investigated for inorganic and organic thin-film analysis with high depth resolution (in the nanometer range), due to the high mass spectra acquisition rate of this instrument.[35,40] It should be highlighted that this instrument has been introduced very recently into the market by Horiba Jobin Yvon, under the denomination of Pulsed Profile (PP)-TOFMS.[41–43]

In Europe, the development of GD techniques and the inter-relationship in the glow-discharge community and therefore also in the GDMS community was mainly maintained by national user groups and by the *VG 9000* user groups. In 1994 the European Working Group on Glow Discharge Spectroscopy (EW-GDS) was established as an additional community combining industrial users with academic research and training institutions, resulting in a network project "Glow Discharge Spectrometry for Spectrochemical Analysis" (Contract No SMT4-CT-98-7517). From this project, a need for the development of a new GDMS sector field type instrument was initiated and finally conducted in the project "Automated Glow Discharge Mass Spectrometry" (Contract No G6RD-CT 2000-00170) resulting in the development of the *Element GD*.[44]

## 13.2 Technical Background

Independent of the manufacturer, the different GDSFMS instruments consist of common parts, including: an ion source, a sampling interface, an electrostatic lens system, a magnetic sector field, an electric sector field, a flight tube, an entrance and an exit slit (whose combination defines the resolution), transfer, zoom and/or filter optics, a detection system and a vacuum system. Most of these components are identical to those discussed in Chapter 12 for scanning ICP-SFMS, except for the ion source and the detection system. Therefore, only these components will be discussed in this chapter in more detail.

### 13.2.1 Glow Discharge Ion Source

The name glow discharge comes from the bright light emission from the plasma generated close to the cathode (negative glow). Glow discharge plasmas are generated by applying a potential difference of up to 2 kV between two (metallic) electrodes (anode and cathode), which are inserted into a noble gas, generally Ar, at a reduced pressure (50–700 Pa). This plasma is generated by fast electrons emitted from the cathode, ionising and exciting the noble gas. The free electrons will then be accelerated to the positive electrode (or anode) giving rise to additional collisions with other gas atoms that produce further ionisation processes. Once the plasma is established, the potential distribution in the GD source (Figure 13.3) changes significantly, and most of the applied discharge voltage drops between the plasma edge and the cathode surface in a region denominated "cathode fall".

Therefore, gas ions that reach the cathode-fall region are accelerated towards the cathode producing the sputtering of atoms from the cathode surface (sample), and the emission of secondary electrons, which are needed to sustain the plasma. Heating effects are observed during the sputtering process. Thus cooling of the electrode becomes essential for a higher discharge power. If thermal effects predominate, the glow discharge might change suddenly into an arc discharge, which is accompanied by a sudden drop of the discharge voltage and a strong increase of the discharge current. Secondary ions, if ejected, are retained by the strong electric field, causing self-sputtering.

Usually different emission and dark zones are observed in the GD plasma if the distance between the electrodes is sufficiently large. However, the number of zones can be reduced to a single one, the negative glow, when the

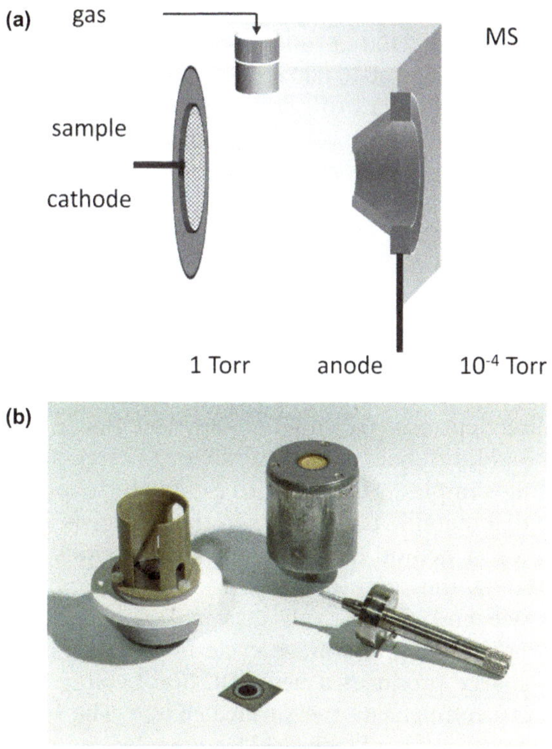


**Figure 13.3** GD ion source: (a) Schematics (b) Photo. (Source (a) Ondrej Hanousek, (b): Photo Cornel Venzago.)

electrode interdistance is short. Moreover, the negative glow is the zone of major analytical interest for emission or mass spectrometry. (For more information about glow discharges please visit the following homepage. *www.glow-discharge.com*[45])

The neutral atoms sputtered from the sample surface get diffused into the glow discharge plasma where they are partly ionised by different mechanisms. In this sense, it is important to highlight that the atomisation and ionisation processes take place locally and temporally separated, and thus are decoupled from each other to a certain extent. This decoupling is an important prerequisite to achieve low matrix effects. Nevertheless, glow discharge plasmas are not completely free from matrix effects as the sample acts also as the cathode of the discharge. Of course, sample properties, such as conductivity, sputtering rate or secondary electron emission yield, have a direct impact on the discharge characteristics, for instance on the current *versus* voltage curves.[29,46–48]

The gas temperature is relatively low (usually below 1000 K) for most of the analytical glow discharges, operated with voltages of up to 2 kV and currents of up to 100 mA. Although the gas temperature is low, the electron temperature is relatively high and not uniform. In particular, two groups of electrons have been observed: the thermalised electrons whose temperature has been measured by Langmuir probe measurements and by Thomson scattering in the range of 2300 to 7000 K and the secondary electrons, whose temperature is in the range of 40 000 to 75 000 K.[49–51] Moreover, the electron densities cover a range from $1\times10^{12}$ cm$^{-3}$ to $4\times10^{14}$ cm$^{-3}$, depending on the discharge conditions (pressure, voltage or current).

Electron impact ionisation is one of the most important and best-known processes in the glow discharge. It is the essential process in the self-sustained plasma. An additional source of ionisation is the Penning ionisation based on an impact with metastable argon atoms. These excited levels at 11.55 eV or 11.72 eV provide sufficient energy for the ionisation of those elements with first ionisation potentials lower than these energy levels. Furthermore, asymmetric charge transfer might also occur between the neutral analyte atoms and the ionised gas species.

Beside the direct current (DC) mode, analytical glow discharges can also be operated in a radio-frequency (RF) mode, most often applied for analysis of nonconducting samples. The power to establish the plasma is capacitive coupled by use of a metallic plate to which the voltage is applied and that is usually arranged just behind the sample. In the analysis of nonconducting materials by RF GD, the sample is polarised by the applied voltage like a dielectric interface and alternatively acts as cathode or anode. A charge builds up during the positive voltage cycle by negative electron bombardment and this charge produces a new "DC-bias" voltage that then further accelerates ions to compensate the surface charge. The typically used radio frequencies are in the range of technical frequencies, from 3.5 to 13.56 MHz.

In order to further improve the analytical figures of merit of GDMS, modifications of the gas flow pattern inside the GD source were evaluated. In

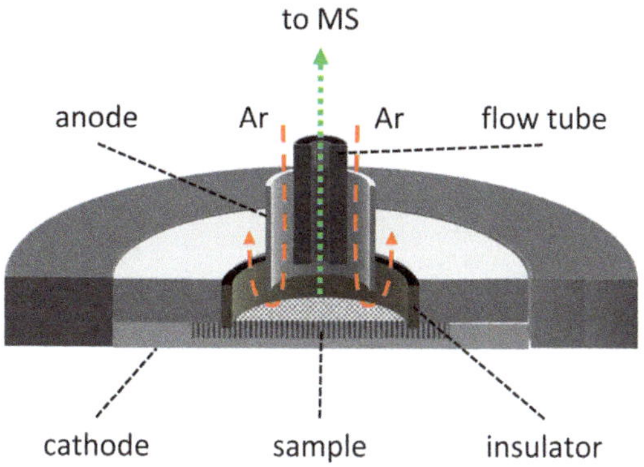


**Figure 13.4** Fast flow glow discharge ion source with internal flow tubes. (Source: Ondrej Hanousek after Jakubowski *et al.*[52]).

particular, the use of discharge gas flows ($\sim$mL min$^{-1}$) directed towards the surface of the sample was investigated, making use of internal flow tubes (see Figure 13.4).

Using such a configuration, it was possible to achieve a gas jet-assisted sputtering that, among other effects, improves the transport of the sputtered atoms into the plasma region where the excitation and ionisation processes take place, as well as the transport of ions into the mass analyser.[53–57] Moreover, due to the low residence time of the different species in the plasma region, this configuration was denominated "fast flow" source. Additional theoretical modelling studies supported the hypothesis that improved transport efficiency is achieved using the fast flow Grimm-type GD source, resulting in enhanced sensitivity and reduced number of polyatomic species.[58]

Table 13.3 lists different types of source designs and power supplies used on prototypes and commercial GDSFMS instruments. It is observed that the first prototypes, in which hollow-cathode DC and RF GDs were coupled to SFMS, were developed by Harrison *et al.*, in the 1970s.[59,60] Since then, there has been a tremendous improvement in technology resulting in the development of several powerful and versatile instruments that are nowadays commercially available. Each of these instruments has a particular design of the GD ion source and power supply. For instance, the main characteristic of the *AstruM* GD ion source is that it is based on a low-pressure static discharge built in tantalum similar to the *VG 9000*. In contrast, the *Element GD* instrument is operated using a fast flow ($\sim$hundreds of mL min$^{-1}$), high-power ($\sim$500 W) DC GD ion source that allows a rapid sputtering rate ($\sim$100 nm s$^{-1}$). This source can also be operated in the pulsed-DC mode, in that case the applied power is much lower (3–4 W) allowing the analysis of low-melting point materials with high sensitivity, as well as allowing depth profile analysis with high depth resolution ($\sim$tens of nm). Furthermore, in this instrument, the sample is placed in a vacuum chamber in order to avoid any risk of leaks between the sample and the GD ion source. On the

**Table 13.3** Different types of source designs and power supplies used on prototypes and commercial GD-SFMS instruments.

| Instrument | Power supply[a] | Source design[b] | Reference |
|---|---|---|---|
| Noncommercial GD-SFMS (*MS702*) | DC GD and RF GD | Hollow cathode | 59, 60 |
| *VG 9000* | DC GD | Pin and flat cathodes | 36, 61 |
| Noncommercial GD-SFMS (*Element*) | RF GD | Flat cathode | 62 |
| Noncommercial GD-SFMS (*Axiom*)[c] | DC GD and RF GD | Flat cathode | 63 |
| Noncommercial On an *Element 2* (ICP-MS) | Interchangeable Hollow-cathode DC GD ion source | Pin sample | 64 |
| *Element GD* | DC GD and μs-pulsed-DC GD | Pin and flat cathodes | 65 |
| *Autoconcept GD90* | DC GD and RF GD | Pin and flat cathodes | 66 |
| *AstruM* | DC GD | Pin and flat cathodes | 67 |

[a]DC = direct current; RF = radio-frequency.
[b]Flat cathode source design mainly refers to Grimm-type design (except *VG 9000*).[7]
[c]*Axiom* is a SFMS instrument with Nier–Johnson configuration from VG Elemental C.[57]

other hand, the GD source of the *Autoconcept GD90* instrument allows a fully automated system with simple changeover between DC and RF, opening new opportunities and increasing the application areas of this technique.

It is known that the strength of GD spectroscopies, including optical emission spectroscopy and mass spectrometry, is the direct analysis of solid samples, including bulk and depth-profile analysis.[35,68] Therefore most of the strategies used for sample preparation are identical for both methods.

As mentioned previously, the GDSFMS instruments demand a sample in pin or flat form. Therefore, samples are machined to the final shape, if required, and additionally polished by grinding, if appropriate. Moreover, powder samples can be pressed with a binder in a suitable form, and even nonconducting powdered samples can be analysed either directly or after mixing with a conducting host matrix; for instance with a powder or very pure ductile solid, such as graphite, gallium or indium.[69–73]

### 13.2.2 Sampling Interface

Due to the difference of working pressure conditions in the GD source ($\sim 10^2$ Pa) and the mass analyser ($10^{-6}$ to $10^{-7}$ Pa), an interface is required for differential pumping, being assisted by a roughing pump. The sampling interface is a crucial part of the mass spectrometer and it has to be optimised to allow a high ion transmission from the ion source to the mass analyser. Typically, the sampling interface on GDMS instruments consists of a cone (skimmer cone), which are hat-shaped metal or flat extraction plates

**Figure 13.5** Photo of the *Element GD* interface, where the skimmer cone and the backside of the GD source can be seen.
(Source: Photo Matthias Balski, Courtesy of Dr. Matthias Balski (BAM).)

plates with a small orifice (0.4–1.2 mm) in the centre in order to sample the central part of the ion beam generated in the plasma. Nevertheless, as the pressure at the GD is relative low, the interface might consist of only one cone. For instance Figure 13.5 shows the interface of the *Element GD* that only contains one cone, equivalent to the skimmer cone in an ICP-MS. Once the ions pass the interface region they are accelerated by a voltage ranging from 4000 to 10 000 V – depending on the particular instrument. In this sense, the high voltage can be applied to the interface and GD, keeping the analyser at ground potential, or *vice versa*.

### 13.2.3 Magnetic and Electric Sectors

As it is described in Chapter 4, the combination of radius and angle of the magnetic and electric sectors need to fulfil certain conditions to guarantee double-focusing conditions. In particular, the differences in ion energies are compensated to a certain extent in the electric sector. Nonetheless, the energy spread of ions originating from GD plasma is much lower than that observed for ions originating from, *e.g.* an ICP plasma, which could be as high as 1–20 eV.

### 13.2.4 Detection System

An extended dynamic range is required in GDSFMS for the determination of all analytes in each run of the sequential mass analyser, including matrix elements (mass%), traces ($\mu g\ g^{-1}$), and ultratraces ($ng\ g^{-1}$) and below. For this purpose, the combination of different detectors, such as Faraday cups and electron multipliers (both discrete dynode detectors and channeltrons), is necessary.[74] The fast Faraday detector used in the *Element GD* does not lengthen the analysis time, due to its fast detection electronics. Moreover, it does not suffer decay or response times and allows integration times down to 1 ms. Also, switching between secondary electron multipliers (SEM) and Faraday modes is fast and below 1 ms.

### 13.2.5 Instrumentation

Only about 160 GDMS instruments have been sold, so far. The commercially available instruments will be presented shortly in the following section.

***VG 9000*** **(VG Elemental)**
The first commercial glow discharge instrument, the *VG 9000* (VG Elemental, Winsford, UK), was based on an organic SFMS (Figure 13.6). It was introduced in the market in 1985 and the first analytical results appeared in 1986. Even though the instrument has been discontinued from 2005, it still represents more than half of the instruments currently in operation. The *VG 9000* was often operated with the sample being a pin. A schematic diagram of the standard ion source for pin samples is shown in Figure 13.7(a).

The source consists of different parts, including two electrodes, the conducting sample used as cathode and the anode with the exit aperture. The pin-type sample is fixed in a special holder and introduced into the ion source by a solid-probe arrangement. The sample is operated as the cathode and is isolated from the anode source body by a ceramic cover. The anode body is cryogenically cooled to reduce the amount of residual gases

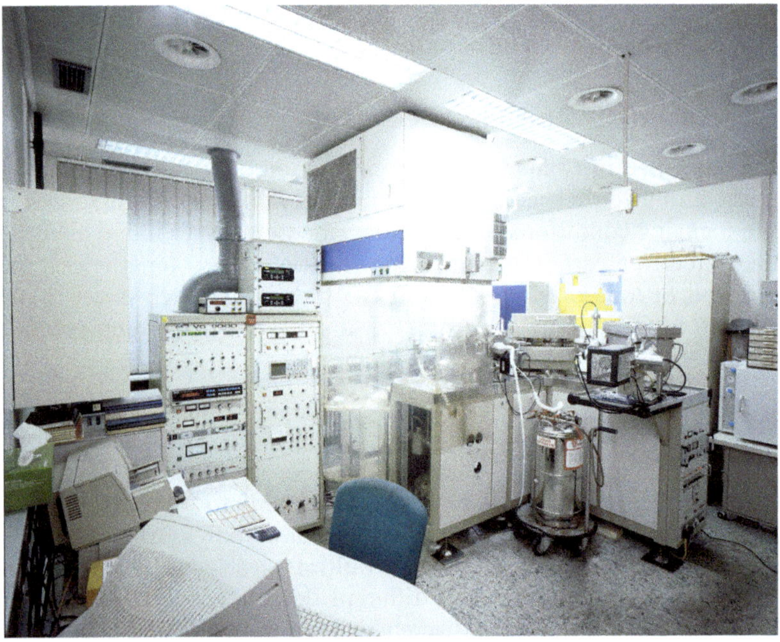

**Figure 13.6** *VG 9000.*
(Source: Photo Cornel Venzago.)

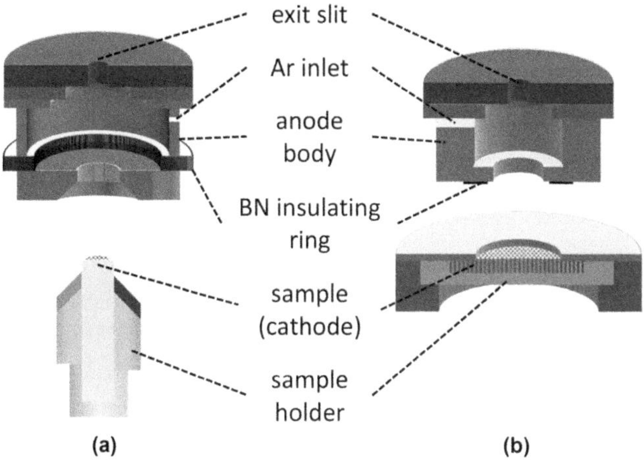


**Figure 13.7** (a) Pin Cell *VG 9000*; (b) Flat Cell *VG 9000*.

significantly. The whole source is floating at the acceleration voltage of the sector field device. A GD source was also commercially available for flat samples (see Figure 13.7(b)), allowing for depth-profiling analysis.

The *VG 9000* instrument is a double-focusing sector field device in a reverse Nier–Johnson geometry. This instrument can be operated at different mass resolutions by modifying the slit width at the entrance and exit of the mass analyser manually. In particular, this instrument can be operated with a maximum mass resolution of $R = 9000$, which was the origin of the name of this instrument. Moreover, the *VG 9000* is equipped with two separate detectors. For high ion intensities of the matrix, a Faraday cup is used, whereas for trace and ultratraces a Daly detector operated in an ion-counting mode is applied.[75,76] Automatic switching between both detectors had been provided so that a dynamic range of more than 9 orders of magnitude could be guaranteed in a single scan.

***Element GD* (Thermo Fischer Scientific)**

Thermo Fisher Scientific introduced a GDSFMS instrument, the *Element GD*, in 2005[77,78] (Figure 13.8).

The *Element GD* combines a new fast flow DC GD ion source with the hardware of the mass spectrometer from the *Element 2* ICP-SFMS instrument. In this GD source, shown in Figure 13.9, the anode and the cathode plate are separated by an insulator. The flat sample is pressed by a holder against the cathode body. Moreover, a flow tube is used to direct the gas flow perpendicular to the sample surface and by application of high flow rates, the ion transport towards the sampling orifice is improved.

The sample handling is optimised for a high sample throughput. Gas flow rates of a few hundred mL min$^{-1}$ are used for operation of the glow discharge, resulting in discharge powers of up to 100 W. The latter are

**Figure 13.8** *Element GD.*
(Source: © Thermo Fisher Scientific.)

advantageous to reduce the preburning phase to reach sputter equilibrium already after a few minutes, which is a relatively short period of time in comparison with the 30 to 60 min required for the pin-type samples in the *VG 9000* source. Moreover, different sets of cones, flow tubes and end-caps are available. For instance, graphite pieces are mainly used for trace-metal analysis while stainless-steel pieces are used for elements such as carbon. Figure 13.10 shows two sets of cone, flow tube and end-cap, made in graphite and stainless steel, respectively.

Additionally, an electric (Peltier) sample cooling (or heating) is applied, eliminating the need for cooling with cryogenic gases (as is the case in the *VG 9000*). The sample is hosted in a vacuum environment, which is different to the design criteria of a Grimm-type source. Therefore, there is a reduced requirement for perfectly flat samples because the sample is not used as a vacuum seal. A separate vacuum manifold employing an optimised sealing mechanism guarantees proper vacuum conditions. This fact enables the measurement of the nonmetals such as oxygen, nitrogen and carbon even at low concentrations. Besides the mostly used flat sample geometries, also pin samples can be analysed using a dedicated pin sample holder for a sample with up to 3 mm diameter. Alternative GD source designs have also been constructed for the *Element GD* instrument. For instance, Gusarova *et al.*, developed a flat and a pin GD source for the analysis of pressed powder samples (see Figure 13.11).[79]

Concerning ion detection a significant difference to the *Element2* ICP-SFMS should be mentioned here. By incorporation of a Faraday cup into the ion-detection module (the same design as used in the *Element XR*, (see

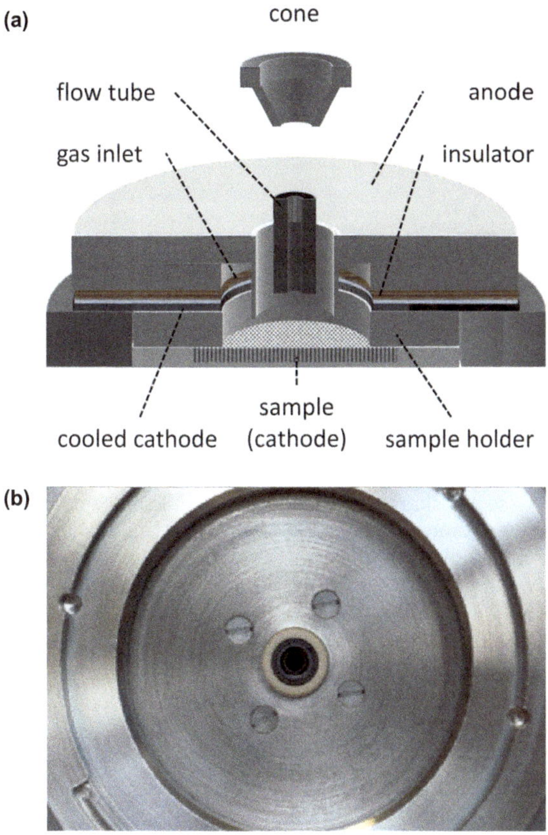


**Figure 13.9** (a) Schematic draw of the fast flow GD ion source. (b) Photo of the *Element GD* glow discharge ion source, showing the ceramic insulator, the anode and the internal flow tube.
(Source: (a) Ondrej Hanousek (b) Photo: Dr. Matthias Balski (BAM).)

Chapter 12), the dynamic range of this instrument is extended to more than 12 orders of magnitude. This is essential for bulk analysis, where the signal intensities for the measured trace elements are always related to the matrix ion intensity, serving as internal standard to overcome possible drift effects, as well as to perform quantification. Except for the ion source and the interface flange, most of the other hardware and also software features are similar to those of the *Element2* or *Element XR*.

***Astrum* (Nu instruments)**

The *Astrum* is a double-focusing, high-resolution GDSFMS, based on the hardware of the Nu *AttoM* ICP-SFMS instrument and was introduced in 2010 (Figure 13.12). The instrument was developed in conjunction with commercial and academic users of GDMS. Following consultation with these users, it became clear that there was a demand for an instrument with

**Figure 13.10** Two sets of skimmer cone, flow tube and end-cap employed in the *Element GD* instrument: (left) made of graphite; (right) made of stainless steel. (Source: Photo Courtesy of Dr. Matthias Balski (BAM).)

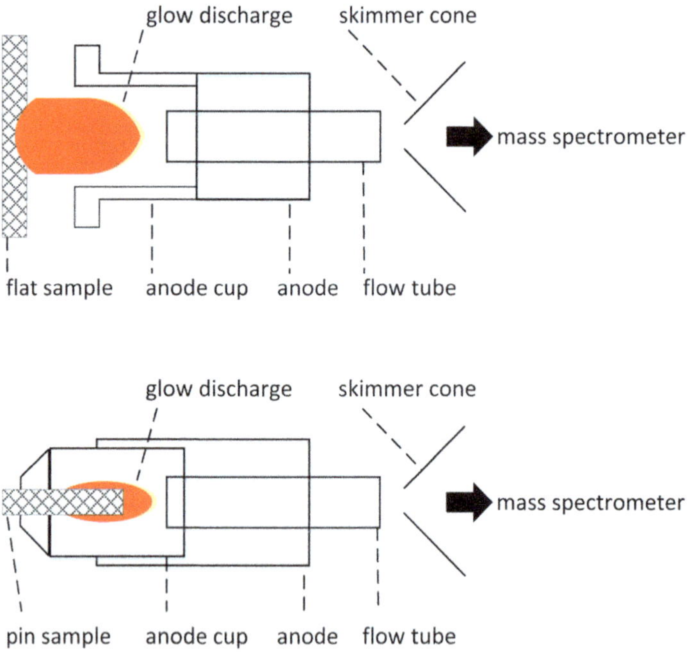


**Figure 13.11** Flat and a pin GD source for the analysis of pressed powder samples (*Element GD*).
(Source: Ondrej Hanousek.)

similar performance characteristics and usability as the *VG 9000*. As a result of this, some design features from the *VG 9000* were adopted for the *Astrum*, particularly with reference to the cell layout and cryogenic cooling capability.

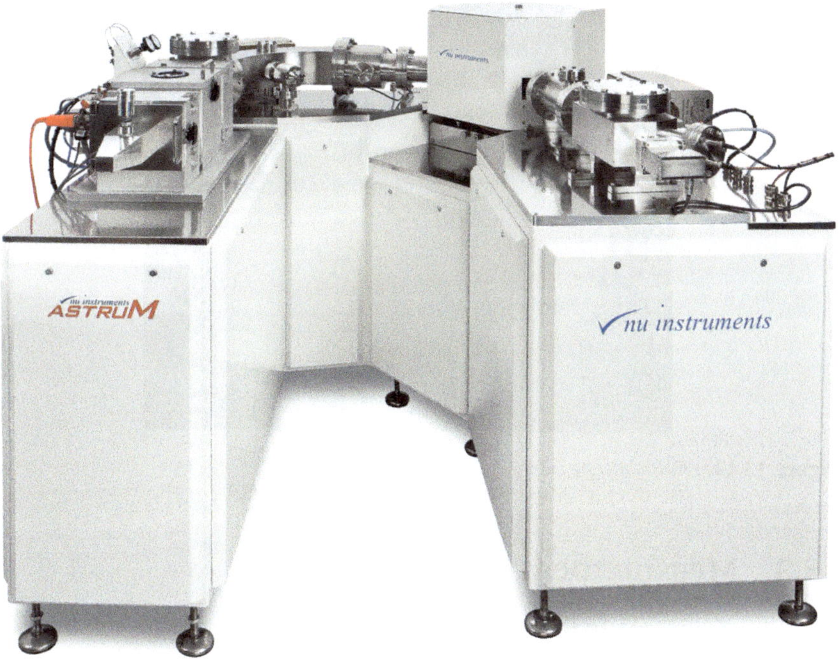


**Figure 13.12** *Astrum.*
(Source: © Nu Instruments.)

The glow discharge source is based on a low-pressure static discharge and consists of a tantalum cell allowing both pin and flat sample configurations. The cryogenic cooling option allows analysis of samples with low melting points and minimises gas backgrounds. The combination of an electron multiplier and a Faraday detector enables a linear dynamic range of 12 orders of magnitude. In October 2010 the first instrument was shipped to Evans Analytical Group (Liverpool, USA) who undertook a detailed evaluation of the instrument.[67]

***Autoconcept GD90*** **(Mass Spectrometry Instruments Inc.)**
The instrument (Figure 13.13) has a Nier–Johnson geometry mounted on a solid framework that optionally has a three-point antivibration support system, resulting in a stable instrument. It provides different GD ion sources that allow both flat and pin samples of different dimensions to be analysed. The ion optics is designed to provide a mass resolution greater than or equal to 160 000 using the 10% valley definition. A mass resolution of 4000 is achieved routinely for inorganic analysis. The *GD90* ion source is a slow-flow design with the DC mode as standard. The RF-GD system for the analysis of nonconducting material is also available. Additionally, an enhanced dynamic range is obtained in this instrument using both a Faraday cup and an electron multiplier that can be used in both analogue and ion-counting modes.

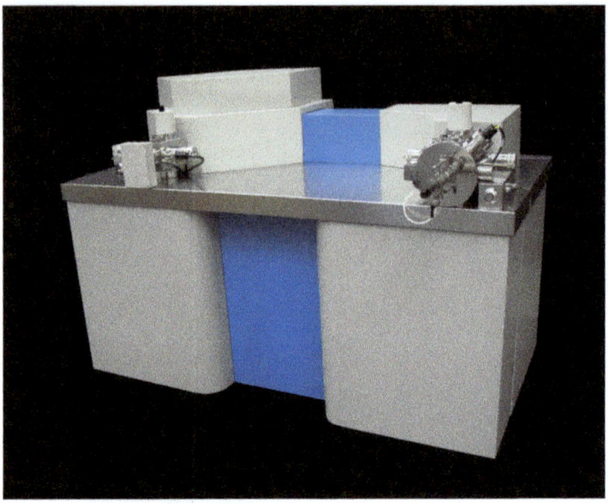

**Figure 13.13** *Autoconcept GD90.*
(Source: © Mass Spectrometry Instruments.)

# 13.3 Measurement Considerations

## 13.3.1 Spectral Interferences

GDMS spectral interferences can be subdivided into several groups similar to ICP-MS spectral interferences, as they can also be attributed to the presence of isobaric atomic ions, multiply charged ions and molecular ions of various origins. For each element, except for indium, there is one isotope that is free from isobaric overlap, although in most cases this is not the most abundant one. Multiply charged ions might be formed in the low-pressure GD, and then found in the mass spectrum at a position *m*/*z*. In particular, doubly and multiply charged ions of the main constituents of the plasma such as components of the gas and the sample matrix are most often observed in GDMS. Molecular ions may consist of atoms of the discharge gas and its contaminants (*e.g.* water, hydrogen, oxygen, nitrogen and noble gases), and/or components of the matrix (*e.g.* cluster ions) (see also Chapter 7).

In contrast to ICP-MS, the most intense ion signal in recent GDMS systems usually comes from the matrix isotope, being even more intensive than any signal from the discharge gas species, including $^{40}Ar^{+}$ or $^{41}ArH^{+}$.[57] For example, Figure 13.14 shows the influence of the gas flow rate (interval between 100 and 240 mL min$^{-1}$) on the discharge current and on some selected ions signals ($^{56}Fe^{+}$, $^{40}Ar^{+}$, $^{40}Ar^{1}H^{+}$, $^{40}Ar_2^{+}$, $^{16}O^{1}H^{+}$) measured for a Fe-matrix SRM using a DC GDMS at a fixed voltage of 600 V.

It is observed that with increasing gas flow rate the discharge current is almost linearly increasing, whereas a much stronger increase is noticed for the ion signals, in particular for $^{56}Fe^{+}$, $^{40}Ar^{+}$ and $^{40}Ar^{1}H^{+}$. In former GD sources, operated with a gas flow rate of less than 1 mL min$^{-1}$, ion-signal ratios between discharge gas species and analyte species (*e.g.* $^{40}Ar^{+}/^{56}Fe^{+}$ ~40) were higher than those observed using high flow rate GD sources (*e.g.*

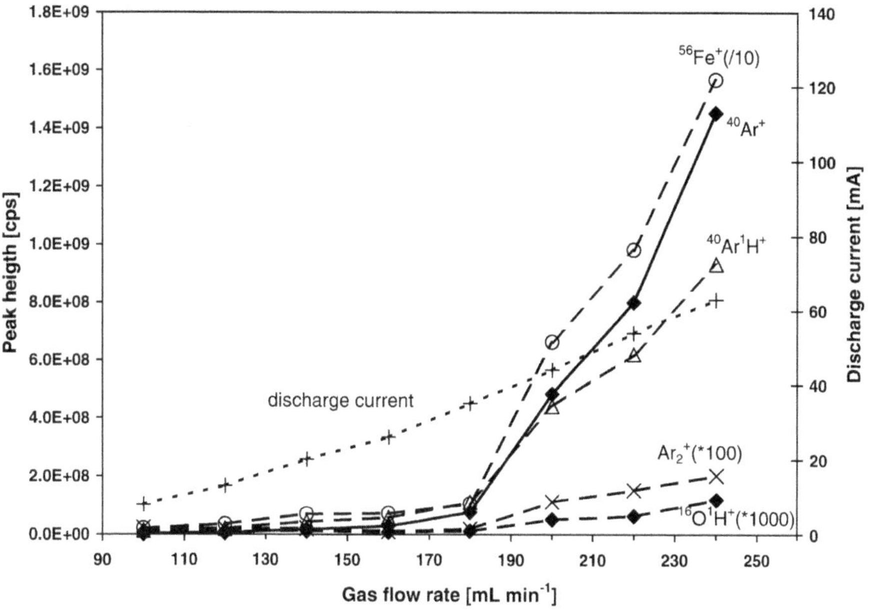


**Figure 13.14** Influence of the Ar gas flow rate on the ion signal intensity of different analytes in a Fe matrix sample and on the discharge current when operating at a constant voltage of 600 V.

$^{40}Ar^{+}/^{56}Fe^{+}$ ~0.1). These gas and background ions have been significantly reduced, but still play an important role and should be carefully considered. So far, it is still not clear if this improvement by two orders of magnitude might be attributed to a selective loss of Ar ions and/or to a transport improvement of analyte ions by the much higher gas flow rates.[80–82]

The extremely high ion signal that the detector is able to collect for the matrix elements, which can be as high as $10^{12}$ cps, might restrict its capabilities to determine low signals from trace elements with similar mass to charge ratios. In this case, the abundance sensitivity, which is typically defined as the ratio of a strong signal at a specified *m*/*z*, to the contribution that same signal makes one unit above or below the same *m*/*z*, is becoming a significant issue. The signal tailing might be due to the collision of ions with residual gas molecules in the mass spectrometer, causing a loss of ion energy and giving rise to a "tail" on the low mass side of a peak. In order to improve the abundance sensitivity, some instruments include a retardation filter lens, which only transmits ions with appropriate kinetic energy. For instance, the abundance sensitivity measured for a Cu matrix at mass 62 is calculated to be <200 ng $g^{-1}$ relative to mass 63, using a mass resolution >4000 in the *Astrum* GDMS,[83] and <10 ng $g^{-1}$ on the *Element GD* (Figure 13.15(a) and (b)).

Many approaches to overcome polyatomic spectral interferences have been described in the literature, including the use of mathematical correction procedures.[85] The sector field MS allows the resolution of polyatomic

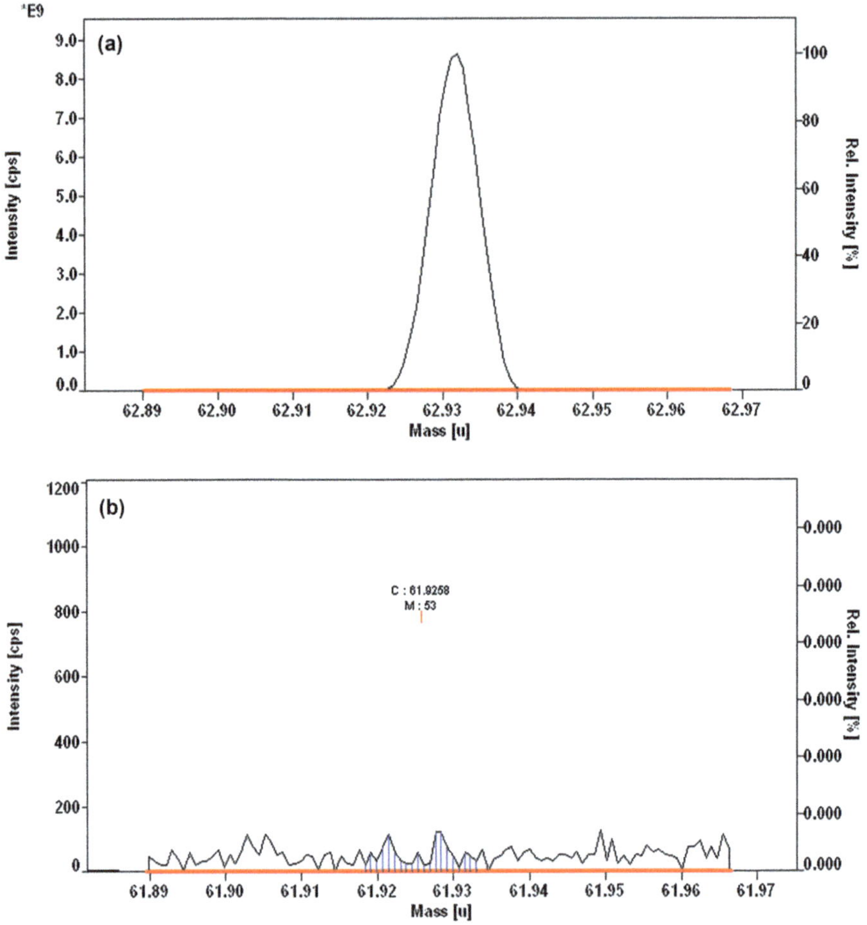


**Figure 13.15** Example of abundance sensitivity of the *Element GD*, where a $^{63}Cu$ matrix signal of $\sim 10^{10}$ cps (a) causes a background of $\sim 50$ cps on the neighbour mass (b). The calculated abundance sensitivity is $\sim 6$ ppb.[84]

interferences by applying high mass resolution (see Chapter 5). In Table 13.4 some examples of typical interferences observed in GDMS are compiled. Most examples given in this table demonstrate that a mass resolution *R* of less than 4000 is sufficient to overcome most interferences in practice.

However, a great number of polyatomic molecular ions, including hydrides, argides and oxides of the matrix components in a mass range above 100 u, require a mass resolution close to or above the upper end of the achievable range of commercial instrumentation. Similarly, isobaric interferences require a really high mass resolution: *e.g.* to resolve $^{40}Ar^+$ from $^{40}Ca^+$ a mass resolution of about 190 000 is necessary. Additionally, the price to be paid for high mass resolution concerns the loss in sensitivity, which is inversely proportional to *R*.

**Table 13.4** Selected typical polyatomic interferences and required mass resolution to separate the nuclide of interest from the interfering polyatomic ion (reproduced after Jakubowski *et al.*[86]).

| Nuclide | Polyatomic ion | Resolution |
|---|---|---|
| $^{28}Si^+$ | $^{14}N_2^+$ | 958 |
| $^{31}P^+$ | $^{14}N^{16}O^1H^+$ | 968 |
| $^{31}P^+$ | $^{15}N^{16}O^+$ | 1458 |
| $^{52}Cr^+$ | $^{35}Cl^{16}O^1H^+$ | 1671 |
| $^{32}S^+$ | $^{16}O_2^+$ | 1801 |
| $^{64}Zn^+$ | $^{32}S^{16}O_2^+$ | 1952 |
| $^{54}Cr^+$ | $^{40}Ar^{14}N^+$ | 2031 |
| $^{54}Fe^+$ | $^{40}Ar^{14}N^+$ | 2088 |
| $^{56}Fe^+$ | $^{40}Ar^{16}O^+$ | 2502 |
| $^{48}Ti^+$ | $^{32}S^{16}O^+$ | 2519 |
| $^{51}V^+$ | $^{35}Cl^{16}O^+$ | 2572 |
| $^{44}Ca^+$ | $^{28}Si^{16}O_2^+$ | 2688 |
| $^{64}Zn^+$ | $^{32}S_2^+$ | 4261 |
| $^{75}As^+$ | $^{40}Ar^{35}Cl^+$ | 7775 |

## 13.3.2 Analytical Figures of Merit

It is a basic advantage of GDMS for the direct analysis of solids to overcome most of the limitations of dissolution and liquid analysis by ICP-MS. Additionally sample preparation is much more straightforward and often simple. Due to the fast sample preparation, a short overall analysis time is achieved and any risk of contaminations from sample pretreatment is avoided. Finally, direct analysis avoids a loss of sensitivity due to dissolution of the sample and therefore considerably lower detection limits can be achieved. Moreover, presputtering of the sample is used to monitor the evolution of the analyte ion signals until stable values are obtained that might represent the bulk analysis data, avoiding the surface contamination.

The analytical figures of merit, which can be achieved by GDMS, strongly depend on the matrix, the sample preparation, operational conditions of the source and the instrument itself. For the matrices Al, Cu or Au total ion matrix currents of $1\times10^{-10}$ to $5\times10^{-9}$ A are achieved for the *VG 9000* and for the *AstruM* measured with $R=4000$. Hence, sensitivities are obtained allowing a detection limit in the range of 1 ng g$^{-1}$ or below. The noise level is typically less than 1 cps, except if large matrix/interference signals are in the vicinity of the analyte isotope.

### Reproducibility

In GDMS, for the definition of the reproducibility it must be distinguished between an external reproducibility, which includes sample preparation and sample changing, and an internal reproducibility, which is calculated from repeated scans during one analysis. Both are expressed by the precision calculated from the standard deviations of intensity measurements. Both internal and external precision depends on the homogeneity and content of the element in the sample. Therefore, most often data for relative standard deviations (RSD) are given in the literature, which depend less on concentration. Internal precision measured on CRM SS-487/1 (Institute for Reference Materials and Measurements (IRMM) is one of the seven institutes of the Joint Research Centre (JRC), a Directorate-General of the European Commission (EC), Geel, Belgium) making three repeated measurements range from $<3\%$ RSD for major matrix components to $<10\%$ for traces and $>10\%$ for ultratraces. The latter are limited by the counting statistics only. Moreover, typical values of the external reproducibility might be of the order of 15%.[63]

**Sample Throughput**

Depending on the analytical requirements regarding elements to be analysed and number of elements to be determined, the measurement time in GDMS is typically between 10 min and 300 min including presputtering times (2–100 min), which depend mostly on the analytical task and the surface quality of the sample. Therefore a typical sample throughput cannot be estimated without additional information. The *Element GD* has the advantages of an instrument without a high-vacuum sample chamber. Therefore sample-changing times can be reduced from 10 min to around 3 min. Moreover, sputtering times of the *Element GD* are usually shorter compared to the other instruments, since a higher plasma power, which results on higher sputtering rates, is applied.

The sample throughput of the *Element GD* instrument is relatively high due to the fact that the GD ion source does not operate at ultrahigh-vacuum conditions. In this sense, in the current commercial instruments, optimum operating pressure conditions at the GD ion source can be achieved in a few minutes after each sample exchange. Furthermore, presputtering of the sample is in many cases accomplished in order to obtain a stable signal by removing some surface contamination and obtaining a sputtering equilibrium.

**Robustness**

Robustness is a measure for the dependence of sensitivities on discharge or instrumental conditions. Although the absolute intensities strongly depend on discharge conditions, the Ion Beam Ratio (IBR) or Relative Sensitivity Factors (RSFs) are generally less dependent. The introduction of the fast flow source brought additional problems regarding robustness since small changes in pressure conditions and flow rates produce significant changes of ion intensities especially in the low- or high-mass ranges. Then, discriminating effects between high and low masses are also observed.

**Limits of Detection[87] (see also Chapter 10)**

In GDMS it is difficult to apply the standard definition of limits of detection (LOD) based on the $3\sigma$ concept ($\sigma =$ standard deviation of the blank). This is due to the fact that a sample containing a low blank is required. Therefore ultrapure samples are necessary to calculate the LOD, but most often these samples are not available. As a result, a so-called instrumental blank (background noise) in combination with the sensitivity of an element is used to calculate the detection limit, providing a more or less theoretical LOD.

The limits of detection for the different elements depend on the matrix of the sample. For instance, Table 13.5 shows the LOD calculated for several elements on Zn-matrix and on Cu-matrix materials using synthetic standards.

Additionally, Table 13.6 shows the LOD determined for several elements on an aluminium-based alloy. In particular, this study was performed using the *VG 9000* instrument with the pin-type GD source geometry, at 5 mA and 1.25 kV. With an integration time of less than 1 min per isotope, LODs of less than 1 ng $g^{-1}$ can be achieved.

Moreover, elements that are relevant for the electronics industry such as Th and U can be measured down to 0.01 ng $g^{-1}$ by increasing the integration time to values between 1 and 5 min per isotope. As a result, a total analysis time of an hour or more is needed for a multielement coverage.[88] Limits of

**Table 13.5** Comparison of limits of quantification (LOQ), determined after a calibration with synthetic pin-shaped standards ($LOQ_{cal}$), or by typical evaluation performed on a *VG 9000* ($LOQ_{VG}$). Standard deviations of approximately 30 or 100 were taken into consideration for $LOQ_{cal}$ values of a Cu- or Zn-matrix, respectively.[79]

| LOQ in Zn-matrix, μg $kg^{-1}$ | | | LOQ in Cu-matrix, μg $kg^{-1}$ | | |
|---|---|---|---|---|---|
| Analyte | $LOQ_{cal}$ | $LOQ_{VG}$ | Analyte | $LOQ_{cal}$ | $LOQ_{VG}$ |
| $^{109}Ag$ | 1500 | 2300 | $^{109}Ag$ | 450 | 2400 |
| $^{11}B$ | 110 | 60 | $^{11}B$ | 310 | 12 |
| $^{9}Be$ | 2 | 20 | $^{9}Be$ | 2 | 12 |
| $^{209}Bi$ | 8 | 20 | $^{114}Cd$ | 280 | 45 |
| $^{140}Ce$ | 13 | 30 | $^{161}Dy$ | 60 | 55 |
| $^{161}Dy$ | 7 | 150 | $^{167}Er$ | 11 | 50 |
| $^{167}Er$ | 5 | 120 | $^{153}Eu$ | 5 | 20 |
| $^{153}Eu$ | 6 | 50 | $^{69}Ga$ | 9 | 20 |
| $^{157}Gd$ | 6 | 180 | $^{157}Gd$ | 15 | 60 |
| $^{165}Ho$ | 1 | 20 | $^{165}Ho$ | 4 | 9 |
| $^{139}La$ | 3 | 20 | $^{95}Mo$ | 800 | 70 |
| $^{7}Li$ | 10 | 30 | $^{23}Na$ | 960 | 12 |
| $^{175}Lu$ | 1 | 20 | $^{146}Nd$ | 18 | 65 |
| $^{146}Nd$ | 920 | 1400 | $^{141}Pr$ | 6 | 12 |
| $^{58}Ni$ | 30 | 100 | $^{45}Sc$ | 2 | 12 |
| $^{141}Pr$ | 1 | 20 | $^{147}Sm$ | 19 | 75 |
| $^{87}Rb$ | 1200 | 111 | $^{159}Tb$ | 3 | 9 |
| $^{187}Re$ | 2 | 45 | $^{232}Th$ | 8 | 12 |
| $^{149}Sm$ | 8 | 210 | $^{205}Tl$ | 75 | 15 |
| $^{159}Tb$ | 1 | 20 | $^{51}V$ | 40 | 12 |
| $^{169}Tm$ | 1 | 20 | $^{172}Yb$ | 25 | 50 |
| $^{51}V$ | 25 | 30 | | | |
| $^{172}Yb$ | 6 | 120 | | | |

**Table 13.6** LOD determined for several elements on an aluminium-based alloy using a *VG 9000*.[88] In the original paper this table with LOD determined in AlSi1%Cu0.5% was shown with the erroneous concentration unit μg $g^{-1}$.

| Element | LOD (ng/g)* | Element | LOD (ng/g)* |
|---|---|---|---|
| Ag | 6.3 | La | 0.5 |
| As | 4.1 | Li | 1.5 |
| Au[a] | 100[c] | Mg | 1.6 |
| B | 1.3 | Mn | 2.9 |
| Ba | 0.8 | Mo[a] | 1.8 |
| Be | 1 | Na | 1.5 |
| Bi | 3.2 | Ni | 4.6 |
| Ca[a] | 14 | P | 2.3 |
| Cd | 15 | Pb | 3.1 |
| Ce | 0.7 | Pd | 4.5 |
| Cl | 1.4 | Pt | 3.3 |
| Co | 0.8 | S | 3 |
| Cr | 2.1 | Sb | 5.6 |
| Cs | 1.2 | Sn[a] | 21 |
| F | 5 | Ti | 0.5 |
| Fe | 1.7 | Th[b] | 0.2 |
| Ga | 9.9 | U[b] | 0.2 |
| Ge[a] | 28 | V | 0.5 |
| Hg[a] | 2.9 | W | 6.6 |
| In | 4.6 | Zn | 14 |
| K | 6.4 | Zr | 5.1 |

[a]Repetitive scans: 3.
[b]Repetitive scans: 10.

detection have been also calculated for different elements, including light elements and nonmetals, on high-purity silicon wafers using the *VG 9000* GDSFMS instrument.[89]

From the first publications of detection limits, no general progress in improving the detection limits has been achieved with the introduction of new generations of instruments. One significant improvement of the *Element GD versus* the *VG 9000*, however, was the possibility to significantly improve analysis speed. This is due to the fact that the *Element GD* has no high-vacuum chamber for the ion source and no cryocooling. Therefore, sample exchange can be done much faster. Also higher magnet scanning speed and lower switching times between detectors brought some advantage.

### 13.3.3 Mass Accuracy

As defined by Balogh:[90] "...mass accuracy is the ability to measure or calibrate the instrument response against a known entity. Usually expressed in parts per million (μg $g^{-1}$ or ppm), the measurement indicates the deviation of the instrument response from a known monoisotopic calculated mass...".

High robustness of the GDMS instrument and low matrix effects, which are nonspectroscopic interferences related to the influence of the matrix constituents on the analyte ion signal, are prerequisites to achieve accurate results. In particular, it is known that matrix effects are low in GDMS, because the steps of atomisation and ionisation are separated in time and space. In this case, the matrix should not have a great influence on the ionisation process of the already atomised gas-phase analytes. In other analytical techniques, where the atomisation and ionisation processes take place simultaneously, as is the case in secondary ion mass spectrometry (SIMS), the matrix effects play an important role. On the other hand, many physical and chemical factors – for instance related to the sample properties – plasma characteristics or to the mass spectrometer, might have an influence on the analyte sensitivity, affecting the accuracy of the analysis. As the sample acts as an electrode in the GD source, properties such as sample temperature[91] (*e.g.* cooling of the sample), surface structure (*e.g.* roughness) or cleanliness of the sample surface have an influence on the sputtering and plasma conditions. Additional parameters that could affect the accuracy are related to the GD operating conditions. For example, the pressure in the source modifies not only the working conditions such as the current–voltage relationship, but also the diffusion and extraction of ions, in particular in the fast-flow GD sources.

## 13.3.4 Quantification Strategies (see also Chapter 8)

The accuracy of the determined concentrations of an analyte element in a certain matrix is defined by the concept of calibration. Similar to other methods of direct analysis of solid materials (such as XRF, LIBS or LA-ICP-MS) GD spectroscopy is not an absolute method and therefore calibration is a prerequisite for accurate analysis. In comparison to GDOES, quantification in GDMS is conceptually easier due to the relatively low numbers of peaks in a mass spectrum compared to the large number of lines in an emission spectrum. GDMS mass spectra are mainly based on the well-known stable isotopes with a well-known natural abundance. Of course, interferences affect the identification and detection of different analytes. In particular, these spectral interferences could be a major problem for the determination of certain ultratrace elements.

**Ion Beam Ratio**

The most common approach for quantification is based on the denominated "ion beam ratio" (IBR) procedure. It can be applied not only for the analysis of pure materials, where one element, which represents the matrix (m), is estimated to have a concentration of 100%, but also for the analysis of nonpure materials (*e.g.* alloys). In this latter case, the concentration of a major element is needed in order to correct its ion signal intensity by its concentration, or alternatively the ion signal intensities of the major elements, whose total concentration is about 100%, are added. Considering that the detected number of ions is proportional to the number of atoms contained in the sample, the ion signal ratio of the ion signals ($I$) (measured on the isotopes under consideration of their abundances) for any analyte (x) with respect to the matrix element represents the number of atoms of this element relative to the atoms of the matrix. Therefore, the atomic concentration ($C$) of an element (x) is obtained from:

$$\mathrm{IBR} = (I_x/I_m) = C_x/C_m \tag{13.1}$$

IBR: ion beam ratio
$I_x$: ion signal of an element x
$I_m$: ion signal of matrix element m
$C_x$: atomic concentration of an element x
$C_m$: atomic concentration of a matrix element m

It should be mentioned that a prerequisite for this procedure is that a sputtering equilibrium has been reached so the intensity ratio remains constant during the time needed for analysis. Moreover, this is a semi-quantitative approach, because it is assumed that the relationship between the intensity of the element x and its concentration (sensitivity of element x) is very close to the relationship between the intensity of the matrix element m and its concentration. Usually, the differences in sensitivity, caused amongst others by differences in ionisation and ion transport in the GD source, can be as high as a factor of 10 and in some cases like N, O and F even more. Therefore, the concentration values calculated by the IBR procedure could be relatively inaccurate, although for many applications, such as ultratrace analyses, this accuracy is considered to be good enough and thus it is often applied for routine analysis. The main advantage of the IBR strategy is its universality. Therefore, IBR can be used even for analytes and matrices where due to the absence of calibration standards and estimated relative sensitivity factors (see below), no other quantification approach can be applied.

**Calibration Curves and Relative Sensitivity Factors**
For a quantitative analysis a calibration procedure with relevant standards or certified reference materials (CRM) is essential. However, the main limitation for calibration is the lack of CRM that could be used for the determination of many elements at trace and ultratrace concentration level. If appropriate CRMs are available, the calibration curve of an element (x) can be obtained by plotting the intensities measured for such element as a function of its certified mass content on the different CRMs. The calibration curve might be obtained using a single CRM or using a set of CRMs that cover a certain interval of analyte concentrations. The slope of the calibration curve is then defined as the sensitivity $S_x$ of the element x.

Moreover, it is possible to define the relative sensitivity factor of an element x ($\mathrm{RSF}_x$) as the ratio between the sensitivity of a reference element, denominated the internal standard IS ($S_{IS}$), and the sensitivity of the element x ($S_x$):

$$RSF_x = S_{IS}/S_x = (I_{IS}/C_{IS})/(I_x/C_x) \qquad (13.2)$$

$RSF_x$: relative sensitivity factor of an element x
$S_{IS}$: sensitivity of a reference element (*i.e.* internal standard IS)
$S_x$: sensitivity of the element x
$I_x$: ion signal of an element x
$I_{IS}$: ion signal of a reference element (*i.e* internal standard IS)
$C_x$: atomic concentration of an element x
$C_{IS}$: atomic concentration of a reference element (*i.e.* internal standard IS)

Therefore, the mass content $\omega$ of an element x is calculated as:

$$\omega_x = ((I_x/I_{IS}) * RSF_x) * \omega_{IS} \qquad (13.3)$$

$RSF_x$: relative sensitivity factor of an element x
$I_x$: ion signal of an element x
$I_{IS}$: ion signal of a reference element (*i.e* internal standard IS)
$\omega_x$: mass content of an element x
$\omega_{IS}$: mass content of a reference element (*i.e.* internal standard IS)

If the internal standard is the matrix element of a pure sample, whose mass content is assumed to be 1 ($\omega_{IS} = 1$), then the equation is:

$$\omega_x = IBR * RSF_x \qquad (13.4)$$

$\omega_x$: mass-content of an element x
IBR: ion beam ratio
$RSF_x$: relative sensitivity factor of an element x

Many different physical processes, such as the sputtering, ionisation, extraction and transport of ions, contribute to the RSF. Therefore, it is recommended that these factors are calculated for a selected set of working conditions.

For instance, Figure 13.16(a) and (b) shows the influence of gas flow rate and discharge current on the RSF calculated for different materials in a Cu-matrix sample. Nevertheless, it is observed that the range of values of the RSFs is relatively limited, from about 0.1 to 4.[79]

Additionally, Figure 13.17 shows that these values are reproducible when measured on different days. Furthermore, RSFs are transferable between instruments once the ion source hardware and plasma discharge conditions are comparable.[79]

It has been noticed that RSFs are very similar for some matrices if identical internal standards are chosen. For instance, a library of RSF values, calculated using the *VG 9000* GDMS system, is commonly used to improve

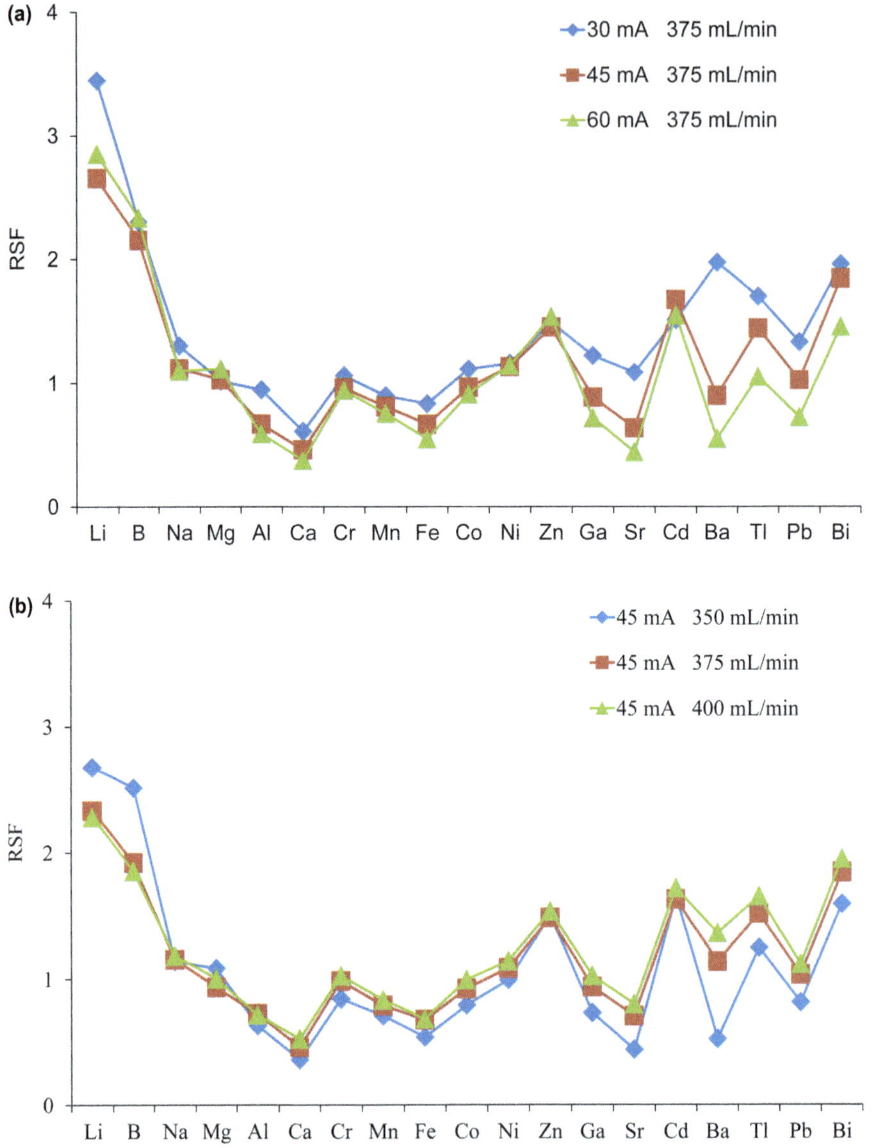


**Figure 13.16** Influence of (a) gas flow rate and (b) discharge current on the RSF calculated for different materials in a Cu-matrix sample.

the accuracy on the quantitative analysis of matrices for which no reference materials or standards are available. Nevertheless, it has been found that RSF values could be matrix specific: For example, RSFs for a silicon matrix differ significantly from those for metallic conductor matrices even under identical instrumental parameters.[65]

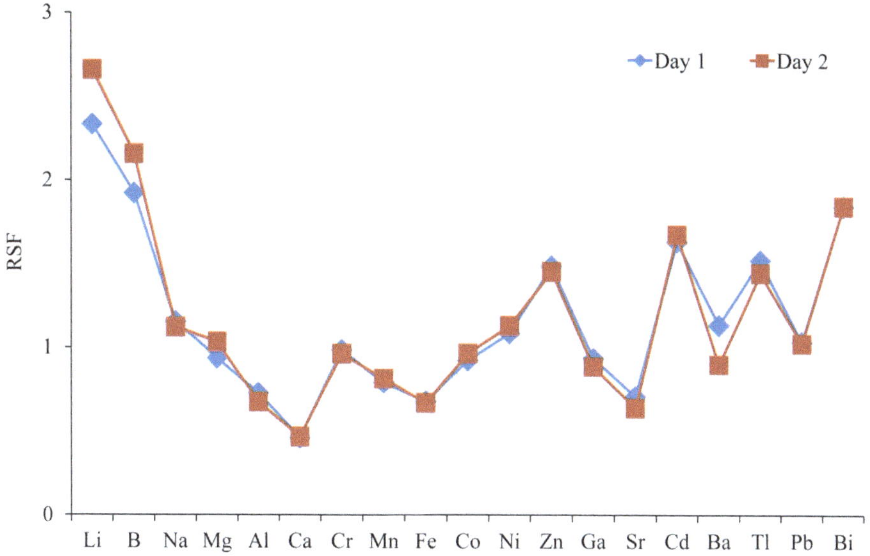


**Figure 13.17** Reproducibility of the RSF measured on two different days in a Cu-matrix sample.[79]

**Standard Relative Sensitivity Factors**

As mentioned above, calibration standards for many matrices and analytes are lacking, resulting in a restricted number of matrix-specific RSF factors. In order to overcome this limitation the denominated Standard RSF quantification strategy was developed. Standard RSF is usually defined as the RSF of an analyte normalised to the RSF of Fe (as one of the most common chemical elements present in analysed samples). An example is used to illustrate the quantification procedure based on Standard RSFs:[29] Different matrix-matched reference material, for example aluminium and iron in nickel and copper and iron in aluminium, may be used for the determination of RSFs:

$$RSF_{Al/Ni} = I_{Ni}/I_{Al} * \omega_{Al/Ni} \quad (13.5a)$$

$$RSF_{Fe/Ni} = I_{Ni}/I_{Fe} * \omega_{Fe/Ni} \quad (13.5b)$$

$$RSF_{Cu/Al} = I_{Al}/I_{Cu} * \omega_{Cu/Al} \quad (13.5c)$$

$$RSF_{Fe/Al} = I_{Al}/I_{Fe} * \omega_{Fe/Al} \quad (13.5d)$$

RSF: relative sensitivity factor of a specific element
*I:* ion signal of a specific element/isotope
$\omega_{x/y}$: certified mass content of element x in matrix y

a) The standard RSF (StdRSF) are calculated. In this sense, the normalisation of the analyte and matrix element specific RSF is performed as follows:

$$\mathrm{StdRSF_{Al}} = \mathrm{RSF_{Al/Ni}}/\mathrm{RSF_{Fe/Ni}} \quad (13.6a)$$

$$\mathrm{StdRSF_{Cu}} = \mathrm{RSF_{Cu/Al}}/\mathrm{RSF_{Fe/Al}} \quad (13.6b)$$

StdRSF: relative sensitivity factor of a specific element
RSF: standard relative sensitivity factor of a specific element

b) The standard RSF values, calculated for many elements in a large number of reference materials and matrices, are stored in the software of the GDMS instrument, defining a standard RSF set.

c) The standard RSF are applied as follows for the quantification of an analyte in any sample (*e.g.* determination of Al in Cu).

$$\omega_{\mathrm{Al/Cu}} = I_{\mathrm{Al}}/I_{\mathrm{Cu}} {}^{*}\mathrm{StdRSF_{Al}}/\mathrm{StdRSF_{Cu}}) \quad (13.7)$$

StdRSF: relative sensitivity factor of a specific element
$I$: ion signal of a specific element/isotope
$\omega$: certified mass content of element x in matrix y

Several extensive round-robin analyses have demonstrated that the standard RSF*s* produce an acceptable level of quantification. This concept works well for survey analyses in various matrices and delivers accuracy of results within a factor of 2 from "true" values for many elements. However, for some elements such as nonmetals the uncertainty can exceed a factor of 10. Nevertheless, standard RSF quantification is often applied in cases where no matrix-specific calibration is available as a more accurate approximation than IBR.

**Calibration with Synthetic Standards**

With the aim of achieving highly accurate quantification with GDMS for rare elements and matrices, the calibration strategy with self-prepared synthetic standards was developed by Matschat *et al.*[92] and then further improved, including evaluation of uncertainties, within the PhD thesis of Gusarova.[79] A similar concept has been applied to LA-ICP-MS quantitative analysis, as well.[93,94]

Synthetic standards are self-prepared matrix-matched samples, made of high-purity matrix powder doped with standard stock solutions of analytes, homogenised and pressed under high pressure into pellets within steel rings as can be seen in Figure 13.18. Figure 13.19 shows that the pressure used for the production of the synthetic calibration samples is a key parameter regarding the accuracy.[79] In this sense, the results indicate that high pressure is recommended.

This approach was shown to be advantageous as it provides matrix-adapted calibration, even at ultratrace ng $g^{-1}$ level. Validation of the trace-element mass fractions of the synthetic standards was carried out using ICP-MS and additional crosschecking was performed using available CRMs

**Figure 13.18** (left) Steel ring (centre) steel ring with pressed powder (right) sputtered sample.

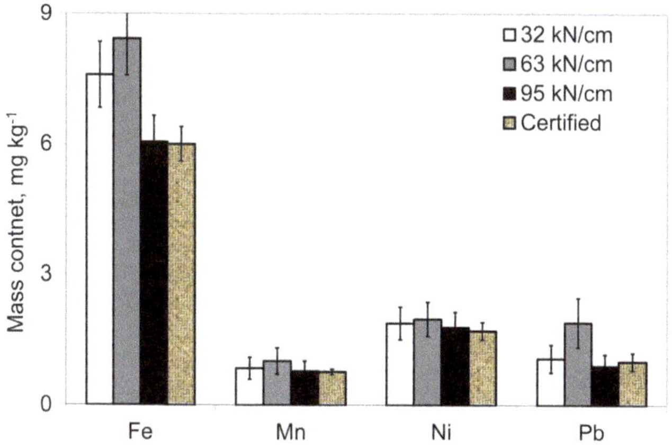


**Figure 13.19** Mass concentration of Fe, Mn, Ni and Pb calculated using calibrations based on three different pressure regimes (32, 63, 95 kN $cm^{-2}$, respectively) for the production of the synthetic calibration samples.

to obtain the calibration curves for some elements. The results showed that accurate results with small uncertainties were achieved for almost 50 trace elements. The other advantage of this concept is that the GDMS calibration can be made traceable to SI in the shortest possible way as synthetic standards are prepared from high-purity metal powders gravimetrically doped with traceable stock solutions. The feasibility of calibration with synthetic standards as a quantification strategy for GDMS was successfully demonstrated in the interlaboratory comparison CCQM P 107 for high-purity zinc samples. Another advantage of this calibration concept is, that synthetic samples can be produced covering the concentration range of the required analyte concentration, including very low concentrations that are not always available using solid calibration materials. For instance Figure 13.20 shows the calibration curves obtained for $^{149}Sm$ and for $^{24}Mg$ using synthetic standards.

Moreover, it should be noted that using this calibration approach it is possible to achieve uncertainty values below 30% for most elements. Table 13.7 lists the measured concentrations with their associated

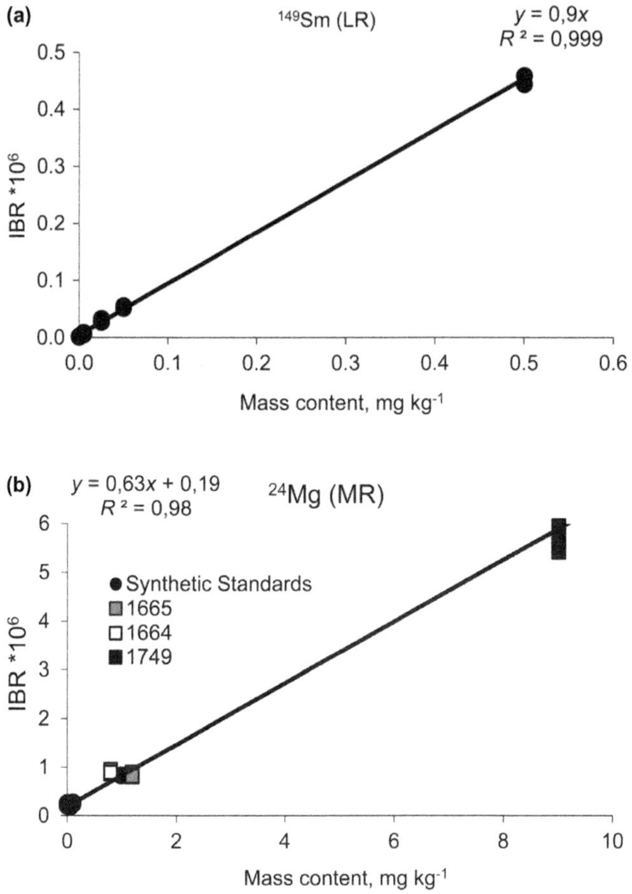


**Figure 13.20** Calibration curves obtained for (a) $^{149}$Sm and (b) for $^{24}$Mg using synthetic standards.[79]

uncertainties for different elements in copper-CRMs (BAM M382, BAM M383, BAM-M384).[79] Pin samples can also be pressed using a different pressing device. However, a significant difference in RSF could be observed when flat or pin-type geometries were used. Figure 13.21 shows the RSF calculated for different elements in samples prepared using different geometries.

Initially, the concept of the synthetic standard calibration was developed to calibrate multielemental trace analysis by GDMS for the certification of high-purity metals as primary standards or for standard materials used in metrological thermometry by National Metrology Institutes. Such calibration can also be used in industrial analytical work for quality control of high-purity metals producers. This concept is more flexible and can be applied to different matrices and a large number of analyte elements and concentration ranges providing accurate results traceable to SI. However the limiting factor for the applicability is the availability of high-purity metal powders in order

**Table 13.7** Measured and certified mass contents $\omega$ of analytes in copper-CRM (BAM M382, BAM M383, BAM-M384) in mg/kg, and relative deviation of the measured from the certified values ($D = 100\times(w_m\ w_z)/w_z$) in per cent. The expanded uncertainties $U_m$ (expanded uncertainty of the measurement) and $U_z$ (expanded uncertatiny of the certified value) were calculated from the combined uncertainty with a coverage factor of $k = 2$.[79]

| | BAM–M382 | | | | | BAM–M383 | | | | | BAM–M384 | | | | |
|---|---|---|---|---|---|---|---|---|---|---|---|---|---|---|---|
| | measured | | certified | | | measured | | certified | | | measured | | certified | | |
| | $w_m$ | $U_m$ | $w_z$ | $U_z$ | $D$, % | $w_m$ | $U_m$ | $w_z$ | $U_z$ | $D$, % | $w_m$ | $U_m$ | $w_z$ | $U_z$ | $D$, % |
| Ag | 2.1 | ± 0.3 | 1.8 | ± 0.2 | 16 | 5.3 | ± 0.4 | 4.7 | ± 0.2 | 12 | 10.7 | ± 0.7 | 10.3 | ± 0.4 | 4 |
| As | n.c.[a] | | n.c.[a] | | | 2.1 | ± 0.5 | 1.9 | ± 0.2 | 7 | 4.3 | ± 0.5 | 5.0 | ± 0.4 | −14 |
| Bi | 0.5 | ± 0.1 | 0.53 | ± 0.03 | −2 | 1.3 | ± 0.1 | 1.02 | ± 0.09 | 24 | 3.3 | ± 0.1 | 3.3 | ± 0.2 | −0.4 |
| Cd | 0.9 | ± 0.6 | 0.90 | ± 0.09 | −4 | 1.8 | ± 0.6 | 1.5 | ± 0.2 | 20 | 4.1 | ± 0.3 | 4.0 | ± 0.1 | 3 |
| Co | 0.6 | ± 0.2 | 0.73 | ± 0.07 | −15 | 1.3 | ± 0.2 | 1.37 | ± 0.05 | −6 | 3.4 | ± 0.2 | 3.0 | ± 0.2 | −12 |
| Mn | 0.7 | ± 0.1 | 0.76 | ± 0.06 | −8 | 1.2 | ± 0.1 | 1.24 | ± 0.05 | −0.8 | 5.8 | ± 0.1 | 6.9 | ± 0.2 | −16 |
| Ni | 1.8 | ± 0.5 | 1.7 | ± 0.2 | 7 | 4.2 | ± 0.6 | 3.6 | ± 0.2 | 16 | 6.2 | ± 0.7 | 5.7 | ± 0.4 | 8 |
| Pb | 1.2 | ± 0.2 | 1.0 | ± 0.2 | 22 | 2.2 | ± 0.2 | 1.3 | ± 0.2 | 68 | 5.5 | ± 0.2 | 5.7 | ± 0.5 | −4 |
| Te | 0.8 | ± 0.3 | 0.61 | ± 0.06 | 24 | 1.8 | ± 0.3 | 1.4 | ± 0.2 | 28 | 7.9 | ± 0.4 | 7.0 | ± 0.5 | 13 |

[a]n. c. – not certified.

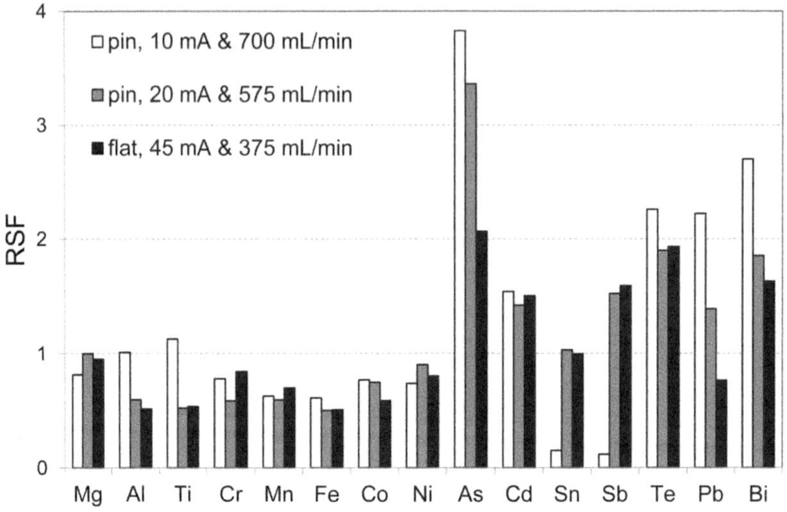


**Figure 13.21** RSF calculated for different elements in samples prepared using different geometries.[79]

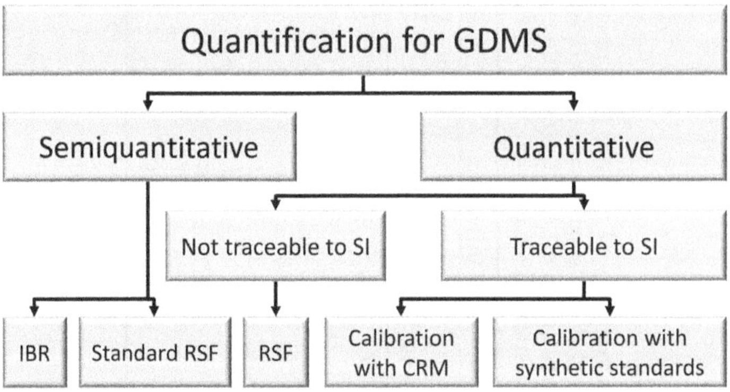


**Figure 13.22** Summary of the main quantification approaches used for GD-MS.[87]

to produce the mixtures with standard stock solutions as well as the chemical compatibility of the metal powders with the stock solutions. Besides, the preparation of such standards required considerable effort and many precautions should be taken in order to avoid losses or contaminations.

Figure 13.22 summarises the main quantification approaches used for GDMS. In particular, quantitative analysis today still requires the determination of matrix-matched element-specific RSF, which can be calculated in 3 different ways:

(1) Using certified reference material (CRM). However, due to the lack of availability, this is only possible for a very limited number of analyte/matrix combinations.[95]

(2) Using quantitative analytical methods, such as wet chemical or other calibrated methods (*e.g.* ICP-MS or ICP-OES after digestion, rapid neutron activation analysis (RNAA), combustion or hot-gas extraction analysis for C, H, N, O, S).[61]
(3) Using prepared synthetic standards, such as high-purity powders doped with CRM-traceable standard stock solutions.[79,87]

From a pragmatic point of view, the accuracy of GDMS has been assessed in several round-robin analyses. For instance, Venzago *et al.* organised a round-robin analysis, where 10 different laboratories in Europe participated in the analysis of aluminium samples with a purity of 4N (99.99%).[61] In particular, it was shown that for 23 elements no significant bias between GDMS results and accepted reference values, determined from RNAA and wet chemistry, was observed.

Another round robin was organised on Cu CRMs (BCR-CRM 075 and BCR-CRM 074) (IRMM Institute for Reference Materials and Measurements (IRMM) is one of the seven institutes of the Joint Research Centre (JRC), a Directorate-General of the European Commission (EC), Geel, Belgium) with the participation of 14 laboratories (see Table 13.8) that employed different instruments: *VG 9000* (GDSFMS) with pin and flat GD source geometry, *VG Axiom* (SFMS from VG Elemental with Nier–Johnson geometry) coupled to a GD source (prototype GDSFMS), *TS Sola* (GDQMS, Turner Scientific Ltd, UK), *LECO GDA750* (GDOES, LECO, St. Josef MI, USA), *Philips PV 8055* (GDOES, Philips, Eindhoven, Netherlands).[95] This round robin showed much higher deviations between laboratories, which is not surprising, since different source geometries, quantification models, detection instruments were used.

Another interlaboratory comparison was performed in order to compare two commercial GDSFMS systems, the *Element GD* and the *VG 9000*, respectively (see Table 13.9).[87] The calibration of both GDSFMS instruments was carried out using standard RSF and self-prepared synthetic compressed powder standards, from a zinc matrix. Validation of the trace-element mass fractions of the synthetic standards was carried out using ICP-MS and additional crosschecking was performed using available CRMs to obtain the calibration curves for some elements. The results showed that accurate results with small uncertainties were achieved for almost 50 trace elements. Nevertheless, the quantification approach based on the use of standard RSF was proven to be inaccurate in a few cases for the *VG 9000* but much more often for the *Element GD*. This is most probably related to the fact that on the *VG 9000*, at the beginning of the lifetime of the instrument, the determination of standard RSF was carried out using many different calibration materials. In contrast, on the *Element GD* the standard RSF were determined only using iron reference materials, combining flat sample steel reference materials with spiked iron pellets.[92] Additionally, the determination of the standard RSF on the *Element GD* was accomplished at an early stage of the instrument development, while the currently used plasma conditions such as flow rates and pressure are slightly different.

**Table 13.8** Round robin organised on Cu CRMs (BCR-CRM 075 and BCR-CRM 074) with a participation of 14 laboratories that employed different instruments: *VG 9000* (GD-SFMS) with pin and flat GD source geometry, *VG Axiom* with GD source (GD-SFMS), *TS Sola* (GD-QMS), *LECO GDA750* (GD-OES), *Philips PV 8055* (GD-OES).[95]

| **Test #** | 1 | 2 | 3 | 4 | 5 | 6 |
|---|---|---|---|---|---|---|
| **Lab #** | 1 | 2 | 3 | 4 | 6 | 7 |
| **Instrument (type):** | GDMS – *VG 9000* | GDMS – *VG 9000* | GDMS – *VG 9000* | GDMS – *VG 9000* | GDMS – *VG 9000* | GDMS – *VG 9000* |
| **Mechanical prep.** | Y | Y | Y | – | – | – |
| **Dilute HNO3 etching** | Y | Y | Y | Y | Y | Y |
| **DI water rinsing** | Y | Y | Y | Y | Y | Y |
| **Alcohol rinsing** | Y | Y | Y | Y | Y | Y |
| **Geometry** | Pin | Pin | Pin | Pin | Pin | Pin |
| **Cell type** | Mega cell | Mega cell | Mega cell | – | Mega Cell | Mega Cell |
| ***U* [kV]** | 1–1.05 | 1 | 1 | 0.96–1.1 | 1.1 | 1.1 |
| ***I* [mA]** | 3–3.4 | 4 | 3 | 1–1.2 | 3 | 3.5 |
| **Matrix intensity [A]** | 3.2–6 E–10 | 1×E–9 | 7.9–9.8 E–10 | 1–3 E–11 | 4 E–10 | 5 E–10 |
| **Calibration/ RSF set** | Standard set | Modified set | Standard set | Standard set | Standard set | Modified set |
| **Point in scan** | 70 | 70 | 50 | 140 | 60 | 30 |
| **DAC step** | 5 | 7 | 6 | 4 | 6 | 10 |
| **Faraday int. time [ms]** | – | 160 | 160 | 160 | 160 | 20 |
| **Daly int. time [ms]** | 300–500 | 100 | 100–1000 | 100 | 400–1200 | 50–1500 |
| **Number of reported elements (sample A/B)** | 28/28 | 29/28 | 30/29 | 31/30 | 31/30 | 31/31 |

## 13.3.5 Analysis of Nonconducting Materials

Nonconducting materials cannot be directly analysed by DC GDSFMS, because it is not possible to sustain the voltage drop at the sample surface (*e.g.* there are no free electrons in the sample) and the plasma cannot be ignited or is extinguished after a short period of time. On the other hand, semiconducting samples might be analysed but in that case the analytical performance of DC GDSFMS is not as good as for the analysis of conducting materials. Different approaches have been investigated in order to analyse nonconducting and semiconducting materials with improved analytical figures of merit, including the use of secondary electrodes, mixing these materials with conductive binders, or using radio-frequency power supplies that are able to generate a negative DC-bias even at the surface of nonconducting samples.

| | | | | | | |
|---|---|---|---|---|---|---|
| 8 | 9 | 10 | 11 | 12 | 13 | 14 |
| 2 | 3 | 5 | 8 | 9 | 9 | 10 |
| GDMS – *VG 9000* | GDMS – *VG 9000* | GDOES – *Philips PV 8055* | GDOES – *GDA750* | GDMS – *TS Sola low resolution* | GDOES – *GD750* | GDMS – *GD/Axiom high resolution* |
| N | Y | – | N | – | N | Y |
| Y | N | N | N | N | N | N |
| Y | Y | N | N | N | N | Y |
| Y | Y | Y | Y | N | N | Y |
| Flat | Flat | Flat | Flat (dia 4mm) | Flat (dia 8mm) | Flat (dia 8mm) | Flat |
| Mega Flat cell | Flat cell | | GRIMM | Modified GRIMM | GRIMM | Modified GRIMM |
| 1 | 1 | 0.6 | 1 | 1.1 | 1 | 0.6 |
| 4 | 6 | 60 | 35 | 100 | 93 | 35–45 |
| 1×E–9 | 6.7–7.8 E–10 | – | – | 1.8E10cps | – | 2E9 cps |
| Modified set | Modified set | Linear calibration | Pure Cu samples | Linear calibration | Linear calibration | Linear calibration |
| 70 | 50 | | | | | |
| 7 | 6 | | | | | |
| 160 | 160 | | | | | |
| 100 | 100–1000 | 100 | – | 200 | 100 | 0.6 |
| 29/28 | 28/26 | 10/0 | 18/0 | 9/9 | 3/3 | 10/4 |

**Secondary Electrode**

The approach of using a conducting mask on the cathode surface as secondary cathode was introduced some time ago.[96] It was employed for some specific application using the *VG 9000* instrument, and it has been recently reintroduced using new commercial instruments.[97,98] In particular, applications have been presented for the analysis of particulate matter,[71] iron ores,[99] Macor[100] and Zr-oxide.[101] Moreover, the optimisation of cathode thickness was described in detail by Schelles and Van Grieken.[102]

**Mixing with Conductive Material**

Mixing of nonconducting powders with conducting powders used as a binder – a technique well known from SSMS – is a more universal approach.

**Table 13.9** Comparison of some RSF values for *VG 9000* and *Element GD* obtained by calibration with synthetic standards ($RSF_{cal}$) and from Standard RSF values ($RSF_{StdRSF}$) respectively. The relationship of the both RSF values is displayed in bold if the uncertainty of Standard RSF concept of a factor of two does not cover the discrepancy, *i.e.* their ratio exceeds two or falls below 0.5.[79]

| | *VG 9000* | | | *Element GD* | | |
|---|---|---|---|---|---|---|
| Isotope | $RSF_{cal}$ | $RSF_{StdRSF}$ | $RSF_{StdRSF}$/ $RSF_{cal}$ | $RSF_{cal}$ | $RSF_{StdRSF}$ | $RSF_{StdRSF}$/ $RSF_{cal}$ |
| $^{7}$Li[a] | 0.22 | 0.29 | 1.3 | 2.7 | 0.81 | **0.3** |
| $^{9}$Be | 0.24 | 0.26 | 1.1 | 2.92 | 1.32 | **0.5** |
| $^{11}$B | 0.08 | 0.22 | **2.9** | 2.25 | 1.69 | 0.8 |
| $^{23}$Na | 0.39 | 0.35 | 0.9 | 1.28 | 0.25 | **0.2** |
| $^{24}$Mg | 0.29 | 0.27 | 0.9 | 1.08 | 0.39 | **0.4** |
| $^{27}$Al | 0.3 | 0.24 | 0.8 | 1.13 | 0.33 | **0.3** |
| $^{31}$P | 0.41 | 0.55 | 1.3 | 4.02 | 0.96 | **0.2** |
| $^{39}$K[b] | 0.44 | 0.21 | **0.5** | 0.27 | 0.15 | 0.6 |
| $^{44}$Ca | 0.07 | 0.1 | 1.3 | 0.52 | 0.12 | **0.2** |
| $^{45}$Sc | 0.22 | 0.07 | **0.3** | 1.02 | 0.15 | **0.2** |
| $^{51}$V | 0.24 | 0.1 | **0.4** | 0.85 | 0.14 | **0.2** |
| $^{52}$Cr | 0.55 | 0.38 | 0.7 | 0.84 | 0.33 | **0.4** |
| $^{55}$Mn | 0.4 | 0.25 | 0.6 | 0.8 | 0.26 | **0.3** |
| $^{56}$Fe | 0.34 | 0.17 | 0.5 | 0.61 | 0.26 | **0.4** |
| $^{59}$Co | 0.31 | 0.19 | 0.6 | 0.86 | 0.27 | **0.3** |
| $^{60}$Ni[a] | 0.46 | 0.26 | 0.6 | 0.86 | 0.39 | **0.5** |
| $^{63}$Cu | 1.31 | 0.89 | 0.7 | 1.46 | 0.64 | **0.4** |
| $^{75}$As[a] | 1.11 | 0.88 | 0.8 | 2.81 | 1.34 | **0.5** |
| $^{77}$Se[b] | 1.3 | 0.56 | **0.4** | 2.22 | 0.98 | **0.4** |
| $^{88}$Sr[a] | 0.36 | 0.11 | **0.3** | 1.23 | 0.14 | **0.1** |
| $^{89}$Y[a] | 0.52 | 0.09 | **0.2** | 1.77 | 0.14 | **0.1** |
| $^{109}$Ag | 0.97 | 0.75 | 0.8 | 1.35 | 1.01 | 0.7 |
| $^{115}$In[a] | 0.93 | 0.68 | 0.7 | 1.51 | 0.38 | **0.3** |
| $^{133}$Cs | 1.54 | 0.17[c] | **0.1** | 3.19 | 0.26 | **0.1** |
| $^{139}$La[a] | 1.22 | 0.13 | **0.1** | 3.76 | 0.17 | **0.04** |
| $^{140}$Ce | 1.87 | 0.13 | **0.1** | 3.59 | 0.2 | **0.1** |
| $^{141}$Pr[a] | 1.47 | 0.14 | **0.1** | 3.19 | 0.21 | **0.1** |
| $^{147}$Sm | 0.8 | 0.13 | **0.2** | 2.18 | 0.19 | **0.1** |
| $^{153}$Eu[a] | 0.66 | 0.14 | **0.2** | 2.45 | 0.17 | **0.1** |
| $^{157}$Gd[a] | 0.87 | 0.12 | **0.1** | 1.18 | 0.18 | **0.2** |
| $^{159}$Tb[a] | 0.88 | 0.12 | **0.1** | 2.49 | 0.22 | **0.1** |
| $^{163}$Dy[a] | 0.74 | 0.13 | **0.2** | 2.33 | 0.2 | **0.1** |
| $^{165}$Ho[a] | 0.72 | 0.14 | **0.2** | 2.27 | 0.21 | **0.1** |
| $^{166}$Er | 0.69 | 0.13 | **0.2** | 1.89 | 0.2 | **0.1** |
| $^{169}$Tm | 0.63 | 0.15 | **0.2** | 1.75 | 0.2 | **0.1** |
| $^{175}$Lu | 0.65 | 0.14 | **0.2** | 1.96 | 0.17 | **0.1** |
| $^{205}$Tl[a] | 1.36 | 0.17[c] | **0.1** | 1.85 | 1.07 | 0.6 |
| $^{208}$Pb[a] | 0.57 | 0.44 | 0.8 | 1.53 | 0.36 | **0.2** |
| $^{209}$Bi[a] | 0.95 | 0.7 | 0.7 | 1.86 | 0.77 | **0.4** |
| $^{232}$Th[a] | 0.95 | 0.1 | **0.1** | 3.03 | 0.24 | **0.1** |
| $^{238}$U[a] | 0.79 | 0.14 | **0.2** | 2.62 | 0.26 | **0.1** |

[a]Was measured in low resolution on *Element GD*.
[b]Was measured in high resolution on *Element GD*.
[c]No StdRSF value was available for Cs and Tl; in this case the StdRSF was set to be equal 1 and divided by the StdRSF value of the matrix element (eqn (13.4)).
The uncertainty of this value corresponds to a factor of five.

In this case, graphite, gallium or indium is used. For example ashed petroleum products[103] or soils were analysed using Al powder as binder.[72,104–106]

A new method has been developed by Qian *et al.*,[107] using a surface coating of molten indium on pin-shaped samples. After about 25 min of stabilisation time, the analyte ion signals from the DC-GDSFMS system were measured with an internal precision better than 10% RSD. Furthermore, between-sample precision could be found in the range of 14% or better. This method was applied to the analysis of $SiO_2$, $Al_2O_3$, $PWO_4$, $ZrO_2$, and $LaTi_2O_7$. Additionally, the accuracy of the determined values in the measurement of NIST 620 (National Institute of Standards and Technology, Gaithersburg, US) oxide standard was reported to be in the range of ±10%. (Tables 13.10 and 13.11).

Nonconducting powders can also be pressed into or onto soft solid metals like indium, lead or tin. An infiltration technique under high pressure similar to the production of metal matrix composites has been demonstrated as suitable for the production of GDMS samples.[73] All these mixing techniques are prone to contamination during sample preparation (mixing, pressing and milling if necessary) and background contaminations from the conducting agent. Prolonged presputter times are necessary to remove near-surface layers containing eventual contamination from pressing devices and possible material inhomogeneity in the near surface area. Additionally, the

**Table 13.10** Typical elemental compositions in $La_2Ti_2O_7$ (relative to Ti) analysed by GDMS using a surface coating method.

| Element | Meas 1 ($\times10^{-6}$) | Meas 2 ($\times10^{-6}$) | Meas 3 ($\times10^{-6}$) | Meas 4 ($\times10^{-6}$) | Average ($\times10^{-6}$) | RSD (%) |
|---|---|---|---|---|---|---|
| Co | 65 | 63 | 63 | 60 | 63 | 2.8 |
| Nb | 409 | 428 | 370 | 381 | 397 | 5.8 |
| Mo | 13 | 14 | 14 | 16 | 14 | 7.6 |
| Sn | 22 | 21 | 19 | 19 | 20 | 6.4 |
| Ba | 495 | 510 | 520 | 533 | 515 | 2.7 |
| Lu | 2.1 | 1.9 | 1.7 | 1.7 | 1.9 | 9.0 |

**Table 13.11** Comparison of DC GDMS results of NIST 620 calibrated by RSF and the certified values (relative to Si).

| Element | Meas 1 (%) | Meas 2 (%) | Meas 3 (%) | Meas 4 (%) | Average (%) | Calibrated value by RSF (%) | Certified value (%) | RSD (%) | Relative error (%) |
|---|---|---|---|---|---|---|---|---|---|
| Na | 18.81 | 17.49 | 16.30 | 19.33 | 17.98 | 29.85 | 31.69 | 7.4 | 5.9 |
| Mg | 3.56 | 2.97 | 3.48 | 3.37 | 3.37 | 6.3 | 6.6 | 7.8 | 4.5 |
| Al | 3.18 | 3.26 | 3.60 | 2.95 | 3.24 | 2.72 | 2.83 | 8.3 | 3.9 |
| S | 0.24 | 0.25 | 0.26 | 0.23 | 0.24 | 0.30 | 0.33 | 5.3 | 9.1 |
| Ca | 10.36 | 10.80 | 10.16 | 9.13 | 10.11 | 16.78 | 15.08 | 7.0 | 0.11 |
| Fe | 0.018 | 0.017 | 0.017 | 0.017 | 0.017 | 0.087 | 0.089 | 2.9 | 2.2 |

(A certified value of each element was directly calculated from the certified value of related oxides ($Na_2O$, MgO, $Al_2O_3$, $SO_3$, CaO and FeO) in the NIST 620.)

addition of a conducting agent provides a dilution of the sample, which additionally lowers sensitivities for trace analyses.

**Radio-Frequency GDSFMS**

A more universal approach for analysis of nonconducting materials is the operation of the GD source with RF voltages.[60,108,109] It should be mentioned that up to now a commercial RF source is only available for the *Autoconcept GD90*. However, no publication is known showing results, so far. The only scientific presentation on this subject was given by Marcus.[110] According to the manufacturer significant progress has been achieved reaching sensitivities down to the low ng $g^{-1}$ range of impurities in alumina [private communication Ekbal Patel, MSI] (Figure 13.23).

Nevertheless Duckworth *et al.*[111,112] used a slightly modified original source of the *VG 9000* and coupled an RF potential *via* a direct insertion probe to a nonconducting sample (*e.g.* a glass sample). They could demonstrate the potential of this approach but were limited by a broadening of the ion-energy distribution, which resulted in a worsening of the ion transmission.

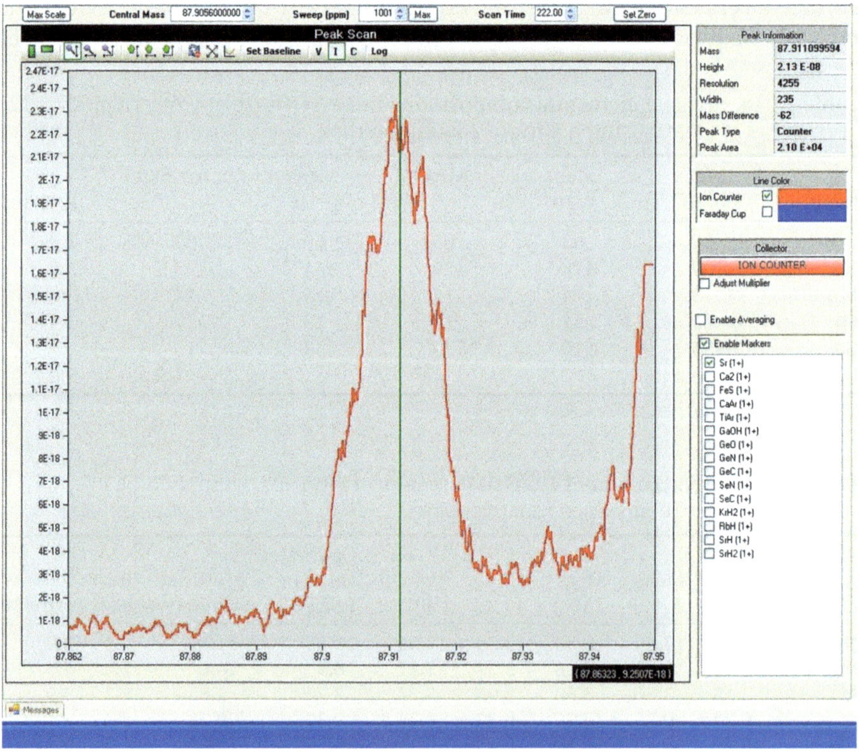


**Figure 13.23** Screen shot of *GD 90* mass spectrum showing the $^{88}Sr$ peak as a trace impurity of 22 ng $g^{-1}$ Sr analysed in fine Alumina Powder CRM 8007-A (National Metrology Institute of Japan, Tsukuba, JP).

The main applications for RF discharges can be seen in the semiconductor industry, for the analysis of glasses and ceramics as well as for polymers. For instance, a home-built RF GD source was developed by Becker and coworkers for the analysis of nonconducting samples,[113] also with a magnetron plasma boost.[114–116] A ring-shaped magnet located behind the sample cathode material direct ions along the magnetic field lines towards the cathode and therefore enhance sputtering and ionisation. A magnetic field strength of several hundred Gauss was applied. They coupled this source to an ICP-SFMS operated with low mass resolution and had achieved detection limits of 10 ng $g^{-1}$ for B in high-ohmic GaAs.[117] Other home-built instruments had been used to analyse Teflon, transition metal oxides, glass samples and ceramics. In all these examples, the composition of the materials investigated was the main interest. In recent times RF GDTOFMS has shown the capability to analyse various non conducting materials such as ceramics glasses and polymers, but detection limits are still some orders of magnitude higher than known from GD-SFMS.[118]

## 13.3.6 Depth Profiling by GDMS

Even though the number of GDMS instruments performing depth profiling and layer analysis is small compared to the bulk metals analysis depth profiling by GDSFMS is still an important method for materials science and development.

The fundamental idea to apply GDSFMS for indepth analysis is based on the fact that a well-defined fraction of atoms sputtered from the sample surface is converted into ions and reaches the detector nearly simultaneously so that in principle the ions detected are representative for the surface composition if the scanning time of the MS is chosen appropriately. It has been shown by Sanderson *et al.*,[119,120] that the commercial sector field instrument operated with a GD source with flat-cell geometry can also be applied for the analytical characterisation of surfaces and near-to-surface layers.

A limiting factor for this concept is the scanning speed of the mass spectrometer. By performing an indepth analysis the intensity time profile is measured. From the knowledge of the sensitivity, the ion intensity can then be converted into the concentration and from the time needed to sputter a particular crater depth the whole concentration-depth-profile can be reconstructed. But this approach can only be applied if the layer or the bulk material provides an internal standard for quantification and if sensitivities or better to say RSF do not change for the bulk and layer material. It is also assumed that the sputter rates do not severely change during the sputtering of the layer so that the depth at each time can be easily calculated from the total crater depth measured after the analysis. However, even in the case when the matrix is changing between layer and substrate either the standard RSF concept of useful yields or normalisation to total ion intensities can be used for quantification. Multielement depth-profiling trace and layer analyses have been important application fields for the characterisation of near-surface contaminations and purity of deposited layers. Quite a high number of papers have been published in this field.[121–135]

By optimisation of sputtering conditions and source geometry, a crater bottom with unevenness of less than 5 μm at a crater depth of more than 70 μm (a sputtering time of 600 min) can be achieved.

Figure 13.24 shows the crater profile achieved on a silicon wafer while reaching detection limits in the low/below ng $g^{-1}$ range. Figure 13.25 shows the trace element depth profile of two different Si-layers grown on SiC substrate.

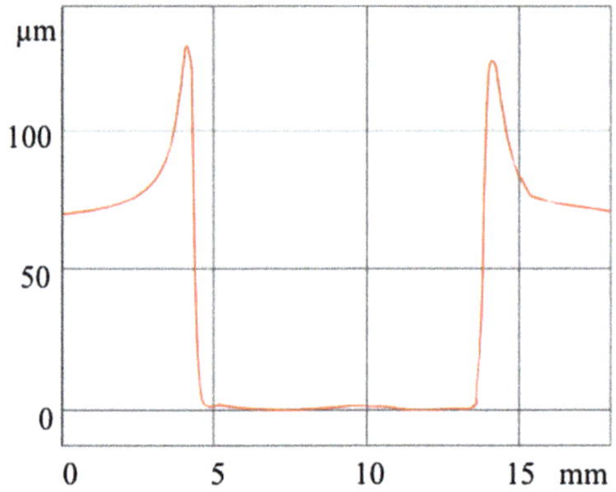


**Figure 13.24** Crater shape of a sputter crater on a silicon wafer. (Source: Cornel Venzago.)

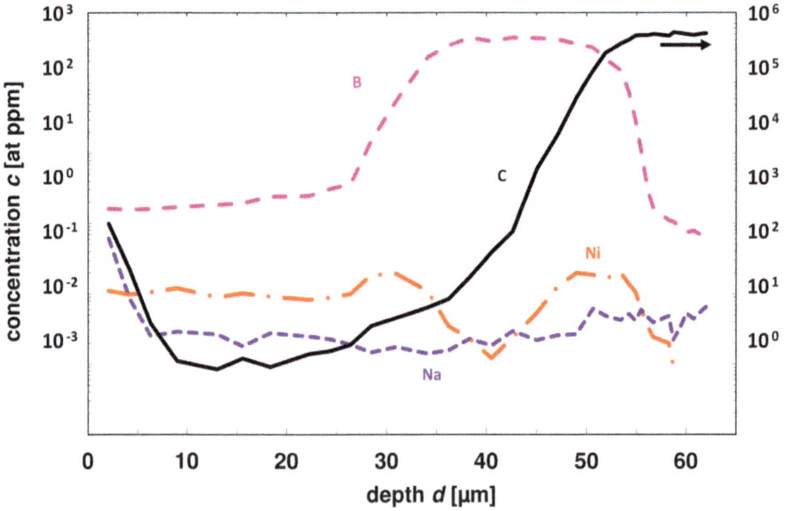


**Figure 13.25** Trace and matrix concentrations of a Si layer system on a SiC substrate (C concentration scaled by the right axis).[136]

The top layer (depth 0–26 μm) shows a low boron dopant level and the bottom layer (depth 26–58 μm) shows a high boron doping level. The impurity profile shows a constant nickel contamination in the top layer and contamination enrichments for several impurities at the layer interfaces, especially at the interface to the SiC substrate.

During the development of the fast flow-GD source it was shown that the ion source can be modified and the working conditions can be adequately optimised to result in flat sputter craters optimised for depth profiling of technical layers. Voronov *et al.*[38,39] applied microsecond DC pulses to the GD source of the *Element GD* and applied time-gated detection by slightly modifying the electronics of the instrument. The addition of microsecond pulsed GD sources to the *Element GD* instrument improved the depth-profiling capabilities for thin layers significantly.

The concept of the μs-pulsed DC glow discharge assembly on a fast-flow high-power source for time resolved analysis in high-resolution mass spectrometry is shown in Figure 13.26. A slightly different approach was established by Churchill *et al.*[137] Finally, Thermo Fischer Scientific adapted this system and introduced their own version with limited capabilities (fixed pulse frequency) in their instrument.

Further improvement could be achieved when detector systems are triggered and synchronised with the pulsing system including gated detection. It was shown that the signal-to-noise ratio could be reduced significantly by setting the noise between pulses to zero.[39] Up to now this has been only realised in early-stage prototype equipment.

Applying a microsecond pulsed glow discharge (μs-pGD) in a fast-flow source, the crater shape and sputtering rate could be adjusted according to the sequence shown in Figure 13.27.

The crater shape with flat bottom and straight edges is ideal for the analysis of thin films. At a low sputtering rate in Si, plasma conditions were

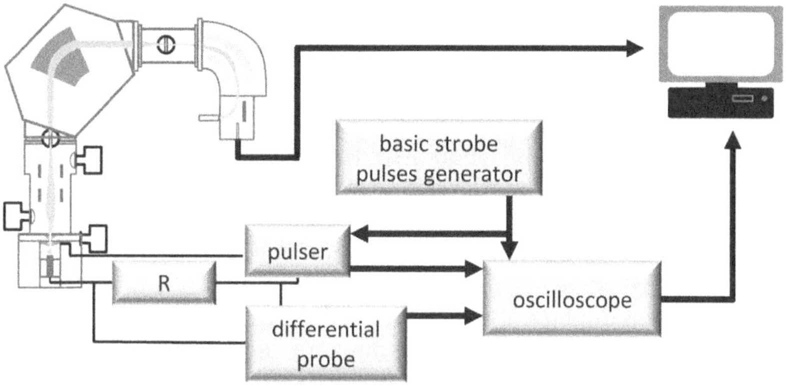


**Figure 13.26** Schematic of the connection of the pulsed power supply. (Source: Ondrej Hanousek after Voronov *et al.*[38])

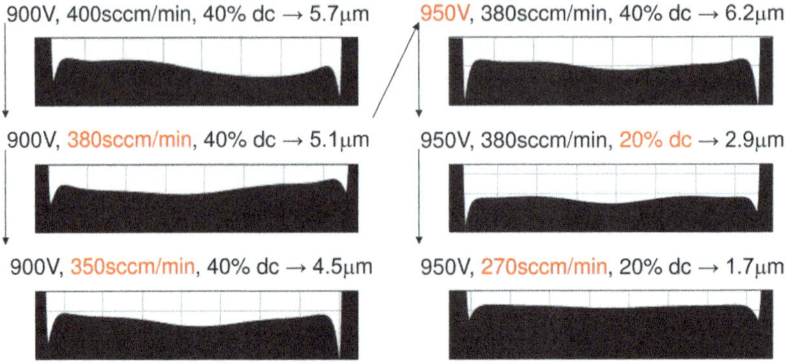


**Figure 13.27** Crater bottoms after 10 min sputtering with the fast flow source (8 mm diameter) in silicon. The plasma conditions for the 2-kHz pulsed mode plasma operation are given, respectively. Lowering the pressure and increasing the voltage balanced the crater shape and reduced the sputtering rate – a low duty cycle (DC) levelled out the crater bottom for homogenous sputtering of the first two μm. Arrows indicate the flow of the sequence to optimise the parameters.[139]

optimised at 950 V operating DC voltage, 270 sccm $min^{-1}$ argon flow and a 2-kHz pulsing frequency with 20% duty cycle.[138] The method was used to characterise various thin films typically used in modern photovoltaic materials research and development showing the potential of GDSFMS for bulk and profiling analysis of thin and ultrathin films for the first time.[13,38,111,138]

## 13.4 Applications

A GD source was used when F.W. Aston built his first mass spectrograph. Therefore, GD sources where involved in the early days already in the research on atomic structures and atomic physics.[3,139] A rediscovery of GD as ion source and the birth of GDMS as an analytical tool (DC and RF) was in the early 1970s.[5] GDMS is used for elemental and isotopic analysis (Figures 13.28(a) and (b)).

Applications for sector field GDMS instruments are now mostly in the nonferrous metals industry, where metals and alloys containing *e.g.* Cu, Al, Si, Ti, Ga, Ni, Cr as well as refractories and precious metals are analysed for their purity. Other applications in the specialised steel industry and regarding nuclear materials, isotopic composition. Environmental samples have been described, as well.

It should be highlighted that many ICP-SFMS instruments are installed in public research institutes, so that the outcome of the research is frequently published. On the contrary, most GDSFMS instruments are operated in private and industrial institutes or labs where publication of results is not required or even unwanted. As a result, the number of citations presented in this chapter is relatively low compared to number of citations such as ICP-SFMS. As a matter of fact, within the last 5 years only very few application papers using sector field GDMS have been published. Already in the first publications for GDMS the possibility for time-resolved measurement and therefore depth-profiling capabilities were described. The recent introduction of μs-pulsed GD supplies have boosted these types of applications especially in the field of thin-film photovoltaic material. Therefore, most of the recent GDMS publications are of time-of-flight mass spectrometry mostly using pulsed RF GD and focused on polymer analysis or on nm-range thin film analyses.

The domain for DC GDSFMS has been for a very long time and is still nowadays the bulk analysis of conducting and semiconducting materials. Discharge parameters, discharge pressure and discharge voltage or current have been optimised for these applications to obtain the highest sensitivity or best stability. A certain presputter phase is required to get stable sensitivities after cleaning the sample surface by sputtering.

Moreover, depth profiling capabilities of GDMS have been demonstrated.[140] Depth profiling and measurement of layer thicknesses have become one of the major applications of GDOES, especially in the steel-coating industry. Depth profiling in GDMS has always been a niche. The optimisation of plasma conditions for depth profiling are mostly in view of the sputter crater shape since homogeneous sputtering over the complete area needs to be achieved in order to obtain decent depth resolution. In sequential scanning sector field mass spectrometry, the relatively low scanning speed of the mass spectrometer has been one major limiting factor in depth resolution. Introducing a μs-pulsed GD supply on a commercial mass spectrometer by reducing sputtering speed without dramatic loss in signal intensity has made new application in thin-film analysis for plasma processes and photovoltaic available.[38,39]

A large number of matrices (Fe, Ni, Cu, Al, Ti) and analyte elements have been analysed in order to produce reliable RSFs.[141] A clear correlation to relative sputtering yields could be found but nevertheless a difference remained, indicating additional processes to Penning ionisation. The possibility for determining RSFs in Fe and Cu using standard stock solution addition to high-purity powders was shown and later developed to a common calibration technique in metrology and industrial analysis.[92] An important factor for the determination of C (RSF = 2.493) and N (RSF = 26.34) in steel is sample preparation and the sputtering conditions in order to obtain low blanks.[142] Al determination in steel showed an RSF of 1.355 analysing 29 iron and steel reference materials.[143] The correlation between RSF and sputtering yields was further investigated by Saka and Inoue.[144] Sample shapes (pin/flat) and discharge conditions were investigated. It could be shown that the normalisation of RSF from different matrices to Fe produced good results and no systematic deviations could be found.

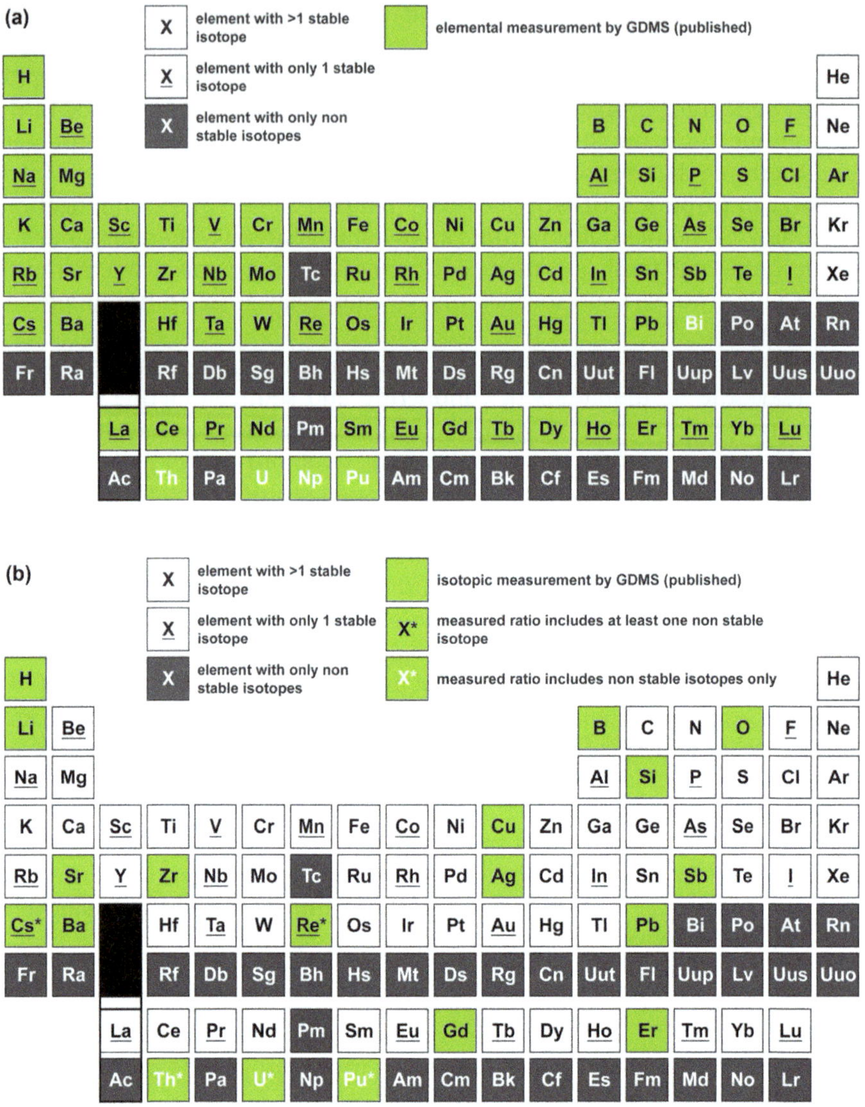


**Figure 13.28** (a) Periodic table of the elements: elements for which elemental data acquired by GDMS has been published (highlighted in lemon green) – also included if used, *e.g.* for interference corrections or internal normalisation. (b) Periodic table of the elements: elements for which isotope ratio data acquired by GDMS has been published (highlighted in lemon green) – also included if measured as inter elemental normalisation standard or in IDMS analysis.

### 13.4.1 Applications in Environmental Sciences

A review on the application of sector field GDMS for environmental samples was presented in 2000 by Baude *et al.*[26] and four years later by Betti and

Aldave de las Heras.[28] Since then, no new work has been published on this subject. This application has never been made into standardised methods or applied to routine analyses. All applications of environmental material have been performed using a secondary cathode or conducting-matrix sample-preparation techniques since most environmental samples are non-conducting. For example, soils, sediments, vegetation were analysed with secondary cathode and host matrix.[145] Automotive exhaust catalysts were analysed using an Al binder, but no superior accuracy or precision data compared to conventional wet chemical analysis could be achieved.[146]

## 13.4.2 Applications in Industrial and Material Research

**High-Purity Metals**

Most instruments in the industry are used for quality control of high-purity metals and alloys. In Table 13.12 a number of measurement standards are listed, which have been produced for industrial applications.

High-purity metals for primary standard reference materials in metrology have been characterised using GDMS. In this context, quantification is a very important item.[92,147,148]

However nuclear methods and isotope dilution using TIMS have been shown to be in certain cases superior to GDMS regarding accuracy and detection limits but they are very time and labour consuming. In neutron activation analysis (RNAA and INAA) a nuclear reactor for the activation of the sample is required as well as a radiochemical laboratory for chemical separation of nuclides and for measurement of activities. Therefore, these methods could never replace the versatility and multielement capabilities of GDMS in routine industrial work.[149,150]

Rapid and accurate process feedback on impurity levels is crucial in the production. A high signal-to-noise ratio is required as a prerequisite for achieving low detection limits. This can be achieved by a careful selection of

**Table 13.12** Selected standards produced for industrial applications.

| Standard | |
|---|---|
| ASTM F1593 - 08 | Standard Test Method for Trace Metallic Impurities in Electronic Grade Aluminum by GDSFMS |
| ASTM F1710 - 08 | Standard Test Method for Trace Metallic Impurities in Electronic Grade Titanium by GDSFMS |
| ASTM F1845 - 08 | Standard Test Method for Trace Metallic Impurities in Electronic Grade Aluminum-Copper, Aluminum-Silicon, and Aluminum-Copper-Silicon Alloys by GDSFMS |
| ASTM F2405 - 04 | Standard Test Method for Trace Metallic Impurities in High Purity Copper by GDSFMS |
| SEMI PV1-0211 | Test Method For Measuring Trace Elements In Silicon Feedstock For Silicon Solar Cells By GDSFMS |
| ISO-TS 15338 | Glow discharge mass spectrometry (GDMS) – Introduction to use |

operational conditions. Additionally, the highest-purity materials for the anode bodies are used, *i.e.* graphite for *Element GD* and Ta for *VG 9000* and *AstruM*. Extraction cone material is also important therefore for ultratrace analysis graphite is used on the *Element GD*) or Ta as exit-slit material on the *VG 9000* and the *Astrum* because these parts are in direct contact with the glow discharge and can contribute to increased blank values. Medium and high mass resolution is required, because the bulk matrix can cause spectral interferences on a number of analytical isotopes.[151]

**Iron and Steel**

Most applications work published from laboratories in Japan deal with the quantification of trace elements in steel, regarding determination of ion formation, *RFS* and the quantification of C, N, and O.[152–155] When analysing Ti traces in high alloy steels, problems arise with interferences and elevated spectral background levels from doubly charged matrix-argides as well as from Mo requiring interference correction.[156]

**Aluminium**

Application of GDMS to the analysis of high-purity aluminium and Al alloys has been a very important field of commercial GDMS even though no recent publication have appeared after the introduction of the method.[61,88,157–161] Further development was published in relation to comparative work on different matrices like new aluminium alloys.[162,163] Some comparative work of different methods has been produced using nuclear methods and isotope dilution using TIMS or ICP-MS. S and O depth profiles on aluminised Ni-superalloy were also recorded by Lee *et al.*[164]

**Copper**

Copper is a matrix that is relatively easy to analyse by GDMS. Therefore, it is investigated in several works especially when very high purity copper qualities are required, *e.g.* for microelectronic applications.[165–167] Several certified reference materials have been used to demonstrate the capabilities of GDMS. An extensive round-robin experiment has been conducted using BCR standards.[92,95]

**Gallium**

Gallium has a melting point of 29.7 °C, which is a challenge for direct solid-state analyses. Very high purity qualities are needed to produce GaAs as well as epitaxial layers of other Ga-containing III-V semiconductors for lasers. It is one of the most difficult matrices for direct solid-state analysis besides the low melting point due to the oxygen affinity as well as due to severe segregation effects during solidification. Already very soon after the market introduction of the *VG 9000*, a so-called Ga cell was developed delivering very good cooling efficiency and using liquid nitrogen sample and source cooling. Therefore, it was made possible to analyse Ga by GDMS as a replacement for the SSMS.[157,168] Nevertheless, direct mass spectrometric methods are not as sensitive as single-crystal resistivity measurements or wet chemical analysis after preconcentration. Therefore, GDMS is not able to distinguish qualities better than 7N (99,99999%) purity. Nonetheless, GDMS is a valuable routine tool for quality control for gallium.

However, to produce reproducible and accurate results sample preparation of Ga is a challenging task, due to the strong segregation effects while solidifying as well as due to the strong oxygen affinity. It can be observed that impurities are collected in the oxide layer on the surface and therefore the material appears purer than the actual quality after sputtering of these layers.[169] Ga is very reactive with water vapour, therefore elevated humidity during sample preparation led to large concentration differences between extremities of the pin-shaped samples. These strong segregation effects produce large concentration ratios as a result of longitudinal heterogeneity to up to several orders of magnitude. A strong decrease of impurity concentration was also shown with sputtering time, showing a large surface bulk inhomogeneity. A rapid quenching technique for sample preparation was introduced by solidification in liquid nitrogen using a thin-walled PTFE tube. A multicrystalline structure with a good homogeneity and low impurity enrichment on the surface could be achieved.[170] Only with the introduction of the μs-pulsed GD supply on the *Element GD* has it become possible to analyse Ga using this instrument as it does not have cryocooling (only Peltier cooling) and a high power input on the sample surface in continuous mode. Table 13.13 shows the measurement details for the analysis of Ga on the μs-pulsed *Element GD*. Table 13.14 shows the achieved results including the relative standard deviation of repeated acquisitions on a Ga-sample. Nonetheless, no results from the *Element GD* or the *AstruM* on Ga have been published so far.

### Titanium

GDMS is described as the method of choice for quality control of titanium.[171] GDMS has been used in comparison to INAA, RNAA and IDMS, were GDMS showed significant differences from the comparative analyses.[172] On the other hand GDMS showed its capability for the determination of Si traces compared to solid-state AAS techniques.[173] The sample-

**Table 13.13** Measurement conditions for μs-pulsed *Element GD* measurement on Ga.

| | |
|---|---|
| (−20) °C Peltier cooling | |
| RSF | STANDARD |
| Voltage | 900 V |
| Frequency | 1 kHz |
| Pulse length | 100 μs |
| Argon flow | 260 sccm |
| Resulting current | 8.4 mA |
| Matrix intensity $^{69}$Ga in medium resolution | 3–4 E+08 cps |
| LOD: 0.001–0.01 μg g$^{-1}$ | |

**Table 13.14** Results of a Ga-sample. (Source: C. Venzago, unpublished data.)

| Element | Mean value [μg g$^{-1}$] | Internal reproducibility RSD |
|---|---|---|
| Fe | 0.08 | 18% |
| Cu | 3.3 | 5% |
| Pb | 0.05 | 12% |

preparation technique for titanium was significantly improved by using $Al_2O_3$ cut off wheels followed by etching with $HNO_3/HF/H_2O_2$.[174]

Al determination in Ti showed an RSF of 2.218 analysing 14 titanium alloys with Al-concentrations from 3 to 8%. However, the RSF in AlTi intermetallic compounds with Al-concentration up to 33.36% was determined with 2.488 showing a significant matrix effect.[143] Sc at mass 45 in Ti is not an easy task to analyse by mass spectrometric methods. Problems include high background from $^{46}$Ti as well as interferences from $^{40}Ar^{50}Ti^{2+}$. Therefore only a limited detection power is shown.[175]

**Refractory Metals**

In the early work of Wilhartitz *et al.*[176] the matrices Mo, W and Cr were investigated with different methods including GDSFMS. It was shown that inhomogeneity can have an influence on the accuracy as the amount of material sputtered (most often less than one mg) is low in comparison to the amount used for solution analysis.[127,177–179]

**Precious Metals**

For the analysis impurity elements in precious metals like platinum, palladium silver and gold delivers good and reproducible results with superior detection capabilities than wet chemical techniques followed by atomic spectrometric methods.[130,166,180–185] The determination of very low mass isotopes like hydrogen and deuterium were also reported.[186]

**Semiconductors and Photovoltaic Materials**

A report from Mykytiuk *et al.*[157] covers the analysis by use of a *VG 9000* of materials mainly used in the semiconductor industry such as CdTe, GaAs, Ga and As. The authors claimed that no standards have been applied for calibration and they pointed out the extremely high dynamic range achieved for these materials of more than 9 orders of magnitude (for the *VG 9000*). Even elements such as C, N and O had been quantified in the matrices mentioned successfully by application of the cryogenic cooling of the GD source.

Within the last few years a new application field has been opened by the photovoltaic industries, even though already in the 1990s GDMS has been applied on photovoltaic materials.[89,121] The boom in the production of photovoltaic materials in the years from 2005 onwards led to an increase in

GDMS applications. GDMS could never obtain a real breakthrough in the semiconductor industry since the sensitivity of GDMS is not sufficient to characterise semiconductor quality material. Therefore mainly indirect methods – like resistance measurements – or specialised single element techniques – like photoluminescence – are applied. The requirements for purity of materials in the photovoltaic industry are not as high as in the microelectronic semiconductor industry. Therefore, GDMS for bulk high-purity silicon for photovoltaic applications has become a major field, *e.g.*, for the *Element GD*.

The accurate determination of dopant elements in photovoltaic silicon is crucial to align the analytical results with the expected physical properties for the material. Some effort has been made to thoroughly determine RSFs for silicon using the *Element GD*. A strong dependence of tuning parameters on the RSFs could be observed, so for reproducible analysis of samples, the determination of the RSF under the identical conditions as the analysis itself is crucial.[65]

GaAs was analysed with the introduction of a radio-frequency discharge on a sector field instrument.[62] ZnSe bulk crystals were grown using the self-seeded technique in horizontal and vertical configurations. By comparing GDMS results from the starting material and the grown crystals, the purification effect of the vapour growth processes as well as the contamination of Si and Al from the fused silica ampoule were observed. The measured photoluminescence (PL) spectra of the starting material and the grown crystal – with no major deep-level emissions – are consistent with the purity levels measured by GDMS. The GDMS and PL results indicate that Al is the most likely donor associated with emission in grown crystals and also imply an increase in Al content and a decrease in Zn vacancy concentration after the growth process.[187,188] A couple of publications deal with a number of specific materials such as ZnO,[189] LiF:Mg,Cu,P,[190] La/B-composition of LaB6 CVD deposited layers,[191] purity of $Al_2O_3$/sapphire,[192] impurity analysis of CdZnTe,[193,194] impurities in REER,[195] and AlCu, Ti, InP, In, GaAs.[196] Moreover, GDMS was shown to be the method of choice to measure differences between unpurified and purified graphite and SiC for nuclear reactors.[197]

Another application is thin-film Al:ZnO (AZO) produced by atomic layer deposition (ALD) as a transparent conductive oxide (TCO).[198] The GDMS instrument was calibrated to determine the Al content of AZO thin films on Si using six calibration samples previously characterised by Rutherford backscattering spectrometry (RBS). Depth-independent average count rates could be used for the calibration of the Al signal with the films c01 to c06 since the stoichiometry of the films turned out to be homogeneous in depth (Figure 13.29).

From the determined $RSF_{AL\ ZNO} = 13.9$, the $\omega_{Al}$ of the two AZO films s01 and s02 of about 150 nm thickness could be determined within a total sputtering time of about 5 min. The method shows that GDMS is a fast alternative to time-consuming techniques such as XPS, SIMS or RBS for the analysis of conductive thin solid films.

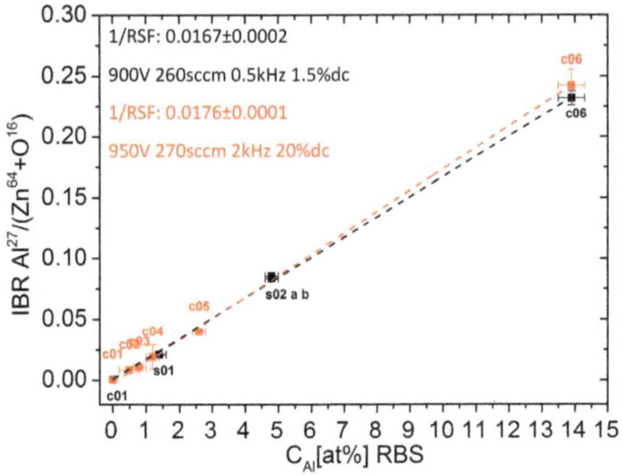


**Figure 13.29** Calibration curves of GD-MS for Al. Black squares indicate the calibration samples linearly fitted by the black line. Red points show the location of the measured samples on the calibration curve. M. Latzel, M. Goebelt, S. W. Schmitt, M. Y. Bashouti, T. Hofmann, C. Venzago and S. H. Christiansen (unpublished results).

Copper indium gallium (di)selenide (CIGS/$CuIn_{1-x}Ga_xSe_2$) thin-film solar cells are the subject of current research since the technology is already reaching 20% efficiency. Absorber layers for CIGS cells are, *e.g.* fabricated by sequential deposition or coevaporation – two methods that allow the engineering of the Ga/In ratio against the absorber depth. The effects of the nonuniform Ga/In ratio in CIGS have major effects on bandgap and device efficiency. Cu deficiency increases the majority (p-) carrier concentration by increasing the number of Cu vacancies. Furthermore, sodium incorporation in CIGS during layer fabrication has various effects on material and device properties. The p-conductivity in CIGS absorbers increases with the sodium amount as well as sodium has positive effects on the stability of CIGS layer growth and the average grain size. All these compositional properties, which are of high importance for CIGS absorber layer development, can be detected by a single GDMS measurement. Measurements on two different CIGS absorbers that were deposited on glass by a coevaporation process (samples provided by T. Rissom, Helmholtz-Zentrum, Berlin, Germany) were performed by GDSFMS using a µs-pulsed GD (900 V, 2.5% duty cycle, 260 sccm min$^{-1}$). Figure 13.30 (left) shows the bulk composition of the two absorber layers.

Both layers show about the same depth independent Cu and Se concentrations, whereas Ga and In profiles are significantly different. The rise in the Mo signal indicates the beginning of the back-contact between CIGS absorber and glass. During the coevaporation process, Na diffuses from the soda-lime glass substrate to the absorber layer. The evaporation process performed at higher temperature shows (Figure 13.30 (right/lower)) an overall higher Na concentration. Since the solubility of Na in the CIGS drops during the cooling process the highest Na concentrations can be found at

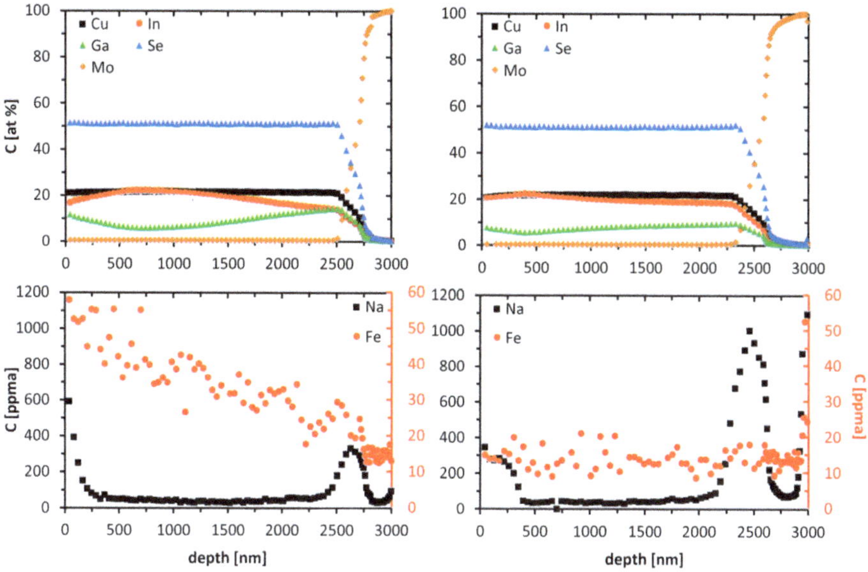


**Figure 13.30** (left) CIGS absorber from a coevaporation process. (right) CIGS absorber from the same coevaporation process at a 10° higher temperature. The upper and lower graph shows bulk and trace element depth profile by HR GDMS, respectively.

the air/CIGS and CIGS/Mo interfaces. The steel housing used for Se targets for deposition is responsible for the Fe contamination found in both layers. The measurements show that the Fe contamination changes as a function of the process parameters (Figure 13.30). From the CIGS/Mo interface the depth resolution of the measurement can be assessed. It can be seen that even in the depth of about 2–3 μm the depth resolution is still as good as about 200 nm.

### 13.4.3 Applications in Geological, Cosmological and Radionuclide Research

The application of GDMS on geological, cosmological and radionuclide research is limited to the addition of electrically conductive binder or on the use of secondary cathodes just as for environmental samples. Therefore, detection capabilities and precisions are limited. Also in order to achieve stable measurement conditions, presputter phases can be very time consuming.[28] For highly radioactive samples it turned out to be necessary to install the sample introduction and the ion-source chamber in a glove box.[199] With this instrument it was possible to determine isotope abundances in highly radioactive samples like Zr-alloys, $U_3O_8$ and other nuclear materials with GDMS using secondary cathode as well as in conducting matrix.[200–205] A precision from 0.1% down to 0.03% and a bias <1% can be achieved on uranium oxide isotope ratios.[186,206,207] The Er isotopic composition in nuclear fuels was investigated in comparison to TIMS with a home-built system from a Mattauch–Herzog SSMS were the spark source had been replaced by a GD source and the photoplate detection system had been replaced by a combined Faraday–SEM system.[208] In another work the isotopic composition of Gd deposits on a Ta target and B in Zr samples were analysed by GDMS and compared to quadrupole ICP-MS.[209] A very good agreement between TIMS and ICP-MS was observed with deviations <1% between the methods for B in Zr. But in GDMS high resolution had to be applied in order to avoid interferences of $^{40}Ar^{4+}$ on $^{10}B$. Good agreement except for the least-abundant isotope could be shown for the comparison of GDMS and TIMS on the Gd sample.

The only paper on cosmochemical application was the measurement of Al, Be, Mg Isotopes in iron meteorites.[210]

# References

1. J. J. Thomson and G. P. Thomson, *Conduction of Electricity through Gases*, Cambridge University Press, Cambridge, UK, 3rd edn., 1928, 1.
2. A. J. Dempster, *Proc. Am. Philos. Soc.*, 1935, **75**, 755.
3. F. W. Aston, *Mass spectra and isotopes*, Longmans, Green & Co.; E. Arnold & Co., New York; London, 2nd edn, 1942, 276.
4. G. Ramendik, J. Verlinden and R. Gijbels, Spark Source Mass Spectrometry, in *Inorganic Mass Spectrometry*, eds. F. Adams, R. Gijbels and R. Van Grieken, John Wiley & Sons, New York, 1988, ch. 2, pp. 17–84.
5. J. W. Coburn, *Rev. Sci. Instrum.*, 1970, **41**, 1219.
6. J. W. Coburn, E. W. Eckstein and E. Kay, *J. Vac. Sci. Technol.*, 1975, **12**, 151.
7. W. Grimm, *Spectrochim. Acta B*, 1968, **23**, 443.
8. N. Jakubowski, D. Stuewer and W. Vieth, *Anal. Chem.*, 1987, **59**, 1825.
9. N. Jakubowski, I. Feldmann and D. Stuewer, *J. Anal. Atom. Spectrom.*, 1997, **12**, 151.
10. N. Jakubowski, I. Feldmann and D. Stuewer, *Spectrochim. Acta Part B*, 1995, **50**, 639.
11. W. W. Harrison, K. R. Hess, R. K. Marcus and F. L. King, *Anal. Chem.*, 1986, **58**, A341.
12. W. W. Harrison, *J. Anal. Atom. Spectrom.*, 1988, **3**, 867.
13. D. J. Hall and P. K. Robinson, *Am. Lab.*, 1987, **19**, 74.
14. F. Adams and A. Vertes, *Fresenius J. Anal. Chem.*, 1990, **337**, 638.
15. R. Gijbels, *Talanta*, 1990, **37**, 363.
16. F. L. King and W. W. Harrison, *Mass Spectrom. Rev.*, 1990, **9**, 285.
17. D. Stuewer, *Fresenius J. Anal. Chem.*, 1990, **337**, 737.
18. W. W. Harrison, C. M. Barshick, J. A. Klingler, P. H. Ratliff and Y. Mei, *Anal. Chem.*, 1990, **62**, A943.
19. W. W. Harrison, *J. Anal. Atom. Spectrom.*, 1992, 7, 75.
20. A. Raith, W. Vieth, J. C. Huneke and R. C. Hutton, *J. Anal. Atom. Spectrom.*, 1992, 7, 943.

21. M. Grasserbauer, *Pure Appl. Chem.*, 1992, **64**, 485.
22. R. Gijbels and A. Bogaerts, *Fresenius J. Anal. Chem.*, 1997, **359**, 326.
23. A. Bogaerts and R. Gijbels, *Spectrochim. Acta Part B: Atom. Spectrosc.*, 1998, **53**, 1.
24. A. Bogaerts and R. Gijbels, *Fresenius J. Anal. Chem.*, 1999, **364**, 367.
25. A. Bogaerts, *J. Anal. Atom. Spectrom.*, 1999, **14**, 1375.
26. S. Baude, J. A. C. Broekaert, D. Delfosse, N. Jakubowski, L. Fuechtjohann, N. G. Orellana-Velado, R. Pereiro and A. Sanz-Medel, *J. Anal. Atom. Spectrom.*, 2000, **15**, 1516.
27. J. S. Becker and H.-J. Dietze, *Int. J. Mass Spectrom.*, 2003, **228**, 127.
28. M. Betti and L. Aldave de las Heras, *Spectrochim. Acta Part B: Atom. Spectrosc.*, 2004, **59**, 1359.
29. V. Hoffmann, M. Kasik, P. K. Robinson and C. Venzago, *Anal. Bioanal. Chem.*, 2005, **381**, 173.
30. C. B. Douthitt, *J. Anal. Atom. Spectrom.*, 2008, **23**, 685.
31. J. Pisonero, B. Fernández and D. Günther, *J. Anal. Atom. Spectrom.*, 2009, **24**, 1145.
32. P. Belenguer, M. Ganciu, P. Guillot and T. Nelis, *Spectrochim. Acta B*, 2009, **64**, 623.
33. A. A. Ganeev, A. R. Gubal, K. N. Uskov and S. V. Potapov, *Russ. Chem. Bull.*, 2012, **61**, 752.
34. N. H. Bings, A. Bogaerts and J. A. Broekaert, *Anal. Chem.*, 2013, **85**, 670.
35. J. Pisonero, N. Bordel, C. Gonzalez de Vega, B. Fernández, R. Pereiro and A. Sanz-Medel, *Anal. Bioanal. Chem.*, 2013, **405**, 5655.
36. K. Robinson and R. Nayler, *Eur. Spectrosc. News*, 1986, **68**, 18.
37. G. M. Hieftje, *J. Anal. Atom. Spectrom.*, 2008, **23**, 661.
38. M. Voronov, T. Hofmann, P. Smid and C. Venzago, *J. Anal. Atom. Spectrom.*, 2009, **24**, 676.
39. M. Voronov, P. Smid, V. Hoffmann, T. Hofmann and C. Venzago, *J. Anal. Atom. Spectrom.*, 2010, **25**, 511.
40. R. Pereiro, A. Solà-Vázquez, L. Lobo, J. Pisonero, N. Bordel, J. M. Costa and A. Sanz-Medel, *Spectrochim. Acta Part B*, 2011, **66**, 399.
41. L. Lobo, J. Pisonero, N. Bordel, R. Pereiro, A. Tempez, P. Chapon, J. Michler, M. Hohl and A. Sanz-Medel, *J. Anal. Atom. Spectrom.*, 2009, **24**, 1373.
42. C. Gonzalez-Gago, J. Pisonero, N. Bordel, A. Sanz-Medel, N. J. Tibbetts and V. S. Smentkowski, *J. Vac. Sci. Technol. A*, 2013, **31**, 06F106.
43. G. Kartopu, A. Tempez, A. J. Clayton, V. Barrioz, S. J. C. Irvine, C. Olivero, P. Chapon, S. Legendre and J. Cooper, *Mater. Res. Innov.*, 2014, **18**, 82.
44. N. Jakubowski, E. Steers and A. Tempez, *J. Anal. Atom. Spectrom.*, 2007, **22**, 715.
45. T. Nelis, Biel-Bienne, Switzerland, www.glow-discharge.com.
46. R. Payling, D. Jones and A. Bengtson, *Glow Discharge Optical Emission Spectrometry*, John Wiley & Sons, Chichester, 1997, 300.

47. R. K. Marcus and J. A. C. Broekaert, eds., *Glow Discharge Plasmas in Analytical Spectroscopy*, John Wiley & Sons, Chichester, England, 2003, 498.
48. M. R. Winchester and R. Payling, *Spectrochim. Acta, Part B*, 2004, **59**, 607.
49. G. Gamez, A. Bogaerts and G. M. Hieftje, *J. Anal. Atom. Spectrom.*, 2006, **21**, 350.
50. G. Gamez, A. Bogaerts, F. Andrade and G. M. Hieftje, *Spectrochim. Acta, Part B*, 2004, **59**, 435.
51. A. Bogaerts, R. Gijbels, G. Gamez and G. M. Hieftje, *Spectrochim. Acta, Part B*, 2004, **59**, 449.
52. N. Jakubowski, R. Dorka, E. Steers and A. Tempez, *J. Anal. Atom. Spectrom.*, 2007, **22**, 722.
53. R. S. Mason, P. D. Miller and I. P. Mortimer, *Phys. Rev. E*, 1997, **55**, 7462.
54. R. S. Mason, D. R. Williams, I. P. Mortimer, D. J. Mitchell and K. Newman, *J. Anal. Atom. Spectrom.*, 2004, **19**, 1177.
55. P. D. Miller, D. Thomas, R. S. Mason and M. Liezers, A New Ion Source For Glow-Discharge Mass Spectrometry, in *Recent Advances in Plasma Source Mass Spectrometry*, ed. G. Holland, BPC Wheatons, Ltd., Exeter, 1995, pp. 91–101.
56. K. Newman, R. S. Mason, D. R. Williams and I. P. Mortimer, *J. Anal. Atom. Spectrom.*, 2004, **19**, 1192.
57. C. Beyer, I. Feldmann, D. Gilmour, V. Hoffmann and N. Jakubowski, *Spectrochim. Acta, Part B*, 2002, **57**, 1521.
58. A. Bogaerts, A. Okhrimovskyy and R. Gijbels, *J. Anal. Atom. Spectrom.*, 2002, **17**, 1076.
59. W. W. Harrison and C. W. Magee, *Anal. Chem.*, 1974, **46**, 461.
60. D. L. Donohue and W. W. Harrison, *Anal. Chem.*, 1975, **47**, 1528.
61. C. Venzago, L. Ohanessian-Pierrard, M. Kasik, U. Collisi and S. Baude, *J. Anal. Atom. Spectrom.*, 1998, **13**, 189.
62. J. S. Becker, A. I. Saprykin and H. J. Dietze, *Int. J. Mass Spectrom. Ion Process.*, 1997, **164**, 81.
63. J. Pisonero, I. Feldmann, N. Bordel, A. Sanz-Medel and N. Jakubowski, *Anal. Bioanal. Chem.*, 2005, **382**, 1965.
64. G. G. Sikharulidze, *Instrum. Exp. Tech.*, 2009, **52**, 242.
65. M. Di Sabatino, A. L. Dons, J. Hinrichs and L. Arnberg, *Spectrochim. Acta, Part B*, 2011, **66**, 144.
66. E. Patel, in conf. proc. of *European Working Group for Glow Discharge Spectroscopy*, Kingston University, p. 14.
67. G. Churchill, D. Barnhart and K. Putyera, in conf. proc. of *European Working Group for Glow Discharge Spectroscopy*, Kingston University, p. 13.
68. J. Pisonero, *Anal. Bioanal. Chem.*, 2006, **384**, 47.
69. S. L. Tong and W. W. Harrison, *Spectrochim. Acta, Part B*, 1993, **48**, 1237.
70. J. C. Woo, N. Jakubowski and D. Stuewer, *J. Anal. Atom. Spectrom.*, 1993, **8**, 881.

71. S. de Gendt, W. Schelles, R. van Grieken and V. Muller, *J. Anal. Atom. Spectrom.*, 1995, **10**, 681.
72. D. C. Duckworth, C. M. Barshick and D. H. Smith, *J. Anal. Atom. Spectrom.*, 1993, **8**, 875.
73. M. Battagliarin, E. Sentimenti and R. Scattolin, *Spectrochim. Acta, Part B*, 1995, **50**, 13.
74. G. D. Schilling, F. J. Andrade, J. H. Barnes IV, R. P. Sperline, M. B. Denton, C. J. Barinaga, D. W. Koppenaal and G. M. Hieftje, *Anal. Chem.*, 2007, **79**, 7662.
75. W. Schütze and F. Bernhard, *Z. Phy.*, 1956, **145**, 44.
76. N. R. Daly, *Rev. Sci. Instrum.*, 1960, **31**, 264.
77. C. Venzago, in conf. proc. of *Winter conference on Plasma Spectrochemistry 2006.*
78. L. Rottmann, W. Schoettker, N. Frerichs, C. Venzago and T. Hofmann, *Performance Characteristics and Applications of a New Glow Discharge Sector Field MS* Application Note: AN30082_E 11/05C, Thermo Electron Corporation, Bremen, Germany, 2006.
79. T. Gusarova, *Wege zur genauen Charakterisierung hochreiner Materialien mit der Glimmentladungs-Massenspektrometrie (GD-MS)*, PhD thesis, BAM Bundesanstalt für Materialforschung und -prüfung, ISBN 978-3-9813346-1-6, 2010, 211.
80. R. S. Mason, D. J. Mitchell and P. M. Dickinson, *PhysChemChemPhys*, 2010, **12**, 3698.
81. R. S. Mason, *PhysChemChemPhys*, 2010, **12**, 3718.
82. R. S. Mason and P. Douglas, *PhysChemChemPhys*, 2010, **12**, 3729.
83. Nu Instruments Limited, Nu Instruments Limited, Wrexham, UK.
84. J. Hinrichs, ThermoFisher Scientific, Bremen, Germany, 2012.
85. S. D. Tanner, V. I. Baranov and D. R. Bandura, *Spectrochim. Acta, Part B*, 2002, **57**, 1361.
86. N. Jakubowski, T. Prohaska, L. Rottmann and F. Vanhaecke, *J. Anal. Atom. Spectrom.*, 2011, **26**, 693.
87. T. Gusarova, T. Hofmann, H. Kipphardt, C. Venzago, R. Matschat and U. Panne, *J. Anal. Atom. Spectrom.*, 2010, **25**, 314.
88. C. Venzago and M. Weigert, *Fresenius J. Anal. Chem.*, 1994, **350**, 303.
89. D. M. P. Milton, R. C. Hutton and G. A. Ronan, *Fresenius J. Anal. Chem.*, 1992, **343**, 773.
90. M. P. Balogh, *Spectroscopy*, 2004, **19**, 34.
91. M. Kasik, C. Michellon and L. C. Pitchford, *J. Anal. Atom. Spectrom.*, 2002, **17**, 1398.
92. R. Matschat, J. Hinrichs and H. Kipphardt, *Anal. Bioanal. Chem.*, 2006, **386**, 125.
93. C. A. Craig, K. E. Jarvis and L. J. Clarke, *J. Anal. Atom. Spectrom.*, 2000, **15**, 1001.
94. D. A. Frick and D. Günther, *J. Anal. Atom. Spectrom.*, 2012, **27**, 1294.
95. M. Kasik, C. Venzago and R. Dorka, *J. Anal. Atom. Spectrom.*, 2003, **18**, 603.

96. R. Jede, O. Ganschow and U. Kaiser, Sputtered neutral mass spectrometry, in *Practical Surf. Analysis – Ion and Neutral Spectroscopy*, eds. D. Briggs and M. P. Seah, John Wiley & Sons Ltd., Chichester, 1983, vol. 2, pp. 495–499.
97. D. M. P. Milton and R. C. Hutton, *Spectrochim. Acta, Part B*, 1993, **48**, 39.
98. W. Schelles, S. De Gendt, V. Muller and R. Van Grieken, *Appl. Spectrosc.*, 1995, **49**, 939.
99. W. Schelles, K. J. R. Maes, S. De Gendt and R. E. Van Grieken, *Anal. Chem.*, 1996, **68**, 1136.
100. W. Schelles, S. De Gendt and R. Van Grieken, *J. Anal. Atom. Spectrom.*, 1996, **11**, 937.
101. V. D. Kurochkin and L. P. Kravchenko, *Pow. Metall. Met. Ceram.*, 2006, **45**, 493.
102. W. Schelles and R. Van Grieken, *J. Anal. Atom. Spectrom.*, 1997, **12**, 49.
103. C. M. Barshick, D. H. Smith, J. H. Hackney, B. A. Cole and J. W. Wade, *Anal. Chem.*, 1994, **66**, 730.
104. C. M. Barshick, D. C. Duckworth and D. H. Smith, *J. Am. Soc. Mass Spectrom.*, 1993, **4**, 47.
105. C. M. Barshick, S. A. Barshick, M. L. Mohill, P. F. Britt and D. H. Smith, *Rapid Commun. Mass Spectrom.*, 1996, **10**, 341.
106. J. Teng, C. M. Barshick, D. C. Duckworth, S. J. Morton, D. H. Smith and F. L. King, *Appl. Spectrosc.*, 1995, **49**, 1361.
107. R. Qian, S. J. Zhuo, Z. Wang and P. K. Robinson, *J. Anal. Atom. Spectrom.*, 2013, **28**, 1061.
108. A. I. Saprykin, F. G. Melchers, J. S. Becker and H. J. Dietze, *Fresenius J. Anal. Chem.*, 1995, **353**, 570.
109. M. J. Heintz, D. P. Myers, P. P. Mahoney, G. Q. Li and G. M. Hieftje, *Appl. Spectrosc.*, 1995, **49**, 945.
110. R. K. Marcus, in conf. proc. of *International Glow Discharge Spectroscopy Symposium IGDSS*.
111. D. C. Duckworth and R. K. Marcus, *Anal. Chem.*, 1989, **61**, 1879.
112. D. C. Duckworth, D. L. Donohue, D. H. Smith, T. A. Lewis and R. K. Marcus, *Anal. Chem.*, 1993, **65**, 2478.
113. A. I. Saprykin, J. S. Becker and H. J. Dietze, *J. Anal. Atom. Spectrom.*, 1995, **10**, 897.
114. A. I. Saprykin, J. S. Becker and H. J. Dietze, *Fresenius J. Anal. Chem.*, 1996, **355**, 831.
115. A. I. Saprykin, J. S. Becker and H. J. Dietze, *Fresenius J. Anal. Chem.*, 1997, **359**, 449.
116. J. S. Becker, J. Westheide, A. I. Saprykin, H. Holzbrecher, U. Breuer and H. J. Dietze, *Mikrochim. Acta*, 1997, **125**, 153.
117. R. Jager, A. I. Saprykin, J. S. Becker, H. J. Dietze and J. A. C. Broekaert, *Mikrochim. Acta*, 1997, **125**, 41.
118. J. Pisonero, J. M. Costa, R. Pereiro, N. Bordel and A. Sanz-Medel, *Anal. Bioanal. Chem.*, 2004, **379**, 658.

119. N. E. Sanderson, E. Hall, J. Clark, P. Charalambous and D. Hall, *Mikrochim. Acta*, 1987, **1**, 275.
120. D. J. Hall and N. E. Sanderson, *Surf. Interface Anal.*, 1988, **11**, 40.
121. C. Venzago, H. von Campe and W. Warzawa, in *11th European Photovoltaic Solar Energy Conference*, Montreux, 1992, pp. 484–486.
122. M. R. Winchester and U. Beck, *Surf. Interface Anal.*, 1999, **27**, 930.
123. R. Snyders, M. Wautelet, R. Gouttebaron, J. P. Dauchot and M. Hecq, *Surf. Coat. Technol.*, 2003, **174–175**, 1282.
124. R. Snyders, M. Wautelet, R. Gouttebaron, J. P. Dauchot and M. Hecq, *Thin Solid Films*, 2003, **423**, 125.
125. D. V. Shenai, M. L. Timmons, R. L. DiCarlo, G. K. Lemnah and R. S. Stennick, *J. Cryst. Growth*, 2003, **248**, 91.
126. L. M. He, K. Putyera, J. D. Meyer, L. R. Walker and W. Y. Lee, *Metall. Mater. Trans. A*, 2002, **33**, 3578.
127. D. M. Wayne, R. K. Schulze, C. Maggiore, D. Wayne Cooke and G. Havrilla, *Appl. Spectrosc.*, 1999, **53**, 266.
128. S. Kumar, V. S. Raju, R. Shekhar, J. Arunachalam, A. S. Khanna and K. G. Prasad, *Thin Solid Films*, 2001, **388**, 195.
129. I. T. Spitsberg and K. Putyera, *Surf. Coat. Technol.*, 2001, **139**, 35.
130. A. Efimov, M. Kasik, K. Putyera and O. Moreau, *Electrochem. Solid State Lett.*, 2000, **3**, 477.
131. V. Vancoppenolle, P. Y. Jouan, M. Wautelet, J. P. Dauchot and M. Hecq, *J. Vac. Sci. Technol. A*, 1999, **17**, 3317.
132. J. Pisonero, J. M. Costa, R. Pereiro, N. Bordel and A. Sanz-Medel, *J. Anal. Atom. Spectrom.*, 2002, **17**, 1126.
133. M. Vazquez Pelaez, J. Pisonero, J. M. Costa-Fernandez, R. Pereiro, N. Bordel and A. Sanz-Medel, *J. Anal. Atom. Spectrom.*, 2003, **18**, 612.
134. L. Aldave de las Heras, O. L. Actis-Dato, M. Betti, E. H. Toscano, U. Tocci, R. Fuoco and S. Giannarelli, *Microchem. J.*, 2000, **67**, 333.
135. L. O. Actis-Dato, L. Aldave de Las Heras, M. Betti, E. H. Toscano, F. Miserque and T. Gouder, *J. Anal. Atom. Spectrom.*, 2000, **15**, 1479.
136. V. Hoffmann, D. Klemm, V. Efimova, C. Venzago, A. A. Rockett, T. Wirth, T. Nunney, C. A. Kaufmann and R. Caballero, Elemental Distribution Profiling of Thin Films for Solar Cells, in *Advanced Characterization Techniques for Thin Film Solar Cells*, 2011, pp. 411–448.
137. G. Churchill, K. Putyera, V. Weinstein, X. Wang and E. B. M. Steers, *J. Anal. Atom. Spectrom.*, 2011, **26**, 2263.
138. S. W. Schmitt, C. Venzago, B. Hoffmann, V. Sivakov, T. Hofmann, J. Michler, S. Christiansen and G. Gamez, *Prog. Photovolt.: Res. Applic.*, 2014, **22**, 371.
139. K. T. Bainbridge and E. B. Jordan, *Phys. Rev.*, 1936, **50**, 282.
140. J. W. Coburn and E. Kay, *Appl. Phys. Lett.*, 1971, **19**, 350.
141. M. Inoue and T. Saka, *Anal. Chim. Acta*, 1999, **395**, 165.
142. S. Itoh, H. Yamaguchi, T. Yoshioka, T. Kimura and T. Kobayashi, *Tetsu-to-Hagane*, 1999, **85**, 666.

143. S. Itoh, H. Yamaguchi, N. Sakuma, T. Hobo and T. Kobayashi, *Bunseki Kagaku*, 2003, **52**, 605.
144. T. Saka and M. Inoue, *Anal. Sci.*, 2000, **16**, 653.
145. M. Betti, S. Giannarelli, T. Hiernaut, G. Rasmussen and L. Koch, *Fresenius J. Anal. Chem.*, 1996, **355**, 642.
146. D. M. Wayne, *J. Anal. Atom. Spectrom.*, 1997, **12**, 1195.
147. H. Kipphardt, R. Matschat, J. Vogl, T. Gusarova, M. Czerwensky, H.-J. Heinrich, A. Hioki, L. A. Konopelko, B. Methven, T. Miura, O. Petersen, G. Riebe, R. Sturgeon, G. C. Turk and L. L. Yu, *Accred. Qual. Assur.*, 2010, **15**, 29.
148. S. Rudtsch, M. Fahr, J. Fischer, T. Gusarova, H. Kipphardt and R. Matschat, *Int. J. Thermophys.*, 2008, **29**, 139.
149. M. Lucic and V. Krivan, *J. Radioanal. Nucl. Chem.*, 1996, **207**, 444.
150. B. Beer and K. G. Heumann, *Fresenius J. Anal. Chem.*, 1992, **343**, 741.
151. J. Hinrichs, M. Hamester and L. Rottmann, *Application Note: 30164*, Thermo Fisher Scientific, Bremen, Germany, 2009.
152. Y. Ishikawa, K. Mimura and M. Isshiki, *Mater. Trans., Jpn. Inst. Met. Mater.*, 2000, **41**, 420.
153. S. Itoh, H. Yamaguchi, I. Hamano, T. Hobo and T. Kobayashi, *Tetsu-to-Hagane*, 2003, **89**, 962.
154. V. D. Kurochkin, *Pow. Metall. Metal Ceram.*, 2008, **47**, 248.
155. T. Takahashi and T. Shimamura, *Anal. Chem.*, 1994, **66**, 3274.
156. S. Itoh, F. Hirose and R. Hasegawa, *Spectrochim. Acta, Part B*, 1992, **47**, 1241.
157. A. P. Mykytiuk, P. Semeniuk and S. Berman, *Spectrochim. Acta Rev.*, 1990, **13**, 1.
158. G. Kudermann, *Fresenius Z. Anal. Chem.*, 1988, **331**, 697.
159. L. F. Vassamillet, *J. Anal. Atom. Spectrom.*, 1989, **4**, 451.
160. G. Kudermann, K. H. Blaufuss, C. Luhrs, W. Vielhaber and U. Collisi, *Fresenius J. Anal. Chem.*, 1992, **343**, 734.
161. J. Hinrichs and M. Hamester, *Application Note: 30142*, Thermo Fisher Scientific, 2011.
162. V. D. Kurochkin, *Anal. Commun.*, 1996, **33**, 381.
163. L. P. Kravchenko and V. D. Kurochkin, *Proceedings of Conference: International Conference "Science for Materials in the Frontier of Centuries: Advantages and Challenges"*, Kyiv, Ukraine, Nov. 4–8 2002, Vol. 2, 702–703.
164. W. Y. Lee, Y. Zhang, I. G. Wright, B. A. Pint and P. K. Liaw, *Metall. Mater. Trans. A: Phys. Metall. Mater. Sci.*, 1998, **29**, 833.
165. Y. Zhu, K. Mimura, Y. Ishikawa and M. Isshiki, *Mater. Trans., Jpn. Inst. Met. Mater.*, 2002, **43**, 2802.
166. Y. Nakamura, S. Maeda, I. Nagai, H. Inoue, M. Ohtaki, M. Yamazaki, M. Hosoi, K. Shinzawa, Y. Sayama and T. Kawabata, *Bunseki Kagaku*, 1991, **40**, 209.
167. M. Uchikoshi, T. Kekesi, Y. Ishikawa, K. Mimura and M. Isshiki, *Mater. Trans., Jpn. Inst. Met. Mater.*, 1997, **38**, 1083.
168. G. Kudermann and K. H. Blaufuss, *Mikrochim. Acta*, 1987, **1**, 269.

169. J. Allègre and B. Boudot, *J. Cryst. Growth*, 1990, **106**, 139.
170. W. Vieth and J. C. Huneke, *Anal. Chem.*, 1992, **64**, 2958.
171. H. W. Rosenberg and J. E. Green, Analyzing high purity titanium, in *Titanium '92, The Science and Technology*, eds. F. H. Froes and I. Caplan, The Minerals & Materials Society, Warrendale, USA, 1993, pp. 2371–2376.
172. D. Wildhagen and V. Krivan, *Anal. Chem.*, 1995, **67**, 2842.
173. H. M. Dong and V. Krivan, *J. Anal. Atom. Spectrom.*, 2003, **18**, 367.
174. D. Fang and P. Seegopaul, *J. Anal. Atom. Spectrom.*, 1992, **7**, 959.
175. A. Held, P. Taylor, C. Ingelbrecht, P. de Bièvre, J. A. C. Broekaert, M. van Straaten and R. Gijbels, *J. Anal. Atom. Spectrom.*, 1995, **10**, 849.
176. P. Wilhartitz, H. M. Ortner, R. Krismer and H. Krabichler, *Mikrochim. Acta*, 1990, **101**, 259.
177. M. Grasserbauer, *Mikrochimica Acta*, 1987, **1**, 291.
178. S. Cherekdjian, M. Ramsbey and M. Anjum, *Nucl. Instrum. Methods Phys. Res., Sect. B*, 1995, **96**, 87.
179. M. Saito, F. Hirose and H. Okochi, *Anal. Sci.*, 1995, **11**, 695.
180. M. v. Straaten, K. Swenters, R. Gijbels, J. Verlinden and E. Adriaenssens, *J. Anal. Atom. Spectrom.*, 1994, **9**, 1389.
181. T. Faahan-Smith and D. Woodford, *Prec. Met.*, 1995, **19**, 205.
182. M. Resano, E. Garcia-Ruiz, K. S. McIntosh, J. Hinrichs, I. Deconinck and F. Vanhaecke, *J. Anal. Atom. Spectrom.*, 2006, **21**, 899.
183. L. Chalumeau, J. Gonzalez, C. Leclere, M. Limayrac, M. Wery and J. Pagetti, in conf. proc. of *AESF SUR/FIN Annual International Technical Conference*, American Electroplaters and Surface Finishers Society, pp. 500–507.
184. D. M. Wayne, T. M. Yoshida and D. E. Vance, *J. Anal. Atom. Spectrom.*, 1996, **11**, 861.
185. G. Beck, H. H. Beyer, W. Gerhartz, J. Hausselt and U. Zimmer, *Edelmetall-Taschenbuch*, Hüthig-Verlag, Heidelberg, 2nd edn., 1995.
186. D. L. Donohue and M. Petek, *Anal. Chem.*, 1991, **63**, 740.
187. C.-H. Su, M. Dudley, R. Matyi, S. Feth and S. L. Lehoczky, *J. Cryst. Growth*, 2000, **208**, 237.
188. M. V. Balarama Krishna, R. Shekhar, D. Karunasagar and J. Arunachalam, *Anal. Chim. Acta*, 2000, **408**, 199.
189. B. Wang, M. J. Callahan, C. Xu, L. O. Bouthillette, N. C. Giles and D. F. Bliss, *J. Cryst. Growth*, 2007, **304**, 73.
190. T. C. Chen and T. G. Stoebe, *Radiat. Protect. Dosim.*, 2002, **100**, 243.
191. S. S. Kher and J. T. Spencer, *J. Phys. Chem. Solids*, 1998, **59**, 1343.
192. X. Jianwei, Z. Yongzong, Z. Guoqing, K. Xu, D. Peizhen and J. Xu, *J. Cryst. Growth*, 1998, **193**, 123.
193. J. P. Tower, S. P. Tobin, M. Kestigian, P. W. Norton, A. B. Bollong, H. F. Schaake and C. K. Ard, *J. Electron. Mater.*, 1995, **24**, 497.
194. A. B. Bollong, F. G., J. P. Tower, S. P. Tobin, M. Kestigian, P. W. Norton, H. F. Schaake and C. K. Ard, *Adv. Mater. Optics Electron.*, 1995, **5**, 87.

195. W. Gao, P. R. Berger, M. H. Ervin, J. Pamulapati, R. T. Lareau and S. Schauer, *J. Appl. Phys.*, 1996, **80**, 7094.
196. R. J. Guidoboni and F. D. Leipziger, *J. Cryst. Growth*, 1988, **89**, 16.
197. P. Grosse, G. Basset, C. Calvat, M. Couchaud, C. Faure, B. Ferrand, Y. Grange, M. Anikin, J. M. Bluet, K. Chourou and R. Madar, *Mater. Sci. Eng. B*, 1999, **61–62**, 58.
198. S. W. Schmitt, G. Gamez, V. Sivakov, M. Schubert, S. H. Christiansen and J. Michler, *J. Anal. Atom. Spectrom.*, 2011, **26**, 822.
199. M. Betti, G. Rasmussen, T. Hiernaut, L. Koch, P. D. M. Milton and C. R. Hutton, *J. Anal. Atom. Spectrom.*, 1994, **9**, 385.
200. M. Betti, *J. Anal. Atom. Spectrom.*, 1996, **11**, 855.
201. M. Betti, G. Rasmussen and L. Koch, *Fresenius J. Anal. Chem.*, 1996, **355**, 808.
202. L. Pajo, G. Tamborini, G. Rasmussen, K. Mayer and L. Koch, *Spectrochim. Acta, Part B*, 2001, **56**, 541.
203. K. Robinson and E. F. H. Hall, *J. Met.*, 1987, **39**, 14.
204. L. Pajo, A. Schubert, L. Aldave, L. Koch, Y. K. Bibilashvili, Y. N. Dolgov and N. A. Chorokhov, *J. Radioanal. Nucl. Chem.*, 2001, **250**, 79.
205. M. Betti, Analysis of Samples of Nuclear Concern with Glow Discharge Atomic Spectrometry, in *Glow Discharge Plasmas in Analytical Spectroscopy*, eds. R. K. Marcus and J. A. C. Broekaert, John Wiley & Sons, Chichester, 2003, pp. 273–292.
206. L. R. Riciputi, D. C. Duckworth, C. M. Barshick and D. H. Smith, *Int. J. Mass Spectrom. Ion Process.*, 1995, **146**, 55.
207. D. C. Duckworth, C. M. Barshick, D. A. Bostick and D. H. Smith, *Appl. Spectrosc.*, 1993, **47**, 243.
208. F. Chartier, M. Aubert, M. Salmon, M. Tabarant and B. H. Tran, *J. Anal. Atom. Spectrom.*, 1999, **14**, 1461.
209. F. Chartier and M. Tabarant, *J. Anal. Atom. Spectrom.*, 1997, **12**, 1187.
210. S. Xue, G. F. Herzog, A. Souzis, M. H. Ervin, R. T. Lareau, R. Middleton and J. Klein, *Earth Planet. Sci. Lett.*, 1995, **136**, 397.

CHAPTER 14

# *Thermal Ionisation Mass Spectrometry*

STEFAN BÜRGER,[a] JOCHEN VOGL,*[b] URS KLOETZLI,[c] LAURIE NUNES[d] AND MARK LAVELLE[e]

[a] International Atomic Energy Agency, Safeguards Analytical Services, Department of Safeguards, Vienna, Austria; [b] BAM Federal Institute for Materials Research and Testing, Berlin, Germany; [c] University of Vienna, Department of Lithospheric Research, Vienna, Austria; [d] Curtin University, Australia; [e] Imperial College, London, UK
*Email: jochen.vogl@bam.de

The principle of TIMS is based on the formation of atom ions on a heated filament and the separation of these ions according to their mass/charge ratio in a mass spectrometer. In general, TIMS is applied for high-precision measurements of isotope ratios. In the basic design of a modern thermal ionisation source, a surface material is used in the shape of a flat ribbon filament. Usually, a solution (separated from the matrix and containing the element of interest in high concentration) is deposited in the centre of the filament as a liquid. Applying a current to the filament will evaporate and ionise the analyte deposited on the filament. In the case of a single-filament geometry, evaporation and ionisation of the analyte is accomplished with one filament, whereas in a double filament design, evaporation and ionisation is performed on two filaments placed parallel to each other. The analyte is evaporated on one filament, with ionisation occurring when the evaporated atoms strike the ionisation filament. A magnetic sector field is used as mass separator.

New Developments in Mass Spectrometry No. 3
Sector Field Mass Spectrometry for Elemental and Isotopic Analysis
Edited by Thomas Prohaska, Johanna Irrgeher, Andreas Zitek and Norbert Jakubowski

Published by the Royal Society of Chemistry, www.rsc.org

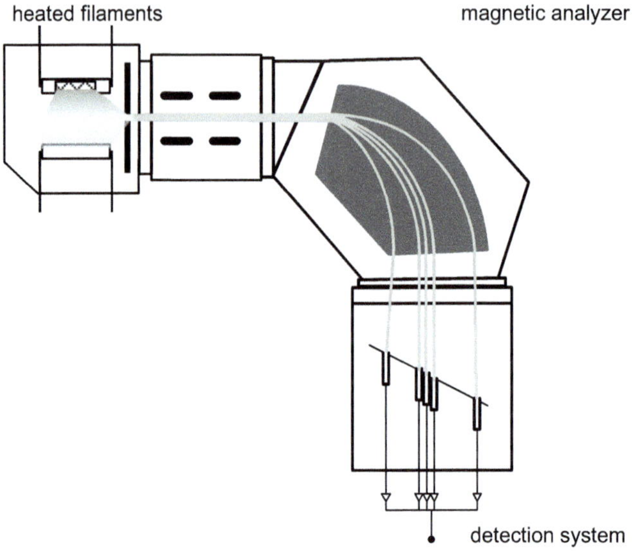


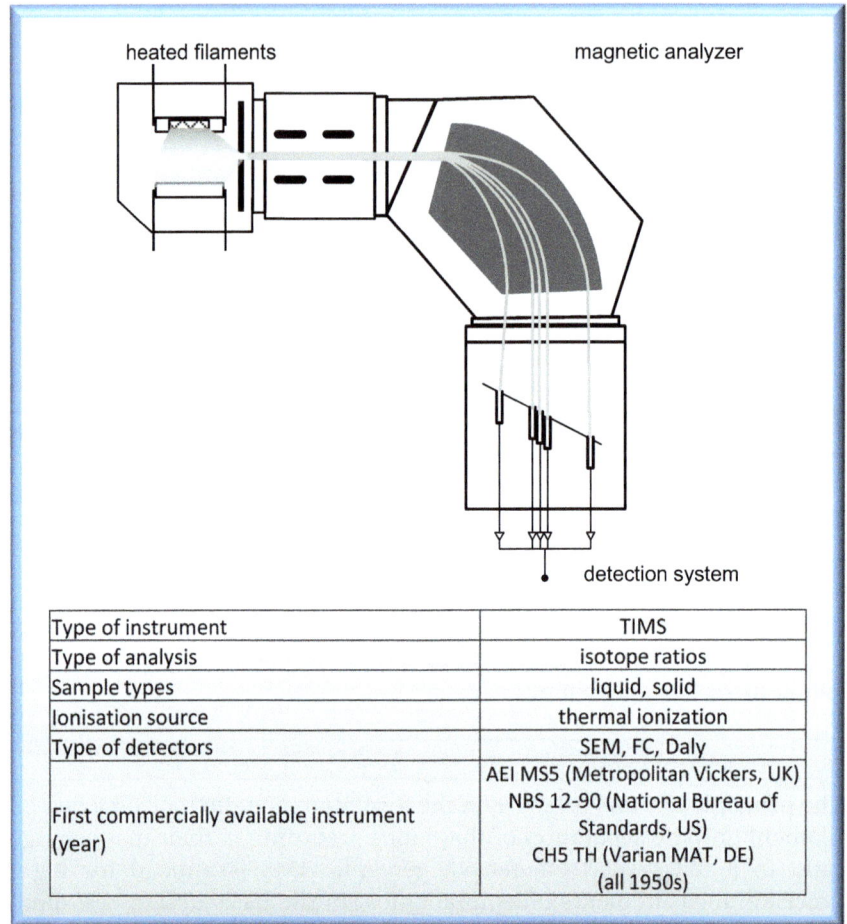

| | |
|---|---|
| Type of instrument | TIMS |
| Type of analysis | isotope ratios |
| Sample types | liquid, solid |
| Ionisation source | thermal ionization |
| Type of detectors | SEM, FC, Daly |
| First commercially available instrument (year) | AEI MS5 (Metropolitan Vickers, UK)<br>NBS 12-90 (National Bureau of Standards, US)<br>CH5 TH (Varian MAT, DE)<br>(all 1950s) |

**Figure 14.1** Features and general schematics of a sector field TIMS. (Source: Ondrej Hanousek)

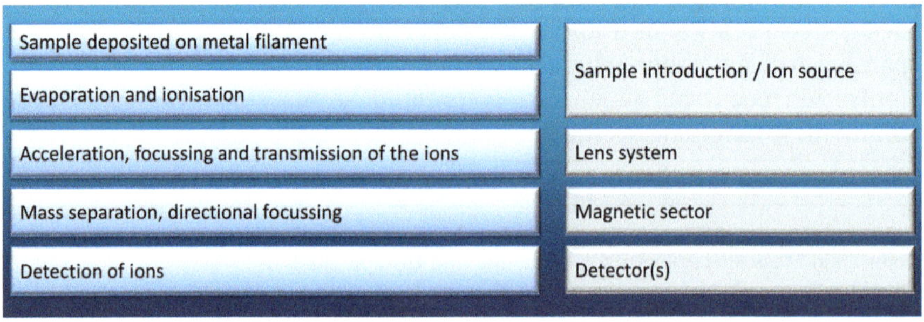

Sample deposited on metal filament
Evaporation and ionisation
Sample introduction / Ion source
Acceleration, focussing and transmission of the ions
Lens system
Mass separation, directional focussing
Magnetic sector
Detection of ions
Detector(s)


**Figure 14.2** Operational flow chart of a TIMS. (Source: Thomas Prohaska.)

**Name and Abbreviation:** The IUPAC conform abbreviation of the technique is TIMS.[1]

## 14.1 General

Take a small piece of metal in the shape of a thin filament. The metal should have a melting point to withstand temperatures as high as 2000 °C, say rhenium or tungsten. Onto the surface of the metal filament place a microscopic amount of a chemical element preferably of low ionisation energy, say an alkali or alkaline-earth metal. Place the filament inside a high vacuum and connect it to a high-voltage power supply. Now, apply a current to the filament and heat it gently. At a specific temperature, the thermal energy of the hot metal filament will cause the chemical element to evaporate and to ionise to some degree. It will emit atomic ions of the element. A technologically simple and, as a consequence, most reliable ion source is at hand. It is not surprising that thermal ionisation, also known as surface ionisation, was recognised and utilised as an ion source in mass spectrometry as early as 1918[2] making it one of the oldest ion sources applied in the field. Usually combined with a single-focusing magnetic sector mass analyser with subsequent ion detection by a multicollector array of Faraday cups and one or more ion counters, these components are the backbone of modern thermal ionisation mass spectrometry (TIMS) instruments (see Figures 14.1 and 14.2). The first commercial instruments were launched as early as the 1960s.

Technologically simple as they may be, thermal ionisation sources provide sufficient ionisation efficiency for the mass-spectrometric analysis of most chemical elements. Indeed, for some elements the ionisation efficiency achievable is close to 100%, turning almost every analyte atom loaded on a filament into an atomic ion. As a consequence, thermal ionisation mass spectrometry has been employed for most chemical elements for isotope ratio determination and/or for amount quantification using isotope dilution,[3] by analysing either positive or negative ions (see Figure 14.3), also referred to as positive (P-TIMS) and negative TIMS (N-TIMS), respectively. Some elements have only one stable or long-lived isotope (including Na, Sc, P, Co, As, Y, Rh, and Cs) or are gaseous or volatile elements like Hg. These elements are usually not measured by sector-field TIMS to determine their isotope ratios. However, organic compounds are measured by surface ionisation organic mass spectrometry (SIOMS) employing thermal ionisation sources in combination with quadrupole, time-of-flight and other mass analysers and – in some cases – with sector-field analysers.[4]

An insight into why the ionisation efficiency differs from element to element is provided by the Saha–Langmuir equation (see also Section 4.2). The Saha–Langmuir equation is based on the Saha equation and work by Langmuir and Kingdom[5] and others. Drawing from the observation that practically all caesium atoms of a caesium vapour that strike a 1200 K hot tungsten surface will be turned into caesium ions, Langmuir and Kingdom modified the Saha equation to explain the experimentally observed ionisation efficiency. The Saha–Langmuir equation states that the ionisation

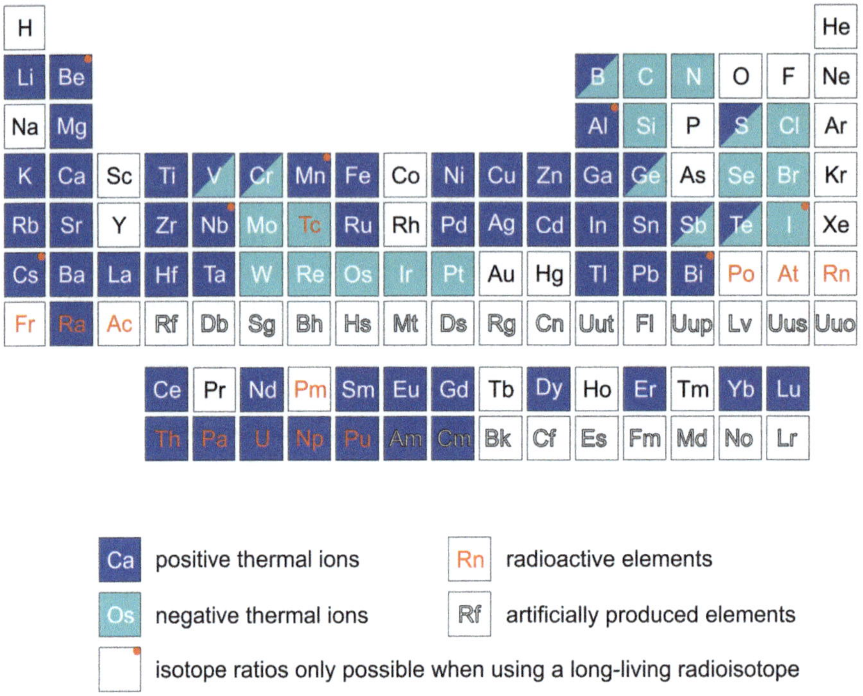


**Figure 14.3** Elements of which isotope ratios can be determined by TIMS (based on an idea by K.G. Heumann[3]).

efficiency $\alpha$ (also known as degree of ionisation) depends on the electron work function $\phi$ of the surface material for electron emission and the equilibrium temperature $T$. Furthermore, it depends on the first ionisation energy $E_i$ (also known as ionisation potential) of the compound in contact with the hot surface in case of positive-ion emission, or the electron affinity $E_{EA}$ in the case of negative-ion emission. The algebraic expression of the Saha–Langmuir equation for positive-ion emission in the absence of an external electric field is:

$$\alpha_+ = \frac{n_+}{n_0} = \frac{g_+}{g_0}\exp\left(\frac{e(\phi - E_i)}{k_B T}\right) \tag{14.1}$$

$\alpha_+$: flux ratio
$n_+$: flux of ions
$n_0$: flux of atoms
$g_+$: statistical weight for the ionic state
$g_0$: statistical weight for the atomic state
$e$: electronic charge constant
$k_B$: Boltzmann constant
$\phi$: work function in eV
$T$: Temperature in K
$E_i$: First ionization energy in eV

As a first approximation, the ratio $g_+/g_0$ in many applications is close to one or to the order of magnitude $10^{-1}$. For more details see, for example.[6–8] Values for the first ionisation energies of selected chemical elements are listed in Table 14.3, work functions of selected surface materials in Table 14.2. It can be concluded from the Saha–Langmuir equation that the ionisation efficiency for the emission of positive ions increases with decreasing first ionisation energy and increasing work function for elements where the difference between the work function and the first ionisation potential is positive and it increases with increasing temperature for elements where the difference between the work function and the first ionisation potential is negative. Thus, the ionisation efficiency for positive ions can be increased by using surface materials with a high work function and a high melting point. As a consequence, rhenium, tungsten, and tantalum are often employed as filament materials as they have high work functions and – in addition – high melting points. As the first ionisation energies of alkali or alkaline-earth metals are low, the ionisation efficiencies are rather high and can approach 100%, as in the case of Cs (see Section 14.2 for more details). For negative-ion emission, high ionisation efficiencies can be realised for elements or compounds with high electron affinities and/or by using a surface material with a high value of its work function. The energy spread of the emitted ions is typically <1 eV. This is low compared to other ion sources used in mass spectrometry, such as inductively coupled plasma or glow discharge. As a consequence, TIMS requires only single-focusing analysers (see Section 14.2.2), resulting in most commercially developed instruments over the past few decades being based on single magnetic sector technology.

The relationship between the filament temperature, rate of analyte evaporation and the ionisation efficiency can be utilised and turned into a major advantage in thermal ionisation. By adjusting the current applied to the filament the surface temperature is changed. This permits control of the ion beam intensity during the measurement. As the ionisation efficiency at a given temperature is different for different chemical compounds, a thermal ionisation source provides elemental selectivity and emission control. This is one reason why interferences are quite rare compared to other ionisation techniques (see Section 14.3.4).

It needs to be pointed out that the Saha–Langmuir equation was derived for substances impinging onto a hot surface, not for substances that are directly placed onto the surface. Indeed, the influence of a monoatomic layer on the surface of the filament, for example a Cs layer on a tungsten filament, was investigated by Langmuir and Kindom, showing that it can significantly decrease the ionisation efficiency. A reduction of the ionisation efficiency with increasing sample sizes has also been pointed out in some cases.[9,10] Furthermore, the equation does not account for the lack of thermal equilibrium, or for chemical processes occurring on the filament surface during heating, for example oxidation. Nonetheless, the Saha–Langmuir equation provides a vital insight into the thermal ionisation source and is useful in predicting the order of magnitude of ionisation efficiency, and in comparing ionisation efficiencies of different elements.

**Table 14.1** Reported total efficiency (ions counted per atoms used), limits of detection using ion-counting detectors, and measurement reproducibility (external standard deviation) using multi-Faraday-cup arrays under favourable conditions in isotope ratio analysis of selected elements using sector-field thermal ionisation mass spectrometry. Elements are measured as positive or negative ions, some as molecular instead of atomic ions. Data from this work and refs. 9,11–26.

| Element | Shape of ionising material | Total efficiency/% | Limit of signal detection/g | Reproducibility for isotope ratio (favourable conditions)/% |
|---|---|---|---|---|
| Li | filament | $10^{-2}$–$10^{-1}$ | | |
| B | filament | $10^{1}$ | | $10^{-3}$–$10^{-2}$ |
| Sr | filament | $10^{-1}$–$10^{0}$ | | $<10^{-2}$, values $<10^{-3}$ reported |
| Cs | filament | $10^{1}$, as high as 50 | | not applicable |
| Nd | filament | $10^{-1}$–$10^{1}$ | | $<10^{-2}$, values $<10^{-3}$ reported |
| Os | filament | as high as $10^{1}$ | | |
| Pb | filament | $10^{-1}$–$10^{0}$ | | $10^{-3}$–$10^{-2}$ |
| Ra | filament | $10^{0}$–$10^{1}$ | $10^{-17}$–$10^{-16}$ | |
| Th | filament | $10^{-2}$–$10^{1}$ | | |
| U | filament | $10^{-2}$–$10^{0}$ | $10^{-14}$–$10^{-17}$ | $10^{-3}$–$10^{-2}$ |
| | cavity | $10^{0}$–$10^{1}$ | | |
| Pu | filament | $10^{-1}$–$10^{1}$ | $10^{-15}$–$10^{-17}$ | $10^{-3}$–$10^{-2}$ |
| | cavity | $10^{0}$–$10^{1}$ | | |

Experimentally observed total efficiencies (ions counted per atoms used) in thermal ionisation mass spectrometry, either using filaments or more sophisticated surface geometries such as in the shape of a cavity are listed for selected elements in Table 14.1. Note that the total efficiency is not exactly equal to the ionisation efficiency but to the ionisation efficiency multiplied by the ion transmission efficiency multiplied by the detector efficiency. The total efficiency is therefore smaller than the ionisation efficiency. Special filament loading techniques can be employed to boost the ionisation efficiency, improve the evaporation behaviour and change the work function (see Section 14.2 for more details). As a result, a large range of ionisation efficiencies can often be reported for a given element; only orders of magnitude are listed here.

Obviously, the reported total efficiency obtainable for Cs is high, reflecting the fact that ionisation efficiency close to 100% can be obtained for Cs as already investigated by Langmuir and Kingdom. Actinides, on the other hand, usually have orders of magnitude smaller ionisation efficiencies due to their rather high first ionisation energies.

Even though the total ionisation efficiencies for actinides are comparatively low, the smallest amounts of substance reported for generating a substantial signal in thermal ionisation mass spectrometry are as low as a few tens of thousand atoms absolute. This sensitivity is a pivotal point in nuclear and earth-science applications, among others. For some elements, pushing the limits of detection is not as crucial as pushing the limits of measurement precision in order to resolve minute differences in their isotope ratios. Foremost among these are measurements of Sr and Nd that lead the way where (relative) measurement reproducibility below 100 parts per million are routinely achieved, and values below 10 parts per million are demonstrated. This is summarised in Table 14.1. This level of precision is largely due to employing a multicollector Faraday-cup array, where all isotopes of interest can be measured simultaneously using a static magnetic field. This renders the precision in the isotope ratio measurement to a large degree independent of the ion-beam fluctuations. It is this precision work, almost unrivalled, that makes thermal ionisation mass spectrometry a powerful tool in research and technology (see Section 14.4). However, it has to be noted that these values are for precision only and do not represent an overall measurement uncertainty. Indeed, for many decades thermal ionisation mass spectrometry has been accepted as the benchmark analytical technique for the precise and accurate measurement of isotopic ratios.[7,27,28] It was the thermal ionisation source – driven by technological advances in sector-field mass analysers, vacuum systems, Faraday-cup detectors, and data processing – that opened the door to a major new area: the detailed study of major geological processes, and the absolute dating of the Earth. In the end, our knowledge about the isotopic composition of the chemical elements and their atomic weights is mainly based on isotope abundance measurements by TIMS, as illustrated by most of the IUPAC recommendations.[29,30]

Today, an estimated 1200 commercial TIMS instruments are installed worldwide.[31]

## 14.2 Technical Background

### 14.2.1 Ion Source

In the basic design of a modern thermal ionisation source, the sample to be analysed is placed onto a flat metal ribbon filament. The filament is soldered to a filament holder for mechanical stability and to form an electrical connection to a high-voltage power supply. A classic double filament structure is shown in Figure 14.4(a) as it is used in one of the commercially available instruments. A drop of the sample solution is deposited in the centre of the filament. By applying a small electrical current, the liquid sample is dried onto the filament before being loaded into the ion source. Once under vacuum, a higher current across the filament causes the sample to evaporate further and partially ionise. In the case of a single-filament geometry, evaporation and ionisation of the analyte occurs on the same filament. With a double-filament geometry, ionisation and evaporation occur on separate filaments arranged parallel to each other: On one filament the analyte is deposited. This filament is referred to as the evaporation filament. The

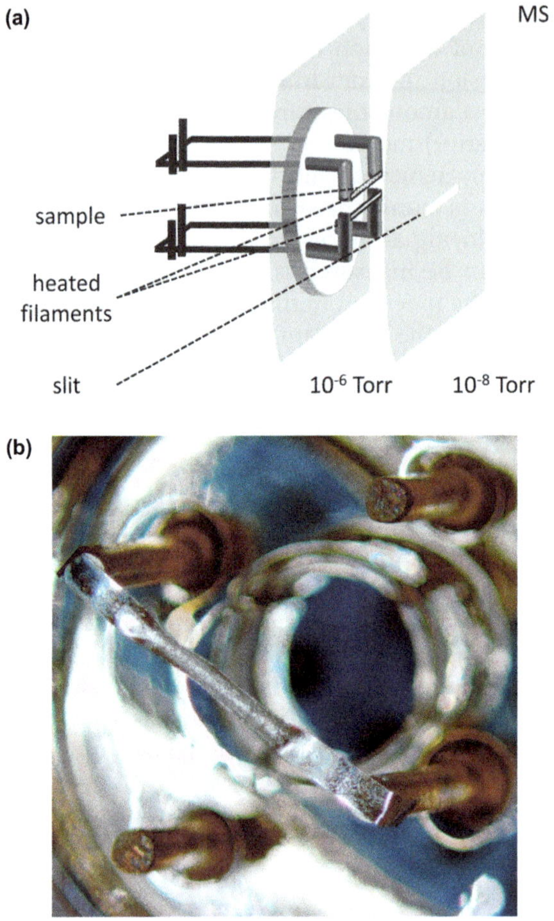


**Figure 14.4** (a) Schematics of a TIMS ion source (b) Glass bead with V-shape filament used for *VG 354* and *Sector 54.*
(Source: (a) Ondrej Hanousek; (b) Photo: Jochen Vogl.)

second filament is referred to as the ionisation filament. When using the double-filament geometry, the analyte is evaporated from the evaporation filament and the analyte atoms or molecules interact with the ionisation filament. Ionisation takes place predominantly on the ionisation filament. In the double-filament geometry, the current applied to both filaments can be regulated independently in order to separate the evaporation and (the potentially much hotter) ionisation process. An extension of the double geometry is the triple-filament geometry.

The filament holders are mounted on a sample turret (also termed filament carousel) (Figure 14.5) and placed inside the ion source housing (also termed ion source chamber) of the instrument. A high voltage ($\sim$10 kV) is applied to extract the ions emitted from the filament and to accelerate them into the flight tube. The ion-source housing is separated from the flight tube

**Figure 14.5** Ion source of a *Sector 54* instrument showing the sample turret inside the opened vacuum housing.
(Source: Jochen Vogl © BAM.)

by a valve that permits venting of the ion-source housing to exchange sample turrets while, at the same time, maintaining the high vacuum in the flight tube.

Significantly improved ionisation efficiencies can be obtained using, for example, V-shaped (also known as canoe or boat-shaped) filaments (see Figure 14.4 (b)) or dimpled filaments[11,26] or the more sophisticated design of a cavity ion source instead of a flat ribbon filament. Boat-shaped filaments are commercially available, dimpled filaments or cavity ion sources, however, are not. The cavity ion source is a solid rod made of, for example, rhenium with a millimetre-size cavity drilled into one of the rod's tips.[11,13,32,33] The analyte is deposited inside the cavity and the tip of the rod is heated in order to evaporate and ionise the analyte. The cavity is a confined geometry and has an increased ratio of surface area to volume compared to flat ribbon filaments, which increases the contact of the analyte with the surface material. As a result, ionisation efficiencies are often higher by several orders of magnitude using cavity ion sources (see Table 14.1).

**Table 14.2** Melting point and (electron) work function for the established surface materials.[35]

| Surface material | Melting point/°C | Work function/eV |
|---|---|---|
| Platinum | ~1800 | ~5.7 |
| Rhenium | ~3200 | 4.96–5.75 |
| Tantalum | ~3000 | 4.00–4.80 |
| Tungsten | ~3400 | 4.30–5.25 |

Platinum, tantalum, tungsten, and rhenium (zone refined or nonzone refined) are usually used as surface materials for the filaments (or cavities) because of their comparatively high melting points and work functions, *i.e.* energy required to remove an electron from the filament surface into the vacuum. Pt shows a very high work function and therefore is often used for negative-ion emission where analytes require a high first ionisation energy. The value of the work function may depend on the crystal structure of the surface material, and it can change if the surface is oxidised or carburised. For example, the work function increases from 5.4 eV to maximum 7.2 eV for oxidised Re and from 5.4 eV to 5.8 eV for carburised Re.[34,35] Table 14.2 summarises melting points and work functions for these surface materials.

A variety of special loading techniques have been developed over the past few decades to influence the ionisation and/or evaporation behaviour of specific elements. As early as in the 1960s a mixture of phosphoric acid and a silica gel activator were used to reduce the volatility of lead compounds in order to permit higher filament temperatures and thus higher ionisation efficiency.[26] Since then, the silica gel technique has been applied to a number of elements with high ionisation potentials.[27] Carbon (*e.g.* as graphite) is another common additive, often used for the analysis of, *e.g.* B, Th, U, and Pu. The overall efficiency can also be improved by using resin bead loading or electroplating instead of liquid-drop loading.[11,34,36] In special applications such as fission-track TIMS or zircon evaporation TIMS, micrometre-size particles are loaded directly onto a filament instead of dissolving each particle and using liquid-drop loading.[37,38] Thermal (surface) ionisation sources are utilised for the chemical analysis of aerosol particles, gases and organic compounds, but usually in combination with time-of-flight, quadrupole or other mass analysers that permit a fast scanning of the entire mass spectrum.[4,39] Sector-field analysers have also been used for the real-time analysis of aerosols. Reported loading techniques and additives as well as the surface materials used for TIMS analysis of selected analyte elements are listed in Table 14.3.

The thermal ionisation source using a flat ribbon filament is characterised by a comparatively small energy spread of the ions, usually below 1 eV.[15,45] A single-focusing geometry is therefore sufficient and indeed, all commercially available TIMS instruments have only a magnetic sector analyser. Another feature of the thermal ionisation source is that predominantly singly charged ions are formed in the thermal ionisation process since the second ionisation energy is considerably larger.[46]

**Table 14.3** Reported filament loading techniques, additives and ions measured for selected elements in isotope ratio analysis using sector-field thermal ionisation mass spectrometry. The first ionisation energies are listed as well.[9,17,19–24,26,40–44]

| Element | First ionisation energy/eV | Filament material | Reported loading techniques | Reported additives | Ions measured |
|---|---|---|---|---|---|
| Li | ~5.4 | Re | Drop loading | Various, *e.g.* phosphoric acid, boric acid, silica gel, Hydrochloric acid | Positive, *e.g.* $Li^+$, $Li_2BO_2^+$, $LiNaBO_2^+$, $Li_2F^+$ |
| B | ~8.3 | Ta | Drop loading | Various, *e.g.* graphite | Positive, *e.g.* $Cs_2BO_2^+$, $Na_2BO_2^+$, $B^+$ |
| | | Pt | Drop loading | Various, *e.g.* graphite, $CaCl_2$, $Ba(OH)_2$ | Negative, *e.g.* $BO_2^-$ |
| Sr | ~5.7 | Re, Ta | Drop loading, resin beads | Various, *e.g.* phosphoric acid, Ta-oxide, $TaF_5$ | $Sr^+$ |
| Cs | ~3.9 | Re | Drop loading, resin beads | | $Cs^+$ |
| Nd | ~5.5 | Re, W | Drop loading, resin beads | Various, *e.g.* phosphoric acid, silica gel, TaCl, $Ta_2O_5$, $TaF_5$ | Positive, $Nd^+$ or Nd-oxide |
| Os | ~ 8.7 | Pt | Drop loading | $Ba(OH)_2$ | Negative, Os-oxides |
| Pb | ~7.4 | Re | Drop loading | Silica gel, phosphoric acid, boric acid | $Pb^+$ |
| Ra | ~5.3 | Re, W, Pt, Ta | Drop loading | Ta-HF, Ta-oxide, phosphoric acid, silica gel | $Ra^+$ |
| Th | ~6.1 | Re | Drop loading | Carbon | Positive, $Th^+$ or Th-oxide |
| U | ~6.2 | Re, W | Drop loading, resin beads, electroplating, particle loading | Carbon, platinum | Positive, $U^+$ or U-oxide |
| Pu | ~6.1 | Re, W | Drop loading, resin beads, electroplating, particle loading | Carbon, platinum | Positive, $Pu^+$ or Pu-oxide |

### 14.2.2 Magnetic Sector and Electric Sector

Modern instruments are predominantly single magnetic sector systems using standard components. The two commercially available TIMS instruments – *Phoenix* (IsotopX) and the *Triton* (Thermo Fisher Scientific) use 90 degree magnetic sectors. However, experimental two-stage and three-stage TIMS instruments have also been built, in an attempt to improve the abundance sensitivity, amongst other features (see Section 14.3).

### 14.2.3 Flight Tube

Modern commercial instruments use flight tubes that are about one to two meters in length and as much as 80 mm in width.

### 14.2.4 Entrance Slit and Exit Slit

Usually, low-resolution slits are used in single magnetic-sector TIMS instruments in order to achieve a mass resolution (at 10% peak height) of several hundred units (at mass 238 u).

### 14.2.5 Transfer, Zoom, and Filter Optics

The *Triton* / *Triton Plus* instruments, for example, are equipped with a set of quadrupole lenses, called "zoom optics". The zoom optics can compensate for slightly different mass dispersions for the different elements. They allow for changing the dispersion of the ion beam after it passes through the magnet, which enables the measurement of rare-earth elements, or uranium and plutonium, within one set of fixed Faraday-cup positions. All commercially available instruments of the present generation (*Triton Plus*, *Phoenix*) and some of the previous instruments (*e.g. Sector 54*) have the option of retarding quadrupole lenses that are employed as energy filters to improve the abundance sensitivity.

### 14.2.6 Detection System

Modern TIMS instruments use predominantly Faraday cups and ion counters such as Daly and secondary electron multipliers arranged in a multi-collector array. The detectors are standard components of sector field mass spectrometers (see Chapter 4).

### 14.2.7 Vacuum System

In modern TIMS instruments, turbo pumps and ion getter pumps are employed for maintaining a high to ultrahigh vacuum in the ion source and flight tube. These are standard components of sector field mass spectrometers (see Chapter 4). Liquid nitrogen-cooled cryogenic traps (also termed cold traps) are utilised to improve the vacuum in the ion source housing, for example to freeze out hydrocarbons and other volatile compounds. They have been used in TIMS since the 1970s.[47]

### 14.2.8 Instrumentation

Thermal ionisation mass spectrometry was used as early as 1918. A. J. Dempster employed a platinum strip to ionise metal salts for the analysis of elements including Na, K, and Ca.[2] The ion separation was achieved in a magnetic field,[2] and the mass resolution approached 100 (half-width maximum). Platinum is still used today as filament material for selected elements.

From the 1950s onwards, going hand in hand with improvements in sector field mass analysers and vacuum technology, various research-built instruments were introduced. To mention just a few, a single-filament source and single-collector magnetic sector field instrument was developed at the National Bureau of Standards (today NIST)[8,36,48] and found use in several laboratories around the world. Also, two- or even three-stage instruments with two magnetic sectors and/or one electrostatic sector were constructed in order to improve abundance sensitivity and to search for previously unknown naturally occurring stable isotopes with abundances in the range of parts per million and below.[27,49–51] Indeed, White *et al.* reported the discovery of Ta-180 and established lower experimental limits for the abundance of a large number of isotopes.[49] Some of these instruments were still in use by the 1990s.[22]

The first commercially available TIMS instruments appeared as early as the end of the 1950s and 1960s, for example the instruments introduced by Metropolitan Vickers, UK, National Bureau of Standards, US or Varian MAT, Germany. Early TIMS instruments were single-collector systems. Isotope ratio measurements performed with a single collector rely on peak jumping between isotopes, which is prone to inaccuracies and imprecisions due to ion-beam fluctuations. This puts a limit on the accuracy and precision achievable in isotope ratio measurements using single-collector systems although drift corrections are applied. To overcome these limitations, multicollector instruments were introduced, being capable of measuring several isotopes simultaneously. This resulted in order of magnitude improvements in precision. Commercial instruments developed in the 1980s and 1990s by Finnigan MAT, Germany, and VG Isotopes, UK, had as many as 9 Faraday cups. Instruments like these were the stepping stones to high-precision geochronology. They enable the resolution of minute difference in – for example – the Sr or Nd isotope ratios employed to determine absolute geological ages using the Rb–Sr or Sm–Nd decay system, respectively. In the early days of multicollector systems, Faraday cups were often arranged with a fixed spacing between them, which limited our ability to analyse certain elements. In later-generation instruments, the Faraday cups were made movable to easily accommodate different isotope systems. From the first multicollector instruments in 1984 onwards, VG instruments and their successors used specific curved magnet entry and exit pole pieces to rotate the focal plane so that the ion beam intersects the focal plane at close to 90 degrees. The advantage is a significant reduction in stray ions, improved ion detection and a simplified mechanism for moving the Faraday detectors.

In order to improve the abundance sensitivity, quadrupole retardation lenses have been employed. In the 1990s, such lenses – named RPQ – were supplied, *e.g.* with the *MAT 261 / 262* instruments and improved the abundance sensitive by one to two orders of magnitude.[45] In the case of the *Sector 54* instruments these same lenses were branded as WARP filters.

The *Triton TIMS*, the successor model of the *MAT 262*, introduced the virtual amplifier for its multicollector Faraday-cup array. The virtual amplifier seeks to eliminate the bias caused by the crosscalibration of the amplifier gains.[15] Unlike the traditional way of using a fixed connection between a Faraday cup and an amplifier, the virtual amplifier works by means of a relay matrix, which permits switching the amplifiers between the different Faraday cups. By rotating the amplifiers during measurement, the uncertainties from the gain calibration cancel out, which improves the achievable measurement uncertainty for isotope ratio determinations. The incorporation of multiple ion counters, for example in the *IsotopX* and predecessor instruments as well as in the *Triton*, were used to improve the signal-to-noise ratios and the sensitivity when compared to Faraday-cup detectors. This enabled low-level measurements of several isotopes, extending the measurement of ion currents to as low as $10^{-19}$ A.

These technological developments combined to produce the state-of-the-art instruments commercially available today. Selected instruments are listed in Table 14.4. As of 2013, there are two TIMS instruments on the market – both sector field multicollector systems. They are the *Phoenix* instrument offered by IsotopX, UK, and the *Triton Plus* instrument by Thermo Fisher Scientific, Germany.

### *Phoenix* (IsotopX (UK))

The *Phoenix* is the latest multicollector TIMS instrument offered by IsotopX, UK (see Figure 14.6). The company IsotopX was formed in 2008 from VG Instruments. The company traces its heritage to the first commercial TIMS instruments, the MM30, launched by VG Micromass in 1973.[12]

The *Phoenix* is a single magnetic sector instrument.[12,54,55] The magnet is a 90 degree sector with a radius of 27 cm. It can be used in positive- and negative-ion mode with an acceleration voltage of plus or minus 8 kV. The multicollector array is fitted with 9 Faraday-cup detectors, one Daly detector or secondary electron multiplier as axial ion counter and additionally six conversion dynode ion counters. This spans a range of $10^{-9}$ A to $10^{-19}$ A in ion current detection. A retarding filter is available for the axial ion counter yielding an abundance sensitivity of better than $1\times10^{-8}$ at $238\pm1$ u. Using the Faraday-cup multicollector array, measurement reproducibilities of 3 parts per million (ppm) and 2 ppm are reported for $^{87}Sr/^{86}Sr$ and $^{143}Nd/^{144}Nd$ isotope ratio measurements, respectively, using several hundred nanograms

**Table 14.4** Selected commercial and research-built sector-field thermal ionisation mass spectrometry instruments (list is not complete; instruments that have been installed only a few times and where no additional data (*e.g.* introduction year or design) are available have not been considered in this list.[12,45,47,51–53]

| Name | Manufacturer[a] | Years of production | Geometry |
|---|---|---|---|
| AEI MS5 | Metropolitan Vickers, UK | 1955–1970 | single-collector magnetic sector |
| CH4-T04 | Varian MAT, Germany | 1958–1962 | single-collector magnetic sector |
| NBS 12–90 | National Burreau of Standards, US | late 1950s | single-collector magnetic sector |
| 12–90-SU | Nuclide Analysis Associates US | 1960s | single-collector magnetic sector |
| AEI 30A | Metropolitan Vickers, UK | 1960s | single-collector magnetic sector |
| CH5 TH | Varian MAT, Germany | 1962–1967 | single-collector magnetic sector |
| TH5 | Varian MAT, Finnigan MAT, both Germany | 1967–1976 | single-collector magnetic sector |
| TSN 206 | Cameca, France | 1968–1980 | single-collector magnetic sector |
| MM30 | VG Micromass, UK | 1973–1977 | single-collector magnetic sector |
| MAT 260 | Finnigan MAT, Germany | 1976–1980 | single-collector magnetic sector |
| VG Isomass 53E | VG Micromass, UK | 1978–1980 | single-collector magnetic sector |
| JMS-05RB | JEOL, Japan | 1980? | single-collector magnetic sector |
| MAT 261 | Finnigan MAT, Germany | 1980–1987 | multicollector magnetic sector |
| THQ | Finnigan MAT, Germany | 1982–1994 | single collector quadrupole instrument |
| VG 354 | VG Micromass, UK | 1984–1987 | multicollector magnetic sector |
| Sector | VG Micromass, UK | 1987–1989 | single-collector magnetic sector |
| MAT 262 | Finnigan MAT, Germany | 1987–1998 | multicollector magnetic sector |
| VG 336 | VG Micromass, UK | 1989–1991 | single-collector magnetic sector |
| VG Sector 54 | VG Isotopes, Fisons Instruments, Micromass, all UK | 1990–1999 | multicollector magnetic sector |
| IsoProbe T | Micromass, GV Instruments, IsotopX, all UK | 1999–2007 | multicollector magnetic sector |
| Triton | Thermo Fisher Scientific, Germany | 1998–2009 | multicollector magnetic sector |
| Phoenix | IsotopX, UK | 2008 to present | multicollector magnetic sector |
| Triton Plus | Thermo Fisher Scientific, Germany | 2009 to present | multicollector magnetic sector |
| Nu TIMS | Nu Instruments | 2014 to present | multicollector magnetic sector |

[a]name of manufacturer at time of release of the instrument.

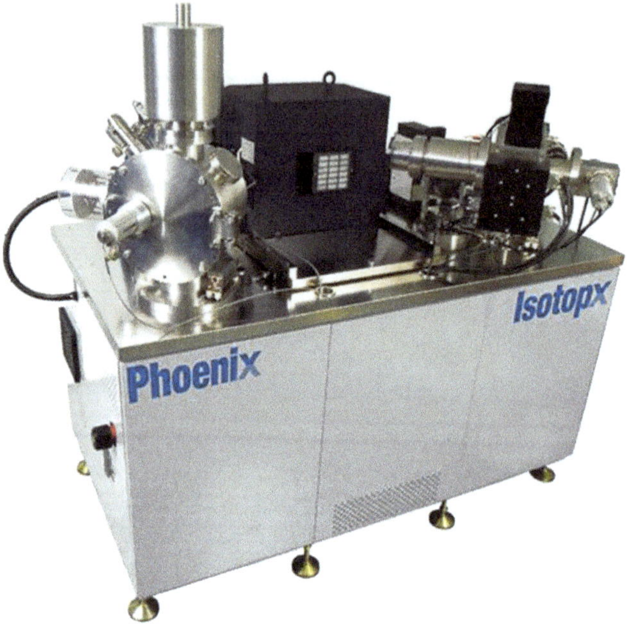


**Figure 14.6** *Phoenix* multicollector TIMS.
(Source: © IsotopX.)

of sample amount per filament. The sample carousel can hold up to 20 samples using triple filament assemblies. The *Phoenix 62* option is available that supports a special detector array that allows simultaneous measurement of $UO_2^+$. A glove box adaptation is also available for handling nuclear materials.

***Triton Plus* (Thermo Fisher Scientific)**

The latest-generation multicollector TIMS instrument supplied by Thermo Fisher Scientific is the TritonPlus, introduced in 2009 (Figure 14.7). It is a modified version of the *Triton* multicollector TIMS instrument, which was launched more than a decade ago. It traces its origin back to the multicollector instruments *MAT 261* and *MAT 262*, which have been produced by Finnigan MAT, Germany, from the 1980s.

The *Triton Plus* is a single magnetic sector instrument. The magnet is a 90 degree sector with a dispersion of 81 cm. It provides positive- and negative-ion modes with an acceleration voltage of 10 kV. The multicollector array can be fitted with up to 9 Faraday-cup detectors, three secondary electron multipliers and eight conversion dynode ion counters. Two of the three SEMs can be fitted with energy filters (retarding potential quadrupoles, RPQs) for high abundance sensitivity ($<2\times10^{-8}$ at mass $238\pm1$ u). This spans a range of $10^{-9}$ A to $10^{-19}$ A in ion current detection. The Faraday cups used are laser machined from solid carbon to achieve uniform

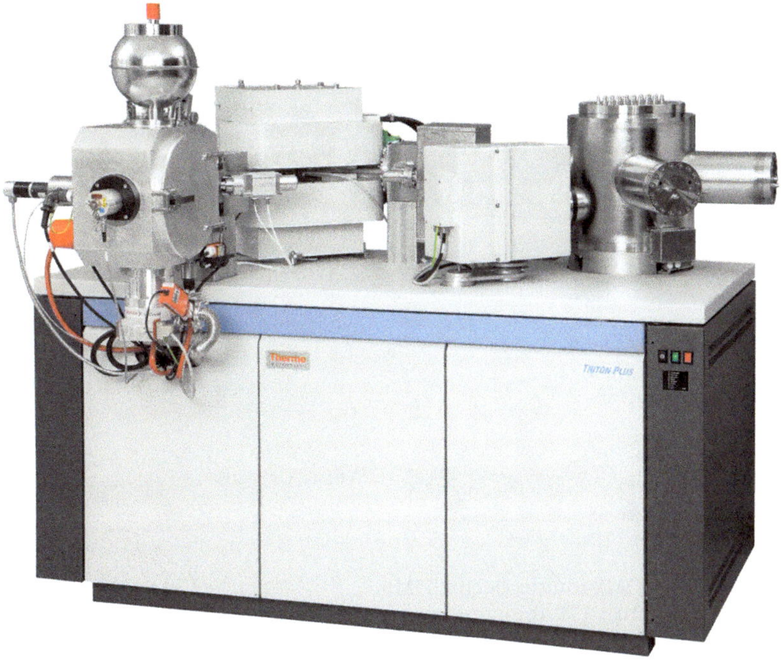

**Figure 14.7** *Triton Plus* multicollector TIMS.
(Source: © Thermo Fisher Scientific.)

response in the cup efficiency factors, high linearity, low noise, and long lifetimes. Thermo Fisher Scientific introduced the virtual amplifier to improve the crosscalibration of individual amplifiers in order to calibrate the differences in the relative amplifier gains between Faraday cups. Using the Faraday-cup multicollector array, a measurement reproducibility of <4 ppm is reported for $^{87}Sr/^{86}Sr$ and $^{143}Nd/^{144}Nd$ isotope ratio measurements using several hundred nanograms of sample amount per filament. It was demonstrated that a measurement reproducibility of better than 200 ppm can be maintained for $^{87}Sr/^{86}Sr$ and $^{143}Nd/^{144}Nd$ isotope ratio measurements analysis even for sample amounts as low as 100 pg by employing $1\times10^{12}$ Ohm resistors.[56] The sample turret can hold up to 21 samples using single or double filament assemblies. Dedicated collector packages for the analysis of Pb, Os, or U isotopes are available. A glove-box adaptation is supported for handling nuclear materials.

***Nu TIMS* (Nu instruments)**

The *Nu TIMS* (released in 2014) has been developed using Nu Instruments field-proven variable-dispersion multicollector technology to improve the versatility and overall performance of this long-established analytical technique. It combines design advances in filament assembly, ion optics and electronics control with ease of use and high precision. The instrument

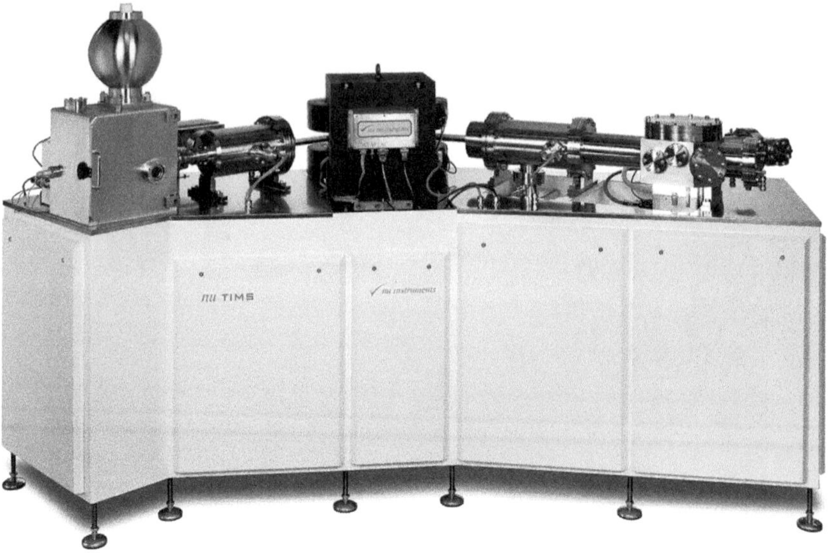


**Figure 14.8** *Nu TIMS* multicollector TIMS.
(Source: © Nu instruments.)

operates at an accelerating potential of 8 kV and is a single-focusing instrument consisting of a 300-mm radius, 60° magnet. The instrument is operational in both positive- and negative-ion modes.

The *Nu TIMS* (Figure 14.8) utilises Nu Instruments unique, patented Zoom lens system to ensure perfect peak alignment without the requirement of moving detectors. The collector array consists of 12 Faraday detectors and multiple multiplier configurations. An ion-counting mini-Daly system can be fitted in place of one of the ion counting electron multipliers. This Daly system consists of a −25-kV conversion dynode that attracts ions onto its surface. These ions cause the ejection of electrons that are in turn repelled onto a scintillator plate. The impact of these electrons produces photons that are detected by a photomultiplier tube that is physically mounted outside of the instrument vacuum envelope. The Daly detector has particularly low dark noise and is both extremely stable and linear over a wide dynamic range. As with other products from Nu Instruments, the *Nu TIMS* can be fitted with deceleration lens filters in front of the ion-counting multipliers and Daly system for improved abundance sensitivity performance.

## 14.3 Measurement Considerations

### 14.3.1 Mass Fractionation and Temperature Profile

Instrumental isotopic fractionation of lighter (*e.g.* $^{6}Li^{+}$) *versus* heavier ions (*e.g.* $^{7}Li^{+}$) in isotope ratio measurements is observed in TIMS, as it is observed in mass spectrometry in general (see Chapter 6). The evaporation and ionisation of the lighter atom is slightly preferred compared to the heavier atom in the thermal ionisation source simply due to the smaller mass of the lighter ion. It results, over the course of a measurement, in a depletion of the lighter atoms relative to the heavier atoms in the analyte reservoir on the filament. This yields positively biased isotope ratios at the beginning and negatively biased isotope ratios towards the end of a measurement. For the element lithium, one of the lightest elements in the periodic system, the mass fractionation in TIMS for $^{6}Li/^{7}Li$ ratio analysis may be several per cent over the duration of a measurement when detecting atomic ions $Li^{+}$. Fractionation will be reduced significantly if molecular ions like $Li_2BO^{+}$ are used, because the relative mass difference for $^{6}Li_2BO^{+}$ *versus* $^{7}Li_2BO^{+}$ is almost one order of magnitude smaller compared to the atomic ions $^{6}Li^{+}$ *versus* $^{7}Li^{+}$.[17–19,57] In contrast, the magnitude of mass fractionation in isotope ratio analysis of the heaviest elements, such as $^{235}U/^{238}U$ or $^{240}Pu/^{239}Pu$, is of the order of half a per cent or less per one mass unit difference.[58,59] The mass fractionation needs to be corrected in order to reduce the introduced bias (see Chapter 6 and Section 14.3.2).

The fractionation observed for isotopes of different elements is in general more pronounced than for isotopes of the same element. This is due to the differences in ionisation and evaporation behaviours of the elements in addition to the mass difference. The temperature applied to the filament can be increased in a stepwise manner over the duration of the measurement to yield a temperature profile with significantly changing ratios between isotopes of different elements. It significantly constrains the utilisation of TIMS for multielement analysis, which is one reason why TIMS is preferentially employed for isotope ratio (and isotope dilution) analysis. This temperature profile, however, can be put to good use, for example, in the identification of interferences (see Section 14.3.4).

## 14.3.2 Calibration and Metrological Traceability

In order to calibrate the isotope ratio measurements against a known reference value, various internal normalisation and external normalisation approaches are used. The calibration is predominately a correction of the mass fractionation effect (see Section 14.3.1 and Chapter 6). The known reference value, against which the isotope ratio measurements are calibrated, is usually the quantity value of reference materials or certified reference materials. In some cases it is a consensus value.

Internal normalisation is used for calibration in several cases. Two prominent examples are $^{87}Sr/^{86}Sr$ or $^{143}Nd/^{144}Nd$ measurements. Here, an internal normalisation is performed using the isotope ratios $^{88}Sr/^{86}Sr$ and $^{146}Nd/^{144}Nd$, respectively. Consensus values of 0.1194 and 0.7219 for the atom amount ratios $^{86}Sr/^{88}Sr$ and $^{146}Nd/^{144}Nd$, respectively, are frequently used in combination with an exponential law or in some cases a Rayleigh-law or power-law model.[42,60–63] (it has to be noted, that the $^{86}Sr/^{88}Sr$ ratio of

0.1194 is a rounding of the certified value of the SRM 987 (NIST, Gaithersburg, US)). These results are regarded as traceable to the consensus value. The addition of a double spike is also used for internal normalisation. This is sometimes employed for the determination of the isotope ratios $^{238}U/^{234}U$ and $^{238}U/^{235}U$ of natural uranium samples, among other applications. Here, the addition of a $^{236}U/^{233}U$ double spike is frequently used[64–66] and the results may be traceable to the SI units, for example, in the case of a certified double-spike reference material. Another example is the use of a $^{205}Pb/^{202}Pb$ double spike for Pb determination.

External normalisation is frequently applied in combination with standard bracketing where either one or more sample measurements are bracketed by measurements of a (certified) reference material. Prominent examples are the isotope ratio measurements of lead, uranium, and plutonium,[59,67,68] often intended to be traceable to the SI units by employing certified reference materials. Various empirical mass fractionation models are employed including exponential, power, and linear law (see Chapter 8).

### 14.3.3 Total Evaporation

The total evaporation (TE) method – also known as the flash evaporation method – in combination with multicollector TIMS has been utilised for decades for the analysis of isotope ratios of uranium and plutonium[69–71] and for elements such as B, Ca, Nd, Eu, and Ra.[20,72,73] In TIMS, the mass fractionation progresses over the duration of a measurement (see Section 14.3.1). The TE method addresses this by measuring the ion signals of all isotopes of the element of interest until exhaustion, *i.e.* a total evaporation of the analyte. This means that the complete sample reservoir is evaporated and measured. The isotopes are detected simultaneously using a multicollector array. The magnetic field and focus settings are usually not changed throughout the duration of the analysis. The isotope ratios are then calculated from the integrated intensities of the individual isotopes. The advantage of TE is that the effect of mass fractionation between the lighter isotope (*e.g.* $^{235}U$) and the heavier isotope (*e.g.* $^{238}U$) is averaged out to a large degree due to the total evaporation of the sample because the entire history of the mass fractionation curve is integrated. The reproducibility of isotope ratio results using the TE technique are to a large degree insensitive to small variations in the analyte amount loaded onto the filament. For example, the mean value of repeated TE measurements for high mass elements like U or Pu exhibit a bias (*i.e.* the difference between the measured value and the certified value) of the order of 0.01% per mass unit or less. However, this bias is not entirely cancelled out in TE but only minimised. This is simply because not one hundred per cent of the analyte atoms loaded on a filament are detected as ions. Therefore, the combination of the timely drift of the ratios from the fractionation (see above) with a possibly nonconstant efficiency can cause biases in total evaporation measurements. The use of TE does not eliminate the necessity of calibrating the isotope ratio measurements (see Section 14.3.2).

A modification to the total evaporation method was introduced in order to overcome the static nature (*i.e.* static magnetic field) of TE.[58,74] By permitting a change in the magnetic field setting at given intervals throughout the TE measurement, the modified total evaporation method (MTE) enables *in situ* determination of the peak-tailing effects that need to be accounted for in the analysis of low abundance isotopes as well as *in situ* intercalibration of ion counters *versus* Faraday cups. It thus addresses some of the disadvantages of the static TE method. The implementation of the MTE method has resulted in significant improvements in the accuracy of analyses of low-abundance isotopes, for example in $^{234}$U and $^{236}$U determination.[59,74]

## 14.3.4 Polyatomic and Molecular Interferences

The first ionisation energy and the evaporation behaviour differ with element and molecular compound and strongly depend on the filament temperature. As a consequence, molecular and polyatomic interferences in TIMS are rare, even less for negative-ion emission compared to positive-ion emission.[46] When raising the filament temperature to the desired point, some isobaric and molecular interferences will evaporate from the filament in the process prior to starting the data acquisition of the analyte. For example, small amounts of Rb can be removed by gently heating the filament prior to the Sr isotope ratio determination. Other potentially interfering elements and compounds may require higher temperatures for ionisation and will thus not overlap with the analyte measurement in the first place. The thermal ionisation source surely offers elemental selectivity to a certain degree. Nonetheless, molecular and isobaric interferences, though rare compared to other mass spectrometric techniques, need to be identified and accounted for. Commercially available TIMS instruments do not offer sufficient mass resolution for resolving potential molecular or even isobaric interferences. Various approaches are in use to account for interferences, as mentioned below. However, the most reliable way of mitigating interferences is to perform a sufficiently good chemical separation of the analyte of interest from the sample matrix prior to the analysis.[46,75]

Isobaric interferences of course can be accounted for by monitoring the signal intensity of a second isotope of the interfering element and applying a correction for the interference. As in the prominent cases of $^{87}$Sr interfered with by $^{87}$Rb or $^{142}$Nd interfered with by $^{142}$Ce, the remaining interference contribution is subtracted by monitoring the isotopes $^{85}$Rb or $^{140}$Ce, respectively, and applying a correction using the known isotopic abundances.[30]

The measurement of oxide ions instead of atomic ions is an approach that has been employed in several cases as the formation of oxide ions varies with the element and can result in less interference. Prominent examples are targeting NdO ions instead of atomic Nd to address Ce and Pr interference, or analysing Re-oxides to minimise the $^{187}$Os interference on $^{187}$Re.[46] The temperature profiles of uranium and plutonium are slightly shifted with Pu ion signals commencing before U ion signals. Here, sequential measurements of first Pu isotope ratios and then U isotope ratios permits a correction for the $^{238}Pu$–$^{238}U$ interference.[76]

Molecular interferences are usually more pronounced in the lower mass range – most notably for the elements Li and B – and different approaches are utilised to address them. For example in boron N-TIMS determination, the measurement of the isotope ratio $^{11}B^{16}O_2/^{10}B^{16}O_2$ suffers from the organic interference $^{12}C^{14}N^{16}O^-$. These interferences can be overcome by preheating the filament. The filament is held at a certain temperature below the temperature needed for boron analysis in order to drive off the more-volatile organic substances and potentially other interfering species. Alternatively, polyatomic boron ions such as $Cs_2BO_2^+$ are analysed in P-TIMS instead to overcome interferences in the low mass range[46] and to minimise mass fractionation. An example of molecular interferences in the high mass range is the formation of $BaPO_2$, which interferes in Pb analysis. Monitoring the ion signal produced by $BaPO_2$ at mass 201 u can be used to correct for the $BaPO_2$ interference to the Pb isotopes.[9]

The temperature profile of the thermal ionisation process can be put to good use for the identification of interferences as the mass fractionation observed for isotopes of different elements is in general more pronounced than for isotopes of the same element. Dramatically shifting isotope ratios throughout the duration of a measurement are a sign of one or more interferences.

## 14.4 Applications

Whether it be the highly precise measurement of trace-element mass fractions at the parts-per-trillion level (*via* multispike isotope dilution), or the analysis of isotope ratios requiring a stable beam (with few molecular and polyatomic interferences), TIMS remains a core analytical technique (see Figure 14.9(a) and (b)). A number of fields of applications can be found similar to MC ICP-MS (see Section 12.5).

### 14.4.1 Applications in Life Sciences

The life sciences represent an area of continuing growth and new method development for thermal ionisation mass spectrometry.

In medicine, the balance and interaction of chemical elements within the body is important to our understanding of health, disease, and basic human physiology. For instance, by employing a novel double-spiking technique to the measurement of sulfur isotopes ($\delta^{34}S$), Mann and Kelly demonstrated an early cancer detection technique.[77] TIMS has also helped us understand how the body metabolises iron. Although standard internal normalisation of TIMS data is capable of producing high-precision Fe isotope ratios ($\pm$0.1 per mil RSD $^{54}Fe/^{56}Fe$), this method removes all trace of any natural, mass-dependent isotopic fractionation in the sample. To measure these variations, Johnson and Beard developed a $^{54}Fe + ^{58}Fe$ double-spike method to correct for instrumental mass fractionation, with associated measurement

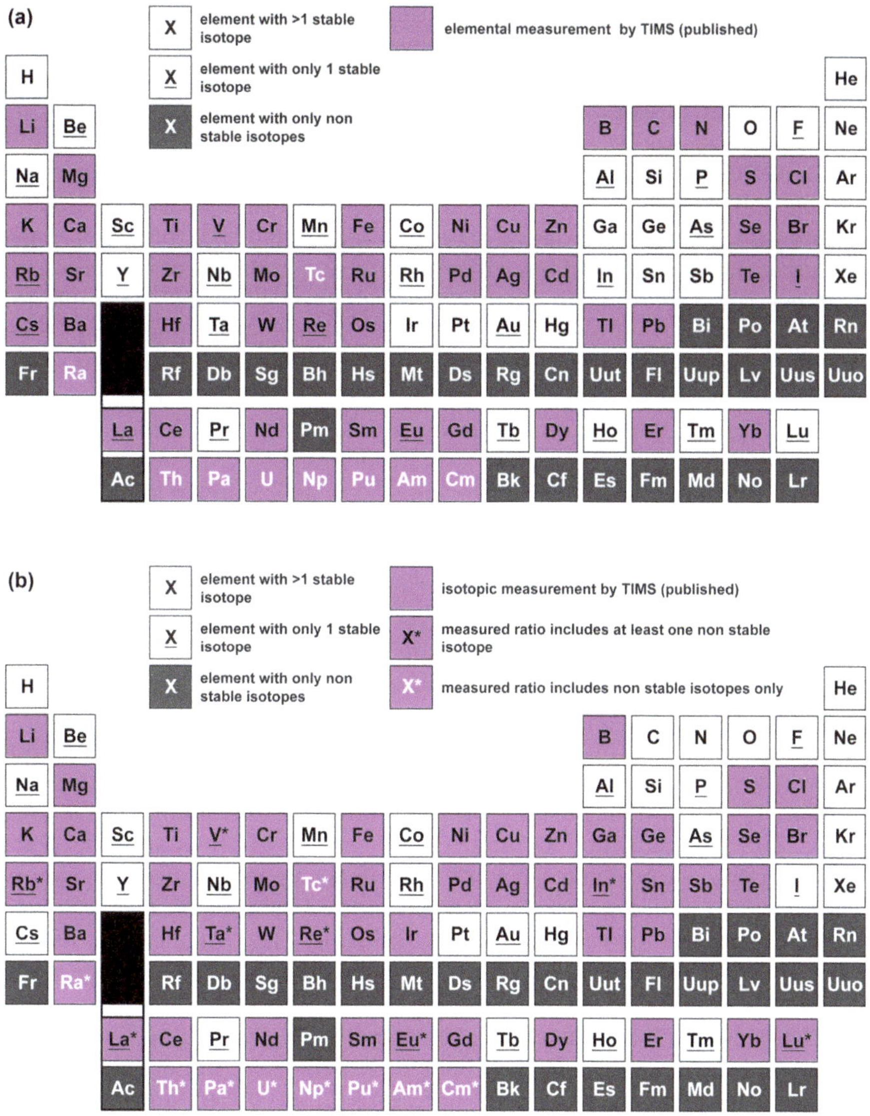


**Figure 14.9** (a) Periodic table of the elements: elements for which elemental data acquired by TIMS has been published (highlighted in lavender lilac) – also included if used, *e.g.* for interference corrections or internal normalisation. (b) Periodic table of the elements: elements for which isotope ratio data acquired by TIMS has been published (highlighted in lavender lilac) – also included if measured as interelemental normalisation standard or in IDMS analysis.

precisions of ± 0.2–0.3 per mil RSD ($^{54}$Fe/$^{56}$Fe).[78] Iron, in addition to its role in oxygen and $CO_2$ transport and storage, is also believed to play a role in various neurodegenerative disorders. Using a stable $^{57}$Fe tracer and negative

TIMS, the fraction of iron transferred from diet to the brain and other organs can be measured with a precision equivalent to 0.1 per mil RSD (standard) and 0.5 per mil RSD (sample).[79]

In addition to work on $^{41}Ca$ tracers,[80] human bone mineralisation/demineralisation studies are supported by TIMS analyses of natural calcium isotope ($^{44}Ca/^{40}Ca$) fractionation in urine.[81] A Monte Carlo error-optimised double-spike method ($^{43}Ca/^{42}Ca$) now allows Ca-isotope TIMS measurements with a long-term external reproducibility of 0.04 per mil (2SD).[82] Also linked to the modelling of bone disorders (plus coronary artery disease), Mg isotopes can now be analysed by TIMS using a gravimetric mix of sample and spike.[83] This unusual technique is necessary, since Mg is made up of three stable isotopes only, and is thus unsuitable for the double-spike method. A general review of the TIMS analysis of additional stable isotopes (such as Zn, Cu, Fe, Ca, Mo) in understanding human and animal metabolism is provided by Turnland.[84]

The continuous and acute (accidental and deliberate) release of radionuclides into the biosphere means TIMS continues to play an important role in many human bioassay procedures.[85] As most radionuclides of concern ($^{90}Sr$, $^{129}I$, $^{137}Cs$, $^{210}Po$, $^{234-238}U$, $^{239-240}Pu$, $^{241}Am$, amongst others) display comparatively large dynamic ranges in isotopic concentrations, the analysis of small tissue and urine samples requires both high abundance sensitivity and ionisation efficiency. In the case of uranium, recent improvements over classical TIMS total evaporation methods for both uncertainty and accuracy now permit stable SEM detection limits as low as $3\times10^{-9}$ for $^{236}U$.[74] Where knowledge is required of ultratrace elements not forming a normal part of human nutrition, their bioassay becomes uniquely difficult. As an example, new TIMS methods for the analysis of subnanogram quantities of the uranium fission product $^{95}Zr$, are now allowing us to measure isotope abundances in human samples at uncertainties $<5\%$.[86]

In the field of plant biochemistry, utilising TIMS to measure natural isotopic variations in key plant nutrients at the submicrogram level (*e.g.* Mg, Ca, Fe, Cu, and Zn), provides a unique insight into water–rock–plant interactions as well as carbon and nutrient cycling.[87] Combining the isotopic analysis of these biologically fractionated elements with standard water, soil, and atmospheric isotopic source indicators (Li, B, Sr, Nd, and Pb), we are better able to assess the reliability of various ecosystem monitors.[88] Similarly, zoologists are using some of these analytical approaches to unravel complex food webs and migration pathways. In this case, measurement of temporal variations in the heavier source-characteristic isotopes such as Sr, Nd, and Pb in bone, feathers, fur, and shell are used to link food sources to geographic location and seasonality.[62,89]

### 14.4.2 Applications in Environmental Sciences

Increased awareness of the impact of anthropogenic activity on the natural environment has become a major concern in recent decades. In turn, this has spurred scientific interest in understanding the processes involved and the potential long-term impact of this activity. Two major facets of this research have subsequently emerged: assessing modern climatic and environmental conditions, relative to conditions prior to the onset of extensive anthropogenic activity and; understanding the extent of global pollution.

TIMS has played an important role in expanding scientific knowledge in these fields of environmental research, particularly through the utilisation of isotopic systems as a method of investigating environmental transport pathways. The fundamental basis of such research lies in the ability to trace the movement of impurities in the natural environment through the use of chemical characteristics that are unique to the specific source locations of those impurities. Isotopic systems such as Pb, Sr, Nd, and U have been shown to reflect the location of geographical source regions, additionally these systems undergo minimal secondary effects during transport that could obscure source-region signatures.[90]

The reconstruction of global paleoclimate and paleoenvironment over glacial cycles requires reliable records of climate tracers (proxies) to evaluate the sensitivity and accuracy of general circulation models. Mineral dust is a major climatic forcing component and glacial ice cores provide an excellent record of past changes in dust levels.[91] The polar ice sheets provide an ideal archive for monitoring past changes in hemispherical scale dust fluxes consequently much research has been focused on the accurate and precise analysis of dust in glacial ice cores. Grousset and coworkers were one of the first groups to utilise isotopic systems for tracing the provenance of dust deposited in glacial ice through the TIMS isotopic analysis of Sr and Nd systems.[92] The results showed strong evidence of a South American, Argentine (Patagonian) loess source for dust reaching Antarctica during the Last Glacial Maximum (LGM). At the time, such analyses required kilogram-sized ice samples in order to obtain milligram amounts of dust for analysis, which limited the resolution capability. However, ongoing TIMS analytical developments have led to the ability to isotopically analyse smaller (microgram) amounts of dust for Sr and Nd,[93–95] with the results similarly showing southern South American dust sources to be dominant during glacial times, with inputs from more proximal (*i.e.* regional Antarctic) sources during Holocene (interglacial) times.

Concurrent to the initial work of Grousset and coworkers on Sr and Nd provenance studies, Rosman and coworkers were developing ultrasensitive TIMS techniques for the analysis of lead (Pb) and its isotopes on much smaller-sized samples, enhancing the analytical resolution capability and utilising an important isotopic tracer. Lead is present at trace levels in crustal material, and the geographical range in isotopic compositions associated with this element makes it an ideal isotopic tracer for the atmospheric movement of crustal material (*i.e.* mineral dust). Consequently, Rosman *et al.* reported the first Pb isotopic measurement of ancient Antarctic ice from the TIMS analysis of East Antarctic ice, similarly identifying a southern South American source.[96] This research spurred significant developments in Pb isotopic analytical methods in the decade to follow. Smaller volume samples, possessing extremely low Pb concentrations (mass fractions <ng $kg^{-1}$), could be accurately and precisely analysed for their isotopic composition by using TIMS, with minimal blank contamination from experimental procedures.[97] Utilising these techniques, studies of climatic fluctuations from glacial to interglacial periods, in the concentration and sources of Pb (hence transport mechanisms) confirmed a strong atmospheric pathway between southern South America (Patagonia) and Antarctica during glacial times.[98,99]

However, the advantages of the ultrasensitive Pb TIMS techniques were most clearly realised in the analysis of interglacial ice. During these climatic periods, global dust levels were at a minimum, hence sensitive analytical techniques are crucial to be able to measure very low concentrations of Pb and the corresponding isotopic compositions. Subsequent studies demonstrated a change in the source contributions of Pb preserved in ancient Antarctic ice during interglacial times as a result of lowered atmospheric dust levels, with a decrease in the strength of the southern South American dust source increasing the prominence of other signals, specifically a background volcanic signal sourced to regional Antarctic volcanic activity and, Australian dust as a secondary Southern Hemisphere dust source.[100–102]

Similarly, isotopic studies of Greenland ice cores have utilised ultrasensitive TIMS techniques, as well, providing information on the past environmental conditions and atmospheric transport pathways in the Northern Hemisphere. In particular, Pb, Sr and Nd isotopes have been used to show atmospheric transport pathways of Asiatic-sourced mineral dust to Greenland, demonstrating the desert regions in this region to be an important climate-forcing component in the Northern Hemisphere both in the past and recent times.[103–105]

Concurrent to the utilisation of Pb isotopes in long-term environmental and climatic-change studies, the analysis of Pb isotopes using TIMS was being successfully utilised to investigate the hemispheric transport of anthropogenic pollution. Lead has been shown to be an important indicator of anthropogenic activity[106] and the isotopic contrast that can exist between crustal Pb and anthropogenically utilised Pb ore, and between global ore bodies, means that this isotopic system is ideal for both tracing the movement of anthropogenic pollution and, providing information on the source of this pollution. Such studies initially focused on the movement of pollution in the Northern Hemisphere, where Greenland ice was used to trace both historical pollution, such as the hemispheric extent of mining and smelting of Spanish Pb during the Roman era[107] and recent pollution linked to North America and Eurasia.[108,109] Later, utilising global mapping of anthropogenic aerosols (similarly obtained from TIMS analysis)[67,110] studies of Antarctic ice similarly showed the extent of Southern Hemispheric pollution. Initial pollution of the continent was linked to Australian lead mining and smelting at the end of the 19th century, later pollution was linked to additional contributions from anthropogenic activity in South

Africa and South America as industrial activities developed in these continental regions.[97,101,102] Lead isotopic analysis (using TIMS) has similarly shown ice core records from continental glaciers in Europe[111,112] and the high-altitude Himalaya[113,114] as useful archives of pollution from the surrounding continental regions. In the case of the latter, the authors demonstrated that by the 1950s, anthropogenic pollution (largely sourced from South Asia) had reached an extent that it was impacting on the remote troposphere.

Lead isotopic analysis of Pb in other environmental media such as soils, sediments, dusts and waters has similarly been shown to be important for tracing the movement of pollution in the environment.[115–118] Additionally, the application of TIMS in the measurement of alternative pollution tracing isotopic systems, such as Sr[115,119,120] and radioactive tracers – Pu and U[121–123] – have also been shown to be successful. The latter being an area of great research interest in the light of the expanding global utilisation of uranium for energy purposes and, the public concern that is often associated with utilisation of radioactive material.

The extent of TIMS contributions to environmental science is by no means limited to either the fields of research or the isotopic systems that have been discussed here. Such techniques have been applied successfully to a number of areas of environmental science including oceanography, ocean circulation and oceanic change (U isotopes,[124] Pb isotopes,[125] Pb and Nd isotopes,[126] Cd isotopes,[127] B isotopes,[128]) and hydrological processes (U isotopes,[129] Sr isotopes,[130–132] Pb and Sr isotopes[133]). Whilst new techniques using alternative instrumentation are being increasingly utilised for environmental isotopic studies (*e.g.* multicollector inductively coupled plasma mass spectrometry, MC ICP-MS, see Section 12.5) it is nonetheless clear that TIMS has played a very important role in the development of many aspects of environmental science for many decades.

### 14.4.3 Applications in Archaeometry and Forensics

From its origins in the nuclear and Earth sciences, TIMS is now often applied to a raft of new research problems across the natural sciences such as applications in archaeology and forensics. Chemical provenancing of a material relies on the principle that many "heavy" elements undergo no significant – or very limited – isotopic fractionation during the normal, low-pressure–temperature processes commonly occurring at the Earth's surface.[134] By measuring isotopic variations in natural and man-made materials of interest, such as human and animal tissues (bone, teeth, hair and feathers), stone tools, ceramic pots, glass, or even nuclear materials, we gain a unique insight into the material's geographic source or production history.

In archaeology, isotopic chemistry can reveal much about early human population mobility. East Polynesia, isolated as it is in the central Pacific, was one of the last regions on Earth to be permanently settled by humans. Collerson and Weisler[135] determined the strontium, neodymium, and lead isotopic composition of prehistoric Polynesian stone adzes by TIMS. In combination with major- and trace-element analyses, the tool blades were linked to distinctive island rock formations occurring around the Pacific. In addition to materials sourced from "local" islands such as the Marquesas and Pitcairn, the stone from an adze recovered on the Polynesian atoll of Napuka was directly sourced to the olivine basalts of Kaho'olawe, Hawaii – 4000 km away across open ocean. The isotopic analyses confirmed a number of local oral histories, indicating postcolonisation trade routes existed across the Pacific as early as about 900 CE.

Isotopic tracing can also be useful beyond the simple first-order correlation of a stone artifact to its source geology. Our understanding of the provenance or technological significance of ancient or antique artifacts such as pottery, ceramics, stonewares, and metallurgical materials or ores, has been significantly expanded by similar "heavy" isotope TIMS studies.[136–141] In a pioneering study, the isotopic composition of Pb in archeological objects was investigated by Brill and Wampler in the 1960s.[142,143] The study showed that the Pb composition analysed by TIMS covered a considerable range and that three distinct groups in the data set correspond to three ancient mining sites in Greece, Britain, and southern Spain. The analysis of Pb isotopes in archaeometallurgy has matured to an indispensable tool and TIMS has played a pioneering role along the way.[141] As elements such as strontium, neodymium, and lead are ingested and incorporated into mammalian tissues with minimal fractionation, the measured fossil isotope ratio can also provide insight into issues such as diet, land use, and migration.[144]

Extending the isotopic provenancing approach to addressing wildlife crime, Vogel *et al.* studied strontium and lead isotopes in tusk and bone from several isolated populations of modern southern African elephant.[145] In identifying and prosecuting illegal poaching, establishing the geographic origin of the seized ivory has been essential. By combining the strontium and lead data with nitrogen and boron isotopes measured by other techniques, the authors were able to match elephant bone and ivory to their distinctive regional "isoscapes". Applying similar techniques to problems as diverse as illegal deer smuggling, the matching and provenancing of comingled human teeth and bone recovered from the Vietnam Conflict, and the separation of local and migrant skeletons from a 14th-century native American pueblo, Beard and Johnson illustrated the expanding role of TIMS across the forensic sciences.[60]

Verifying the origin of the food we eat is important in both the monitoring of human health, and the protection of producers, suppliers and the public from fraud. The isotopic characterisation of relatively high-value products such as wine, cheese, and mineral water has been the focus of a number of recent studies.[146,147] Following a similar approach, West *et al.* used TIMS techniques to show a correlation between the measured strontium isotope values in marijuana from across the USA, and the modelled strontium "isoscape" on which they were grown.[148] In all of these potentially evidential studies, establishing the statistical significance of any multi-isotope approach relies heavily on analytical accuracy and precision – an area where TIMS still surpasses many other techniques.

Radiometric age dating remains an indispensable tool in paleontology and archaeometry. Frumkin *et al.* combined uranium-thorium dating of natural speleothems by TIMS, with the radiocarbon analyses of organic matter, to constrain the date of construction of Jerusalem's Siloam Tunnel.[149] Attributed in the Bible to the achievement of King Hezekiahin (727–698 BCE), the Siloam radiometric ages converge at about 700 BCE, confirming the tunnel as the best independently dated Iron-Age biblical structure currently known. Working along similar lines, but with material significantly older, Barkai *et al.* dated speleothems linked to the first human occupation of the Qesem cave (Israel) at about 400 thousand years ago.[150] The cave archaeology offers a rare insight into the early stages of, what is now considered modern human behaviour in the "out of Africa" corridor: the use of fire; tool production; systematic hunting and butchering and food sharing.[151]

Closely linked to radiometric age dating, the investigation of illicit trafficking in nuclear materials (also known as nuclear forensics) relies on a complex toolbox of analytical techniques and methods to uncover the history of a seized substance. Information pertaining to the radiological hazard, intended use, last legal owner, or smuggling route can all be required to support a criminal prosecution. In cases involving uranium and plutonium, TIMS is frequently used to identify the isotopic composition of the material, the time since last chemical purification, and possible geospatial indicators.[152–154]

### 14.4.4 Applications in Industrial, Nuclear Fuel Cycle and Material Research

The use of ID TIMS for industrial applications is reported when accurate quantifications of element concentrations in complex matrices are required or elements at trace levels have to be determined reliably. Applications can be found, for example, in the field of semiconductor production. They focus on the determination of microelectronically relevant elements such as Cu, Fe, Zn, Ag, Cd, Tl, Pb, U, and Th in high-purity metals including aluminium,[155] titanium,[156] and cobalt[157] and in other high-purity materials such as metal silicides and silicon dioxide[158] at trace levels and below. Detection limits of only a few picograms per gram are reported for the determination of several heavy metals such as Cr, Ni, Cu, Zn, Ag, Cd, Tl, Pb, Th, and U in semiconductor-grade hydrofluoric acid by ID TIMS.[44]

Accurate quantification of low-level sulfur content in gas oils has traditionally been performed using ID TIMS in combination with high pressure asher (HPA) or Carius tube digestion.[159] This is an analytical task of importance as sulfur specifications are legislated by regulatory bodies for fossil fuels such as petrol and diesel because of environmental concerns. Due to its proven accuracy, TIMS has been employed for the certification of industrial relevant reference materials such as "sulfur in petrol"[160,161] and "toxic elements in polymers",[162] among others (see Section 14.4.7).

Boron is used as neutron absorber due to the considerably high neutron cross section of $^{10}B$. This amounts to about $3.8\times10^3$ barns for thermal neutrons, what exceeds almost one million times the neutron cross section of $^{11}B$.[163] In nuclear power plants the neutron flux is controlled by the concentration of boric acid in the cooling circuit. The boric acid absorbs neutrons, hence slowing down the rate of nuclear fission. The concentration of $^{10}B$ needs to be monitored, as the overall amount of $^{10}B$ decreases (converted by capturing of a neutron) and as a consequence, the rate of neutron absorption and its influence on the neutron flux steadily decreases, as well. The amount of $^{10}B$ is determined by the analysis of the boric-acid concentration and the $^{10}B/^{11}B$ isotope amount ratio. This analysis is carried out using various techniques including ICP-MS. Boron isotope reference materials were certified using TIMS procedures[164] to provide an accurate reference for this and other applications.

Another application is the use of boron doped into steel, aluminium, and other alloys for radiation shielding. These metals are used to shield nuclear fuel rods during transportation and storage and therefore the $^{10}B$ concentration in those materials has to follow strict regulations for guaranteeing proper shielding. Therefore, the $^{10}B$ amount and distribution has to be verified by measurements. Such analyses have been carried out for decades, for example, at the Federal Institute for Materials Research and Testing (BAM), Germany. The boron concentration has been analysed by ICP-OES and the $^{10}B/^{11}B$ isotope amount ratio by P-TIMS using the $Na_2BO_2$ technique.[165] The measurement precision achieved for the $^{10}B/^{11}B$ isotope ratio determination was better than 0.05% and the expanded uncertainty ($k=2$) about 0.2% for the corresponding $^{10}B$ isotope abundance, $n(^{10}B)/n(B)$ given in per cent. This analytical performance enables fulfilment of the official requirements.

The isotopic and elemental analysis of actinides and fission products for applications in the nuclear industry and fuel cycle research has been one of the domains of TIMS. A demonstration of the widespread use of TIMS in uranium and plutonium analysis of nuclear materials is found in the results of various interlaboratory comparisons. New Brunswick Laboratory (NBL), US,[166] Institute for Reference Materials and Measurements (IRMM), E.C.,[167] Commission for the Establishment of Analytical Methods (CETAMA), France,[168] and the International Atomic Energy Agency (IAEA), among others, have organised interlaboratory comparisons. Participants represented industry and research laboratories active in various parts of the industrial nuclear fuel cycle, nuclear sciences, and other scientific disciplines. For example, REIMEP-18,[167] one of the recent interlaboratory comparisons organised by IRMM, had 71 participants. Multicollector TIMS and single- or multicollector ICP-MS have been the main analytical techniques used by the participants for uranium and plutonium isotopic analysis of nuclear material samples in these interlaboratory comparisons. Typical differences to the certified values are in the range of 0.1% on major isotope ratio measurements (*e.g.* $^{235}U/^{238}U$ or $^{240}Pu/^{239}Pu$) using MC TIMS, in some cases below 0.05%.

The implementation of the total evaporation method as early as in the 1990s and, more recently, the modified total evaporation method (see Section 14.3.3) in combination with multicollector TIMS has contributed significantly to the widespread use of thermal ionisation mass spectrometry in the analysis of uranium and plutonium nuclear materials. Isotope ratio analyses can span more than eight orders of magnitude.[169] On a routine basis, performing as many as several thousand analyses per year per instrument, expanded uncertainties ($k=2$) on the order of 0.1% and in some cases below 0.05% are maintained on major isotope ratio measurements.[59,74] Expanded uncertainties in the range of 0.3% to 0.5% and below are reported on a routine basis for the determination of minor isotopes, for example the $^{238}U/^{234}U$ ratio in natural uranium. The quantification of the abundance of the minor isotopes $^{233}U$ and $^{236}U$ in uranium materials is of interest as it is produced from $^{232}Th$ and $^{235}U$ by neutron capture, respectively, hence it is a signature of neutron irradiation in nuclear reactors. But even in natural uranium samples such as ores, neutrons from spontaneous fission and other sources produce $^{233}U$ and $^{236}U$, alas at extremely low abundance. TIMS was employed over the past decades for the quantification of the uranium isotope ratios in natural uranium samples with reported $^{238}U/^{236}U$ and for $^{238}U/^{233}U$ isotope ratios down to $10^{-10}$ and $10^{-9}$, respectively.[169,170] (Accelerator mass spectrometry has pushed the limits for $^{233}U$ and $^{236}U$ even further.[171]) Due to its proven analytical performance in nuclear material analysis, TIMS remains pivotal in the certification of nuclear reference materials at leading metrological institutions such as NBL and IRMM (see Section 14.4.8).

Crucial knowledge of the extent of nuclear fission and transmutation in nuclear fuel is gained by analysing irritated or spent nuclear fuel. For this purpose, fuel samples are dissolved and the analytes of interest separated. By analysing the isotopic composition of one or more of the fission products, the burn-up of the fuel can be determined. Examples are the isotopic measurements of Nd,[172] Cs,[173] $^{235}U$ burn-up or the production of Pu, Am, and Cm.[71,173,174] These analytical data are also used for evaluation of calculation codes or nuclear waste disposal.[175]

The discovery of the Oklo-Bangombe natural reactors at various sites in Gabon, Africa, in 1972 was due to an accurate determination of the isotope ratio $^{238}U/^{235}U$ in respective uranium ore samples (reviewed by De Laeter and Hidaka[176]). Nature had anticipated humankind's achievement of sustaining a fission chain reaction by about two billion years. Elemental and isotopic analyses of the uranium ores and surrounding geological environment performed using TIMS and other analytical techniques revealed the reactors' two billion year age, the depletion in the $^{235}U$ abundance compared to natural uranium to due fission, and anomalous rare-earth element contents showing fission product contamination. It also revealed that the duration of nuclear criticality varied widely for the various reactor zones, with as long as several hundred thousand years. The studies have implications to radioactive-waste containment as they showed that some fission products like rare-earth elements have been well retained over two billion years but that partial to almost complete loss has occurred to other fission products.

International nuclear safeguards is responsible for deterring the proliferation of nuclear weapons by detecting early the misuse of nuclear material or technology, and by providing credible assurances that international safeguards obligations are honoured by the States. The analysis of nuclear material samples and environmental samples taken by nuclear safeguards inspectors is a crucial part in this undertaking.[177,178] Multicollector TIMS is one of the main analytical techniques that are traditionally being employed at safeguards laboratories[71,74,169,177,179] for uranium and plutonium analysis of nuclear materials originating from various parts of the industrial nuclear fuel cycle. These instruments are also installed at on-site laboratories at reprocessing facilities such as Rokkasho, Japan[180] and Sellafield, UK[181] to ensure timely analytical response. The results of nuclear safeguards analyses need to meet strict criteria as driven by the so-called International Target Values (ITVs).[182] The ITVs are a set of analytical performance criteria revised periodically by an international group of experts to reflect current practice in safeguards analysis of industrial nuclear and fissile materials using various analytical techniques. In additional to isotopic and elemental quantifications, analytical procedures using TIMS and other analytical techniques have been established permitting the production age (since last purification) of a suspect nuclear material to be unraveled[154,183] or to capitalise on other unique material signatures such as the $^{18}O/^{16}O$ isotope ratio for material attribution.[184]

In the aftermath of the Gulf War in 1991, an IAEA action team undertook inspections of Iraqi nuclear installations and collected environmental samples and bulk construction samples to search for unusual traces of nuclear materials or nuclear activities.[185] TIMS measurements on graphite calutron pieces, for example, revealed that the electromagnetic isotope separation process was used for enriching fissile isotope $^{235}U$. Environmental samples included smear samples taken by swiping surfaces to catch microscopic uranium or plutonium particles. They – in combination with a multitude of other analytical investigations – revealed evidence for a clandestine nuclear program. Due to its efficacy, environmental sampling was established in the following years[177] as an additional tool to support international nuclear safeguards. These environmental samples are now routinely scrutinised for nuclear particles and their tale-telling isotopic composition using TIMS and other analytical techniques such as ICP-MS, SIMS, and radiometry.[37,76,186–189] This is achieved either by performing bulk dissolutions of the smear swipe and subsequent TIMS or ICP-MS analysis, or by micromanipulation of selected particles and subsequent analysis of individual particles, for example, by fission track TIMS or SIMS.

### 14.4.5 Applications in Earth Sciences

Thermal ionisation mass spectrometry has played an indispensable role in the development and application of isotope systematics in the geosciences. Modern geoscience research critically relies on the use of isotope systematics to unravel the fundamental processes and time frames of how Earth and indeed the solar system work. Two fundamental processes thereby provide the grounds and for which TIMS is so eminently important: radioactive decay of isotopes and the ingrowth of the respective radiogenic decay isotopes; and isotope fractionation invoked by geological processes.

Basically, radioactive decay leads to a temporal change of the abundance ratio between the decaying radioactive (parent) isotopes and the accumulating radiogenic (daughter) isotopes. So, radioactive decay constitutes a measure of time and thus gives rise to the possibility of determining the absolute age of Earth materials and of determining the ages and the duration of processes working on these. This is called geochronology and is also known as absolute dating or radiometric age determination. Prominent examples of parent–daughter isotopes related to TIMS are $^{87}Rb$–$^{87}Sr$, $^{147}Sm$–$^{143}Nd$, and $^{238}U$–$^{206}Pb$.[190–193]

Isotope fractionation means that – while a process is working on a material – isotopes of a given element present in the material react in different degrees to the process. So the postprocess isotope abundances of an element (and thus the isotopic composition thereof) can be different to the preprocess isotope abundance. The difference is called isotope shift. By analysing this shift one can identify and quantify the process, which has worked on the material. In geological environments the major factors controlling isotope fractionation are either mass dependent (MDF) or mass independent (MIF). The actual causes of isotope fractionation can be attributed to a large number of different physical parameters, *e.g.* temperature, reaction kinetics, diffusion, pressure, crystal structure, and combinations thereof.[193]

Isotope systematics relevant to the Earth system have been routinely exploited since the 1950s. TIMS analysis of the following elements are routinely used in Earth-system research: Rb, Sr, Sm, Nd, U, Th, and Pb. Other elements analysed by TIMS include Lu, Hf, Re, Os, B, Ca, Cl, and Se. However, the analysis of these elements has been superseded to some extent by (MC) ICP-MS.

Although it is common practice to call the investigated decay systems "geochronometers" extraterrestrial minerals and rocks are also routinely dated. So the term "chronometer" is probably more appropriate. Table 14.5 lists achievable reproducibilities for selected chronometer.

**Absolute Dating – Methodology**

The decay rate of radioactive nuclei after a certain amount of time $t$ is proportional to the number of nuclei $N$ present:[191]

$$-\frac{dN}{dt} = \lambda N \tag{14.2}$$

**Table 14.5** Achievable reproducibilities for selected chronometer.

| System | internal reproducibility ($2\sigma$) | external reproducibility ($2\sigma$) | duration/hours |
|---|---|---|---|
| $^{147}Sm/^{150}Sm$ | 0.005% | 0.002% | 1–5 |
| $^{143}Nd/^{144}Nd$ | 0.002% | 0.001% | 1–5 |
| $^{147}Sm/^{144}Nd$ | 1% | 1% | 1–5 |
| $^{235}U/^{238}U$ | 0.05% | 0.02% | 1 |
| $^{207}Pb/^{206}Pb$ | 0.002% | 0.001% | 1 |
| $^{206}Pb/^{238}U$ | 0.1% | 0.1% | 1 |

where the proportionality is expressed as the decay constant λ.

Integrating

$$-\int \frac{dN}{N} = \lambda \int dt \tag{14.3}$$

leads to

$$-\ln N = \lambda t + C. \tag{14.4}$$

If

$$C = -\ln N_0 \tag{14.5}$$

where $N_0$ is the initial number of atoms/nuclei.

Thus,

$$\ln \frac{N}{N_0} = -\lambda t \tag{14.6}$$

and

$$\frac{N}{N_0} = e^{-\lambda t} \tag{14.7}$$

and reformulating gives the basic equation

$$N = N_0 e^{-\lambda t}. \tag{14.8}$$

Assuming closed-system behaviour we can write

$$D^* = N_0 - N \tag{14.9}$$

where $D^*$ is the number of daughter nuclei formed from the decay of $N_0$ during the time $t$. The asterisk denotes the radiogenic origin of $D$. This equation assumes that $D^* = 0$ at $t = 0$.

Then

$$D^* = N_0 - N_0 e^{-\lambda t} \tag{14.10}$$

and thus

$$D^* = N_0(1 - e^{-\lambda t}). \tag{14.11}$$

In most geological systems $N_0$ is not known independently. But as

$$N_0 = Ne^{\lambda t} \tag{14.12}$$

the above equation becomes

$$D^* = Ne^{\lambda t} - N = N(e^{\lambda t} - 1). \tag{14.13}$$

Now, $D^*$ is only related to quantities that can be determined directly.

Unlike the assumption above, in many geological systems $D^* \neq 0$. Thus, a correction for an initial amount of $D$ has to be made in the form of

$$D = D_0 + D^*. \tag{14.14}$$

This leads to the final age equations

$$D = D_0 + N(e^{\lambda t} - 1) \tag{14.15}$$

and

$$t = \frac{1}{\lambda} \ln\left(\frac{D - D_0}{N}\right) + 1. \tag{14.16}$$

Exemplified using the Rb–Sr system (for simplicity $N(^{87}\mathrm{Sr})$ is abbreviated as $^{87}\mathrm{Sr}$ and accordingly):

$$N(^{87}\mathrm{Sr}) = N(^{87}\mathrm{Rb}) \cdot (e^{-\lambda_{\mathrm{Rb}} t}) \tag{14.17}$$

assuming that $D = 0$

For practical reasons, the absolute amounts of $D$, $D_0$ and $N$ are normalised to a common denominator, *i.e.* in the above example $N(^{87}\mathrm{Sr})$ and $N(^{87}\mathrm{Rb})$ are normalised to $N(^{86}\mathrm{Sr})$ (and using the abbreviations of above):

$$\left(\frac{^{87}\mathrm{Sr}}{^{86}\mathrm{Sr}}\right) = \left(\frac{^{87}\mathrm{Rb}}{^{86}\mathrm{Sr}}\right)(e^{-\lambda_{\mathrm{Rb}} t}) \tag{14.18}$$

$D_0$ then becomes $\left(\frac{^{87}\mathrm{Sr}}{^{86}\mathrm{Sr}}\right)_i$ where $i$ denotes the initial isotopic ratio.

Thus the age equation becomes

$$\frac{^{87}\mathrm{Sr}}{^{86}\mathrm{Sr}} = \left(\frac{^{87}\mathrm{Sr}}{^{86}\mathrm{Sr}}\right)_i + \frac{^{87}\mathrm{Rb}}{^{86}\mathrm{Sr}}(e^{\lambda_{\mathrm{Rb}} t} - 1). \tag{14.19}$$

or solved for $t$

$$t = \frac{1}{\lambda_{\mathrm{Rb}}} \ln\left[\left(\frac{\left(\frac{^{87}\mathrm{Sr}}{^{86}\mathrm{Sr}}\right) - \left(\frac{^{87}\mathrm{Sr}}{^{86}\mathrm{Sr}}\right)_i}{\left(\frac{^{87}\mathrm{Rb}}{^{86}\mathrm{Sr}}\right)}\right) + 1\right]. \tag{14.20}$$

$$\frac{^{87}\mathrm{Sr}}{^{86}\mathrm{Sr}} = \left(\frac{^{87}\mathrm{Sr}}{^{86}\mathrm{Sr}}\right)_i + \frac{^{87}\mathrm{Rb}}{^{86}\mathrm{Sr}}(e^{\lambda_{\mathrm{Rb}} t} - 1) \tag{14.21}$$

is a linear equation describing a line with slope $e^{\lambda_{\mathrm{Rb}} t} - 1$ and the $y$-axis intercept $\left(\frac{^{87}\mathrm{Sr}}{^{86}\mathrm{Sr}}\right)_i$.

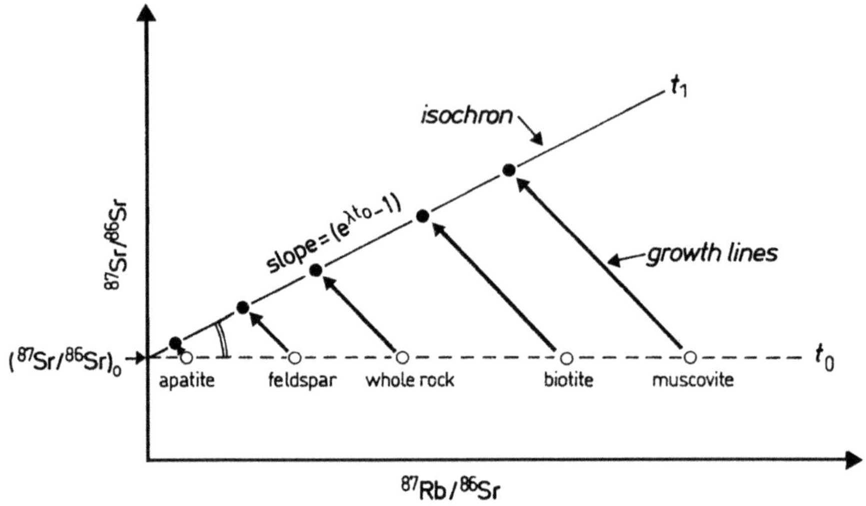


**Figure 14.10** Isochron for Rb–Sr dating.
(Source: Reproduced from Geyh *et al.*[195])

Such a line is called an *isochron*.[194] An example is shown for Rb–Sr dating in Figure 14.10.

By analysing two or more whole-rock samples, minerals or mineral fractions the line $t_1$ can be calculated by more or less sophisticated linear regression algorithms. From the slope of the regression line the age $t_1$ of the system can then be derived. The $y$-axis intercept gives additional isotope geochemical information. The mineral or rock age thus calculated corresponds to the respective system closure, *i.e.* the moment in time when parent and daughter isotopes are no longer separated by, *e.g.* diffusional processes. Obviously, the age system under inspection has to remain closed from the time point of system closure until today in order to be able to derive meaningful ages.

Due to the diffusional properties of the radiogenic decay products the "age" of minerals and rocks corresponds to a certain temperature threshold, called blocking or closing temperature, below which the diffusion of the radiogenic decay products becomes so slow that from an analytical point of view (and also from a geological one) the system can be considered as closed. Thus, derived ages are called cooling ages. Closing temperatures range from >800 °C (U–Pb in zircon) to 70 °C (U/Th–He in apatite), depending on the isotope system, type of mineral or rock, mineral size, cooling rate, stress field and strain rates, presence or absence of fluids (amongst others). It is thus very important to note that under normal conditions in the Earth absolute dating of minerals and rocks does not directly result in mineral growth ages or rock-formation ages but rather in the point in time when a certain mineral and rock cooled below the respective blocking or closing temperature.

On Earth, under crustal conditions ($T < 1000$ °C, $P < 4$ GPa) pressure does not exert any noticeable effect on geochronometers.

**Rb–Sr Geochronology**

Rubidium has two naturally occurring isotopes $^{85}$Rb and $^{87}$Rb. $^{87}$Rb decays to $^{87}$Sr by $\beta^-$ decay with a half-life of about 50 Ga.[196] $^{85}$Rb is stable. Strontium has four naturally occurring stable isotopes $^{84}$Sr, $^{86}$Sr, $^{87}$Sr and $^{88}$Sr. $^{87}$Sr is partly radiogenic due the decay of $^{87}$Rb. As a result, the $^{87}$Sr/$^{86}$Sr ratio varies depending on the initial Rb/Sr ratio and the age. During mineral growth or their subsequent alteration, Rb and Sr substitute in traces for K and Ca, respectively. Thus, K-rich and concomitant Ca-poor minerals such as micas and potassium-feldspars are favourable for dating by the Rb–Sr method.

In order to perform the TIMS analyses, Rb and Sr are separated from the matrix by specialised ion-exchange procedures after complete dissolution of the mineral or rock. Rb and Sr concentrations are determined by ID, for example using a $^{87}$Rb–$^{84}$Sr mixed spike. Filament-loading techniques are reviewed in Table 14.3. Instrumental mass fractionation of Sr is corrected online by assuming, for example $^{88}$Sr/$^{86}$Sr = 8.3753 and applying an empirically derived fractionation law to $^{87}$Sr/$^{86}$Sr (see Chapter 8). The reproducibility on the $^{87}$Sr/$^{86}$Sr ratio is often better than 0.01% (see Table 14.1).

Examples of successful Rb–Sr dating are various: Rb–Sr ages have been derived for a huge number of igneous and metamorphic rocks on Earth, for rocks from the Moon and for meteorites. From these, the initial radiogenic Sr isotopic composition (BABI = basaltic achondrite best initial) of the Earth could be derived ($^{87}$Sr/$^{86}$Sr$_{\text{(BABI, 4'560 Ma)}}$ = 0.69899 ± 0.00005).[192] Other important contributions of the Rb–Sr system to Earth sciences are the recognition of some of the oldest rocks on Earth, the deduction of growth rates of continental crust, mantle evolution and the determination of cooling ages of rocks.[191–193] A typical application of Rb–Sr dating shows a multimineral isochron from a sample of the Great Dyke in Zimbabwe, S Africa (see Figure 14.11).[197] "Clean micas" and leached pyroxenes define an isochron with an age of 2543.0 ± 4.4 Ma. Whole-rock samples and feldspars show an open-system behaviour resulting in a deviation of analyses from the reference isochron.

**Sm–Nd Geochronology**

The Sm–Nd geochronometer was developed in the 1970s and 1980s as a result of the demand for a reliable geochronometer for the dating of mafic and ultramafic rocks ($SiO_2 < 55$ weight%), which are not readily dated by Rb–Sr as they are generally K- and Rb-poor and Ca- and Sr-rich.

Both Sm and Nd are rare-earth elements (REEs). Sm has 7 natural isotopes ($^{144}$Sm, $^{147}$Sm, $^{148}$Sm, $^{149}$Sm, $^{150}$Sm, $^{152}$Sm, $^{154}$Sm) of which $^{147}$Sm (abundance: 15.00 mole%) decays to $^{143}$Nd by $\alpha$ - decay with a half-life of 106 Ga, corresponding to a decay constant of $6.54 \times 10^{-12}$ a$^{-1}$. Nd has also 7 natural isotopes ($^{142}$Nd, $^{143}$Nd, $^{144}$Nd, $^{145}$Nd, $^{146}$Nd, $^{148}$Nd, $^{150}$Nd). Due to the

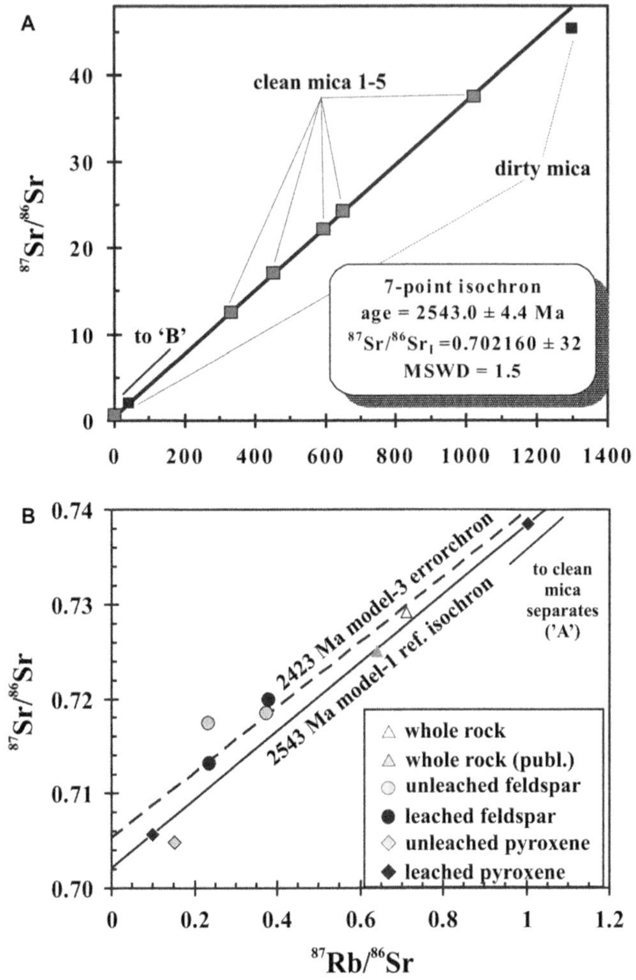


**Figure 14.11** Rb–Sr isochron of seven separates marks the time of isotope closure of the Rb–Sr system. Nonleached separates show significant scatter and yield apparently no geological significant age information. (Source: Reproduced from Nebel and Mezger.[197])

radiogenic ingrowth of $^{143}Nd$, today's $^{143}Nd/^{144}Nd$ for the bulk Earth is 0.512638 but varies for individual minerals and rocks between 0.506754 and ~ 0.518.[192]

As for dating by the Rb–Sr method, minerals and rocks should be strongly enriched in Sm and relatively depleted in Nd. Useful minerals are garnet, pyroxene and amphiboles. Plagioclase is strongly enriched in Nd and is often used as a nonradiogenic counterpart. But also highly REE-enriched accessory minerals as apatite, monazite, xenotime, zircon, bastnaesite, zoisite, allanite, sphene, epidote are of interest. On average, Sm mass fractions can reach several 10 mg kg$^{-1}$ (0.1–1000 mg kg$^{-1}$), corresponding $^{147}Sm/^{144}Nd$

ratios in minerals are between 0.1 and 2 (0.01–5). Whole rock $^{147}Sm/^{144}Nd$ is between 0.1 and 0.3 (0.4). $^{143}Nd/^{144}Nd$ ratios are 0.506754 to 0.518.

Sm and Nd are separated from the matrix by specialised ion-exchange procedures after complete dissolution of the mineral or rock. Sm and Nd concentrations are determined by ID commonly using a $^{147}Sm$–$^{150}Nd$ mixed-spike. This is added to the sample prior to dissolution in order to achieve complete sample-spike homogenisation. After separation both elements are loaded as nitrates onto Re filaments. Loading quantities are <100 ng and >1 ng. Analyses follow normal TIMS procedures depending on the MS brand. Instrumental mass fractionation of Nd is corrected online by assuming $^{146}Nd/^{144}Nd = 0.7219$ and applying an empirically derived exponential fractionation law to $^{143}Nd/^{144}Nd$.

Similar to the afore-mentioned Rb–Sr system Sm–Nd ages have been derived for a huge number of igneous and metamorphic rocks on Earth, for rocks from the Moon and for meteorites. Earth's radiogenic Nd isotopic composition today (CHUR = Chondritic uniform reservoir) could be derived ($^{143}Nd/^{144}Nd_{(CHUR,\ today)} = 0.512638 \pm 0.000005$) again allowing to decipher the Sm–Nd evolution of Earth and the solar system to be deciphered. Other important contributions of the Sm–Nd system to Earth Sciences are the deduction of Earth's mantle evolution and the dating of eclogites, for instance.[191–193]

The example shows the Sm–Nd dating of iron-rich eclogites from the Ötztal region in W Austria (Figure 14.12).[198] Eclogites are rocks formed by high-pressure (>1.5 GPa) metamorphism in subduction zones. In the Sm–Nd isochron diagram two mineral fractions (cpx = clinopyroxene, ga = garnet, in

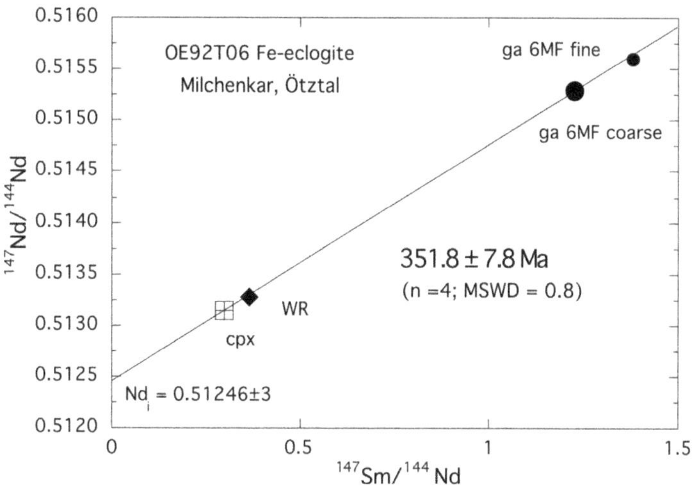


**Figure 14.12** Sm–Nd mineral isochrone ages (Grt, Cpx, Zo, Amp, wr) constrain the eclogite-facies metamorphism in the central Ötztal metabasite suite at (347 ± 9) Ma.
(Source: Reproduced from Miller and Thöni.[198])

2 subfractions) and the whole rock (WR) define an isochron, whose slope corresponds to an age of 351.8 ± 7.8 Ma. This age is interpreted as reflecting the time of system closure during eclogite formation c. 352 Ma ago.

**U–Pb and Th–Pb Geochronology**

Applications of the U–Pb and Th–Pb geochronometers have probably played the most important role in the development of geochronology. In fact, the U–Pb decay system was the first chronometer to deliver accurate and precise age estimates of 4.57 Ga for the Earth and meteorites.[199]

Among all chronometers based on radioactive decay the U–Pb system is unique as two isotopes of one mother element (U) decay to two isotopes of one daughter element (Pb). This makes possible internal consistency testing, which allows recognising system disturbances due to open-system behaviour, mixing effects and others. This is a fundamental advantage over other chronometers. Useful age data can be gained from one single analysis and not from at least two, as is needed for the other geochronometers.

Uranium has 4 natural isotopes $^{234}U$, $^{235}U$, $^{236}U$ and $^{238}U$. All are radioactive. Relevant for the present discussion are $^{238}U$ (abundance: 99.7 mole%), which decays to $^{206}Pb$ by a decay chain with an overall half-life of 4486.3 Ma (decay constant: $1.55125 \times 10^{-10}$ $a^{-1}$) and $^{235}U$ (abundance: 0.3 mole%) that decays to $^{207}Pb$ by another decay chain with an overall half-life of 703.8 Ma (decay constant: $9.8485 \times 10^{-10}$ $a^{-1}$). Th has one long-lived isotope, $^{232}Th$ that decays to $^{208}Pb$ with a half-life of 14.0 Ga (decay constant: $4.94752 \times 10^{-11}$ $a^{-1}$). Pb has 5 naturally occurring isotopes, $^{204}Pb$, $^{206}Pb$, $^{207}Pb$, $^{208}Pb$, $^{210}Pb$. $^{204}Pb$ and $^{210}Pb$ are radioactive, $^{206}Pb$, $^{207}Pb$ and $^{208}Pb$ are in parts radiogenic. Thus, natural Pb isotopic abundances are highly variable. For instance, the least radiogenic ratio $^{206}Pb/^{204}Pb = 18.5$ (Canon Diablo troilite), whereas in minerals it can be in the range of $10^5$.

Uranium mass fractions in minerals are between 1 mg/kg and 5000 mg/kg. Corresponding Pb mass fractions are therefore 0.1 mg/kg to >500 mg/kg. As for dating by the other methods, minerals should be strongly enriched in U and relatively depleted in Pb. In this respect, useful minerals are zircon, monazite, xenotime, apatite, sphene, rutile, allanite, uranite, uranothorite, thorite. Zircon is by far the most often dated mineral. Whole-rock samples are not dated.

U and Pb are separated from the matrix by specialised ion-exchange procedures after complete dissolution of the mineral under investigation. U and Pb concentrations are determined by ID commonly using a $^{233}U$–$^{205}Pb$ mixed spike. This spike is added to the sample prior to dissolution in order to achieve complete sample-spike homogenisation. After separation, both elements are loaded as chlorides on Re filaments using Si-gel as an activator. Loading quantities for both elements are <10 ng and >10 pg. Analyses follow normal TIMS procedures depending on the instrument type. Instrumental mass fractionation of U and Pb are corrected by applying externally derived mass fractionation. Also in use are $^{233}U$–$^{235}U$ and $^{202}Pb$–$^{205}Pb$ double-spike techniques.

U–Pb age data is plotted in a so-called concordia diagram. This shows both U–Pb decay systems simultaneously (see Figure 14.13), thus allowing a comparison between the two. The concordia line (blue line in Figure 14.13) represents all points of identical $^{206}Pb/^{238}U$ and $^{207}Pb/^{235}U$ ages from 0 Ma to 4560 Ma. Any analysis lying within analytical error on the concordia is interpreted to represent an undisturbed sample, the age thus being geologically meaningful. Any analysis not lying on the concordia is called discordant with the discordance implying some system disturbance by open-system behaviour, lead loss or mixing, for instance. With only radiogenic Pb in a system, the $^{207}Pb/^{206}Pb$ ratio is only depending on the age of the system. This allows under certain assumptions to derive meaningful ages solely from the $^{207}Pb/^{206}Pb$ ratio.

The example (Figure 14.13) shows zircon analyses from the Winnebach migmatite from the Ötztal region in W. Austria.[200] The crystal forms of the analysed zircons (6 crystals A–M) are shown schematically. Crystal A is concordant at 490 ± 9 Ma (see inset). The age is thus interpreted as defining the time of migmatisation. The other more elongated crystals (B–M) are all

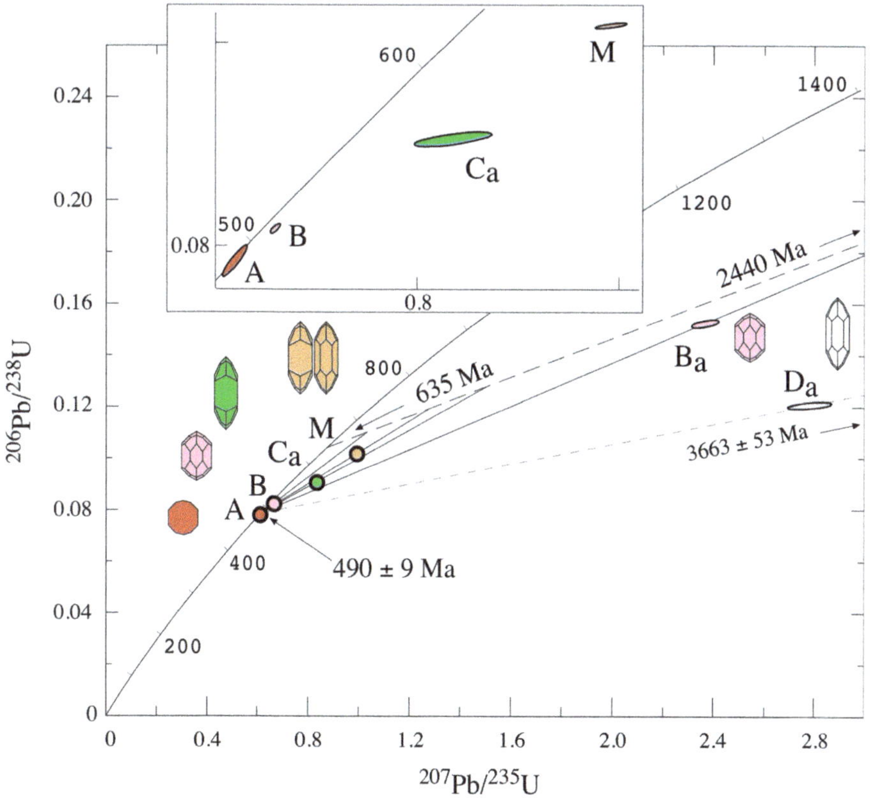


**Figure 14.13** Concordia diagram of zircons from the Winnebach migmatite. (Source: Reproduced from Klötzli-Chowanetz.[200])

analytically discordant. They define sublinear arrays indicative of mixing of differently aged zircon components. Numbers along the concordia are ages in Ma. Ages with arrows point to intersection points of discordia line with the concordia. Analytical errors are smaller than the data point symbols. The inset shows the true 2-sigma error ellipses for four crystals. Beware that not the complete concordia but only the part from 0 to 1400 Ma is shown.

**Other Geochronometers**

A number of other geochronometers partially making use of TIMS analyses are currently in use in Earth sciences. These are, for instance, Re–Os, and the U-series methods. Lu–Hf was developed using TIMS, but is now practically exclusively measured by MC-ICP-MS. The same holds true for Re–Os.

### 14.4.6 Applications in Geology – Radiogenic Isotope Geochemistry

The radiogenic ingrowth of Sr, Nd, Pb, Hf and Os isotopes has also important geochemical bearings. Topics, which employ radiogenic isotope systematics, are, *e.g.* mantle and crustal reservoir characteristics, mineral and whole rock model ages or palaeotemperatures and -salinity. Only two examples out of a huge number of applications illustrating the use of radiogenic isotopes are shown.

**Radiogenic Sr Isotope Geochemistry**

One classical application is the variation of $^{87}Sr/^{86}Sr$ in ocean water with time, which reflects differential input of Sr derived from continental crust *via* the runoff from the continents and Sr input from the mantle by oceanic rift systems. The $^{87}Sr/^{86}Sr$ of marine sediments is thereby taken as proxy for the ocean water $^{87}Sr/^{86}Sr$ ratio. This thus allows investigating ocean water Millions of years in age, which obviously cannot be sampled directly anymore. Highly radiogenic $^{87}Sr/^{86}Sr$ ratios reflect major Sr contributions from continents due to favourable weathering conditions and continent configuration. Low radiogenic $^{87}Sr/^{86}Sr$ ratios, on the contrary, reflect a predominance of Sr input from the mantle due to strong rifting activities and the formation of large igneous provinces. An example is given in Figure 14.14 showing the secular curve for seawater $^{87}Sr/^{86}Sr$ variations during the Phanerozoic based on the analyses of marine carbonate, evaporate rocks and fossils.

Similar approaches and arguments are taken by using radiogenic Nd, Os, Hf and Pb isotope systematics. Other Sr isotope TIMS applications deal with, *e.g.* the isotopic composition of Earth's mantle and its variation or the initial Sr isotopic composition of plutonic and volcanic rocks.

**Combined Sr–Nd–Hf–Pb; Radiogenic Isotope Geochemistry**

The combination of different isotope systems to the same minerals and rocks allows for a deeper understanding of the fundamental processes

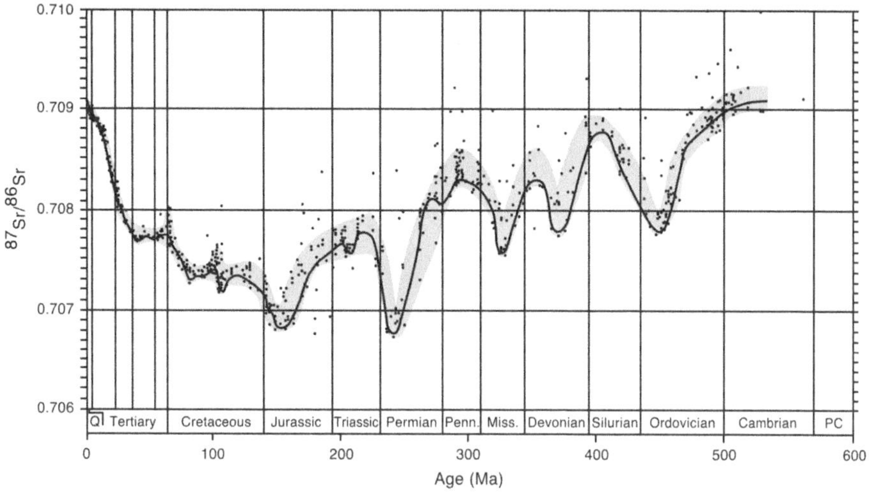


**Figure 14.14** The Burke *et al.*[201] secular curve for seawater $^{87}Sr/^{86}Sr$ variations during the Phanerozoic based on analyses of marine carbonate and evaporate rocks and fossils. Increasing scatter with age is due to greater diagenetic alteration of orignila marine phases and increased effects of high Rb–Sr phases such as silicates in older samples and greater uncertainties in age assignments for older samples. The shaded area around the curve is the error envelope of best estimate seawater value by Burke *et al.*[201] Results are plotted here relative to a $^{87}Sr/^{86}Sr$ value of 0.71034 for NIST $SrCO_3$ Standard Reference Material 987 (SRM 987).
(Source: Reproduced from Banner.[202])

governing the respective genesis and evolution. An important aspect of such multi-isotope applications is to understand the meaning and cause of the coupling or the decoupling of isotope systems. This is illustrated in Figure 14.15, which shows the correlation of the Sr, Nd, Hf and Pb isotope systems in very young basaltic rocks from different oceans. Basaltic rocks are formed by partial melting of the Earth's mantle and are interpreted to provide primary mantle isotope signatures. Clearly different isotopic compositions can be correlated with different geographic regions thus implying that the mantle has geographically different compositional domains. Most of the shown fields can be interpreted as stemming from bimodal or trimodal mixing of different compositional end members, mixing being of different degrees, which leads to the often-observed quasilinear compositional trends.

## 14.4.7 Applications in Metrology

Some of the main metrological applications of TIMS are isotope dilution mass spectrometry (IDMS) for the characterisation of reference materials (RMs), and the determination of absolute isotope ratios of elements and

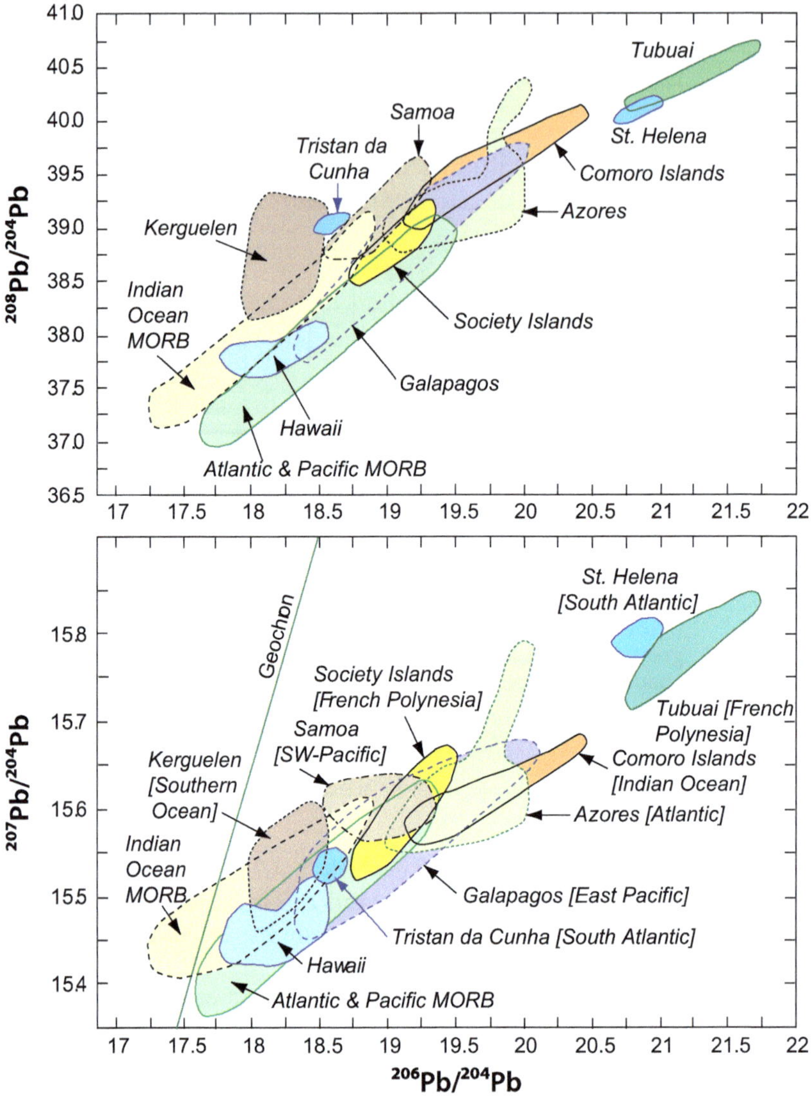


**Figure 14.15** Sr, Nd and Pb isotopic composition of oceanic basalts. (Reproduced from Stosch.[203])

their relative atomic weights. In the past, a main activity was the determination of physical constants like radioactive half-lives and nuclear cross sections.

## Isotope Dilution Mass Spectrometry

Isotope dilution mass spectrometry is a calibration technique where an accurately weighed amount of the sample is mixed with an accurately weighed amount of a so-called spike. The spike contains the analyte element in an isotopically enriched form. By measuring the isotope amount ratio in this mixture (also termed blend) the analyte concentration in the sample can be determined (see Chapter 8 and the literature[204–206]). Based on the principle of isotope dilution mass spectrometry an equation can be set up, describing the complete analytical procedure with all necessary corrections and input quantities (*e.g.* masses, measured isotope ratios, spike concentration, calibration). With this equation the analyte mass fraction in the sample can be calculated and subsequently an uncertainty evaluation with uncertainty budget performed. The characteristics of isotope dilution mass spectrometry make it a reference method in elemental analysis, which is, due to its proven records in reference materials characterisation, internationally accepted as such. And it is of particular importance as it is a method for trace analysis in complex matrices having the potential to be a primary method of measurement (see Chapter 8).

The combined use of isotope dilution mass spectrometry with TIMS (ID TIMS) was triggered in the late 1940s / early 1950s by the broad availability of enriched stable isotopes provided by the US Atomic Energy Commission.[207] Since the early 1970s ID TIMS has been increasingly applied to the characterisation of reference materials by the National Bureau of Standards (NBS), now National Institute for Standards and Technology (NIST).[208] A number of metal elements such as Cr, Ni, Pb, U, and V were quantified at NIST in a variety of matrices such as ores, oils, bovine liver, and citrus leaves by using positive TIMS (P-TIMS). After the introduction of negative TIMS (N-TIMS), isotope dilution was used in the late 1980s to characterise nonmetallic elements such as B, I, and Se in biological, food, and environmental reference materials.[209]

From the 1970s to the 1990s metrological laboratories such as NIST and the Institute for Reference Materials and Measurements (IRMM, formerly Central Bureau for Nuclear Measurements), among others, used ID TIMS for the characterisation of reference materials. A part of this work has been published in a series of reviews[208–210] whereas the complete work is only visible through the numerous certification reports and reference materials (*e.g.* BCR-680/681 "Toxic Elements in Polyethylene"[211] or ERM-EF211/212/213 "Sulfur in Petrol"[212]). From the 1990s onwards the use of isotope dilution mass spectrometry for reference material characterisation has increased continuously due to the increased application of ICP-MS, while the use of ID TIMS steadily decreased.[213–215] Comparative studies on the characterisation of reference materials by ICP-MS and TIMS using isotope dilution procedures showed that both are equally valid, when multicollector instruments were used, and that similar measurement uncertainties can be achieved.[216,217] These developments and related applications have been described in a detailed review paper.[207]

Improvements to the IDMS procedure can be achieved by employing the double or triple IDMS approach (see Chapter 8 for more details). For example, the Federal Institute for Materials Research and Testing (BAM, Germany) implemented improvements to the double IDMS procedure.[206]

In this way the relative expanded uncertainties obtained with ID-TIMS were reduced to below 0.2% for the quantification of Cd, Cr, and Pb in polymer materials, where typical relative expanded uncertainties employing ID ICPMS range from 0.5% to 2%.[162] A further improvement was the application of triple IDMS to elemental analysis.[218] The major advantage of triple IDMS is that the isotope amount ratio of the spike does not need to be determined, as it is represented by a third blend in the IDMS equation. Therefore, triple IDMS is useful when highly enriched spikes are used or when strong memory effects occur. Additionally, lowest expanded uncertainties (<0.08%) can be obtained as demonstrated for the quantification of Cd in acrylonitrile-butadiene-styrene by ID TIMS.[162]

**Absolute Isotopic Composition, Atomic Weights and Isotope Reference Materials**

Along with some of the first instruments, TIMS has been used to determine the isotopic composition and from this the atomic weight of chemical elements. For some of the first atomic weight determinations, *e.g.* the atomic weight of Mg,[219] simply the observed isotope ratios were used. Very soon, however, the analysts recognised that there is a bias in the observed isotope ratio due to mass fractionation effects. The problem of determining mass fractionation without having an isotope standard of well-known isotope ratios was solved by Nier, who used synthetic isotope mixtures that allowed him to accurately determine the atomic weight of Ar.[220] At least two samples containing highly enriched but different isotopes of the same element were produced or purchased, for example one sample with enriched $^{40}$Ar and one sample with enriched $^{36}$Ar. If necessary, they were chemically purified to remove impurities from other elements. The enriched isotopes still contain other isotopes of the same element as the isotopic abundance of the enriched isotope is not 100% and usually does not exceed 99%. Subsequently, the isotopic abundance of each enriched isotope is determined by TIMS instrument without knowledge of the extent of mass fractionation and thus without correction thereof. After this, mixtures of the enriched isotopes are prepared under full gravimetric control. With the isotope ratios of the enriched isotopes and the gravimetric data of the mixtures the isotope ratios in the mixtures can now be calculated. These calculated values are still biased by the mass fractionation as the measured isotope ratios of the enriched isotopes are biased as mentioned above. To solve this, the mixtures are now measured with the same TIMS instrument. From the observed isotope ratios, being biased as well, and the calculated isotope ratios a so-called mass fractionation factor (K-factors) can be calculated. This is a first approximation of the extent of the mass fractionation in the TIMS instrument. This first approximation of the K-factor can be used to correct the initial measurements of the isotope ratios in the enriched isotopes prior to mixing. Now the K-factor can be recalculated. This iteration has to be repeated several times until the K-factor converges to a constant value expressing the full extent of the mass fractionation of the instrument and the unbiased isotope ratios of the mixtures. Now the TIMS instrument is "calibrated" and can be used to determine absolute isotope ratios, especially for characterising primary isotope reference materials.

This elegant approach has been used since then for the determination of atomic weights and is still capitalised today for the analysis of the isotopic composition of the elements and for calculating the corresponding atomic weight in terrestrial reference samples or reference materials. In 1954 NBS (now NIST) set up a programme to certify isotope reference materials by using this gravimetrical isotope mixtures approach in combination with TIMS. In the following years – the 1960s to the 1980s – NIST certified a significant number of isotope reference materials, which still make up the largest part of today's isotope reference materials in use. In the 1960s CBNM (now IRMM) started a similar programme on isotope reference materials. Until 1993 much progress has been made and isotope reference materials have been certified for 20 elements by applying TIMS instruments calibrated by synthetic isotope mixtures.[210] At the beginning of the new millennium work in this field was taken up, for example, at the BAM and since 2004 BAM and IRMM isotope reference materials are available as European Reference Materials (ERMs). Since 1993 work on isotope reference materials declined due to the low visibility and acceptance of this work. On the contrary, the needs for isotope reference materials increased due to the increased application of ICP-MS instruments for isotope analysis. A concise historical review of this development is given by De Laeter *et al.* in the "Atomic weights of the elements: Review 2000",[221] the present situation is reviewed by Vogl and Pritzkow[222] and the need for new isotope reference materials is described by Vogl *et al.*[223] Currently, isotope reference materials are available for more than two dozens of elements. The measurement uncertainties for isotope ratio values between 0.1 and 10 obtained by TIMS were improved to as low as 0.008%. Nowadays TIMS is used, for example, to certify secondary isotope reference materials such as the boron isotope reference materials ERM-AE101 to AE104 and ERM-AE120 to AE124.[164,222,223] Recently, uranium double-spike IRMM-3636 with a $^{233}U/^{236}U$ ratio of unity was made available and NBL CRM 112-A, a well-known reference material used substantially in geochronology as well as in the nuclear community, was recertified using TIMS.

There is increasing emphasis on establishing metrological traceability in sample analyses by a rigorous use of certified reference materials for instrument calibration and a careful evaluation of the associated measurement uncertainties. An example is the field of nuclear safeguards where a significant proportion of the accountancy of fissile isotopes U-235 and Pu-239 in nuclear material samples is performed using multicollector TIMS and metrological traceability of the results to the SI units is recognised.[59,74]

## 14.4.8 Metrological Applications in Nuclear Science

TIMS, mostly in combination with IDMS, has often been applied to determine a variety of physical constants such as half-lives, fission yields, or branching ratios crucial in nuclear science. Work in this field until 1988 has been reviewed by De Laeter[224] and this topic is discussed to some extent in Section 15.4.4. Work from 1988 onwards in this field is discussed briefly here.

The half-life of a radioactive decay can be determined in at least two ways: by determining the decay of the parent nuclide, or by determining the ingrowth of the daughter nuclide. The parent decay method is usually applied for short half-lives, because a relatively short time duration of the experiment already results in a significant decrease in the parent nuclide concentration. The daughter ingrowth method, however, is preferably used for long half-lives, as the ingrowth of a small number of nuclides can readily be observed. The latter technique, for example, has been applied for the direct determination of the half-life of $^{187}$Re, which decays to $^{187}$Os by β-decay.[225] Lindner *et al.*[225] determined the ingrowth of $^{187}$Os in an Os-free perrhenic acid solution over 4 years by ID-TIMS and subsequently calculated the half-live of $^{187}$Re as $(4.35 \pm 0.13)\times 10^{10}$ years. Especially in geochronology the $^{187}$Re half-life is important for using the Re–Os clock. The first method was used for the determination of the $^{241}$Pu half-life where the isotopic abundance of $^{241}$Pu was determined periodically.[226] Half-lives of other nuclides, such as $^{41}$Ca, $^{126}$Sn, $^{151}$Sm, $^{230}$Th, $^{234}$U, $^{235}$U, and $^{238}$U were determined in the past years by applying ID-TIMS and can be found in the literature.[227–231]

Nuclides can decay by different decay types. In the case of $^{64}$Cu, for example, a $\beta^-$ decay leading to $^{64}$Zn can occur as well as a $\beta^+$ decay leading to $^{64}$Ni. Both decays occur with a fixed ratio, the so-called branching ratio. This branching ratio was determined by Wermann *et al.* applying ID-TIMS.[232] Solid copper with low Zn and Ni impurities was irradiated in a nuclear reactor. The ingrowth of $^{64}$Ni and $^{64}$Zn, caused by the decay of the $^{64}$Cu produced from stable $^{63}$Cu by irradiation with thermal and epithermal neutrons, was determined by ID-TIMS. With the so-obtained results the $\beta^-$ branching ratio of $^{64}$Cu could be calculated as $(38.06 \pm 0.30)\%$, with an expanded uncertainty that is smaller than those from previous studies.

First ionisation potentials were determined using thermal ionisation mass spectrometry as early as the 1960s for elements important to the nuclear fuel cycle such as Th, Np, Pu, and Cm.[233,234]

## References


1. M. L. Gross, R. M. Caprioli, S. Houk, D. Beauchemin and D. E. Matthews, eds., *The Encyclopedia of Mass Spectrometry - Volume 5: Elemental, Isotopic & Inorganic Analysis by Mass Spectrometry*, Elsevier Science & Technology Books, 2010, 5, 1065.
2. A. J. Dempster, *Phys. Rev.*, 1918, **11**, 316–325.
3. K. G. Heumann, *Fresenius Z. Anal. Chem.*, 1986, **325**, 661–666.
4. T. Fujii, *Eur. J. Mass Spectrom.*, 1996, **2**, 91–114.
5. I. Langmuir and K. H. Kingdon, *Proc. Roy. Soc. Lond. Series A*, 1925, **107**, 61–79.

6. M. J. Dresser, *J. Appl. Phys.*, 1968, **39**, 338–339.
7. K. G. Heumann, S. M. Gallus, G. Rädlinger and J. Vogl, *J. Anal. Atom. Spectrom.*, 1998, **13**, 1001–1008.
8. C. M. Barshick, Glow discharge mass spectrometry, in *Inorganic Mass Spectrometry: Fundamentals and Applications (Practical Spectroscopy)*, eds. C. M. Barshick, D. C. Duckworth and D. H. Smith, Marcel Dekker Inc, New York, 2000, vol. 23 , ch. 2, pp. 31–66.
9. Y. Amelin and W. J. Davis, *J. Anal. Atom. Spectrom.*, 2006, **21**, 1053–1061.
10. M. Edwards, X. Tang and R. Shakeshaft, *Phys. Rev. A*, 1987, **35**, 3758–3767.
11. S. Bürger, L. R. Riciputi, D. A. Bostick, S. Turgeon, E. H. McBay and M. Lavelle, *Int. J. Mass Spectrom.*, 2009, **286**, 70–82.
12. IsotopX, http://www.isotopx.com, (accessed 11.06.2014).
13. Z. Li-hua, D. Hu, W. Guan-yi, L. Zhi-ming, W. Chang-hai, L. Xue-song, Z. Guo-qing, S. Yong-yang and Z. Zi-bin, *Int. J. Mass Spectrom.*, 2011, **305**, 45–49.
14. M. G. Watrous, J. E. Delmore and M. L. Stone, *Int. J. Mass Spectrom.*, 2010, **296**, 21–24.
15. Thermo Scientific, http://www.thermoscientific.com/TritonPlus, (accessed 11.06.2014).
16. J. L. Birck, *Geostand. Newslett.*, 2001, **25**, 253–259.
17. R. H. James and M. R. Palmer, *Chem. Geol.*, 2000, **166**, 319–326.
18. S. K. Sahoo and A. Masuda, *Anal. Chim. Acta*, 1998, **370**, 215–220.
19. N. Jabeen, E. Rehman and S. Ahmed, *J. Radioanal. Nucl. Chem.*, 2003, **258**, 427–430.
20. T. Yokoyama and E. Nakamura, *J. Anal. Atom. Spectrom.*, 2004, **19**, 717–727.
21. B. Ghaleb, E. Pons-Branchu and P. Deschamps, *J. Anal. Atom. Spectrom.*, 2004, **19**, 906–910.
22. A. M. Volpe, J. A. Olivares and M. T. Murrell, *Anal. Chem.*, 1991, **63**, 913–916.
23. A. S. Cohen, N. S. Belshaw and R. K. O'Nions, *Int. J. Mass Spectrom. Ion Process.*, 1992, **116**, 71–81.
24. J. J.-S. Shen and C.-F. You, *Anal. Chem.*, 2003, **75**, 1972–1977.
25. T. Nakano and E. Nakamura, *Int. J. Mass Spectrom.*, 1998, **176**, 13–21.
26. A. E. Cameron, D. H. Smith and R. L. Walker, *Anal. Chem.*, 1969, **41**, 525–526.
27. J. R. de Laeter, *Mass Spectrom. Rev.*, 1998, **17**, 97–125.
28. T. Walczyk, *Anal. Bioanal. Chem.*, 2004, **378**, 229–231.
29. M. E. Wieser, N. Norman Holden, T. B. Coplen, J. K. Böhlke, M. Berglund, W. A. Brand, P. De Bièvre, M. Gröning, R. D. Loss, J. Meija, T. Hirata, T. Prohaska, R. Schoenberg, G. O'Connor, T. Walczyk, S. Yoneda and X. K. Zhu, *Pure Appl. Chem.*, 2013, **85**, 1047–1078.
30. M. Berglund and M. E. Wieser, *Pure Appl. Chem.*, 2011, **83**, 397–410.
31. C. Douthitt, Thermo Fisher Scientific, personal communication.

32. P. G. Johnson, A. Bolson and C. M. Henderson, *Nucl. Instrum. Methods*, 1973, **106**, 83–87.
33. Y. Duan, R. Danen, X. Yan, R. Steiner, J. Cuadrado, D. Wayne, V. Majidi and J. Olivares, *J. Am. Soc. Mass Spectrom.*, 1999, **10**, 1008–1015.
34. D. H. Smith, W. H. Christie and R. E. Eby, *Int. J. Mass Spectrom. Ion Phys.*, 1980, **36**, 301–316.
35. W. D. Davis, *Environ. Sci. Technol.*, 1977, **11**, 587–592.
36. D. J. Rokop, R. E. Perrin, G. W. Knobeloch, V. M. Armijo and W. R. Shields, *Anal. Chem.*, 1982, **54**, 957–960.
37. C.-G. Lee, D. Suzuki, F. Esaka, M. Magara and T. Kimura, *Talanta*, 2011, **85**, 644–649.
38. U. S. Klötzli, *Analyst*, 1997, **122**, 1239–1248.
39. M. Svane, T. L. Gustafsson, B. Kovacevik, J. Noda, P. U. Andersson, E. D. Nilsson and J. B. C. Pettersson, *Aerosol Sci. Technol.*, 2009, **43**, 653–661.
40. D. R. Lide, ed., *CRC Handbook of Chemistry & Physics*, CRC Press, Boca Raton, FL, 73 edn., 1992, 2660.
41. S. A. Kasemann, A. B. Jeffcoate and T. Elliott, *Anal. Chem.*, 2005, **77**, 5251–5257.
42. Z. Chu, F. Chen, Y. Yang and J. Guo, *J. Anal. Atom. Spectrom.*, 2009, **24**, 1534–1544.
43. T. Ishikawa and K. Nagaishi, *J. Anal. Atom. Spectrom.*, 2011, **26**, 359–365.
44. M. Horn and K. G. Heumann, *Fresenius J. Anal. Chem.*, 1994, **350**, 286–292.
45. M. E. Wieser and J. B. Schwieters, *Int. J. Mass Spectrom.*, 2005, **242**, 97–115.
46. K. G. Heumann, S. Eisenhut, S. Gallus, E. H. Hebeda, R. Nusko, A. Vengosh and T. Walczyk, *Analyst*, 1995, **120**, 1291–1299.
47. K. M. Downard and J. R. de Laeter, *J. Mass Spectrom.*, 2005, **40**, 1123–1139.
48. *NBS Technical Note No. 426*, NBS/NIST Technical Notes, NBS/NIST, 1967.
49. F. A. White, T. L. Collins and F. M. Rourke, *Phys. Rev.*, 1956, **101**, 1786–1791.
50. D. H. Smith, W. H. Christie, H. S. McKown, S. L. Walker and G. R. Hertel, *Int. J. Mass Spectrom. Ion Phys.*, 1973, **10**, 343–351.
51. *IAEA Bulletin: The Safeguards Analytical Laboratory – Its Functions and Analytical Facilities V*, **19**, IAEA, Vienna, 1977, 38–47.
52. C. Brunnée, *Rapid Commun. Mass Spectrom.*, 1997, **11**, 694–707.
53. ThermoScientific, *High Resolution Multicollector Mass Spectrometers*, http://www.thermoscientific.com, 2008.
54. IsotopX, *Phoenix Instrument Specifications, Nuclear Installation, Product Information* **2**, http://www.isotopx.com, 2012.
55. IsotopX, *Phoenix Site Planning Guide* http://www.isotopx.com, 2011.
56. J. M. Koornneef, C. Bouman, J. B. Schwieters and G. R. Davies, *J. Anal. Atom. Spectrom.*, 2013, **28**, 749–754.

57. S. K. Aggarwal and H. C. Jain, *Int. J. Mass Spectrom. Ion Process.*, 1995, **141**, 149–160.
58. S. Richter and S. A. Goldberg, *Int. J. Mass Spectrom.*, 2003, **229**, 181–197.
59. S. Bürger, S. D. Balsley, S. Baumann, J. Berger, S. F. Boulyga, J. A. Cunningham, S. Kappel, A. Koepf and J. Poths, *Int. J. Mass Spectrom.*, 2012, **311**, 40–50.
60. B. L. Beard and C. M. Johnson, *J. Foren. Sci.*, 2000, **45**, 1049–1061.
61. T. Tanaka, S. Togashi, H. Kamioka, H. Amakawa, H. Kagami, T. Hamamoto, M. Yuhara, Y. Orihashi, S. Yoneda, H. Shimizu, T. Kunimaru, K. Takahashi, T. Yanagi, T. Nakano, H. Fujimaki, R. Shinjo, Y. Asahara, M. Tanimizu and C. Dragusanu, *Chem. Geol.*, 2000, **168**, 279–281.
62. L. Font, G. M. Nowell, D. G. Pearson, C. J. Ottley and S. G. Willis, *J. Anal. Atom. Spectrom.*, 2007, **22**, 513–522.
63. S. Bürger, R. M. Essex, K. J. Mathew, S. Richter and R. B. Thomas, *Int. J. Mass Spectrom.*, 2010, **294**, 65–76.
64. L. A. Dietz, G. A. Land and C. F. Pachucki, *Anal. Chem.*, 1962, **34**, 709–710.
65. S. Richter, A. Alonso-Munoz, R. Eykens, U. Jacobsson, H. Kuehn, A. Verbruggen, Y. Aregbe, R. Wellum and E. Keegan, *Int. J. Mass Spectrom.*, 2008, **269**, 145–148.
66. J. Hiess, D. J. Condon, N. McLean and S. R. Noble, *Science*, 2012, **335**, 1610–1614.
67. A. Bollhöfer and K. J. R. Rosman, *Geochim. Cosmochim. Acta*, 2000, **64**, 3251–3262.
68. K. Mathew, P. Mason, A. Voeks and U. Narayanan, *Int. J. Mass Spectrom.*, 2012, **315**, 8–14.
69. J. H. Chen, R. Lawrence Edwards and G. J. Wasserburg, *Earth Planet. Sci. Lett.*, 1986, **80**, 241–251.
70. E. Callis and R. Abernathey, *Int. J. Mass Spectrom. Ion Process.*, 1991, **103**, 93–105.
71. R. Fiedler, *Int. J. Mass Spectrom. Ion Process.*, 1995, **146**, 91–97.
72. S. Wakaki, S.-N. Shibata and T. Tanaka, *Int. J. Mass Spectrom.*, 2007, **264**, 157–163.
73. S. Mialle, A. Quémet, A. Ponvienne, A. Gourgiotis, M. Aubert, H. Isnard and F. Chartier, *Int. J. Mass Spectrom.*, 2012, **309**, 141–147.
74. S. Richter, H. Kuhn, Y. Aregbe, M. Hedberg, J. Horta-Domenech, K. Mayer, E. Zuleger, S. Burger, S. Boulyga, A. Kopf, J. Poths and K. Mathew, *J. Anal. Atom. Spectrom.*, 2011, **26**, 550–564.
75. I. T. Platzner, A. J. Habfast and A. G. Walder, *Modern Isotope Ratio Mass Spectrometry*, John Wiley & Sons, Chichester, 1997, 157.
76. Y. Saito-Kokubu, D. Suzuki, C.-G. Lee, J. Inagawa, M. Magara and T. Kimura, *Int. J. Mass Spectrom.*, 2012, **310**, 52–56.
77. J. L. Mann and W. R. Kelly, *Rapid Commun. Mass Spectrom.*, 2010, **24**, 2673–2679.
78. C. M. Johnson and B. L. Beard, *Int. J. Mass Spectrom.*, 1999, **193**, 87–99.

79. J.-H. Chen, S. Shahnavas, N. Singh, W.-Y. Ong and T. Walczyk, *Metallomics*, 2013, **5**, 167–173.
80. C. Hennessy, M. Berglund, M. Ostermann, T. Walczyk, H.-A. Synal, C. Geppert, K. Wendt and P. D. Taylor, *Nucl. Instrum. Methods Phys. Res. Sect. B: Beam Interact. Mater. Atoms*, 2005, **229**, 281–292.
81. A. Heuser and A. Eisenhauer, *Bone*, 2010, **46**, 889–896.
82. G. O. Lehn, A. D. Jacobson and C. Holmden, *Int. J. Mass Spectrom.*, 2013, **351**, 69–75.
83. G. Chew and T. Walczyk, *Anal. Chem.*, 2013, **85**, 3667–3673.
84. J. R. Turnlund, *J. Animal Sci.*, 2006, **84**, E73–E78.
85. S. F. Boulyga, *Int. J. Mass Spectrom.*, 2011, **307**, 200–210.
86. M. B. Greiter, V. Höllriegl and U. Oeh, *Int. J. Mass Spectrom.*, 2011, **304**, 1–8.
87. A.-D. Schmitt, N. Vigier, D. Lemarchand, R. Millot, P. Stille and F. Chabaux, *Compt. Rend. Geosci.*, 2012, **344**, 704–722.
88. P. Stille, A.-D. Schmitt, F. Labolle, M.-C. Pierret, S. Gangloff, F. Cobert, E. Lucot, F. Guéguen, L. Brioschi, M. Steinmann and F. Chabaux, *Compt. Rend. Geosci.*, 2012, **344**, 297–311.
89. A.-F. Maurer, S. J. Galer, C. Knipper, L. Beierlein, E. V. Nunn, D. Peters, T. Tütken, K. W. Alt and B. R. Schöne, *Sci. Total Environ.*, 2012, **433**, 216–229.
90. F. E. Grousset and P. E. Biscaye, *Chem. Geol.*, 2005, **222**, 149–167.
91. B. Delmonte, I. Basile-Doelsch, J. R. Petit, V. Maggi, M. Revel-Rolland, A. Michard, E. Jagoutz and F. Grousset, *Earth-Sci. Rev.*, 2004, **66**, 63–87.
92. F. E. Grousset, P. E. Biscaye, M. Revel, J.-R. Petit, K. Pye, S. Joussaume and J. Jouzel, *Earth Planet. Sci. Lett.*, 1992, **111**, 175–182.
93. A. Bory, E. Wolff, R. Mulvaney, E. Jagoutz, A. Wegner, U. Ruth and H. Elderfield, *Earth Planet. Sci. Lett.*, 2010, **291**, 138–148.
94. B. Delmonte, P. S. Andersson, H. Schöberg, M. Hansson, J. R. Petit, R. Delmas, D. M. Gaiero, V. Maggi and M. Frezzotti, *Quat. Sci. Rev.*, 2010, **29**, 256–264.
95. B. Delmonte, C. Baroni, P. S. Andersson, H. Schoberg, M. Hansson, S. Aciego, J.-R. Petit, S. Albani, C. Mazzola, V. Maggi and M. Frezzotti, *J. Quat. Sci.*, 2010, **25**, 1327–1337.
96. K. J. R. Rosman, W. Chisholm, C. F. Boutron, J. P. Candelone and C. C. Patterson, *Geophys. Res. Lett.*, 1994, **21**, 2669–2672.
97. P. Vallelonga, K. Van de Velde, J. P. Candelone, C. Ly, K. J. R. Rosman, C. F. Boutron, V. I. Morgan and D. J. Mackey, *Anal. Chim. Acta*, 2002, **453**, 1–12.
98. P. Vallelonga, P. Gabrielli, K. Rosman, C. Barbante and C. F. Boutron, *Geophys. Res. Lett.*, 2005, **32**, L01706.01701–L01706.01704.
99. P. Vallelonga, P. Gabrielli, E. Balliana, A. Wegner, B. Delmonte, C. Turetta, G. Burton, F. Vanhaecke, K. Rosman and S. Hong, *Quat. Sci. Rev.*, 2010, **29**, 247–255.
100. A. Matsumoto and T. K. Hinkley, *Geochem. J.*, 1997, **31**, 175–181.

101. P. Vallelonga, K. Van de Velde, J.-P. Candelone, V. Morgan, C. Boutron and K. Rosman, *Earth Planet. Sci. Lett.*, 2002, **204**, 291–306.
102. L. J. Burn-Nunes, P. Vallelonga, R. D. Loss, G. R. Burton, A. Moy, M. Curran, S. Hong, A. M. Smith, R. Edwards, V. I. Morgan and K. J. R. Rosman, *Geochim. Cosmochim. Acta*, 2011, **75**, 1–20.
103. A. J. M. Bory, P. E. Biscaye, A. Svensson and F. E. Grousset, *Earth Planet. Sci. Lett.*, 2002, **196**, 123–134.
104. A. J. M. Bory, P. E. Biscaye and F. E. Grousset, *Geophys. Res. Lett.*, 2003, **30**, 1167.
105. G. R. Burton, K. J. R. Rosman, J.-P. Candelone, L. J. Burn, C. F. Boutron and S. Hong, *Earth Planet. Sci. Lett.*, 2007, **259**, 557–566.
106. M. Murozumi, T. J. Chow and C. Patterson, *Geochim. Cosmochim. Acta*, 1969, **33**, 1247–1294.
107. K. J. R. Rosman, W. Chisholm, S. Hong, J.-P. Candelone and C. F. Boutron, *Environ. Sci. Technol.*, 1997, **31**, 3413–3416.
108. K. J. R. Rosman, W. Chisholm, C. F. Boutron, J. P. Candelone and U. Gorlach, *Nature*, 1993, **362**, 333–335.
109. K. J. R. Rosman, W. Chisholm, C. F. Boutron, J. P. Candelone, J. L. Jaffrezo and C. I. Davidson, *Earth Planet. Sci. Lett.*, 1998, **160**, 383–389.
110. A. Bollhöfer and K. Rosman, *Geochim. Cosmochim. Acta*, 2001, **65**, 1727–1740.
111. K. J. R. Rosman, C. Ly, K. Van de Velde and C. F. Boutron, *Earth Planet. Sci. Lett.*, 2000, **176**, 413–424.
112. M. Schwikowski, C. Barbante, T. Doering, H. W. Gaeggeler, C. Boutron, U. Schotterer, L. Tobler, K. Van de Velde, C. Ferrari, G. Cozzi, K. Rosman and P. Cescon, *Environ. Sci. Technol.*, 2004, **38**, 957–964.
113. K. Lee, S. D. Hur, S. Hou, L. J. Burn-Nunes, S. Hong, C. Barbante, C. F. Boutron and K. J. R. Rosman, *Sci. Total Environ.*, 2011, **412–413**, 194–202.
114. S. Hong, K. Lee, S. D. Hur, S. Hou, L. J. Burn-Nunes, C. F. Boutron and C. Barbante, in conf. proc. of *15th International Conference on Heavy Metals in the Environment (ICHMET-15)*, Chemical Faculty, Gdańsk University of Technology, 2012, pp. 345–357.
115. M. Lahd Geagea, P. Stille, F. Gauthier-Lafaye and M. Millet, *Environ. Sci. Technol.*, 2008, **42**, 692–698.
116. F. Thevenon, N. D. Graham, M. Chiaradia, P. Arpagaus, W. Wildi and J. Poté, *Sci. Total Environ.*, 2011, **412**, 239–247.
117. B. Gulson, M. Korsch, W. Winchester, M. Devenish, T. Hobbs, C. Main, G. Smith, K. Rosman, L. Howearth, L. Burn-Nunes, J. Seow, C. Oxford, G. Yun, L. Gillam and M. Crisp, *Environ. Res.*, 2012, **112**, 100–110.
118. A. K. Mackay, M. P. Taylor, N. C. Munksgaard, K. A. Hudson-Edwards and L. Burn-Nunes, *Environ. Pollut.*, 2013, **180**, 304–311.
119. G. R. Burton, K. J. R. Rosman, K. P. Van de Velde and C. F. Boutron, *Earth Planet. Sci. Lett.*, 2006, **248**, 217–226.

120. F. Guéguen, P. Stille, M. Lahd Geagea, T. Perrone and F. Chabaux, *Chemosphere*, 2012, **86**, 641–647.
121. S. K. Sahoo, H. Isobe, T. Sato, K. Fujimoto and Y. Nakamura, *J. Radioanal. Nucl. Chem.*, 2002, **252**, 241–245.
122. S. K. Sahoo, Y. Nakamura, K. Shiraishi and A. Masuda, *Int. J. Environ. Anal. Chem.*, 2004, **84**, 919–926.
123. R. Jakopic, S. Richter, H. Kuhn and Y. Aregbe, *J. Anal. Atom. Spectrom.*, 2010, **25**, 815–821.
124. D. Delanghe, E. Bard and B. Hamelin, *Marine Chem.*, 2002, **80**, 79–93.
125. W. Abouchami and M. Zabel, *Earth Planet. Sci. Lett.*, 2003, **213**, 221–234.
126. H.-F. Ling, S.-Y. Jiang, M. Frank, H.-Y. Zhou, F. Zhou, Z.-L. Lu, X.-M. Chen, Y.-H. Jiang and C.-D. Ge, *Earth Planet. Sci. Lett.*, 2005, **232**, 345–361.
127. A. D. Schmitt, S. J. G. Galer and W. Abouchami, *J. Anal. Atom. Spectrom.*, 2009, **24**, 1079–1088.
128. J. Trotter, P. Montagna, M. McCulloch, S. Silenzi, S. Reynaud, G. Mortimer, S. Martin, C. Ferrier-Pagès, J.-P. Gattuso and R. Rodolfo-Metalpa, *Earth Planet. Sci. Lett.*, 2011, **303**, 163–173.
129. R. C. Roback, T. M. Johnson, T. L. McLing, M. T. Murrell, S. Luo and T.-L. Ku, *Geol. Soc. Am. Bull.*, 2001, **113**, 1133–1141.
130. M. J. Grove, P. A. Baker, S. L. Cross, C. A. Rigsby and G. O. Seltzer, *Palaeogeo., Palaeoclimatol., Palaeoecol.*, 2003, **194**, 281–297.
131. A. Guendouz, A. Moulla, W. Edmunds, K. Zouari, P. Shand and A. Mamou, *Hydrogeol. J.*, 2003, **11**, 483–495.
132. K. M. Frei and R. Frei, *Appl. Geochem.*, 2011, **26**, 325–340.
133. P. de Caritat, D. Kirste, G. Carr and M. McCulloch, *Appl. Geochem.*, 2005, **20**, 767–787.
134. T. D. Bullen and A. Eisenhauer, *Elements*, 2009, **5**, 349–352.
135. K. D. Collerson and M. I. Weisler, *Science*, 2007, **317**, 1907–1911.
136. P. Degryse, J. Schneider, N. Kellens, M. Waelkens and P. H. Muchez, *Archaeometry*, 2007, **49**, 75–86.
137. M. Renzi, I. Montero-Ruiz and M. Bode, *J. Archaeol. Sci.*, 2009, **36**, 2584–2596.
138. M. S. Walton and M. S. Tite, *Archaeometry*, 2010, **52**, 733–759.
139. B.-P. Li, J.-X. Zhao, A. Greig, K. D. Collerson, Z.-X. Zhuo and Y.-X. Feng, *Nucl. Instrum. Methods Phys. Res. Sect. B: Beam Interact. Mater. Atoms*, 2005, **240**, 726–732.
140. B.-P. Li, J.-X. Zhao, A. Greig, K. D. Collerson, Y.-X. Feng, X.-M. Sun, M.-S. Guo and Z.-X. Zhuo, *J. Archaeol. Sci.*, 2006, **33**, 56–62.
141. F. Cattin, B. Guenette-Beck, M. Besse and V. Serneels, *Archaeol. Anthropol. Sci.*, 2009, **1**, 137–148.
142. R. H. Brill and J. M. Wampler, *Am. J. Archaeol.*, 1967, **71**, 63–77.
143. J. De Laeter, *Mass Spectrom. Rev.*, 2011, **30**, 757–771.
144. S. R. Copeland, M. Sponheimer, D. J. de Ruiter, J. A. Lee-Thorp, D. Codron, P. J. le Roux, V. Grimes and M. P. Richards, *Nature*, 2011, **474**, 76–78.

145. J. Vogel, B. Eglington and J. Auret, *Nature*, 1990, **346**, 747–749.
146. S. Kelly, K. Heaton and J. Hoogewerff, *Trends Food Sci. Technol.*, 2005, **16**, 555–567.
147. J. Montgomery, J. A. Evans and G. Wildman, *Appl. Geochem.*, 2006, **21**, 1626–1634.
148. J. B. West, J. M. Hurley, F. Ö. Dudás and J. R. Ehleringer, *J. Foren. Sci.*, 2009, **54**, 1261–1269.
149. A. Frumkin, A. Shimron and J. Rosenbaum, *Nature*, 2003, **425**, 169–171.
150. R. Barkai, A. Gopher, S. Lauritzen and A. Frumkin, *Nature*, 2003, **423**, 977–979.
151. A. Gopher, A. Ayalon, M. Bar-Matthews, R. Barkai, A. Frumkin, P. Karkanas and R. Shahack-Gross, *Quat. Geochronol.*, 2010, **5**, 644–656.
152. K. Mayer, M. Wallenius and I. Ray, *Analyst*, 2005, **130**, 433–441.
153. L. Pajo, K. Mayer and L. Koch, *Fresenius J. Anal. Chem.*, 2001, **371**, 348–352.
154. M. Wallenius and K. Mayer, *Fresenius J. Anal. Chem.*, 2000, **366**, 234–238.
155. B. Beer and K. G. Heumann, *Fresenius J. Chem.*, 1992, **343**, 741–745.
156. B. Beer and K. G. Heumann, *Anal. Chem.*, 1993, **65**, 3199–3203.
157. B. Beer and K. G. Heumann, *Fresenius J. Anal. Chem.*, 1993, **347**, 351–355.
158. P. Herzner and K. G. Heumann, *Anal. Chem.*, 1992, **64**, 2942–2944.
159. R. Hearn, M. Berglund, M. Ostermann, N. Pusticek and P. Taylor, *Anal. Chim. Acta*, 2005, **532**, 55–60.
160. W. Pritzkow, J. Vogl, R. Koppen and A. Ostermann, *Int. J. Mass Spectrom.*, 2005, **242**, 309–318.
161. T. Linsinger, W. Andrzejuk, A. Bau, J. Charoud-Got, P. De Vos, H. Emteborg, R. Hearn, A. Lamberty, A. Oostra, W. Pritzkow, C. Quetel, G. Roebben, I. Tresl, J. Vogl and S. Wood, *Energy Fuels*, 2007, **21**, 2240–2244.
162. J. Vogl, M. Koenig, W. Pritzkow and G. Riebe, *J. Anal. Atom. Spectrom.*, 2010, **25**, 1633–1642.
163. D. R. Lide, ed., *CRC Handbook of Chemistry & Physics*, CRC Press, Boca Raton, 82nd edn., 2001.
164. J. Vogl and M. Rosner, *Geostand. Geoanal. Res.*, 2012, **36**, 161–175.
165. *Standard operating procedures 7, 8 & 9*, BAM Federal Institute for Materials Research and Testing, Division 1.1, Berlin, Germany, 2013.
166. P. Mason, *An Evaluation of Uranium Measurement Capabilities and Comparison to State-of-the-Practice Target Values, including an Examination of Historical Performance* MEASUREMENT EVALUATION ANNUAL REPORT, New Brunswick Laboratory, Argonne, US, 2011.
167. S. Richter, A. Alonso, J. Truyens, H. Kühn, A. Verbruggen and R. Wellum, *Int. J. Mass Spectrom.*, 2007, **264**, 184–190.
168. G. Granier, F. Pointurier and F. Chartier, *J. Radioanal. Nucl. Chem.*, 2009, **279**, 875–884.
169. S. Richter, A. Alonso, W. De Bolle, R. Wellum and P. Taylor, *Int. J. Mass Spectrom.*, 1999, **193**, 9–14.

170. D. Rokop, D. Metta and C. Stevens, *Int. J. Mass Spectrom. Ion Phys.*, 1972, **8**, 259–264.
171. M. Paul, D. Berkovits, I. Ahmad, F. Borasi, J. Caggiano, C. N. Davids, J. P. Greene, B. Harss, A. Heinz and D. J. Henderson, *Nucl. Instrum. Methods Phys. Res. Sect. B: Beam Interact. Mater. Atoms*, 2000, **172**, 688–692.
172. L. W. Green, C. H. Knight, T. H. Longhurst and R. M. Cassidy, *Anal. Chem.*, 1984, **56**, 696–700.
173. J. S. Kim, Y. S. Jeon, S. D. Park, S. H. Han and J. G. Kim, *J. Nucl. Sci. Technol.*, 2007, **44**, 1015–1023.
174. L. Green, N. Elliot, F. Miller and J. Leppinen, *J. Radioanal. Nucl. Chem.*, 1989, **131**, 299–309.
175. K. Joe, Y.-S. Jeon, S.-H. Han, C.-H. Lee, Y.-K. Ha and K. Song, *Appl. Radiat. Isotopes*, 2012, **70**, 931–936.
176. J. De Laeter and H. Hidaka, *Mass Spectrom. Rev.*, 2007, **26**, 683–712.
177. D. Donohue, *J. Alloys Compd.*, 1998, **271**, 11–18.
178. *The IAEA Safeguards Analytical Laboratories, The Science Essential to Verifying the Peaceful Use of Nuclear Material, Fact Sheets*, Information Series – Division of Public Information, International Atomic Energy Agency, Vienna, 2012.
179. O. P. de Oliveira, W. D. Bolle, S. Richter, A. Alonso, H. Kühn, J. E. S. Sarkis and R. Wellum, *Int. J. Mass Spectrom.*, 2005, **246**, 35–42.
180. K. Raptis, G. Duhamel, R. Ludwig, S. Balsley, S. Bürger, V. Mayorov, A. Koepf, S. Hara, Y. Itoh and K. Yamaguchi, *J. Radioanal. Nucl. Chem.*, 2013, **296**, 585–592.
181. L. Duinslaeger, P. van Belle, K. Mayer, K. Casteleyn, S. Abousahl, P. Daures, H. Eberle, T. Enright, A. Guiot and M. Hild, *Atw. Int. Zeit. Kernenergie*, 2004, **49**, 420–426.
182. *International Target Values 2010 for Measurement Uncertainties in Safeguarding Nuclear Materials*, International Atomic Energy Agency, Vienna, Austria, 2010.
183. M. Wallenius, A. Morgenstern, C. Apostolidis and K. Mayer, *Anal. Bioanal. Chem.*, 2002, **374**, 379–384.
184. L. Pajo, G. Tamborini, G. Rasmussen, K. Mayer and L. Koch, *Spectrochim. Acta, Part B*, 2001, **56**, 541–549.
185. D. Donohue and R. Zeisler, *Anal. Chem.*, 1993, **65**, 359A–368A.
186. S. Usuda, K. Yasuda, Y. Saito-Kokubu, F. Esaka, C.-G. Lee, M. Magara, S. Sakurai, K. Watanabe, F. Hirayama and H. Fukuyama, *Int.J. Environ. Anal. Chem.*, 2006, **86**, 663–675.
187. T. Shinonaga, F. Esaka, M. Magara, D. Klose and D. Donohue, *Spectrochim. Acta, Part B: Atom. Spectrosc.*, 2008, **63**, 1324–1328.
188. R. Jakopič, S. Richter, H. Kühn, L. Benedik, B. Pihlar and Y. Aregbe, *Int. J. Mass Spectrom.*, 2009, **279**, 87–92.
189. M. Kraiem, S. Richter, H. Kühn, E. A. Stefaniak, G. Kerckhove, J. Truyens and Y. Aregbe, *Anal. Chem.*, 2011, **83**, 3011–3016.
190. R. H. Steiger and E. Jäger, *Earth Planet. Sci. Lett.*, 1977, **36**, 359–362.

191. G. Faure, *Principles of Isotope Geology*, John Wiley & Sons, New York, 2nd edn., 1986.
192. A. P. Dickin, *Radiogenic Isotope Geology*, Cambridge University Press, Cambridge, 2005.
193. C. J. Allègre, *Isotope Geology*, Cambridge University Press, Cambridge, 2008.
194. L. Nicolaysen, *Ann. New York Acad. Sci.*, 1961, **91**, 198–206.
195. M. A. Geyh and H. Schleicher, Absolute age determination: Physical and Chemical Dating Methods and Their Application, Springer, Berlin, 1990, vol. 1, p. 514.
196. O. Nebel, E. E. Scherer and K. Mezger, *Earth Planet. Sci. Lett.*, 2011, **301**, 1–8.
197. O. Nebel and K. Mezger, *Precamb. Res.*, 2008, **164**, 227–232.
198. C. Miller and M. Thöni, *Chem. Geol.*, 1995, **122**, 199–225.
199. C. Patterson, *Geochim. Cosmochim. Acta*, 1956, **10**, 230–237.
200. E. Klötzli-Chowanetz, *Migmatite des Ötztalkristallins: Petrologie und Geochronologie*, Universität Wien, Wien, 2001, 156.
201. W. Burke, R. Denison, E. Hetherington, R. Koepnick, H. Nelson and J. Otto, *Geology*, 1982, **10**, 516–519.
202. J. L. Banner, *Earth-Sci. Rev.*, 2004, **65**, 141–194.
203. H. G. Stosch, *Einführung in die Isotopengeochemie*, Universität Karlsruhe, Institut für Mineralogie und Geochemie, 2012.
204. K. G. Heumann, *Int. J. Mass Spectrom. Ion Process.*, 1992, **118**, 575–592.
205. P. De Bievre and H. S. Peiser, *Fresenius J. Anal. Chem.*, 1997, **359**, 523–525.
206. J. Vogl and W. Pritzkow, *Mapan-J. Metrol. Soc. India*, 2010, **25**, 135–164.
207. J. Vogl, *J. Anal. Atom. Spectrom.*, 2007, **22**, 475–492.
208. L. J. Moore, H. M. Kingston, T. J. Murphy and P. J. Paulsen, *Environ. Int.*, 1984, **10**, 169–173.
209. K. G. Heumann, *Mass Spectrom. Rev.*, 1992, **11**, 41–67.
210. P. de Bièvre, J. R. de Laeter, H. S. Peiser and W. P. Reed, *Mass Spectrom. Rev.*, 1993, **12**, 143–172.
211. A. Lamberty, W. Van Borm and P. Quevauviller, *The Certification of Mass Fraction of As, Br, Cl, Cr, Hg, Pb and S in Two Polyethylene CRMs—BCR-680 and BCR-681*, bcr information – REFERENCE MATERIALS, EUR 19450 EN, European Commission, Luxembourg, 2001.
212. W. Andrzejuk, A. Bau, J. Vharoud-Got, P. de Vos, H. Emteborg, R. Hearn, A. Lamberty, T. Linsinger, A. Oostra, W. Pritzkow, C. Quétel, G. Roebben, I. Tresl, J. Vogl and S. Wood, *Certification of the sulfur mass fraction in three commercial petrol materials – Certified Reference Materials ERM-EF211, ERM-EF212, ERM-EF213*, European Commission, Institute for Reference Materials and Measurements, Luxembourg, 2007, 37.
213. K. E. Murphy, E. S. Beary, M. S. Rearick and R. D. Vocke, *Fresenius J. Anal. Chem.*, 2000, **368**, 362–370.
214. K. E. Murphy, S. E. Long and R. D. Vocke, *Anal. Bioanal. Chem.*, 2007, **387**, 2453–2461.
215. Y. Aregbe and P. Taylor, *Metrologia*, 2003, **40**, 08001.

216. P. Klingbeil, J. Vogl, W. Pritzkow, G. Riebe and J. Muller, *Anal. Chem.*, 2001, **73**, 1881–1888.
217. J. Vogl, P. Klingbeil, W. Pritzkow and G. Riebe, *J. Anal. Atom. Spectrom.*, 2003, **18**, 1125–1132.
218. J. Vogl, *Rapid Commun. Mass Spectrom.*, 2012, **26**, 275–281.
219. A. J. Dempster, *Science*, 1920, **52**, 559.
220. A. O. Nier, *Phys. Rev.*, 1950, **77**, 789–793.
221. J. R. De Laeter, J. K. Bohlke, P. De Bievre, H. Hidaka, H. S. Peiser, K. J. R. Rosman and P. D. P. Taylor, *Pure Appl. Chem.*, 2003, **75**, 683–800.
222. J. Vogl and W. Pritzkow, *J. Anal. Atom. Spectrom.*, 2010, **25**, 923–932.
223. J. Vogl, M. Rosner and W. Pritzkow, *Anal. Bioanal. Chem.*, 2013, **405**, 2763–2770.
224. J. R. De Laeter, *Mass Spectrom. Rev.*, 1988, **7**, 71–111.
225. M. Lindner, D. A. Leich, R. J. Borg, G. P. Russ, J. M. Bazan, D. S. Simons and A. R. Date, *Nature*, 1986, **320**, 246–248.
226. R. Wellum, A. Verbruggen and R. Kessel, *J. Anal. Atom. Spectrom.*, 2009, **24**, 801–807.
227. G. Jorg, Y. Amelin, K. Kossert and C. L. von Gostomski, *Geochim. Cosmochim. Acta*, 2012, **88**, 51–65.
228. F. Oberli, P. Gartenmann, M. Meier, W. Kutschera, M. Suter and G. Winkler, *Int. J. Mass Spectrom.*, 1999, **184**, 145–152.
229. M. He, H. T. Shen, G. Z. Shi, X. Y. Yin, W. Z. Tian and S. Jiang, *Phys. Rev. C*, 2009, **80**, 064305.
230. H. Cheng, R. L. Edwards, J. Hoff, C. D. Gallup, D. A. Richards and Y. Asmerom, *Chem. Geol.*, 2000, **169**, 17–33.
231. J. M. Mattinson, *Chem. Geol.*, 2010, **275**, 186–198.
232. G. Wermann, D. Alber, W. Pritzkow, G. Riebe, J. Vogl and W. Görner, *Appl. Radiat. Isotopes*, 2002, **56**, 145–151.
233. D. H. Smith and G. R. Hertel, *J. Chem. Phys.*, 1969, **51**, 3105–3107.
234. D. H. Smith, *J. Chem. Phys.*, 1971, **54**, 1424–1425.

CHAPTER 15

# *Secondary Ion Mass Spectrometry*

LAURE SANGELY,*[a] BERNARD BOYER,[b] EMMANUEL DE CHAMBOST,[c] NATHALIE VALLE,[d] JEAN-NICOLAS AUDINOT,[d] TREVOR IRELAND,[e] MICHAEL WIEDENBECK,[f] JÉRÔME ALÉON,[g] HARALD JUNGNICKEL,[h] JEAN-PAUL BARNES,[i] PHILIPPE BIENVENU[j] AND UWE BREUER[k]

[a] International Atomic Energy Agency, Austria; [b] Université Montpellier 2, France; [c] formerly CAMECA, France; [d] Centre de Recherche Public - Gabriel Lippmann, Luxembourg; [e] The Australian National University, Australia; [f] Helmholtz-Zentrum Potsdam Deutsches GeoForschungsZentrum GFZ, Potsdam, Germany; [g] Centre de Sciences Nucléaires et de Sciences de la Matière (CSNSM), University Paris Sud laboratory in Orsay, France; [h] Federal Institute for Risk Assessment, Department of Product Safety, Berlin, Germany; [i] Commissariat à l'énergie atomique et aux énergies alternatives, DEN Cadarache, France; [j] Commissariat à l'énergie atomique et aux énergies alternatives, XXX, France; [k] Forschungszentrum Jülich GmbH, Juelich, Germany
*Email: L.Sangely@iaea.org

The basic principle of SIMS is that a focused beam of energetic ions (so-called primary ions) is targeted onto the surface of a solid sample. Primary ions dissipate their energy, leading to the sputtering and ionisation of the outmost atoms of the sample surface. The resulting secondary ions are accelerated and transferred to a magnetic analyser.

New Developments in Mass Spectrometry No. 3
Sector Field Mass Spectrometry for Elemental and Isotopic Analysis
Edited by Thomas Prohaska, Johanna Irrgeher, Andreas Zitek and Norbert Jakubowski

Published by the Royal Society of Chemistry, www.rsc.org

Figure 15.1 displays the general layout of a SIMS instrument. Figure 15.2 gives an overview of its analytical flow chart. SIMS is applicable to the determination of the isotopic and trace, minor (and to some limited extent, major) element composition across the entire periodic table for any solid material compatible with high-vacuum conditions. Elemental and isotopic analysis can be performed i) locally at the micrometre scale; ii) in depth along so-called depth profiles; iii) over 2D surfaces (ion imaging) or within 3D volumes (ion tomography) with a resolution range between 50 nm and 5 μm. SIMS is used in a variety of application fields: geochemistry and cosmochemistry, material sciences, microelectronics, life sciences, and nuclear safeguards.

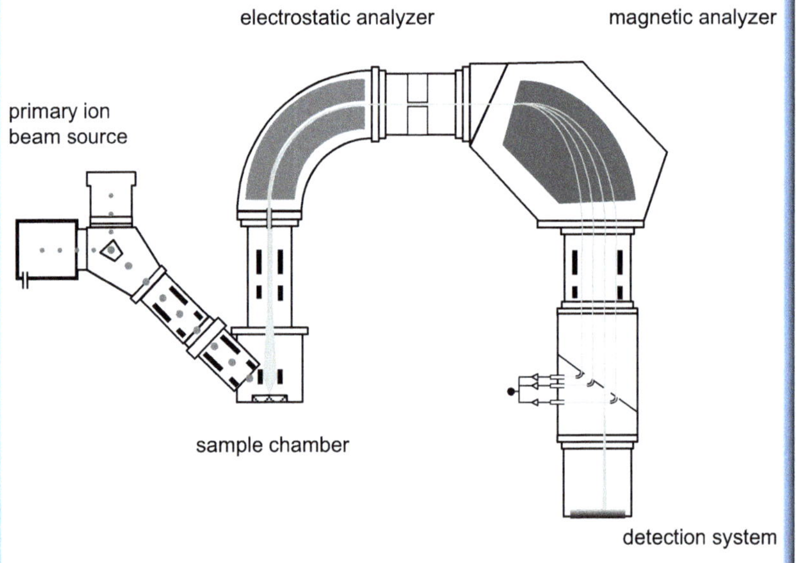

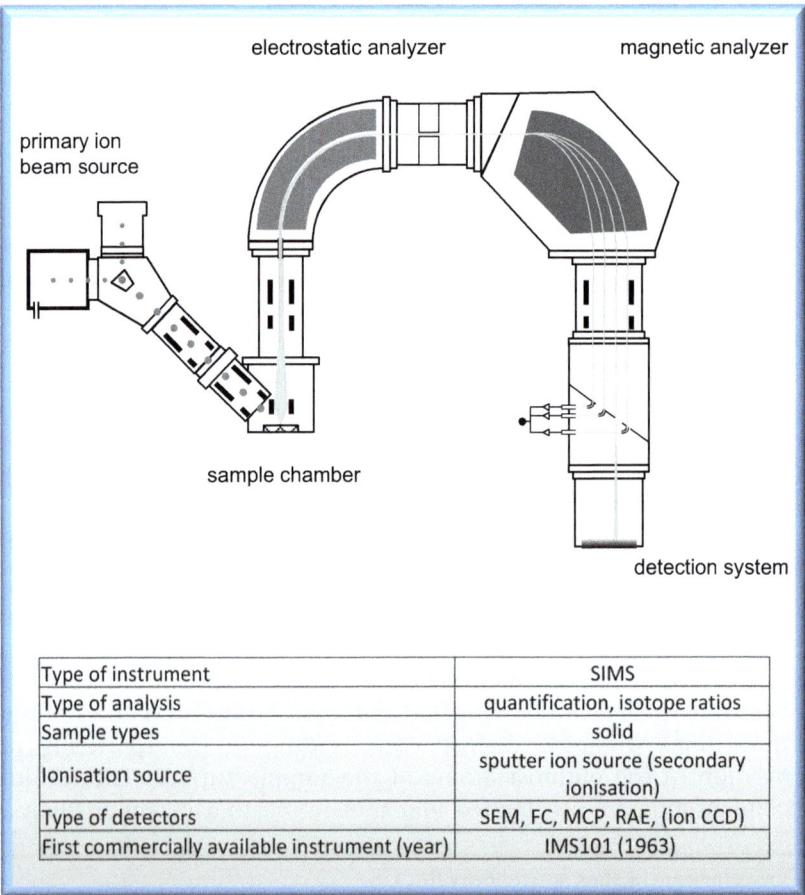

| | |
|---|---|
| Type of instrument | SIMS |
| Type of analysis | quantification, isotope ratios |
| Sample types | solid |
| Ionisation source | sputter ion source (secondary ionisation) |
| Type of detectors | SEM, FC, MCP, RAE, (ion CCD) |
| First commercially available instrument (year) | IMS101 (1963) |

**Figure 15.1** Features and general schematics of a magnetic SIMS (based on *CAMECA IMS 7f*).
Source: Ondrej Hanousek.

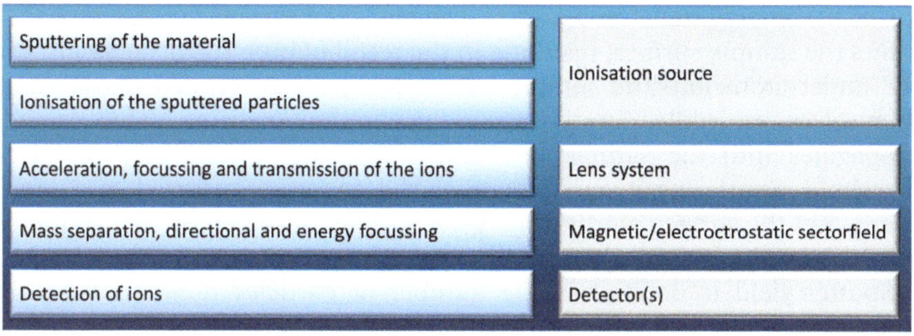

Sputtering of the material
Ionisation of the sputtered particles
Ionisation source
Acceleration, focussing and transmission of the ions
Lens system
Mass separation, directional and energy focussing
Magnetic/electroctrostatic sectorfield
Detection of ions
Detector(s)


**Figure 15.2** Operational flow chart of a SIMS.
Source: Thomas Prohaska.

**Name and Abbreviation:** The IUPAC conform abbreviation of the technique is SIMS according to the 2013 recommendations on the definitions of terms relating to mass spectrometry.[1]

# 15.1 General

Under ultrahigh-vacuum conditions, the bombardment of energetic ions (the so-called primary ions) onto the surface of a solid sample results in the ejection of atoms, molecules, and clusters from the uppermost sample layers (typically within the first 0.1–2 nm). This process is known as sputtering. A small fraction of the sputtered material escapes as positively or negatively charged species, namely secondary ions. This overall process (the emission of secondary ions from a solid sample surface as a result of an interaction with energetic primary ions) is referred to as secondary ionisation. Even if the complete mechanisms of secondary ionisation still have to be understood, it has been empirically established for decades that the composition of secondary ions reflects to some extent the isotopic, elemental, and – in some cases – the molecular properties of the target sample. As a result, the development in the late 1960s of the SIMS technique (by definition, the combination of a secondary ionisation chamber and a mass spectrometer) opened up an outstanding range of possibilities for the analysis of surfaces and solids.

## 15.1.1 Fundamentals of Secondary Emission

Benninghoven *et al.*[2] and Vickerman *et al.*[3] are recognised by the SIMS community as the major references regarding *inter alia* the chemical and physical principles involved in the process of secondary emission. The information provided in this section was generally drawn from these books and references therein.

### Sputtering

The accepted basic principle for the sputtering process is based on a series of binary collisions, referred to collectively as "collision cascade." A primary ion of keV energy transfers a part of its energy to a sample target atom when it hits the sample surface, resulting in the recoil of both the primary ion and the target atom into the sample lattice. Now, these recoiled particles are themselves projectiles in the second-generation collision. This cascade propagates into the sample until the recoil energy falls below a given threshold. Sputtering occurs whenever the energy transferred to a target overcomes the surface binding energy. It is generally accepted that sputtered particles originate from the very first nm of a surface.

Sputter yield is defined as the number of particles removed from the sample per impacting primary ions. It typically ranges between 5 and 25 and depends on both sample properties (*e.g.* mass of atomic constituents, binding energy of target atoms and crystal orientation) and the properties of primary ions (*e.g.* mass, energy and incidence angle). The total sputter yield in multicomponent materials equals the sum of the partial sputter yields of the constituent atoms. Due to a preferential sputtering phenomenon, the ratio between partial sputter yields for two different species does not necessarily equal the concentration ratio of this element at the active surface. As a consequence, the near-surface region tends to be depleted in elements, which have higher sputter yields, while it is enriched in elements that have lower sputter yields.

If sputtered particles originate from the very first nm of a given surface, the effect of primary ions/surface interactions can be traced up to a few dozen nm into the surface. In this so-called damaged volume, the structure and composition of the sample is strongly altered due to the implantation of primary ions and the transport of atoms from one layer of the sample into another. The depth of the damaged volume depends on the sample's properties (mass of atomic constituents, crystal orientation, and density) as well as primary beam properties (mass, energy, and incidence angle).

**Ionisation**

As mentioned before, the vast majority of sputtered material consists of neutral atoms. For a given secondary-ion species, the ionisation rate (or ionisation probability) is defined as the ratio between the number of ions formed and the number of atoms (or molecules) of the same species initially present in the sample volume which is sputtered. The ionisation rate is usually around a few tenths of a per cent up to a few per cent and is critical for the sensitivity of the SIMS analysis. The ionisation rate is strongly affected by the atomic properties of the analysed element (a positive ionisation rate tends to decrease with the ionisation potential, while a negative ionisation rate tends to increase with electron affinity). The choice of the primary ion species is also of utmost importance. The use of oxygen as primary ions is known to enhance positive ionisation, while cesium primary ions favour negative ionisation.

For primary oxygen ion bombardment, the implantation of primary ions leads to the formation of metal–oxygen bonds. When the sputtering occurs and metal-oxygen bounds break, the oxygen atom captures one electron due to its higher electron affinity, while the sample constituent is emitted with a positive charge. As the cesium implantation proceeds, the work function of the surface tends to decrease and more secondary electrons are excited over the potential barrier of the surface. When sputtering occurs, the sample constituent emitted is likely to capture one of the available excited electrons.

Ionisation efficiency also depends on the chemical environment of sputtered species. For example, the presence of oxygen as a sample constituent is known to enhance positive ionisation. The use of a microleak of dioxygen right above the sample surface results in a similar effect.[4]

## 15.1.2 Pre-Equilibrium and Equilibrium State

As mentioned above, the interaction between energetic primary ions and sample surface atoms results in the strong alteration of the chemical composition and structure of the sample within the first dozens of nanometres from the bombarded surface. When the bombardment starts, the concentration of implanted primary ions at the active surface increases drastically from zero up to a saturation level that is reached when the flux of implanted primary ions leaving the surface due to resputtering compensates for the flux of newly implanted primary ions. In addition – as a result of preferential sputtering – the relative proportions between sample elements and isotopes at the active surface progressively changes from that of the fresh sample up to a point of maximum alteration. This point is reached once the active surface reaches a depth range that exceeds the initial damaged volume. This time span – called pre-equilibrium (or transition) state – is characterised by transient sputter, ionisation rates and sample surface composition that compromise the use of common calibration methods (see Section 15.4).

Two kinds of strategies can be adopted to overcome this limitation:

(i) The first one corresponds to what is called static SIMS. The primary ion fluence is kept below a given threshold, ensuring that each emitted ion is generated from an area of the surface that was not affected by a former sputtering event. As a result, the active surface does not exceed the topmost atomic layer of the solid surface. Static SIMS is a technique of choice for *stricto sensu* surface analysis especially in adsorbed molecular layers. It will not be further discussed in this chapter since time-of-flight analysers are generally preferred for such applications.

(ii) In dynamic SIMS, the primary ion beam density (*i.e.* the primary ion current normalised to the bombarded area) is high enough to result in a layer-by-layer etching of the sample surface. In this mode, the analysis is restricted to the layers exposed at a depth where the equilibrium regime prevails (typically between a few nm and a few hundred μm from the surface). During the equilibrium state, sputter yield and ionisation rates are constant and the chemical/isotopic composition of the sputtered material is considered to match the composition of the fresh sample. Under these circumstances, the basic assumptions underlying the common standardisation methods are met (see Section 15.4). On a macroscopic level, dynamic SIMS results in the formation of a crater that deepens at a constant rate (namely, the sputter rate) of 0.5 to 5 nm $s^{-1}$. The sputter rate depends on both the primary ion density and sputter yield. Since the sputter yield varies as a function of the crystal orientation, roughening usually occurs at the bottom of the craters in polycrystalline materials due to differential sputter rates.

### 15.1.3 Characteristics of Secondary Ions

In dynamic SIMS, the secondary emission spectra are generally dominated by atomic (single-charged) ions rather than complex molecular ions. This effect is even more pronounced if atomic species are employed as primary ions (*e.g.* $Ar^+$, $O^-$, $Cs^+$) compared to polyatomic species (*e.g.* $SF_5^+$, $C_n^+$, $Au_n^+$ or $Bi_n^+$). However, a limited number of elements are more prone to be emitted as di- or triatomic ions in combination with a major element from the matrix or a primary ion, rather than as atomic ions. This is, for instance, the case for rare-earth elements, uranium and thorium that readily form positive oxide and dioxide ions under oxygen bombardment. It is also the case for nitrogen in carbonaceous matrices that is preferentially emitted as a negative cyanide ion.

Secondary ions are emitted from the sample surface in a variety of trajectory angles and translational kinetic energies, which range from 0 to several hundred eV. Polyatomic ions generally exhibit a narrower translational energy distribution than atomic ions. The reason for this is that the kinetic energy transferred to the emitted polyatomic ions is distributed between internal (vibrational and rotational) and translational modes, whereas atomic ions only possess translational modes.

As mentioned above, the elemental/isotopic composition of the material sputtered under the equilibrium regime is considered to reflect the composition of the fresh sample. However, this is not reflected by the composition of secondary ions. As pointed out by Shimizu and Hart,[5] this fact is commonly attributed to the ionisation aspects of secondary emission. Ionisation rates vary from element to element and between isotopes of a given element. In addition, they vary independently one from another as a function of sample properties (resulting in what is generally called matrix effects). As a consequence, the measurement of trace element and isotopic compositions using SIMS is subject to the availability of reference materials that match the sample matrix, for which the composition of each analyte has been certified or measured using a different technique with a level of uncertainty similar to (or lower than) the uncertainty expected for the SIMS results.

In addition to secondary emission processes, the differential transmission of elements and isotopes through the flight tube contributes (but to a minor extent) to the instrumental mass fractionation. Since the Earth's magnetic field is responsible for a slight dispersion of ion trajectories as a function of their mass, the way each slit and diaphragm crosscuts the secondary beam will influence the instrumental mass fractionation. For example, a shift between $^{29}Si$ and $^{28}Si$ trajectories by a few micrometres is expected at the point of the entrance slit. It is likely to cause a fractionation in the per mil range. As a consequence, achieving high-precision isotopic measurements requires all parameters that influence the secondary beam and aperture/slit widths and their relative positions to be kept strictly identical over the period of time when measurements on the samples and the related reference materials are performed (this includes the local geometry and distribution of potential at the sample surface).

### 15.1.4 Samples and Analytes

Virtually any solid matrix can be analysed using SIMS, provided that the surface of interest is flat, conductive and compatible with ultrahigh-vacuum conditions (on top of that, high-precision isotopic analysis usually requires a mirror-polished surface). In the case of nonconductive materials, the analysis is only possible using a negative/positive combination for primary/secondary polarities and if the sample surface is coated with a thin layer of gold, aluminium or carbon (a few dozen nm). However, the analysis of insulators using other primary/secondary polarity combinations is possible pending the use of an electron gun to compensate the positive charges that accumulate over the rastered area. The size of the sample to be introduced in the specimen chamber is limited in practice by the dimensions of the sample holders.

In principle, all elements and isotopes from hydrogen to plutonium can be analysed using SIMS. Practically, good performance across the whole periodic table implies the use of several types of primary ion sources (*e.g.* a cesium source for electronegative elements and an oxygen source for metals). In some cases, further requirements may apply. For example, the measurement of hydrogen requires enhanced vacuum conditions, while memory effects may hinder the measurement at the trace level of some elements that were present as major constituents in a matrix previously analysed. In this case, cleaning the specimen chamber and transfer optics is the only way to limit background levels.

### 15.1.5 Analytical Capability

All SIMS designs include a primary column, which is fitted between the primary ion sources and the specimen chamber. It enables focusing of the primary beam at the sample plane, thus forming a spot with either a uniform or a 2D Gaussian density distribution and ranges from a few to a few hundreds of micrometres in diameter. The primary beam can be either static or rastered over a square area up to a few tenths of millimetres. The primary ion beam intensity can be selected within a range from a few μA to a few μA. This flexibility in primary ion beam conditions enables the user to switch between three main analysis modes, namely microanalysis, imaging and depth profiling.

**Imaging: Isotopic and Elemental Mapping**

The secondary ions extracted from the sample surface reach the entrance of the analyser, through the so-called transfer optics. On several instrument models (*e.g. CAMECA IMS 7f*, *IMS 1280-HR* and predecessors), both segments (transfer and analyser) were designed and combined such as to form a stigmatic optical system, capable of transmitting an image of the sample surface, in which the ions of the selected mass that were emitted from a given object point converge into a single image point. The location of the image point only depends of the location of the emitting point in the sample plane (regardless the emission angle, initial energy and mass of the ions). As a result, when a sample is illuminated with a broadly defocused or a rastered primary beam, a mass-filtered ion image of the sample surface can be projected at the exit of the mass spectrometer onto a microchannel plate (MCP) detector that is fitted together with a phosphorescent screen or on a 2D solid state digitiser (*e.g.* the CAMECA resistive anode encoder system, RAE). This method of imaging, called ion microscopy, is commonly used for elemental and isotopic mapping. An ultimate lateral resolution of around 2 μm can be achieved for analytes present in concentrations down to 100 g $g^{-1}$. In so-called microscope mode, the lateral resolution mainly depends on the transfer settings (thus, it has to be balanced with respect to the transmission but is independent of the primary beam cross section). As compared to scanning element maps produced by electron microprobe, the SIMS approach is some 10 to 100 times faster, but suffers in that the data are, at best, only semiquantitative in nature.

On all SIMS instruments, an additional method of imaging is available. In this mode called microprobe, a focused primary beam is rastered over the sample surface. At the exit of the analyser, the time sequence of secondary-ion detection by the electron multipliers in converted into a real-time 2D intensity distribution, called scanned ion images. A square area up to 500 μm×500 μm can be transmitted to the analyser thanks to a dynamic transfer system. It enables the synchronisation between primary and secondary beam deflections relative to the optical axis, so that off-axis abberations can be minimised. In microprobe mode, the lateral resolution depends mostly on the primary beam cross section at the sample plane. It has to be mainly balanced with the sputter rate but is independent of the transfer settings (an ultimate lateral resolution down to 50 nm can be achieved). Stacking repeated images in either microscope or microprobe mode allows a 3-dimensional "tomographic" ion image to be constructed.

On recent *SHRIMP* instruments, ion image maps can be acquired using a sample stage stepping system. Here, the spatial resolution is defined by both the primary beam cross section and the increment and precision of the stage mechanical stepping. This approach, similar to what is available on ToF-SIMS for large-area imaging, is expected to emerge in the magnetic SIMS literature in the near future.

**Depth Profiling**

Under specific primary beam conditions, sputtering results in flat-bottomed craters with sharp edges and a layer-by-layer etching of the sample. SIMS therefore enables the continuous, indepth analysis of the sample with a depth resolution in the nm range (the so-called depth profiles), provided that a "gate" is created to exclude the ions emitted by the edges of the crater from the integrated signal. All measures taken for limiting the extent of the damaged volume can be used to optimise the depth resolution (especially reducing the primary ion impact energy). Depth profiles usually display concentrations expressed in atoms per $cm^3$ as a function of depth. The relationship between sputtering time and depth has to be determined independently by profilometry, while the relationship between secondary-ion intensity and concentration is usually established by running a depth profile on a reference sample. The reference sample consists of a matrix similar to the matrix of the sample, where a known quantity of the analyte has been implanted. In depth profiling, the detection limit for a given element is typically 4 to 5 orders of magnitude less than the maximum concentration measured across the profile and can be as low as $5\times10^{13}$ atoms $cm^{-3}$ ($\sim$1 ppb) in favourable cases.

**Trace and Isotopic Microanalysis**

When a static and low intensity primary ion beam is used, it is possible to sputter a sample volume of a few $\mu m^3$ within a few minutes. Under these conditions, the SIMS technique enables the *in situ* microanalysis of trace elements with detection limits in the ppm to ppb range. Isotopic micro-analysis can also be performed at an uncertainty level in a range between a few per cent to a few tenths of ‰.

## 15.2 Technical Background

The following description of a typical sector field SIMS is based on the *CAMECA IMS 7f* (Figure 15.3), the latest in the series of *CAMECA IMS* instruments, which started with the *IMS 3f*.[6] The current generation *IMS 7f* along with *IMS Wf*, *SC Ultra* and *IMS 1280-HR* are the beneficiaries of advances in mechanical engineering, vacuum system improvements, electronic, software speed and power breakthroughs. Also, they are the response to new demands from industry and from material science. A common basic design will be described in this chapter.

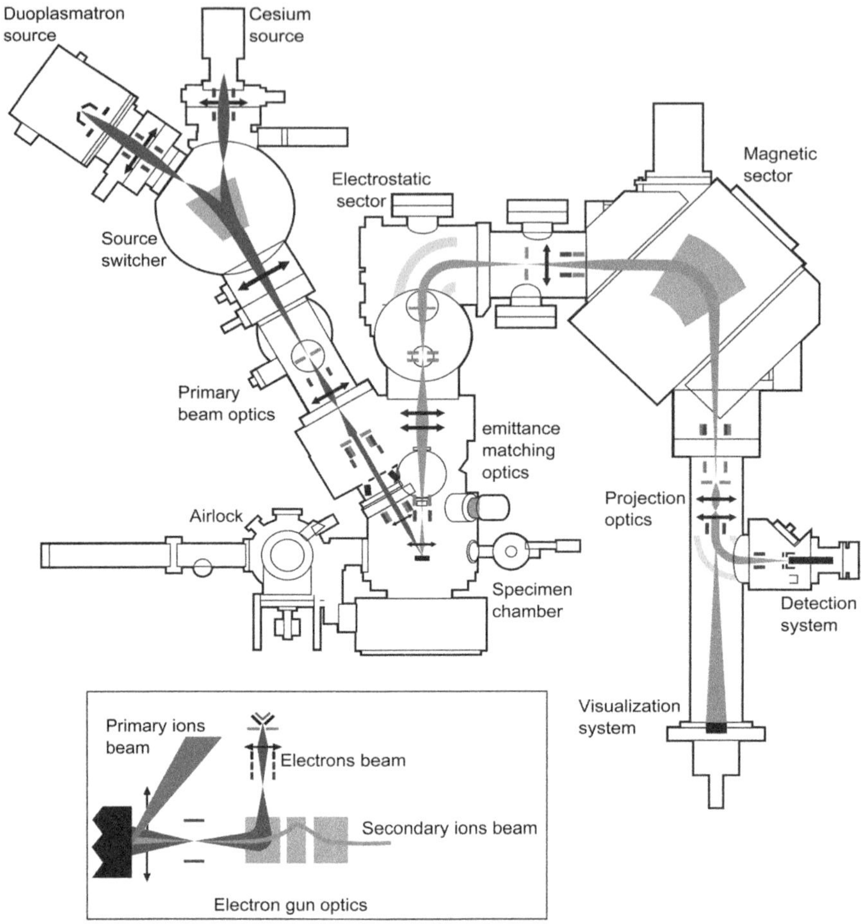


**Figure 15.3** The schematic of a *CAMECA IMS7F*. (Source: © CAMECA.)

## 15.2.1 Sample Introduction

Due to the high sensitivity of SIMS, one must treat a sample with extreme caution for cleanliness before analysing it with this technique. The analytical chamber must be kept under high or ultrahigh vacuum. Before its introduction to the high-vacuum system, the sample is placed in an airlock, where the sample is pumped down to a suitable vacuum level for transfer (Figure 15.4). The airlock may have an additional local baking system to support surface decontamination. Some fully automated SIMS instruments (see Section 15.3.2) are equipped with a robot, which transfers samples in the airlock. The sample size is commonly in the cm range (*e.g.* up to a 300-mm wafer can be placed into a *CAMECA Wf* and *SC ULTRA*).

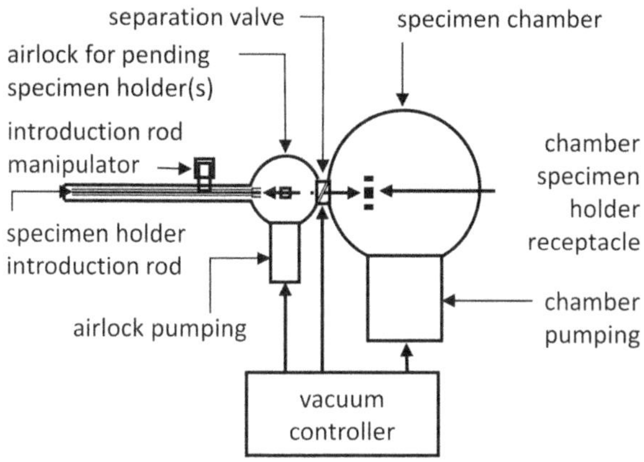


**Figure 15.4** Schematics of an airlock.
(Source: © CAMECA courtesy of CAMECA.)

## 15.2.2 Ion Source

In general, two or three ion sources are available. The choice of the source, ion polarity and primary ion impact energy on the sample surface depends on the nature of the sample and the kind of secondary ions to be analysed.

**The Duoplasmatron Source**

The duoplasmatron (Figure 15.5) consists of three electrodes: a pure nickel-made cylindrical cathode; an iron-made intermediate (Z) conical electrode with a 2-mm hole and a flat anode with a 400 μm hole. The three electrodes are aligned along a solenoid axis. The cathode and the Z electrode can be moved along a radial path of about 2 mm.

The gas (usually oxygen or argon) is introduced at low pressure through the cathode while a potential of −1000 V is applied to the cathode relative to the anode. In these conditions an oxygen or argon plasma is created (Paschen's law). This plasma is confined in the space between these electrodes. The magnetic field created with the solenoid helps to keep the plasma along the duoplasmatron axis. Once the plasma is ignited it is sustained by the collision between electrons extracted from the cathode surface and the gas molecules, which continuously feed this volume. Ionised (+) molecules are attracted by the cathode, and their impact produces new electrons.

The positive ions are confined in the centre of the plasma. Negative ions (for oxygen) are present in the outer plasma. When the primary column HV potential is positive relative to the extraction electrode and the Z and the cathode are close to the duoplasmatron axis, positive ions ($O_2^+$, $Ar^+$) are extracted from the compressed plasma through the anode hole through which pumping is done. If the Z electrode and the cathode are off-centre and the primary column HV is negative, one can extract negative ions ($O^-$) from the plasma. HV extraction potential can switch between a few kV to ±15 kV.

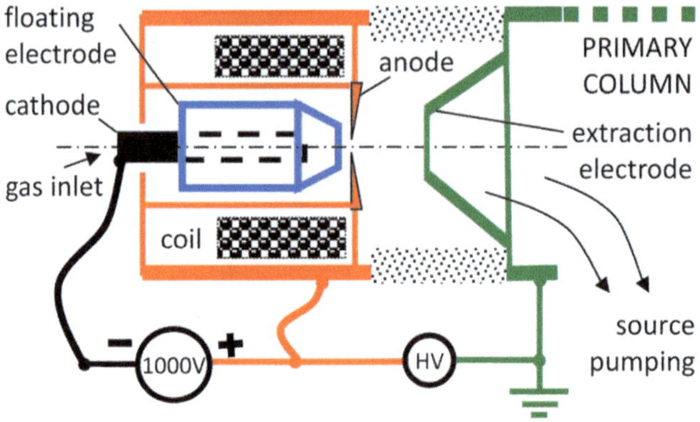


**Figure 15.5** Schematics of a duoplasmatron source.
(Source: © CAMECA courtesy of CAMECA.)

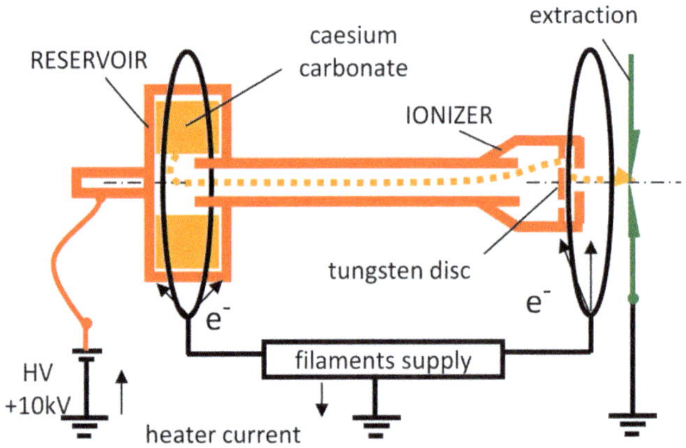


**Figure 15.6** Schematics of a cesium source.
(Source: © CAMECA courtesy of CAMECA.)

Since 2010, a high-brightness radio-frequency (RF) plasma gas source, developed by Oregon Physics LCC, has been available on *SC Ultra* intruments for $O_2^+$ primary ions.[7] Compared to a conventional duoplasmatron, the RF plasma source enables substantially higher beam densities (by approximately a factor 10). In parallel, an RF source for $O^-$ ions that could be fitted on a *NanoSIMS* is currently tested at CAMECA.

**The Cesium Source**

A cesium ion source is shown in Figure 15.6. A cesium carbonate tablet is heated to about 150 °C in a reservoir. Cesium vapour is evacuated to an ioniser where it is ionised at the contact with a holed flat tungsten washer. The tungsten washer temperature in the ioniser can reach 1100 °C. Both reservoir and ioniser are heated by cathode heating. These temperatures and the choice of tungsten material enhance cesium ionisation yield.[8] Since the reservoir and ioniser assembly is biased at +10 kV relative to the extraction electrode, $Cs^+$ ions are accelerated through the primary column.

## 15.2.3 Ion Source Projection Column (Figure 15.7)

The instrument may use a separate column for each source or only one column equipped with a source switcher system (IMS 7f). For the latter, ions extracted from the chosen ion source are deviated and filtered by a magnetic prism followed by an optical adjustable mass-selection aperture. The filtering/mass selection function is particularly useful. When oxygen ions are used, traces of gas contaminant ($N_2$; $H_2O$) and duoplasmatron electrodes releases may contaminate the sample surface currently being sputtered by the primary ion beam. For primary columns with a monosource, the beam-filtering function is done by a device like a Wien filter.[9]

After the primary ion beam is selected, it is focused onto the sample surface by using Einzel-type electrostatic lenses.[10] The beam must pass through the lens axes using electrostatic deflectors. The beam shape can be adjusted by the use of electrostatic quadrupoles and an adjustable aperture may partly remove beam aberrations inherent to this kind of ion optical setup. The beam position on the sample is controlled by electrostatic deflection plates. In most cases – during analysis or column alignment – deflection plates are used to raster the beam over the sample surface.

## 15.2.4 Analytical Chamber

In the analytical chamber, the sample surface is biased relative to the extraction electrode, which is usually grounded (Figure 15.8). The difference of the potentials between the sample surface and the primary ion extraction gives impact energy of $10^2$ to $10^3$ eV, which is sufficient to sputter the target surface. The implantation of primary ions determines the volume of

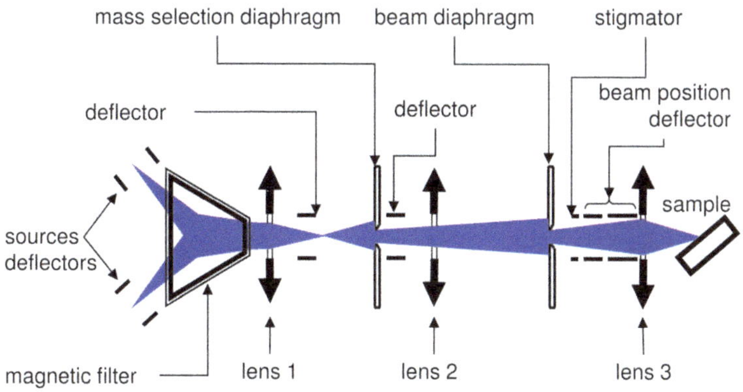


**Figure 15.7** The ion source projection column.
(Source: © CAMECA courtesy of CAMECA.)

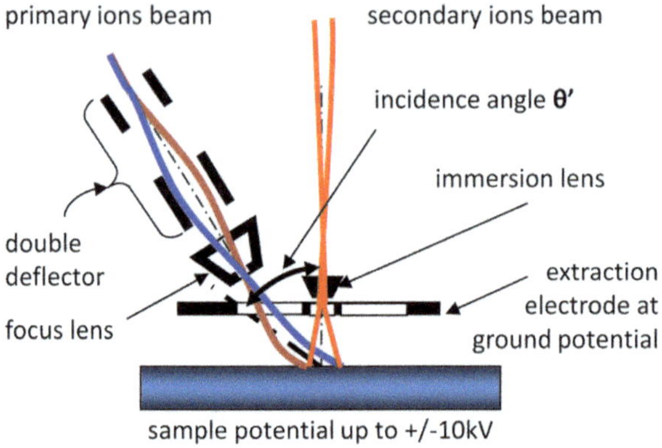


**Figure 15.8** Ion trajectories in the extraction area over the sample surface. (Source: © CAMECA courtesy of CAMECA.)

**Table 15.1** Incidence angle for different accelerating voltages (after Peres *et al.*[13]).

| Accelerating voltages (in kV) Primary Ions | Secondary Ions | Incidence angle (°) |
|---|---|---|
| +8 | +5 | 52 |
| +5 | +3 | 50 |
| +3 | +2 | 56 |
| +1.15 | +0.9 | 81 |
| +2 | -0.6 | 26 |

interaction with sample atoms and therefore the indepth resolution of a SIMS profile depends on this impact energy for a given ion-source type.

Sample surface preparation is crucial for SIMS analysis quality due to the electrostatic field and possible charge migration induced by unbalanced primary-ion arrival and secondary-ion departure. A perfect sample is flat, polished, conductive or semiconductive. Insulators may be analysed if the surface is coated with a metallic layer (carbon, gold, *etc.*) and even then the use of a normal incidence electron flood gun may be necessary to compensate for charges.[11,12] It has to be noted that the primary beam is bent in the sample vicinity due to the local electrical field (Figure 15.8). The primary beam incidence angle therefore needs to be taken into account since surface roughness may be generated and/or amplified during a SIMS analysis.

Table 15.1 shows examples of incidence angle *versus* primary and secondary voltages setting.

### 15.2.5 Transfer Optics

The optic assembly between the sample surface and the spectrometer consists of electrostatic lenses and optical apertures. This assembly has two functions: (i) to match the sample surface secondary-ion emittance with the

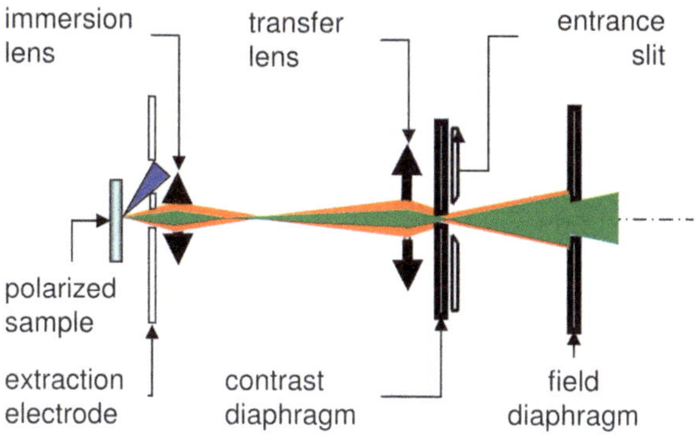


**Figure 15.9** The emittance matching optic from sample surface to spectrometer. (Source: © CAMECA.)

acceptance of the secondary ion spectrometer optics and (ii) to transfer a real image of sample surface secondary emission through the whole device (in the case of a *CAMECA IMS 7f*).

Figure 15.9 shows schematically this "transfer" optics. Coupling the immersion lens with the transfer lens produces:

(i) A real image of the emission crossover in the plane of the contrast aperture. The contrast aperture limits the ion beam angle due its energy dispersion and hence limits chromatic aberrations. The smaller the aperture, the better the ion image contrasts.
(ii) An image of the sample surface emission in the plane of the field aperture. The field aperture is an optical limiter of the analysed area. The ion beam angle, which emerges from this aperture, has to be optimised for a given spectrometer mass-resolving power.

The counterpart of better image quality and mass-resolving power is loss of secondary ion beam intensity.

### 15.2.6 The Spectrometer

A double-focus SIMS spectrometer consists of an electrostatic sector that can focus the secondary beam relative to the ion energy and a magnetic sector that can focus the secondary beam relative to the ion mass (Figure 15.10). With such a spectrometer, a mass-resolving power above $m/\Delta m$ 5000 at 10% of the maximum peak definition was reached in earlier 1977.[14]

The electrostatic sector is made of two spherical shape electrodes between which a potential is applied to bend the secondary beam at 90° for ions extracted at the sample voltage. This potential is usually optimised for an instrument and then fixed. Although sample surface potential is fixed, secondary ion emission is spread over a wide spectrum of energies: The role of

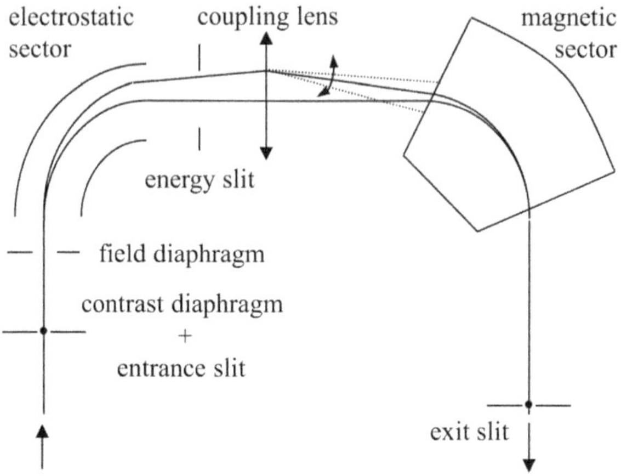


**Figure 15.10** Cancelling the energy dispersion with the ESA, the magnetic sector and the coupling lens.
(Source: © CAMECA.)

the electrostatic sector is to disperse the secondary beam relative to the ion energy (see Chapter 4).

The magnetic sector separates the ions as for a given sample voltage, the ion curvature radius depends on its mass to electric charge ratio. The magnetic field is controlled by a computer to select the mass to be analysed. But owing to the energy spectra of sample surface extracted ions, actual mass separation is *de facto* attenuated by aberrations. This needs to be improved and is discussed in the following paragraph: In Figure 15.10 the coupling lens controls the beam aperture angle of the ions, which emerge from the electrostatic sector. This lens adjusts the ion-trajectory aperture to match the magnet entrance in a way that the energy dispersion analysed by the electrostatic sector leads to a single crossover at the exit-slit plane: ions of the same mass to charge ratio converge at the crossover in the exit-slit plane.

The initial beam energy spectra is somewhat compensated for but not completely. The noncompensated part of the spectra may be reduced by adjusting the energy slit opening. By moving this slit off axis or by applying an offset to the accelerating voltage, the ion beam energy spectrum may be run through. It is a means for separating simple ions from isobaric cluster ions. Figure 15.11 shows an energy spectrum of these two types of ions. Perfect alignment of the spectrometer is obtained by adjusting the electrostatic sector potentials and the energy slit position in such a way that the ion beam trajectory coincides with the coupling lens axis for a given secondary-ion extraction voltage. In these conditions the image of the crossover in the exit-slit plane is the image of the crossover in the entrance slit.

### 15.2.7 The Projection Optics

The electrostatic projection lenses can focus on a virtual image of the field aperture (*i.e.* the sample surface) located in the magnetic sector, or on the

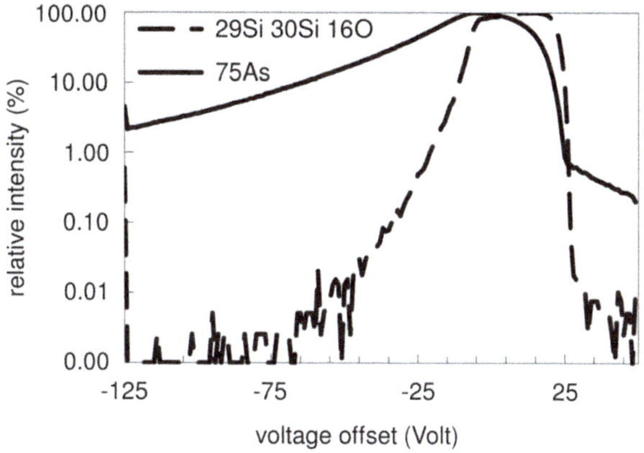


**Figure 15.11** Energy spectrum of a single ion ($^{75}As^+$) compared to a cluster ion ($^{29}Si^{30}Si^{16}O^+$). Sample voltage swings from −125 to +30 V around the secondary nominal voltage (*i.e.* +5 kV). The central energy window is set at about 30 eV *via* the energy-slit opening.[15]

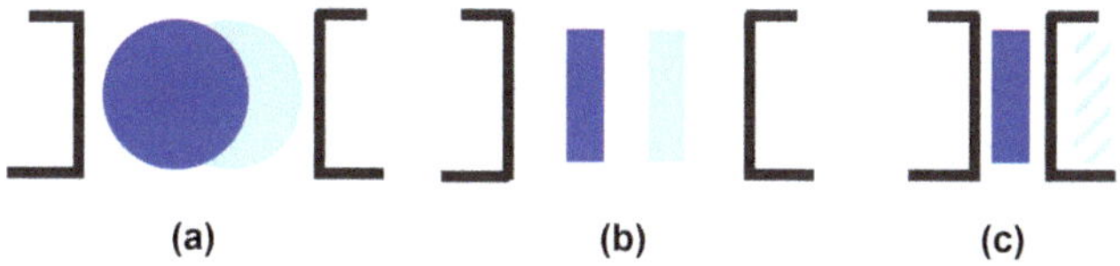


**Figure 15.12** Setting the mass-resolving power of the spectrometer (a) low mass resolution (b) separation of the ion beams by a closed entrance slit and (c) selection of the ion beam by an closed exit slit. (Source: Bernard Boyer after de Chambost *et al.*[15])

real image at the exit slit (the sample emission crossover). The projections optics is used to achieve the alignment of the ion beam described in all the previous paragraphs. After the spectrometer and the exit slit, the secondary ion beam is ideally constituted of only one isotope of one kind of chemical species. But actually, on the image of the contrast aperture, the ion beam of interest and potential isobaric interferences may be superimposed when the spectrometer is set in low mass resolution (Figure 15.12(a)). The separation of the beam of interest from its interference is accomplished by closing the entrance slit (Figure 15.12(b)). Then, the selection of the ion of interest is accomplished by closing the exit slit over the interference (Figure 15.12(c)) (see also Chapter 5).

## 15.2.8 The Detection System

The alignment of SIMS is accomplished by viewing the secondary beam image on the screen of a special ion/photon transducer, the channel plate (Figure 15.13). Every 10-μm size channel of about 500 μm length plays the

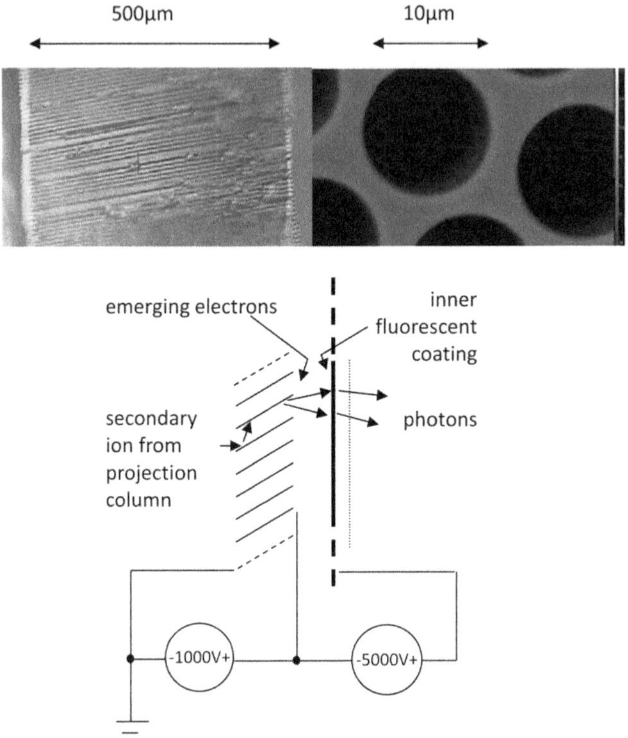


**Figure 15.13** Channel plate: photos of cross section and zoomed channel and schematic drawing of a channel circuit.
(Source: Bernard Boyer after de Chambost *et al.*[15])

role of a microcontinuous electron multiplier for each incoming ion, and the resulting electrons burst produces photons on the fluorescent screen, which hence shows a direct isotopic image of the sample surface or of the emission crossover. The transducer sensitivity can be enhanced in the multiple device stack mounting.

Once the alignment is accomplished, the secondary ion beam may be projected onto other types of detectors. The resistive anode encoder (RAE) uses a MCP or multiple MCPs stack to convert ions to electrons. The fluorescent screen is replaced by a position-sensitive charge detector. The coordinates of incoming ions are calculated, stored in a memory and therefore an ion image can be recorded. Usually, a RAE is a retractable assembly fitted on the ion trajectory just before the conventional MCP. Typical count rate for RAE is in the $10^4$ cps range.

The electron multiplier (EM) consists of multiple electrodes called dynodes, specially designed to produce secondary electron pulse signal under the impact of incoming particles (*e.g.* secondary ions). It is the most sensitive detector in use on the *IMS 7f*, it is fast (low dead time between pulses) and has extremely low background (typically $<0.001$ cps). The main drawbacks of an EM counting channel (*i.e.* the EM itself and the associated electronics) are:

i) Its inability of operation at high count rate. The maximum count rate depends on the type of EM and is $\ll 10^6$ cps.
ii) The dependence of its gain upon the types and energy of incoming ions (the so-called discrimination). However, this effect may be corrected thanks to an adequate data postprocessing.
iii) The yield drift (*i.e.* the variation of the ratio between output pulse rate and number of incoming ions due to the device natural aging).

Nevertheless, very precise measurement is allowed when a pulse height analyser (PHA) routine is used to determine the optimal high voltage for the EM and signal threshold for the associated electronics.

The Faraday cup (FC) consists of a hollow electrode and is charge sensitive. It is used when the secondary-ion intensity is higher than the EM count rate limit (ideally count rate should range between $10^4$ cps and $10^8$ cps). Modern electronics associated with FCs features temperature-stabilised charge or current amplifiers with low thermal noise. As a result, FCs counting channels are preferentially used for isotopic measurement, provided that high count rate is possible. Its main advantages are:

i) The absence of an ion-discrimination effect.
ii) Its longer lifetime and lower cost compared to EM.

### 15.2.9 The Vacuum System

Ultrahigh vacuum (UHV) is absolutely indispensable for SIMS analysis as various interferences may come from desorption off the instrument walls and/or sample. Thus, the pumping system has to be efficient for light gases ($N_2$, $H_2O$) present in the ambient atmosphere and unintentionally introduced in the analysis chamber during sample transportation through the airlock. These molecules tend to adsorb on the chamber walls and to desorb continuously. It is necessary to bake the entire instrument at least once a year during normal operation to accelerate desorption in order to keep a "clean" vacuum. The vacuum in the sources, the primary column(s), the airlock and sample chamber is achieved by turbomolecular pumps connected to a primary dry pump. The sources can be isolated by a valve for maintenance. The spectrometer is often pumped by ion pumps and can be isolated from the sample chamber by means of a valve. An efficient but temporary light-molecule pumping is performed by a Ti-sublimation pump. The resulting vacuum in the analytical chamber can get to below $7\times10^{-8}$ Pa.

# 15.3 Instrumentation

## 15.3.1 History of Magnetic SIMS Instrumentation

The history of SIMS can be traced back to 1949.[16] At that time, in Vienna, Austria, Professor R.F.K. Herzog (who had proposed the famous "Mattauch–Herzog" spectrometer a few years earlier) and his student, F.P. Viehböck constructed an instrument with the principle of using primary ions to spray a sample. Some of the atoms, which were sputtered off the sample were ionised and further accelerated into a mass spectrometer. This setup combining secondary ion emission and (magnetic sector field) mass spectrometry was actually the first published implementation of SIMS. At the turn of the 1960s, two teams independently achieved instrumental investigations, which led to marketed instruments that used this new analytical technique, named secondary ion mass spectrometry (SIMS): Castaing–Slodzian in France and Herzog-Liebl in USA.

Table 15.2 provides an overview of instruments marketed until 2013.

**Castaing, Slodzian and CAMECA (1959)**

It seems that in 1956, Professor Castaing and his student Slodzian were not aware of the 1949 Herzog paper and discovered secondary ion emission by chance while investigating secondary electron emission. Castaing is considered to be the father of EPMA (electron probe microanalysis), which he developed within the frame of his thesis at the turn of the 1950s. He seized the opportunity to make an ion microscope that could give quantitative

**Table 15.2** Marketed instruments (until 2013).

| Manufacturer | Model | Period of availability |
|---|---|---|
| GCA | *IMS 101* | 1963–1965 |
| ARL | *IMMA* | 1965–1978 |
| CAMECA | *SMI 300* | 1968–1977 |
| CAMECA | *IMS 3F* | 1977–1985 |
| ANUTECH/ASI | *Shrimp I* | 1982–2010 |
| CAMECA | *IMS 4f* | 1985–1991 |
| VG | *Ionex IX70S* | 1990–1993 |
| VG | *Isolab* | 1989–1990 |
| CAMECA | *IMS 5f* | 1992–1994 |
| ANUTECH/ASI | *Shrimp II* | since 1992 |
| CAMECA | *IMS 1270* | 1993–2003 |
| KRATOS | *S1030* | 1992–1993 |
| ANUTECH/ASI | *Shrimp RG* | 1995–2010 |
| CAMECA | *IMS 6f* | 1995–2002 |
| CAMECA | *NanoSIMS 50* | 1999–2004 |
| CAMECA | *IMS Wf/ IMS SC-Ultra* | since 2003 |
| CAMECA | *IMS 7f* | since 2003 |
| CAMECA | *IMS 1280* | 2004–2009 |
| CAMECA | *NanoSIMS 50L* | since 2005 |
| CAMECA | *IMS 1280-HR* | since 2009 |

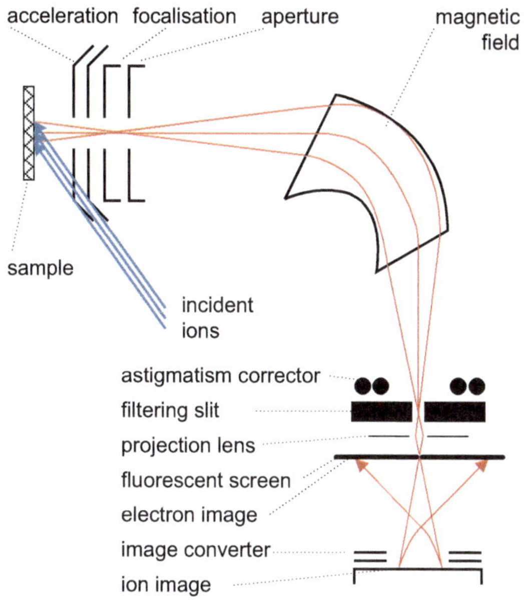


**Figure 15.14** Schematic of the Castaing–Slodzian ion microscope (1960). (Redrawn by Ondrej Hanousek after Castaing and Slodzian.[17])

information on the elemental chemical composition, as well as information about the electron microprobe, but which offered the possibility of producing directly a complete image of the sample. The apparatus of Slodzian's thesis consisted of a primary ion column, a sample holder immersed in an accelerating lens, a stigmatic magnet and a kind of Möllenstedt converter for converting the ion images – mass filtered – into an image displayed on a phosphorus screen (Figure 15.14). Beautiful images were published as early as 1960 and the complete "ion microscope" was presented at the 1962 display of the French Physical Society.[17]

Slodzian paid special attention to the so-called "extraction" lens and the matching of extraction conditions with the magnetic sector. The truly innovative component in the Castaing–Slodzian draft is the idea, that it is possible to take advantage of the magnet stigmatic properties for achieving microscopic imaging and to match the instrumental toolbox of electron microscopy (the electrostatic lenses) with the toolbox of mass spectrometry (magnets and electrostatic sectors).[14]

## Liebl, *IMS 101* and *IMMA* (1963–1968)

At the same time, Herzog teamed up with Liebl within the framework of the American company GCA (Geophysics Corp. of America) at Bedford near Boston, to develop an ion probe based on the principle of scanning microscopy. This project was funded by NASA.[2] Liebl and Herzog were at least one year behind the Castaing–Slodzian project. Their first publication presenting

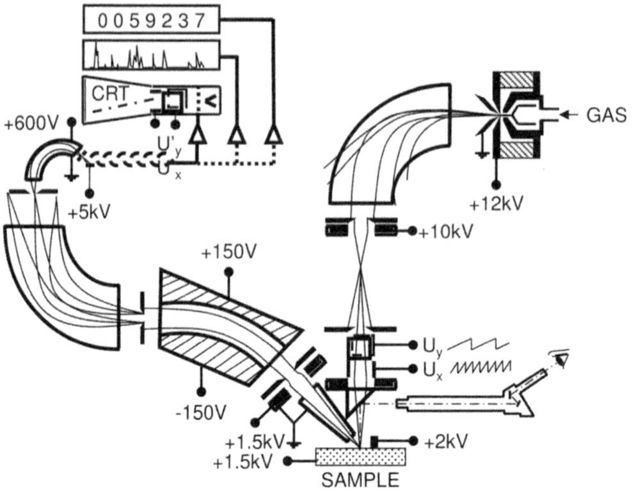


**Figure 15.15** Schematic of the ARL *IMMA* (1967).
(Source: © ARL.)

the *IMS101* dates back to 1963,[18] while the first images of Slodzian were obtained in 1960. Compared to the Castaing–Slodzian instrument, the *IMS101* had the big advantage that it included a double-focusing spectrometer using both a magnet and an electrostatic sector, while the French instrument was equipped with only a magnet. The primary ion source of the *IMS101* was a duoplasmatron, which was invented by Von Ardenne and published in 1956.[19]

Helmut Liebl defended his thesis in Munich in 1956 and joined the Geophysics Corp. of America (GCA) in 1959. Around 1964, Liebl left GCA for the Californian Applied Research Lab (ARL) and presented in 1967 a new instrument, the *IMMA* (Figure 15.15).[20] The capability of selecting the fastest ions (energy filtering) allowed Andersen and Hinthorne to solve with an *IMMA* the interference issue for measuring lead isotope ratios in zircons.[21] Afterwards, in 1968, Liebl returned to the Max-Planck Institut in Munich.[22]

### The *SMI 300* (1968)

The weakness of the first Castaing–Slodzian design was its lack of energy focusing. The energy was focused within the ion image plane but not within the filtering slit, leading to a very poor mass resolution. To overcome this deficiency, Castaing proposed a new design where the energy compensation was achieved by the so-called Castaing–Henry filter (Figure 15.16). It includes both a magnet and an electrostatic mirror. The mirror is a plane surface biased at the same voltage as the sample. Depending on the excitation of the magnet, only a given mass passes through the device. The more energetic ions are less deflected by the first part of the filter, but they are also less deflected in the second half.

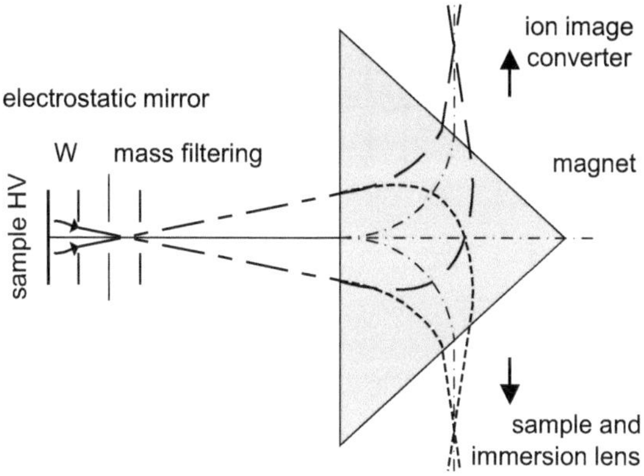


**Figure 15.16** The Castaing–Henry filter.
(Source: Ondrej Hanousek after an original © CAMECA courtesy of CAMECA.)

This device was inserted first into a vertical ion microscope, which was reproduced by CAMECA in 1965 and later also used in the *SMI 300*. It achieved energy focusing, which is not illustrated in the sketch (Figure 15.16). On the basis of this sketch, the French company CAMECA, which was already manufacturing the Castaing electron probe, developed the *SMI 300*, an energy-compensated magnetic SIMS presented in 1968.[23] 37 *SMI 300* instruments were produced from 1968 to 1978. The *SMI 300* was an ion microscope that was capable of transferring the entire ion image, whereas the Liebl instrument had only scanning imaging capabilities.

**CAMECA *IMS 3f* (1978)**

There was a fierce competition between CAMECA and ARL in the 1970s and this even included a patent war. At the beginning of the 1970s Cameca was not in an advantageous position in this competition because it was clear to the scientific community that the *SMI 300* was unnecessarily complicated. It turned out that the ion microscope structure of the instrument was far better suited for depth profiling than for the ion probe. Under the guidance of CAMECA CTO Rouberol, a more conventional double-focusing spectrometer, of the Nier–Johnson type, was designed and developed. Both the electrostatic sector analyser (ESA) and the magnet were stigmatic, implying that the ESA was of a spherical type. The *IMS 3f* was launched in 1977 and included a microchannel plate device which replaced the Möllenstedt converter for the microscope image capability (Figure 15.17).[6] The *IMS 3f* included a so-called transfer optics (emittance matching optics), as well, proposed by Slodzian and Figueras: a set of electrostatic lenses, which – in combination with extraction lenses – allows the operator to vary the ion microscope magnification.[24] Later, the instruments were readily equipped with a cesium source, in addition to the duoplasmatron.

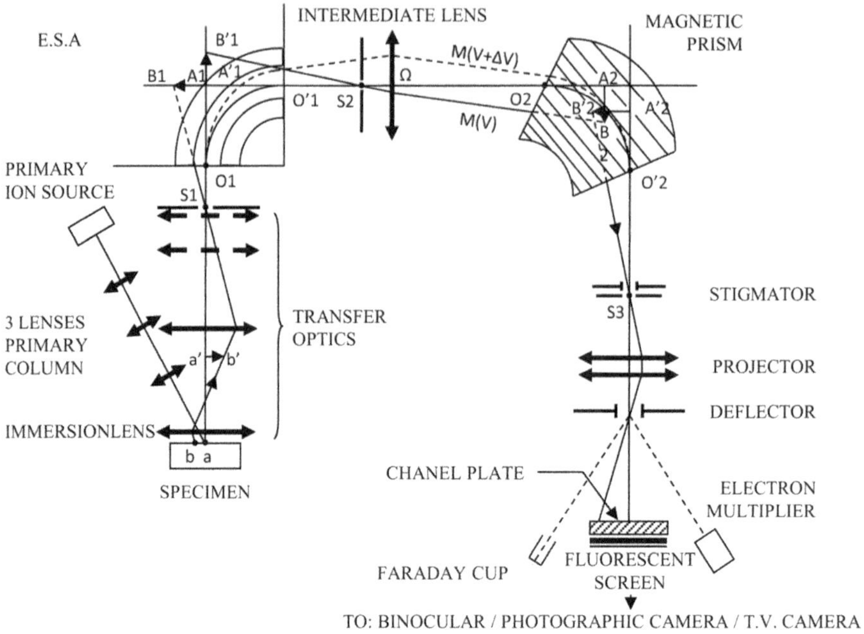


**Figure 15.17** Schematic of the IMS 3f (1977). (© CAMECA courtesy of CAMECA.)

Even with these improvements, the mass resolution of both the *SMI 300* and the *IMMA* could not exceed 3000. The *IMS 3f* could achieve a mass resolution of 5000 on a routine basis, thus enabling new applications such as phosphorus in silicon depth profiling or trace element and isotope ratio analyses in the field of geochemistry to be run.[25] The key component for meeting such a requirement was a magnet supply based on a Hall sensor, which could exhibit high stability and switching performance.[14]

Between 1978 and 1985, some 100 *IMS 3f* instruments were delivered around the world. CAMECA gained monopoly over the magnetic SIMS and ARL was eliminated as a competitor.

A normal electron gun (NEG) was designed by Slodzian for dealing with the issue of insulating samples.[11] It operated on the principle of flooding the samples with electrons in the axis of the accelerating lens. This technology was incorporated on the *IMS 4f*, which was the successor of the *IMS 3f* in 1985.[12]

**Large-Radius Probes: the *SHRIMP*, the *IMS 1270/1280/1280-HR* (1981–2013)**

Since the invention by Aston in the early 1920s, the mass spectrometer has been a powerful tool in geochemistry labs, well suited for measuring the chemical or isotopic abundances of geological material. Whenever a sample cannot be assumed to be homogeneous, a localised analysis is relevant, and the ion microprobe is the right tool to do the job. Such is the case of uranium-lead analysis used for dating zircons. Professor Bill Compston, from the Australian National University of Canberra, initiated the project of a dedicated SIMS instrument, the so-called SHRIMP (sensitive high-resolution ion microprobe) for running such analyses. Because of the very low concentration of lead and of the spurious peaks at the vicinity of the $^{206}Pb$ peak, the instrument was required to deliver both high mass resolution and high transmission rates. The key components for providing such performances were a very large magnet radius (1000 mm – to be compared to the 117 mm for the *IMS 3f*) and a spectrometer designed by Matsuda, which was free from second-order aberrations. The design was based on a cylindrical electrostatic sector, a quadrupole lens and some magic deflection angles for the electrostatic sector and the magnet (Figure 15.18).[26] The design and fabrication took seven years and the first analytical results were obtained in 1982. A large number of geochemists came to Canberra to analyse their samples in the SHRIMP, which led to the improvement and duplication of the instrument in 1992 for other Australian labs. SHRIMP was finally commercialised abroad in 1995.[27]

The SHRIMP created a market for large-radius SIMS instruments dedicated to geochronology in the Earth and space sciences. In the earliest 1990s CAMECA proposed for this market the *IMS 1270*, which featured a magnet of radius 585 mm and coarsely the same sketch of the ion microscope part as in the *IMS 3-4f*, with 90° angles, stigmatic ESA and magnet and the capability of transferring the sample ion image to a microchannel plate device. Reducing the second-order aberrations while keeping a 90° geometry was obtained by using hexapoles and introducing anamorphosis in the ion-image transfer. Furthermore, for this instrument, CAMECA developed a multicollector, which allowed the simultaneous measurement of Pb isotopes, but also

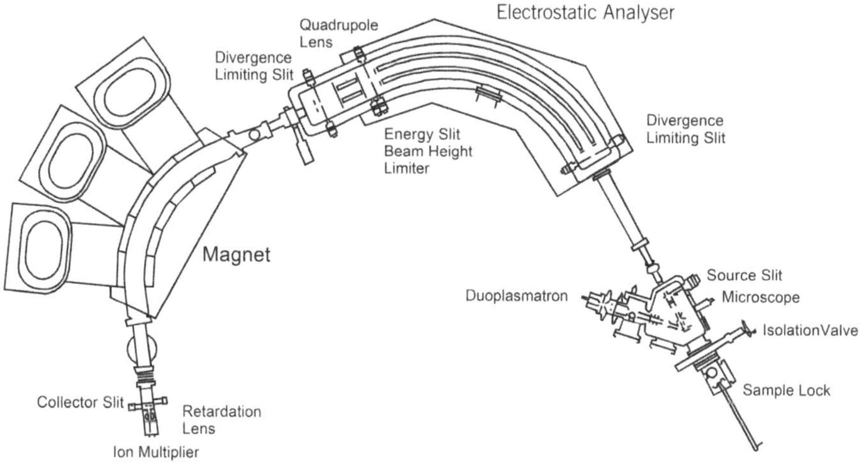


**Figure 15.18** Schematic of the *SHRIMP* (1977).
(Source: Trevor Ireland simplified version as published by Ireland *et al.*[28])

low-mass ion isotopes, down to lithium. This multicollector could be equipped with electron multipliers or Faraday cups. The normal electron gun made the *IMS 1270* well suited for stable isotope analysis. The instrument used oxygen ions and thus had positive primary ions and negative secondary ions. A reproducibility better than 0.2‰ was routinely obtained for oxygen isotope ratio analyses.[29] Second-order aberrations were completely eliminated by the successor of *IMS 1270*, the *IMS 1280-HR* proposed by CAMECA in 2009. This model enabled the implementation of applications requiring a mass resolution better than 25 000.[30]

In 2000, it was demonstrated that the large-radius SIMS was quite suitable for nuclear-safeguards applications, *i.e.* the analysis of micrometre-size particles from nuclear-handling facilities for the purpose of searching for nondeclared activities.[14] This application requires dedicated imaging software for particle detection and measurement.[31]

### High Spatial Resolution SIMS: the *NanoSIMS 50* (1999–2013)

It is known from the optics of charged particles that a short working distance is the key for reducing spherical aberrations of a lens. The fact that the sample is immersed within the secondary extraction lens, generally leads the designer to move the primary last lens away. In order to bypass this problem, in 1974, Liebl proposed a design where the same electrostatic objective lens was used for both primary and secondary ions (Figure 15.19).[22]

Some years later, the same coaxial principle was included in the ion microprobe project started in 1980 by Slodzian at the French aeronautic research institute ONERA. The setup included a Mattauch–Herzog spectrometer, which allowed the simultaneous measurement of different masses in the range of 1 to 20.[32] This project was eventually transferred to CAMECA and the first manufactured instrument was renamed to *NanoSIMS 50*. It was delivered in 1999 to the Harvard Medical School laboratory of professor Lechêne. With a primary ion probe as small as 50 nm, this instrument proved to be well suited for applications in cell biology and microbiology, but also for analysing small inclusions in extraterrestrial samples.[33]

The magnet supply developed by CAMECA for both the *NanoSIMS 50* and the *IMS 1280* at the turn of the 21st century combined a Hall-sensor control and a nuclear magnetic resonance (NMR) control.

### High Depth Resolution: The *IMS Wf* and the *SC Ultra* (2000–2013)

Around 1994, it appears that with the continuous reduction in the size of circuits, the junction depth was no longer in the range of a few hundred nanometres, but only a few tens of nanometres. These junctions that are very close to the surface are called "shallow junctions" or "ultrashallow junctions". In order to measure profiles of shallow junctions with good accuracy, it is important that the instrument provides a good depth resolution, and this – in turn – requires a low impact energy. The first successful experiments of low impact energy were achieved using a quadrupole SIMS in which the primary column is "floating" and where the primary ion impact energy

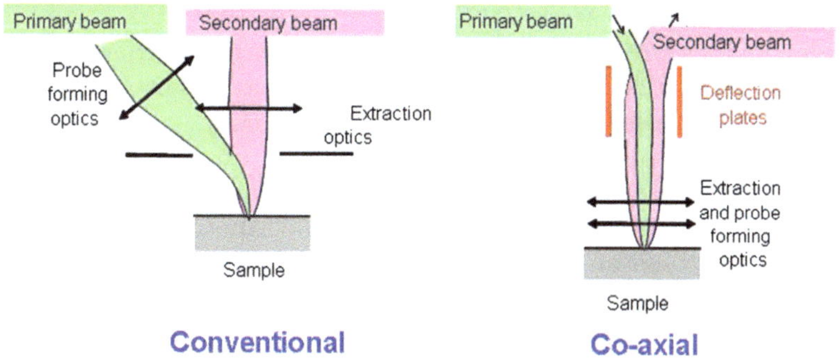


**Figure 15.19** Sketch of the *NanoSIMS* coaxial principle. (© CAMECA, courtesy of CAMECA.)

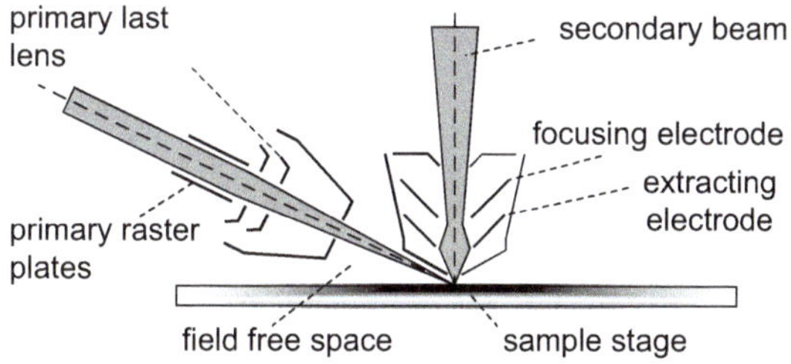


**Figure 15.20** CAMECA *IMS Wf* sampling region. (Source: Ondrej Hanousek after original © CAMECA, courtesy of CAMECA.)

was less than 500 eV.[34] In fact, a primary floating column is mandatory when primary and secondary ions are of the opposite polarity, which is the case in analysing phosphorus with primary $Cs^+$.

The CAMECA *IMS Wf* and *SC Ultra* were equipped with a primary floating column. The design improvements were concentrated mainly on the extraction region. The issue was to prevent the extraction field at the sample plane from strongly disturbing the low-energy incident ions. A very low extraction field leads to huge Recknagel aberrations, and therefore to a very poor transmission. The best compromise was to include in the extraction lens a shield electrode at the same voltage as the sample and the last primary lens electrode (Figure 15.20).[15] As the optimum incidence angle was not the same for cesium (60°) and for oxygen (<45°), both the *IMS Wf* and the *SC Ultra* were equipped with two columns where one was dedicated to cesium and the other to oxygen primary sources. As the *IMS Wf* accepts full 300 mm wafers, SIMS was introduced in semiconductor cleanrooms in 2002.[35] At the end of the 2000s, the so-called EXLIE-SIMS (extreme low impact energy - SIMS) was developed, and this instrument provided primary impact energies down to 100 eV.[7]

**SIMS with Accelerator: *MegaSIMS* (2005)**

In order to measure oxygen isotopes on the samples retrieved from the Genesis Mission, a team at the University of California, Los Angeles (UCLA) led by McKeegan designed and set up the *MegaSIMS*, an instrument combining half of a CAMECA *IMS 6f* a and a tandem accelerator mass spectrometer (AMS, see section 15.5.1 – Cosmochemistry). The first oxygen isotope ratio measurements were run in 2005.[36] Another SIMS-Accelerator instrument was built at the Naval Research Lab (NRL).[37]

## 15.3.2 Instruments Currently on the Market

***IMS 7f* (CAMECA)**

The CAMECA *IMS 7f* (Figure 15.21) was launched on the market in 2004. Its physical design is generally based on the proven *IMS 3*, *4*, *5*, *6f* models that were commercialised in 1978. As for its predecessors, the main selling point for this compact monocollector instrument is its versatility. Sometimes referred to as a universal SIMS, the *IMS 7f* was specified to address a variety of applications, which include depth profiling, trace element/isotopic microanalysis and imaging.

The *IMS 7f* is fitted with a duoplasmatron in addition to a cesium source. The primary column is fitted at an angle of 30° to the sample plane normal. The sample high voltage of the *IMS 7f* can be continuously adjusted between −10 to +10kV, pending on the desired secondary ion polarity. Since the source high voltage is also adjustable to some extent, it is theoretically possible to optimise both the primary ion incidence angle and impact energy. Practically, such optimisation is limited due to the loss of primary-beam density and secondary-beam transmission resulting from lower primary and secondary high voltage, respectively.

**Figure 15.21** *IMS 7f.*
(Source: Photo © CAMECA.)

The "accel–decel"[38] and "postacceleration" options were designed in order to overcome this limitation, which is critical for high depth resolution indepth analysis. The "accel–decel" option enables a deceleration of primary ions in the vicinity of the sample, whereas the duoplasmatron source is operated at high voltage. The "postacceleration" option enables an acceleration of secondary ions at the level of the electron multiplier first dynode. In order to further enhance its capability for high depth resolution, the *IMS 7f* can be fitted with a fully eucentric rotating stage, which minimises the surface roughening created through sample sputtering.

The latest developments in the scope of the *IMS 7f* were focused on additional kits of accessories. Two instrument submodels were designed (the *IMS 7f-GEO* in 2006 and the *IMS 7f-Auto* in 2012) in order to enhance the instrument capability for high-precision stable isotope microanalysis (*IMS 7f-GEO*) and high-throughput automated operation (*IMS 7f*-Auto).

For high-precision measurements performed in monocollection mode, it is critical to minimise the waiting times between signal integration spans and to monitor the parameters, which reflect short-term instrumental instabilities.[13] On the *IMS 7f-GEO*, the primary beam intensity is monitored quasicontinuously (every 80 ms) during the measurement enabled by a primary Faraday cup working in charge mode and a fast primary-beam blanking system. The *IMS 7f-GEO* also features a double Faraday-cup detection system. Alternatively, using two distinct detectors spares the waiting time component usually allocated for Faraday cup settling between two successive integration spans. In addition, a Hall probe was fitted onto the magnetic sector in order to minimise the waiting time needed to stabilise the exit beam after switching the magnet. The Hall probe measures the magnetic field in real time, its output signal is used to apply an electrostatic feedback to the magnet flight tube, such as to compensate the gap between the instantaneous and set point values of magnetic field.

The *IMS 7f-Auto* features a fully redesigned primary column (which provides easier tuning and enhanced primary beam stability) in addition to a fully motorised storage chamber, which enables automatic loading/unloading of sample holder to/from the analysis chamber. This latter option combined with a new capability for optical centring and chained analysis is an asset for applications requiring high analysis throughput with minimal or remote user's intervention.

### *IMS 6fR* (CAMECA)

The application of secondary ion mass spectrometry to the analysis of highly radioactive samples requires the implementation of specific devices to protect operators as well as critical parts of the instrument from contamination and irradiation effects. The first designs in this field were made at the Paul Scherrer Institute at the end of the 1970s on a quadrupole ATOMIKA instrument[39] before the introduction of a shielded sector field SIMS by CAMECA about 20 years later. Based on an *IMS 6f* scheme, this instrument was installed first at the Commissariat à l'Energie Atomique et aux Energies Renouvelables (CEA) Cadarache[40] and then at the Institute for Transuranium Elements (ITU) Karlsruhe[41] for the characterisation of irradiated nuclear fuels.

The main features of radiological protections used in shielded SIMS consist of a glove-box with handling equipment for sample transfer and loading into the machine, surrounded by a hot cell that encloses the complete introduction system and the specimen chamber. With 10-cm thick lead walls, such a protection enables to analyse nuclear samples with activity levels as high as 75 GBq with a resultant dose below 5 $\mu$Sv $h^{-1}$ at the operator position. It also limits the influence of $\beta$- and $\gamma$-rays on the detection system background (*i.e.* the electron multiplier used for quantitative measurements and the microchannel plate – fluorescent screen assembly, the resistive anode encoder and the video camera used for direct imaging of the sample).

Additional adaptations have also been made on the machine itself to minimise manual operations and to ensure safe interventions for maintenance and servicing. Major modifications made on the basic CAMECA *IMS 6fR* consist of the automation of instrument functions such as the procedure of sample introduction and the movement of slits and diaphragms on the primary and secondary columns, which are remotely controlled from the computer station. Special attention is also paid to dispose dedicated tools for inspection, cleaning and exchange of contaminated pieces.

### *IMS Wf/SC Ultra*

The CAMECA *IMS Wf* and *SC Ultra* (Figure 15.22) were marketed in 2000 to meet the semiconductor industry's new analytical needs facing the miniaturisation of their components. (*Wf* and *SC* stand for "Wafer" and "Semi-Conductor"), respectively. These two instruments are magnetic sector SIMS especially designed for depth profiling at low impact energy without compromising the mass-resolution power and the transmission.

Both the airlock chamber and the analysis chamber of the *IMS Wf* were sized to enable the analysis of whole 300 mm wafers. An automated system for the load of the wafers makes possible successive fully automated analysis. The *IMS Wf* is used in semiconductor cleanrooms close to production lines, whereas the *SC Ultra* is used in the research laboratory. The airlock chamber of the *SC Ultra* was designed to accept a holder for smaller samples.

The *IMS Wf* and the *SC Ultra* are fitted with two primary columns mounted at 60° and 36° with respect to the sample surface normal. The first column is a floating primary column equipped with a cesium source and/or an oxygen source. This floating column can be biased at any voltage allowing tuning the impact energy independently of the primary acceleration voltage and the sample voltage. In this way, the impact energy can be adjusted from 10 keV down to 150 eV in both secondary polarities (negative and positive). The second optional primary column is equipped with a duoplasmatron mounted at an angle of 36° (optimised configuration). Since 2010, the duoplasmatron source, which usually runs in the acceleration–deceleration

**Figure 15.22** *IMS Wf.*
(Source: Photo © CAMECA.)

mode on both columns, is replaced by a RF plasma ion source with a higher brightness. It is then possible to carry out low and ultralow impact energy depth profile analyses within short analysis times.

The immersion-lens region was specially designed to avoid the deviation of the low-energy primary ions in the extraction field. The analysis of non-conducting samples is assisted with an electron gun for the charge compensation.

Both instruments can be equipped with a eucentric rotating stage and an oxygen-flooding option to minimise the sample-surface roughness created by the primary ion impact. These features combined with the low impact energy capability allow achieving a depth resolution down to 0.7 nm/decade.

***IMS 1280-HR*** **(CAMECA)**

The *IMS 1280*-HR (Figure 15.23) is the successor of the *IMS 1270* and of the *IMS 1280*. This large-radius instrument is dedicated to applications that require a high transmission and a high reproducibility of isotope ratio measurements. It can be equipped with two ion sources – oxygen and cesium – that can be switched by means of a so-called primary beam magnet filter (PBMF). The primary column provides the sample with either a Gaussian beam formed by the demagnification of the ion source or a shaped beam formed by the demagnification of an aperture. The duoplasmatron source can provide both negative ($O^-$, $O_2^-$) and positive ($O_2^+$, $O^+$) primary ions.

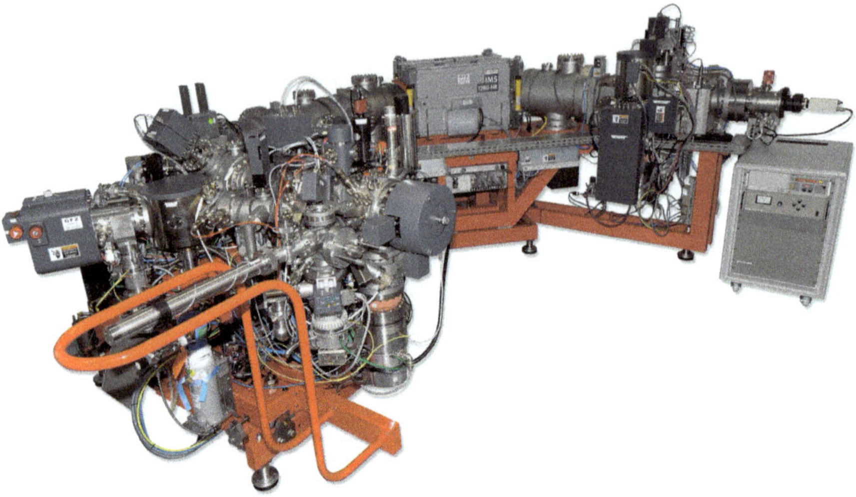

**Figure 15.23** CAMECA *IMS 1280-HR*.
(Source: © Elisabeth Glanz, Helmholtz Centre Potsdam.)

The sample is biased by the acceleration voltage (positive or negative) depending on the secondary ions polarity. The secondary ions are therefore accelerated to energies of 10 keV *via* a the voltage drop to a grounded extraction plate. The sample chamber is equipped with an oxygen-flooding leak, which enhances the secondary ion emission, especially for lead in the case of U–Pb dating applications. In the case of a negative secondary voltage, a normal electron gun (NEG) is used for charge compensation of insulating materials.

The *IMS 1280-HR* double-focusing spectrometer is stigmatic as is the *IMS 7f*. High transmission is achieved by a combination of a large-radius magnet (585 mm) and the cancellation of the full set of second-order aberrations obtained by using correction hexapoles and planar lenses located between the electrostatic sector and the magnet.[30,42] The sample image is transferred through the spectrometer and can be displayed at the end of the instrument on a microchannel plate device (MCP). A pair of lenses located between the immersion lens and the entrance slit can be used in a zoom mode for matching the sample emission with the acceptance of the spectrometer.

The *IMS 1280-HR* is equipped with a multicollector consisting of five moveable collector units.[43] Either an electron multiplier or a Faraday cup can be mounted onto each unit. The magnetic field of the electromagnet is controlled by both a Hall probe and multiple NMR (nuclear magnetic resonance) probes.

***NanoSIMS 50L* (CAMECA)**

The *NanoSIMS 50L* (Figure 15.24) was marketed in 2005 and succeeded the *NanoSIMS 50* commercialised in 1999. "*50*" is related to the lateral resolution of this type of SIMS instrument, which can be as low as 50 nm, whereas "*L*" stands for the large size of the magnet radius. These two machines are essentially dedicated to high-resolution SIMS imaging and

**Figure 15.24** CAMECA *NanoSIMS 50L*. (Source: Photo: © CAMECA.)

have a broad field of applications in microbiology and cell biology but also geology, space science and in material science.

The *NanoSIMS 50L* is equipped with two types of sources: a Cs – source and a duoplasmatron. Unlike the other instruments only $O^-$ ions selected through a Wien filter can continue their way into the primary column of the instrument when the oxygen source is used. Contrary to the *IMS 7f*, the *IMS 1280-HR* and the *SC-Ultra*, the *NanoSIMS* analyses are always carried out at the same impact energy of 16 keV ($\pm$8kV for primary accelerating voltage and $\mp$ 8kV for the sample high voltage).

The primary column was designed to achieve a reduction of the image of the source by a factor of 1000. The incident beam reaches the sample surface with normal incidence allowing reduction of the shadow effects commonly encountered with an oblique incidence for the measurement of a rough sample or a mutiphase material (*IMS 7f*, *IMS 1280-HR*). The trajectory of the primary ions is then coaxial with the secondary beam in the objective part forcing primary and secondary ions to be of different polarities. This implies the use of $Cs^+$ ions to investigate electronegative elements (*e.g.* H, C, O, S) and that of $O^-$ ions for electropositive elements (*e.g.* Na, K, Ca).

The immersion lens is used both to focus the primary ion beam and to collect the secondary ions. The distance between the sample surface and the optical assembly is decreased to 400 μm (instead of a few mm in the other SIMS instruments) to greatly reduce the optical aberrations. This system facilitates the tuning of intense small probes focused down to 50 nm. A high extraction field applied to the sample associated with a reduced system dispersion improves the collection efficiency of secondary ions. The secondary beam is then routed to the mass spectrometer with a minimum of energy dispersion. The spectrometer is a magnetic sector of Mattauch–Herzog geometry, wherein the mass spectrum of secondary ions is circulated in the focal plane of the magnet (radius of 680 mm). This configuration allows the use of a multicollection system composed of seven electron multiplier detectors, six of which are mobile in this focal plane. The short distance between the different detectors makes possible the simultaneous measurement of the three oxygen isotopes (16, 17 and 18) and the three silicon isotopes (28, 29 and 30). Up to the mass 58, the different isotopes of Cr, Ti and Fe can be measured during a single acquisition. Beyond the mass 58, two adjacent masses separated by one unit (for instance $^{60}$Ni and $^{61}$Ni) must be recorded during two acquisitions. The images produced with the *NanoSIMS 50L* can be recorded with a resolution from 16×16 pixels up to 2048×2048 pixels with a lateral resolution of 100 nm under standard conditions but can reach 50 nm when necessary.

***SHRIMP* (The Australian National University; Australian Scientific Instruments)**

The *SHRIMP I* design was based on the concept of a geological mass spectrometer and an ion microprobe: The primary beam is focused to a static spot on the target and the secondary ions that are generated are transferred to the mass spectrometer with as little beam loss as possible.[28]

The mass spectrometer was made as large as thought practically possible, with consideration of both manufacturing and cost. The *SHRIMP* magnet turning radius was designated at 1000 mm, and the rest of the Matsuda (1974) design scaled accordingly.

The primary beam was generated with a General Ionex duoplasmatron, with replacement of the hot filament with a cold cathode. The beam formation was produced by two main Einzel lenses demagnifying the extraction aperture that could produce a 10-μm spot on a target. This was subsequently modified to include a third Köhler Einzel lens to transfer the ion beam to an aperture located at the focal point of the condensor Einzel lens immediately before the target. This arrangement produces an analogue of the optical Köhler illumination,[27] whereby any aberrations in the ion beam arriving at the Köhler aperture are not transferred to the final lens system. This produces a spot with extremely even illumination that is the demagnified (factor of four) image of the Köhler aperture. A Wien filter was added to allow separation and selection of $O_2^-$ vs $O^-$ primary beams. The incidence angle of the primary beam on to the surface is 45°.

The *SHRIMP* primary column is kept at +10 kV for negative primary ion beams. A power supply for the duoplasmatron is kept on column ground but is held at −10 kV so that the duoplasmatron is actually close to real ground potential. This configuration was used when a positive primary – positive secondary configuration was required. The sample also sits at close to column ground with a small bias to accelerate the ion beam to the acceleration lens of the secondary column. The acceleration lens is held close to real ground potential to extract positive ions at 10 keV.

The acceleration gap produces a unity magnification beam transfer to the beam-matching system consisting of three slit Einzel lenses. These lenses shape the ion beam in terms of size and divergence so that the beam passing through the mass spectrometer is maximised. At the end of the source chamber is the source (entrance) slit of the mass analyser. In the secondary ion extraction column is a Schwarzschild mirror system allowing direct imaging of the sample during analysis. This optical system also configures the secondary ion system because when the sample is in focus, primary ion and secondary ion beam axes are also coincident at the target.

The mass analyser consists of the electrostatic analyser (1272 mm turning radius, 85° turning angle), a quadrupole lens at the end of the ESA chamber and the magnet (46° turning angle). The total beam length of the mass spectrometer is approximately 7 m. An energy crossover between the ESA and magnet can be used for energy selection. This design incorporates a quadrupole lens between electrostatic and magnetic sectors to minimise second-order aberrations,[44] with a magnification of 0.44 and a mass dispersion of $7.73\times10^5$. The original single-collector was replaced with a multiple collector chamber in 1985. This enabled Faraday-cup measurement as well as multiple ion counters (up to 8 channels). *SHRIMP I* was finally decommissioned in 2011.

*SHRIMP II* maintained the same mass analyser ion optic design but changed elements of the primary column and secondary extraction to increase performance. The primary column was configured to make Köhler illumination the prime method of primary beam formation and used a stronger defocusing condensor lens (factor of 7 demagnification), resulting in higher primary beam intensities. The slit-Einzel system in *SHRIMP I* was replaced by a quadrupole triplet that has lower aberrations and better beam transmission. *SHRIMP II* was fitted with a Cs gun and electron gun system in the mid-1990s. The Kimball Physics Cs gun was attached directly to the duoplasmatron extraction gap allowing a simple interchange. The Cs gun has a Cs zeolite charge and can be stored unevacuated and at room temperature when the duoplasmatron is used. An electron gun focuses an electron beam to the target with a final energy of around 1.2–1.5 keV and causes neutralisation of the charge build up associated with the Cs primary beam.

A multiple collector fitted in the mid 1990s allows three Faraday-cup channels and up to 5 ion-counter channels. An ion-counter triplet is used for unity mass spacing for the Pb isotopes and can be moved to unity mass spacings appropriate for Hf to Pu.

*SHRIMP IIe* was added as a commercial instrument by Australian Scientific Instruments in 2008 and features improved electronic control.

A reverse geometry variant of *SHRIMP* was produced in the mid-1990s, the *SHRIMP RG*. The design uses the same primary column – source chamber configuration as *SHRIMP II*, while the mass analyser is based on one of the ion optic solutions of Matsuda[45] for minimisation of third-order aberrations.

The mass analyser configuration has the magnet and ESA in reverse order as compared to the forward-geometry *SHRIMP II CQH* design. The reverse-geometry design also includes a chamber containing a quadrupole doublet that causes lateral spread in the ion beam but preserves the vertical profile (hence transmission) through the magnet (Figure 15.25). The magnet turning radius is 1000 mm through an angle of 46°. The ESA chamber contains the electrostatic plates (turning radius 751 mm, angle 88.65°), as well as an entry quadrupole lens and a larger projection lens at the exit. The magnification of the *SHRIMP RG* mass analyser is 0.32 and the dispersion is $3\times10^6$, which results in an approximately four-fold improvement in the mass-resolution capability compared to *SHRIMP II*.[46] The mass spectrometer only passes a very narrow mass range and hence operates only as a single collector.

*SHRIMP SI* is based on the *SHRIMP II* design with changes in the source chamber and in the multiple collector.[28] The source chamber has been machined out of a single piece of stainless steel in order to optimise the vacuum. A double vacuum lock also allows sample introduction and interchange without venting the whole interlock system. The primary column is configured using a 5-keV Cs gun as the only source. *SHRIMP SI* uses a column at ground potential with an immersion lens (asymmetric potentials) as the final probe-forming lens to bring the final primary ion energy to 15 keV. Both Köhler illumination and critical illumination are supported through three aperture Einzel lenses, each with a demagnification power of 10. Critical illumination allows spot sizes down to an aberration limit of 0.6 μm. The multiple-collector array was designed to enable the use of 3 Faraday cups or 3 large ETP multipliers at mass spacings up to 1/40, but has been reconfigured subsequently to allow multiple collection of all four S isotopes.[47] The development of charge-mode electrometers allows count rates down to 50 000 cps to be measured on the electrometer system.[48]

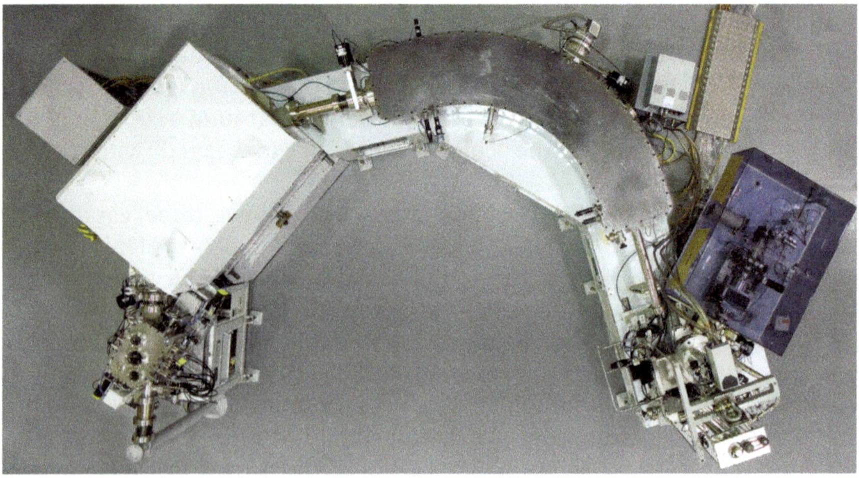

**Figure 15.25** *SHRIMP IIe* ion microprobe at Australian Scientific Instruments. (Source: Photo: courtesy of B. Ferguson, ANU.)

## 15.4 Measurement Considerations

Dealing with the nuts and bolts of SIMS depth-profile and trace element analysis in a variety of materials (mostly relevant to the field of microelectronics), one reference is of special note: Wilson *et al.*[49] Regarding the specificities of isotopic analyses (mostly relevant to geochemistry and cosmochemistry), the reader is invited to refer to Ireland (2004) and Ireland (2014).[47,50]

### 15.4.1 Mass Resolution Power

As described in Chapter 5, the mass resolution power is defined by the value of $m/\Delta m_{(x\%)}$, where $m$ is the mass of the analyte and $\Delta m_{(x\%)}$ is the full mass peak width at $x$% of the peak maximum (both expressed in mass units). In the SIMS literature, the mass-resolution power is usually specified at 50% or 10% and more occasionally at 1%. Under usual working conditions, the mass-resolution power is determined by the exit-slit width and the intensity profile of the secondary ion beam at the exit-slit plane (the so-called exit beam). This latter corresponds to the image of the entrance slit through the analyser plus the associated chromatic and aperture aberrations. Since most SIMS applications require flat-top mass peaks, it is common practice to set the entrance slit so that the exit beam width is at least half the exit-slit width. Under these conditions, the value of $m/\Delta m_{(50\%)}$ only depends on the exit-slit width according to:

$$m/\Delta m_{(50\%)} = K_{\mathrm{m}}/W_{\mathrm{exit}} \tag{15.1}$$

$m$: mass
$\Delta m_{(50\%)}$: full mass peak width at 50% of the peak maximum
$K_{\mathrm{m}}$: mass dispersion coefficient of the magnet in the exit-slit plane
$W_{\mathrm{exit}}$: width of the ext slit

The value of $m/\Delta m_{(x\%)}$, for $x < 50\%$, depends not only on the exit-slit width but also on the exit-beam profile. In addition to the entrance-slit width, this latter depends on a combination of parameters; namely the size of the field aperture, the size of the energy slit, and the optic transfer magnification. It has been demonstrated that an optimal trade-off between mass resolution and transmission can be achieved when the energy slit, the field aperture and the transfer optics magnification are set so that the spread of aberrations is comparable to the size of the entrance-slit image. Under these conditions, the exit-beam profile generally follows a truncated Gaussian distribution and the values of $m/\Delta m_{(10\%)}$, $m/\Delta m_{(1\%)}$, and $m/\Delta m_{(1‰)}$ approximately equals to 0.8, 0.7, and 0.65 times the value of $m/\Delta m_{(50\%)}$, respectively (assuming a flat-top peak configuration). For $x < 1‰$, the value of $m/\Delta m_{(x\%)}$ is also significantly influenced by the vacuum conditions prevailing in the coupling and collection optics (residual gas molecules being responsible for scattering the ion trajectories).

In most SIMS applications, small-radius sector instruments are operated at a $m/\Delta m_{(50\%)}$ that ranges between 300 and 10 000. The level of transmission obtained at a $m/\Delta m_{(50\%)}$ of 2000 and 5000 is approximately 10 and 50 times less than the full transmission (corresponding to a $m/\Delta m_{(50\%)}$ of 300). Due to their higher mass-dispersion coefficients (*e.g.* 1215 mm for the *CAMECA IMS 1280* against 250 mm for the *CAMECA IMS 7f*), large-radius sector instruments can be operated at a much higher $m/\Delta m_{(50\%)}$ (namely between 2000 and 50 000). The level of transmission obtained at a $m/\Delta m_{(50\%)}$ of 5000 and 10 000 is approximately 1.5 and 5 times less than the full transmission (corresponding to a $m/\Delta m_{(50\%)}$ of 2000), respectively.

## 15.4.2 Interferences

Besides monoatomic ions, the secondary ionisation process produces a tremendous variety of molecular ions that coming from all possible combinations between the atomic constituents inherent to the sample, the residual gas molecules, and the primary ion species, independently of their respective valencies. The potential for interferences is primarily determined by the configuration of the analyser in most SIMS applications. There are two types of methods for resolving interferences: (i) is based on the mass-resolution power, (ii) is based on energy filtering. In both cases, interference resolution can only be achieved at the expense of the transmission.

As a rule of thumb, a given interference is considered as resolved when the value of $m/\Delta m$ ($m$ being the mass of the analyte and $\Delta m$ the mass difference between the analyte and the interfering species) is similar to the mass-resolution power defined at half-maximum, $m/\Delta m_{(50\%)}$. Under these circumstances, the contribution of the interfering species to the signal is less than 1‰ (assuming identical peak maxima for the analyte and the interfering species). However, a less-conservative approach is also possible in order to minimise the loss of transmission. If an interference contribution of 1% and 1‰ is acceptable with respect to the other terms of the uncertainty budget, a mass-resolution power $m/\Delta m_{(50\%)}$ of approximately 0.7 to 0.75 times $m/\Delta m$ can be considered as sufficient.

In the energy-filtering method, the energy band pass is shifted to a higher range (by typically 80 eV) in order to enhance the selectivity towards monoatomic ions (see Section 15.1). In favourable cases, the relative contribution of the interfering species to the signal can be reduced by 4 orders of magnitude. Despite its selectivity, the energy-filtering method results in a transmission loss for monoatomic ions by a factor 50 (which corresponds to what is encountered at a $m/\Delta m_{(50\%)}$ of 5000 on a small radius sector instrument).

Given their enhanced mass resolution *vs.* transmission trade-off, the energy filtering is generally not a method of choice on large-radius sector instruments. However, it has been widely used for decades on small-radius instruments to resolve monoxide interferences from a variety of elements in a mass range > 60 amu (*e.g.* light rare-earth element monoxides *vs.* heavy rare-earth elements, $Si_2O$ *vs.* As). Dealing with non-oxide interfering dimers, the energy-filtering method is virtually more favourable in terms of transmission over a mass range from 60 to 120 amu.

However, this method turns out to be inefficient in resolving interferences related to some dimers that have a very low vibrational/rotational energy component (such as hydride ions). If mass resolution has been widely used to resolve hydride interferences in the low-mass range, it is practically out of reach in the high-mass range, even for large-radius sector instruments.

For an interference that cannot be resolved by any of the techniques mentioned above, it is common practice to strip its contribution from the peak of interest by monitoring a molecular ion of similar chemical composition at a different nominal mass (*e.g.* the contribution of $^{235}U^1H$ to the $^{236}U$ signal is usually estimated by monitoring $^{238}U^1H$).

In SIMS, the level of vacuum that is generally achieved at the level of the analyser results in very low abundance sensitivity. In uranium isotopic analysis, the contribution of the tail of the major isotope $^{238}U$ to the signal recorded at the mass of the minor isotope $^{236}U$ is typically a few hundreds of ppb.

## 15.4.3 Calibration

Due to the dependence of the isotopic and interelemental fractionations on the sample properties and the analytical conditions (see Section 15.1), the calibration of measurements is strictly subjected to the availability of an adequate set of external standards. Standards must share the chemical and physical properties of the samples and have to be analysed in exactly the same analytical conditions as the unknown sample. The range of SIMS applications is so wide in terms of analytes and sample composition, that the availability of certified reference materials is generally too limited to fulfil those requirements. Under these circumstances, the SIMS community has been developing and validating a wealth of synthetic and natural secondary standards (such as implants for the calibration of depth profiles or large and homogeneous minerals for elemental and isotopic microanalysis).

**Calibration of Trace-Element Analysis**

In most SIMS applications, the analysis of trace elements is based on a double – internal and external – calibration. An external standard is used to determine the relative response factors (usually called within the SIMS community, relative sensitivity factors, RSFs) for the different analytes with respect to an internal standard (the major element in the sample is usually chosen). The RSF is defined for an element a relative to an internal standard IS by the relationship:

$$\mathrm{RSF}_{\mathrm{a,IS}} = \frac{y_{\mathrm{a,std}}/x_{\mathrm{a,std}}}{y_{\mathrm{IS,std}}/x_{\mathrm{IS,std}}} \tag{15.2}$$

$y_{a,std}$ and $y_{IS,std}$ are the concentrations of the analyte and the internal standard that are known for the external standard, respectively. $x_{a,std}$ and $x_{IS,std}$ correspond to the signal measured on the external standard for the analyte and the internal standard, respectively.

The analyte to internal standard concentration ratio $y_{a,spl}/y_{IS,spl}$ is calculated from the signal ratios measured on the sample using the relationship (Eq. 15.3).

$$y_{a,spl}/y_{IS,spl} = \mathrm{RSF}_{a,IS} \times x_{a,spl}/x_{IS,spl} \quad (15.3)$$

$x_{a,spl}/x_{IS,spl}$: analyte to internal standard signal ratio measured on the sample.

An absolute quantification of trace elements is accessible using this one-step calibration method, provided that the absolute concentration of the internal standard in the sample is known *a priori* (as in the case of simple stoichiometric matrices) or when it can be determined using an independent microanalysis method (such as the electron microprobe analysis or the micro-Fourier transform infrared microscopy).

The calibration method based on RSF is generally applied for the determination of impurities in concentration up to 1%. Above this range, the chemical composition of the sample is not considered as matching that of the external standard (where the analyte is present at trace level) and potential matrix effects are likely to compromise the accuracy of measurements (see Section 15.4).

**Calibration of Isotopic Analysis**

SIMS isotopic analysis is generally calibrated using a set of external standards spanning a range in isotopic composition that covers the isotopic variability expected in the samples. For a given isotopic ratio $y_1/y_2$, the instrumental mass fractionation (IMF) factor $\alpha_{inst}$ is determined by the relationship:

$$\alpha_{inst} = \frac{(x_1/x_2)_{std}}{(y_1/y_2)_{std}} \quad (15.4)$$

$(x_1/x_2)_{std}$: signal ratio measured on the external standard
$(y_1/y_2)_{std}$: true value (*i.e.* the certified value or a value determined on the standard using an independent method)
$\alpha_{inst}$: instrumental mass fractionation

On samples, the corrected isotopic ratios $(y_1/y_2)_{spl}$ are calculated from the measured signal ratios $(x_1/x_2)_{spl}$ using the relationship:

$$(y_1/y_2)_{spl} = (x_1/x_2)_{spl}/\alpha_{inst} \quad (15.5)$$

The uncertainty related to the calibration is captured by the variability of $\alpha_{inst}$ values over the set of standards. In the scope of high-precision measurements (in the sub-‰ range), the variability of $\alpha_{inst}$ can virtually be minimised thanks to a very strict reproduction of the analytical conditions from analysis to analysis and the assessment of potential matrix effects (see Section 15.4.4).

### 15.4.4 Matrix Effects

As mentioned in Section 15.1, variations in the sample chemical composition are likely to cause variations in ionisation rates that are not uniform between the different elements or isotopes of a given element. The resulting variations in RSFs and IMF factors are referred to as "matrix effect" in the SIMS literature. Strong matrix effects are usually related to differences in the concentration of ionisation-enhancing species, like oxygen and halogens in the case of positive ionisation or alkali metals in the case of negative ionisation. As the mechanism for secondary ionisation itself, the mechanism for matrix effects is still a matter of debate. One of the accepted models suggests that differences in the sample composition are primarily responsible for differences in sputter yields. As sputter yield controls the concentration of implanted primary ions at the active sample surface, this may alter in turn the ionisation rates for the different elements or isotopes.

Matrix effects have been investigated for a range of SIMS applications, where variations in the sample chemical composition are commonly encountered from one analysis point to another (in microanalysis) or between the successive sample layers (in depth profiling). Most challenging cases include the analysis of trace elements in nonstoichiometric matrices (like $Al_xGa_{1-x}As$ and $Al_xGa_{1-x}N$ materials) or isotopic analysis in samples that represent a compositional continuum (such as glasses and minerals forming solid-solution series).

As matrix effects represent a limitation for the accuracy of SIMS measurements, it is common practice to adapt the calibration procedures described above, so as to calibrate the RSFs and IMF factors as a function of the sample composition. The main challenge is to establish an empirical relationship between the RSFs or IMF factors and one variable characteristics of the compositional property responsible for the matrix effect, using a set of external standards covering the compositional range expected in the samples. Provided that the empirical variable is monitored on the samples at the same location and with comparable lateral or depth resolution as the analytes, accurate RSFs and IMF factors can be determined despite sample chemical heterogeneity. In microanalysis, the empirical variable used for the determination of RSFs and IMF factors is usually based on offline measurements of major and minor sample constituents (such as EPMA). In depth profiling, the empirical variable is usually based on SIMS analysis and is simultaneously monitored together with the analyte signals as the successive sample layers are sputtered away (*e.g.* for the determination of trace impurities in $Al_xGa_{1-x}N$ matrices, RSFs need to be calibrated against $AlN^-/GaN^-$ signal ratios).

In some specific applications, it is possible to minimise the amplitude of matrix effects. Under $Cs^+$ bombardment, the determination of minor and major constituents in complex matrices appears to be much less sensitive to matrix effects when $MCs^+$ and $MCs_2{}^+$ secondary ions are measured (M being a sample constituent and Cs a primary ion) instead of $M^-$ secondary ions.[51] It has been proposed that $MCs^+$ and $MCs_2{}^+$ result from the combination above the sample surface between neutrals and primary ions, whereas the production of $M^-$ ions would be concurrent with the sputtering event. Under oxygen bombardment, all measures favouring the complete oxidation of the emitting surface have been shown to minimise the matrix effects related to initial differences in oxidation states. This can be achieved by using a microleak of gaseous dioxygen above the sample surface or by decreasing the incidence angle of primary ions with respect to the surface normal. Another approach to minimise matrix effects is to shift the energy bandpass for secondary ions to a higher range, since secondary ions leaving the sample surface with high initial kinetic energy are generally less affected by matrix effects compared to low-energy ions.

## 15.5 Applications

As for the Castaing electron probe, the earliest SIMS instruments found their first applications in the field of metallurgy, but some geological applications were achieved with the ARL IMMA. The first analysis of lead in zircon was conducted by Andersen in 1969.[52] These analyses were not yet performed at high resolution. Since the earliest 1970s, a new application enhanced the SIMS manufacturing: measurement of dopant implantation profiles in semiconductors, or, in other words, depth profiling. SIMS is widely used for elemental and isotopic analysis for most of the elements of the PSE (Figures 15.26(a) and (b)).

### 15.5.1 Applications in Geochemistry and Cosmochemistry

SIMS is one of the most powerful methods available to the analytical geochemist. For more detailed discussions, the reader is referred to publications by Hinton and Fayek.[53,54] For more recent reviews of trends in SIMS applications in the geosciences the reader is referred to Wiedenbeck[55] and Wiedenbeck *et al.*[56] A recent publication by Hoppe *et al.*[33] reviews the capabilities of *NanoSIMS* in the fields of biogeoscience and cosmochemistry.

**Geochemistry**

A key feature of analytical geochemistry is that minerals and natural glasses tend to be highly complex, commonly containing most elements at very varying concentration. As *in situ* measurement by SIMS eliminates the need for any pretreatment or chemical separation, the analyst must commonly deal with highly complex spectra, which commonly contain multiple isobaric species derived from molecular and/or multiply charged ions. A second consideration when analysing geomaterials by SIMS is that ion yields are strongly influenced by the chemical composition of the

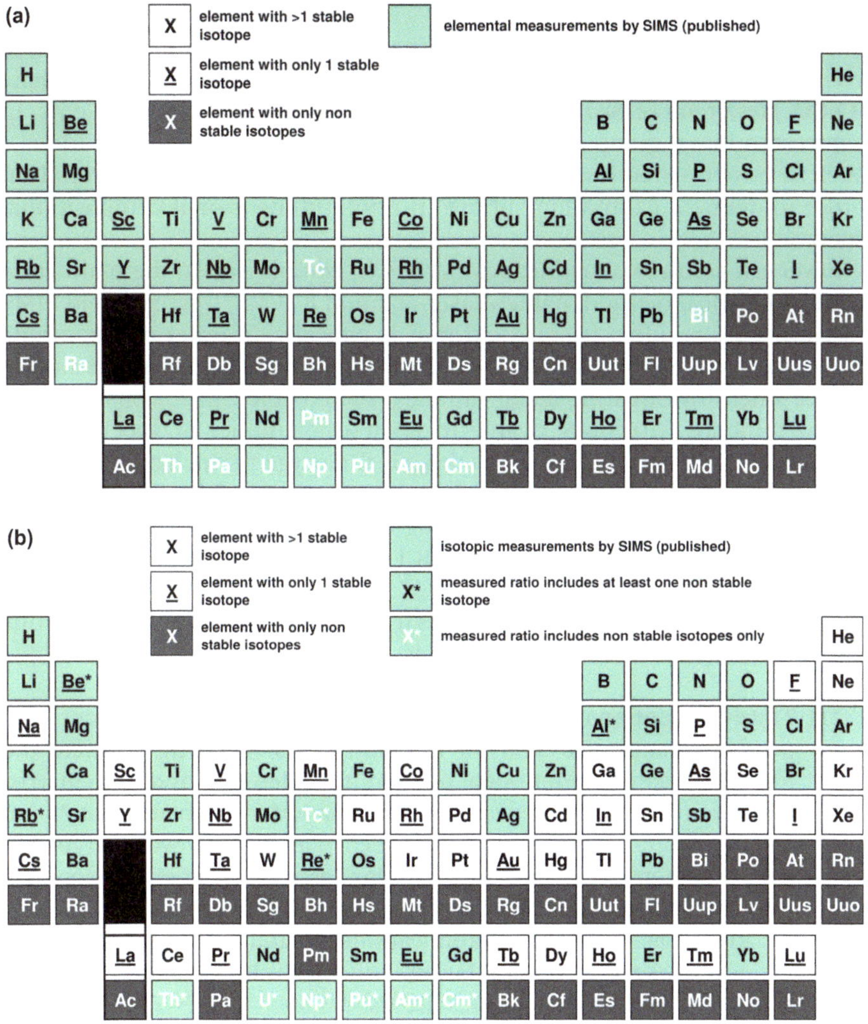


**Figure 15.26** (a) Periodic table of the elements: elements for which elemental data acquired by SIMS has been published (highlighted in ocean green) – also included if used, *e.g.* for interference corrections or internal normalisation. (b) Periodic table of the elements: elements for which isotope ratio data acquired by SIMS has been published (highlighted in ocean green) – also included if measured as interelemental normalisation standard or in IDMS analysis.

matrix being analysed. Ion yields can vary significantly as a function of the chemistry of the sample, such that reliable data can only be obtained when ion yields are calibrated using reference materials (RM) that are closely matched to the samples under investigation. This prerequisite for matrix matching between RM and sample applies to trace element[57,58] as well as isotopic analyses.[59,60] The requirement for matrix matching is a serious challenge for SIMS in the geosciences due to the large number of different minerals. In general, those mineral systems, which have little or no variations in their major element compositions, such as quartz ($SiO_2$), calcite ($CaCO_3$) or zircon ($ZrSiO_4$), present the least severe calibration problems. In contrast, mineral groups that accept multiple solid solution substitutions – such as garnets – may necessitate the use of numerous reference materials in order to define a calibration surface in a multidimensional compositional space.[61]

SIMS can claim two distinct advantages in comparison to the main competing analytical method, namely laser-ablation ICP-MS: (i) sector field SIMS instruments provide a useful yield that is typically an order of magnitude better than laser-ablation systems. This means that SIMS is the method of choice when investigating small samples, such as melt inclusions or finely zoned minerals. (ii) SIMS is able to analyse a number of geochemically critical elements, including both hydrogen and oxygen, which are difficult or impossible to assess with laser-ablation ICP-MS. In contrast, LA-ICP-MS has the advantages of high sample throughput, less pronounced matrix dependence and ease of use, making LA-ICP-MS the method of choice in many geochemical investigations.

In the case of dynamic SIMS there are five general categories for which this technology is applied in the geosciences:

(1) Trace-element analysis. In general, sector field SIMS instruments can provide quantitative data for concentrations $<100$ ng/g, where the limit of quantification depends on the element and the matrix. Modern instruments provide a spatial resolution of about 10 to 25 μm for such studies. Data quality and speed of analysis between large- and small-geometry machines are similar, especially when the energy-filtering method is applied for interference separation. Over the past decade many trace-element applications previously addressed by SIMS are now pursued by laser ablation, although if high spatial resolution is required ion probe facilities are used.

(2) Isotopic analysis. There are numerous isotopic systems that provide crucial records of geochemical fluxes at both the local and global scales. Some elements show broad variations in nature (*e.g.* hydrogen, boron, carbon, sulfur), that total analytical uncertainties in the order of $\pm 1‰$ are sufficient to address important topics in fields such as ore petrology, the tracing of fluids or deep Earth chemical cycling.[62] Such analytical uncertainties are within the reach of small-geometry SIMS instruments that have been available for the past three decades. Other key geochemical systems – and in particular the determination of the $\delta^{18}O$ values – require much better analytical uncertainties, which can only be provided by the large-geometry instruments. Oxygen isotope data were reported with uncertainties of better than $\pm 0.2‰$ (1 SD) from both *1280*[63] and *SHRIMP*[64] instruments. With a small sampling diameter of $<10$ μm and rapid sample throughput

of up to 200 analyses per day, SIMS has become an essential tool for many geochemical investigations ranging from volcanology to climate research.

(3) Age determination. The development of SIMS-based U–Th–Pb age determinations of the mineral zircon marked a key milestone in analytical geochemistry, obviating the need to dissolve this refractory mineral in extremely powerful acids. Such *in situ* dating of minerals only became possible with the advent of the large-geometry instruments.[65] Although the reported age uncertainties of SIMS data remain larger than those currently reported by wet chemical techniques, the high spatial resolution and speed of analyses mean that SIMS is an important tool for geochronologists. In contrast, studies requiring large volumes of data – such as the study of broad age populations in clastic sediments – are often better suited to laser-ablation ICP-MS analyses. U–Pb zircon geochronology is difficult to pursue with a small-geometry instrument, where only samples with ages in excess of several billion years can be readily assessed.[66]

(4) Depth profiling. Although depth profiling is the most commonly used SIMS mode in material sciences, it has found only limited use in geochemical research.[67–70]

As depth resolution is generally limited only by sample surface roughness, diffusion studies based on isotopically or elementally doped films deposited on highly polished crystalline substrates are well suited for SIMS. One of the major challenges of this strategy can be increased surface roughness resulting from the diffusion experiment run at high temperatures and pressures, necessitating the need to introduce a roughness correction to the resulting profiles.[71] Depth profiling has also been used in conjunction with ion-implant reference materials where no other means of calibrating relative ion yields were possible.[72]

(5) Imaging and 3D ion tomography. Although potentially very powerful, imaging and 3D tomography have found only limited application in geochemistry until now.[73] This is of special note that excellent imaging resolution provided by the *NanoSIMS* instrument is yet to be widely taken up by the geochemical community.[74]

**Cosmochemistry**

Magnetic sector SIMS and cosmochemistry have been engaged in a half-century long love story ever since the invention of SIMS in the 1960s. One of the first two SIMS prototypes was built at the Applied Research Laboratories under the joint sponsorship of NASA and GCA Corp. for the specific analysis of the lunar rocks returned by the Apollo 11 mission in 1969. In those early days U, Th, Pb and REE abundances were measured in the Apollo and Luna samples, notably with the aim to determine the age of the lunar crust.[21] The first generation of magnetic sector SIMS from CAMECA was used to study extraterrestrial rocks from the Moon and from the fall of the Allende meteorite all along the 1970s, which allowed a milestone discovery in early solar system science: the presence of extinct $^{26}$Al in the oldest solar system rocks, the so-called refractory inclusions from Allende.[75] This discovery, which is related to the galactic context of solar system formation is still far from being completely understood. It was conducted with the *PANURGE IMS 3f* just installed at the California Institute of Technology.

Noble-gas isotopes have revealed since the early 1960s that an interstellar component was kept hidden in primitive meteorites. After a search that lasted 25 years, the almost simultaneous discovery of interstellar nm-sized diamonds and µm-sized silicon carbides by a joint team of scientists from the Chicago and Saint Louis universities, was made possible by the *IMS 3f* analysis of individual silicon carbide grains in residues extracted from meteorites after a harsh acid treatment.[76] This discovery led the Saint Louis SIMS group to influence the development of the *NanoSIMS* instrument by Georges Slodzian and CAMECA by continuously pushing and demanding for better precision and lateral resolution. Since the commercial release of the *NanoSIMS* in the late 1990s, an increasing variety of submicrometre interstellar materials formed before the birth of the solar system have been discovered in primitive meteorites. The exotic isotopic compositions of these presolar grains record the nuclear reactions that took place in the interior of stars older than the solar system. This notably includes presolar silicates, which were previously destroyed by acid treatment and only discovered by systematic isotopic imaging campaigns with µm or sub-µm spatial resolution.[77,78] Amongst the most primordial solar-system materials, isotopic imaging revealed the existence of extreme excesses in D and $^{15}$N in µm-sized domains within interplanetary dust particles and organic matter extracted from primitive meteorites for which an interstellar origin previously accepted is now under debate.[79,80] Apart from searching the needle in the haystack, magnetic sector SIMS has been constantly used to study meteorites and an important improvement in analysis came in the 1990s with the release of large sector instruments such as the CAMECA *IMS 1270*. These instruments are now routinely used to study extraterrestrial rocks and to characterise their trace element and isotopic heterogeneities at the micrometre scale with precision that reaches a fraction of a per mil. Amongst key discoveries, one should note several extinct radionuclides, $^{26}$Al but also $^{10}$Be,[81] and $^{60}$Fe,[82] – discovered by magnetic sector SIMS. Finally – because of its high spatial resolution, sensitivity and versatility in analytical modes (*i.e.* high mass resolution, imaging, depth profiling) – magnetic sector SIMS has become the number one technique for the isotopic analysis of scarce and precious samples returned by space missions (*e.g.* Stardust mission at comet Wild2,[83] Hayabusa mission at asteroid Itokawa.[84] It demonstrated that comets were not primarily made of interstellar dust but were rather typical solar-system objects related to the carbonaceous chondrite meteorites. It also allowed the isotopic characterisation of solar wind implanted in lunar soils and in collectors from the Genesis mission.[85,86] In that respect, perhaps

**Figure 15.27** UCLA *MegaSIMS*. Distortion in the picture is due to the size of the instrument. Chairs are shown for scale (circle). The main parts of the instruments are shown by arrows and are listed following the path of secondary ions. *IMS6F*: body of the Cameca *IMS6F* with $Cs^+$ source, primary beam mass filter, sample chamber and introduction system, transfer optics, normal incidence electron gun, field aperture and entrance slit. *IR*: Isotope Recombinator to select all three oxygen isotopes. *TA*: Tandem Accelerator. *ES*: Electrostatic Sector. *MS*: 1.25 m radius Magnetic Sector. *FP*: Focal Plane, three electron multipliers are placed along the focal plane in the multicollector system. (Source: © Kevin McKeegan, UCLA.)

the most important result of the last 10 years is the measurement of the oxygen isotopic composition of the Sun by McKeegan and coworkers, which was conducted using their University of California at Los Angeles home-made ultralarge instrument, the *MegaSIMS*, which combines a *CAMECA IMS 6F* with a tandem accelerator and a ~2 m multicollector system placed after a 1.25 m radius magnet for the analysis of secondary ions in the MeV range (Figure 15.27). The tandem accelerator can eliminate the gigantic OH interference at mass 17 and is able to carefully analyse a handful of $^{17}O$ atoms. The most recent cosmochemical advance, which uses magnetic sector SIMS as a leading technique, is the study of trace amount of water on the Moon and its isotopic composition. Many first-order publications were produced on this topic in the last 2 years, which point to (i) an elevated water content in the early Moon, which is difficult to reconcile with the giant impact hypothesis for the formation of the Moon and (ii) an early delivery most likely by carbonaceous chondrite impactors.[87]

### 15.5.2 Applications in Material Science

In the years since its development, the SIMS technique has been applied so frequently in the field of materials science that it is impossible to draw up an exhaustive list of its use in the following brief presentation. Only selected applications related to some of the main characteristics of the technique are presented in here. In particular, the application in microelectronics and nuclear sciences – whose development has undeniably imposed the successive technical improvements of SIMS – are further described. Further information can also be found in different review papers.[88,89] In the future it seems that magnetic SIMS instruments will continue to play an important role in characterisation for microelectronics even if some application will necessarily be taken over by other techniques such as atom probe or SSRM for measuring 3D doping distributions.

Materials science investigates the relationships between the microstructure of materials and their macroscopic properties. An element present in a material – even in a very low concentration – can drastically modify its properties according to the way it is distributed within its microstructure. For instance, the addition of boron in steel in amounts of tens of parts per million is enough to increase the hardenability of this material if boron is present in a solid solution. Nevertheless, the precipitation of boron into carbides hinders this positive effect. By combining high sensitivity and high lateral resolution imaging capability, the SIMS technique offers the possibility of investigating the distribution of the two light elements, boron and carbon, in steel,[90] whilst the use of electron-based techniques remains very limited due to poor detection limits. It is therefore possible to explain and control the steel microstructure through SIMS measurements. In the particular case of a trace-element measurement, the relation between SIMS intensities and the concentrations of a given element is linear, making the SIMS analysis semiquantitative and even quantitative provided that the relative sensitivity factor can be determined.[89] This is partly illustrated by Christien *et al.*[91] who have recently shown the feasibility of an accurate quantitative analysis of equilibrium grain boundary with nanoanalysis SIMS.

Among the light elements, hydrogen plays a significant role in the corrosion mechanisms of materials, thus it is primordial for this role to be identified. Because few techniques are able to measure hydrogen, the measurement using SIMS is of particular interest for studies related for instance to the corrosion of glasses or hydrogen embrittlement of steels. Using SIMS depth profiles of hydrogen and the Fick's law of diffusion, diffusion coefficients of water were determined in buried medieval stained glasses that had been altered by natural weathering conditions for hundreds of years.[92] Thus, new insights were, *e.g.* gained for predicting the durability of nuclear-waste glasses. Similar SIMS data was also successfully obtained for evaluating hydrogen concentrations around the fatigue crack tip and estimating a hydrogen apparent diffusion coefficient in stainless steels.[93]

Another strength of SIMS applied to materials science is its capability to analyse isotopes for studying transport processes in solids using isotopic tracing. The method employs one of the stable minor isotopes of the element of interest (similar atomic radii and chemical properties) as tracer. For instance, oxidation studies of materials are two-stage experiments. The first stage consists of the production of an oxide layer at the surface of the material. It takes place in a media containing the three oxygen isotopes $^{16}O$, $^{17}O$ and $^{18}O$ according to their respective natural abundances 99.76, 0.04 and 0.20%. During the second step, the diffusion of oxygen through the oxide layer formed is visible by artificially increasing the concentration of $^{18}O$ ($^{17}O$ is also possible but more expensive) in the media where the sample is corroded. Then, the diffusion coefficient of $^{18}O$ can be deduced from SIMS depth profiles giving the $^{18}O$ intensities as a function of the depth in the oxide layer by using Fick's laws of diffusion. In this kind of experiment, it is possible to differentiate between bulk and grain-boundary diffusion in polycrystalline materials.[94] Additionally, isotopic tracing allows us to better understand corrosion or ageing mechanisms.[88,95]

The development of new materials has often relied on SIMS characterisations. For example, the latest generations of thin film solar cells (*e.g.* CIGS, copper indium gallium selenide) have been analysed with SIMS in order to understand how the composition of layers affects the performances of the cells.[96] In particular, SIMS enables the assessment of the concentration gradients of the constituents, the study of the possible elemental diffusion between layers, the identification of contaminations or trace elements with excellent detection limits (from ppm down to ppb depending on the element investigated) or the highlighting of the reaction mechanism involved in the growth process of layers.

**Microelectronics**

SIMS has been used for over 35 years for analysis related to microelectronics.[97] It has adapted to the reduction in dimensions by going to increasingly low impact energies – a challenge that is still present. SIMS must now cope with new challenges: the diversity of materials used today and the development of 3D devices. In a broader sense SIMS must also meet challenges in display technology, memory applications and solar-cell technology posing similar challenges. This chapter can only hope to scratch the surface of magnetic SIMS applications in microelectronics but aims to show the diversity of the applications through the following examples:

Ultrashallow junctions: These are not a new concern to SIMS analysts, but one that has become increasingly complicated to address as the implantation depth decreases bringing with the necessity to quantify high levels of dopant, close to the surface with sub-nm depth resolution.[98] It is necessary to decrease the impact energy low enough in order that the decay in the measured signal is representative of the steepness of the dopant profile and not for the SIMS depth resolution. Differences in ion yield and sputter rate in the transient region and surface oxide need to be accounted for as a large fraction of the dose may be contained here. It is thus necessary to have a good knowledge of the oxide thickness, but also to use other techniques to verify the SIMS profile near the surface. This has been accomplished by various groups using techniques such as medium-energy ion scattering (MEIS) or high-resolution Rutherford backscattering (HR-RBS).[99] One group has even proposed using reactive ion etching to further reduce impact energy.[100] With the advent of FinFET and other 3D devices, techniques such as atom probe or scanning spreading resistance microscopy (SSRM) may be more suited for dopant characterisation as the planar implant profile is no longer representative of doping on sidewalls of devices.[101]

Contamination: Although this domain is often addressed by ToF-SIMS, the magnetic sector instrument has advantages in terms of detection limit for inorganic contaminants. When coupled with a vapour-phase decomposition procedure extremely good lower detection limits can be obtained.[102]

Gate stack: This is a particularly pertinent example of the type of thin multilayer film that SIMS is particularly good at characterising but is still challenging due to the different sputtering characteristics of the high-k and metal gate layers, typically $HfO_2$- and TiN-based respectively. Again, the use of other techniques such as transmission electron microscopy (TEM), X-ray photoelectron spectroscopy (XPS), MEIS, atom probe as well as analysing samples in a "backside" configuration are important to overcome sputtering and ion-yield artefacts.[103]

Cesium cluster "MCs" analysis: the analysis of $Cs^+$ or $Cs_2^+$ ions associated with the element of interest is often used for the characterisation of II-VI and III-V multilayers[104] for optoelectronic applications. This is mainly due to reduced matrix effects and easier quantification, but can also be used on magnetic instruments that are not equipped with a floating Cs column to obtain relatively low impact energies (typically in the 500 eV to 4 keV range).

**Nuclear Sciences**

SIMS is considered to be a major analytical technique for nuclear-fuel characterisation and actually is a complimentary tool to the other techniques used in this field. The first advantage of SIMS, when compared to EPMA notably, is its capability to measure the isotopes of fission products and actinides. Other strengths include the achievement of low detection limits for several elements of interest, including light elements and low yield fission products, the high precision and accuracy of isotope ratio measurements and the possibility to acquire depth profiles inside fuel samples. In contrast, the main limitation of SIMS lies in the difficulty to obtain quantitative results, especially in irradiated fuels for which no reference material is available and for which mass spectra are very complex and thus interferences can be hard to resolve.

A review of the literature on nuclear-fuel studies reveals two main fields of investigation where SIMS analyses turned out to provide original and remarkable results: (i) the chemical behaviour of fission products during normal and off-normal (power transient) irradiation conditions and (ii) the isotopic abundance evolution due to nuclear reactions.

Concerning the behaviour of fission products, the application of SIMS through ionic imaging has enabled the study of the migration of volatile elements such as Rb, Cs, Te and I inside fuel pellets but also of their release outside the fuel during power ramps, with potential effects on cladding corrosion.[105–107] Thanks to the possibilities offered by depth profiling, significant results have been obtained on the behaviour of fission gases, mainly Xe and Kr, from quantitative estimations of their local retention and their partitioning between the fuel matrix and the bubbles formed during irradiation.[106,108,109] In parallel, more fundamental studies conducted with the use of SIMS after ion implantation have made possible the calculation of diffusion coefficient values for several fission products in $UO_2$.[110]

Concerning the evolution of isotopic abundances, one of the most attractive features of SIMS for postirradiation examinations is the determination of local burn-up values from the measurement of rare-earth fission products, *e.g.* Nd isotopes, to $^{238}U$ ratios.[106,111] Another field of fruitful applications is the inventory of actinides produced in irradiated $UO_2$ and MOX fuels.[106,112] Research in this field has enabled the study of the evolution of major U and Pu isotopes as well as minor actinide isotopes (Np, Am, Cm) in relation to nuclear reactions (fission, neutron capture and decay reactions) and has fed and validated neutron and fuel behaviour codes developed for industrial nuclear reactors.

More specific studies dealing for instance with the variations of impurity element isotope ratios (Li, B) with fuel burn-up[113] or with the influence of "burnable absorbers" (Gd) on fuel performances[114] are worth mentioning in this review to illustrate the large range of SIMS applications for nuclear-fuel characterisation.

### 15.5.3 Applications in Nuclear Safeguards

A network of 7 SIMS laboratories has been qualified to provide analytical support to the International Atomic Energy Agency (IAEA) in the scope of the strengthened safeguards measures implemented under the agreements of the Treaty on the Non-Proliferation of Nuclear Weapons and its Additional Protocol. In that respect, the main task for SIMS is i) to detect all the micrometre-sized particles containing uranium in the presence of a very large proportion of particles originating from the environment and ii) to individually analyse a selection of particles for uranium isotopes.

Particles of interest are collected, mostly on cotton swipes, in the environment of nuclear sites for the purpose of drawing conclusions regarding the absence of undeclared nuclear material or activities. This type of sampling (called environmental sampling within the safeguards community) is based on the premise that every nuclear process, no matter how leak tight, emits trace amounts of process material to the environment.[115]

The quantities of nuclear material involved are well below concern from a health physics and safety standpoint but they are generally detectable using environmental sampling in association with high-sensitivity mass-spectrometry techniques. The isotopic fingerprint of these uranium traces as revealed by particle analysis, is indicative of the process from which they derive. SIMS has been a technique of choice for this application due to its combined capability for high-sensitivity, large-field isotopic mapping (up to 500 μm by 500 μm), and *in situ* isotopic microanalysis.

To characterise the entire $^{235}U$ enrichment range in a given sample, it is critical to perform an exhaustive survey of the sample mount (a signature of interest may be featured by less than 1:10 000 of the uranium particles present). This was made possible by means of a dedicated piece of software compatible with the *IMS 4f* electronics and developed in the late 1990s by Charles Evans and his associates: P-Search.[116] An automated and "slightly destructive" screening is performed overnight thanks to a series of chained image acquisitions. The associated algorithm for image processing enables the exhaustive recognition of particles and the determination of their respective isotopic composition. Though the derived results are of limited precision given the small amount of material sputtered, they provide valuable hints to select critical particles for subsequent destructive microanalysis.[117]

The main limitation for particle analysis using the *IMS 4f* (and successors) is related to the contribution of interference stemming from elements collocated with uranium, which compromises the measurement of minor isotopes (given the limited amount of analyte, instruments are generally operated at full transmission, which can only be achieved at the expense of mass-resolution power on compact instruments).

The implementation of the *CAMECA IMS 1280* in replacement of *IMS 4-7f* instruments in 2009 was a milestone for the safeguards community.[118,119] This turning point followed the development by CAMECA of an *IMS 1280*, *IMS 6f* and *IMS 7f* – compatible piece of software for the purpose of particle screening: APM (automated particle measurement).[31,120]

Most of the *IMS 1280* and *1280-HR* instruments dedicated to nuclear-safeguards applications feature a multicollector system including 5 electron multipliers. The gap between uranium isotope trajectories at the detector plane is so small that the detectors have to be brought together as close as possible. Even so, it is necessary to use a stigmator device to deflect the trajectories apart from each other, so that they all fit into their respective detector. The capability of the *IMS 1280* to simultaneously collect all uranium isotopes and its enhanced mass-resolution power *vs.* transmission trade-off resulted in an increase in sensitivity by a factor of 10, while interference contribution related to common inorganic species is generally kept at a negligible level. In addition, multicollection enabled a significant gain in sample throughput compared to the peak-jumping mode in use on compact instruments. Currently, a total of *CAMECA IMS 1280* and *1280-HR* instruments is fully or partially dedicated to IAEA safeguards applications.

### 15.5.4 Applications in Life Sciences

Modern life sciences are in urgent need of hyphenating analytical techniques, which can contribute to the challenges of modern society in a globalised world. New approaches are also needed in medicine and toxicology, especially in the fields of personalised medicine or the TOX-21c programme.[121] Imaging mass spectrometry and especially *NanoSIMS* are well suited to answer hot topics, questions and give insight into the quantitative subcellular spatial distribution of environmental chemicals like heavy metals, endocrine-disrupting compounds, carcinogens like polycyclic aromatic hydrocarbons as well as the effect of pharmaceuticals in combination with their metabolites after cellular uptake.[122] *NanoSIMS* has the ability to discern questions surrounding the quantitative subcellular metabolic flux analysis of various biomolecules like DNA, nucleic acids, proteins, sugars and lipids.[123]

Single-cell distribution and quantification of biomolecules and environmental toxicants is the ultimate challenge not only for a better understanding of the molecular backgrounds for a variety of diseases such as cancer, Parkinson's disease, Alzheimer's disease and coronary heart disease.[33,124–128] The resulting insights will enhance drug discovery and enable new ways for targeted drug delivery. Additional information could be gathered from mass spectral imaging to monitor time-dependent pharmacological processes and drug-treatment progression and therefore significantly enhance modern medicine. The exact assessment of how a pharmacological drug treatment affects the intracellular distribution of administered pharmaceuticals and drug-related metabolites in relation to a comprehensive subcellular quantification is a must.[126] In combination with population subgroup analysis, imaging mass spectrometry is a strong candidate to become part of the tool-box needed for the development of modern analytical equipment needed for the modernisation and evaluation of new medical treatments based on personalised medicine.

Visualisation and quantification of the intracellular metabolomics in subcellular structures as well as the exact location of environmental toxicants after cellular uptake enables the exact assessment of intracellular metabolic and signalling pathways for life sciences and especially for modern toxicology.[129] Distribution pattern of chemicals within cells and tissues allows the biochemical characterisation of uptake and release mechanisms like pinocytosis, phagocytosis or ion-transport channel activation. Modern mass-spectrometry-based imaging techniques such as *NanoSIMS*, which combine high mass resolution with high spatial resolution in the nanometre range will contribute considerably to answering a variety of questions resulting from the demands in those fields. Furthermore, the *NanoSIMS* can especially contribute to the identification of spatially resolved intracellular metabolomic biomarkers not only for disease diagnosis and early medical treatment successes, but also for the specific regulation of metabolomic and toxicological pathways within subcellular structures.

Modern quantitative imaging techniques use multi-isotope labelling imaging mass spectrometry in combination with 3D imaging.[130–132] Stable isotope-labelled molecules or minerals are quantified within single cells and cellular substructures such as the nucleus, mitochondria, lysosomes and the Golgi apparatus.[123,127,133–135] The method can also be used to characterise the exact location of cell-membrane-bound compounds or to visualise participating lipids in plaque formation within cell membranes.[136] A three-dimensional reconstruction of the acquired data allows for the exact quantitative localisation of biomolecules within single cells and cellular substructures. Hereby stable isotope labelling is used in magnetic sector SIMS analysis for the exact quantification and localisation of nucleic acids, DNA molecules, amino acids, to quantify and visualise membrane lipids in single cells as well as for the of characterisation and quantification of nanomaterials for toxicological studies after cellular uptake.[129,137–141]

Moreover, stable-isotope labelling has significant advantages: the molecules do not decay nor do they change their 3D chemical structure and therefore influence the size of the molecule, in comparison to other labelling techniques.

SIMS is therefore well suited to meet new and emerging challenges spanning a multitude of topics from the development of personalised drug treatment regimes to the elucidation of toxicological signalling pathway regulation for a better environmental health risk assessment.

# References

1. K. K. Murray, R. K. Boyd, M. N. Eberlin, G. J. Langley, L. Li and Y. Naito, *Pure Appl. Chem.*, 2013, **85**, 1515–1609.
2. A. Benninghoven, F. G. Rüdenauer and H. W. Werner, *Secondary Ion Mass Spectrometry: Basic Concepts, Instrumental Aspects, Applications, and Trends*, John Wiley & Sons Inc., New York 1st edn, 1987, 1264.
3. J. C. Vickerman, N. M. Reed and A. Brown, *Secondary Ion Mass Spectroscopy: Principles and Applications (Retroactive Coverage)*, Oxford University Press, Oxford, 1989, 341.
4. G. Slodzian and J. F. Hennequin, *Cr. Acad. Sci. B Phys.*, 1966, **263**, 1246–&.
5. N. Shimizu and S. R. Hart, *J. Appl. Phys.*, 1982, **53**, 1303–1311.
6. J. M. Rouberol, M. Lepareur, B. Autier and J. M. Gourgout, in *conf. proc. of VIII International Congress on X-ray Optics and Microanalysis* pp. 322–328.
7. A. Merkulov, P. Peres, S. Choi, F. Horreard, H. U. Ehrke, N. Loibl and M. Schuhmacher, *J. Vac. Sci. Technol. B*, 2010, **28**, C1C48–C41C53.
8. G. Slodzian, B. Daigne, F. Girard, F. Boust and F. Hillion, *Biol Cell*, 1992, **74**, 43–50.
9. A. Galejs and C. E. Kuyatt, *J. Vac. Sci. Technol.*, 1978, **15**, 865–867.
10. A. Adams and F. Read, *J. Phys. E: Sci. Instrum.*, 1972, **5**, 150.
11. G. Slodzian, H. Chaintreau and R. Dennebouy, in *ICXOM II Lecture*, Cameca News, London, Canada, 1986.

12. H. N. Migeon, M. Schuhmacher and G. Slodzian, *Surf. Interface Anal.*, 1990, **16**, 9–13.
13. P. Peres, E. de Chambost and M. Schuhmacher, *Appl. Surf. Sci.*, 2008, **255**, 1472–1475.
14. E. de Chambost, *A History of Cameca (1954–2009)*, in Advances in Imaging and Electron Physics, ed. P. W. Hawkes, Elsevier, Amsterdam, The Netherlands, 1st edn., 2011, **vol. 167**, **ch. 1**, pp. 1–121.
15. E. de Chambost, B. Boyer, B. Rasser and M. Schuhmacher, in *Secondary Ion Mass Spectrometry SIMS XII*, eds. A. Benninghoven, P. Bertrand and H. N. Migeon, Elsevier, Brussels, Belgium, 1999, p. 1058.
16. R. F. K. Herzog and F. P. Viehböck, *Phys. Rev.*, 1949, **76**, 855–856.
17. R. Castaing and G. Slodzian, *J. Microsc. (Oxford)*, 1962, **1**, 395–410.
18. H. J. Liebl and R. F. K. Herzog, *J. Appl. Phys.*, 1963, **34**, 2893-&.
19. M. Van Ardenne, *Tabellen der Elektronenphysik, Ionenphysik und der Übermikroskopie*, VEB Deutscher Verlag der Wissenschaften, Berlin, 1956.
20. H. Liebl, *J. Appl. Phys.*, 1967, **38**, 5277–5283.
21. C. Andersen and J. Hinthorne, *Earth Planet. Sci. Lett.*, 1972, **14**, 195–200.
22. H. Liebl, *Anal. Chem.*, 1974, **46**, 22A–30a.
23. J. M. Rouberol, J. Guernet, P. Deschamps, J. P. Dagnot and J. M. Guyon de la Berge, in *conf. proc. of V. Internationaler Kongress für Röntgenoptik und Mikroanalyse*, Springer, pp. 311–318.
24. G. Slodzian and A. Figueras, in *conf. proc. of 8th Intern. Congress on X-Rays Optics and Microanalysis*, Pendell, pp. 659–665.
25. N. Shimizu, *Nature*, 1981, **289**, 575–577.
26. S. W. J. Clement, W. Compston and G. Newstead, in *Proceedings of the 1st international conference on secondary ion mass spectrometry* ed. A. Bennighoven, John Wiley & Sons Inc., Münster, Germany, 1977, pp. 12–17.
27. T. R. Ireland, Ion microprobe mass spectrometry: techniques and applications in cosmochemistry, geochemistry, and geochronology, in *Advances in Analytical Geochemistry*, eds. M. Hyman and M. Rowe, Elsevier, Oxford, 1995, vol. 2, pp. 1–118.
28. T. R. Ireland, S. Clement, W. Compston, J. J. Foster, P. Holden, B. Jenkins, P. Lanc, N. Schram and I. S. Williams, *Austral. J. Earth Sci.*, 2008, **55**, 937–954.
29. N. T. Kita, T. Ushikubo, B. Fu and J. W. Valley, *Chem. Geol.*, 2009, **264**, 43–57.
30. Patent FR2942072, 2009.
31. P. M. L. Hedberg, P. Peres, J. B. Cliff, F. Rabemananjara, S. Littmann, H. Thiele, C. Vincent and N. Albert, *J. Anal. Atom. Spectrom.*, 2011, **26**, 406–413.
32. E. de Chambost, M. Schuhmacher, G. Lövestam and S. Claesson, in *Secondary Ion Mass Spectrometry, SIMS X*, John Wiley & Sons Inc., 1997, pp. 1003–1006.

33. P. Hoppe, S. Cohen and A. Meibom, *Geostand. Geoanal. Res.*, 2013, **37**, 111–154.
34. M. G. Dowsett, N. S. Smith, R. Bridgeland, D. Richards, A. C. Lovejoy and P. Pedrick, in *conf. proc. of Secondary ion mass spectrometry, SIMS X*, John Wiley & Sons Inc., 1997, p. 367.
35. M. Juhel and F. Laugier, *Appl. Surf. Sci.*, 2004, **231**, 698–703.
36. K. D. McKeegan, C. D. Coath, P. H. Mao, G. Jarzebinski and D. Burnett, in *35th Lunar and Planetary Science Conference*, 2004.
37. C. Cetina, K. Grabowski and D. Knies, in *40th Lunar and Planetary Institute Science Conference*, 2009.
38. M. Schuhmacher, B. Rasser and F. Desse, , *J. Vac. Sci. Technol. B*, 2000, **18**, 529–532.
39. G. Bart, T. Aeme, U. Flückner and E. Sprunger, *Nucl. Instrum. Methods*, 1981, **180**, 109–116.
40. B. Rasser, L. Desgranges and B. Pasquet, *Appl. Surf. Sci.*, 2003, **203**, 673–678.
41. S. Portier, S. Brémier and C. T. Walker, *Int. J. Mass Spectrom.*, 2007, **263**, 113–126.
42. E. de Chambost, M. Schuhmacher, G. Lövestam and S. Claesson, in *Secondary Ion Mass Spectrometry, SIMS X*, John Wiley & Sons Inc., 1997, pp. 1003–1006.
43. E. de Chambost, P. Fercocq, F. Fernandes, E. Deloule and M. Chaussidon, in *Secondary Ion Mass Spectrometry, SIMS XI*, John Wiley & Sons Inc., 1998, pp. 727–730.
44. H. Matsuda, *Int. J. Mass Spectrom. Ion Phys.*, 1974, **14**, 219–233.
45. H. Matsuda, *Nucl. Instrum. Methods Phys. Res. A*, 1990, **298**, 199–204.
46. T. Ireland and M. Bukovanska, *Geochim. Cosmochim. Acta*, 2003, **67**, 4849–4856.
47. T. R. Ireland, Ion Microscopes and Microprobes, in *Treatise on Geochemistry*, eds. H. H.D. and T. K.K., Elsevier, Oxford, 2nd edn., 2014, vol. 15, pp. 385–409.
48. T. R. Ireland, *Rev. Sci. Instrum.*, 2013, **84**, 011101–011121.
49. R. G. Wilson, F. A. Stevie and C. W. Magee, *Secondary Ion Mass Spectrometry: A Practical Handbook For Depth Profiling and Bulk Impurity Analysis* John Wiley & Sons Inc., New York, 1989, 384.
50. T. R. Ireland, SIMS Measurement of Stable Isotopes, in *Handbook of Stable Isotope Analytical Techniques*, ed. P. A. De Groot, Elsevier, 2004, vol. 1, ch. 30, pp. 652–691.
51. Y. Marie, Y. Gao, F. Saldi and H. N. Migeon, *Surf. Interface Anal.*, 1995, **23**, 38–43.
52. C. Andersen, *Int. J. Mass Spectrom. Ion Phys.*, 1970, **3**, 413–428.
53. R. W. Hinton, Ion Microprobe Analysis in Geology, in *Microprobe Techniques in the Earth Sciences*, eds. P. J. Potts, J. F. Bowles, S. J. Reed and R. Cave, Mineralogical Society of Great Britain and Ireland, 1995.

54. M. Fayek, *Secondary Ion Mass Spectrometry in the Earth Sciences: Gleaning the Big Picture from a Small Spot* Mineralogical Association of Canada, Toronto, 2009, 41, 160.
55. M. Wiedenbeck, *Geostand. Geoanal. Res.*, 2010, **34**, 387–394.
56. M. Wiedenbeck, R. Bugoi, M. J. M. Duke, T. Dunai, J. Enzweiler, M. Horan, K. P. Jochum, K. Linge, J. Kosler, S. Merchel, L. F. G. Morales, L. Nasdala, R. Stalder, P. Sylvester, U. Weis and A. Zoubir, *Geostand. Geoanal. Res.*, 2012, **36**, 337–398.
57. E. Deloule, O. Paillat, M. Pichavant and B. Scaillet, *Chem. Geol.*, 1995, **125**, 19–28.
58. R. Hinton, *Chem. Geol.*, 1990, **83**, 11–25.
59. C. Rollion-Bard and J. Marin-Carbonne, *J. Anal. Atom. Spectrom.*, 2011, **26**, 1285–1289.
60. L. Sangely, M. Chaussidon, R. Michels and V. Huault, *Chem. Geol.*, 2005, **223**, 179–195.
61. F. Z. Page, N. T. Kita and J. W. Valley, *Chem. Geol.*, 2010, **270**, 9–19.
62. M. Chaussidon and A. Jambon, *Earth Planet. Sci. Lett.*, 1994, **128**, 731–733.
63. M. J. Whitehouse and A. A. Nemchin, *Chem. Geol.*, 2009, **261**, 31–41.
64. R. B. Ickert, J. Hiess, I. S. Williams, P. Holden, T. R. Ireland, P. Lanc, N. Schram, J. J. Foster and S. W. Clement, *Chem. Geol.*, 2008, **257**, 114–128.
65. W. Compston, I. S. Williams and S. W. Clement, in 30$^{th}$ annual conference on mass spectrometry and applied topics, 1982, pp. 593–595.
66. M. Wiedenbeck and J. Goswami, *Geochim. Cosmochim. Acta*, 1994, **58**, 2135–2141.
67. S. Chakraborty and D. C. Rubie, *Contrib. Mineral. Petrol.*, 1996, **122**, 406–414.
68. A. Dimanov and M. Wiedenbeck, *Eur. J. Mineral.*, 2006, **18**, 705–718.
69. R. Milke, M. Wiedenbeck and W. Heinrich, *Contrib. Mineral. Petrol.*, 2001, **142**, 15–26.
70. J. A. Van Orman, T. L. Grove, N. Shimizu and G. D. Layne, *Contrib. Mineral. Petrol.*, 2002, **142**, 416–424.
71. H. Z. Fei, C. Hegoda, D. Yamazaki, M. Wiedenbeck, H. Yurimoto, S. Shcheka and T. Katsura, *Earth Planet. Sci. Lett.*, 2012, **345**, 95–103.
72. H. Keppler, M. Wiedenbeck and S. S. Shcheka, *Nature*, 2003, **424**, 414–416.
73. M. A. Kusiak, M. J. Whitehouse, S. A. Wilde, A. A. Nemchin and C. Clark, *Geology*, 2013, **41**, 291–294.
74. N. McLoughlin, D. Wacey, C. Kruber, M. R. Kilburn, I. H. Thorseth and R. B. Pedersen, *Chem. Geol.*, 2011, **289**, 154–162.
75. T. Lee, D. Papanastassiou and G. Wasserburg, *Astrophys. J.*, 1977, **211**, L107–L110.
76. E. Zinner, M. Tang and E. Anders, *Nature*, 1987, **330**, 730–732.
77. S. Messenger, L. P. Keller, F. J. Stadermann, R. M. Walker and E. Zinner, *Science*, 2003, **300**, 105–108.

78. A. N. Nguyen and E. Zinner, *Science*, 2004, **303**, 1496–1499.
79. E. Zinner, K. D. McKeegan and R. M. Walker, *Nature*, 1983, **305**, 119–121.
80. H. Busemann, A. F. Young, C. M. O. Alexander, P. Hoppe, S. Mukhopadhyay and L. R. Nittler, *Science*, 2006, **312**, 727–730.
81. K. D. McKeegan, M. Chaussidon and F. Robert, *Science*, 2000, **289**, 1334–1337.
82. S. Tachibana and G. R. Huss, *Astrophys. J. Lett.*, 2003, **588**, L41–L44.
83. K. D. McKeegan, J. Aleon, J. Bradley, D. Brownlee, H. Busemann, A. Butterworth, M. Chaussidon, S. Fallon, C. Floss, J. Gilmour, M. Gounelle, G. Graham, Y. B. Guan, P. R. Heck, P. Hoppe, I. D. Hutcheon, J. Huth, H. Ishii, M. Ito, S. B. Jacobsen, A. Kearsley, L. A. Leshin, M. C. Liu, I. Lyon, K. Marhas, B. Marty, G. Matrajt, A. Meibom, S. Messenger, S. Mostefaoui, S. Mukhopadhyay, K. Nakamura-Messenger, L. Nittler, R. Palma, R. O. Pepin, D. A. Papanastassiou, F. Robert, D. Schlutter, C. J. Snead, F. J. Stadermann, R. Stroud, P. Tsou, A. Westphal, E. D. Young, K. Ziegler, L. Zimmermann and E. Zinner, *Science*, 2006, **314**, 1724–1728.
84. H. Yurimoto, K. Abe, M. Abe, M. Ebihara, A. Fujimura, M. Hashiguchi, K. Hashizume, T. R. Ireland, S. Itoh, J. Katayama, C. Kato, J. Kawaguchi, N. Kawasaki, F. Kitajima, S. Kobayashi, T. Meike, T. Mukai, K. Nagao, T. Nakamura, H. Naraoka, T. Noguchi, R. Okazaki, C. Park, N. Sakamoto, Y. Seto, M. Takei, A. Tsuchiyama, M. Uesugi, S. Wakaki, T. Yada, K. Yamamoto, M. Yoshikawa and M. E. Zolensky, *Science*, 2011, **333**, 1116–1119.
85. K. D. McKeegan, A. P. A. Kallio, V. S. Heber, G. Jarzebinski, P. H. Mao, C. D. Coath, T. Kunihiro, R. C. Wiens, J. E. Nordholt, R. W. Moses, D. B. Reisenfeld, A. J. G. Jurewicz and D. S. Burnett, *Science*, 2011, **332**, 1528–1532.
86. B. Marty, M. Chaussidon, R. C. Wiens, A. J. G. Jurewicz and D. S. Burnett, *Science*, 2011, **332**, 1533–1536.
87. A. E. Saal, E. H. Hauri, J. A. Van Orman and M. J. Rutherford, *Science*, 2013, **340**, 1317–1320.
88. D. S. McPhail, *J. Mater. Sci.*, 2006, **41**, 873–903.
89. E. Chatzitheodoridis, G. Kiriakidis and I. Lyon, Secondary ion mass spectrometry and its application to thin film characterisation, in *Handbook of Thin Film Materials: Characterisation and Spectroscopy of Thin Films*, ed. H. S. Nalwa, Academic Press, 2002, vol. 2, ch. 13, pp. 637–683.
90. J. Drillet, N. Valle and T. Iung, *Metall. Mater. Trans. A*, 2012, **43**, 4947–4956.
91. F. Christien, C. Downing, K. L. Moore and C. R. M. Grovenor, *Surf. Interface Anal.*, 2012, **44**, 377–387.
92. J. Sterpenich and G. Libourel, *J. Non-Cryst. Solids*, 2006, **352**, 5446–5451.
93. N. Saintier, T. Awane, J. M. Olive, S. Matsuoka and Y. Murakami, *Int. J. Hydrogen Energy*, 2011, **36**, 8630–8640.
94. R. A. de Souza and M. Martin, *Phys. Status Solidi C*, 2007, **4**, 1785–1801.

95. N. Valle, A. Verney-Carron, J. Sterpenich, G. Libourel, E. Deloule and P. Jollivet, *Geochim. Cosmochim. Acta*, 2010, **74**, 3412–3431.
96. K. Durose, S. E. Asher, W. Jaegermann, D. Levi, B. E. McCandless, W. Metzger, H. Moutinho, P. D. Paulson, C. L. Perkins, J. R. Sites, G. Teeter and M. Terheggen, *Prog. Photovolt.*, 2004, **12**, 177–217.
97. H. Oechsner, H. Schoof and E. Stumpe, *Surf. Sci.*, 1978, **76**, 343–354.
98. W. Vandervorst, *Appl. Surf. Sci.*, 2008, **255**, 805–812.
99. C. W. Magee, R. S. Hockett, T. H. Buyuklimanli, I. Abdelrehim and J. W. Marino, *Nucl. Instrum. Methods Phys. Res. Sect. B-Beam Interact. Mater. Atoms*, 2007, **261**, 594–599.
100. N. Vanhove, P. Lievens and W. Vandervorst, *Surf. Interface Anal.*, 2011, **43**, 159–162.
101. P. Eyben, W. Vandervorst, D. Alvarez, M. Xu and M. Fouchier, Probing Semiconductor Technology and Devices with Scanning Spreading Resistance Microscopy, in *Scanning Probe Microscopy: Slectrical and Slectromechanical Phenomena at the Nanoscale* eds. S. V. Kalinin and A. Gruverman, Springer, New York, 2007, vol. 1, pp. 31–87.
102. M. Juhel, C. Trouiller, D. Guiheux, C. Arsac, N. Drogue, S. Couvrat and C. Grosjean, *Surf. Interface Anal.*, 2011, **43**, 582–585.
103. M. J. P. Hopstaken, D. Pfeiffer, M. Copel, M. S. Gordon, T. Ando, V. Narayanan, H. Jagannathan, S. Molis, J. A. Wahl, H. Bu, D. K. Sadana, L. Czornomaz, C. Marchiori and J. Fompeyrine, *Surf. Interface Anal.*, 2013, **45**, 338–344.
104. C. J. Gu, F. A. Stevie, C. J. Hitzman, Y. N. Saripalli, M. Johnson and D. P. Griffis, *Appl. Surf. Sci.*, 2006, **252**, 7228–7231.
105. L. Desgranges, B. Pasquet, T. Pujol, I. Roure, T. Blay, J. Lamontagne, T. Martella, B. Lacroix, O. Comiti and L. Caillot, in *International Seminar on Pellet/Clad Interactions with Water Reactor Fuels*, CEA Cadarache/DEN/DEC, Aix en Provence, France, 2004, pp. 241–252.
106. C. T. Walker, S. Bremier, S. Portier, R. Hasnaoui and W. Goll, *J. Nucl. Mater.*, 2009, **393**, 212–223.
107. L. Desgranges, C. Riglet-Martial, I. Aubrun, B. Pasquet, I. Roure, J. Lamontagne and T. Blay, *J. Nucl. Mater.*, 2013, **437**, 409–414.
108. L. Desgranges, C. Valot, B. Pasquet, J. Lamontagne, T. Blay and I. Roure, *Nucl. Instrum. Methods Phys. Res. Sect. B-Beam Interact. Mater. Atoms*, 2008, **266**, 147–154.
109. S. Portier, S. Bremier, R. Hasnaoui, O. Bildstein and C. T. Walker, *Appl. Surf. Sci.*, 2008, **255**, 1323–1326.
110. M. Saidy, W. H. Hocking, J. F. Mouris, P. Garcia, G. Carlot and B. Pasquet, *J. Nucl. Mater.*, 2008, **372**, 405–415.
111. C. Valot, L. Desgranges, B. Pasquet, J. Lamontagne, J. Noirot, T. Blay and I. Roure, in *conf. proc. of European Working Group on Hot Laboratories and Remote Handling*, Petten, The Netherlands, 2005.
112. L. Desgranges, B. Pasquet, C. Valot and I. Roure, *J. Nucl. Mater.*, 2009, **385**, 99–102.

113. D. C. Gerlach, J. B. Cliff, D. E. Hurley, B. D. Reid, W. W. Little, G. H. Meriwether, A. J. Wickham and T. A. Simmons, *Appl. Surf. Sci.*, 2006, **252**, 7041–7044.
114. H. U. Zwicky, E. T. Aerne, G. Bart, F. Petrik and H. A. Thomi, *Radiochim. Acta*, 1989, **47**, 9–12.
115. E. Kuhn, D. Fischer and M. Ryjinski, in *conf. proc. of Symposium on international safeguards: Verification and nuclear material security*, Vienna, Austria, 2001.
116. D. S. Simons, G. Gillen, C. J. Zeissler, R. H. Fleming and P. J. McNitt, in *Secondary ion mass spectrometry SIMS XI*, eds. G. Gillen, R. Lareau, J. Bennett and F. Stevie, John Wiley & Sons Inc., New York, 1st edn., 1998, pp. 59–62.
117. P. M. L. Hedberg, K. Ingeneri, M. Watanabe and Y. Kuno, in *conf. proc. of 46th Institute of Nuclear Materials Management (INMM) Annual Meeting*, Arizona, USA, 2005.
118. Y. Ranebo, P. M. L. Hedberg, M. J. Whitehouse, K. Ingeneri and S. Littmann, , *J. Anal. Atom. Spectrom.*, 2009, **24**, 277–287.
119. J. Poths, L. Sangely, O. Bildstein, T. Kitao, A. Schwanhaeusser, M. Hosoya and T. Tanpraphan, in *conf. proc. of 33rd ESARDA Conference*, Budapest, Hungary, 2011.
120. P. Peres, P. M. L. Hedberg, S. Walton, N. Montgomery, J. B. Cliff, F. Rabemananjara and M. Schuhmacher, , *Surf. Interface Anal.*, 2013, **45**, 561–565.
121. T. Ramirez, M. Daneshian, H. Kamp, F. Y. Bois, M. R. Clench, M. Coen, B. Donley, S. M. Fischer, D. R. Ekman, E. Fabian, C. Guillou, J. Heuer, H. T. Hogberg, H. Jungnickel, H. C. Keun, G. Krennrich, E. Krupp, A. Luch, F. Noor, E. Peter, B. Riefke, M. Seymour, N. Skinner, L. Smirnova, E. Verheij, S. Wagner, T. Hartung, B. van Ravenzwaay and M. Leist, *Altex-Alternat. Animal Exp.*, 2013, **30**, 209–225.
122. S. Behrens, A. Kappler and M. Obst, *Environ. Microbiol.*, 2012, **14**, 2851–2869.
123. M. L. Steinhauser, A. P. Bailey, S. E. Senyo, C. Guillermier, T. S. Perlstein, A. P. Gould, R. T. Lee and C. P. Lechene, *Nature*, 2012, **481**, 516–519.
124. C. W. Mueller, P. K. Weber, M. R. Kilburn, C. Hoeschen, M. Kleber and J. Pett-Ridge, in *Advances in Agronomy*, ed. L. S. Donard, Academic Press, 2013, vol. 121, pp. 1–46.
125. J. L. Guerquin-Kern, T. D. Wu, C. Quintana and A. Croisy, *Biochim. Biophys. Acta-General Sub.*, 2005, **1724**, 228–238.
126. L. E. Wedlock, M. R. Kilburn, R. Liu, J. A. Shaw, S. J. Berners-Price and N. P. Farrell, *Chem. Commun.*, 2013, **49**, 6944–6946.
127. S. E. Senyo, M. L. Steinhauser, C. L. Pizzimenti, V. K. Yang, L. Cai, M. Wang, T.-D. Wu, J.-L. Guerquin-Kern, C. P. Lechene and R. T. Lee, *Nature*, 2013, **493**, 433–436.
128. C. Quintana, S. Bellefqih, J. Y. Laval, J. L. Guerquin-Kern, T. D. Wu, J. Avila, I. Ferrer, R. Arranz and C. Patino,*J. Struct. Biol.*, 2006, **153**, 42–54.

129. A. Georgantzopoulou, Y. L. Balachandran, P. Rosenkranz, M. Dusinska, A. Lankoff, M. Wojewodzka, M. Kruszewski, C. Guignard, J. N. Audinot, S. Girija, L. Hoffmann and A. C. Gutleb, *Nanotoxicology*, 2012, 7, 1168–1178.
130. C. P. Lechene, G. Y. Lee, J. C. Poczatek, M. Toner and J. D. Biggers, *Plos One*, 2012, 7, e42267–e42270.
131. D. S. Zhang, V. Piazza, B. J. Perrin, A. K. Rzadzinska, J. C. Poczatek, M. Wang, H. M. Prosser, J. M. Ervasti, D. P. Corey and C. P. Lechene, *Nature*, 2012, **481**, 520–U137.
132. D. Duday, F. Clément, E. Lecoq, C. Penny, J. N. Audinot, T. Belmonte, K. Kutasi, H. M. Cauchie and P. Choquet, *Plasma Process. Polym.*, 2013, **10**, 864–879.
133. C. Lechene, F. Hillion, G. McMahon, D. Benson, A. M. Kleinfeld, J. P. Kampf, D. Distel, Y. Luyten, J. Bonventre, D. Hentschel, K. M. Park, S. Ito, M. Schwartz, G. Benichou and G. Slodzian, *J. Biol.*, 2006, **5**, 20.
134. P. Gorzelak, J. Stolarski, P. Dubois, C. Kopp and A. Meibom, *J. Struct. Biol.*, 2011, **176**, 119–126.
135. K. E. Smart, J. A. C. Smith, M. R. Kilburn, B. G. Martin, C. Hawes and C. R. Grovenor, *Plant J.*, 2010, **63**, 870–879.
136. M. M. Lozano, Z. Liu, E. Sunnick, A. Janshoff, K. Kumar and S. G. Boxer, *J. Am. Chem. Soc.*, 2013, **135**, 5620–5630.
137. H. A. Klitzing, P. K. Weber and M. L. Kraft, *Methods Molec. Biol. (Clifton, N.J.)*, 2013, **950**, 483–501.
138. J. P. Piret, D. Jacques, J. N. Audinot, J. Mejia, E. Boilan, F. Noel, M. Fransolet, C. Demazy, S. Lucas, C. Saout and O. Toussaint, *Nanoscale*, 2012, **4**, 7168–7184.
139. J. N. Audinot, A. Georgantzopoulou, J. P. Piret, A. C. Gutleb, D. Dowsett, H. N. Migeon and L. Hoffmann, *Surf. Interface Anal.*, 2013, **45**, 230–233.
140. T. Eybe, J. N. Audinot, T. Bohn, C. Guignard, H. N. Migeon and L. Hoffmann, *J. Appl. Microbiol.*, 2008, **105**, 1502–1510.
141. T. Eybe, J. N. Audinot, T. Udelhoven, E. Lentzen, B. El Adib, J. Ziebel, L. Hoffmann and T. Bohn, *Chemosphere*, 2013, **90**, 1829–1838.

CHAPTER 16

# *Gas Source Isotope Ratio Mass Spectrometry (IRMS)*

WILLI A. BRAND,[a] CHARLES B. DOUTHITT,[b] FRANCOIS FOUREL,[c] RODRIGO MAIA,[d] CARLA RODRIGUES,[e] CRISTINA MAGUAS[d] AND THOMAS PROHASKA*[f]

[a] Max-Planck-Institute for Biogeochemistry, Jena, Germany; [b] Thermo Fisher Scientific, US; [c] Laboratoire de Géologie de Lyon, CNRS-UMR 5276, Université Claude Bernard Lyon 1, Ecole Normale Supérieure de Lyon, France; [d] University of Lisbon, Portugal; [e] Diverge Grupo Nabeiro Innovation Centre, R&D Projects, Portugal; [f] University of Natural Resources and Life Sciences Vienna (BOKU), Department of Chemistry, Division of Analytical Chemistry, VIRIS Laboratory for Analytical Ecogeochemistry, Tulln, Austria
*Email: thomas.prohaska@boku.ac.at

Gas source isotope ratio mass spectrometry is usually referred to as isotope ratio mass spectrometry (IRMS) or stable-isotope ratio mass spectrometry (SIRMS). The method describes a mass spectrometer used specifically for the measurement of stable-isotope ratios of a limited number of elements (C, H, N, O and S) after transfer into a gaseous species. Si, Cl, Br and Se can be added to the list even though their applications are limited compared to the other isotope systems. IRMS, in addition to measuring isotope ratios, can also be used to measure ratios of gas species (*e.g.* $O_2/N_2$, $Ar/N_2$), isotopologues (*e.g.* $^{15}N$ positions in $N_2O$, "clumped" isotopes), or even elemental concentrations (from an elemental analyser, EA) or molecular abundances (from a gas chromatograph,

New Developments in Mass Spectrometry No. 3
Sector Field Mass Spectrometry for Elemental and Isotopic Analysis
Edited by Thomas Prohaska, Johanna Irrgeher, Andreas Zitek and Norbert Jakubowski

Published by the Royal Society of Chemistry, www.rsc.org

GC). (***Note***: The noble-gas mass spectrometer (also referred to as static gas mass spectrometer or static isotope ratio mass spectrometer (static IRMS)) shares a number of characteristics with IRMS and will be explained in more detail in Chapter 17.)

Since the elements for which isotope ratios are measured are introduced into the mass spectrometer as a gaseous molecule, samples (if not already in gaseous form) must be converted into gaseous form (*e.g.* $CO_2$, $H_2$, $N_2$, $N_2O$, CO, $O_2$, $SO_2$, $SF_6$, $CH_3Cl$, $CH_3Br$, or $SiF_4$). In stable gas source mass spectrometry, the gas is continuously introduced into the ion source (whereas in static IRMS the gas is fed into the source/analyser system once and left there without further supply or pumping).

The gas molecules are ionised in an electron impact (EI) ion source and the major isotopologues of the molecular ion are measured (in some special cases, also fragment ions are the target of analysis). After acceleration by a static electric field, ions are separated in a momentum

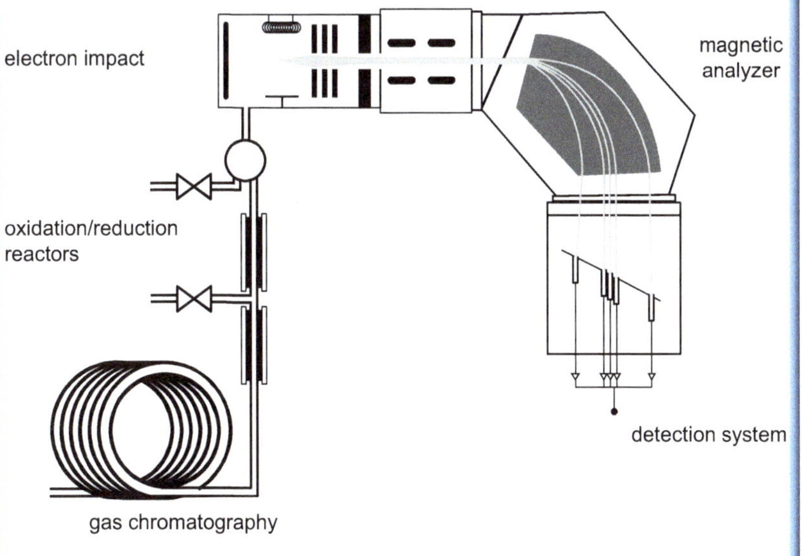

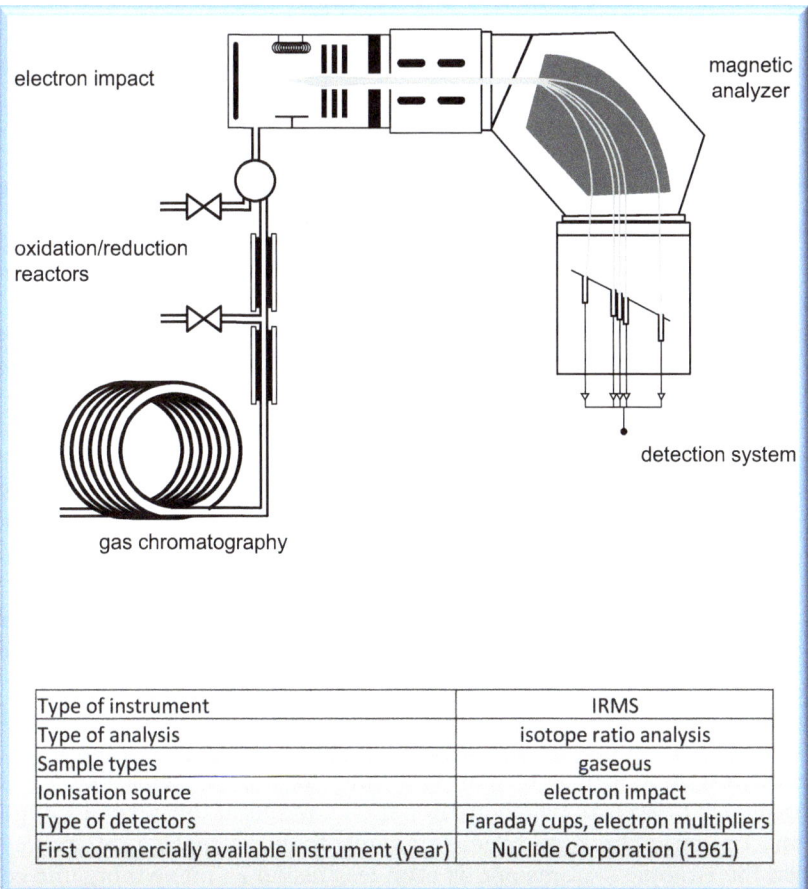

| | |
|---|---|
| Type of instrument | IRMS |
| Type of analysis | isotope ratio analysis |
| Sample types | gaseous |
| Ionisation source | electron impact |
| Type of detectors | Faraday cups, electron multipliers |
| First commercially available instrument (year) | Nuclide Corporation (1961) |

**Figure 16.1** Features and general schematics of an IRMS. (Source: Ondrej Hanousek.)

separator, usually an electromagnet, although permanent magnets have also been used occasionally. The ions are detected in a set of detectors (mostly Faraday cups), which simultaneously collect the ions of interest. The software converts the measured intensity ratios of the isotopologues into an isotopic composition, which is usually reported as an isotope delta value relative to a certified reference material (CRM). Because of the (usually) small isotopic differences reporting is made by multiplying the delta value by a factor of 1000, thus expressing the difference in per mill (‰) "units". For tracer studies, where higher isotopic enrichments are used, Excess Atom Fraction (former notation atom per cent excess (APE)) are in frequent use as well (see Figures 16.1 and 16.2).

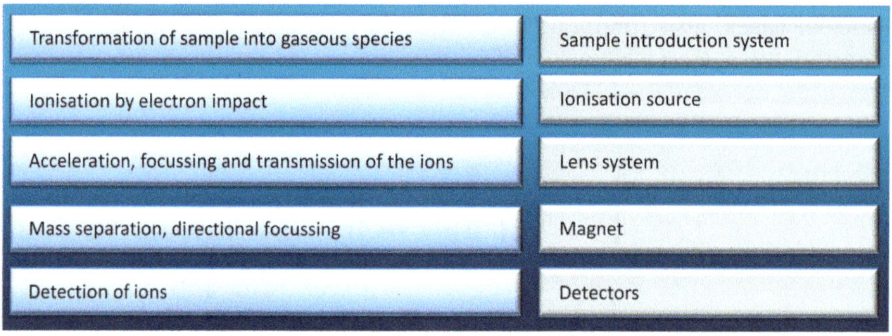

Transformation of sample into gaseous species
Sample introduction system
Ionisation by electron impact
Ionisation source
Acceleration, focussing and transmission of the ions
Lens system
Mass separation, directional focussing
Magnet
Detection of ions
Detectors


**Figure 16.2** Operational flow chart of an IRMS.

**Name and Abbreviation**

According to a recent IUPAC publication, IRMS is defined as "Measurement and study of the relative abundances of the different isotopes of an element in a material using a mass spectrometer".[1]

Following this definition, the term IRMS is not limited to the analysis of light-element stable isotopes using a gas source mass spectrometer. The analysis of heavier-element isotope ratios applying various ionisation techniques may also be referred to using the term "IRMS". Nonetheless, throughout this chapter, we will use the term "IRMS" for gas source isotope ratio mass spectrometry only.

In the context of IRMS, the term "light", "stable" or the combination "light stable" isotopes is generally used for stable nuclides of the bioelements H, C, O, N and S. Even though the use of the term "light stable" is well established in the scientific community, it is misleading because it does not account for the isotopic systems of Li and B. Moreover, the applications of the systems described here have expanded to other systems like, *e.g.* the halogens, Si or even U, which are not at all covered in the generally applied terms. The term "bio-" and "geo-" is rarely found to make a difference between the isotopic systems and is even less useful as many inorganic commodities consist of H, C, O and N and isotopic systems generally investigated in geosciences will appear in biological systems, as well. As a consequence, none of these used terms can be recommended.

(***Note***: Even though elemental analysis can be accomplished with mass spectrometers dedicated to the measurement of gaseous species, this section will focus only on the determination of isotope ratios.)

**Notation of Data in IRMS**

**Isotope Scales and the Delta Notation in IRMS (see also Chapter 8)**

IRMS analyses are performed as relative measurements. A sample gas is compared to a calibration gas (also termed "reference gas" or "working gas"). The calibration gas has a well-defined isotopic composition, *i.e.* the gas is carefully calibrated using an international isotopic reference material. The deviation between sample and reference gas is expressed by using the isotope $\delta$-notation.[2–5] Because these deviations are usually very small, $\delta$-values are reported in per mil (‰). The resulting isotope delta values are dimensionless quantities and their associated value can be negative or positive. This comparison technique allows compensation of instrumental effects such as drift or mass discrimination, which may change over time and vary from instrument to instrument. In dual-inlet system measurements, the "change-over" valve allows for an alternating admission of sample and reference gases to the ion source. In continuous-flow measurements, a pulse of a calibration gas is introduced before or after the sample gas. Reference and sample materials should be chemically as similar as possible and should have the same "chemical history" (chemical conversion and sample separation steps) following the "principle of identical treatment".[6,7] Isotopic reference materials are available from international resources like NIST, USGS or IAEA. They are used to define delta scales[3,8–11] and to position local working-reference materials on the respective scales.[12] A comprehensive compilation of primary and secondary reference materials for IRMS and their respective $\delta$-value history was published recently.[13]

**Atom Per Cent Excess**

When enriched stable isotopes are applied in, *e.g.* tracer studies, it is popular to report the level of isotopic abundance above a given background reading in units of atom per cent excess (APE). APE is calculated by subtracting the background reading in atom per cent (of the isotope under investigation) from the experimental level. However, according to the rules for the international SI system (see 8th SI brochure[14]), both atom per cent and APE should be replaced with "atom fraction" and "excess atom fraction". (For details see Coplen.[2])

## 16.1 General

(Gas source) isotope ratio mass spectrometry is a conventional method for measuring isotope ratios and has benefited from more than 65 years of research and development. Modern mass spectrometers are all based on sector field mass separators. More recently, the development of high-resolution sector field devices has added a new dimension to IRMS. Modern instruments achieve a high sample throughput, which is a prerequisite, *e.g.* for ecosystem studies where usually a large number of samples is analysed in a given study with high precision, typically around 0.1 ‰ (carbon).

(***Note***: In contrast to IRMS, isotope abundance ratio measurements by emerging laser techniques employ direct photoabsorption of infrared light by the different isotopologues of $H_2O$, $CO_2$, $N_2O$, and $CH_4$. This promising new technology, which is rapidly expanding and finding applications in new fields, does not employ mass spectrometry and will therefore not be discussed here further.[15])

A brief history of the stable-isotope technology and selected applications has been assembled, *e.g.* by Habfast.[16] Already in the early years of mass spectrometry, the isotopic variability of light elements (H, C, O) were studied[17] and quickly extended to nitrogen and sulfur. These pioneers founded the field of "conventional stable-isotope studies" dedicated to the "light life-science elements". The major technique of measuring the isotope ratios for those elements has been gas source isotope ratio mass spectrometry ever since. This technique was started by Nier's original mass spectrometer (Figure 16.3) developed during the 1940s.[18–20] Today, the same basic principles still apply, albeit with numerous technical improvements, not the least being full automation under computer control.

**1950 Breakthrough: The Dual-Inlet System**

The first sustained use of stable-isotope mass spectrometry was made at the laboratories of Harold C. Urey at the University of Chicago. A new sample

**Figure 16.3** A.O. Nier "tuning" one of his early mass spectrometers. (Courtesy of IsoPrimeLtd.)

inlet system was added (dual viscous flow inlet system, McKinney *et al.*[21]) to the instrument designed by Nier. The inlet system was based on a "changeover valve" developed by Murphey in 1947 for thermal diffusion studies.[22] During the 1950s Epstein and coworkers at the California Institute of Technology in Pasadena,[23–27] Urey in Chicago,[4,28,29] and Harmon Craig at Scripps, worked on improvements of the initial techniques and developed the use of stable-isotope variations for investigations in geosciences, thus founding "stable-isotope geochemistry". Stable-isotope methods and IRMS systems in Russia were developed in the 1940s and applied with a specific focus on sulfur. The investigation of sulfur isotopic composition started with the pioneering works of Trofimov and Thode at the end of the 1940s.[30,31]

The field of stable-isotope ratio mass spectrometry grew considerably during the 1960s and 1970s: commercial mass spectrometers became available and many new applications, mainly in the geosciences, were developed. With the discovery of the $^{13}C/^{12}C$ isotope ratio difference between the C-3 and C-4 photosynthetic cycles,[32–34] stable-isotope studies entered the world of food and agricultural science, as well, and found applications in food authenticity and food control. The method was extended soon to the investigation of food webs, biosynthesis and ecosystem studies (see for instance the pioneering work of DeNiro and Epstein[35–37]).

**New Tools: irm and ConFlo**

The world of stable-isotope instrumentation saw a major change and breakthrough in applications during the 1980s. Hyphenated techniques evolved, in which chromatographic separations using a helium carrier gas were coupled with IRMS. This technique enabled precise isotopic measurements from gas-chromatography effluents and other systems, which also made use of a carrier gas. When made on a GC-IRMS, such measurements were referred to as "isotope ratio monitoring" (irm).[38–44] When performed using an elemental analyser-IRMS, the term "continuous flow" (ConFlo) was more common.[45–48] The development of these techniques led to an explosion of stable-isotope studies.

**Essential: Sample Preparation**

Before the introduction of irm/ConFlo, the basic steps of isotope ratio analysis were as follows: Starting from the original sample, which can be a solid, liquid or a gaseous form, the element(s) of interest has to be chemically transformed into a simple gaseous form prior to isotopic analysis (*e.g.* $CO_2$, $H_2$, $N_2$, $N_2O$, CO, $O_2$, $SO_2$, $SF_6$). Quantitative conversion is necessary to avoid isotopic fractionation. In some cases, fractionation is inherent in the reaction. In this case quantification of the fractionation is required and the establishment of techniques for keeping the fractionation factors constant is needed. An example is the oxygen exchange reaction between water samples and $CO_2$ gas, which is introduced into the mass spectrometer for the actual isotopic analysis of the water sample. At room temperature, the oxygen in the $CO_2$ – when in equilibrium with the liquid water sample – is enriched in $^{18}O$

by about 40‰. As a consequence, the temperature has to be kept constant to within ± 0.1 °C for the water sample and likewise for the reference water measured in the same sequence. After equilibration, both (sample and reference) $CO_2$ gases are introduced into the dual-inlet system of the IRMS, thereby preventing the alteration of the isotopic composition of the $CO_2$ to influence the final isotope ratio result.

Until the mid-1980s most sample preparation steps were performed off-line in chemical laboratories equipped with dedicated, homemade vacuum lines. The gas produced from the sample was stored in glass vials or in sealed tubes for later admission to the mass spectrometer. The dual-inlet system usually had dedicated reference and sample gas inlets. The reference gas was often the same for most sample gases – a daily-fresh aliquot from a larger gas reservoir. This reference gas had been calibrated previously using internationally available reference gases or materials like carbonates, from which the calibration gas had to be generated. This process often resulted in a significant measurement bias. In addition, it was not easy to identify all possible sources of uncertainties, making the accurate measurement of very subtle isotopic signatures cumbersome, sometimes even impossible.

**Computer Control and Automation**

The classical dual-inlet techniques were extended by the development of automated preparation systems for isotopic measurements from carbonates[49–51] and water, by multiport inlet systems for automated analysis of multiple samples and by cryogenic trapping systems, which could be used to analyse combustible organics.[52,53]

**Open Split Inlet**

With the invention of the new irm and ConFlo techniques in the 1980s, a new area of stable-isotope ratio research emerged: The online coupling of automated preparation systems (elemental analysers, gas chromatographs, characterised by helium as carrier gas) to stable-isotope mass spectrometers operating at ultrahigh vacuum in general was achieved *via* an "open split". This device had already been in use for coupling gas chromatography to organic mass spectrometers,[54] but had been abandoned in favour of more sensitive direct coupling devices. The open split turned out to be the ideal coupling device for IRMS. In its simplest form, it consists of a tube-in-tube arrangement, where the larger tube is the exhaust of the preparation device and the smaller tube acts as the flow-limiting conduit to the IRMS ion source. The open split allowed for a simple reference injection by positioning a reference gas outlet capillary upstream from the sniffing point, as well.[41]

**irm: Beyond Carbon Isotopes**

The commercial introduction of irm/ConFlo systems permitted a simplification of sample preparation for (automated) stable-isotope measurements, thereby also reducing analysis costs and fostering migration of stable-isotope techniques to new research areas as well as commercial laboratories, *e.g.* for quality control in industrial applications. The initial ConFlo isotope systems allowed $^{13}C/^{12}C$ and $^{15}N/^{14}N$ analysis[46,55] with an elemental analyser. Sulfur $^{34}S/^{32}S$ determination was added in the 1990s.[56] Applications of early irm GC-IRMS systems were largely focused on $^{13}C/^{12}C$ determinations. Nitrogen $^{15}N/^{14}N$ analyses were added 3–4 years later,[57] almost together with the first $^{18}O/^{16}O$ measurement capabilities using CO ($m/z = 30/28$) as the analysis gas.[58] Using GC-IRMS systems, measuring carbon isotopes became routine work in many laboratories, whereas nitrogen and oxygen analysis remained a specialty of only a few laboratories for quite some time. Obstacles as encountered during formation of CO from oxygen-bearing compounds were also observed in attempts to analyse $^{2}H/^{1}H$ ratios using irm/ConFlo arrangements. However, these problems only became visible after having solved a more serious problem: Deuterium is recorded as $^{2}HH^{+}$ on the mass-3 channel, close to where $He^{+}$ ($m/z = 4$) would be measured. The low-mass (energy-loss) tail of the $He^{+}$ ion current is large enough for the sensitive cup at the position $m/z = 3$ cup to be saturated in a single-focusing mass spectrometer. The state-of-the-art in $\delta^{2}H/^{1}H$ isotope measurements was discussed in 1996 in detail by Brand *et al.*[59] The tail problem at $m/z = 4$ was finally solved by the introduction of a suitable energy filter – either as a saddle-point retarding lens filter[60] or *via* addition of an electrostatic sector[61] in front of the Faraday cup at the position $m/z = 3$.

In 2000, irm/ConFlo commercial systems were more or less "complete"; they were suitable for carbon, hydrogen, oxygen, nitrogen and sulfur (for elemental analyser only) isotope ratio measurements. Coupling with gas chromatographs included the commercially available purge-and-trap options, cryogenic preconcentration and focusing and even carbon isotope analysis from dissolved organic and inorganic carbon, provided that the final analysis gas could be carried in a helium gas stream and presented to the mass spectrometer at the open-split interface.

After two moderately successful attempts to build an interface for measuring carbon isotope ratios from HPLC effluents using a moving-wire system,[62,63] a breakthrough came with a new approach, incorporating a postchromatography chemical oxidation step into the liquid phase and purging the formed $CO_2$ gas using a gas-permeable membrane and helium as the carrier gas.[64,65] Because of the requirement for total conversion of C in the HPLC effluent, it is evident that the solvent cannot contain any combustible C other than that in the analyte. Therefore, the solvent is restricted to ultrapure water with usable gradients in inorganic salt content or temperature. While this restricts the system to a limited number of separations, it has proven to be very useful for analysing sugars and sugar derivatives, amino acids, short-chain fatty acids, DNA and other biocompounds that are soluble in aqueous solutions as well as for analysis of dissolved organic and inorganic carbon (DOC and DIC).

**The New Dream Machine: Double Focusing and High Mass Resolution**
In the early 2000s a new stable-isotope research tool called "clumped isotopes" was introduced, in which the abundance of doubly substituted isotopologues were quantified.[66] The ion beams are weak, because these species are not abundant. Therefore, detection of and subsequent correction for isobaric interferences on minor ion beams has become a crucial issue and can best be done by a double-focusing mass spectrometer operating at medium to high resolution.[67] As a consequence, high mass resolution IRMS instruments have been developed and became commercially available, recently.

## 16.2 Technical Background

The first methods for stable-isotope ratio analysis were extremely time consuming in order to achieve the basic requirements of high precision and accuracy. Compared to modern performance assessment using reference materials, the achieved precision was rather moderate, so that only larger isotopic fractionation phenomena could be studied. The original IRMS had only one inlet system: the dual viscous flow inlet system, used for comparing an unknown sample to a reference gas. All samples had to be prepared offline. Manual or semiautomated glass lines were common in stable-isotope laboratories. Gases were sealed in tubes for subsequent introduction into the variable bellows volumes of the dual-inlet system of the mass spectrometer. The dual-inlet system could be enhanced with microvolume inlets in order to enable small samples to be analysed. The first automation steps used multiple inlet ports and automated tube crackers for unattended overnight analysis of multiple samples. A large variety of homemade automated sample-preparation devices focused on the isotopic analysis of carbonates and water samples.

The increased availability of fully automated, high-precision analytical systems changed the situation. Nowadays, isotope ratio mass spectrometry is used in many different fields. Samples can be solids, liquids, plain gases, or trace gases. They can be studied for their bulk isotopic signatures or the signatures may be broken down to the compound-specific level. However, one general rule still holds in spite of all advances that have been made: a given sample always has to be transformed into something that can be manipulated, separated and detected by the mass spectrometer. The ability to analyse the contents of a given sample is facilitated when the sample complexity is reduced. The easiest way to do this is *via* separation of the individual chemical components before the actual measurement. This principle has paved the way for the commercial success of preparation systems hyphenated with isotope ratio mass spectrometers. The most common valuable sample preparation is the oxidation of an organic material inside a CHN analyser, thereby producing $N_2$, $CO_2$ (and $H_2O$). Today, the instrumental approach most often found in stable-isotope laboratories for measuring carbon and nitrogen isotope ratios is the direct coupling to an IRMS *via* an open-split interface. Other online systems with special high-temperature (>1350 °C) reactors filled with glassy carbon are successfully operated for hydrogen and oxygen isotope ratio measurements. In all cases, the analyte gases reach the mass spectrometer as a transient pulse entrained in a helium carrier gas.

The stable-isotope ratio mass spectrometer is characterized by a dual inlet system, an electron impact ion source, an ion separation device and a detector array for sampling the ion beams and quantifying their current intensities with high precision. All modern IRMS mass spectrometers are magnetic sector field devices.

The inlet system for IRMS can be a vacuum vessel (or two in the dual-inlet system) or an open-split device for introducing effluents from a chromatographic separation. The inlet systems are designed to handle pure gases – principally $CO_2$, $N_2$, $H_2$, and $SO_2$ (but also $N_2O$, $O_2$, $SF_6$, CO and more).

The molecules are introduced *via* the inlet system into the ion source, where they are ionised by electron impact (EI) and accelerated by several kilovolts, separated by a magnetic field and detected in Faraday cups. The essentials of the mass spectrometric design have already described and discussed in detail in the literature; see for instance the *Handbook of Stable Isotope Analytical Techniques*.[68] This Handbook, edited by De Groot, is one of the major references for stable-isotope analytical techniques.[69] Further instrumental reviews have been published and are recommended for further reading in order to gain a deeper instrumental insight.[70–75] In the following sections, only a brief overview of the basic principles is given.

## 16.2.1 Sample Preparation

Either the sample is already in its gaseous form or needs to be transformed into a gaseous form ($CO_2$, $H_2$, $N_2$, $N_2O$, CO, $O_2$, $SO_2$, $SF_6$, $CH_3Cl$, $CH_3Br$, $SiF_4$, ($UF_6$)).

**Elemental Analyser (EA)**

Solid samples (*e.g.* soil, tissues or plant material) or liquids are combusted prior to further analysis. Generally, there are two types of elemental analyser (EA): Either the sample is combusted in a helium atmosphere enriched with oxygen (EA for C and N resulting in $CO_2$ and $N_2$) or the sample undergoes a high-temperature thermal conversion (TC/EA for H and O resulting in $H_2$ or CO). The solid or nonvolatile liquid sample is introduced into the EA using tin (for C and N analysis) or silver (for H and O analysis) capsules. Liquids with limited viscosity can be introduced directly using a liquid inlet system (like a syringe).

The combustion of C and N takes place in an $O_2$-enriched atmosphere, where $CO_2$, $NO_X$ and $H_2O$ are produced. The reactor can contain $Cr_2O_3$ and $Co_3O_4 + Ag$ for binding sulfur and halogens. The temperature in the reactor is at 950–1050 °C. Owing to the combustion of the tin, the reaction temperature is briefly increased up to 1800 °C. The removal of excess of $O_2$

and the reduction of $NO_X$ to $N_2$ takes place in a subsequent quartz reactor containing Cu (with possible variations) at a temperature of 600–650 °C. The water is removed from the analyte stream using a chemical trap usually filled with $Mg(ClO_4)_2$. When only N-isotopes are measured, an additional chemical trap removes $CO_2$. Finally, $CO_2$ and $N_2$ are separated by means of an isothermal packed GC column or alternatively "purge and trap" systems are applied. The latter consist of short GC columns (adsorption tubes), which collect the gases and further release them *via* electrical heating.

High-temperature conversion takes place in a carbon-containing reactor, which is operated between 1350 and 1450 °C; H- and O-bearing compounds are converted to $H_2$ and CO. The system usually consists of concentric tubes. The outer tube is made of $Al_2O_3$ (or SiC) and the inner tube is made of glassy carbon and filled with glassy carbon chips. Silver wool inside the inner tube bind halogens. The gases are separated subsequently using an isothermal packed GC column - capable of separating $N_2$ and CO. $N_2$ forms an isobaric interference with CO and can affect the ionisation of $H_2$. Additional traps (*e.g.* charcoal, $Mg(ClO_4)_2$ or $P_4O_{10}$) can be used to remove further reaction product gases.

**GC-IRMS**

Gas chromatography IRMS (GC-IRMS) techniques have been developed together with online systems for the EA-IRMS coupling.[38,39,44,47,76,77] In essence, the GC-IRMS systems are the multiple-peak version of the EA systems. In combustion mode, the GC effluent passes through a microcombustion reactor to convert all eluting organic compounds to $CO_2$, $N_2$ and $H_2O$. Water removal is usually accomplished using a semipermeable Nafion® membrane (Perma Pure, Toms River, NJ 08755 USA), sometimes complemented by an additional dry-ice trap mounted in front of the open split. A liquid-nitrogen trap for $CO_2$ removal is employed for $^{15}N/^{14}N$ analyses.

Instead of an oxidation reactor, a high-temperature reduction furnace kept at >1350 °C can be used for compound-specific hydrogen isotope analysis from GC effluents. While now routine for $\delta^2H$ determination, analysis of $\delta^{18}O$ has turned out to be more difficult.[78] As in EA/IRMS, the analyte gas is admitted to the IRMS ion source *via* a low-flow open split, thereby carefully optimising the split efficiency for sensitive detection from capillary chromatography. For more detailed descriptions we refer to the vast amount of specialised and review literature[79–82] and citations therein.

**LC-IRMS**

LC-IRMS coupling is mostly used for measurements of C isotope ratios in organic substances after separation by an HPLC. The direct coupling of a LC system provides the solution to the IRMS system *via* an interface - mostly a wet-chemical oxidation interface[64,83] or a moving-wire interface.[84] In the wet-chemical interface, the carbon-containing organic substances in the liquid phase are directly transformed in a capillary oxidising reactor to $CO_2$ by means of an oxidising reagent such as $(NH_4)_2S_2O_8$ in combination with

$H_3PO_4$ and $AgNO_3$ as catalysts. The resulting $CO_2$ is transferred *via* a permeable membrane into a counter purge stream of helium, further dried using a Nafion® membrane and transferred to the mass spectrometer *via* an open split. In order to avoid interferences, the mobile phase must not contain organic substances.

**LA-IRMS**

Laser ablation has been used to release the sample by heating (decarbonation of $CaCO_3$) and to combust samples *in situ* (combustion of S to SO) with high spatial resolution.[85] The direct solid sampling allows, *e.g.* fluorination of O in silicates and spinels (to $O_2$).[86]

**Fluorination ($SF_6$, $UF_6$, $SiF_4$, $SeF_6$)**

Fluorine and fluorine-bearing compounds ($BrF_5$, $ClF_3$, $BF_3$) with or without HF have been used with IRMS as a method to transfer the analytes of interest into gaseous fluorine-containing species or for studying the triple oxygen isotopes of oxygen. Fluorination techniques have also been successfully extended to other elements like S, Se and U. Besides conventional techniques (fluorination in high-temperature furnaces), laser-induced heating techniques have been applied. For a more detailed description of reduction and fluorination techniques we refer the reader to Taylor's chapter in the "Handbook of Stable Isotope Analytical Techniques".[87]

## 16.2.2 Sample Introduction

Inlet systems for IRMS are usually straightforward consisting of inert materials such as, *e.g.* stainless steel or glass compounds (*e.g.* capillaries and valves) and the design tends to avoid cavities as these are major sources of cross contamination. The inlet system for IRMS can be a vacuum vessel (or two in the dual-inlet system) or an open-split device for introducing effluents from a chromatographic separation. An important aspect of the inlet system lies in the use of capillary tubes. Within capillary tubes, viscous flow ensures that there is no time-dependent isotopic fractionation affecting the relative measurement of the gases. Most IRMS instruments are operated in dynamic mode, whereas noble-gas spectrometers are generally operated in static mode, even though modifications can be used (see Chapter 17).

**Dynamic IRMS**

An interface is needed for the introduction of the gaseous species into the mass analyser. The interface has to limit the amount of gas entering the ion source and also enables the introduction of working gases and the introduction of He for its dilution. Dynamic IRMS is operated in two basic types of IRMS: either in continuous flow (CF) or as dual inlet (DI), with the latter providing improved precision (Figure 16.4).

In a continuous-flow system, sample and standard gases are carried in a helium stream allowing pulse injections of the sample gas thus reducing

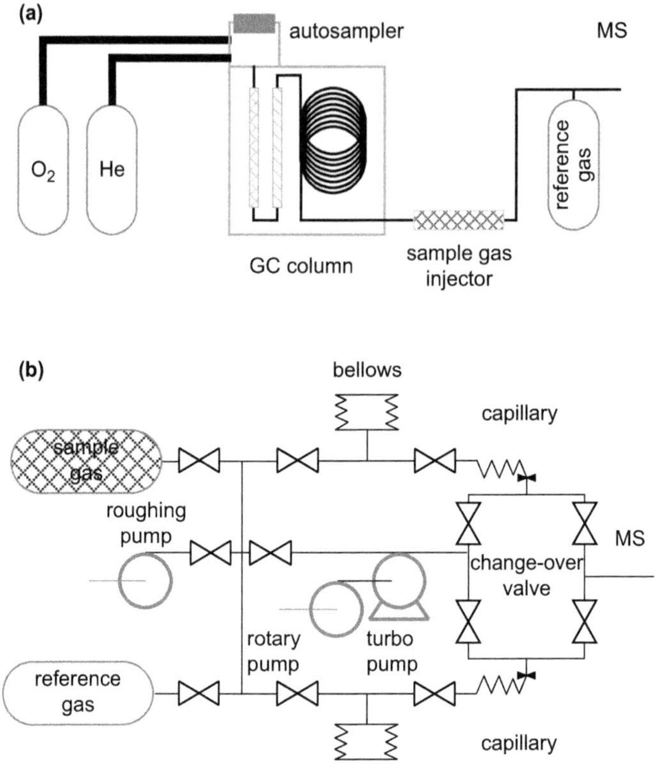


**Figure 16.4** IRMS interface (a) continuous flow and (b) dual inlet.

sample size and consumed amounts. Continuous-flow inlet systems can be configured with a laser-ablation system for direct measurement of transient signals of solid samples.

In a dual-inlet system, the sample gas is introduced into a variable-volume reservoir (also known as bellows) where it is adjusted to match the pressure of the reference gas (or vice versa). The sample gas flows through a capillary to the changeover valve which directs the fluxes to the mass spectrometer or to the waste reservoir, respectively. This system maintains a constant flow of both gases and allows direct and repeated measurements of the sample and standard under identical temperature and pressure conditions as required for accurate isotope ratio measurements.

### 16.2.3 Ion Source

Most IRMS systems employ "electron impact" ionisation (EI), which has the least isotopic discrimination and generates in general an overall intense ion beam. Electrons are produced from a heated filament and are accelerated through a potential of 50–120 eV into the source chamber. The electrons are confined into a beam using a small magnetic field induced by permanent magnets. The electrons hit the molecules (M) producing positive ions according to:[88]

$$e^- + M \rightarrow M^{+\bullet} + 2e^-$$

Two electrons exit from the collision complex, leaving a positively charged radical cation, the "molecular ion" ($M^{+\bullet}$). Most ions produced by EI are single charged. Doubly charged ions are observed as well, albeit at a much lower intensity. The triplet at $m/z = 22$, 22.5 and 23 in the $CO_2$ mass spectrum is a good example. These are the doubly charged molecular ions ($CO_2{}^{2+}$), which arrive at the collector plane at half the nominal masses of the $CO_2$ isotopologues.

The ions are extracted from the ion source and shaped into a beam, which enters the analyser through the alpha – or source – slit. Usually, beam-steering plates allow the beam to be guided optimally through the flight tube and magnet gap.

IRMS usually use Nier-type ionisation sources (Figure 16.5 and Figure 16.6) including various modifications.

The Nier source (as shown in Figure 16.5) uses a tungsten filament, which emits electrons that pass through the ionisation volume and are received by a trap measuring the electron current (*I*). The source is regulated so that (*I*) is kept constant as the electron beam has a significant impact on instrumental mass fractionation and sensitivity and must therefore not fluctuate. A small magnetic field (*B*) in the Nier source keeps the electrons confined to a small cross-sectional area from which the newly formed ions are extracted. The ion extraction is achieved by the voltages applied to the repeller (if present) and to the draw out plates.

Sensitivity as well as instrumental mass fractionation depend on the gas pressure in the source (*i.e.* the amount of gas). Thus, standards and sample should be similar in composition (*e.g.* purity) and pressure. On the other hand, modern instruments are designed for lowest dependence of the

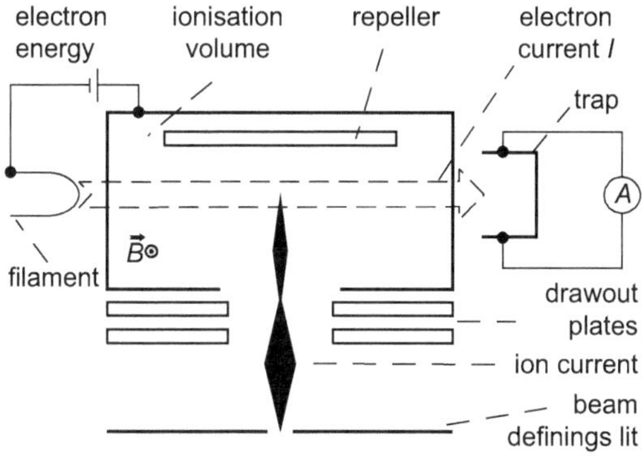


**Figure 16.5** Schematics of a Nier source.
(Source: Ondrej Hanousek adapted from Burnard *et al.*[89])

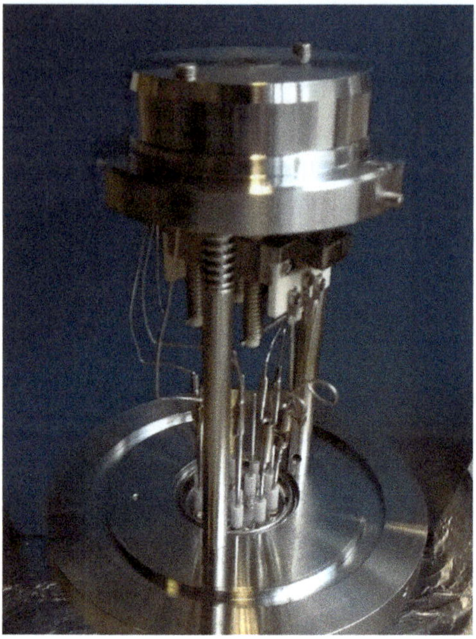

**Figure 16.6** Photograph of a Thermo-Fisher MAT Delta plus EI source. (Source: Willi A. Brand, © MPI-BGC Jena.)

measured ion current ratios upon signal size. Typically, the current commercially available instruments often have less than 0.01 ‰/nA signal difference at $m/z=$ 44 ($CO_2$).

## 16.2.4 Sector Field Mass Analyser

IRMS use - almost without exception - sector field mass separators. Generally, an IRMS only needs a magnet for the separation of the ions and directional focusing as the ion source provides ions with a narrow energy distribution. Most modern IRMS systems are equipped with electromagnets scanning from $m/z=2$ to $m/z=70$ or to even higher masses ($m/z=150$). In (older) systems with a permanent magnet, the mass range is limited. Mass selection is made by variation of the acceleration voltage alone. Hence, the sensitivity for different masses is not constant. Moreover, low masses (especially $m/z=18$, water) are difficult to reach without changing the magnetic field strength. Therefore, electromagnets have become the preferred option. They are designed to provide full coverage of the ion path when working with widely separated masses, *e.g.* $m/z=2$ and 3.

Even though a single magnetic sector has been considered adequate for IRMS instrumentation, recent developments have introduced a double-focusing design for achieving high mass resolution and abundance sensitivity.[67] This design enables high drawout fields to be employed and, thus high ratio linearity without compromising the mass spectrometric peak shape. Moreover, small isobaric interferences can be detected and, sometimes, eliminated.

### 16.2.5 Interface and Vacuum System

Most instruments use dual-stage vacuum systems consisting of a primary vacuum stage provided by rotary pumps and a secondary vacuum stage using turbomolecular pumps. Oil diffusion pumps can still be encountered on older systems. Yet, modern instruments are exclusively equipped with turbomolecular pumps. Some systems use differential pumping with one turbomolecular pump connected to the source of the instrument and one connected to the collector part of the flight tube. Most IRMS systems are equipped with vacuum gauges to control the analyser vacuum. Usually, IRMS instruments operate at vacuum levels around $1–5\times10^{-8}$ mbar ($10^{-6}$ Pa) in dual-inlet mode and $2–4\times10^{-6}$ mbar ($10^{-4}$ Pa) when in continuous flow mode – the latter representing the noncorrected reading of the carrier-gas pressure.

### 16.2.6 Detection System

Nearly all IRMS use Faraday collectors in their detection system. After separation according to their $m/z$ ratios, the ion beams are focused into a Faraday cup. Modern systems are equipped with an expanded dynamic analysis range and variable resistor values. Some IRMS systems are available with moveable collectors, but most commercial systems have a set of fixed collectors with a different width to accommodate the dispersion of the various isotopic multiplets. Zoom-lens technology is an exception to this rule.

Supplementary techniques for suppressing, *e.g.* the low-energetic $He^+$ tail in online $\delta^2H$ measurements (see earlier discussion) vary considerably among the different manufacturers. In hydrogen, the abundance of natural deuterium is typically below 160 ppm, therefore its ion current at $m/z=3$ is rather weak. When analysing hydrogen isotope abundances in a He carrier gas system, the neighboring peak at $m/z=4$ is rather high and has a pronounced peak tailing, interfering with the $m/z=3$ ion current measurement. Several technical options have been chosen by various manufacturers to solve this problem, *e.g.* applying electrostatic filters or retarding lens arrangements. These improvements benefit from the fact that the helium ions at $m/z=4$ have an energy deficit, when they are able to enter the $m/z=3$ Faraday-cup collector.

## 16.3 Commercially Available Instruments

This section provides a list of currently commercially available gas source MS instruments with a significant market share.

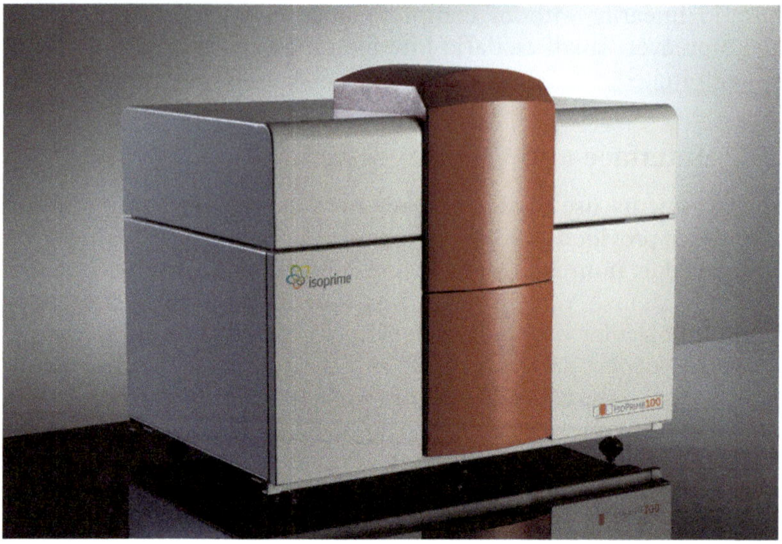


**Figure 16.7** *IsoPrime 100* IRMS.
(Source: © IsoPrime.)

***IsoPrime 100 IRMS* (IsoPrime Elementar)**

The *IsoPrime 100* is a magnetic sector mass spectrometer with a 90° geometry and a 9-cm radius electromagnet. In its basic configuration $\delta^2H$, $\delta^{13}C$, $\delta^{18}O$, $\delta^{15}N$ and $\delta^{34}S$ can be measured (Figure 16.7).

The instrument is fitted with a "universal triple collector" as standard for the analysis of $\delta^{13}C$, $\delta^{18}O$, $\delta^{15}N$ and $\delta^{34}S$. An additional Faraday cup may be installed to allow for $\delta^2H$ analysis, complete with an electrostatic filter for $H^{3+}$ correction. The instrument has the capacity for detecting and measuring up to 10 ion beams simultaneously and can be used for special applications such as analysis of chlorine and bromine isotopes, $N_2O$ isotopomers, and $SO_2$–SO fractions. The *IsoPrime 100* is compatible with several automated preparation systems for stable-isotope determinations available from the same manufacturer.

***Delta V* (Thermo Fisher Scientific)**

The *Delta V* mass spectrometer is the most recent instrument in a series of "*Delta*" IRMS manufactured at Thermo Fisher Scientific in Bremen, starting in the early 1980s with the "*Delta E*". These instruments share a 3-kV acceleration and 18-cm dispersion geometry. Mass ranges from $m/z = 28$ to $m/z = 46$ can be measured simultaneously. The hydrogen isotopes are measured in the same image plane where the "mass-3 Faraday cup" has a deceleration filter for $He^+$ stray ion removal. The instrument shares the same family of peripheral devices for automated sample preparation and analysis with the *MAT-253* and the *MAT-253* Ultra. The instruments are operated using the same software (Figure 16.8).

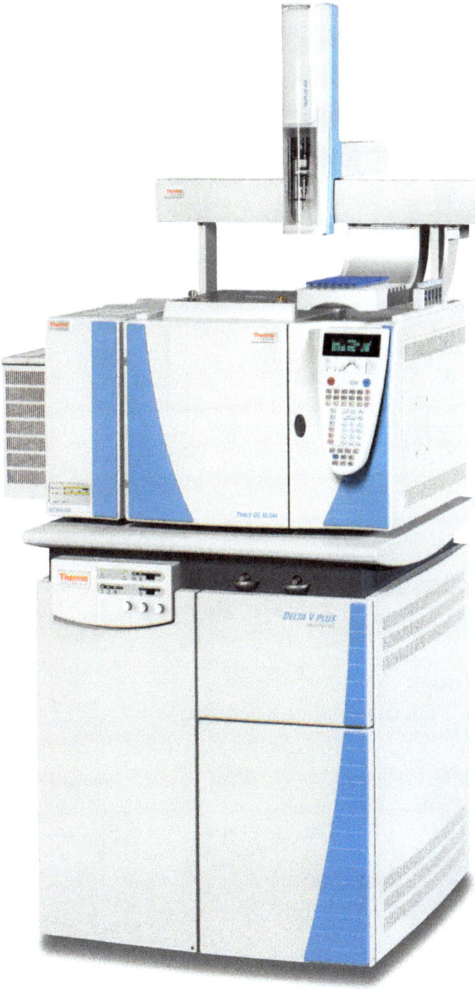

**Figure 16.8** *Delta V* IRMS.
(Source: © Thermo Fisher Scientific.)

***MAT 253* (Thermo Fisher Scientific)**

The 10-kV MAT 253 IRMS is the fourth generation of a series of mass spectrometers that started in 1978 with the MAT 250. The large-dispersion (42 cm) flight path is able to get a high sensitivity and wide flat top of the respective mass peaks at the same point in time (Figure 16.9).

***MAT 253 Ultra* (Thermo Fisher)**

The *MAT 253 Ultra* is a completely new double-focusing mass spectrometer, further developed from the TIMS mass spectrometer (*Triton*) and the MC ICP-MS machine (*Neptune*) (Figure 16.10). Instead of the ICP torch and ion optics, the *Ultra* features a *MAT-253 EI* gas ion source. The instrumental

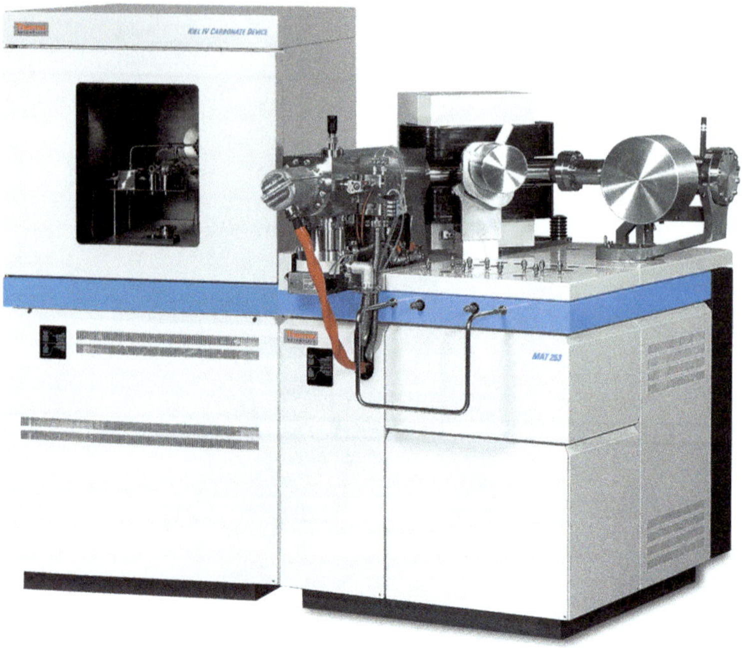

**Figure 16.9** *MAT 253* IRMS.
(Source: © Thermo Fisher Scientific.)

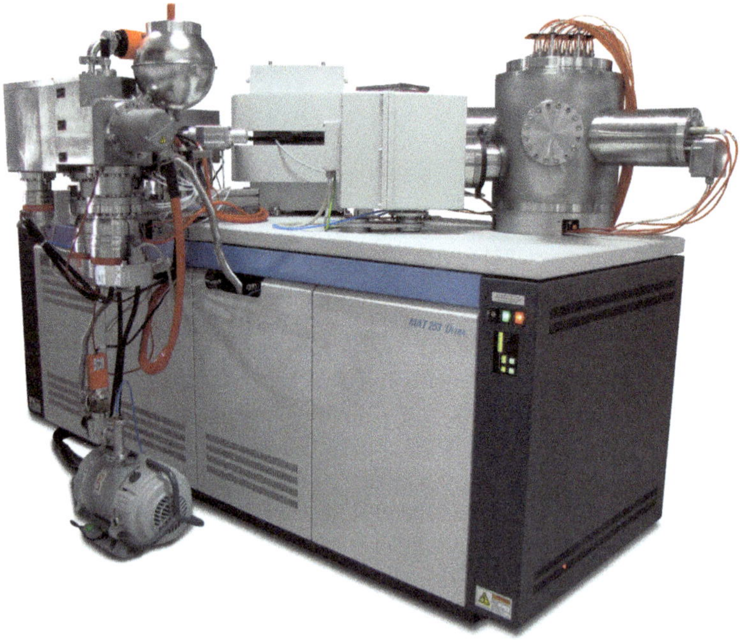

**Figure 16.10** *MAT 253 Ultra* IRMS.
(Source: © Thermo Fisher Scientific.)

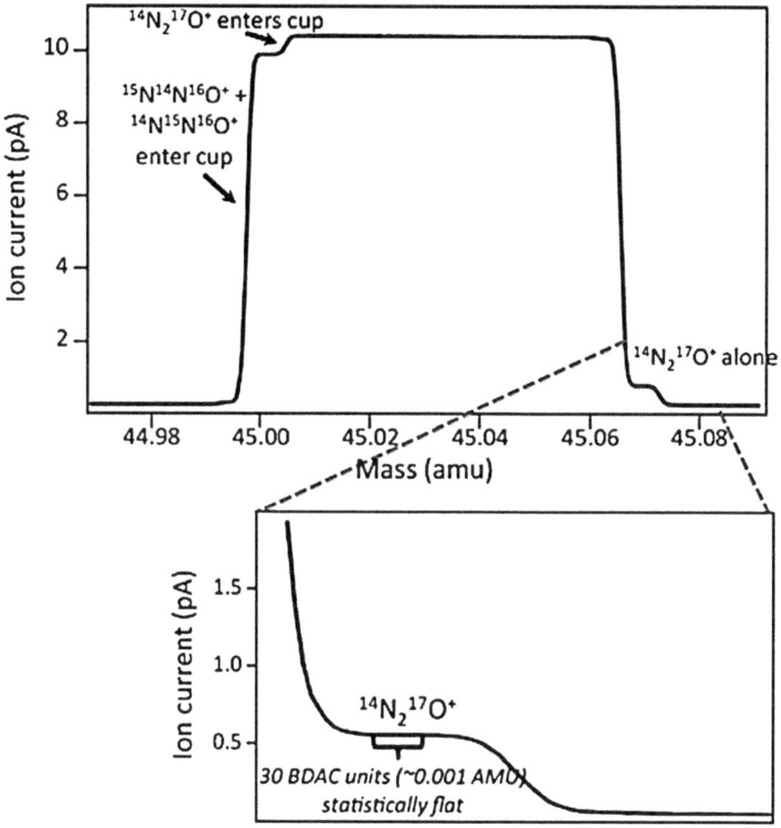


**Figure 16.11** $N_2O$ at $m/z = 45$; ion current at $m/\Delta m = 17\,000$ resolution. (Source: courtesy A. Hilkert, Thermo Fisher Scientific.)

development was triggered by new demands for sensitivity and stability in "clumped" isotope ratio analysis.[66] Consequently, the first instrument found its home in the Caltech laboratories of Eiler.[67] A striking feature is the high mass resolution (up to $m/\Delta m = 25\,000$), as exemplified by the $N_2O$ ion beam at $m/z$ 45 depicted in Figure 16.11.

### *SerCon 20-22* (and *Integra-2*) (SerCon)

The *SerCon 20-22* IRMS system is a magnetic sector mass spectrometer with a 120° geometry and a 11-cm radius electromagnet[90] (Figure 16.12). Depending on the options chosen, the *SerCon 20-22* can operate in dual-inlet and irm/ConFlo mode. It is designed to measure $\delta^2$H, $\delta^{13}$C, $\delta^{18}$O, $\delta^{15}$N and $\delta^{34}$S and can be configured as well for $\delta^{37}$Cl and $\delta^{81}$Br isotopic measurements. The *SerCon 20-22* is compatible with an automated preparation system for stable-isotope determinations (DI water analyses, DI carbonate analyses, EA, GC, atmospheric trace gases). Using the same instrument, SerCon also offers a combined elemental analyser–IRMS system called *Integra-2* with a small footprint.

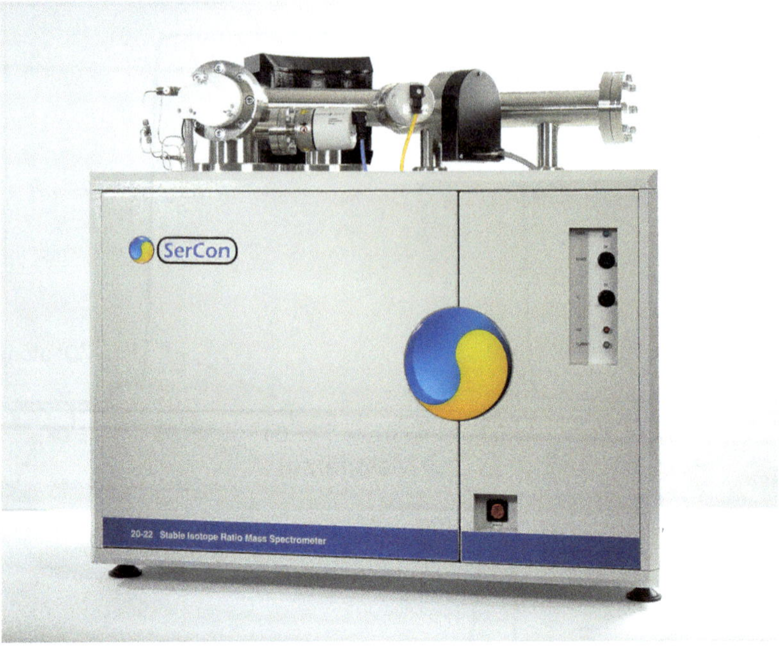


**Figure 16.12** *SerCon 20-22* IRMS.
(Source: © SerCon)

***SerCon MIRA* (SerCon)**

In 2012, SerCon introduced a new research-type stable-isotope ratio mass spectrometer, the SerCon MIRA (Figure 16.13). This instrument is a large-dispersion 120° deflection single-focusing mass spectrometer specifically designed for "clumped" isotope research.

***IDmicro* (Compact Science Systems Ltd.)**

The *IDmicro* is a small IRMS with a fixed magnet for routine $\delta^{13}C$, $\delta^{15}N$, $\delta^{18}O$, and $\delta^{34}S$ measurements. The system can work remotely and has low power consumption as it is based on the system, which was designed for the European Space Agency. The *IDmicro* is coupled to a common elemental analyser interface allowing the user to choose the EA most suitable to their application. The *IDmicro Breath* is dedicated to breath-gas analysis.

***Isologger* (Compact Science Systems Ltd.)**

The basic idea was the development of a portable but robust system with a simple design and small footprint. One version of this instrument was deployed even in outer space on the Mars Express in 2003.

***Nu Horizon* (Nu Instruments)**

The *Nu Horizon* (Figure 16.14) is a 5-kV magnetic sector mass spectrometer based on a 90° geometry with a large mass dispersion (30 cm effective

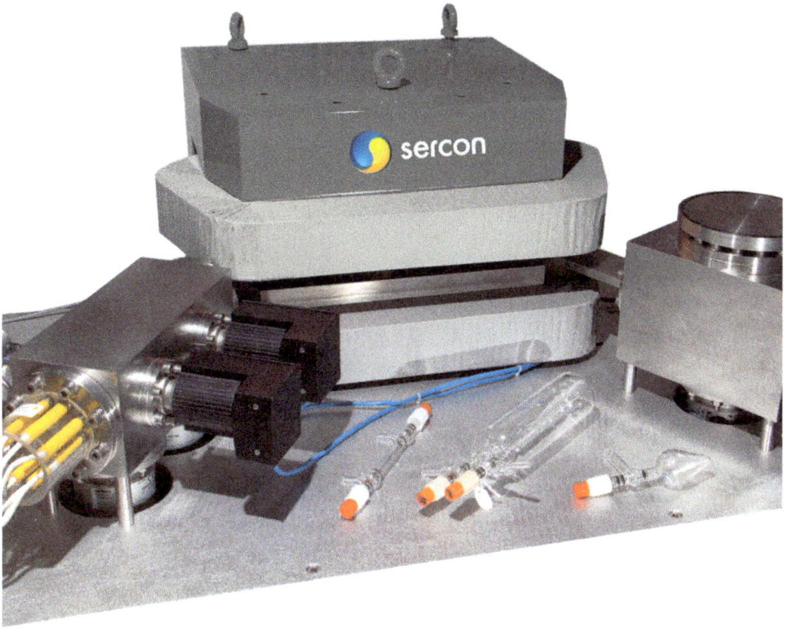


**Figure 16.13** *SerCon MIRA* IRMS.
(Source: © SerCon.)

**Figure 16.14** *Nu Horizon* IRMS.
(Source: © Nu Instruments.)

magnetic deflection radius). With its configuration up to 6 collectors, the *Nu Horizon* is able to measure $\delta^2$H, $\delta^{13}$C, $\delta^{18}$O, $\delta^{15}$N and $\delta^{34}$S and can be configured as well for $\delta^{37}$Cl and $\delta^{81}$Br isotopic measurements. The variable zoom optics allows the use of small-entry Faraday cups for all isotope triplets. The *Nu Horizon* is compatible with automated sample-preparation systems for stable-isotope determinations (DI water analyses, DI carbonate analyses, EA, GC, atmospheric trace gases).

***Nu Perspective* (Nu Instruments)**

The *Nu Perspective* (Figure 16.15) is an 8-kV single-focusing magnetic sector mass spectrometer based on a 90° geometry with an extra-large 60-cm effective mass dispersion. Like the smaller *Nu Horizon*, the zoom optics enables most isotope triplets to be measured with small Faraday cups. Up to 12 such detectors can be positioned along the focusing plane for simultaneous ion current monitoring. With a standard setup, $\delta^2$H, $\delta^{13}$C, $\delta^{18}$O, $\delta^{15}$N, $\delta^{34}$S, $\delta^{37}$Cl and $\delta^{81}$Br isotopic measurements can be made. The *Nu Perspective* is compatible with automated preparation systems for stable-isotope determinations (DI water analyses, DI carbonate analyses, EA, GC, atmospheric trace gases).

***Nu Panorama* (Nu Instruments)**

The *Nu Panorama* (Figure 16.16) is a newly designed double-focusing magnetic sector mass spectrometer with a 4×5 m footprint. The instrument is designed as a high-resolution double-focusing mass spectrometer and its

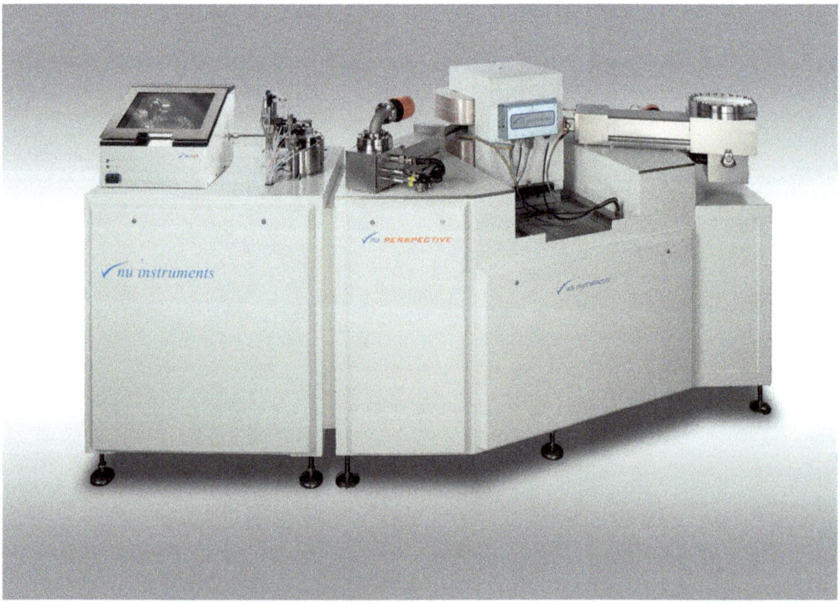


**Figure 16.15** *Nu Perspective* IRMS.
(Source: © Nu Instruments.)

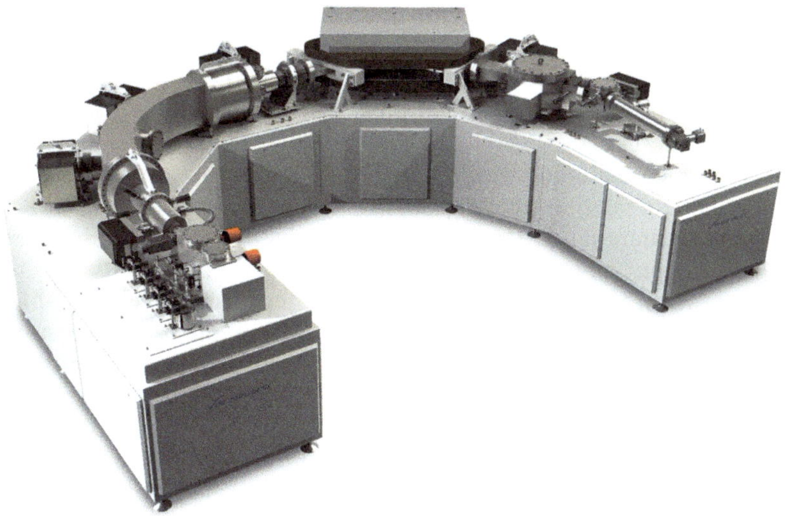

**Figure 16.16** *Nu Panorama* IRMS.
(Source: © Nu Instruments.)

major intended use is to reliably measure the very small signatures of "clumped" isotopes.

## 16.4 Measurement Considerations

IRMS is typically applied for measuring isotopic compositions at low enrichment and natural abundance levels. Small variations in very low amounts of the heavier (or less-abundant) isotopes are quantified with utmost precision when in the presence of large amounts of the lighter isotope. Moreover, the low detection limit of IRMS for measuring isotopic enrichments is increasingly applied for the use of small amounts in isotopic tracer studies, *e.g.* in medical studies, as isotopically labelled compounds are usually expensive.

### 16.4.1 Hydrogen

Hydrogen isotope ratios have been measured predominantly from water in the past (see above). Today, they can be measured reliably with TC/EA-IRMS techniques as well, avoiding the conversion to water. The system was already described briefly above for water samples. Like for oxygen isotope analysis, solid or liquid samples need to be wrapped in Ag foil (or Ag capsules). The major obstacles to a wider application up to now are two-fold:

(1) International reference materials available are NBS 22 (oil) and IAEA-CH7 (PE foil). Both materials have a similar $\delta^2H_{VSMOW}$ value (−117‰[91] and −103.5‰,[92] respectively).

(***Note***: VSMOW is the Vienna Standard Mean Ocean Water reference standard used for defining this delta scale. Find out more about reference standards for delta scales in Brand.[68])

In contrast – owing to their biosynthesis – natural (organic) samples in general are depleted in $^2H$ considerably, often beyond −300‰. Two reference materials with a wide isotopic distance are necessary for a reliable analysis. A corresponding effort to produce a set of new reference materials is currently being made.

(2) Hydrogen isotopes in organic materials appear in two forms. One form is stable over time and another form is exchangeable with water in the environment. The nonexchangeable fraction is normally the target of interest. In order to separate the two, an equilibration with moisture for several days is needed. However, matrix effects can severely affect this process and care needs to be taken to select the right analytical procedure for the matrix under investigation.[5,91,93–95]

## 16.4.2 Oxygen

Oxygen isotope ratios from organic materials can be analysed in a similar arrangement using a high-temperature reduction technique (also Thermal-Conversion/EA, "TC/EA").[96–99] The reactor, typically an $Al_2O_3$ or SiC tube filled with glassy carbon chips and (sometimes) nickelised carbon, is kept at temperatures above 1350 °C. Oxygen- and hydrogen-containing materials (*e.g.* biological materials, water, oxyanions, phyllosilicates) or oxyanions (*e.g.* such as nitrate or sulfate) are encapsulated in silver foil and are converted to CO and $H_2$. The CO and/or $H_2$ are analysed in the IRMS. The masses $m/z = 28$ and $m/z = 30$ are monitored for CO. Compared to plain oxidation with $O_2$, the high-temperature reduction technique is chemically less well constrained.[100–102] The reactions of sample material containing nitrogen are not well understood and postconversion reactions like recombination between CO and $H_2$ may occur. Numerous gaseous byproducts (*e.g.* HCN, $(CN)_2$) have been identified even at 1450 °C and the analytical implications have been discussed.[103] As a consequence, the nature of the reference materials interspersed with the samples in the sample carousel is critical and should be matched. A number of reference materials have been evaluated thoroughly and are available from USGS, IAEA and NIST.[101] Moreover, direct comparison with water has been difficult when using Ag capsules. More recently, water samples encapsulated in crimped Ag tubules have become available.[104]

Currently, there are two methods for generating $O_2$ from sulfates: (1) Ni-tube barite fluorination, where barite is directly fluorinated in a conventional Ni tube with a yield of about 50% and (2) $CO_2$ fluorination, where $CO_2$ derived from the graphite-reduction method is converted to nearly 100% into $O_2$ by fluorination in a Ni tube at 800 °C for 45 h. Recent developments employ a $CO_2$-laser system to generate $O_2$ from barite in a $BrF_5$ atmosphere.

### 16.4.3 Water

The isotopic systematic of water has been one of the early fields of stable-isotope investigation. Isotopic fractionation occurs during phase shifts, *i.e.* evaporation from the oceans and from soils, transpiration through membranes and small orifices (*e.g.* leaves) and condensation (clouds, precipitation). In addition, water is formed by combustion processes, where the oxygen stems from the atmosphere. Relative to ocean water, atmospheric oxygen is enriched by +24‰ (the "Dole effect"[105,106]).

Both hydrogen and oxygen isotopic determinations of water are performed applying several techniques: for $\delta^{18}O$, the most widely method used involves equilibration of the water samples with $CO_2$.[26] In this process, the $\delta^{18}O$ signature of the water sample is acquired by the $CO_2$ gas in the vial headspace, albeit with a large isotopic fractionation of roughly +40‰. Subjecting reference and sample water to the same conditions and keeping the temperature constant allows for comparison of $^{18}O/^{16}O$ isotope ratios with precisions approaching ± 0.02‰. Similarly, determinations of $^{2}H/^{1}H$ ratios can be achieved when a suitable catalyst (Pt) is present in the head space.[107,108] In spite of a large fractionation of more than −700‰, $\delta^{2}H$ values can be determined with a precision of better than 0.5‰. Today, water equilibration is mostly accomplished in automated preparation systems both in Dual-Inlet and continuous-flow mode. More recently, reduction at high temperatures (>1350 °C) in the presence of carbon in continuous-flow mode has achieved precisions close to the dual-inlet techniques as described above.[109] This technique has enabled dual-isotope (H, O) analysis from the same sample in the same preparation setup. Moreover, samples other than water can be analysed directly, thus avoiding the conversion to water, which has led to a considerable increase in throughput and availability of $\delta^{18}O$ and $\delta^{2}H$ data from such samples.

### 16.4.4 Organic Carbon and Nitrogen

Today, the most common sample-preparation apparatus connected to an IRMS is the elemental analyser (EA). In a helium carrier gas atmosphere a Dumas combustion generates the gases of interest ($CO_2$ and $N_2$), which are separated chromatographically, entrained in a helium carrier-gas stream and admitted directly to the IRMS source *via* an open-split interface. This technique allows a high sample throughput.

The dried and homogenised sample material is weighed into tin capsules and loaded into a multisample carousel on the EA. During oxidation, the whole sample – including the tin capsule – is combusted in a high-temperature (~1050 °C) quartz tube with a pulse of $O_2$ added. Downstream, surplus oxygen is scavenged from the gas stream and nitrous oxides are reduced to elemental nitrogen ($N_2$) in a quartz glass tube filled with copper and maintained at a temperature of usually 600 °C.

To remove the water vapour generated during combustion, a water trap (usually $Mg(ClO_4)_2$ or $P_4O_{10}$; sometimes also a Nafion trap (Perma Pure, Toms River, NJ 08755 USA)) is required. Before admitting the combusting gases to the ion source of the MS it is essential to remove the water arising from the combustion. Any such water residue would lead to protonation of $CO_2$ (thus producing $HCO_2^+$),[110] an isobaric interference of $^{13}CO_2^+$ at $m/z = 45$.[39,81] The analyte gases ($CO_2$, $N_2$) resulting from combustion are entrained in the helium carrier gas stream. The gases are usually separated on a gas chromatography column (typically a $\frac{1}{4}$'' Porapak packed column[111]) before admission to the IRMS *via* an open-split interface.[112] Both $\delta^{13}C$ and $\delta^{15}N$ values can be obtained from the same sample (with some compromise in precision) using a diluter system in order to account for the differences in concentration for N and C. In adherence to the "principle of identical treatment", it is advised to include working references, quality control references and blank capsules in the daily sequences. Ideally, the working reference materials should be similar in elemental composition to the samples of interest.[6,113]

### 16.4.5 Measurement of Carbonates

Historically, the first major applications of IRMS were the isotopic analysis of carbonates and water. Carbonates often arise from the calcite-generating microfauna. In these, calcite formation shows a temperature-dependent isotopic fractionation, in particular for oxygen. Thus, paleo-temperatures can be inferred from studying $\delta^{18}O$ variations in microfossils (foraminifera). The basic experimental principles for analysis of carbonates were developed in the early days of IRMS and are still in use today. $CO_2$ is liberated from the carbonate sample material by a reaction with phosphoric acid ($H_3PO_4$). After drying, the $CO_2$ gas is introduced into the IRMS instrument, where the ion currents of the isotopologues at $m/z = 44$, 45 and 46 are measured. After correcting the 45/44 ion current ratio for the $^{17}O$ contribution,[3,114–116] the $\delta^{13}C$ of the sample is obtained (see Section 16.4.10 on isobaric interferences). Similarly, the $(m/z = 46)/(m/z = 44)$ ion current ratio gives access to the $\delta^{18}O$ value of the sample. In the past, the phosphoric acid reaction was carried out in specially built glass reaction vessels. Automated commercial preparation lines based on the same principle have replaced these. Progress has been made mainly in reducing the sample size to (single foraminifers can be analysed in some cases) and in analytical precision. Some of the systems are compatible with dual-inlet systems, generating the most sensitive and precise data from carbonate samples. Other systems work in continuous-flow mode, thus enabling higher analysis throughput.

## 16.4.6 Clumped Isotopes

The early 2000s witnessed the advent of a new stable-isotope systematics called "clumped isotopes"[66] The basic idea is that the isotopic information extracted from different isotopologues of the same compound is not identical.

(***Note***: $CO_2$ gas consists of different isotopologues covering the mass range from $m/z = 44$ ($^{12}C^{16}O^{16}O$) up to $m/z = 49$ ($^{13}C^{18}O^{18}O$), with every mass position in between occupied by several isotopologues.)

Instead, the less-abundant stable isotopes tend to "clump" together over the stochastic – *i.e.* statistically expected – mean value. This tendency to clump is a fundamental property of these molecules, which have different zero-point energies for different isotopes. Hence, the clumping effect is temperature dependent and largely independent of the isotopic composition.[117–120] As a consequence, the temperature of the formation processes can be inferred – provided, that the original isotopologue distribution survives in the respective environment. In order to detect this subtle signature and calibrate the temperature information, the isotope ratio information from the often very minor multiply labeled species (*e.g.* $^{13}C^{16}O^{18}O$, $m/z = 47$) needs to be measured with very high precision. This requires extremely sensitive and precise instrumentation with a very low and clean baseline as well as a thorough calibration strategy. Detection of and correction for isobaric interferences on these minor ion beams is important, which are accomplished preferably by using a double-focusing mass spectrometer operating at medium to high resolution.[67]

## 16.4.7 Compounds with Sulfur

Various methods are available for sulfur isotopic analysis based on the same general principles as for N and C isotopic determinations. There are several challenges in sulfur isotope analysis: (1) Upon oxidation, $SO_2$ and $SO_3$ are formed. The reactor packing needs to be optimised to convert $SO_3$ to $SO_2$ quantitatively. (2) $SO_2$ dissolves in water on surfaces including the combustion water formed during sample oxidation. Therefore, all tubing should be made of hydrophobic material. (3) $SO_2$ is a "sticky" gas. It tends to adsorb on the transfer line and ion source walls and generate memory effects. (4) The oxygen and sulfur isotopes cannot be separated. Therefore, sample and reference material conversion must occur in the same oxidising environment to ensure identical oxygen isotopic composition of the produced gases. Thus, $SO_2$ gas from a cylinder alone is not a suitable reference. $SO_2$ can be analysed on the molecular ion masses at $m/z = 64$, 65, 66 and 68 and SO at $m/z = 48$, 49 and 50.

Conversion to $SF_6$ is the preferred method for high-accuracy measurements. Such analyses are made on dual-inlet systems and mass spectrometers capable of measuring the high $SF_5^+$ fragment masses at $m/z = 127$, 128, 129 and 131. $SF_6$ has the advantage of being entirely inert – almost like a noble gas. Hence, memory effects do not compromise the accuracy of the results.

### 16.4.8 Halogens

Halogens (X = Cl and Br) are usually introduced by transforming them into gaseous forms after precipitation as AgX. Different methods have been used to produce gaseous species such as $Cl_2$ or $COCl_2$. Hoering and Parker[121] used gaseous HCl to measure chlorine isotopic ratios. HCl was chosen as the source material, as it could be prepared quantitatively from small samples. However, HCl has the disadvantages of being corrosive and creating a long-term mass-spectrometric memory effect.

In a more common approach, the halogens (X = Cl, Br) are precipitated as AgX, further converted to $CH_3X$ and – after separation – introduced into the mass spectrometer resulting in $CH_3X^+$ ions for further measurements.[122–125] As an alternative approach, Taylor and Grimsrud[126] used negative ions of $CH_3X^-$ for the isotope ratio measurements.

### 16.4.9 Silicon, Selenium, Germanium, Uranium

All these elements can be fluorinated to gaseous species in order to provide high-precision isotope ratio measurements by gas source MS because of the monoisotopic character of F.

Si isotope analysis by IRMS is usually accomplished by conversion of silicon to $SiF_4$ after fluorination *e.g.* using $BrF_5$ in vacuum at temperatures around 500 °C,[127] or using laser heating in a fluorination chamber. De La Rocha *et al.* used an improved vacuum fluorination system to liberate silicon from the $SiO_2$ by laser heating under pure $F_2$ to produce $SiF_4$ gas for further analysis.[128]

There is only limited literature on the investigation of Se monitored as $SeF_6$[129] as well as on germanium, which can be converted *via* $BaGeF_6$ or *via* direct fluorination to $GeF_4$ and measured by IRMS.[130]

The $UF_6$ gas ion source allows an unprecedented level of reproducibility for isotope ratio measurements and has therefore been chosen as the reference measurement method in nuclear science. For this purpose, a new GSMS instrument, called *URANUS* from Thermo-Fisher, was installed at the European Commission Joint Research Center IRMM (EC-JRC-IRMM) in 2010. The instrument is equipped with a Faraday multicollector detection system ("triple cup"), enabling the minor isotopes $^{234}U$ and $^{236}U$ to be measured simultaneously with the major isotopes $^{235}U$ and $^{238}U$. To mount the three Faraday cups in the same housing without the possibility of a mutual shielding – *e.g.* by a metal plate – has been developed because the relative mass differences between the masses of the $UF_5^+$ ions at $m/z = 329$, 330, and 331 are very small. As a consequence, ion scattering and crosstalk effects between the three cups had to be investigated carefully.[131]

### 16.4.10 Corrections for Isobaric Interferences

Interfering ion currents on the mass positions of interest are often encountered. (see also Chapter 7) An example has already been described for $m/z = 45$ of $N_2O$. To distinguish the two different isotopologues comprising the $m/z = 45$ ion current ($^{15}N^{14}N^{16}O^+$ and $^{14}N_2{}^{17}O^+$) – in this particular case – a mass resolution $m/\Delta m$ of about 20 000 is required, which is clearly beyond the capabilities of most IRMS systems.

A more prominent example is the $^{17}O$ contribution to the $m/z = 45$ ion current of $CO_2$, which would require an even higher mass resolution of more than 60 000. Hence, one has to take advantage of other properties to correct for these isobaric interferences. To remove the (about 7%) contribution to the $m/z = 45$ ion current, one makes use of the (more or less) fixed ratio between $^{17}O$ and $^{18}O$ in the relevant world water pool. By measuring the $^{18}O$ deviation, the $^{17}O$ signature can be established and subtracted from the 45/44 ratio. The procedure has first been described by Craig[3] and was later refined by Santrock *et al.*[132] and Assonov and coauthors.[104,114,115]

Corrections for the oxygen isotopic composition of $SO_2$ in sulfur isotopic measurements have been presented by Leckrone and Ricci[133] in Volume I of the Handbook edited by de Groot.[69] Continuous-flow measurements of sulfur isotopes are accomplished using the mass spectrum of the $SO_2$. From a combined measurement of the 66/64 and 50/48 ratios, the 34/32 ratio deviation can be inferred.

The spectral interferences do not necessarily arise from the measured species itself. Often, they arise from ion–molecule reaction inside the ion source, mainly leading to protonated $(m+1)$ molecular ions. Protonation reactions can be avoided by reducing the possible proton-donating precursor (*e.g.* $H_2O$, $CH_4$ or other organics, *e.g.* from pump oil) by optimising the vacuum conditions. However, this is not possible for hydrogen. In an autoprotonation reaction, $H_2{}^+$ reacts with neutral $H_2$ to form $H_3{}^+$ (and a hydrogen atom). The reaction is proportional to both partners ($H_2$ and $H_2{}^+$). As a consequence, the observed ratio is a (linear) function of the $H_2$ pressure (or intensity) in the ion source. In order to measure the net $^2$HH content in the gas, a correction of the ratio dependence ("$H_3{}^+$ factor") must be established and applied to every sample analysis.

## 16.5 Selected Applications

Because a large number of reviews and books are available, we have in the following focused on a condensed overview with some selected examples throughout the different disciplines. For more information we warmly recommend further reading of the Handbook of Stable Isotope Analytical Techniques.[135] General principles and the isotopic variation of various elements are given in the book by Faure and Mensing,[136] and Chapters 10 to 14 in Beauchemin and Matthews.[75] Reviews on compound specific isotope analysis and its applications in a wide area of research are available in a book by Jochmann and Schmidt[137] and in a paper by Lichtfouse.[138]

Therefore, the intention of the following section is neither to be comprehensive nor complete but to allow for catching a glimpse of the fascinating world of stable-isotope mass spectrometry (see Figure 16.17).

## 16.5.1 Applications in Earth and Geosciences, Paleoclimate Research and Cosmochemistry

The intensive investigation of stable isotopes in geological and geochemical research has developed into the field of "stable-isotope geochemistry", as an aspect of geology based upon study of the natural variations in the relative abundances of isotopes of various elements arising from mass-dependent isotopic fractionation. Comprehensive overviews on the applications of isotopes in geochemistry are given – besides the books already mentioned – in Hoefs[139] or Sharp.[140]

**Geology and Geochemistry**

Geochemists have used isotopes of H, C, N, O and S for studying geochemical and global element cycles, tracing rock sources or ore deposits, enhancing the knowledge about the origin of gases and water bodies, revealing information about the ages of rocks, hydrothermal vent systems or past climatic conditions. Carbon isotopes (amongst others) are used in, *e.g.* petrochemical research to monitor oil/oil, oil/gas and oil/gas/source correlations, as well. Besides these "classic elements", Si has gained increasing interest in geochemical research. Mass-spectrometric studies of silicon isotope variations in natural environments started in the 1950s.[141] In 1982, Douthitt pioneered the investigation of a large number of the Earth's samples showing that the largest variation of silicon isotopes in the Earth's samples can be found in surface environments.[142] A number of important geological applications of silicon isotope studies, such as tracing the source of materials, characterising sedimentary environments and identifying the genesis of rocks and ores, were proposed, *e.g.* by Ding and Tong[143] and Wu *et al.*[144]

**Paleoclimate Research**

The $^{13}C/^{12}C$ ratio has been used widely as indicator of paleoclimate: a change of the ratio in the residues of plants indicates a change in the amount of photosynthetic activity and thus, in how favourable the environment was for the plants.[136] More recently, Eiler presented a review on the application of clumped isotope analysis for the reconstruction of paleoclimate.[145]

**Hydrology**

Isotopes have been applied in catchment research in the last few decades to trace sources of waters and solutes. The data were evaluated by mixing

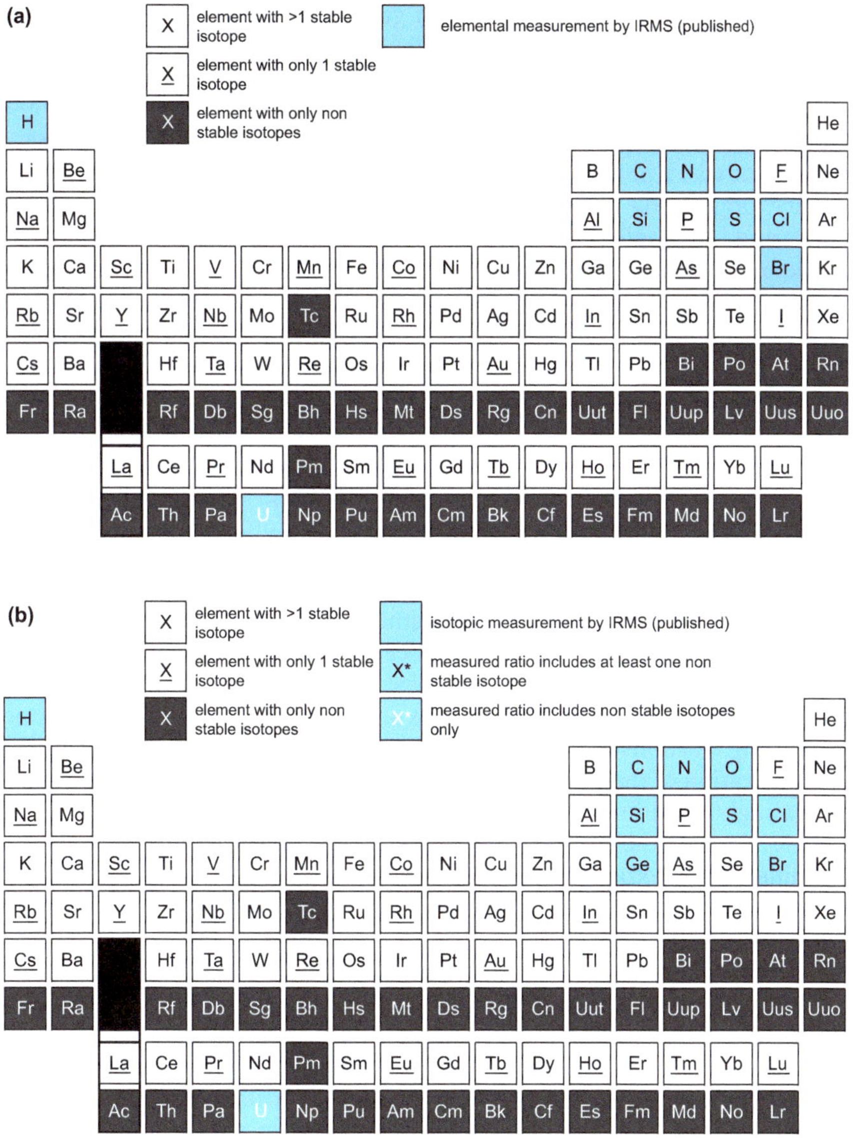


**Figure 16.17** (a) Periodic table of the elements: elements for which elemental data acquired by IRMS has been published (highlighted in sky blue) – also included if used, *e.g.* for interference corrections or internal normalisation. (b) Periodic table of the elements: elements for which isotope ratio data acquired by IRMS has been published (highlighted in sky blue) – also included if measured as interelemental normalisation standard or in IDMS analysis.

models to determine the impact of multiple (assumed) constant-composition sources. Isotope hydrology addresses the application of the measurements of H and O isotopes. They are ideal tracers for detecting water origin and migration in particular in groundwater systems.[146] As a major advantage, stable hydrogen and oxygen isotopes of water do not interact heavily with oxygen and hydrogen in the organic and geologic materials as they move through a catchment resulting in a negligible effect on the isotope ratios. The main processes that can alter the isotopic composition of catchment water are: (1) phase changes occurring at or close to the ground surface originating from evaporation, condensation, melting, or sublimation) and (2) simple mixing of water bodies close to the surface. Therefore, stable oxygen and hydrogen isotopes can be used, *e.g.* to investigate hydrologic processes in various catchment types (rain-dominated temperate and tropical catchments, snowmelt-dominated catchments, arid basins, and lake-dominated systems, respectively) or to determine the contributions of old and new water to a stream (and to other components of the catchment).[147] In addition, isotopes of constituents that are dissolved in the water (*e.g.* S, N, C) can be investigated. The ratios of such solute isotopes can be changed significantly when reacting with biological and/or geological materials upon movement of water through the catchment. Therefore, the variation of these isotopes can be used to trace, *e.g.* water sources and flow paths. Solute isotopes provide information on the reactions that are responsible for their presence in the water, as well, and possible flow paths implied by their presence. Finally, solute isotopes in catchment research can be applied for tracing the solute sources themselves. Besides the extensive use of C, N and S isotopes in forest and agricultural studies, solute isotopes are increasingly used for determining weathering reactions and sources of solutes in water catchments.[148] Further to the application of the "classical set of isotopes", Cl and Br stable isotopes are increasingly used in hydrology research, especially to trace groundwater sources. Eggenkamp provides a comprehensive bibliography of chlorine and bromine stable isotopes on his webpage.[149]

**Cosmochemistry**

In order to identify the chemical composition in outer space, space industries have designed light-weight mass spectrometer to be used in outer space. The ASPERA-3 Ion Mass Analyser was sent in 2003 on the Mars mission to identify ions mainly of H and He. Particles entering the analyser through an outer grid are deflected into the electrostatic analyser (ESA) and further separated in a cylindrical magnetic field set up by permanent magnets. The ions finally hit a microchannel plate (MCP) and are detected simultaneously regarding both direction and mass per charge.[150]

## 16.5.2 Applications in Environmental Sciences

Stable isotopes are widely used in ecology and environmental sciences and thus found their way into numerous reviews and books.[150]

Ecologists and environmental chemists have investigated the use of isotopes to study plants and animals. Isotopic signatures can be used to find patterns and mechanisms down to the single-organism level: they are used to trace food webs and to follow whole ecosystem nutrient cycling in both terrestrial and aquatic ecosystems. As an example, the study of the natural abundance of carbon isotopes in the environment and the use of stable carbon isotope tracers have proved to be very useful in investigating the soil carbon cycle and soil trophic relationships.[151] Moreover, the variation of stable isotopic compositions has been applied to trace the origin and migration of wildlife.[152]

Isotope ratios in plant or animal tissues are the traces left by significant physiological or ecological processes. The timescales of such integrative processes depend on the element turnover rate in the tissue or pool.[153] Moreover, stable isotope variations can result from key ecological processes.[154–156] The presence or absence of such processes and their magnitude can studied by observing stable-isotope ratios relative to known background levels. Biological responses to changing environmental conditions can also leave their trace by altered isotope ratios. In cases where substances or residues accumulate in an incremental fashion, such as tree rings, animal hair and ice cores, isotope ratios can be used as a record of system response to changing environmental conditions or a proxy record for environmental change. The $\delta^{13}C$ of organic and gas samples varies during the processing of carbon in the bio- and geosphere. The basis for much of the observed variation rests on the two major metabolic processes, photosynthesis and respiration[157] with additional variation being expressed during biosynthetic, anabolic, or catabolic reactions that rely on carbon-based substrates.[158] Carbon stable isotopes in plants are fractionated primarily by photosynthesis.[136] During photosynthesis, organisms using the C-3 pathway show different enrichments compared to those using the C-4 pathway, allowing organic matter to be distinguished from abiotic carbon, to trace sources and identify the photosynthetic pathway of the organic matter.

Plants – although relying on atmospheric $CO_2$ for photosynthesis – are "depleted" in $^{13}C$ relative to the atmosphere. Enzymatic and physical processes that discriminate against $^{13}C$ in favour of $^{12}C$ cause this depletion. In plants, water availability can be important in inducing changes in stomatal physiology and/or biochemical discrimination that in turn are expressed in the $\delta^{13}C$ of photosynthetic products as well as tissues and compounds synthesised from these products.[159] Particularly in the case of C-3 plants, $CO_2$ is assimilated from the atmosphere, entering and diffusing out of the leaf through stomata. While diffusing into the intercellular space within the leaf, $CO_2$ is fractionated in favour of carbon-12 by $\sim 4.4‰$. Owing to the different mobilities of the $CO_2$ isotopologues this effect largely depends on factors such as temperature or vapour pressure.[160] During C-3 photosynthesis, the internal $CO_2$ ($c_i$) is combined enzymatically with leaf water to produce sugars *via* carboxylation. This results in a biochemical fractionation against the heavy isotope, with a total fractionation of $\sim -18‰$ against the atmospheric source.[161] An change in stomatal conductance will affect the renewal of the $CO_2$ inside the leaf. Thus, when external factors exert a direct influence on either of these controls, the plant will integrate such influences, which become apparent in photosynthesis and are ultimately expressed as a change in the $\delta^{13}C$ of the resulting products.[162] When sugars are used in the synthesis of the different components of the plant, there are additional fractionations in various compounds, such that cellulose and lignin exhibit lower $\delta^{13}C$ than the sugars formed in the leaf.[163] While the nature of these fractionation processes have been studied in detail at the leaf level,[164] a more precise characterisation for other plant tissues (*e.g.* seeds) is still an open task. For instance, direct solid $\delta^{13}C$ analysis of single tree rings was conducted by Vaganov *et al.*[165] using laser-ablation combustion GC-IRMS. The authors presented a significant carry-over effect, based on correlations between $\delta^{13}C$ of early and the preceding late wood.

The large range of $\delta^{13}C$ values found in biological and geological materials[166] is indicative of the variety of processes which allows to trace carbon in the environment.[167–169] As environments change in space and time, variations in the $\delta^{13}C$ of different types of materials occur, and can thus serve as an indicator of the change.[170]

Besides the use of the natural variation of C isotopes, labelling methods especially with $^{13}C$ have found wide acceptance in ecosystem research, especially in soil chemistry.[171,172]

Bruneau *et al.* determined rhizosphere $^{13}C$ pulse signals in thin soil – sections by laser-ablation isotope ratio mass spectrometry, tracking carbon flow from roots into the soil *in situ.*[173] A good example of the value of $^{13}C$ labelling in food-web studies investigated the pathway of carbon from benthic algae to bacteria, nematodes and macrofauna, which enabled the length of time of initial and maximal $^{13}C$ incorporation up the food chain to be measured.[174]

At plant root level, there is no observable fractionation of O and H isotopes during water uptake.[175] As a consequence, the isotopic composition of water in roots and stems retains the isotopic signature of water available to the plant. The two major sources are water taken up from the deep, groundwater reservoir, or water from recent precipitation. The isotopic composition of ground water is the reult of a long-term integration of precipitation from the area.[176] While there is no fractionation of isotopes during water uptake by plants, fractionation does occur during evapotranspiration. The isotopic composition of the source water is overprinted by the evapotranspirative signal in the leaf, which is largely influenced by vapour-pressure deficit (relative humidity).[162] Compounds, which incorporate H or O during biosynthesis can alter their isotopic signatures, which are "recorded" in the organic molecules.[170] Many of these fractionation effects are among the largest known in biological systems, leading to a significant oxygen isotopic enrichment in organic matter. In recent years the processes that lead to such enrichment of the heavy O isotopes in organic matter are becoming better understood and such materials are becoming increasingly more valuable as "biomarkers" of ecological change, largely because they are known to record temperatures, water sources, and even levels of relative humidity at the time of synthesis. The hydrogen-isotope story in living matter is even more complicated. On top of the water fractionation in the leaves before photosynthesis, a large number of fractionation effects occur during biosynthesis of the various compounds (sugars, fatty acids, amino acids, peptides, *etc.*). In general, enzymatic processes discriminate against the heavier isotopes, resulting in a large variety of negative $\delta^2$H values.[177] The H and O stable-isotope analysis of this broad suite of organic molecules also provides useful archives of ecological change.[178]

Another intensively studied biogeochemical cycle, yet not fully understood from a stable-isotope point of view, is the nitrogen cycle.[154] In terrestrial ecosystems plant growth is often limited by nitrogen available to the plant in the form of nitrates, ammonia or amino acids.[179] As studies that have used $\delta^{15}$N data in the areas of physiology, ecology, and biogeochemistry increased, it became clear that new challenges had to be faced in applying nitrogen isotopes to trace, integrate, or record certain process.[154] Of utmost importance is the challenge of characterising the many and varied fractionation factors associated with the transformation, utilisation, and immobilisation of N substances as they move through the nitrogen cycle. Several authors state that plant $\delta^{15}$N is not a tracer of nitrogen source; instead, it provides a synthesis of the $\delta^{15}$N of the nitrogen source, of the fractionation events that occur during nitrogen absorption, of fractionations originating from different mycorrhizal associations, and during assimilation, allocation and loss of nitrogen from the plant.[180] However, as nitrogen demands frequently exceed nitrogen supply in natural systems, it has been suggested that plant $\delta^{15}$N is still a good approximation of $\delta^{15}$N of the available nitrogen source(s), under most field conditions.[181]

Recently, the great potential of S isotopes for detecting and understanding the nature and magnitude of ecological processes and its change has been exploited. The isotopic signature of $SO_2$ in the atmosphere can be altered by marine or terrestrial sources, which in turn have natural as well as anthropogenic components.[182] However, sulfur is not only a source of atmospheric pollution. It is also an essential nutrient in all ecosystems. For most plants, the normal source for sulfur is sulfate, taken up by the roots. The plant's assimilatory sulfate reduction pathway, providing "organic sulfur" from sulfate, proceeds without major sulfur isotopic fractionation.[183] According to general experience, the bulk plant sulfur is depleted by only 1–2‰ relative to its primary sources, soil and sea spray sulfate or $SO_2$ from the atmosphere.[183] This element exhibits natural variations in the isotopic ratio due to fractionations that occur during chemical, physical, and certain biological processes. Hence, the "isotopic signature" can be used to trace the origins of sulfur-bearing compounds.[184]

Within the last several years, silicon has gained new interest in monitoring the global Si biogeochemical cycle for tracing the origin of dissolved silicon transferred from various soil–plant systems towards rivers[185] and in order to investigate implications for climate change and human perturbation.[186,187] In this context, especially the role of Si isotope fractionation by diatoms was exploited as indicators of past oceanic change.[188]

### 16.5.3 Applications in Archaeometry and Forensics

The analysed elements have the potential to provide specific isotopic fingerprint information of a sample, which can be used in forensic science to prove the link of evidence to a crime, to ensure provenance and authenticity or to detect illegal substances or doping.[189,134,190–193] Even human-body parts (*e.g.* hair and fingernail) can be assigned to a geographic origin by means of H and O isotopes.[194] Isotope ratio mass spectrometry plays also a key role in particular for the analysis of explosives, ignitable liquids and illicit drugs, as well, such as outlined in Benson *et al.*[190] A comprehensive overview and guide to forensic stable-isotope analysis is given in the book of Meier-Augenstein.[122] All these efforts are well recognised and resulted in the establishment of a network developing forensic applications of isotope ratio mass spectrometry, called the Forensic Isotope Ratio Mass Spectrometry (FIRMS), in 2002.

IRMS and the application of stable isotopes have also found their way into doping control for identifying steroid abuse. The $\delta^{13}C$ value of urinary steroids is measured by IRMS and the results can confirm their exogenous origin based on the abnormal $\delta^{13}C$ result.[195]

Stable isotopes have been used in "environmental forensics", as well in order to trace back the origin of contaminants. Besides the light stable isotopes, Cl isotopes have been used to quantify the flux of man-made chlorinated compounds[196] but also proved to be useful to detect the origin chlorinated organic compounds of groundwater contaminants.[197]

Stable isotopes (predominantly C, N and O) are used in anthropology as a tool for studying demography or life history of ancient populations,[198] paleodietary pattern and paleoclimatic changes.[199,200]

In addition, the potential of $\delta^{34}S$ values of human and faunal bone collagen was shown to provide paleodietary information and to indicate residence or migration.[201]

Recent instrumental developments make use of laser-ablation systems coupled to an IRMS in order to investigate small fragments or incrementally grown layers (*e.g.* teeth) for their C and O isotopic composition.[202,203]

### 16.5.4 Applications in Life Sciences

Biological processes are generally good examples of kinetic isotope reactions. Usually, the lighter isotopic species are preferred during biochemical reactions, owing to the lower energy associated with breaking the bonds in molecules containing the lighter isotopes. This results in significant fractionations between the substrate (heavier) and the biologically mediated product (lighter). The size of the fractionation depends on the details of the reaction pathway and the relative energies of the chemical bonds altered by the respective reaction. In general, reactions near the energetic threshold are slow and exhibit a larger isotopic fractionation than faster steps that happen with a larger amount of available energy and, thus, are less selective. Hence stable isotopes can be used in metabolic pathways of, *e.g.*, microbes.[203]

Due to the recent instrumental developments, most isotope ratio studies in the life sciences are based on molecular mass spectrometry since they provide molecular information along with the isotopic pattern.

In analogy to paleodietary studies, stable isotopes of C and N can be used to reconstruct the diets of modern animals [paleodietary studies, stable isotopes of C and N can be used to reconstruct the diets of modern animals, Michener].[213] The determination of the isotopic composition of Amino acids requires chemical derivatisation to allow GC separation prior to GC-IRMS.

Medical and biochemistry research make use of labelling techniques (mainly C, N, H, and S) in order to provide information on biochemical reactions or metabolic pathways. In this case, the metabolites are synthetised using enriched isotopes. Again, the isotopic composition is determined preferentially by molecular mass spectrometry, instead of IRMS. Labelled $^{13}C$ substances are also applied in medicine in combination with highly sensitive mass spectrometers as an alternative to radioactive markers. In this context, labelled isotopes are administered for determining metabolic products that develop during metabolic pathways, *e.g.* by bacteria. A prominent example is the $^{13}C$ breath test for the identification of Helicobacter pylori.[204] A recent study examined carbohydrate digestion and glucose absorption by measuring plasma glucose after oral intake of $^{13}C$-enriched carbohydrates.[205] Another biochemical investigation used a blood portion to measure the fractional synthesis rate of glutathione (GSH) after infusion of (1-$^{13}C$)-glycine.[206]

In addition to total isotopic composition, it has become feasible in selected cases to obtain site-specific isotopic information of the molecules of interest. Nuclear magnetic resonance (NMR) has been employed as major technique for this purpose, especially in the context of food authentication.[207] NMR is attractive because its output is sensitive to the local chemical environments within the molecules, so that positional isotopic information is relatively easy to obtain. Isotope ratio mass spectrometry provides a far more precise measurement technique but since samples are normally prepared by combustion techniques, site-specific information is lost, which can only be overcome partially by chemical treatment or adapted pyrolysis techniques.[208]

### 16.5.5 Applications in Industry, Nuclear Research and Material Research

One of the major impacts of IRMS in modern industry is quality control and authentication of goods and products *via* the stable-isotope pattern. Many ecological processes produce a unique isotopic fingerprint, which may be used to discriminate the geographical origin of plant or animal tissues.[155,163,209] Therefore the method is increasingly applied in the determination of the provenance or authenticity of food or drugs.[134]

IRMS has found a niche application in high-precision measurements of uranium isotopes. Gaseous $UF_6$ is one of the primary forms of uranium in the nuclear fuel cycle, in particular in the isotope enrichment process, which has to be controlled by accurate isotopic measurements of $UF_6$ samples.[131] (***Note***: Furthermore, the benefit of applying stable-isotope measurements of xenon and krypton to supplement existing techniques for nuclear-safeguards verification has been demonstrated and recognised by the International Atomic Energy Agency (IAEA).[210]) Another important impact can be seen at the determination of highly accurate isotope data for Si within the Avogadro project for the redefinition of the Avogadro constant,[211,212] and thus of the kilogram.

# References


1. K. K. Murray, R. K. Boyd, M. N. Eberlin, G. J. Langley, L. Li and Y. Naito, *Pure Appl. Chem.*, 2013, **85**, 1515–1609.
2. T. B. Coplen, *Rapid Commun. Mass Spectrom.*, 2011, **25**, 2538–2560.
3. H. Craig, *Geochim. Cosmochim. Acta*, 1957, **12**, 133–149.
4. H. C. Urey, *Science*, 1948, **108**, 602–603.
5. W. A. Brand and T. B. Coplen, *Isotopes Environ. Health Stud.*, 2012, **48**, 393–409.
6. R. A. Werner and W. A. Brand, *Rapid Commun. Mass Spectrom.*, 2001, **15**, 501–519.
7. J. F. Carter and B. Fry, *Anal. Bioanal. Chem.*, 2013, **405**, 2799–2814.
8. T. B. Coplen, *J. Res. Natl. Inst. Stand. Technol.*, 1995, **100**, 285–285.
9. T. B. Coplen, W. A. Brand, M. Gehre, M. Gröning, H. A. J. Meijer, B. Toman and R. M. Verkouteren, *Rapid Commun. Mass Spectrom.*, 2006, **20**, 3165–3166.
10. T. B. Coplen, W. A. Brand, M. Gehre, M. Gröning, H. A. J. Meijer, B. Toman and R. M. Verkouteren, *Anal. Chem.*, 2006, **78**, 2439–2441.
11. G. Hut, in conf. proc. of *Consultants' group meeting on stable isotope reference samples for geochemical and hydrological investigations*, International Atomic Energy Agency, 1987, p. 42.
12. T. B. Coplen, J. K. Böhlke, P. De Bievre, T. Ding, N. E. Holden, J. A. Hopple, H. R. Krouse, A. Lamberty, H. S. Peiser, K. Revesz, S. E. Rieder, K. J. R. Rosman, E. Roth, P. D. P. Taylor, R. D. Vocke and Y. K. Xiao, *Pure Appl. Chem.*, 2002, **74**, 1987–2017.
13. W. A. Brand, T. B. Coplen, J. Vogl, M. Rosner and T. Prohaska, *Pure Appl. Chem.*, 2014, **86**, 425–467.
14. BIPM, *The International System of Units (SI), Bureau international des poids et mesures*, Sèvres Cedex, France, 2006.
15. E. Kerstel and L. Gianfrani, *Appl. Phys. B*, 2008, **92**, 439–449.
16. K. Habfast, *Int. J. Mass Spectrom. Ion Process.*, 1983, **51**, 165–189.
17. A. O. Nier and E. A. Gulbransen, *J. Am. Chem. Soc.*, 1939, **61**, 697–698.
18. A. O. Nier, *J. Am. Chem. Soc.*, 1939, **61**, 697.

19. A. O. Nier, *Rev. Sci. Instrum.*, 1940, **11**, 212–216.
20. A. O. Nier, *Rev. Sci. Instrum.*, 1947, **18**, 398–411.
21. C. R. McKinney, J. M. McCrea, S. Epstein, H. A. Allen and H. C. Urey, *Rev. Sci. Instrum.*, 1950, **21**, 724–730.
22. B. F. Murphey, *Phys. Rev.*, 1947, **72**, 834–837.
23. S. Epstein, R. Buchsbaum, H. A. Lowenstam and H. C. Urey, *Geol. Soc. Am. Bull.*, 1951, **62**, 417–426.
24. S. Epstein, R. Buchsbaum, H. A. Lowenstam and H. C. Urey, *Geol. Soc. Am. Bull.*, 1953, **64**, 1315–1325.
25. S. Epstein, E. Degens and D. L. Graf, *J. Geophys. Res.*, 1962, **67**, 1636–&.
26. S. Epstein and T. Mayeda, *Geochim. Cosmochim. Acta*, 1953, **4**, 213–224.
27. S. Epstein, R. P. Sharp and I. Goddard, *J. Geol.*, 1963, **71**, 698–720.
28. H. C. Urey, *Proc. Natl. Acad. Sci. USA*, 1952, **38**, 351–363.
29. H. C. Urey, H. A. Lowenstam, S. Epstein and C. R. McKinney, *Geol. Soc. Am. Bull.*, 1951, **62**, 399.
30. (a) A. Trofimov, *Doklady Akad. Nauk SSSR*, 1949, **66**. (b) H. Thode, J. Monster and H. Dunford, *Geochim. Cosmochim. Acta*, 1961, **25**, 159–174.
31. V. I. Vinogradov, *Rol'osadochnogo tsikla v geokhimii izotopov sery (The Role of Sedimentary Cycle in Sulfur Isotope Geochemistry)*, Nauka, Moscow, 1980.
32. M. M. Bender, *Radiocarbon*, 1968, **10**, 468–&.
33. M. M. Bender, *Phytochemistry*, 1971, **10**, 1239–&.
34. M. M. Bender, I. Rouhani, H. M. Vines and C. C. Black, *Plant Physiol.*, 1973, **52**, 427–430.
35. M. J. DeNiro and S. Epstein, *Science*, 1977, **197**, 261–263.
36. M. J. DeNiro and S. Epstein, *Geochim. Cosmochim. Acta*, 1978, **42**, 495–506.
37. M. J. DeNiro and S. Epstein, *Geochim. Cosmochim. Acta*, 1981, **45**, 341–351.
38. D. E. Matthews and J. M. Hayes, *Anal.Anal. Chem.*, 1978, **50**, 1465–1473.
39. W. A. Brand, *J. Mass Spectrom.*, 1996, **31**, 225–235.
40. A. Barrie, J. Bricout and J. Koziet, *Biomed. Mass Spectrom.*, 1984, **11**, 583–588.
41. D. A. Merritt, W. A. Brand and J. M. Hayes, *Org. Geochem.*, 1994, **21**, 573–583.
42. M. P. Ricci, D. A. Merritt, K. H. Freeman and J. M. Hayes, *Org. Geochem.*, 1994, **21**, 561–571.
43. K. H. Freeman, J. M. Hayes, J. M. Trendel and P. Albrecht, *Nature*, 1990, **343**, 254–256.
44. J. M. Hayes, K. H. Freeman, B. N. Popp and C. H. Hoham, *Org. Geochem.*, 1990, **16**, 1115–1128.
45. T. Preston and A. Barrie, *American Laboratory*, 1991, **23**, H32–&.
46. T. Preston and N. J. P. Owens, *Analyst*, 1983, **108**, 971–977.
47. T. Preston and N. J. P. Owens, *Biomed. Mass Spectrom.*, 1985, **12**, 510–513.
48. R. A. Werner, B. A. Bruch and W. A. Brand, *Rapid Commun. Mass Spectrom.*, 1999, **13**, 1237–1241.

49. H. Erlenkeuser, 1987.
50. M. J. Nadeau, M. Schleicher, P. M. Grootes, H. Erlenkeuser, A. Gottdang, D. J. W. Mous, J. M. Sarnthein and H. Willkomm, *Nucl. Instrum. Methods Phys. Res. Sect. B: Beam Interact. Mater. Atoms*, 1997, **123**, 22–30.
51. P. K. Swart, S. J. Burns and J. J. Leder, *Chem. Geol.*, 1991, **86**, 89–96.
52. B. Fry, W. Brand, F. J. Mersch, K. Tholke and R. Garritt, *Anal. Chem.*, 1992, **64**, 288–291.
53. B. Fry, R. Garritt, K. Tholke, C. Neill, R. H. Michener, F. J. Mersch and W. Brand, *Rapid Commun. Mass Spectrom.*, 1996, **10**, 953–958.
54. D. Henneberg, U. Henrichs and G. Schomburg, *Chromatographia*, 1975, **8**, 449–451.
55. T. Preston and D. C. McMillan, *Biomed. Environ. Mass Spectrom.*, 1988, **16**, 229–235.
56. A. Giesemann, H. J. Jager, A. L. Norman, H. P. Krouse and W. A. Brand, *Anal. Chem.*, 1994, **66**, 2816–2819.
57. D. A. Merritt and J. M. Hayes, *J. Am. Soc. Mass Spectrom.*, 1994, **5**, 387–397.
58. W. A. Brand, A. R. Tegtmeyer and A. Hilkert, *Org. Geochem.*, 1994, **21**, 585–594.
59. W. A. Brand, H. Avak, R. Seedorf, D. Hofmann and T. Conradi, *Isotopes Environ. Health Stud.*, 1996, **32**, 263–273.
60. A. W. Hilkert, C. B. Douthitt, H. J. Schluter and W. A. Brand, *Rapid Commun. Mass Spectrom.*, 1999, **13**, 1226–1230.
61. J. Morrison, T. Brockwell, T. Merren, F. Fourel and A. M. Phillips, *Anal. Chem.*, 2001, **73**, 3570–3575.
62. W. A. Brand and P. Dobberstein, *Isotopes Environ. Health Stud.*, 1996, **32**, 275–283.
63. R. J. Caimi and J. T. Brenna, *Anal. Chem.*, 1993, **65**, 3497–3500.
64. M. Krummen, A. W. Hilkert, D. Juchelka, A. Duhr, H. J. Schluter and R. Pesch, *Rapid Commun. Mass Spectrom.*, 2004, **18**, 2260–2266.
65. E. Hettmann, W. A. Brand and G. Gleixner, *Rapid Commun. Mass Spectrom.*, 2007, **21**, 4135–4141.
66. J. M. Eiler and E. Schauble, *Geochim. Cosmochim. Acta*, 2004, **68**, 4767–4777.
67. J. M. Eiler, M. Clog, P. Magyar, A. Piasecki, A. Sessions, D. Stolper, M. Deerberg, H.-J. Schlueter and J. Schwieters, *Int. J. Mass Spectrom.*, 2013, **335**, 45–56.
68. W. A. Brand, Mass Spectrometer Hardware for Analyzing Stable Isotope Ratios, in *Handbook of Stable Isotope Analytical Techniques (Chapter 38)*, ed. P. A. de Groot, Elsevier Science, Amsterdam, 2004, vol. 1, pp. 835–856.
69. P. A. De Groot, *Handbook of Stable Isotope Analytical Techniques*, Elsevier Science, 2004, 1.
70. M. Wendeberg and W. A. Brand, Isotope ratio mass spectrometry (IRMS) of light elements (C, H, O, N, S): The principles and

characteristics of the IRMS instrument, in *Encyclopedia of Mass Spectrometry; Elemental and Isotope Ratio Mass Spectrometry*, eds. D. Beauchemin and D. E. Matthews, Elsevier, Amsterdam, 2010, vol. 5, pp. 739–749.
71. K. Habfast, Advanced isotope ratio mass spectrometry I: Magnetic isotope ratio mass spectrometers., in *Modern Isotope Ratio Mass Spectrometry*, ed. I. T. Platzner, John Wiley & Sons, Chichester, UK, 1997, pp. 11–82.
72. M. E. Wieser and W. A. Brand, Isotope Ratio Studies using Mass Spectrometry, in *Encyclopedia of Spectroscopy and Spectrometry*, eds. J. C. Lindon, G. E. Tranter and J. L. Holmes, Academic Press, Amsterdam, 1999, vol. 1, pp. 1072–1086.
73. H. R. Krouse, Isotope Ratio Mass Spectrometry, in *Encyclopedia of Analytical Chemistry*, John Wiley & Sons, Ltd, 2006.
74. W. G. Mook and P. Grootes, *Int. J. Mass Spectrom. Ion Phys.*, 1973, **12**, 273–298.
75. D. Beauchemin and D. E. Matthews, *Encyclopedia of Mass Spectrometry ; Elemental and Isotope Ratio Mass Spectrometry*, Elsevier, Amsterdam, 2010, 5.
76. W. A. Brand, Isotope Mass Spectrometry: Precision from Transient Signals, in *Advances in Mass Spectrometry*, eds. E. J. Karjalainen, A. E. Hesso, J. E. Jalonen and U. P. Karjalainen, Elsevier Science, Amsterdam, 1998, vol. 14, pp. 661–686.
77. M. Sano, Y. Yotsui, H. Abe and S. Sasaki, *Biol. Mass Spectrom.*, 1976, **3**, 1–3.
78. M. Zech, R. A. Werner, D. Juchelka, K. Kalbitz, B. Buggle and B. Glaser, *Org. Geochem.*, 2012, **42**, 1470–1475.
79. Y. Zhang, H. J. Tobias, G. L. Sacks and J. T. Brenna, *Drug Test. Anal.*, 2012, **4**, 912–922.
80. W. Meier-Augenstein, *J. Chromatogr. A*, 1999, **842**, 351–371.
81. W. Meier-Augenstein, *Handbook of Stable Isotope Analytical Techniques*, ed. P. De Groot, *Elsevier*, Amsterdam, The Netherlands, 2004, 1, 153.
82. M. Elsner, M. A. Jochmann, T. B. Hofstetter, D. Hunkeler, A. Bernstein, T. C. Schmidt and A. Schimmelmann, *Anal. Bioanal. Chem.*, 2012, **403**, 2471–2491.
83. J. P. Godin, L.-B. Fay and G. Hopfgartner, *Mass Spectrom. Rev.*, 2007, **26**, 751–774.
84. A. L. Sessions, S. P. Sylva and J. M. Hayes, *Anal. Chem.*, 2005, 77, 6519–6527.
85. T. E. Cerling and Z. D. Sharp, *Palaeogeogr, Palaeoclimatol, Palaeoecol*, 1996, **126**, 173–186.
86. Z. D. Sharp, *Geochim. Cosmochim. Acta*, 1990, **54**, 1353–1357.
87. B. E. Taylor, Fluorination methods in stable isotope analysis, in *Handbook of Stable Isotope Analytical Techniques*, ed. P. De Groot, Elsevier, Amsterdam, The Netherlands, 2004, vol. 1, pp. 400–472.

88. A. G. Clark, M. B. Eisen, D. R. Smith, C. M. Bergman, B. Oliver, T. A. Markow, T. C. Kaufman, M. Kellis, W. Gelbart, V. N. Iyer, D. A. Pollard, T. B. Sackton, A. M. Larracuente, N. D. Singh, J. P. Abad, D. N. Abt, B. Adryan, M. Aguade, H. Akashi, W. W. Anderson, C. F. Aquadro, D. H. Ardell, R. Arguello, C. G. Artieri, D. A. Barbash, D. Barker, P. Barsanti, P. Batterham, S. Batzoglou, D. Begun, A. Bhutkar, E. Blanco, S. A. Bosak, R. K. Bradley, A. D. Brand, M. R. Brent, A. N. Brooks, R. H. Brown, R. K. Butlin, C. Caggese, B. R. Calvi, A. B. de Carvalho, A. Caspi, S. Castrezana, S. E. Celniker, J. L. Chang, C. Chapple, S. Chatterji, A. Chinwalla, A. Civetta, S. W. Clifton, J. M. Comeron, J. C. Costello, J. A. Coyne, J. Daub, R. G. David, A. L. Delcher, K. Delehaunty, C. B. Do, H. Ebling, K. Edwards, T. Eickbush, J. D. Evans, A. Filipski, S. Findeiss, E. Freyhult, L. Fulton, R. Fulton, A. C. L. Garcia, A. Gardiner, D. A. Garfield, B. E. Garvin, G. Gibson, D. Gilbert, S. Gnerre, J. Godfrey, R. Good, V. Gotea, B. Gravely, A. J. Greenberg, S. Griffiths-Jones, S. Gross, R. Guigo, E. A. Gustafson, W. Haerty, M. W. Hahn, D. L. Halligan, A. L. Halpern, G. M. Halter, M. V. Han, A. Heger, L. Hillier, A. S. Hinrichs, I. Holmes, R. A. Hoskins, M. J. Hubisz, D. Hultmark, M. A. Huntley, D. B. Jaffe, S. Jagadeeshan, W. R. Jeck, J. Johnson, C. D. Jones, W. C. Jordan, G. H. Karpen, E. Kataoka, P. D. Keightley, P. Kheradpour, E. F. Kirkness, L. B. Koerich, K. Kristiansen, D. Kudrna, R. J. Kulathinal, S. Kumar, R. Kwok, E. Lander, C. H. Langley, R. Lapoint, B. P. Lazzaro, S. J. Lee, L. Levesque, R. Q. Li, C. F. Lin, M. F. Lin, K. Lindblad-Toh, A. Llopart, M. Y. Long, L. Low, E. Lozovsky, J. Lu, M. H. Luo, C. A. Machado, W. Makalowski, M. Marzo, M. Matsuda, L. Matzkin, B. McAllister, C. S. McBride, B. McKernan, K. McKernan, M. Mendez-Lago, P. Minx, M. U. Mollenhauer, K. Montooth, S. M. Mount, X. Mu, E. Myers, B. Negre, S. Newfeld, R. Nielsen, M. A. F. Noor, P. O'Grady, L. Pachter, M. Papaceit, M. J. Parisi, M. Parisi, L. Parts, J. S. Pedersen, G. Pesole, A. M. Phillippy, C. P. Ponting, M. Pop, D. Porcelli, J. R. Powell, S. Prohaska, K. Pruitt, M. Puig, H. Quesneville, K. R. Ram, D. Rand, M. D. Rasmussen, L. K. Reed, R. Reenan, A. Reily, K. A. Remington, T. T. Rieger, M. G. Ritchie, C. Robin, Y. H. Rogers, C. Rohde, J. Rozas, M. J. Rubenfield, A. Ruiz, S. Russo, S. L. Salzberg, A. Sanchez-Gracia, D. J. Saranga, H. Sato, S. W. Schaeffer, M. C. Schatz, T. Schlenke, R. Schwartz, C. Segarra, R. S. Singh, L. Sirot, M. Sirota, N. B. Sisneros, C. D. Smith, T. F. Smith, J. Spieth, D. E. Stage, A. Stark, W. Stephan, R. L. Strausberg, S. Strempel, D. Sturgill, G. Sutton, G. G. Sutton, W. Tao, S. Teichmann, Y. N. Tobari, Y. Tomimura, J. M. Tsolas, V. L. S. Valente, E. Venter, J. C. Venter, S. Vicario, F. G. Vieira, A. J. Vilella, A. Villasante, B. Walenz, J. Wang, M. Wasserman, T. Watts, D. Wilson, R. K. Wilson, R. A. Wing, M. F. Wolfner, A. Wong, G. K. S. Wong, C. I. Wu, G. Wu, D. Yamamoto, H. P. Yang, S. P. Yang, J. A. Yorke, K. Yoshida, E. Zdobnov, P. L. Zhang, Y. Zhang, A. V. Zimin, J. Baldwin, A. Abdouelleil, J. Abdulkadir, A. Abebe, B. Abera, J. Abreu, S. C. Acer, L. Aftuck, A. Alexander, P. An, E. Anderson, S. Anderson, H. Arachi, M. Azer, P.

Bachantsang, A. Barry, T. Bayul, A. Berlin, D. Bessette, T. Bloom, J. Blye, L. Boguslavskiy, C. Bonnet, B. Boukhgalter, I. Bourzgui, A. Brown, P. Cahill, S. Channer, Y. Cheshatsang, L. Chuda, M. Citroen, A. Collymore, P. Cooke, M. Costello, K. D'Aco, R. Daza, G. De Haan, S. DeGray, C. DeMaso, N. Dhargay, K. Dooley, E. Dooley, M. Doricent, P. Dorje, K. Dorjee, A. Dupes, R. Elong, J. Falk, A. Farina, S. Faro, D. Ferguson, S. Fisher, C. D. Foley, A. Franke, D. Friedrich, L. Gadbois, G. Gearin, C. R. Gearin, G. Giannoukos, T. Goode, J. Graham, E. Grandbois, S. Grewal, K. Gyaltsen, N. Hafez, B. Hagos, J. Hall, C. Henson, A. Hollinger, T. Honan, M. D. Huard, L. Hughes, B. Hurhula, M. E. Husby, A. Kamat, B. Kanga, S. Kashin, D. Khazanovich, P. Kisner, K. Lance, M. Lara, W. Lee, N. Lennon, F. Letendre, R. LeVine, A. Lipovsky, X. H. Liu, J. L. Liu, S. T. Liu, T. Lokyitsang, Y. Lokyitsang, R. Lubonja, A. Lui, P. MacDonald, V. Magnisalis, K. Maru, C. Matthews, W. McCusker, S. McDonough, T. Mehta, J. Meldrim, L. Meneus, O. Mihai, A. Mihalev, T. Mihova, R. Mittelman, V. Mlenga, A. Montmayeur, L. Mulrain, A. Navidi, J. Naylor, T. Negash, T. Nguyen, N. Nguyen, R. Nicol, C. Norbu, N. Norbu, N. Novod, B. O'Neill, S. Osman, E. Markiewicz, O. L. Oyono, C. Patti, P. Phunkhang, F. Pierre, M. Priest, S. Raghuraman, F. Rege, R. Reyes, C. Rise, P. Rogov, K. Ross, E. Ryan, S. Settipalli, T. Shea, N. Sherpa, L. Shi, D. Shih, T. Sparrow, J. Spaulding, J. Stalker, N. Stange-Thomann, S. Stavropoulos, C. Stone, C. Strader, S. Tesfaye, T. Thomson, Y. Thoulutsang, D. Thoulutsang, K. Topham, I. Topping, T. Tsamla, H. Vassiliev, A. Vo, T. Wangchuk, T. Wangdi, M. Weiand, J. Wilkinson, A. Wilson, S. Yadav, G. Young, Q. Yu, L. Zembek, D. Zhong, A. Zimmer, Z. Zwirko, D. B. Jaffe, P. Alvarez, W. Brockman, J. Butler, C. Chin, S. Gnerre, M. Grabherr, M. Kleber, E. Mauceli, I. MacCallum, D. G. Consor, B. I. G. Sequencing and B. I. W. G. Ass, *Nature*, 2007, 450, 203–218.

89. P. Burnard, L. Zimmermann and Y. Sano, History and Background, in *The Noble Gases as Geochemical Tracers*, Springer, Berlin Heidelberg, 2013, pp. 1–15.
90. S. J. Prosser, *Int. J. Mass Spectrom. Ion Process.*, 1993, **125**, 241–266.
91. T. B. Coplen and H. P. Qi, *Rapid Commun. Mass Spectrom.*, 2010, **24**, 2269–2276.
92. R. Gonfiantini, W. Stichler and K. Rozanski, *IAEA TECDOC*, 1995, **825**, 13–29.
93. J. R. Ehleringer, G. J. Bowen, L. A. Chesson, A. G. West, D. W. Podlesak and T. E. Cerling, *Proc. Natl. Acad. Sci.*, 2008, **105**, 2788–2793.
94. T. B. Coplen and H. P. Qi, *Foren. Sci. Int.*, 2012, **214**, 135–141.
95. H. P. Qi and T. B. Coplen, *Rapid Commun. Mass Spectrom.*, 2010, **25**, 2209–2222.
96. M. Gehre and G. Strauch, *Rapid Commun. Mass Spectrom.*, 2003, **17**, 1497–1503.
97. B. E. Kornexl, M. Gehre, R. Hofling and R. A. Werner, *Rapid Commun. Mass Spectrom.*, 1999, **13**, 1685–1693.
98. J. Koziet, *J. Mass Spectrom.*, 1997, **32**, 103–108.

99. R. A. Werner, B. E. Kornexl, A. Rossmann and H. L. Schmidt, *Anal. Chim. Acta*, 1996, **319**, 159–164.
100. F. Accoe, M. Berglund, B. Geypens and P. Taylor, *Rapid Commun. Mass Spectrom.*, 2008, **22**, 2280–2286.
101. W. A. Brand, T. B. Coplen, A. T. Aerts-Bijma, J. K. Böhlke, M. Gehre, H. Geilmann, M. Gröning, H. G. Jansen, H. A. J. Meijer, S. J. Mroczkowski, H. P. Qi, K. Soergel, H. Stuart-Williams, S. M. Weise and R. A. Werner, *Rapid Commun. Mass Spectrom.*, 2009, **23**, 999–1019.
102. G. B. Hunsinger and L. A. Stern, *Rapid Commun. Mass Spectrom.*, 2012, **26**, 554–562.
103. G. B. Hunsinger, C. A. Tipple and L. A. Stern, *Rapid Commun. Mass Spectrom.*, 2013, **27**, 1649–1659.
104. H. Qi, M. Gröning, T. B. Coplen, B. Buck, S. J. Mroczkowski, W. A. Brand, H. Geilmann and M. Gehre, *Rapid Commun. Mass Spectrom.*, 2010, **24**, 1821–1827.
105. M. Dole, *J. Am. Chem. Soc.*, 1935, **57**, 2731–2731.
106. N. Morita and T. Titani, *Bull. Chem. Soc. Jpn.*, 1936, **11**, 414–418418.
107. J. Horita, *Chem. Geol.*, 1988, **72**, 89–94.
108. T. B. Coplen, J. D. Wildman and J. Chen, *Anal. Chem.*, 1991, **63**, 910–912.
109. M. Gehre, H. Geilmann, J. Richter, R. A. Werner and W. A. Brand, *Rapid Commun. Mass Spectrom.*, 2004, **18**, 2650–2660.
110. V. G. Anicich, *J. Phys. Chem. Ref. Data*, 1993, **22**, 1469–1570.
111. G. Castello and G. D'Amato, *J. Chromatogr. A*, 1980, **196**, 245–254.
112. R. A. Werner, B. A. Bruch and W. A. Brand, *Rapid Commun. Mass Spectrom.*, 1999, **13**, 1237–1241.
113. M. Teece and M. L. Fogel, Preparation of ecological and biochemical samples for isotope analysis, in *Handbook of Stable Isotope Analytical Techniques*, ed. P. De Groot, Elsevier, Amsterdam, The Netherlands, 2004, vol. 1, pp. 177–202.
114. S. S. Assonov and C. A. M. Brenninkmeijer, *Rapid Commun. Mass Spectrom.*, 2003, **17**, 1007–1016.
115. S. S. Assonov and C. A. M. Brenninkmeijer, *Rapid Commun. Mass Spectrom.*, 2003, **17**, 1017–1029.
116. W. A. Brand, S. S. Assonov and T. B. Coplen, *Pure Appl. Chem.*, 2010, **82**, 1719–1733.
117. J. M. Eiler, P. Ghosh, H. Affek, E. Schauble, J. Adkins, D. Schrag and P. Hoffman, *Geochim. Cosmochim. Acta*, 2005, **69**, A127–A127.
118. C. N. Garzione, G. D. Hoke, J. C. Libarkin, S. Withers, B. MacFadden, J. Eiler, P. Ghosh and A. Mulch, *Science*, 2008, **320**, 1304–1307.
119. P. Ghosh, J. Adkins, H. Affek, B. Balta, W. F. Guo, E. A. Schauble, D. Schrag and J. M. Eller, *Geochim. Cosmochim. Acta*, 2006, **70**, 1439–1456.
120. P. Ghosh, C. N. Garzione and J. M. Eiler, *Science*, 2006, **311**, 511–515.
121. T. C. Hoering and P. L. Parker, *Geochim. Cosmochim. Acta*, 1961, **23**, 186–199.

122. W. Meier-Augenstein, *Stable Isotope Forensics: An Introduction to the Forensic Application of Stable Isotope Analysis*, Wiley-Blackwell, United Kingdom, 2010, 3.
123. R. Kaufmann, A. Long, H. Bentley and S. Davis, *Nature*, 1984, **309**, 338–340.
124. A. Long, C. J. Eastoe, R. S. Kaufmann, J. G. Martin, L. Wirt and J. B. Finley, *Geochim. Cosmochim. Acta*, 1993, **57**, 2907–2912.
125. J. F. Willey and J. W. Taylor, *J. Am. Chem. Soc.*, 1980, **102**, 2387–2391.
126. J. W. Taylor and E. P. Grimsrud, *Anal. Chem.*, 1969, **41**, 805–810.
127. T. Ding, Analytic methods for silicon isotope determinations, in *Handbook of Stable Isotope Analytical Techniques*, ed. P. A. De Groot, Elsevier, Amsterdam, The Netherlands, 2004, vol. 1, pp. 523–539.
128. C. L. De La Rocha, M. A. Brzezinski and M. J. DeNiro, *Anal. Chem.*, 1996, **68**, 3746–3750.
129. H. Krouse and H. Thode, *Can. J. Chem.*, 1962, **40**, 367–375.
130. H. Kipphardt, S. Valkiers, F. Henriksen, P. De Bièvre, P. Taylor and G. Tolg, *Int. J. Mass Spectrom.*, 1999, **189**, 27–37.
131. S. Richter, H. Kühn, J. Truyens, M. Kraiem and Y. Aregbe, *J. Anal. Atom. Spectrom.*, 2013, **28**, 536–548.
132. J. Santrock, S. A. Studley and J. M. Hayes, *Anal. Chem.*, 1985, **57**, 1444–1448.
133. K. Leckrone and M. Ricci, Oxygen isotope Corrections in Continuous-Flow Measurements of SO2, in *Handbook of Stable Isotope Analytical Techniques*, ed. P. A. De Groot, Elsevier, Amsterdam, The Netherlands, 2004, vol. 1.
134. Z. Muccio and G. P. Jackson, *Analyst*, 2009, **134**, 213–222.
135. P. A. De Groot, ed., *Handbook of Stable Isotope Analytical Techniques*, Elsevier Science & Technology, Amsterdam, The Netherlands, 2004, 1.
136. G. Faure and T. Mensing, *Isotopes: principles and applications*, Wiley, Hoboken, NJ, USA, 3rd edn, 2005.
137. M. A. Jochmann and T. C. Schmidt, *Compound-specific Stable Isotope Analysis*, Royal Society of Chemistry, 2012, 376.
138. E. Lichtfouse, *Rapid Commun. Mass Spectrom.*, 2000, **14**, 1337–1344.
139. J. Hoefs, *Stable Isotope Geochemistry*, U.S. Government Printing Office, 6th edn, 2009.
140. Z. Sharp, *Principles of Stable Isotope Geochemistry*, Pearson Education, Upper Saddle River, NJ, 2007.
141. J. H. Reynolds and J. Verhoogen, *Geochim. Cosmochim. Acta*, 1953, **3**, 224–234.
142. C. Douthitt, *Geochim. Cosmochim. Acta*, 1982, **46**, 1449–1458.
143. T. Ding and L. Tang, *Silicon Isotope Geochemistry*, Geological Publishing House, Beijing, China, 1996.
144. S. Wu, T. Ding, X. Meng and L. Bai, *Chin. Sci. Bull.*, 1997, **42**, 1462–1465.
145. J. M. Eiler, *Quat. Sci. Rev.*, 2011, **30**, 3575–3588.
146. T. D. Bullen, D. P. Krabbenhoft and C. Kendall, *Geochim. Cosmochim. Acta*, 1996, **60**, 1807–1821.

147. C. Kendall and E. A. Caldwell, Fundamentals of isotope geochemistry, in *Isotope Tracers In Catchment Hydrology*, eds. C. Kendall and J. J. McDonnel, Elsevier, Amsterdam, 1998, pp. 51–86.
148. C. Kendall, M. G. Sklash and T. D. Bullen, Isotope Tracers of Water and Solute Sources in Catchments, in *Solute modelling in catchment systems*, ed. S. T. Trudgill, John Wiley & Sons, New York, 1995, pp. 261–303.
149. H. G. M. Eggenkamp, http://www.eggenkamp.info/halogen/, (accessed 16.06.2014).
150. S. Barabash, R. Lundin, H. Andersson, K. Brinkfeldt, A. Grigoriev, H. Gunell, M. Holmström, M. Yamauchi, K. Asamura, P. Bochsler, P. Wurz, R. Cerulli-Irelli, A. Mura, A. Milillo, M. Maggi, S. Orsini, A. J. Coates, D. R. Linder, D. O. Kataria, C. C. Curtis, K. C. Hsieh, B. R. Sandel, R. A. Frahm, J. R. Sharber, J. D. Winningham, M. Grande, E. Kallio, H. Koskinen, P. Riihelä, W. Schmidt, T. Säles, J. U. Kozyra, N. Krupp, J. Woch, S. Livi, J. G. Luhmann, S. McKenna-Lawlor, E. C. Roelof, D. J. Williams, J. A. Sauvaud, A. Fedorov and J. J. Thocaven, The Analyzer of Space Plasmas and Energetic Atoms (ASPERA-3) for the Mars Express Mission, in *The Mars Plasma Environment*, ed. C. T. Russell, Springer New York, 2007, ch. 6, pp. 113–164.
151. P. L. Staddon, *Trends Ecol. Evol.*, 2004, **19**, 148–154.
152. K. A. Hobson, *Oecologia*, 1999, **120**, 314–326.
153. T. E. Cerling, G. J. Bowen, J. R. Ehleringer and M. Sponheimer, The Reaction Progress Variable and Isotope Turnover in Biological Systems, in *Terrestrial Ecology*, eds. E. D. Todd and R. T. W. Siegwolf, Elsevier, 2007, vol. 1, pp. 163–171.
154. T. E. Dawson, S. Mambelli, A. H. Plamboeck, P. H. Templer and K. P. Tu, *Annu. Rev. Ecol. Syst.*, 2002, **33**, 507–559.
155. G. J. Bowen, L. I. Wassenaar and K. A. Hobson, *Oecologia*, 2005, **143**, 337–348.
156. T. E. Dawson and R. Siegwolf, *Stable Isotopes as Indicators of Ecological Change*, Academic Press, 2011, 1.
157. E. Brugnoli and G. D. Farquhar, Photosynthetic fractionation of carbon isotopes, in *Photosynthesis*, Springer, 2000, pp. 399–434.
158. O. Schmidt, J. Quilter, B. Bahar, A. Moloney, C. Scrimgeour, I. Begley and F. Monahan, *Food Chem.*, 2005, **91**, 545–549.
159. G. Gleixner, H.-J. Danier, R. A. Werner and H.-L. Schmidt, *Plant Physiol.*, 1993, **102**, 1287–1290.
160. G. Farquhar and J. Lloyd, Carbon and oxygen isotope effects in the exchange of carbon dioxide between terrestrial plants and the atmosphere, in *Stable Isotopes and Plant Carbon-Water Relations*, eds. J. R. Ehleringer, A. E. Hall and G. D. Farquhar, Academic Press, San Diego, 1993, vol. 40, pp. 47–70.
161. M. Barbour and G. Farquhar, *Plant, Cell Environ.*, 2000, **23**, 473–485.
162. N. J. Loader, D. McCarroll, M. Gagen, I. Robertson and R. Jalkanen, *Terrestr. Ecol.*, 2007, **1**, 25–48.

163. M. M. Barbour, T. J. Andrews and G. D. Farquhar, *Funct. Plant Biol.*, 2001, **28**, 335–348.
164. G. D. Farquhar, J. R. Ehleringer and K. T. Hubick, *Annu. Rev. Plant Biol.*, 1989, **40**, 503–537.
165. E. A. Vaganov, E.-D. Schulze, M. V. Skomarkova, A. Knohl, W. A. Brand and C. Roscher, *Oecologia*, 2009, **161**, 729–745.
166. T. W. Boutton, II. Atmospheric, terrestrial, marine, and freshwater environments, in *Stable Carbon Isotope Ratios of Natural Materials*, eds. D. C. Coleman and B. Fry, Academic Press, 1991, vol. 1, pp. 173–244.
167. M. Lakatos, B. Hartard and C. Máguas, The stable isotopes delta13C and delta18O of lichens can be used as tracers of microenvironmental carbon and water sources, in *Stable Isotopes As Indicators of Ecological Change*, eds. T. Dawson and R. W. Siegwolf, Elsevier, Oxford, 2007, pp. 73–88.
168. C. Màguas and H. Griffiths, *Prog. Bot.*, 2003, 472–505.
169. D. C. Coleman, ed., *Carbon Isotope Techniques*, Academic Press, 1991.
170. T. E. Dawson and R. T. Siegwolf, *Terrestr. Ecol.*, 2007, **1**, 1–18.
171. N. Ostle, P. Ineson, D. Benham and D. Sleep, *Rapid Commun. Mass Spectrom.*, 2000, **14**, 1345–1350.
172. B. A. Hungate, R. B. Jackson, C. B. Field and F. S. Chapin III, *Plant Soil*, 1995, **187**, 135–145.
173. P. Bruneau, N. Ostle, D. A. Davidson, I. C. Grieve and A. E. Fallick, *Rapid Commun. Mass Spectrom.*, 2002, **16**, 2190–2194.
174. J. J. Middelburg, C. Barranguet, H. T. Boschker, P. M. Herman, T. Moens and C. H. Heip, *Limnol. Oceanogr.*, 2000, **45**, 1224–1234.
175. M. M. Barbour, *Func. Plant Biol.*, 2007, **34**, 83–94.
176. L. Flanagan and J. Ehleringer, *Func. Ecol.*, 1991, 270–277.
177. A. L. Sessions, T. W. Burgoyne, A. Schimmelmann and J. M. Hayes, *Organ. Geochem.*, 1999, **30**, 1193–1200.
178. J. B. West, H. W. Kreuzer and J. R. Ehleringer, Approaches to plant hydrogen and oxygen isoscapes generation, in *Isoscapes*, Springer, 2010, pp. 161–178.
179. R. D. Evans, *Trends Plant Sci.*, 2001, **6**, 121–126.
180. D. Robinson, *Trends Ecol. Evolut.*, 2001, **16**, 153–162.
181. J. D. Marshall, J. R. Brooks and K. Lajtha, Sources of variation in the stable isotopic composition of plants, in *Stable Isotopes in Ecology and Environmental Science*, Blackwell Publishing, Oxford, 2007, pp. 22–60.
182. M. A. Wadleigh and D. M. Blake, *Environ. Pollut.*, 1999, **106**, 265–271.
183. N. Tanz and H.-L. Schmidt, *J. Agricul. Food Chem.*, 2010, **58**, 3139–3146.
184. M. A. Wadleigh, *Environ. Pollut.*, 2003, **126**, 345–351.
185. J.-T. Cornelis, B. Delvaux, R. Georg, Y. Lucas, J. Ranger and S. Opfergelt, *Biogeosciences*, 2011, 8.
186. C. L. De La Rocha, M. A. Brzezinski and M. J. DeNiro, *Geochim. Cosmochim. Acta*, 2000, **64**, 2467–2477.
187. K. Ziegler, O. Chadwick, E. Kelly and M. Brzezinski, *Geochim. Cosmochim. Acta*, 2002, **66**, A881.

188. C. De La Rocha, M. A. Brzezinski, M. DeNiro and A. Shemesh, *Nature*, 1998, **395**, 680–683.
189. T. Prohaska and J. Draxler, *Encyclopedia of Analytical Chemistry*, Wiley On-line Library, 2014, pp. 1–30.
190. S. Benson, C. Lennard, P. Maynard and C. Roux, *Foren. Sci. Int.*, 2006, **157**, 1–22.
191. N. Gentile, L. Besson, D. Pazos, O. Delémont and P. Esseiva, *Foren. Sci. Int.*, 2011, **212**, 260–271.
192. F. Roelofse and U. E. Horstmann, *Foren. Sci. Int.*, 2008, **174**, 64–67.
193. K. Pye, S. J. Blott, D. J. Croft and J. F. Carter, *Foren. Sci. Int.*, 2006, **163**, 59–80.
194. W. Meier-Augenstein, Stable isotope fingerprinting, in *Forensic Human Identification: An Introduction*, eds. T. Thompson and S. Black, CRC Press, Boca Raton, Florida, 2006, pp. 29–54.
195. A. T. Cawley and U. Flenker, *J. Mass Spectrom.*, 2008, **43**, 854–864.
196. N. Tanaka and D. M. Rye, *Nature*, 1991, **353**, 707.
197. E. Van Warmerdam, S. Frape, R. Aravena, R. Drimmie, H. Flatt and J. Cherry, *Appl. Geochem.*, 1995, **10**, 547–552.
198. M. A. Katzenberg, Stable isotope analysis: a tool for studying past diet, demography, and life history, in *Biological Anthropology of the Human Skeleton*, eds. M. A. Katzenberg and R. S. Shelley, 2nd edn, 2007, ch. 13, pp. 411–441.
199. S. Jim, S. H. Ambrose and R. P. Evershed, *Geochim. Cosmochim. Acta*, 2004, **68**, 61–72.
200. B. J. Mac Fadden, *Paleobiology*, 1998, **24**, 274–286.
201. M. P. Richards, B. T. Fuller, M. Sponheimer, T. Robinson and L. Ayliffe, *Int. J. Osteoarchaeol.*, 2003, **13**, 37–45.
202. B. H. Passey and T. E. Cerling, *Chem. Geol.*, 2006, **235**, 238–249.
203. T. E. Cerling and Z. D. Sharp, *Palaeogeogr., Palaeoclimatol., Palaeoecol.*, 1996, **126**, 173–186.
204. D. Schilling, R. Jakobs, U. Peitz, M. Sulliga, M. Stolte, J. F. Riemann and J. Labenz, *Digestion*, 2001, **63**, 8–13.
205. K. P. Rembacz, K. N. Faber and F. Stellaard, *Rapid Commun. Mass Spectrom.*, 2007, **21**, 3169–3174.
206. H. Schierbeek, F. te Braake, J. Ä. Godin, L. Ä. Fay and J. B. Van Goudoever, *Rapid Commun. Mass Spectrom.*, 2007, **21**, 2805–2812.
207. I. J. Colquhoun and M. Lees, Nuclear Magnetic Resonance Spectroscopy, in *Analytical Methods of Food Authentication*, eds. P. R. Ashurst and M. J. Dennis, Springer, Glasgow, 1998, ch. 3.
208. M. Dennis, P. Wilson, S. Kelly and I. Parker, *J. Anal. Appl. Pyrol.*, 1998, **47**, 95–103.
209. G. J. Bowen, J. R. Ehleringer, L. A. Chesson, A. H. Thompson, D. W. Podlesak and T. E. Cerling, *Am. J. Phys. Anthropol.*, 2009, **139**, 494–504.
210. Y. Aregbe, S. Valkiers, K. Mayer, M. Varlam and R. Wellum, *ESARDA Bull.*, 2007, **37**, 30–34.

211. P. De Bièvre, S. Valkiers and H. Peiser, *J. Res. Natl. Inst. Stand. Technol.*, 1994, **99**, 201–202.
212. Bureau International des Poids et Mesures, http://www.bipm.org/en/scientific/mass/avogadro/, (accessed 16.06.2014).
213. R. Michener and K. Lajtha, eds. *Stable Isotopes in Ecology and Environmental Science*, John Wiley & Sons, 2008.

CHAPTER 17

# Gas Source Isotope Ratio Mass Spectrometry for the Analysis of Noble Gases

SERGEY ASSONOV[a] AND THOMAS PROHASKA*[b]

[a] International Atomic Energy Agency, Vienna, Austria; [b] University of Natural Resources and Life Sciences Vienna (BOKU), Department of Chemistry, Division of Analytical Chemistry, VIRIS Laboratory for Analytical Ecogeochemistry, Tulln, Austria
*Email: thomas.prohaska@boku.ac.at

Gas source mass spectrometers used in static vacuum mode are mostly applied for the analysis of the isotopic composition of noble gases and thus usually referred to as static or noble-gas mass spectrometry.

Noble-gas mass spectrometry is using a similar type of ion source and magnetic sector field mass spectrometer as described in the previous chapter (IRMS) for the major difference that this type of systems is used in static mode, where the gas is fed once into the whole source/analyser system and left there without further supply or pumping. This technique has found specific use in the isotopic analysis of noble gases (He, Ne, Ar, Kr, Xe) as well as of $H_2$ and $N_2$ (in terrestrial minerals and meteorites). All gases are introduced in gaseous form. A particular property is to use ultrahigh-vacuum (UHV) materials, UHV pumping and also chemical

New Developments in Mass Spectrometry No. 3
Sector Field Mass Spectrometry for Elemental and Isotopic Analysis
Edited by Thomas Prohaska, Johanna Irrgeher, Andreas Zitek and Norbert Jakubowski

Published by the Royal Society of Chemistry, www.rsc.org

getters maintaining static vacuum during measurements. Sample gas preparation is rather specific. Noble gases are prepared by extracted physical extraction (*e.g.* melting and stepwise heating for solids, crushing of solids in vacuum, degassing of waters) followed by chemical purification and cryogenic separation.

The mass spectrometer usually consists of an electron impact ion source and a magnetic sector field analyzer. Ions are detected either in a peak-jump mode by a single detector (often a secondary electron multiplier) or in a set of multiple detectors (mostly Faraday cups) collecting the ions of interest simultaneously. Results are reported as isotope ratios, elemental ratios or concentrations (Figure 17.1 and Figure 17.2).

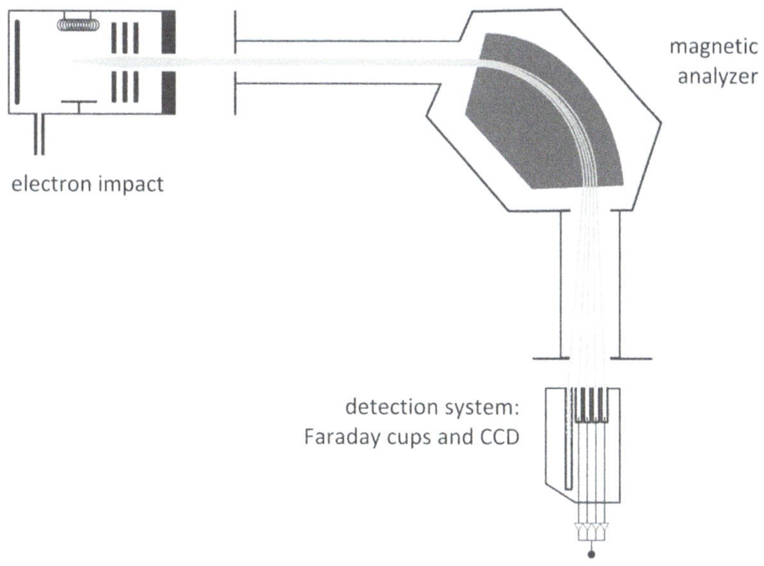

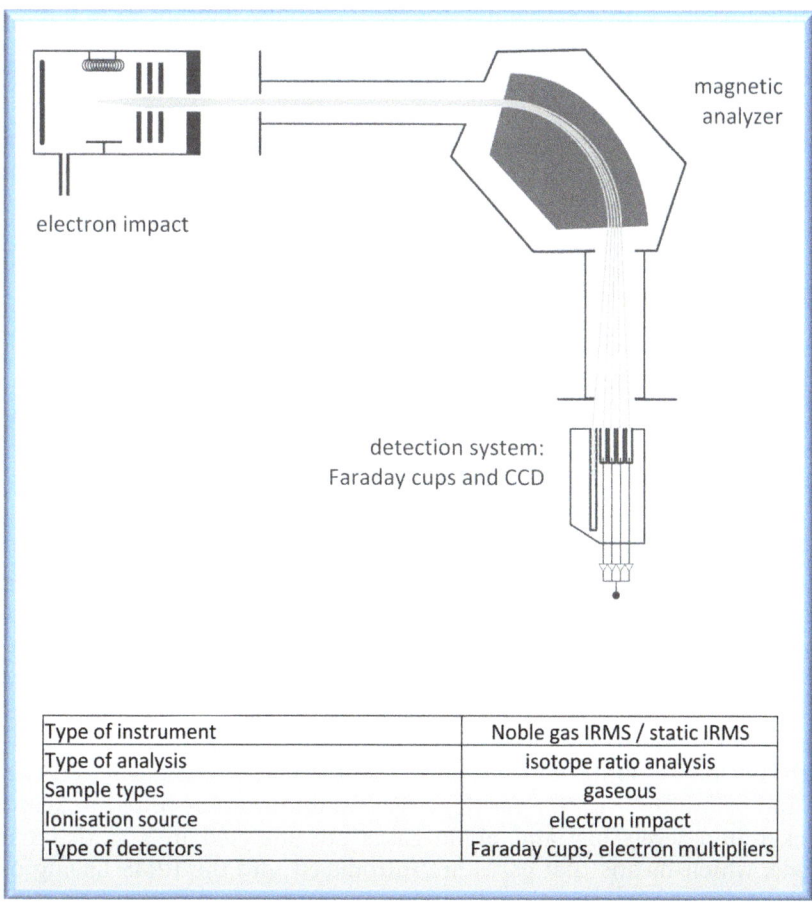

| | |
|---|---|
| Type of instrument | Noble gas IRMS / static IRMS |
| Type of analysis | isotope ratio analysis |
| Sample types | gaseous |
| Ionisation source | electron impact |
| Type of detectors | Faraday cups, electron multipliers |

**Figure 17.1** Features and general schematics of a static (noble gas) IRMS. (Source: Ondrej Hanousek.)

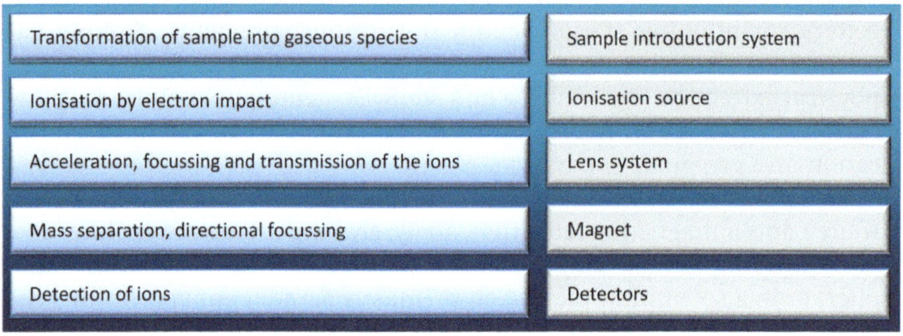
Transformation of sample into gaseous species
Sample introduction system
Ionisation by electron impact
Ionisation source
Acceleration, focussing and transmission of the ions
Lens system
Mass separation, directional focussing
Magnet
Detection of ions
Detectors

**Figure 17.2** Operational flow chart of an IRMS.

**Name and Abbreviation**

This technique uses a gas source mass spectrometer with a number of similarities as in IRMS (Chapter 16) and is usually referred to as noble-gas mass spectrometry, noble-gas IRMS or static IRMS without a common abbreviation used, so far. Therefore, we introduce the abbreviation NG IRMS for noble-gas mass spectrometry.

(***Note***: we have omitted the hyphen as NG is an additional description of the instrument and not a hyphenation of an additional technique (see also Chapter 3).)

## 17.1 General (Peculiarities of Noble-Gas Mass Spectrometry)

Noble-gas measurements include isotope ratio, elemental ratio and concentration determinations of He, Ne, Ar, Kr and Xe. Similar to stable isotopes of light elements, noble gases are analysed by using magnetic sector field gas source mass spectrometers. However, NG IRMS encompasses a number of very specific aspects of IRMS (*e.g.* hardware and techniques in use, referencing approaches and above all, sample preparation).

As noble gases are mostly available in small quantities, sample preparation, purification and separation are of crucial importance. It has to be noted that all noble gases are present in the atmosphere and Ar, Kr, Xe are adsorbed on surfaces. Thus, thorough degassing is a prerequisite. Removing adsorbed gases from surfaces of samples, gas-extraction lines and of inside the mass spectrometer is usually achieved by baking at 150–300 °C. To deal with very low amounts of NG available for analysis (*e.g.* about $10^{-7}$ cc STP for Ar and below $10^{-11}$ cc STP for Xe) UHV conditions are required to keep the background levels low and noble gases are analysed in a static-vacuum mode, which means that gases are introduced into the IRMS having been separated from the pumping. Ion sources are mostly optimised for the analysis of all noble gases. In this context, most NG IRMS systems are equipped with electromagnets as the investigated mass range lies between $m/z = 3$ and

$m/z = 136$. Moreover, double-focusing conditions provide flat-top peaks and the potential for high mass resolution as an asset for noble-gas mass spectrometry in order to separate potential spectral interferences. (Most critical isobaric interferences include $HD^+$ and $H_3^+$, which interfere with $^3He^+$. Thus a mass resolution of $R = 650$ to 700 and higher is needed.) To cover a large range of isotopic abundances, various combinations of Faraday cups with high ohmic resistors ($10^{12}$ Ω) and secondary electron multipliers are used to cover isotope ratios $^3He/^4He$ and $^{40}Ar/^{36}Ar$ ranging by several orders of magnitude.

**Historical Aspects**

The investigation of the isotopic variations of noble gases resulting from both radiogenic and nucleogenic processes were of central importance already within the early years of mass spectrometry. The first reported spectrum of a mass spectrometer was the mass spectrum of several elements consisting of mass to charge lines. We cite: "We find, for example, that the atom of mercury, the heaviest atom I have tested, can have as many as 8 charges, krypton can have as many as 5, argon 3, neon 2, and so on."[1] That was the discovery of the isotopes. Within the following 10 years most isotopes of the noble gases were determined except for $^3He$, following only in 1939 when Alvaraz and Cornog showed that $^3He$ is a stable isotope of He by using a cyclotron mass spectrometer.[2] Later Aldrich and Nier continued studies on $^3He$ isotope abundance in the atmosphere and of different He sources.[3,4] They also detected radiogenic $^{40}Ar$ and laid the foundations for K–Ar dating.[5] A mass spectrometer with an electron impact source was built for this purpose.[6] Most modern mass spectrometers for noble-gas isotope ratio measurements are based on the Nier-type ion source. A major impact in the field of noble-gas mass spectrometry was the development of the static mass spectrometer by Reynolds in 1956 that appears to have introduced the use of electron multiplier.[7] Later, in 1980, Hohenberg described several developments resulting in a high sensitivity NG IRMS.[8] Major improvements included signal detection by secondary electron multiplier (SEM) in the ion-counting mode and an ion source of high transmission (more than 90%). The latter is an electron impact source of cylindrical geometry, with exceptional low pressure dependence and low mass discrimination based on the Baur ion source.[9] For a long time, a large number of homemade machines had been used throughout as well as instruments from now-extinct companies with the Nier-type ion source.

Similar developments – both in measurement techniques and scientific findings – were accomplished by the Russian/Soviet noble-gas school. The work was in many cases published in Russian only, simultaneously to or sometimes earlier than papers in internationally appearing journals. Many works and findings were thus not accessible to the international community as Soviet scientific literature remained mostly untranslated. The Russian school initiated by V.I. Vernadsky and A.E. Fersman focused initially on uranium minerals and uranium decay, helium and other daughter elements.

In 1922, Vernadsky initiated the Radium Institute (St. Petersburg, Russia) specialising on nuclear physics, radio- and geochemistry, U and Th minerals, migration of radioactive elements, radiogenic He and on dating by using radioactive decay. That resulted in the Russian noble-gas school headed by E.K. Gerling and coworkers (in the 1940s, 1950s and later). They largely used home-built isotope mass spectrometers – created also in collaboration with physical research institutes and – later – commercial instruments (mostly IRMS aimed for $UF_6$ measurements) were adapted for noble gases. Important scientific findings include works by E.K. Gerling on K–Ar dating, which were accomplished in 1948 in parallel with Aldrich and Nier.[5] In the 1950s and early the 1960s numerous publications appeared in Russian on the results of K–Ar dating and geochronology. In a review of dating methods in 1965,[10] it was noted that the first K–Ar dating was accomplished by Gerling and Yashchenko[11] "who obtained nine argon ages (mainly on microcline) and who reported satisfactory agreement between argon ages and those of associated minerals determined by the lead and helium methods."

One specific focus was set on the isotopic geochemistry of He. Thus, a high resolution ($R \sim 2000$) mass spectrometer for $^3He$ /$^4He$ was developed[12,13] applying a resonance modulation of the $^3He^+$ ion beam in the magnetic field. This gave a full resolution of $^3H^+$ from all interferences. Based on this development and the specifically designed instrument, several pioneering works on $^3He$ and $^3He/^4He$ geochemistry were accomplished (for example the precise determination of the $^3He/^4He$ ratios in the atmosphere,[14] helium isotope studies on volcanic gasses, rocks and thermal springs[15–17] and developing new Tritium/$^3He$ dating method for ground waters,[18] which is widely used nowadays worldwide). An overview on He isotope studies in Russia is given, *e.g.* by Mamyrin and Tolstikhin.[19]

Helium isotope geochemistry stimulated further developments of "traditional" sector-type NG IRMS. In the late sixties, Clarke *et al.*[20] described a 10-inch radius NG IRMS with high resolution on $^3He^+$ ($R = 625$) and an improved peak shape with the aim to study Tritium by $^3H$ accumulation and also to redetermine the atmospheric $^3He/^4He$. (Results were reported as delta-value *vs.* the atmospheric $^3He/^4He$.)

In 1976, Takaoka[21] described a low-blank high resolution ($R = 500$) NG IRMS aimed at the analysis of all noble gases ($^3He$ to Xe) in meteorites. The IRMS was able to resolve most of the interference peaks such as $H^{3+}$ and $HD^+$ from $^3He^+$, $H_2{}^{18}O^+$, $^{20}NeH^+$and $^{40}Ar^{2+}$ on masses of Ne, $C_3H_2{}^+$ from $^{38}Ar^+$ and hydrocarbon ions from Kr and Xe isotopes. A whole metal, low blank gas extraction system with high-temperature furnace was optimised as well, and the IRMS and gas extraction system were equipped with UHV components. That work actually included a number of developments accomplished by the group of Hintenberger at the Max Planck Institute for Chemistry in Mainz, DE (see, *e.g.* references in Takaoka[21]).

In the 1960s, K–Ar dating was further developed and $^{40}Ar$–$^{39}Ar$ dating proposed by Turner and colleagues,[22,23] which resulted in creating a specialised Ar mass spectrometer in 1966.[24] The instrument was used for

$^{40}Ar$–$^{39}Ar$ dating of Apollo lunar samples (*e.g.* Turner *et al.*[25]). The *MS-1* instrument equipped initially with a Nier-type ion source and an ion-getter pump, was updated with the Baur source in 1985 and is still in operation up to the present.[24]

It should be noted that the noble gas-school in Manchester has developed and introduced many technical and hardware solutions, *e.g.* the laser extraction for $^{40}Ar$–$^{39}Ar$,[26,27] crushing in vacuum,[28] mass-spectrometer with cryogenic gas concentrator and laser resonance ionisation for Kr and Xe.[29–32]

Numerous developments and technical solutions optimised on home-built instruments as well as an increased demand for NG IRMS resulted in commercial IRMS products. Several companies have been created and existed in the Manchester area. (Ownerships and company names have been changed several times since then.) In the early 1980s, the *MM-3000* was commercialised by VG Instruments. The instrument included two detectors (Faraday and electron multiplier) and was able to well resolve $^{3}He^{+}$ and $HD^{+}$ (*e.g.* Hooker *et al.*[33]). All commercial NG IRMS were equipped with the Nier ion source. Later the *VG-3600* was produced. The successor of the *MM-3000* and the *VG-3600*, the *VG-5400* was reported in 1986. The *VG-5400* is a statically operated, double-focused, 90° sector field mass spectrometer with 57-cm extended geometry, adjustable magnet poles, equipped with one Faraday-cup detector and one electron multiplier. Later the *VG-5440* was introduced, which is a modified version of *VG-5400* with a wide-angle flight tube, five fixed Faraday collectors and one Daly collector.

MAP (Mass Analyzer Products, Manchester, UK) was started in 1982 and shortly after, the *MAP-215-50* was introduced to the market. Later, the *MAP 216* was introduced as a successor of the *MAP 215*. The VG and MAP instruments have long been the most important workhorses for noble-gas isotope measurements for the following 20 years when development in noble-gas mass spectrometer came almost to a stop.

The new generation of mass spectrometer marks the age of multicollector noble-gas instruments as single-collector instruments became more and more obsolete. Finally in 2004, two new instruments were announced to the market: The *Helix Split Flight Tube* (*Helix SFT*) was originally announced by the company HtGX, which became part GVI before being bought by Thermo, where the model was finally finished. The second instrument, the *Noblesse*, was launched by Nu instruments. The *Noblesse* continues the VG tradition of using existing MS technology taking the optics and flight tube from the Nu Plasma. The *Aurora* (also introduced by Nu Instruments) was a NG IRMS with a simplified detector arrangement (one Faraday cup and one single ion counting multiplier).

Today, there is a proliferation of instruments: Thermo Scientific has three different purpose built models, the *Argus*, the *Helix SFT* and the *Helix MC*. Nu Instruments has currently one model, the *Noblesse*, and IsotopX has launched a new instrument (the *NGX*) (see Section 17.3).

The development of figures of merits in noble-gas mass spectrometry published by Burnard[34] show that the precision has improved by almost three orders of magnitude and the sample size has been reduced by almost 6 orders of magnitude from the early measurements in the 1940s up today.

Nonetheless – similar to light stable-isotope IRMS – alternative methods for analysing noble-gas isotope ratios have been exploited such as, *e.g.* laser spectroscopic methods on He isotopes[35] using laser light absorption and resonance ionisation mass spectrometry of Kr and Xe cryogenically concentrated in the ion source.[29–32]

## 17.2 Technical Background

The stable-isotope ratio mass spectrometer is comprised of an inlet system, an ion source, an analyser for ion separation and detectors for ion collection and quantification. All modern noble-gas mass spectrometers are magnetic sector field devices. As similar considerations as compared to IRMS instruments can be applied, the basic components will not be described in detail. The most crucial point is in any case the sample preparation.

### 17.2.1 Sampling, Extraction and Purification

As already mentioned, sample preparation is one of the most crucial steps in NG IRMS. Noble gases are mostly available in very low quantities and in a matrix of other accompanying gases such as $CO_2$, $N_2$, $O_2$ or $H_2O$. As these gases are chemically reactive, purification of noble gases is often accomplished by using Ti–Zr getters, which chemically absorb reactive gases. As there is a dependence of measured NG isotope ratios on the total pressure in IRMS, gases are separated into fractions (He, Ne or He + Ne, Ar, Kr + Xe), which are analysed separately.

**Sampling**

Most samples are solids (minerals, rocks, meteorites, *etc.*) and waters. In some cases air samples are analysed (*e.g.* firn air, air trapped in ice cores or soil gases). For solid samples there are no specific sampling requirements. The only recommendation maybe to avoid mineral / density separation in heavy liquids as these may increase the level of organic interferences.

Difficulties in collecting samples without gas loss and/or contamination have been an obstacle in accurate He isotope ratio measurement ever since. Water samples (with the aim of measuring noble-gas isotope ratios and/or performing Tritium/$^3$He dating requires collection of water *in situ*, with the major concerns (i) to avoid contamination by air and (ii) ensure no gas losses during storage. Therefore, water flowing directly from a well or from a pump is sampled in Cu tubes with metal clamps for crimping. Crimping creates a cold welding of the tube, so that samples can be stored and shipped without further precautions. In the lab, the water is degassed and gases are transferred to a noble-gas extraction/purification line.

Taking air samples requires avoiding contamination by air and hydrocarbons (*e.g.* oil-fee pumps) and storing air samples in clean stainless steel containers with a valve or in a Cu tube; using glass should be avoided, as He can be partly lost by diffusion through glass. Ice core samples need to be taken and stored in the same way as used for analysis of trapped air in ice.

**Noble-Gas Extraction Methods**

In general, the concentrations of noble gases in geological materials and fluids are very low. Thus, a central part of the instrumental considerations for measuring noble gases is usually focusing on various techniques to release noble gases from solid samples or fluids, whereby each sample type requests its individual challenging approach in sample preparation before analysis. Extraction of noble gases from fluids is mostly achieved by phase separation. Thermal extraction or crushing under vacuum is used to release noble gases present in minerals/solids either within the mineral lattice or within fluid inclusions. A furnace or a laser is applied for releasing the noble gases either by stepwise heating and/or complete melting. Values obtained by stepwise heating mostly reflect different phases, though values (mostly elemental ratios) may be somewhat affected by diffusion. Up to now, most laboratories use homemade gas extraction and purification systems in combination with commercial components like lasers or cryogenic systems for gas separation. Selecting the appropriate gas extraction method and its parameters, or combination of methods is often the only means to address a certain geochemical question and separate NG components accordingly.

Often, volcanic samples with gas–fluid inclusions are crushed in vacuum followed by complete melting thus separating gases in inclusions and radiogenic gases components. (For all methods maintaining low and extra-low blank levels is critical.)

**Noble-Gas Extraction by Heating**

Since the very beginning of noble-gas research, heating furnaces were the most common method. The most critical component was to create a high-temperature furnace with a low blank, long lifetime of the reactor and heating element and being robust in operation. In 1976, Takaoka[21] described a low-blank metal gas extraction system based on the developments by the group of Hintenberger (*e.g.* Hintenberger *et al.*[36]) in Mainz, DE. This system included a double-vacuum high-temperature furnace with a heating element in the outer vacuum volume, a reaction tube made of Ta with the Ta insert aimed to prevent reactor from corrosion, with reactive silicate and metal sample melts. In 1978 Staudacher *et al.*[37] described further refinements of the furnace for noble-gas extraction namely by step heating (temperature control with a thermocouple) and melting. Similar to Takaoka,[21] the furnace included a double vacuum for the heating element, a Ta-reactor tube with the insert, a thermocouple and a low-ohmic heating element requiring up to 120 A. Very low blank levels were achieved. This development has been like a "golden standard" for noble-gas mass-spectrometric techniques.

Another principle is a single-vacuum furnace where the heating element and sample are located in the same vacuum volume. Though this is a simple technical solution, the major disadvantage is that degassing of the heater in the sample volume cannot be avoided, thus leading to increased blanks. Recently, the use of a single vacuum furnace was proposed for cosmogenic $^{3}$He dating.[38] Due to very low $^{3}$He levels in the atmosphere and also due to miniaturisation of the heating element (located in the same vacuum as the sample) and eliminating the large reactor tube, the authors have reached $^{3}$He blank levels lower than in "classical" double-vacuum furnaces.

**Noble-Gas Extraction by Laser Techniques**

Gas extraction by laser heating, first proposed for noble gases in meteorites,[39] has been developed for $^{40}$Ar–$^{39}$Ar,[26,27,40,41] later for nitrogen and noble gases in meteorites[42,43] and for lunar samples and terrestrial rocks (Hashizume and Marty[44] and references therein). A wide range of lasers (Nd:YAG, $CO_2$, IR and UV lasers) have been applied, which allow local heating with the consequence of very low blanks (especially critical for Ar and $N_2$) and also enable stepwise heating of the sample. Moreover, a laser allows for selective spot sampling and also for opening fluid inclusions (laser decrepitating, *e.g.* Bernard *et al.*[45]). The extraction system consists of a UHV chamber with an optical window (compatible with laser wavelengths) connected to UHV pumping, a noble-gas purification system, a laser, a binocular microscope and often a video camera. Advantages of laser extraction include, *e.g.* much lower blanks compared to resistance furnace with a large reactor, negligible memory and the possibility to analyse single grains or to make profiles over a polished section.

**Noble-Gas Extraction by Crushing in Vacuum**

Extracting NG from fluid inclusions is typically done by crushing in vacuum. Initially mechanical, screw-operated crushers were constructed from stainless steel vacuum valves (*e.g.* Stuart and Turner[28]). Later, the design has been optimised by using magnetically activating (external solenoid) piston moving inside the crushing volume. The method is commonly used to extract NG from fluid inclusions in basalts and magmatic rocks and later also in stalagmites (to determine paleotemperatures based on noble gases dissolved in formation water in caves).[46,47] The most critical issue is to remove heavy noble gases adsorbed both on surface of the sample and of the crusher itself, what is accomplished by preheating of the crusher loaded with a sample under vacuum.

**Noble-Gas Extraction by Acid Etching of Lunar Samples in Vacuum**

Noble-gas extraction by sample etching in vacuum was developed by the Zürich group in order to extract solar-wind noble gases implanted in lunar minerals. In the first publication[48] the system was made of glass with Teflon valves and samples were exposed initially to $HNO_3$ vapours and then to

$HNO_3$ liquid, with etching steps lasting from 1 h to several days. This method permitted two components to be recognised – solar wind and solar energetic particles (implanted with higher energy, to higher depth) – and to obtain isotope and elemental ratios without element or isotope fractionation due to diffusion as occurs during stepwise heating. In 1986 Wieler *et al.*[48] obtained values for He and Ne. They reduced blanks by making the system of gold and platinum and by using HF. Later values for He–Ne–Ar[49] and for Kr–Xe[50] were obtained. These findings were an important step in solar-system research.

**Noble-Gas Extraction from Water Samples**

He from water samples needs to be extracted quantitatively for Tritium/$^3$He dating. (We refer here to the system created and optimised at the IAEA (*e.g.* Stolp *et al.*[51]).) Water samples taken in Cu tubes with metal clamps are transferred in larger stainless steel volumes and heated, so that NGs are released in the head space. Then the head space is connected to a smaller transfer volume cooled by liquid nitrogen *via* a capillary. Due to the pressure gradient (very low pressure above ice), water vapours are directed to the transfer volume, transferring noble gases like a diffusion pump. The overall efficiency of degassing and gas transfer is $\sim$99%. The gas in the transfer volume contains only a small amount of water and is ready for purification, which is performed on a NG purification line. All steps (NG purification, separation and isotope measurements on *VG-5400*) are performed under computer control, in a reproducible way.

**Noble-Gas Purification and Separation**

As a single and pure noble gas is required for the subsequent analysis, in a further step, the noble gases are purified and separated *via* their physical and chemical properties. Purification is mostly accomplished by "chemical separation" making use of the chemical inertness of the gases using, *e.g.* metal surfaces as "getter". Commonly, Ti–Zr and Al–Zr getters (SAES getters, Lainate, Italy) are used. (Before, CaO was also used.[48]) During these steps, gases are either transferred from one volume to another by expansion (He and Ne) or by freezing on a sorbent (cleaned under vacuum-activated charcoal or molecular sieves). In an additional step, cryoseparation can be applied for the separation of purified mixtures of noble gases (in many cases all gases, from He to Xe are present). Thereby, the noble gases are adsorbed on a sorbent (*e.g.* activated charcoal) at low temperatures and the different noble gases are released by stepwise increasing the temperature. For more information on the sample preparation we refer to textbooks, which describe these techniques in more detail (*e.g.* Burnard[34]).

## 17.2.2 Sample Introduction

In contrast to stable-isotope IRMS operated under "dynamic vacuum mode" (constant introduction of the gas into the MS while the MS is pumped), noble-gas mass spectrometers are operated under "static vacuum mode". This means that the complete amount of gas is transferred (by expansion) to the NG IRMS having the pumps isolated.[7] In the static mode, much higher sensitivity can be reached, as gas is more effectively used. For a comparison, typical noble-gas amounts are $n\times10^{-7}$ cc STP for Ar or $n\times10^{-11}$ cc STP (or even less) for Xe, whereas IRMS in the dual-inlet mode (see Chapter 16) need about 0.3–0.5 cc STP gas (down to 0.03 with the microvolume trap). Several injections are needed for IRMS operated in the continuous-flow mode as the gas amounts for $\sim2\times10^{-5}$ cc STP (*e.g.* 10 ng of C) in a single injection. Moreover, very low backgrounds of the NG IRMS are required as the background rises as soon as the MS is isolated from the pump. This is reached by using Ti–Zr getters in the NG IRMS.

**Nontraditional Sample-Introduction Systems**

For many years, sample introduction was based on separation of noble gases from solid or water sample, followed by gas purification, separation in fraction and introduction into NG-IRMS by gas expansion, see the section on sample preparation. However, in recent years there is a tendency for a synergy between noble-gas approaches and stable-isotope approaches. Below we outline some examples.

**Noble-Gas Measurements on Unseparated Air Samples by Using Conventional IRMS**

Isotope and elemental ratios (in particular $Ar/N_2$ and $Kr/N_2$ elemental ratios and Ar isotope ratios) of firn air and air trapped in ice (ice cores) are used for paleoenvironmental applications, in order to evaluate the past temperature gradients. Severinghaus *et al.*[52] describe the method developed for precise measurements of $^{40}Ar/^{36}Ar$ and Kr/Ar ratios of air trapped in ice by using the DI-mode of the *MAT252* and *MAT253* with sample and reference gas filled in dual inlet (DI-mode) bellows. Air extracted from ice is purified by a Zr-Al (SAES) getter in order to remove $N_2$ and other gases and then purified noble gases are diluted with pure $N_2$ gas. Later, Kobashi *et al.*[53] described further developments of measurements of $^{40}Ar/^{36}Ar$ ratios based on the DI-mode of the *MAT253*. The air sample is treated with heated copper (500 °C) in order to remove molecular oxygen and thus eliminating the isotopomer $^{18}O^{18}O$, which gives an isobaric interference on $^{36}Ar$ during measurements. In this work, argon isotopic ratios $^{40}Ar/^{36}Ar$, $^{86}Kr/^{40}Ar$, $^{14}N^{15}N/^{14}N^{14}N$ nitrogen isotopic and $^{14}N_2/^{40}Ar$ ratio are expressed as delta values normalised to the present atmosphere.

**Ar-Isotope Measurements in a Continuous Flow of Helium**

Ar-isotope measurements in a continuous flow of helium were suggested by Ignatiev *et al.*,[54] with the aim of performing K–Ar and $^{40}Ar$–$^{39}Ar$ geochronology. Argon is extracted by a laser in a chamber within multiple samples. Then, argon is purified by a chromatographic capillary column in a flow of helium and directed to IRMS. Measurements of the $^{36}Ar$, $^{38}Ar$ and $^{40}Ar$ are performed by a *MAT 253* operated in continuous flow mode, with Faraday cups for $^{36}Ar$, $^{38}Ar$ and $^{40}Ar$. No high-vacuum extraction and purification line is required. The $^{40}Ar$ blank of the entire system (including the high-vacuum line of laser argon extraction and the interface of the high-vacuum line to helium flow GC and *MAT-253*) is better than $3.4\times10^{-9}$ ccSTP, with a certain part of the blank caused by Ar traces in the He carrier gas. The potential improvements include the maximisation of the efficiency of the open split and the use of an electron multiplier for $^{36}Ar$ measurement as well reducing $^{40}Ar$ levels in the He tank.

**A Membrane Inlet System for Noble Gases in Gas and Water Samples**
Visser *et al.*[55] described a noble-gas membrane inlet system developed to measure noble-gas elemental ratios at natural abundances in gas and water samples. The inlet system consists of a membrane inlet, a water trap, a carbon-dioxide trap, getters and a gate valve. As a detector, they use a quadrupole mass spectrometer, which is sufficient for noble-gas elemental ratios at natural abundances. The system measures five noble-gas isotopes ($^{4}He$, $^{22}Ne$, $^{38}Ar$, $^{84}Kr$ and $^{132}Xe$) every 10 s, providing high time resolution. Water(s) equilibrated with air at certain temperatures can be used as reference.

## 17.2.3 Ion Source (see also Section 16.3)

Similar to IRMS, commercial NG IRMS instruments apply the Nier-type ion source (see Section 16.3) where the "electron impact" ionisation is accomplished by electrons emitted by a hot filament. The Nier-type ion source is robust and of high sensitivity. The disadvantage is that the magnetic field in the source results in instrumental isotopic fractionation. This is usually approximated by a linear function *vs.* the mass difference. The correction factor can be averaged over several isotope ratios (*e.g.* for Kr and Xe) measured on a gas standard.

In order to eliminate this effect, the Baur–Singer source was developed.[9] The source uses a circular filament providing a focused conical ion beam without requiring magnets. The source was successfully used by Hohenberg in a high-sensitivity high-transmission NG IRMS.[8] This source is known to result in lower instrumental mass discrimination but have lower sensitivities as compared to the Nier-type source.

In order to increase sensitivity, Baur[56] utilised a turbopump compressing gases from the NG-IRMS to the ion source region, thus increasing the partial pressure in the source and improving the sensitivity by about two orders of magnitude for He, with further increase for heavy gases (*e.g.* see Figure 5 in Ott *et al.*[57]). Another group tested this design with a commercially available turbopump, which resulted in increased levels of isobaric interferences, presumably due to pump oil and viton sealings in the turbo pump.[58] The signal gain was found to be smaller and mass dependent, from $\sim$2 for He, to $\sim$22 Kr and $\sim$50 for Xe.

Sensitivity and instrumental isotopic fractionation of the Nier-type sources depend on the gas pressure in the source (*i.e.* the amount of gas in the NG IRMS), most probably due to space-charge effects. Therefore, standards and samples should be similar in composition (*e.g.* He/Ne ratios) and in total pressure. For this reason, noble gases are usually separated and analysed in fractions (He, Ne, Ar, Kr and Xe) separately (see above).

Ar, Kr and Xe suffer from a memory (pumping) effect in the NG-ion source (see above) resulting in a gradual decrease of ion beam intensities. Normally the effect is rather small/negligible for He and Ne and increases with mass (from Ar to Xe). Therefore, ion-beam intensities have to be extrapolated to the moment of gas introduction into the MS ("zero time").

### 17.2.4 Sector Field Mass Analyser

Sector field mass spectrometers are the dominant mass separators in noble-gas mass spectrometry. Modern noble-gas machines have a high accelerating potential of several kilovolts and accomplish peak switching by changing the magnetic field of the electromagnet. The mass range to be covered ranges from $m/z = 3$ (He) to $m/z = 136$ (Xe).

In order to achieve high resolving power, a peak shape with flat peak top and – at the same time – double focusing is often used – to obtain low contribution of the $^{4}He^{+}$ tail to $m/z = 3$. This is achieved by ion-beam focusing (requirement of a high-voltage ion source) and by using a magnet with adjustable entrance and exit poles, thus optimising beam focusing by the magnet by optimising ion-beam entrance and exit angles. In order to further reduce the $^{4}He^{+}$ tailing to $m/z = 3$, split flight tubes separating $^{4}He^{+}$ and $^{3}He^{+}$ beams in space and electrostatic filters are applied (*e.g. Noblesse, Helix SFT*). (A very specific NG IRMS for $^{3}He/^{4}He$ measurements, with resolving power of $R \sim 2000$ was developed in Russia in the 1960–1970s (see above).)

Many older instruments used a low accelerating potential of a few hundred volts and a small permanent-field magnet. The accelerating potential was then switched (peak-jump mode) to focus different nuclides into the collector. This type of machine suffered to a lesser degree from memory effects but provided poorer precision for isotope ratio measurements.

### 17.2.5 Interface and Vacuum System

Low amounts of NGs require very low background levels in NG IRMS. To achieve this, UHV pumping is needed. In earlier years diffusion pumps (often Hg diffusion pumps) with liquid $N_2$-vapour traps were in use. Nowadays, sample preparation lines are pumped by turbopumps backed with preferably oil-free fore-vacuum pumps (*e.g.* pumps with oil filters or diaphragm pumps). A combination of turbopumps and ion-getter pumps is used for pumping the mass spectrometer. Turbopumps are used during baking out and for pumping large amounts of gases, while ion-getter pumps (giving vacuum of $10^{-9}$ mbar and better) are used to maintain UHV and low backgrounds.

Ion-getter pumps are rather specific: Gases to be pumped are ionised by a discharge between Ti-electrode plates being at high voltage, ions are accelerated and then implanted in titanium electrodes. Some amount of titanium is sputtered by ions and then redeposited. While chemically active gases ($O_2$, $CO_2$, $N_2$) are chemically trapped in the form of compounds formed with Ti, noble gases are trapped only by implantation and redeposition of Ti. Though ion-getter pumps aimed to pump noble gases have an optimised geometry of Ti electrodes (in order to increase sputtering), pumps can be saturated with most abundant noble gases (*i.e.* with Ar), which results in a burst of gas. This explains why large gas amounts need to be pumped by turbopumps.

All components of the vacuum system and sample-preparation lines have to be made of high vacuum purity materials (high-quality stainless steel, often electropolished and preferably annealed in vacuum, is used both for the inner parts of the mass spectrometer as well as for sample preparation lines). The system should have metal to metal UHV flanges (*e.g. Conflat*®) flanges with Cu gaskets) and be bakeable up to 300 °C. UHV valves with Cu seals are in use, either manually and/or pneumatically operated.

Despite all these measures, a certain amount of hydrogen *via* hydrocarbons is build up, which results in spectral interferences. In particular, hydrogen gives ${}^{1}H^{2}D^{+}$ and ${}^{1}H_3{}^{+}$ interfering with ${}^{3}He^{+}$. In order to resolve ${}^{3}He^{+}$ from ${}^{1}H^{2}D^{+}$ and $H_3{}^{+}$ peaks, a resolving power of $R>650$ is needed. In addition, a low contribution of the ${}^{4}He^{+}$ tailing to $m/z=3$ is required.

Vacuum as well as gas pressure ought to be measured during pumping, gas extraction, gas purification and gas separation. Penning and ionisation gauges pump ionised gas in a way similar to ion-getter pumps and can therefore not be used. Pirani gauges and Baratrone sensors are often in use as these do not ionise and do not pump gases.

### 17.2.6 Detection System

Many older machines had one detector (multiplier) measuring multi-isotopes in peak-jump mode whereas modern machines have a multicollector array for simultaneous detection of several isotopes (see description Section 17.3). NG isotope ratios often cover several orders of magnitude, in particular for ${}^{3}He/{}^{4}He$ ranging from $n\times10^{-4}$ to $n\text{–}10^{-8}$. ${}^{40}Ar/{}^{36}Ar$ ratios cover the range from 295.5 up to $n\times1000$ (for terrestrial samples) and about 0.5 and lower for extraterrestrial samples. To cover this large range, combinations of Faraday cup(s) with high-ohmic resistors ($10^{12}$ Ω) and secondary electron multiplier(s) are applied. (For instance when ${}^{3}He/{}^{4}He$ ratios or high ${}^{40}Ar/{}^{36}Ar$ ratios are measured a SEM is used for the lower abundant isotope (${}^{3}He$ or ${}^{36}Ar$) and a Faraday cup for ${}^{4}He$ or ${}^{40}Ar$.)

As a consequence of a multicollector approach, all detectors need to be intercalibrated with utmost care. That is particular critical for multicollector NG-IRMS aiming ${}^{40}Ar$–${}^{39}Ar$ dating, as ${}^{39}Ar$ is not present in the atmospheric argon (typically used as reference gas) and calibration by using another isotope in a peak-jumping mode may not be easy. A gas mixture containing $^{40}Ar$, $^{39}Ar$ and $^{38}Ar$ produced by mixing gasses derived from neutron-irradiated sanidine with the $^{38}Ar$ spike was suggested[59] to calibrate all collectors by combing measurements on this mixture and on atmospheric Ar.

## 17.3 Instrumentation

Within this section, currently commercially available instruments are presented and briefly described.

***ARGUS VI* (Thermo Fisher Scientific)**

The *ARGUS VI* (Figure 17.3) multicollector NG-IRMS is a 4.5-kV, small volume (~700 cc) instrument with a magnetic sector field of 13 cm radius and 90 deg angle, bakeable to 300 °C. Sensitive Nier type ion source ($10^{-3}$ A / Torr), five fixed Faraday detectors and optionally, a compact secondary electron multiplier (optimised slim version of SEM) are mounted next to the low-mass Faraday cup. The instrument can be used to analyse all noble gases but it is specially designed to measure simultaneously all five Ar isotopes and can be operated in three different modes. In the multicollector mode, masses $m/z = 36, 37, 38, 39$ and 40 are collected on a Faraday collector array (resistors from $10^{10}$ to $10^{12}$ Ohm) for samples with sufficient amounts of Ar available. The peak-jumping mode is applied like in a single-collector MS using the

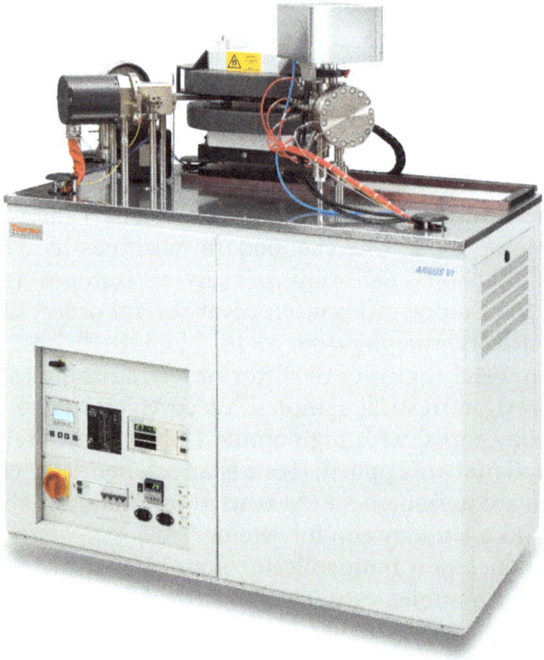


**Figure 17.3** *ARGUS VI.*
(Source: © Thermo Fisher Scientific.)

electron multiplier in ion-counting mode for the analyses of (very) small gas fractions. In a combined mode, all masses will be measured simultaneously with the lowest isotope measured by the electron multiplier on mass $m/z = 36$. Even though the *ARGUS VI* has a fixed collector setup, an electronic steering plate placed in front of each collector is responsible for exact positioning of the peaks, even for other gases such as Kr. *ARGUS VI* can be connected to a number of noble-gas preparation systems provided by the company.

***Helix MC* (Thermo Fisher Scientific)**

The *Helix MC Plus*™ (Figure 17.4) is a magnetic sector field multicollector noble-gas MS with a 350 mm / 120 deg magnet and allows measurements at high mass resolution of up to five isotopes at the same time. The detectors allow a mass resolution of 750 (10% valley definition) whereas the (optional) high-resolution detector allows mass resolutions of up to 1500.

***Helix SFT* (Thermo Fisher Scientific)**

The *Helix SFT* (Figure 17.5) is a split flight tube noble-gas mass spectrometer, which is similar to the *Helix MC* and allows high-precision analysis of all noble-gas isotopes to be performed. The major difference is due the split flight tube design, which allows measurement of $^4He^+$ on the Faraday cup, whereas $^3He^+$ is directed to an electron multiplier. The $^3He^+$ channel includes a 90 deg energy filter to reduce $^4He^+$ peak tailing and to improve abundance sensitivity to $< 1$ ppb (tailing contribution from $^4He$ to $m/z = 3$). The mass resolution is sufficient to separate $^3He^+$ from $^1H^2D^+$ and $^1H_3{}^+$ interferences.

**Figure 17.4** *Helix MC*.
(Source: © Thermo Fisher Scientific.)

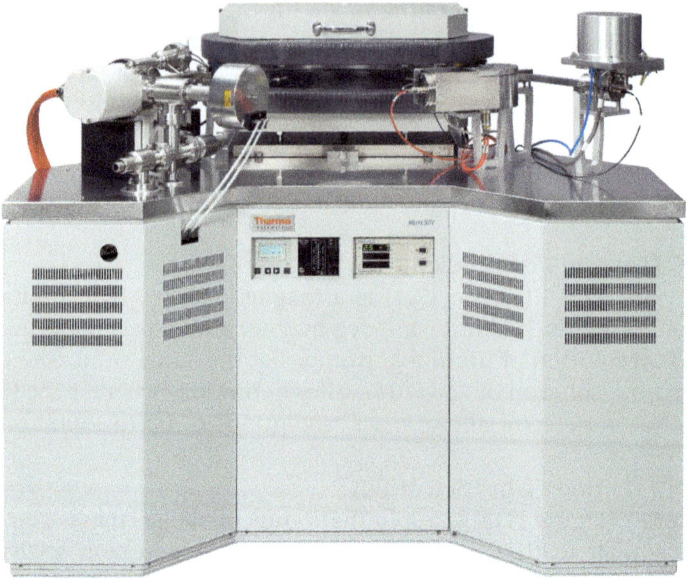

**Figure 17.5** *Helix SFT.*
(Source: © Thermo Fisher Scientific.)

*Helix MC* and *Helix SFT* share the same software and several other technical solutions with ARGUS VI; Thermo Scientific also provides a line of sample preparation units being compatible with all their NG-IRMS and software.

***Noblesse* (Nu Instruments)**

The *Noblesse* (Figure 17.6) multicollector mass spectrometer is designed for the analysis of all noble gases. The instrument has a Nier source, a single-focusing 24-cm radius / 75 deg angle magnet and can be baked up to 300 C. The fixed multicollector array (in combination of up to 6 Faraday cups and up to 4 multipliers) with variable-dispersion zoom optics allowing instantaneous switching and exact peak positioning for several sets of isotopes (previously optimised for the MC ICP-MS *Nu-Plasma*). An all-metal sample preparation line produced by Nu Instruments (*e.g.* a furnace, gas pipetting, laser system, getter purification) can be interfaced to the mass spectrometer.

***NGX* (IsotopX)**

The *NGX* (Figure 17.7) is a multicollector machine maintaining a very low volume. The instrument uses an 8-kV Nier-type ion source, a sector field magnet with adjustable entrance and exit poles, asymmetrical optical geometry, whereas the magnet shortens the image length of the ion beam. This allows the use of a smaller flight tube for reducing the internal volume and lowering the instrument footprint. The instrument uses a 270-mm / 90 deg electromagnet based on the design from the *Phoenix* TIMS instrument (IsotopX). The NGX uses a fixed multicollector geometry

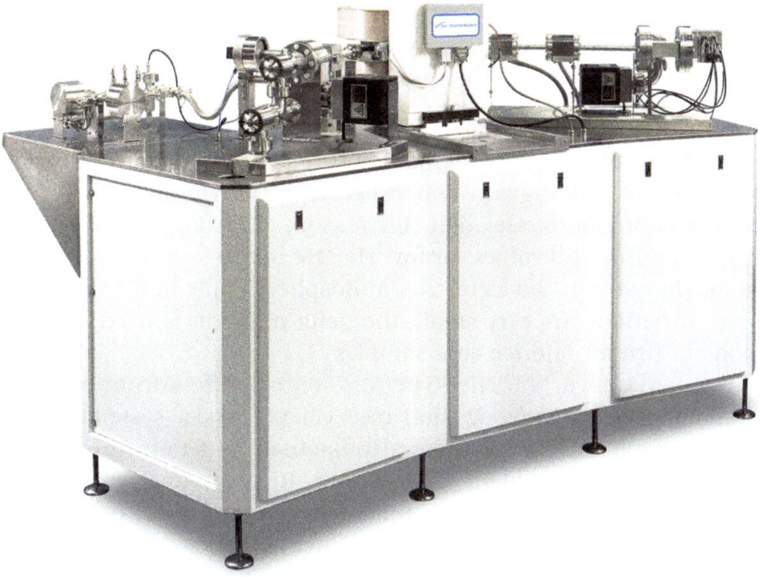

**Figure 17.6** *Noblesse.*
(Photo: © Nu Instruments.)

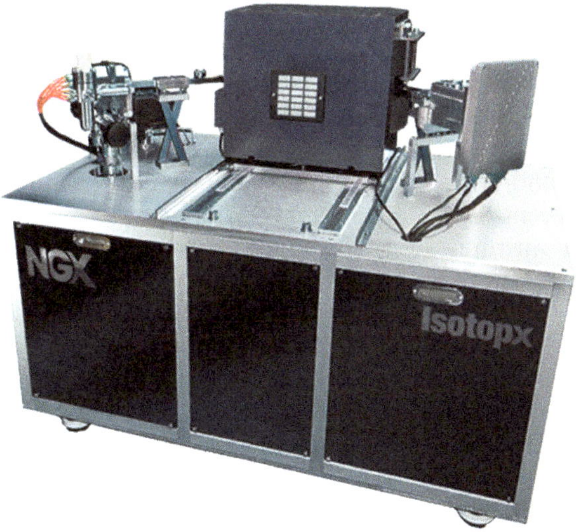


**Figure 17.7** *NGX.*
(Photo: © IsotopX.)

specified for specific isotopes at the time of purchase. In addition, a fully automated gas-extraction line with a furnace and purification of noble-gas samples and gas standards is available.

# 17.4 Measurement Considerations

## 17.4.1 Referencing Strategies and Data Reporting

Noble gases in the terrestrial atmosphere reservoir are used as international (scale-defining) reference material. In contrast to isotope ratios of light elements (H, C, N, O and S) demonstrating very small natural variations, isotope ratios of noble gases and especially $^3He/^4He$ and $^{40}Ar/^{36}Ar$ cover several orders of magnitudes. For this reason, NG isotope ratios are mostly expressed as numerical values, or for $^3He/^4He$ in the form of $n \times R_A$ where $R_A$ stands for the ratio in the terrestrial atmosphere. Only in a few cases when noble-gas variations are very small, the delta notation is used (*e.g.* Ar ratio variations in firn air and ice-core samples).

As discussed above, the Nier-type ion source suffers from a systematic instrumental isotopic fractionation (see Chapter 6), a systematic bias in measured isotope ratios, which may also fluctuate depending on, *e.g.* on ion-source tuning or the ambient temperature. In order to determine the instrumental isotopic fractionation (mass-discrimination factor) and also to calibrate / reference NG isotope and elemental ratios, aliquots of one or more laboratory standards are measured. By comparing ratios measured on a standard with tabulated ratios (*e.g.* for the atmospheric reservoir), the instrumental isotopic fractionation factor is determined and later used for correcting sample measurements. Notably, this correction may play a substantial role in the overall uncertainty (see also Chapters 6, 8 and 10).

Usually, NG lab reference gases (Lab-Ref gases) are kept in stainless steel reservoirs (of a few L) filled with purified atmospheric noble gases, with 2 dosing valves (gas pipette) thus permitting to aliquot the Lab-Ref gas in a reproducible way. The volume of a gas pipette needs to be unchanged over time. For the best practice, the use of a dynamometric wrench (for manual dosing valves) and/or pneumatically operated values (for automated systems) as well as well-controlled temperature conditions are recommended. In some cases, isotopic mixtures are prepared and used, in particular for He, which isotope ratios in nature cover several orders of magnitude. Often, He–Ne mixtures with the $^4He/^{20}Ne$ ratio different from the atmospheric ratio are used (due to accumulation of radiogenic $^4He$, $^4He/^{20}Ne$ exceed the ratio in the atmosphere in many samples $^4He/^{20}Ne$ exceed the ratio in the atmosphere). All isotope ratios in gas mixtures are calibrated by using the atmospheric NG reservoir.

All above considerations stress the importance that isotope ratios of atmospheric noble gases are well known. Absolute determinations of NG isotope ratios in the atmosphere have been accomplished by several researchers, with the first determinations for atmospheric Ne, Ar, Kr and Xe done by Nier[60,61] who calibrated the mass spectrometer by using an Ar isotope mixture. For many years, these have been commonly accepted values for $^{40}Ar/^{36}Ar$ and $^{38}Ar/^{36}Ar$. Several re-redeterminations were performed recently for Ar, exemplifying the need for accurate values. Namely, Lee *et al.*[62] measured atmospheric argon (from La Jolla air) by using gravimetrically prepared mixtures of highly enriched $^{36}Ar$ and $^{40}Ar$ for calibration, and obtained $^{40}Ar/^{36}Ar = 298.56 \pm 0.31$ about 1% higher than the Nier[61] value of $295.5 \pm 0.5$).

Valkiers *et al.*[63] reported Ar isotope ratios for commercial high-purity Ar, *via* calibration by gravimetric mixtures of $^{36}Ar$, $^{38}Ar$ and $^{40}Ar$. A potential problem arises as Valkiers *et al.*[63] did not analyse atmospheric Ar directly (which is the primary scale-defining source) but exclusively commercially available high-purity Ar. The latter is produced by separation of air. As a consequence, isotope fractionation cannot be excluded. Recently, Mark *et al.*[64] analysed both four different cylinders of high-grade Ar (Air Products) and also atmospheric Ar purified by chemical way (getters) from air aliquots. The two sets of values were found to be in good agreement. By intercomparison of all available values on the $^{40}Ar/^{36}Ar$ *vs.* $^{38}Ar/^{36}Ar$ plot, they notice a mass fractionation trend with the most deviating value reported by Valkiers *et al.*[63] Mark *et al.*[64] proposed to re-evaluate the values commonly in use and noticed the value by Lee *et al.*[62] as the most accurate. Using accurate value appears to be important for $^{40}Ar/^{39}Ar$ geochronology especially on young samples[65] and also for geochemical studies, as even a small increase from the atmospheric $^{40}Ar/^{36}Ar$ indicates a presence of radiogenic $^{40}Ar$.

The first determinations for atmospheric Ne, Kr and Xe done by Nier[60] are still in use. Redeterminations done by Valkiers and coworkers for Kr, and Xe[66,67] cannot be directly traced to the primary/international reference material – which is the atmospheric reservoir of noble gases – as commercial gases were used for redetermination. Since the production process of these gases includes cryogenic separation of liquefied atmospheric gases, systematic deviations in the isotope composition of commercial gas from the atmospheric-gas reservoir can be expected, as discussed for Ne.[68]

Many geochemical studies are focused on $^{3}He/^{4}He$ in terrestrial rocks, gases and fluids. The variation of the ratio may be caused by, *e.g.* radiogenic $^{4}He$, the presence of primordial He (solar-type He) or atmospheric gases recycled by subduction (means – sediments recycling). One of the first and most precise determination of atmospheric $^{3}He/^{4}He$ was done by Mamyrin *et al.*,[14,69] who prepared calibration mixtures based on pure (high enrichment) $^{3}He$ and natural He with very low $^{3}He/^{4}He$ (radiogenic He). Their value $(1.399 \pm 0.013)\times10^{-6}$ ($1\sigma$) agrees with redeterminations performed later (*e.g.* Clarke *et al.*[70] redetermined as $(1.384 \pm 0.006)\times10^{-6}$ ($1\sigma$) and then $(1.343 \pm 0.013)\times10^{-6}$ by Sano *et al.*[71]). In 1990 Davidson and Emerson[72] provided an additional estimate of the atmospheric $^{3}He/^{4}He$ of $(1.393 \pm 0.014)\times10^{-6}$ ($1\sigma$). Thus, the value by Mamyrin *et al.*[14,69] stands valid up to now.

The use of atmospheric air as a high-quality international standard and its stability over time is particularly important for He as it has low abundance. In 1989 a 10-year decrease of the atmospheric $^{3}He/^{4}He$ value by ~1.7% was reported by Sano *et al.*[71] Anthropogenic release of radiogenic $^{4}He$ due to natural oil and gas production was discussed as the source of this change. Later, Lupton and Evans[73] measured 30-year old atmospheric air samples and concluded that the possible decrease of $^{3}He/^{4}He$ ratio (their value is within uncertainty and much less than the evaluation by Sano *et al.*[71]) is insignificant to prevent using atmospheric air as valid international standard for $^{3}He/^{4}He$. This was confirmed 10 years later by the same authors. Systematic measurements performed on Cape Grim Air Archive (years 1978–2011) provides the evidence for the stability of the atmospheric Ne, Ar, Kr and Xe compositions.[97] This work estimated the decrease in $^{3}He/^{4}He$ ratio as 0.23 to 0.30 permil per year which does not agree with measurements by Sano *et al.*[74]

Besides isotope ratios, NG concentrations/amounts in a sample and/or NG elemental ratios need to be determined, or an amount of a certain NG component has to be calculated. Applications requiring NG amounts to be determined include various types of dating (K–Ar and $^{40}Ar–^{39}Ar$ dating, U-Th/He dating, Tritium/$^{3}H$ dating on natural waters, exposure age determinations based on accumulation of spallogenic NG), determination of paleotemperatures based on dissolved noble gases and others. Quantities of noble gases are reported either as $cm^3$ at standard temperature and pressure (cc STP, defined as 273.15 K and 101325 Pa) or moles per gram of rock or L of water.

NG amounts are mostly determined based on a comparison between a sample and a known amount of laboratory reference gas from a large stainless steel reservoir with a calibrated volume between two valves (gas pipette). In order to keep constant volume of the pipette, pneumatically or manually (tightened with a dynamometric wrench) operated valves are in use. The amount of gas in a pipette is also to be corrected for the number of pipettes taken since gas reservoir refilling.

In order to account for an amount-dependent sensitivity of a NG mass-spectrometer – specifically for amount determination – several pipettes can be introduced into a mass spectrometer and/or several gas pipettes with different volumes can be used. Besides, it is a common practice to maintain several laboratory reference gas reservoirs, with different gases (*e.g.* with Ar, He, Ne, Kr and Xe), or variable elemental amount ratios. All these "home-made" mixtures are calibrated against atmospheric noble gases.

Another way to determine the $^{40}Ar$ amount in a sample aimed at K–Ar dating is based on isotope dilution with a $^{38}Ar$ tracer. The tracer is added (by pipetting) to the reactor volume during sample melting, so that the tracer accompanies all the steps along with the sample gas (such as, *e.g.* gas transfer by freezing on a sorbent, gas purification and introduction to MS). This permits accounting for any incompleteness of gas transfer and/or fluctuations in the IRMS sensitivity among runs.

When elemental NG ratios (*e.g.* $^{4}He/^{36}Ar$) are of interest, these are frequently reported as relative noble-gas abundances, reported relative to $^{36}Ar$ and normalised to air, commonly known as *F* values:

$$F(^{i}E) = \frac{(^{i}E\,/\,^{36}\mathrm{Ar})_{\mathrm{sample}}}{(^{i}E\,/\,^{36}\mathrm{Ar})_{\mathrm{air}}} \tag{17.1}$$

*F*: *F* value
$^{i}E$: abundance of the isotope *i* of an element *E*
$^{36}Ar$: abundance of $^{36}Ar$

Often, the results are expressed relative to the ratios of the solar component estimated earlier by Anders and Grevesse.[75] Similarly to elemental ratios, $^3He/^4He$ isotope ratios are often reported relative to that of the atmosphere:

$$n = \frac{R}{R_A} \tag{17.2}$$

$R$ : $^3He/^4He$ in the sample
$R_A$ : $^3He/^4He$ in the atmosphere

### 17.4.2 Noble-Gas Components (He, Ne, Ar, Kr and Xe) and Component Deconvolution

Before considering noble-gas applications, we briefly overview noble-gas components found in terrestrial and extraterrestrial environments.

**Atmospheric Noble Gases**
Terrestrial atmosphere contains all noble gases (He to Xe), with Ar contributing to 0.93% of the dry atmosphere. The atmospheric component may be present in all type of samples and often needs to be corrected for.

**Radiogenic Components**
$^4He$ is produced by alpha decay of U and Th. Radiogenic $^{40}Ar$ is produced by electron capture of $^{40}K$ ($^{40}K$ is also decayed by beta-decay giving $^{40}Ca$). Some samples may contain "excess $^{40}Ar$" – that is radiogenic $^{40}Ar$ being not produced by $^{40}K$ *in situ* in the mineral but trapped from surrounding fluids where it could have been "lost" by other minerals and rocks. A correction for "excess $^{40}Ar$" has to be applied. This can be done by the graphical "isochrone" method for K–Ar and by stepwise heating for $^{40}Ar/^{39}Ar$.

**Neutron-Induced Components**
These components comprise, for instance, Ar isotopes produced during irradiation in a nuclear reactor as a result of reactions on Cl, K and Ca. Kr and Xe isotopes can be produced during irradiation in a nuclear reactor as reactions on U and Th.

**Trapped NG Components**
This is a collective term indicating noble gases trapped in terrestrial rocks and/or meteorites. These contain all isotopes of all noble gases, sometimes named as a primary component(s).

**Solar Component**
Solar-wind corpuscular irradiation contains all elements, with noble gases being the best studied. Solar components contain all isotopes of all noble gases and can be found in solar-wind-irradiated meteorites and lunar soils.

**Fissiogenic Component(s)**

Natural spontaneous fission decay of $^{238}U$ and $^{232}Th$ results in several isotopes of Kr and Xe. Neutron-induced fission of $^{235}U$ results in another fissiogenic component, with a different isotopic composition of Kr and Xe compared to the spontaneous-fission. One more component, spontaneous fission Xe, produced by the decay of extinct $^{244}Pu$ is found in some meteorites and lunar samples.

**Cosmogenic (Spallogenic) Components**

These components are produced by spallation of various nuclei by high-energy cosmic irradiation and can be found in meteorites and the surface layer of terrestrial rocks exposed to cosmic irradiation. They constrain isotopes of all noble gases such as $^{3}He$, $^{4}He$, $^{20}Ne$, $^{21}Ne$, $^{22}Ne$, $^{36}Ar$, $^{38}Ar$ and isotopes of Kr and Xe. The isotopic composition depends on the chemical composition of the irradiated material and also on the shielding depth. In meteorites and lunar soil, neutron capture by Br and I can result in additional $^{80}Kr$, $^{82}Kr$ and $^{128}Xe$.

**Nucleogenic Components**

Nucleogenic components have been produced by nuclear reactions during supernova events and are found in specific gas carriers in primitive meteorites.

**Component Deconvolution**

As noble gases occur as a mixture of several components, often one needs to decouple NG components and calculate the amount of each component in a sample. That can be done graphically, *e.g.* on a 3-isotope plot, or numerically, based on a known/specific isotope composition of the components. For example, the amount of radiogenic $^{40}Ar$ in a sample can be calculated from the total $^{40}Ar$ corrected for the atmospheric $^{40}Ar$ as follows:

$$[^{40}Ar_{rad}] = [^{40}Ar_{total}] - 295.5 \times [^{36}Ar] \tag{17.3}$$

$[^{i}Ar]$ : amount of the Ar isotope $i$ ($i = 40$ or 36); rad = radiogenic; total = total]
295.5 : $^{40}Ar/^{36}Ar$ ratio in the atmospheric component
(***Note***: Here it is assumed that $^{36}Ar$ is only fom atmospheric Ar.)

Often, stepwise heating is used to release the atmospheric component in low temperature steps and to get some temperature fractions (often at high $T$) being more enriched in certain components. Other selective methods include: crushing in vacuum in order to release gases trapped in fluid inclusions; surface etching of lunar soils in vacuum to release solar wind noble gases implanted in the very surface of lunar soils and minerals or stepwise combustion of insoluble organics and SiC which are chemically released from meteorites.

### 17.4.3 Tritium Determination by $^3$He Accumulation

Tritium is typically determined by preconcentration (water electrolysis) followed by counting method. In order to determine Tritium by NG mass spectrometry – which significantly increases sensitivity and improves detection limits (*e.g.* Clarke *et al.*[70]), the sample is completely degassed in order to remove all dissolved gases (including $^3$He) and then the water sample is stored in order to allow for accumulation of $^3$He from Tritium decay, which is – again – degassed from the sample. Hereafter, the extracted gases with the accumulated $^3$He are further purified and measured by NG IRMS. The technique is in use for measurements of very-low to ultra-low levels of Tritium in natural waters.

### 17.4.4 Isobaric Interferences and Memory Effects

Even though there are a number of potential isobaric interferences,[34] the most critical ones are $HD^+$ and $H_3{}^+$ on $m/z = 3$, $^{40}Ar^{2+}$ on $m/z = 20$, $CO_2{}^{2+}$ on $m/z = 22$, $C_3{}^+$ on $m/z = 36$, and the family of hydrocarbons and fragment-ion peaks on masses of Kr and Xe. Even when using high-vacuum stainless steel and getters inside the NG IRMS chamber, a certain amount of hydrogen of hydrocarbons is build up that interferes with the NG mass spectrum, resulting in isobaric interferences. In particular, hydrogen gives $HD^+$ and $H_3{}^+$, which interfere with $^3He^+$. In order to resolve $^3He^+$ from $HD^+$ and $H_3{}^+$ peaks, a resolving power of 650–700 and higher is needed. In addition, a low contribution of the $^4He^+$ tailing to $m/z = 3$ is required.

Other typical interferences include double-charged $^{40}Ar^{++}$ ($m/z = 20$) and sometimes $CO_2{}^{++}$ ($m/z = 22$), which interfere with $^{20}Ne^+$ and $^{22}Ne^+$ correspondingly. By using a liquid-nitrogen-cooled finger with a sorbent (activated charcoal or molecular sieves) connected to NG IRMS during Ne analysis, levels of Ar can be significantly reduced. $CO_2$ is reduced by using both getters and a liquid-nitrogen-cooled finger.

Another specific aspect of NG IRMS is the presence of memory effects. These include both the ion-source pumping effect and gas adsorption/desorption. The effect is small to negligible for He and Ne, with the magnitude increasing from Ar to Xe. Given high-vacuum materials in use (*i.e.* low adsorption on the surface), memory in NG IRMS is mostly due to ions implanted in the ion-source slits (*e.g.* Verkouteren *et al.*[76] and references therein). This results in decreased beam intensities of a running sample as well as a gradual release (washing-out) of a previous sample. In order to correct for the memory, beam intensities (or isotope ratios) are extrapolated to the moment of sample-gas introduction to NG IRMS. A memory due to NG adsorption/desorption is remarkable for gas-extraction lines, being significant for heavy gases (Ar, Kr and Xe). As a rule, it is not recommended to analyse samples with very different isotope compositions and/or very different Ar, Kr and Xe amounts in a sequence. From time to time, the NG IRMS needs to be baked out to remove implanted noble gases. In order to minimise or eliminate this memory, a high-transmission ion source is critical (due to better focusing much less gas implanted in ion-source slits). It was known for years that minimising the emission current and thus reducing the ion-formation area, one could reduce the ion-implantation region. In recent years, IRMS with high-transmission ion sources become available.

## 17.5 Selected Applications

Recently, a book on the use of noble gases as geochemical tracers was published.[34] Therefore, the intention of the following section is neither to be comprehensive nor complete but to present a glimpse of the fascinating world of noble-gas isotope mass spectrometry (Figure 17.8).

### 17.5.1 Applications in Earth and Geosciences, Paleoclimate Research and Cosmoscience

Noble gases possess unique properties, which make them important in isotope geology. Thus they have shown ever since a great potential of contributing to the fundamental understanding of the formation and evolution of our Earth. Zartman *et al.*[77] described in their 1961 paper the basic principles behind the use of noble gases. The isotopic signature of noble gases can give important information, *e.g.* about the origin and history of a rock or meteorites, lunar materials, fluid samples, about radiogenic ages and redistribution of components. In many cases, He isotopes are used as powerful tracers of a number of processes pertaining to geophysics and geochemistry. In this context, helium isotope geochemistry of MORB basalts and samples from subduction regions have to be mentioned. For more details, we refer to the book on noble-gas geochemistry, which was published by Ozima and Podosek[78] or Burnard[34] and other specialised publications. Below, we list a few selected applications.

**Dating Methods and Geochronology**

K–Ar dating and later its advanced versions, the $^{40}Ar$–$^{39}Ar$ dating, are widely used in geochronology for determining the age and timing of a large variety of geological processes as well as alteration and diagenesis of minerals – from meteorite samples as old as the Earth (4.5 billion years) to the age of historical events (*e.g.* eruption of volcanoes). The method utilises the natural decay of radioactive $^{40}K$ to $^{39}Ar$ and can be applied to any rocks and minerals containing K and preserving radiogenic Ar.

The Tritium –$^{3}He$ dating of ground waters was pioneered by Tolstikhin and coworkers.[18,19] The method is based on accumulation of –$^{3}He$ due to –$^{3}He$/Tritium decay. Both $^{3}He$ and –$^{3}He$/Tritium concentrations have to be determined. A correction for dissolved atmospheric He is based on dissolved $^{4}He$ and heavier noble gases. As atmospheric –$^{3}He$/Tritium is mostly due to nuclear bomb tests in 1960s, the range of water ages to be determined is limited.

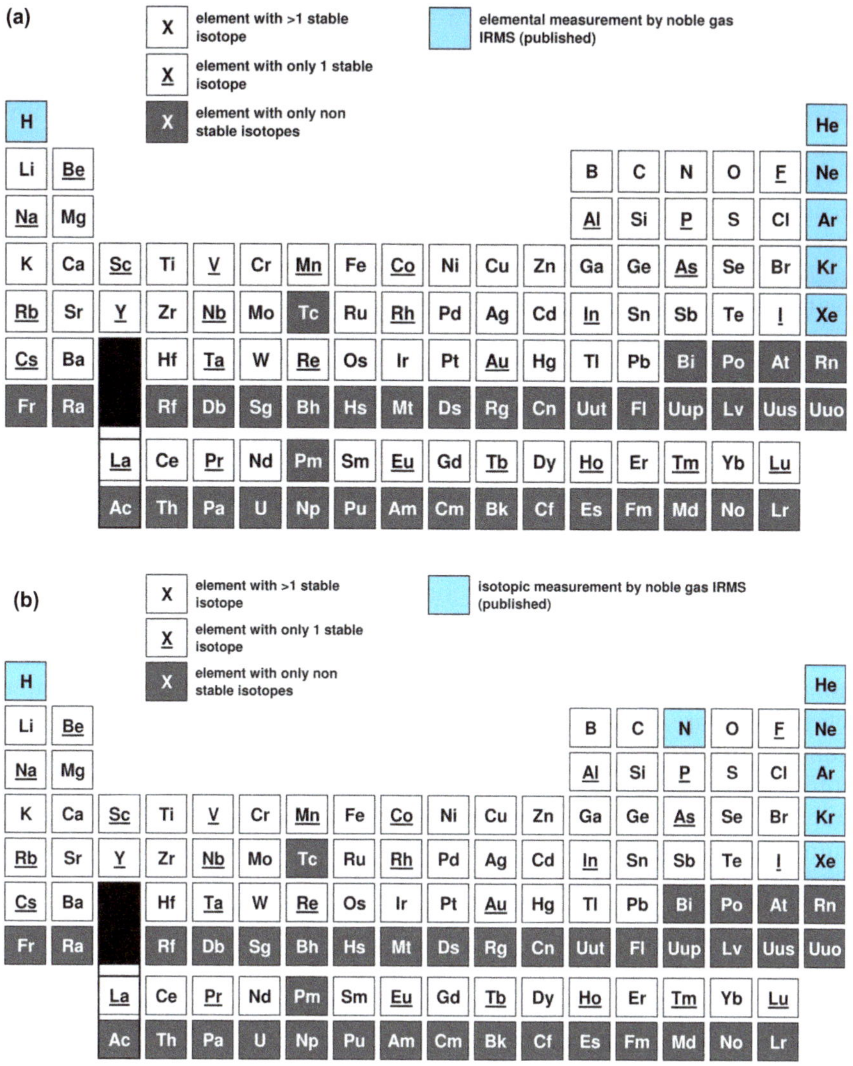


**Figure 17.8** (a) Periodic table of the elements: elements for which elemental data acquired by NG IRMS has been published (highlighted in sky blue). (b) Periodic table of the elements: elements for which isotope ratio data acquired by NG IRMS has been published (highlighted in sky blue).

A short-lived $^{129}$I ($t_{1/2}$ about 17 Ma y) was present in meteorites in the early solar system. $^{129}$I decayed to $^{129}$Xe. By measuring $^{129}$Xe, a "meteorite formation age" can be determined, that is the I–Xe dating method.

One more interesting geochronology application is based on Xe isotopes. Xe isotopes can be produced by spontaneous fission of $^{238}$U, spontaneous and neutron-induced fission of $^{235}$U so that Xe can be used as age function. The Xe-U and later Xe-Xe dating method developed by Shukolyukov *et al.*[79]

The method is similar to the $^{40}Ar$–$^{39}Ar$ method – the amount of natural fission Xe (daughter element) is proportional to the age and the amount of U, which is the parent element. In order to determine U content by mass spectrometry, samples are neutron irradiated in a reactor together with a standard of known U-content.

There are several publications by Gerling and later by Shukolyukov and coauthors (mostly in Russian), who developed the Xe isotope geochemistry. For instance, Shukolyukov *et al.*[80] found that migration of short-lived precursors of Xe isotopes (these include isotopes of Sn, Sb, Te and I) produced by the $^{235}U$ fission in the natural nuclear reactor Oklo (Gabon) resulted in anomalous isotope composition of Xe remaining there in uranium minerals. As one of the precursors, $^{129}I$ is intermediate-lived (about 17 Ma y), it was hypothesised (in the 1980s and 1990s) that migration of the $^{129}I$ - the precursor of $^{129}Xe$ – might have resulted in some isotope peculiarities of Xe on the early Earth.[80]

**Hydrology and Palaeoclimate Research**

Noble gases have been widely used as tracers for the origin and the transport of fluids as they are chemically inert and volatile (thus a having strong tendency to partition between phases). Noble gases were applied, *e.g.* in seawater as tracers for physical and biogeochemical ocean processes.[34] The concentration of dissolved atmospheric noble gases in water is determined by temperature, salinity and atmospheric pressure. Hence, the noble-gas concentrations of a water body isolated from the atmosphere can provide information about the conditions prevailing during gas exchange with the atmosphere.[81,82] Based on noble gases dissolved in waters one can determine paleotemperatures from groundwaters[81] or from fluid inclusions in speleothem carbonates.[46] This application is based on the difference in temperature-dependent solubility factors for different noble gases.

In palaeoclimate research, noble gases bear the potential to be used as indicators of the Earth's history[34] and as a paleothermometer.[83] For instance, the measured concentrations of atmospheric noble gases in old groundwater are used to calculate paleotemperatures.[83–85]

The temperature dependence in solubility of different noble gases can be applied for the determination of paleotemperatures of groundwaters (*e.g.* Aeschbach-Hertig *et al.*[81]) or of fluid inclusions, *e.g.* in speleothem carbonates (*e.g.* Kluge *et al.*[46]).

**Cosmochemistry**

Noble gases and stable isotopes have been widely studied in lunar soils and lunar rocks samples delivered by American and Russian space missions as well as in meteorites. There are numerous publications by Lunar Planetary Science conferences and numerous special publications (*e.g.* Wieler and Heber[86]).

Moreover, the use of simplified isotope mass spectrometers for space missions has been one important research tool to study planetary atmospheres. These were miniaturised magnetic field and quadrupole mass spectrometers. To mention a few initial works, measurements of the Lunar atmosphere were accomplished by Apollo 17 using a small magnetic field mass spectrometer (*e.g.* Hodges and coworkers[87,88]) followed later by the measurement of the Mars atmosphere and other mission projects.

## 17.6 Stable Isotopes by Using Noble-Gas Static Vacuum Approach

Use of the noble-gas static vacuum approach to measure ng amounts of light elements was primarily developed by Pilinger and coworkers. First, this technique was proposed for hydrogen isotopes[89] further for carbon isotopes[90] and nitrogen isotope ratios.[91,92] The use of static mode mass-spectrometry for methane was also developed and tested in the 1990s for atmospheric methane.[93,94] Signal ratios at $m/z = 17$ / $m/z = 16$ ratio were determined so that the calculation of $\delta^{13}C(CH_4)$ required the value of $\delta D(CH_4)$ to be known/assumed. Nowadays, the method is not used and methane isotope determinations are performed by CF IRMS on methane preconcentrated from air samples (see Chapter 16).

Nowadays, the static vacuum technique is still widely used for low-level nitrogen isotope measurements in rocks, meteorites and lunar samples, with several methods of gas extraction, *e.g.* by laser (Hashizume and Marty[44] and references therein) or by crushing basaltic rocks (Barry *et al.*[95], also by step-heating and step-combustion.[96] The advantage is that both very low amounts of nitrogen and noble-gas isotopes can be measured from the same sample. In all methods, the released gases have to be carefully purified – namely for nitrogen by oxidation on CuO followed by cryogenic purification (removing $H_2O$ and $CO_2$) and for noble gases by getters – then introduced into the static vacuum NG IRMS and analysed. Specifically, a correction for CO (also $m/z = 28$, 29, 30) building up in NG IRMS during static vacuum measurements of $N_2$ has to be applied for or CO has to be resolved.[95]

## References


1. J. J. Thomson, *Proc. Roy. Soc. Lond. Series A*, 1913, **89**, 1–20.
2. L. W. Alvarez and R. Cornog, *Phys. Rev.*, 1939, **56**, 379–379.
3. L. T. Aldrich and A. O. Nier, *Phys. Rev.*, 1946, **70**, 983–983.
4. L. T. Aldrich and A. O. Nier, *Phys. Rev.*, 1948, **74**, 1225–1225.
5. L. T. Aldrich and A. O. Nier, *Phys. Rev.*, 1948, **74**, 876–877.
6. A. O. Nier, *Rev. Sci. Instrum.*, 1947, **18**, 398–411.
7. J. H. Reynolds, *Rev. Sci. Instrum.*, 1956, **27**, 928–934.
8. C. M. Hohenberg, *Rev. Sci. Instrum.*, 1980, **51**, 1075–1082.
9. H. Baur, *Numerische Simulation und praktische Erprobung einer rotationssymmetrischen Ionenquelle für Gasmassenspektrometer.*, PhD, ETH, Zürich, Switzerland, 1980.
10. L. H. Ahrens, *Rep. Prog. Phys.*, 1956, **19**, 80–106.

11. E. K. Gerling and M. L. Yashchenko, *Proc. USSR Adad. Sci.*, 1952, **83**, 901–902.
12. B. A. Mamyrin and B. N. Shustrov, *Sov. Phys.-Tech. Phys.*, 1957, **2**, 1246–1254.
13. B. A. Mamyrin, B. N. Shustrov, G. S. Anufriev, Boltenko.Bs, V. A. Zagulin, Kamenski.Il, I. N. Tolstikhin and L. V. Khabarin, *Prib. Tekh. Eksp.*, 1972, 148–150.
14. B. A. Mamyrin, G. S. Anufriev, Kamenski.Il and Tolstikh.In, *Dokl. Akad. Nauk SSSR*, 1970, **195**, 188–&.
15. B. A. Mamyrin, G. S. Tolstikhin, Anufriev and I. Kamenski, *Dokl. Akad. Nauk SSSR*, 1969, **184**, 1197–&.
16. E. K. Gerling, I. N. Tolstikhin, S. S. Yakovlev and B. A. Mamyrin, *Geochem. Int. USSR*, 1971, **8**, 755–&.
17. B. A. Mamyrin, I. N. Tolstikhin, G. S. Anufriev and I. I. Kamenski, *Geokhimiya*, 1972, 1396–1396.
18. I. N. Tolstikhin and I. Kamenski., *Geochem. Int. USSR*, 1969, **6**, 810–&.
19. B. A. Mamyrin and I. N. Tolstikhin, *Helium Isotopes in Nature*, Elsevier Scientific Publishing Amsterdam, The Netherlands, 1984.
20. W. B. Clarke, M. A. Beg and H. Craig, *Earth. Planet. Sci. Lett.*, 1969, **6**, 213–220.
21. N. Takaoka, *Mass Spectrosc.*, 1976, **24**, 73–86.
22. G. Turner, J. A. Miller and R. L. Grasty, *Earth. Planet. Sci. Lett.*, 1966, **1**, 155–157.
23. C. Merrihue and G. Turner, *J. Geophys. Res.*, 1966, **71**, 2852–2857.
24. G. Turner, http://www.seaes.manchester.ac.uk/our-research/researchgroups/isotopegeochemistryandcosmochemistry/facilities/noblegaslabs/ms-1, (accessed 19.06.2014).
25. G. Turner, J. C. Huneke, F. A. Podosek and G. J. Wasserburg, *Earth. Planet. Sci. Lett.*, 1971, **12**, 19–35.
26. R. Burgess, G. Turner, M. Laurenzi and J. W. Harris, *Earth. Planet. Sci. Lett.*, 1989, **94**, 22–28.
27. G. Turner, R. Burgess, M. Laurenzi, S. Kelley and J. Harris, *Chem. Geol.*, 1988, **70**, 142–142.
28. F. M. Stuart and G. Turner, *Chem. Geol.*, 1992, **101**, 97–109.
29. S. A. Crowther, R. K. Mohapatra, G. Turner, D. J. Blagburn, K. Kehm and J. D. Gilmour, *J. Anal. Atom. Spectrom.*, 2008, **23**, 938–947.
30. J. D. Gilmour, I. C. Lyon, W. A. Johnston and G. Turner, *Rev. Sci. Instrum.*, 1994, **65**, 617–625.
31. Y. Iwata, C. Ito, H. Harano and T. Aoyama, *Int. J. Mass Spectrom.*, 2010, **296**, 15–20.
32. I. Strashnov, D. J. Blagburn and J. D. Gilmour, *J. Anal. Atom. Spectrom.*, 2011, **26**, 1763–1772.
33. P. J. Hooker, R. Bertrami, S. Lombardi, R. K. Onions and E. R. Oxburgh, *Geochim. Cosmochim. Acta*, 1985, **49**, 2505–2513.
34. P. E. Burnard, ed., *The Noble Gases as Geochemical Tracers*, Springer-Verlag, Berlin Heidelberg, 2013, 391.

35. L. B. Wang, P. Mueller, R. J. Holt, Z. T. Lu, T. P. O'Connor, Y. Sano and N. C. Sturchio, *Geophys. Res. Lett.*, 2003, **30**, 1592.
36. H. Hintenberger, H. W. Weber, H. Voshage, H. Wanke, F. Begemann, E. Vilscek and F. Wlotzka, *Science*, 1970, **167**, 543–545.
37. T. Staudacher, E. K. Jessberger, D. Dorflinger and J. Kiko, *J. Phys. E-Sci. Instrum.*, 1978, **11**, 781–784.
38. L. Zimmermann, P. H. Blard, P. Burnard, S. Medynski, R. Pik and N. Puchol, *Geostand. Geoanal. Res.*, 2012, **36**, 121–129.
39. G. H. Megrue, *Science*, 1967, **157**, 1555–1556.
40. S. P. Kelley, *Geochim. Cosmochim. Acta*, 2006, **70**, A311.
41. S. P. Kelley, N. O. Arnaud and S. P. Turner, *Geochim. Cosmochim. Acta*, 1994, **58**, 3519–3525.
42. S. J. Norris, P. W. Brown and C. T. Pillinger, *Meteoritics*, 1981, **16**, 369–369.
43. I. A. Franchi, I. P. Wright, E. K. Gibson and C. T. Pillinger, *J. Geophys. Res.-Solid Earth*, 1986, **91**, D514–D524.
44. K. Hashizume and B. Marty, Nitrogen isotopic analyses at the sub-picomole level using an ultra-low blank laser extraction technique, in *Handbook of Stable Isotope Analytical Techniques*, ed. P. D. Groot, Elsevier, Amsterdam., 2004, vol. 1, ch. 17, pp. 361–375.
45. P. G. Burnard, F. M. Stuart, G. Turner and N. Oskarsson, *J. Geophys. Res.-Solid Earth*, 1994, **99**, 17709–17715.
46. T. Kluge, T. Marx, W. Aeschbach-Hertig, C. Spotl and D. K. Richter, *Chem. Geol.*, 2014, **368**, 54–62.
47. W. Aeschbach-Hertig and D. K. Solomon, Noble gas thermometry in groundwater hydrology, in *The Noble Gases as Geochemical Tracers*, Springer, 2013, pp. 81–122.
48. R. Wieler, H. Baur and P. Signer, *Geochim. Cosmochim. Acta*, 1986, **50**, 1997–2017.
49. J. P. Benkert, H. Baur, P. Signer and R. Wieler, *J. Geophys. Res.-Planets*, 1993, **98**, 13147–13162.
50. R. Wieler and H. Baur, *Meteoritics*, 1994, **29**, 570–580.
51. B. J. Stolp, D. K. Solomon, A. Suckow, T. Vitvar, D. Rank, P. K. Aggarwal and L. F. Han, *Water Resour. Res.*, 2010, 46.
52. J. P. Severinghaus, A. Grachev, B. Luz and N. Caillon, *Geochim. Cosmochim. Acta*, 2003, **67**, 325–343.
53. T. Kobashi, J. P. Severinghaus and K. Kawamura, *Geochim. Cosmochim. Acta*, 2008, **72**, 4675–4686.
54. A. V. Ignatiev, T. A. Velivetskaya and S. Y. Budnitskiy, *Rapid Commun. Mass Spectrom.*, 2009, **23**, 2403–2410.
55. A. Visser, M. J. Singleton, D. J. Hillegonds, C. A. Velsko, J. E. Moran and B. K. Esser, *Rapid Commun. Mass Spectrom.*, 2013, **27**, 2472–2482.
56. H. Baur, *EOS, Trans. Am. Geophys. Union*, 1999, **46**, F1118.
57. U. Ott, P. Hoppe and G. W. Lugmair, *New Astron. Rev.*, 2004, **48**, 165–169.
58. T. Matsumoto, J. Matsuda, I. Yatsevich and M. Ozima, *Geochem. J.*, 2010, **44**, 167–172.

59. M. A. Coble, M. Grove and A. T. Calvert, *Chem. Geol.*, 2011, **290**, 75–87.
60. A. O. Nier, *Phys. Rev.*, 1950, **79**, 450–454.
61. A. O. Nier, *Phys. Rev.*, 1950, 77, 789–793.
62. J.-Y. Lee, K. Marti, J. P. Severinghaus, K. Kawamura, H.-S. Yoo, J. B. Lee and J. S. Kim, *Geochim. Cosmochim. Acta*, 2006, **70**, 4507–4512.
63. S. Valkiers, D. Vendelbo, M. Berglund and M. de Podesta, *Int. J. Mass Spectrom.*, 2010, **291**, 41–47.
64. D. F. Mark, F. M. Stuart and M. de Podesta, *Geochim. Cosmochim. Acta*, 2011, **75**, 7494–7501.
65. P. R. Renne, W. S. Cassata and L. E. Morgan, *Quat. Geochronol.*, 2009, **4**, 288–298.
66. Y. Argebe, S. Valkiers, J. Poths, J. Norgaard, H. Kipphardt, P. De Bievre and P. D. P. Taylor, *Int. J. Mass Spectrom.*, 2001, **206**, 129–136.
67. S. Valkiers, Y. Aregbe, P. D. P. Taylor and P. De Bievre, *Int. J. Mass Spectrom.*, 1998, **173**, 55–63.
68. F. Pavese, B. Fellmuth, D. I. Head, Y. Hermier, K. D. Hill and S. Valkiers, *Anal. Chem.*, 2005, 77, 5076–5080.
69. B. A. Mamyrin, G. Anufriye, I. Kamenski and Tolstikhin, *Geochem. Int. USSR*, 1970, 7, 498.
70. W. B. Clarke, W. J. Jenkins and Z. Top, *Int. J. Appl. Radiat. Isotopes*, 1976, **27**, 515–522.
71. Y. Sano, H. Wakita, Y. Makide and T. Tominaga, *Geophys. Res. Lett.*, 1989, **16**, 1371–1374.
72. T. A. Davidson and D. E. Emerson, *J. Geophys. Res.: Atmos. (1984, 2012)*, 1990, **95**, 3565–3569.
73. J. Lupton and L. Evans, *Geophys. Res. Lett.*, 2004, **31**, L13101.
74. J. Lupton and L. Evans, *Geophys. Res. Lett.*, 2013, **40**, 6271–6275.
75. E. Anders and N. Grevesse, *Geochim. Cosmochim. Acta*, 1989, **53**, 197–214.
76. R. M. Verkouteren, S. Assonov, D. B. Klinedinst and W. A. Brand, *Rapid Commun. Mass Spectrom.*, 2003, **17**, 777–782.
77. R. Zartman, G. Wasserburg and J. Reynolds, *J. Geophys. Res.*, 1961, **66**, 277–306.
78. M. Ozima and F. A. Podosek, *Noble Gas Geochemistry*, Cambridge University Press, Cambridge, UK, 2002.
79. J. Shukoljukov, T. Kirsten and E. Jessberger, *Earth. Planet. Sci. Lett.*, 1974, **24**, 271–281.
80. Y. A. Shukolyukov, E. K. Jessberger, A. P. Meshik, D. Vu Minh and J. L. Jordan, *Geochim. Cosmochim. Acta*, 1994, **58**, 3075–3092.
81. W. Aeschbach-Hertig, F. Peeters, U. Beyerle and R. Kipfer, *Water Resour. Res.*, 1999, **35**, 2779–2792.
82. R. Kipfer, W. Aeschbach-Hertig, F. Peeters and M. Stute, *Rev. Mineral. Geochem.*, 2002, **47**, 615–700.
83. M. Stute and P. Schlosser, *Clim. Change Cont. Isotopic Rec.*, 1993, 89–100.
84. E. Mazor, *Geochim. Cosmochim. Acta*, 1972, **36**, 1321–1336.
85. U. Beyerle, R. Purtschert, W. Aeschbach-Hertig, D. M. Imboden, H. H. Loosli, R. Wieler and R. Kipfer, *Science*, 1998, **282**, 731–734.

86. R. Wieler and V. S. Heber, *Space Sci. Rev.*, 2003, **106**, 197–210.
87. R. R. Hodges Jr, J. H. Hoffman and F. S. Johnson, *Icarus*, 1974, **21**, 415–426.
88. J. H. Hoffman and R. R. Hodges, Jr., *Moon*, 1975, **14**, 159–167.
89. I. P. Wright and C. T. Pillinger, *Meteoritics*, 1983, **18**, 425–426.
90. R. H. Carr, I. P. Wright, A. W. Joines and C. T. Pillinger, *J. Phys. E-Sci. Instrum.*, 1986, **19**, 798–808.
91. I. P. Wright, S. R. Boyd, I. A. Franchi and C. T. Pillinger, *J. Phys. E-Sci. Instrum.*, 1988, **21**, 865–875.
92. S. R. Boyd, I. P. Wright and C. T. Pillinger, *Meas. Sci. Technol.*, 1993, **4**, 1000–1005.
93. S. M. Jackson, G. H. Morgan, A. D. Morse, A. L. Butterworth and C. T. Pillinger, *Rapid Commun. Mass Spectrom.*, 1999, **13**, 1329–1333.
94. A. D. Morse, G. H. Morgan, A. L. Butterworth, I. P. Wright and C. T. Pillinger, *Rapid Commun. Mass Spectrom.*, 1996, **10**, 1743–1746.
95. P. H. Barry, D. R. Hilton, S. A. Halldorsson, D. Hahm and K. Marti, *Geochem. Geophys. Geosyst.*, 2012, **13**, doi:10.1029/2011GC003878.
96. A. B. Verchovsky, M. A. Sephton, I. P. Wright and C. T. Pillinger, *Earth. Planet. Sci. Lett.*, 2002, **199**, 243–255.
97. M. S. Brennwald and N. Vogel, *et al.*, *Earth Planet. Sci. Lett.*, 2013, **366**, 27–37.

CHAPTER 18

# *Outlook*[†]

THOMAS PROHASKA,* JOHANNA IRRGEHER* AND ANDREAS ZITEK*

University of Natural Resources and Life Sciences Vienna (BOKU), Department of Chemistry, Division of Analytical Chemistry, VIRIS Laboratory for Analytical Ecogeochemistry, Tulln, Austria
*Email: thomas.prohaska@boku.ac.at; johanna.irrgeher@boku.ac.at; andreas.zitek@boku.ac.at

Thrilling developments of analytical capabilities and technological novelties of the 20th century have also fostered the field of elemental and isotopic mass spectrometry. Computer control catapulted the method into modern times and – still – the tremendously increasing amount of data is almost impossible to handle with basic modern computer technology. After the first generation of mass spectrometers had been dedicated to pure science, the instruments have faced a steady development and can nowadays not only be found in almost every scientific laboratory but have also entered many fields of routine analysis. Continued enhancements in performance allow for the investigation of areas that have not been dreamt of before. Smaller sample sizes, increased measurement precision, detection limits heading towards the detection of single atoms, combination of simultaneous measurements of sample properties – to name a

[†]*With contributions from* Jean Nicolas Audinot, Jean-Paul Barnes, Willi A. Brand, Uwe Breuer, Emanuel de Chambost, Charles B. Douthitt, Trevor Ireland, Urs Klötzli, Cristina Máguas, Jorge Pisonero, Peter Roos, Lothar Rottmann, Laure Sangely, Nathalie Valle, Frank Vanhaecke, Cornel Venzago, Jochen Vogl and Michael Wiedenbeck.

New Developments in Mass Spectrometry No. 3
Sector Field Mass Spectrometry for Elemental and Isotopic Analysis
Edited by Thomas Prohaska, Johanna Irrgeher, Andreas Zitek and Norbert Jakubowski

Published by the Royal Society of Chemistry, www.rsc.org

few – and the race is not over, yet. Even though it is illusory and impossible to predict the future – which is by the way the intrinsic beauty of the future – we allow a glimpse into the crystal ball – or better to say, we summarise the rainbow wishes and lilac dreams of analytical scientists for the future.

*"...the good thing is that there is a treasure at both ends of the rainbow."* (Prohaska)

Sector field mass spectrometers – even described as bulky, clumsy and expensive, and expected to disappear sooner or later from the market – have kept their position and importance as could be highlighted within the previous chapters. Above that, there is also a bright future for this kind of instruments, which will be hardly replaceable as such.

In addition, the fast instrumental developments in analytical sciences demand for three utmost important criterions:

(1) We require a sound understanding of the science behind the measurement and as a consequence, a sound metrological basis to avoid conclusions being drawn from analytical artefacts or by simply neglecting the science behind an analytical process.
(2) The demand for highest metrological quality goes along with the demand for reliable certified reference standards in elemental and isotopic analysis.
(3) We require well-trained and educated scientists in the field of analytical sciences in order to cope with the high requirements and to deal with challenging developments in science.

*"Analytical sciences have to stand their field in mastering the ongoing transdisciplinary challenges they face in the continuous development of techniques and adequate application to burning questions in science or to sound routine measurements with highest metrological quality."* (Prohaska)

**Elemental and Isotopic Analysis**

Elemental and isotopic analysis cover a large field of mass spectrometry and still, there is the demand for lower detection limits and improved instrumental precision. In addition, higher sensitivity, interference-free measurements and improvement in instrumental isotopic and elemental fractionation have been on the wish list of scientists for many years and will not disappear in the near future.

The field of chemical imaging has brought new dimensions in the visualisation of chemical information and the race is continuing. Improved sensitivity of detectors and decreasing dwell/integration times to measure ultrafast transient signals are required to increase the potential spatial resolution of solid sampling techniques (SIMS and LA-ICP-MS).

The field of isotopic measurements has been connected to the development of magnetic sector field mass spectrometry ever since.

*"Although other techniques are also capable of providing information on the isotopic composition of target elements, mass spectrometry is playing a leading role in isotopic analysis."* (Vanhaecke)

The development of instruments allowing for isotope ratio measurements with increasing precision and increased sensitivity has allowed for the analysis of the natural variation of elements for which a stable isotopic composition had been assumed for a long time as it was hidden behind the instrumental performance. It is now realised that all elements with two or more isotopes show natural variation in their isotopic composition due to isotope fractionation effects accompanying physical processes and (bio)-chemical reactions and it has been demonstrated that on top of mass-dependent fractionation, mass-independent isotope fractionation also occurs for some elements.

This journey is not over, yet. Still, more isotopic systems are challenged and investigated for their use in geochemistry and cosmochemistry but also entering the fields of, *e.g.* ecosystem research, biomedicine, biology, materials science, forensics or archaeometry.

*"It is to be expected that the application of isotopic analysis in this and other less-traditional fields will be intensified in the next decade."* (Vanhaecke)

*"Once believed to be constants in nature, we know today that the isotopic composition of the elements is not irrevocable. Thus, future surprises and wonders are guaranteed as there will fortunately be new elements to be isolated and additional isotopes to be discovered that require to be explored by mass spectrometry."* (Irrgeher)

**Instrumental Developments**

*"Generally, my opinion is that nowadays instruments reached a performance level that satisfies all our needs. What we often need is not an improvement in terms of sensitivity or precision, but an improvement in the lab, e.g. lower blank or alternative resin materials."* (Vogl)

Even though it is widely accepted that mass-spectrometric techniques have become mature and entered routine laboratories, we still face new developments proving that analytical science is a true transdisciplinary field imprinting disciplines with new analytical achievements and challenge analytical developments for solving upcoming questions in a wide area of research disciplines.

*"We are still on a wild ride exploring new capabilities. Every new instrument should allow new measurements. The companies that make scientific*

*instruments just to make them cheaper do not do the community a service, because pushing the boundaries of instrument development requires an honest and active interchange between the community and the company as well as healthy companies.*" (Douthitt)

There is still a demand for improving detectors with respect to sensitivity, linear dynamic range, fast handling of signals and high linearity. In addition, simultaneous measurement of the whole mass range with high precision and sensitivity can be put in the basket.

"*There is still a desire for multiple collection at moderate count rates either with multiple Faraday cups or ion counters or combinations for isotope ratio measurements.*" (Ireland)

Even the magnet as the core of a sector field does not remain untouched from new developments. Faster magnets with stable magnetic fields can be found in almost every modern sector field mass spectrometer – and still, there is a demand for even faster and more stable magnets.

It is not only the standalone feature of an instrument that makes the impact. Besides, we face a continuous combination of techniques in order to enhance the knowledge of our samples tremendously. This development covers the fact that the full information spectrum of a sample needs to be assessed simultaneously in, *e.g.*, transient signals, and that valuable samples do not allow for the sequential measurement by different techniques. Therefore, not only are matrix and trace elements measured simultaneously making use of the large dynamic range but also the combination of elemental and species information along with molecular information or chemical surface data in combination with other surface properties (*e.g.* pH, optical properties or reactivity) are required. Pushing the limits, single-cell analysis for multiple parameters is on the top of the wish list in life sciences.

"*We will see more combinations including two or more analysers in series and in parallel, which will require much better software than we currently have.*" (Douthitt)

"*There is a need to develop and improve strategies for the spatially distinct integration of data from different sources (e.g. SIMS and LA-ICP-MS, structural information, etc.) and the metrologically sound analysis of spatially arranged data*" (Zitek)

Improved mass resolution without loss in sensitivity can be found on many notes when requesting the demands of scientists in mass spectrometry. Different geometries have proven their capabilities and from time to time, old abandoned technologies are facing a revival, such as the Mattauch–Herzog geometry.

Besides, the instrumental periphery is facing developments. Following this, there is a need for automated sample preparation, *e.g.*, in analyte separation. It is desirable to reach a level comparable to stable isotope analysis with techniques such as EA-IRMS coupling. This means an online separation system, which can be completely coupled to, *e.g.*, an MC ICP-MS and that has an additional channel for standard introduction. This should be fully automated, including a fully automated calibration mode, and controlled by one type of software, which of course also includes all possible calculation modules.

> *"Pushing the limits is one thing – we also need reliable systems, which we have to start to fully understand."* (Prohaska)

**ICP-MS**

Simultaneous multielement analysis, fast transient signals from laser-ablation or chromatographic systems coupled to an ICP-MS require ultrafast handling of data. This is also true for the investigation of nanoparticles, still a hot topic lacking adequate methods. To improve these capabilities, faster magnets and faster signal handling need to be developed – also for potential multiisotope ratio measurements.

"Less, smaller and faster" – single atoms in nm spatial resolution at ns time resolution can be seen as the limits, which need to be achieved. These are the demands for fields of science where elemental and isotopic analysis has entered within the last years.

> *"With respect to ICP-MS applications in biological sciences, proteomic characterisation of single cells will be an inviting goal to achieve a better understanding of carcinogenetic processes, as single cells are the origin of a later tumour and the protein interplays define the cells' properties and are fingerprints for differentiation, dedifferentiation and malignant transformation. Higher spatial resolution and absolute quantification are the major challenges."* (Roos)

Isotopic measurements by MC ICP-MS have enabled the assessment of isotopic systems, which were out of reach for decades. Nonetheless, even though competitive, the different types of instruments have their clear features and distinct fields of application.

> *"The evolution of the MC ICP-MS has been wonderful to watch, from the early very limited view that it would be a replacement for TIMS for radiogenic isotopes, to what it is today. And we are far from the end, because unlike TIMS, we can hyphenate the technique with chromatography and laser ablation to open up new dimensions."* (Douthitt)

**GDMS**

GD-SFMS is a well-established technique mainly in the quality control of high-purity materials as well as in materials research and will stay in this position.

*"Future improvements in pulsed direct current and radio-frequency glow discharge ion sources and in ultrafast sector field mass spectrometers will further increase the analytical potential of this technique."* (Pisonero)

Different instrumental configurations have been developed and introduced into the market. Today, conventional DC-GD coupled with SFMS does not expect much further improvement, since these instruments (*e.g. VG9000*) stand out due to their robustness and reliability. An extension of the application range was realised through the addition of μs – pulsed power supplies (*e.g. Element* GD). Even if most research papers nowadays in GDMS are related to RF GDTOFMS, a significant commercialisation has not taken off yet.

*"One can be curious if an addition of RF could bring a new rise GD-SFMS by extending the application range of GD-SFMS more to nonconducting materials."* (Venzago)

Quantification in GDMS can still be improved, for example, by producing more synthetic calibration materials. The advantage of semiquantitative calibration strategies like standard RFSs are most common and make the GDSFMS very flexible. Besides, the isotopic ratio measurement capabilities of GDMS are not exploited, yet.

*"Another development could a multicollector GDMS for isotope analysis in industrial applications, e.g. boron isotopes in steel."* (Vogl)

**TIMS**

Even though TIMS has been seen in many fields as the "gold standard" for isotopic measurements, there are a lot of fundamental processes that are not fully exploited, yet, and still improvements are made also with respect to the fundamental chemical processes taking place on the TIMS filaments. In addition, the instrumental demands are similar as compared to other techniques described here.

*"...demand for faster magnets and "useful" ion counters, more stable and more increased linear dynamic ranges of more than seven orders of magnitude."* (Klötzli, translated from German.)

**SIMS**

Ongoing improvements of existing and developments of new and alternate primary ion sources in SIMS are needed in order to enhance specific performances (*e.g.* high-brightness sources to gain lateral resolution or cluster primary ion species in order to improve the depth resolution, to lower divergence or increase the degree of ionisation).

*"With the increasing diversification of materials used in advanced electronics in the future, magnetic SIMS instruments will need to cope with highly*

*insulating and organic materials. New ion beams (such as argon clusters) and new charge-compensation schemes will need to be considered."* (Barnes)

*"The general trend of magnetic SIMS is an improvement of both accuracy and reproducibility. In the near past, dramatic improvements were obtained by implementing fully automated calibration and alignment routines. In the near future, improvements are expected from the availability of brighter primary negatives ion sources that will make possible, for example, to perform zircon-dating analyses at the micrometre scale."* (*de Chambost*)

Sound quantification is the ultimate challenge in SIMS as matrix effects strongly hamper both accuracy and precision in measurements. Therefore, future developments towards improvements in measurement and evaluation processes along with the provision of adequate standards are required.

*"For the future, emphasis should be placed on the SIMS quantitative measurements that up to now have mainly been restricted to semiconductor applications. Needless to say, it is not a straightforward process, but in this way, the SIMS technique will inevitably broaden its field of application."* (Valle)

*"The use of SIMS in geochemistry is greatly hampered by the extremely wide range of possible matrices that may be studied. Thus, there is a pressing need to find a means of defining or calibrating the relative sensitivities between elements or isotopes that is free of the need to possess matrix matched reference materials."* (Wiedenbeck)

**IRMS**

The well-established field of stable isotope ratio mass spectrometry has seen a large number of new developments in the last 25 years. These are mainly related to the introduction of carrier-gas systems, so that stable-isotope ratios could be directly analysed from chromatographic effluents with very high precision. The very recent introduction of new double focusing mass spectrometers has improved peak shapes and sensitivity, and the increase in mass resolution to working resolutions in excess of 25 000 has made detection and quantification of mass interferences a routine tool, allowing systematic biases in stable isotope results to be reduced. Hence smaller, yet robust, isotope signatures in nature like the "clumped isotopes" can be addressed with good success.

*"It is expected that such high-resolution records of isotope signatures, with new analytical tools, will allow further development of isotope-based models in basic research areas that aims to understand the complexity of relations between geochemical cycles and both ecosystem and atmosphere processes."* (Máguas)

**Figure 18.1** "Il prossimo libro lo scriviamo insieme." – Andreas Zitek, Thomas Prohaska, Johanna Irrgeher and Ondrej Hanousek. (Source: Photo: unknown photographer; © Irrgeher.)

Furthermore, laser-based technologies have started to challenge mass spectrometry.

> *"During the last years, laser techniques have become precise enough to rival the mass-spectrometric dominance in the field. It is expected that this role will further increase with new capabilities, which probably will include applicability to chromatographic techniques as well."* (Brand)

A speciality of these new laser photoabsorption techniques is their high isotopologue specificity. As an example, position-specific $^{15}N$ analysis in atmospheric $N_2O$ has become possible. With these techniques, the door to more general intramolecular stable isotope distribution analysis seems to open, which will mark the start of an entire new field of stable isotope applications.

What More to Say.......? (see Figure 18.1)

> *"There is a need for professors and a need for R&E directors, but they need to be less god-like than they were and more voices need to be listened to."* (Douthitt)

> *"There is a clear need for the development of transdisciplinary research centres focusing on the integration of high-end mass spectrometric analysis with different scientific disciplines."* (Zitek)

> *"The future is very exciting. We just have to stay alive to get there. I see that you are running races, that is good, that will help you live longer."* (Charles B. Douthitt, personal communication to Thomas Prohaska)

# *Epilogue*

Editing a book proved to be a race against the mounted souls with giant wings arising from other dimensions and – if you are not aware – the venom creeps into every single corner of your life – when you are left without leisure time, social contacts and sleep, you know that you are at the end of the editing process.

Editing this book brought me close to my mountain bike experience in Tibet: At a height of about 4000–5000 metres above sea level, it's hard for the lungs to find any oxygen – you fight with every kick into your pedals. Then, there is the permanent wind, trying to kick you over the edge of the gravel road, always head wind, whatever direction you go, no escape, even in the night, the wind visits you through every chink of your tent. And not to forget the dust, there is plenty of it, impeding breathing, chafing your eyes. And for the rest, there are stones – you have a ball riding over stone chippings with a bike for hundreds of miles, every jump a bump in your rump. . .but this is not what I wanted to tell you about Tibet, it is more the wisdom that comes out of that all:

". . .behind every hilltop, there is. . .another hilltop." So, this is editing this book. . . . . .a never-ending challenge. But challenges are what we need to grow.

The story of this book started some 15 years ago, when a certain Charles Douthitt approached two novices in the field, namely Christopher Latkoczy and myself trying to convince us that there is an urgent need to provide the scientific community with a book about sector field mass spectrometry. It was too early to succeed and the project never lifted off. It took some attempts by others, but not successful at their times, remaining damp squibs. We note the year 2011 when – after a successful finishing of a review together on sector field ICP-MS and some unpublished contributions to the previous mentioned attempts – Norbert Jakubowski approached me to convince me to edit a book on sector field mass spectrometry – naïve to agree

New Developments in Mass Spectrometry No. 3
Sector Field Mass Spectrometry for Elemental and Isotopic Analysis
Edited by Thomas Prohaska, Johanna Irrgeher, Andreas Zitek and Norbert Jakubowski

Published by the Royal Society of Chemistry, www.rsc.org

without knowing the consequences and the dusty road of pain to follow. In 2013, the project almost failed, nondelivering authors in combination with a steadily changing concept brought us at the edge of no return, we seemed to vanish in quicksand.

Somewhere new energy arouse and after Johanna Irrgeher and Andreas Zitek joined and reformed our editing team, light at the end of the tunnel appeared to be visible, in the hope these are not only some blown synapses of the shop-worn brain. But the lunacy continued, once one hole was plugged there was more water from another crack. . .you start to see dwarfs and imps spinning round and – yes – they are grabbing some paragraphs there, rearranging pictures and leaving a full mess. You read through a text and find it fine; you read it again and throw it in the trash, rearranging everything again. There is light but no exit. Amazingly, the patience of our publisher proved to be almost without limits and their never-ending faith gave us confidence in reaching our goals. With the help of the VIRIS caboodle we were finally able to fight the virtual giant octopus trying to haul us down with long tentacles covered with sticky printing ink. Even though a lilac cloud of lunacy covered our sight and with birds spinning round our brains making us looking for the end of a rainbow in this virtual lunatic world with whispers all around making you lose your senses and control, we finally succeeded, uploading the first version almost 15 years later.

My greatest thanks to Johanna, for her endless *joie de vivre* bringing this opus together, to Andreas for his limitless confidence standing restlessly his ground besides me and to my wife Sabine for her unbelievable understanding and her boundless love.

This book is dedicated to confidence, passion and love in any colour.

With this I close the book by opening a new chapter.

Thomas Prohaska

# Subject Index